University of California Publications

ENTOMOLOGY
Volume 116

A Bibliography of Ant Systematics

Philip S. Ward, Barry Bolton,
Steven O. Shattuck, William L. Brown, Jr.

University of California Press

A BIBLIOGRAPHY OF ANT SYSTEMATICS

A Bibliography of Ant Systematics

Philip S. Ward
University of California, Davis

Barry Bolton
The Natural History Museum, London

Steven O. Shattuck
CSIRO, Canberra

William L. Brown, Jr.
Cornell University, Ithaca

UNIVERSITY OF CALIFORNIA PRESS
Berkeley • Los Angeles • London

UNIVERSITY OF CALIFORNIA PUBLICATIONS IN ENTOMOLOGY

Volume 116
Issue Date: August 1996

UNIVERSITY OF CALIFORNIA PRESS
BERKELEY AND LOS ANGELES, CALIFORNIA

UNIVERSITY OF CALIFORNIA PRESS, LTD.
LONDON, ENGLAND

PRINTED IN THE UNITED STATES OF AMERICA

Library of Congress Cataloging-in-Publication Data

A bibliography of ant systematics / Philip S. Ward . . . [et al.].
p. cm. — (University of California publications in entomology ; v. 116)
ISBN 0-520-09814-5 (pbk : alk. paper)
1. Ants—Bibliography. 2. Ants—Classification—Bibliography.
I. Ward, Philip S., 1952- . II. Series.
Z5858.A6B53 1996
[QL568.F7]
016.59579'6—dc20

96-23721
CIP

The paper used in this publication meets the minimum requirements of American National Standard for Information Sciences—Permanence of Paper for Printed Library Materials, ANSI Z39.48-1984.

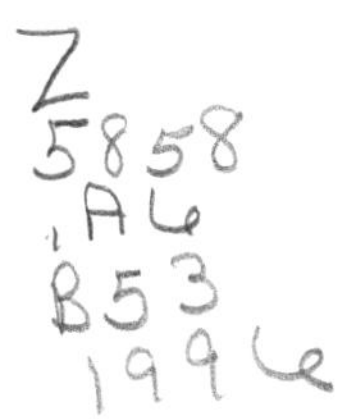

Contents

Preface

Begun ten years ago, this project was initially motivated by the need to bring together and provide consistent citations for the scattered descriptive literature in ant taxonomy, but the effort soon expanded to include published references on the phylogeny, evolution, geographical distribution, and morphology of ants, as well as comparative studies of ant biochemistry, behavior, and ecology. In the course of trawling for relevant literature, we accumulated a database of more than 16,000 references covering all aspects of ant biology.

The systematics bibliography presented here is a specialized subset of records that have been carefully reviewed for content and accuracy of citation. The process of checking (and often correcting) the citation details of each reference proved to be far more time-consuming than the mere accumulation of the records themselves. We have used the software program dBASE IV® to organize information on the literature, with the publications on ant systematics and related topics being held in a single file named ANTBIB. At the time of writing (April 1996) ANTBIB contained about 8,000 references.

Principal responsibility for the maintenance of this database and for the verification of the literature citations lies with the senior author. Shattuck (now at CSIRO, Division of Entomology, Canberra) was involved primarily in the early stages of this project and wrote most of the dBASE programs used to manage the database. Bolton (Department of Entomology, Natural History Museum, London) and Brown (Department of Entomology, Cornell University) each provided substantial input from their extensive files on ant literature. Bolton additionally checked the publication dates and citation details of many of the articles listed here. While we have striven to make the bibliography as complete as possible, errors and omissions doubtless remain. We would be grateful to receive comments, corrections, or additions to this bibliography. Such information should be sent to Philip S. Ward, Department of Entomology, University of California, Davis, CA 95616, U.S.A.

Our thanks are due to many people who assisted in this project. We are especially grateful to the library staff at the University of California at Davis, University of California at Berkeley, and the Natural History Museum, London. We are also indebted to Wojciech J. Pulawski (California Academy of Sciences) for an exceptionally thorough manuscript review, and to Rose Anne White (University of California Press) for expert editorial guidance. For additional assistance on matters bibliographical and linguistic we thank Gary Alpert, Joaquin Baixeras, Jacinto Barquín, Janne Bengtsson, Melissa Bennett, Carlos R. F. Brandão, Byeong-Moon Choi, Fabianna Cuezzo, Chris Darling, Terezinha Della Lucia, Mark DuBois, Xavier Espadaler, Neal Evenhuis, Fernando Fernández, Brian Fisher, Michael Fisher, David Furth, Masaki Kondoh, Jehoshua Kugler, Bede Lowery, Susumu Maeda, Michael Ohl, Sanford Porter, Alexandr Radchenko, Alexandr Rasnitsyn, Zhanna Reznikova, E. Ocete Rubio, Roy Snelling, Karel Spitzer, Alberto Tinaut, Georges Tohmé, Petr Werner, and librarians at the California Academy of Sciences, Center for Research Libraries (Chicago), Clark University (Worcester, Massachusetts), Museum of Comparative Zoology (Harvard University), National Agricultural Library (Beltsville, Maryland), Stanford University, and the University of Southern California.

Introduction

SCOPE AND CONTENT OF THE BIBLIOGRAPHY

This bibliography is intended to be a comprehensive reference source for publications on the taxonomy, evolution, and comparative biology of ants (Hymenoptera: Formicidae). Concerned with ant systematics in the broad sense, it includes not only strictly taxonomic papers (containing descriptions of new taxa, revisions of classifications, and nomenclatural changes), but also those dealing with such related fields as morphology, identification, phylogeny, biogeography, and faunistics (including recent work falling under the rubric of "biodiversity studies"). The bibliography also contains references on ant ecology, social behavior, genetics, physiology, and biochemistry. The major criterion for inclusion of a non-taxonomic publication is that it involves comparisons among related species or higher taxa of ants, or otherwise has systematic or evolutionary implications. In addition we have included single-species studies on taxa of special phylogenetic significance. A final category is that of bibliographic references, i.e., those containing information about the publication details of one or more primary references in the database.

The bibliography covers the period 1758 to 1995. It contains 8,185 entries, of which 8,109 are literature records and the remainder cross-references. These records have been extracted from a computerized database, ANTBIB, maintained at the University of California (Davis) by P. S. Ward. Each literature entry in the bibliography contains information from four fields: **author**, **year** of publication, **title**, and **citation**. In most cases this is followed by the exact **date** of publication, in square brackets after the citation. Additional **comments**, usually concerning the contents of the article or citation details, are sometimes appended to the end of the reference. Thus each entry has up to six items of information associated with it. We have not made use of a keywords field, but efficient searches for words appearing in the author, title, citation, or comments fields can be carried out on an electronic version of the bibliography, in conjunction with software such as EndNote® (see below under "Electronic dissemination of the bibliography").

FORMAT AND CONVENTIONS ADOPTED

Bibliographic entries are listed in conventional alphabetical order, by author and year. All text of the bibliography appears in a single format, i.e., italics, underline, and bold are not used, with the exception that genus and species names are italicized.

The entire bibliography is in the Roman alphabet, and publications using this script are cited in their original language. For papers and books in languages other than English the following combinations of diacritical marks and letters are recognized: á, à, â, ä, å, ã, Ä, Å, ç, Ç, é, è, ê, ë, É, í, ì, î, ï, Î, ñ, Ñ, ó, ò, ô, ö, õ, Ö, ú, ù, û, ü, Ü and ÿ. All other diacritical marks have been ignored. For publications appearing in other scripts (mostly Cyrillic, Chinese or Japanese) the titles have been translated into English, with the original language noted after the title, while the serial names have been transliterated.

Papers or books published in parts are given multiple entries, corresponding to each separately published (and differently dated) part. Unpublished theses and reports are not included in the bibliography.

Author

For ease of use and simplicity of alphabetization, we have standardized the names of all senior authors. For those names with variant spellings (including variable initials and variable transliterations), we have somewhat arbitrarily chosen a single spelling, usually one that corresponds to the most common or familiar usage (Table 1).

All alternative spellings of senior author names are cross-listed in the bibliography. This includes word-order variations in names that do not otherwise vary in spelling (such as "Guillou, E. J. F. le" for "Le Guillou, E. J. F."), as well as the variations listed in Table 1.

The following convention is adopted for Nicolas Kusnezov: papers published before 1945 are assigned to "N. N. Kuznetsov-Ugamsky" because Kusnezov used this compound name (in one form or another) almost exclusively during this period, the only exceptions being a few papers published between 1923 and 1926. For publications appearing after 1945, we use the form adopted by the author after his emigration to Argentina, i.e., "N. Kusnezov."

Year

The year listed after the author's name (or authors' names) is our estimate of the year in which the work was published. Where the actual year of publication is different from the apparent year, the latter is placed in quotes and parentheses after the former, e.g. 1988 ("1987"). This is not an uncommon situation for articles appearing in serial publications. In this case the apparent year is the one which appears prominently on the title page of the relevant issue or volume, while the actual year of publication is established from other, usually more cryptic information, either in the same or subsequent issues of the serial (e.g., from the end-of-volume table of contents) or occasionally from outside sources of evidence (see further discussion under "Determination of publication dates"). Although at odds with Recommendation 22A of the International Code of Zoological Nomenclature (Ride *et al.*, 1985) which recommends that only the actual date of publication be cited, the format adopted here is a useful way of alerting readers to discrepancies between the two dates.

Multiple papers by the same author, or the same permutation of authors, within a year are given different, consecutive lower case suffixes after the year (e.g., 1986a, 1986b), concordant with their chronological sequence of appearance to the extent that this can be determined.

Title

Titles of each article or book are given in their original orthography if in the Roman alphabet; otherwise they have been translated into English, with the language of the original noted. In the latter case the "language of the original" is that in which most of the text

Table 1. Senior author names with variable renditions: standards adopted in this bibliography, and alternatives appearing in the literature.

Standard	Alternative(s)
Acosta Salmerón, F. J.	Acosta, F. J.
Arakelian, G. R.	Arakelyan, G. R.
Arnol'di, K. V.	Arnoldi, K. V.
Atanassov, N.	Atanasov, N.; Atanassow, N.
Billen, J. P. J.	Billen, J.
Boven, J. K. A. van	Boven, J. van
Brandão, C. R. F.	Ferreira Brandão, C. R.
Collingwood, C. A.	Collingwood, C.
Comín, P.	Comín del Río, P.
Dalla Torre, K. W. von	Dalla Torre, C. G. de
De Haro, A.	De Haro Vera, A.
Deyrup, M.	Deyrup, M. A.
Dlussky, G. M.	Dlusskiy, G. M.
Donisthorpe, H.	Donisthorpe, H. S. J. K.
Escalante Gutiérrez, J. A.	Escalante G., J. A.
Espadaler, X.	Espadaler Gelabert, X.
Ezhikov, T.	Ezikov, J.
Fontenla Rizo, J. L.	Fontenla, J. L.
Franch, J.	Franch Batlle, J.
Frauenfeld, G. R. von	Frauenfeld, G.
Gerstäcker, A.	Gerstaecker, A.
Hermann, H. R.	Hermann, H. R., Jr.
Ipinza-Regla, J. H.	Ipinza, J.; Ipinza-Regla, J.
Karavaiev, V.	Karavajev, V.; Karawaiew, W.; Karawajew, W.
Krausse, A. H.	Krausse, A.
Kusnezov, N. (1948-1978)	Kusnezow, N.
Kuznetsov-Ugamsky, N. N. (1923-1930)	Kusnezov, N. N.; Kusnezov-Ugamsky, N.; Kusnezow, N. N.; Kusnezow-Ugamsky, N.; Kuznecov-Ugamskij, N. N.; Kuznetzov-Ugamskij, N. N.
Lattke, J. E.	Lattke, J.
MacKay, W. P.	MacKay, W.
Martínez, M. D.	Martínez Ibáñez, M. D.
Mayr, G.	Mayr, G. L.
Menozzi, C.	Minozzi, C.
Motschoulsky, V. de	Motschulsky, V. de
Mukerjee, D.	Mukerji, D.; Mukherji, D.
Ortiz, F. J.	Ortiz y Sánchez, F. J.
Quiran, E. M.	Quiran, E.
Radchenko, A. G.	Radtchenko, A. G.; Radtschenko, A. G.
Ramos-Elorduy de Conconi, J.	Conconi, J. R. E. de
Reyes, J. L.	Reyes Lopez, J. L.
Rodríguez, A.	Rodríguez González, A.
Ruzsky, M.	Ruzsky, M. D.
Schembri, S. P.	Schembri, S.
Schkaff, B.	Škaff, B.
Schmidt, G. H.	Schmidt, G.
Tinaut, A.	Tinaut, J. A.; Tinaut Ranera, J. A.
Tiwari, R. N.	Tewary, R. N.
Tohmé, G., Tohmé, H.	Thome, G., Thome, H.
Wesselinoff, G. D.	Vesselinov, G.
Weyrauch, W.	Weyrauch, W. K.

appears; summaries may be given in other languages. All title words are uncapitalized except where grammatical use requires otherwise (e.g., German nouns, English proper nouns, etc.).

In cases where there is no formal title, we place pertinent information in square brackets. Book reviews, for example, are titled as "[Review of: ", followed by a full citation of the reviewed work (i.e., author, year, title, place, publisher, number of pages). Untitled contributions appearing in published accounts of society meetings are cited as "[Untitled. Introduced by: ", followed by the first one to several lines of relevant text. Isolated taxonomic descriptions by someone other than the principal author(s) of a work usually have no obvious title. In such cases we cite the name of the new taxon as it is written in the paper.

Citation

The citation contains all the remaining standard bibliographic information that follows after the title. This is usually one of three types: a serial citation, book citation, or nested citation. Since the content of the citation depends on the type, we discuss each of these separately.

For serial citations we give the serial name, volume number, and pages. The issue number within a volume is given only where it is needed for unambiguous identification of the pages (this typically occurs where pagination starts anew with each issue rather than with each volume). In this case the issue number is placed in parentheses after the volume number. For periodicals with multiple series, the number or name of the series is given in parentheses before the volume number. Where individual volumes of a serial are labeled by year rather than by a volume number (e.g., *Proceedings of the Zoological Society of London* for the years 1833 to 1936) we treat the year as the volume number. For serials where each issue is separately numbered and paginated and does not form part of a larger volume (e.g., *Breviora*), the issue number is cited as the volume number.

All serials are abbreviated in accordance with standards set by the ISSN International Centre (1994) and those used by BIOSIS® for its *Zoological Record Serial Sources*™. A complete list of all periodicals cited in this bibliography, in both their abbreviated and full titles, is given in the next chapter ("List of Serials").

The following citation information is given for books: place of publication, publisher (in a few cases the printer, where the publisher is unclear), and the number of pages. This information is given not only in cases where the book itself is the primary reference, but also for articles cited within books.

Nested citations typically begin as "Pp. xx-xx in:" (or "P. xx in:"), followed by citation of the larger work of which they are a part. Examples include book chapters, single-author appendices within multiple-author papers, forewords, footnotes, etc. A frequent kind of nested citation in the bibliography is that of isolated species descriptions by someone other than the principal author(s) of a paper. These have been entered as separate bibliographic records. This practice applies even in cases where the author(s) to whom the descriptions should be attributed forms an overlapping set with the author(s) of the more inclusive work (see, e.g., the following in the bibliography: Wheeler and Mann, 1942a to 1942i; MacKay, 1988; Francoeur, 1985; Menozzi, 1930d, 1930e, 1930f). Such entries inflate the number of references somewhat, but permit more consistent record-keeping in the database.

Date of Publication

We have made a considerable effort to ascertain as accurately as possible the date of publication of each paper or book, particularly those containing nomenclaturally significant events (new taxa, new combinations, synonymy, etc.). The publication date appears in square brackets following the citation and is usually given to the nearest month or day. Where the publication could be dated only to year, the year alone is placed between brackets. The absence of a date signifies that we did not research the publication date with sufficient exhaustiveness to make a statement. The procedures followed in establishing dates of publication are discussed below.

Date information is given in ANSI (yyyy.mm.dd) format. Hence 25 November 1965 would be rendered as [1965.11.25]. For those works dated only to month, the designation contains only six digits, e.g., [1965.11]. Here it is to be understood that the default date for nomenclatural purposes is the last day of the month, just as the default date for publications dated only to year is December 31 of that year (Article 21c, ICZN).

Comments

These are usually brief and are used primarily to give sources for dating or, in papers dealing with a greater variety of taxa, to indicate on which pages ants (Formicidae) are treated. Cross-references to related papers or to translations are also given.

DETERMINATION OF PUBLICATION DATES

General Procedures

The primary source of information for establishing the publication date is the date specified in the work itself, although for papers appearing in serials the relevant information may be found in later issues of the same volume. In all cases we have searched first for explicit evidence about the publication date in the original work. Statements such as "mailing date 31 July 1967," "paru le 6 avril 1952," or "ausgegeben 21 Oktober 1908" can often be found on the front or back covers of the journal issue in which a paper appears. (This highlights the importance of binding issue covers with the rest of the volume, a practice not always followed by libraries.) In many instances journal issues are simply dated to the nearest month, sometimes without any clear statement about publication date. We accept such dates, in the absence of any contradictory evidence. If a range of dates is given (e.g., "January-March, 1924") we use the latest date. For serial publications in which no dating information is provided in the pertinent issue (or for which there is some doubt as to its accuracy), we have examined subsequent issues of the same volume in which the work was published. Sometimes the publication date for a particular issue is given in the next issue of the journal (e.g., *Psyche*), or in the end-of-year table of contents.

For publications produced within the past fifty years in France and in some other French-speaking regions, the "dépôt légal" date, appearing in fine print at the back of most books and serials, provides a good estimate of the date of publication. Most commonly this is given to the nearest trimester, in which case we take the end of the three-month period. Thus, "dépôt légal, 1er trimestre 1974" becomes [1974.03], while "4e trimestre 1974" would be dated as [1974.12].

From about 1930 onwards, works published in Russia, the former Soviet Union, and adjacent countries usually contain the imprimatur date. In the absence of any additional evidence concerning the publication date, we cite this "signed for printing" ("podpisano k pechati", in transliterated Russian) date, while recognizing that the completion of printing and release of the publication may have occurred several days to several months later. To indicate this the "podpisano" dates in the ant systematics bibliography are prefaced with ">" and accompanied by a brief explanatory note. For some works published in Poland a distinction is made between the date on which the text was consigned for printing ("podpisano do druku") and the date on which printing was completed ("druk ukończono"), and in such a case we use the latter date as the publication date.

Some journals published in the nineteenth and early twentieth centuries labeled each issue according to the date of the meeting (séance, sessione, Sitzung) whose activities are recounted therein. Such dates cannot be accepted as publication dates, even if the papers appearing in the issue were read at the meeting. Unless there is a clear indication that printed separates were made available—and in most cases we do not have such evidence—it becomes necessary to seek additional information regarding the date of publication of the issue. For some journals (e.g., *Bulletin de la Société Entomologique de France*) publication typically occurred about two weeks after the meeting; in other instances this did not happen until several months to a year later. Some specific situations that we have dealt with are discussed below under "Case-by-case discussion of serials."

If author's separates *were* distributed before the publication of an article in the journal (or other venue) for which it was intended, the date of distribution of the separates is the publication date of the work (Article 21h, ICZN). In the course of compiling the bibliography of ant systematics, we have encountered few instances of this—examples include papers by Emery (1900b, 1901k), Santschi (1920i), and Spinola (1851)—although there may be additional cases of which we are unaware.

External Evidence

If the original work (or, in the case of serial publications, subsequent issues of the same journal) either does not provide information on the date of publication or is suspected of being inaccurate, it becomes necessary to pursue additional evidence, external to the work itself. For some periodicals we have found it useful to consult multi-year indices since these sometimes provide precise publication dates which are not given in the original issues (or which are lacking in library copies in which the issue covers were discarded). In other cases publication dates of previous issues are announced at irregular intervals in the volumes of a journal.

For older books and monographs, as well as serial publications, less direct evidence may be sought. This includes information about the date that a particular work was received at a public institution, for example the receipt date at a library, or the date on which a book was reviewed in a weekly publication such as *Bibliographie française*. Other lines of investigation may involve the scrutiny of the minutes of society meetings or of letters from publishers. There is a large but scattered literature on the dating of natural history publications, usefully summarized in Griffin, Sherborn, and Marshall (1936) and supplements (Griffin, 1943; Stearn and Townsend, 1953; Goodwin, 1957; Goodwin, Stearn and Townsend, 1962). Other reference sources that have proven helpful in the current study include Sherborn (1922, 1932), Stafleu and Cowan (1976, 1979, 1981, 1983, 1985), Evenhuis *et al.* (1989), and the bibliography in Evenhuis (1994b). Where we have relied on one of these published secondary sources to date a work, we cite it in the comments field. Unpublished findings of our own are also made explicit.

Case-by-case Discussion of Serials

In this section we review dating and citation difficulties that have arisen with particular serials, and our decisions with respect to these.

Anales del Museo Nacional de Historia Natural de Buenos Aires.
Articles in the ant bibliography are from the years 1914 (volume 26) to 1934 (volume 38). Two somewhat contradictory sources of evidence on publication dates are available: the pagebase dates (which often vary slightly within an article) and the dates given for each article in the yearly index to the volume. We accept the latter as publication dates.

Annales de la Société Entomologique de Belgique.
So titled, the journal ran from 1857 to 1925 (volumes 1-64). Ant literature citations begin in volume 21 (1878). Starting with volume 41 (1897), publication dates are available from pagebase dates, except for volume 64 (see below). Prior to 1897, each issue is labeled only according to the monthly meeting that it covers. Publication appears to have occurred within several weeks of the meeting, based on séance and publication dates for volumes 41 to 45. Nevertheless, we have been unable to uncover exact publication dates so we date articles appearing before 1897 as after (">") the meeting date (e.g., > 1892.07.02).

The last volume (volume 64) consists of a single issue and lacks pagebase dates. The volume cover is dated "1924", but on page 179 there is a footnote "ajoutée à l'impression, IV, 1925." Hence we date all articles in this volume as April 1925.

Annales de la Société Entomologique de France.
Two useful indices ("Tables générales de 1861 à 1880 inclusivement," appearing in 1885, and "Tables générales de 1881 à 1890 inclusivement," appearing in 1895) give dates of publication for all issues (fascicles) for the period 1861 to 1890. For most issues appearing prior to 1861, we have used a set of dates based on an unpublished study by Neal Evenhuis (pers. comm.).

Dating of later (20th century) issues of this journal presents no problems insofar as exact publication dates are usually given on the originals. Note that volumes 114 (Année 1945), 115 (1946) and 116 (1947) were published as a single issue, dated "Décembre 1949." The back cover has a dépôt légal date of "1er trimestre 1950," however, so that we date this as March 1950.

See also: *Bulletin Bimensual de la Société Entomologique de France.*

Annali del Museo Civico di Storia Naturale (1870-1915), continued as *Annali del Museo Civico di Storia Naturale "Giacomo Doria"* (1916-present).
For articles appearing in volumes 9 (1877) and later, exact publication dates are available from the pagebases. We have used these except for volume 34, a special memorial issue in honor of Leonardo Fea, whose front and back covers are bordered in black and are both dated "1895." We have therefore dated the Emery article therein to year only (1895), despite its pagebase date of 7 September 1894. The foreword of this volume, by Doria and Gestro, is dated 20 April 1895.

Arquivos do Museu Nacional (Rio de Janeiro)
For Borgmeier's paper in volume 42 ("1951"), there are two contradictory statements about the publication date: one at the end of the article ("publicado em 30 de agôsto de 1951") and the other in the volume index ("distribuído em 7 de abril de 1952"). We have accepted the latter date, while noting that Borgmeier (1971) and Kempf (1972b) date this paper as appearing even later, in 1953.

Bulletin of the Bernice Pauahi Bishop Museum.
Entomological articles in volumes 98, 113, and 114 (also titled "Pacific Entomological Survey Publication" numbers 1, 6, and 7, respectively) were issued individually, and later bound together in the whole volumes of the *Bulletin* (Evenhuis *et al.*, 1989:919-920). We accept the individual issue dates, which can be up to three years before the volume dates.

Bulletin Bimensuel de la Société Entomologique de France.
Between 1873 and 1894 there were two editions of the "Bulletin" of the Société Entomologique de France (MacGillavry, 1936; Evenhuis *et al.*, 1989). One of these, which may be titled the *Bulletin Bimensuel de la Société Entomologique de France*, was a biweekly publication. The other, containing the same articles but usually with different pagination, appeared quarterly as part of the *Annales de la Société Entomologique de France*. Since the *Bulletin Bimensual* came out first, its dates are the relevant ones for nomenclatural purposes. These publication dates have been researched by Neal Evenhuis (Bishop Museum, Honolulu) who kindly sent us a copy of his unpublished notes.

Bulletin de la Société Entomologique de France.
We use this title to refer to the single, independant issuance of the "Bulletin" which began in 1895. It is the successor to the *Bulletin Bimensuel de la Société Entomologique de France*. Precise dates of publication are given in the "Bulletin" from 1902 onwards. For articles appearing between 1895 and 1902 only the séance date is available. We have dated these by extrapolation, based on the pattern (established by Neal Evenhuis when examining the *Bulletin Bimensual*) that each meeting's bulletin was published at least four days prior to the next meeting.

Bulletin de la Société d'Histoire Naturelle de l'Afrique du Nord.
Precise dates of publication are given in volumes 1-43 (1910-1952), but not thereafter. Hence we have dated articles appearing after 1952 to year only, with one exception: the last issue of the 1954 volume (volume 45, no. 7-8), published in 1955, contains a paper by Bernard that obviously predates his other papers of that year including a paper in volume 46 of the same journal. A default date of 31 December 1955 puts this paper out of sequence, so

we have used the receipt date at the UC Berkeley library (29 March 1955) to establish an upper bound on the publication date of Bernard (1955a).

Comptes Rendus (Hebdomadaires) des Séances de l'Academie des Sciences (Paris) (1835-1965); continued, in part, by *Comptes Rendus (Hebdomadaires) des Séances de l'Academie des Sciences. Série D. Sciences Naturelles* (1966-1980).
Citations in the ant systematics bibliography from this weekly serial span the period 1844 (volume 19) to 1972 (volume 275). Each issue of the serial is labeled according to the weekly meeting (séance) whose activities are recorded therein. In early years there is documentation, in the "Bulletin bibliographique" part of the journal, that at each weekly meeting the Academy received the published *Comptes rendus* covering the previous week's meeting. In a testimonial to the publisher, M. J.-A. Gauthier-Villars, published in 1898 (volume 126, page 453), a statement is made implying that such prompt publication was always achieved, at least throughout the nineteenth century: "Depuis 1835, l'imprimerie qu'il dirigeait, après Bachelier et Mallet-Bachelier, n'a pas une seule fois manqué à faire paraître les *Comptes rendus* à l'heure voulue, malgré les difficultés de toute nature ..." Hence we assign a publication date of one week after the séance date for articles appearing in the *Comptes rendus*.

Genera Insectorum
A recent study by Evenhuis (1994a) demonstrates that the title page dates and the "colophon" dates (sign-off dates of the authors) of the fascicles of *Genera Insectorum* cannot be relied upon. We have adopted the dates of publication established by Evenhuis, which are based primarily on receipt dates at various institutions.

Journal and Proceedings of the Linnean Society of London. Zoology (1855-1865) and *Journal of the Linnean Society of London. Zoology* (1867-1968).
Publication dates of articles appearing prior to 1891 are taken from Kappel (1896).

Journal (and *Proceedings*) *of the Royal Society of Western Australia*
Watson (1945) provides publication dates for volumes 1 (1916) to 28 (1944) of this serial. He omits to mention, however, that, beginning with volume 10, individual articles were released as separates ("pamphlets") prior to the whole volume, a fact noted in the annual report for the year 1924 (*J. Proc. R. Soc. West. Aust.* 10:xx). The publication date of each separate is given on the first page of the article. We have used these dates rather than those given in Watson (1945). On the other hand, to date J. Clark's (1924a) paper appearing in volume 9 of this journal, it is necessary to refer to Watson (1945).

La Naturaleza.
For publication dates to the nearest year (for all volumes of this serial) see H. M. Smith (1942).

Memorie della Reale Accademia delle Scienze dell'Instituto di Bologna.
We have been unable to date Emery's articles appearing in this serial between 1891 and 1901 (volumes 1 to 9 of the fifth series) beyond the year. There is a meeting (Sessione) date associated with each paper, which we mention in the comments after each citation, but these meeting dates cannot be taken as publication dates. A similar situation applies to

articles in the *Rendiconti delle Sessioni della Reale Accademia delle Scienze dell'Istituto di Bologna* (see below). It should be noted that Emery's papers in the *Memorie* were also issued as separates, with pagination different from that in the volume, suggesting a publication date preceding that of the whole volume.

Proceedings of the Academy of Natural Sciences of Philadelphia.
Precise dates of publication for each issue are to be found on the reverse of the title page of each volume, from 1871 (volume 23) onwards. Information on the publication dates of earlier issues is fragmentary (Nolan, 1913). For articles appearing before 1871, we have used the pagebase dates except where more accurate dating information is provided in Nolan (1913).

Proceedings of the California Academy of Sciences.
Articles by Pergande in volumes (2)4 and (2)5 are accompanied by pagebase dates, but a review of the evidence suggests that these cannot be relied upon as dates of publication. For example, Pergande's first article ("On a collection of Formicidae from Lower California and Sonora, Mexico" in the Proceedings, volume (2)4, pages 26-36) has a pagebase date of 19 September 1893 but in the annual report for 1893, presented at a meeting of the Academy on 2 January 1894 (see Proceedings, volume (2)4, page 629), this paper is said to be "in press". Even allowing for preparation of the annual report several weeks before the January meeting, publication of this paper is unlikely to have occurred in September 1893. Moreover, beginning with the third series of the Proceedings (which started in 1897), both pagebase dates and explicit publication dates are given for each article, and their lack of equivalency is transparent. Thus, for the twelve articles in the Zoology part of volume (3)1 the discrepancy between the actual publication date and the latest pagebase date—some articles have more than one pagebase date—ranges from 0 to 18 days (mean 6.1 days), with the pagebase dates always preceding the publication dates. If Pergande's articles were issued no less promptly one could conclude that publication occurred within about one or two weeks of the pagebase dates but obviously this is not certain, and it appears not to have been the case for his first paper. An upper bound on the publication date of each Pergande paper is provided by the date of issuance of the whole volume. For further discussion see under Pergande in the bibliography. Exact publication dates are available for later papers in the Proceedings by Wheeler (1919d, 1919e, 1933a, 1934a), and by this time the pagebase dates and publication dates coincide.

Proceedings of the Entomological Society of British Columbia.
Leech (1942) provides publication dates for earlier issues of this journal.

Proceedings of the Entomological Society of Philadelphia.
Information on the publication dates of volumes 1-6 (1861-1867) is given by F. M. Brown (1964).

Proceedings of the Zoological Society of London.
We have followed Waterhouse (1937) and Duncan (1937) for the publication dates of articles appearing in this journal between 1830 and 1926.

Psyche (Cambridge).
This journal contains numerous articles of relevance to ant systematics, from volumes 8 (1899) to 101 (1994-1995). Starting with volume 53 (1946), precise publication dates ("mailing dates") are available for each issue by consulting the inside front cover of the subsequent issue. For papers appearing in volumes 9-52 we have only the apparent month date (e.g., "December 1916" or "June-August 1927") from the front cover of each issue. There is considerable evidence to indicate that these monthly dates are unreliable, but for consistency with usage in Bolton's (1995b) recent world catalog of ants these dates have been cited here. In addition, however, we indicate in the comments field the "stamp date" in the library of the Museum of Comparative Zoology, Harvard University. According to Frank Carpenter (pers. comm. via David Furth), these MCZ stamp dates correspond closely to the mailing dates of the issues. Such receipt dates are present on the MCZ issues as far back as volume 10 (1903). There are irregularities, however, including cases where the stamp date is invariant for two or three consecutive issues, and stamp dates that *postdate* those of the Cornell University (Ithaca, New York) library. Thus the MCZ stamp dates are at best useful guidelines, providing a latest possible publication date.

Rendiconti delle Sessioni della Reale Accademia delle Scienze dell'Istituto di Bologna (continued, in part, as *Rendiconti delle Sessioni della Reale Accademia delle Scienze dell'Istituto di Bologna. Classe di Scienze Fisiche*).
The Rendiconti were released as several issues (fasciculo) per volume. These are labeled according to the months whose meetings they cover, e.g. "Fasciculo 3°, Marzo-Aprile 1897" for the meetings held between 14 March 1897 and 25 April 1897. There is also a year of publication stamped at the bottom of the title page of each fasciculo. This latter date usually agrees with the two-month fasciculo date except for those of "Novembre-Dicembre", for which the title page bears the following year. Thus it is clear that there was some delay in the publication of each fasciculo, but just how much delay is unknown. Taking a conservative course, we have dated articles appearing in this serial (all by Emery) to year only, using the year date stamped on the title page of each issue. In the comments field we indicate the meeting (Sessione) date on which the paper was presented.

Russkoe Entomologicheskoe Obozrenie.
Articles by Karavaiev and Ruzsky in volumes 9-16 (1910-1916) have two sets of dates. The first corresponds to that of the Julian (old Russian) calendar, while the second date, which is placed in parentheses, is that of the Gregorian calendar. We cite the latter dates as the publication dates. See also below under *Trudy Russkago Entomologicheskago Obshchestva*.

Sitzungsberichte der Kaiserlichen Akademie der Wissenschaften in Wien. Mathematisch-Naturwissenschaftliche Classe [*Klasse*]. *Abteilung I*.
The serial was published with this title from 1861 to 1917. Ant systematics papers occur between 1866 (volume 53) and 1907 (volume 116). For most volumes from 1873 (volume 67) onwards we encountered explicitly dated issue covers (dated to month) in the library copies examined by us. For earlier articles we inferred the month of publication from the issue number (Heft 1 = January, Heft 2 = February, etc.) on the basis of statements appearing in the indices (Registers) for volumes 43-50 and 51-60 (Anonymous, 1865, 1870) to the effect that each volume comprises ten issues, one appearing each month except August and September.

This serial was continued as *Sitzungsberichte der Akademie der Wissenschaften in Wien. Mathematisch-Naturwissenschaftliche Klasse. Abteilung I* (1918-1946) and later *Sitzungsberichte der Österreichischen Akademie der Wissenschaften. Mathematisch-Naturwissenschaftliche Klasse. Abteilung I* (1947-), but the issues of these journals in which we have found relevant ant articles are undated beyond the year and we have been unable to uncover evidence on their actual dates of publication. Hence the articles appearing therein, by Finzi and Bachmayer, are dated to year only.

Transactions of the Entomological Society of London.
Starting with volume 1896, exact dates are given for each part of a volume. For earlier issues, including those of the *Journal and Proceedings of the Entomological Society of London*, we have relied upon the dates meticulously researched by G. Wheeler (1912).

Transactions of the Entomological Society of New South Wales.
The years of publication of the first issues (various parts of volumes 1 and 2) are given by Andrewes (1930).

Transactions of the Linnean Society of London.
We consulted Raphael (1970) for publication dates of articles appearing in the first series (1791 to 1875).

Trudy Russkago Entomologicheskago Obshchestva.
For pre-revolutionary issues of this and other Russian publications that have a single exact publication date, we assume that such dates are those of the Julian calendar. We have converted these to Gregorian calendar dates by adding 12 days (19th century) or 13 days (20th century up to 31 January 1918). Publications dated only to the nearest month (e.g., early volumes of *Russkoe Entomologicheskoe Obozrenie*) have not been modified, although technically some of these could have been published in the following month, by Gregorian calendar reckoning.

Verhandlungen der Kaiserlich-Königlichen Zoologisch-Botanischen Gesellschaft in Wien.
We have been unable to find dated issue covers for volumes prior to 1880, hence articles therein are dated to year only. More precise dates of publication are available later, beginning in the ant bibliography with Mayr's (1886c, 1886d) papers.

INFORMATION SOURCES

The sources of information used to compile this bibliography are many and varied. One of the most important has been the *Zoological Record*, for the years 1864 to 1994/1995. This is an extraordinarily useful reference, documenting a wealth of information on systematics and general biology. Nevertheless, mistakes and omissions do occur occasionally, making verification of literature citations necessary. Complementing the *Zoological Record* are the two series of the *Index litteraturae entomologicae* (Horn and Schenkling, 1928-1929; Derksen and Scheidung, 1963-1972; Gaedike, 1975) and Hagen's (1862, 1863) *Bibliotheca entomologica*. These works proved helpful for researching the eighteenth and nineteenth century literature.

Published bibliographies of Borgmeier (Borgmeier, 1971), Creighton (Gregg, 1974), Emery (Ghigi, 1925), Finzi (Menozzi, 1941), Forel (Kutter, 1931), Gallardo (Bruch, 1934), Karavaiev (Paramonov, 1941), Kempf (Lopes, 1978), Kusnezov (Gavrilov, 1964), Mann (Wilson, 1960b), Mayr (Kohl, 1909), Menozzi (Grandi, 1943), Pisarski (Czechowski, 1994), M. R. Smith (D. R. Smith, 1973), Soudek (Kratochvíl, 1936a), Wasmann (Schmitz, 1932), W. M. Wheeler (Evans and Evans, 1970), and other myrmecologists were additional productive starting points. We also gleaned the literature references in general works on myrmecology such as Forel (1874, 1920b), Gösswald (1985, 1989, 1990), Hölldobler and Wilson (1990), Wheeler (1910), and Wilson (1971), and in citation-rich systematic treatments and faunistic catalogs (e.g., Baroni Urbani, 1971c; Bernard, 1967a; Borgmeier, 1955; Brandão, 1991; Dlussky, 1967a; Kempf, 1972b; Kupyanskaya, 1990; Pisarski, 1975; Stitz, 1939; Terayama, Choi and Kim, 1992). Specialized bibliographies devoted to particular aspects of myrmecology were also examined: Buckley, 1982c; Cherrett and Cherrett, 1989; Cotti, 1963; Loureiro and Soares, 1990; Okano, 1989; and Onoyama and Terayama, 1994.

We have included in this bibliography all of the 3,670-odd taxonomic references from Bolton's (1995b) recent world catalog of ants. We have also been able to incorporate additional taxonomic papers and new data on publication dates that were acquired too late for inclusion in that catalog. Where there is a discrepancy between the publication date in Bolton (1995b) and that given for the same paper here, the latter date may be considered more definitive.

Among unpublished sources, the index card files on ant literature assembled by Bolton and Brown over a period of many years were a rich mine of bibliographic information. Many of the initial entries of records into the ANTBIB database came from these card files. They were also useful for checking against references acquired independently by Ward or Shattuck (cross-comparisons revealing errors) and in suggesting leads for resolving problematic citations.

For retrieving recent literature we browsed current journals, requested reprints from colleagues, and used computer-based information systems (BIOSIS®, Current Contents®, MELVYL®, etc.) available through the University of California library system. There is, in fact, an enormous discrepancy between the ease of online retrieval of current literature and the difficulties posed by access to the older literature. It is arguably in the latter area where this bibliography makes its greatest contribution, since it appears unlikely that the main body of older taxonomic literature will be accessible in computerized form in the near future.

We should again emphasize that none of the literature references obtained from second-hand sources was accepted without further scrutiny. We made an attempt to verify each citation by examination of the original work. This procedure, the most time-consuming part of the project, frequently led to emendations or corrections of the citation, as well as to discoveries of additional relevant literature. In those few cases where we were unable to locate a publication, we indicate this and give the source of the citation.

ELECTRONIC DISSEMINATION OF THE BIBLIOGRAPHY

A copy of the database file (ANTBIB) that forms the basis of this bibliography can be obtained from the senior author (Philip S. Ward, Department of Entomology, University of

California, Davis, CA 95616, U.S.A.; e-mail: psward@ucdavis.edu). Efficient literature searches can then be carried out with the use of an appropriate bibliographic or database software program. The ANTBIB file is available in either dBASE® or EndNote® format, but responsibility for obtaining a copy of the relevant software lies with the user. The ANTBIB file also does not include the contents of chapters 1 and 2 of this bibliography, for which the present *University of California Publications in Entomology* volume should be consulted.

Other computerized databases on the Formicidae include FORMIS, a general bibliography of ant literature (available from S. D. Porter, USDA, ARS-MAVERL, P. O. Box 14565, Gainesville, FL 32604, U.S.A.), and ATTINE, a specialized database of leaf-cutting ant literature (Fowler *et al.*, 1995). We anticipate that the ANTBIB records will be incorporated into FORMIS in the future.

LITERATURE CITED

(Note: Ant literature mentioned in the above discussion is cited in the bibliography proper.)

Andrewes, H. E. 1930. *Catalogue of Indian insects. Part 18 - Carabidae.* Calcutta: Government of India Central Publication Branch, xxii + 389 pp.

Anonymous. 1865. *Register zu den Bänden 43 bis 50 der Sitzungsberichte der Mathematisch-Naturwissenschaftlichen Classe der Kaiserlichen Akademie der Wissenschaften. V.* Wien: Karl Gerold's Sohn, 157 pp.

Anonymous. 1870. *Register zu den Bänden 51 bis 60 der Sitzungsberichte der Mathematisch-Naturwissenschaftlichen Classe der Kaiserlichen Akademie der Wissenschaften. VI.* Wien: Karl Gerold's Sohn, 199 pp.

Brown, F. M. 1964. Dates of publication of the various parts of the Proceedings of the Entomological Society of Philadelphia. *Trans. Am. Entomol. Soc.* **89**:305-308.

Derksen, W., Scheidung, U. 1963-1972. *Index litteraturae entomologicae. Serie II: Die Welt-Literatur über die gesamte Entomologie von 1864 bis 1900.* Band I-IV. Berlin: Deutsche Akademie der Landwirtschaftswissenschaften zu Berlin, xii + 697 pp; 678 pp; 528 pp; 482 pp.

Duncan, F. M. 1937. On the dates of publication of the Society's 'Proceedings', 1859-1926. *Proc. Zool. Soc. Lond. Ser. A* **107**:71-77.

Evenhuis, N. L. 1994a. The publication and dating of P. A. Wytsman's *Genera Insectorum*. *Arch. Nat. Hist.* **21**:49-66.

Evenhuis, N. L. 1994b. *Catalogue of the fossil flies of the world.* Leiden: Bachhuys Publishers, 8 + 600 pp.

Evenhuis, N. L., Thompson, F. C., Pont, A. C., Pyle, B. L. 1989. Literature cited. Pp. 809-891 *in*: Evenhuis, N. L. (ed.) Catalog of the Diptera of the Australasian and Oceanian regions. *Bishop Mus. Spec. Publ.* **86**:1-1155.

Fowler, H. G., Schlittler, F. M., Schlindwein, M. N. 1995. ATTINE: a computerized database of leaf-cutting ant (*Atta* and *Acromyrmex*) literature. *J. Appl. Entomol.* **119**:255.

Gaedike, R. 1975. *Index litteraturae entomologicae. Serie II: Die Welt-Literatur über die gesamte Entomologie von 1864 bis 1900.* Band V. Register. Eberswalde: Akademie der Landwirtschaftswissenschaften der Deutschen Demokratischen Republik, 238 pp.

Goodwin, G. H., Jr. 1957. A catalogue of papers concerning the dates of publication of natural history books. Third supplement. *J. Soc. Bibliogr. Nat. Hist.* **3**:165-174.

Goodwin, G. H., Jr., Stearn, W. T., Townsend, A. C. 1962. A catalogue of papers concerning the dates of publication of natural history books. Fourth supplement. *J. Soc. Bibliogr. Nat. Hist.* **3**:5-12.

Griffin, F. J. 1943. A catalogue of papers concerning the dates of publication of natural history books. First supplement. *J. Soc. Bibliogr. Nat. Hist.* **2**:1-17.

Griffin, F. J., Sherborn, C. D., Marshall, H. S. 1936. A catalogue of papers concerning the dates of publication of natural history books. *J. Soc. Bibliogr. Nat. Hist.* **1**:1-30.

Hagen, H. A. 1862. *Bibliotheca entomologica. Die Litteratur über das ganze Gebiet der Entomologie bis zum Jahre 1862.* Erster Band. A-M. Leipzig: W. Engelmann, xii + 566 pp.

Hagen, H. A. 1863. *Bibliotheca entomologica. Die Litteratur über das ganze Gebiet der Entomologie bis zum Jahre 1862.* Zweiter Band. N-Z. Leipzig: W. Engelmann, 512 pp.

Horn, W., Schenkling, S. 1928-1929. *Index litteraturae entomologicae. Serie I: Die Welt-Literatur über die gesamte Entomologie bis inklusive 1863.* Band I-IV. Berlin-Dahlem: published by the senior author, 1426 pp.

ISSN International Centre. 1994. *Liste d'abréviations de mots des titres de publications en série. List of serial title word abbreviations.* Diskette edition. Paris: ISSN International Centre.

Kappel, A. W. 1896. *General index to the first twenty volumes of the Journal (Zoology) and the zoological portion of the Proceedings, November 1838 to 1890, of the Linnean Society of London.* London: Longmans, Green, and Co., viii + 437 pp.

Leech, H. B. 1942. The dates of publication of certain numbers of the Proceedings of the Entomological Society of British Columbia. *Proc. Entomol. Soc. B. C.* **38**:29-36.

MacGillavry, D. 1936. Les Bulletins de la Société entomologique de France. (Bibliographische bijdrage. VII.). *Entomol. Ber.* (Amst.) **9**:204-208.

Nolan, E. J. (ed.) 1913. *An index to the scientific contents of the Journal and Proceedings of the Academy of Natural Sciences of Philadelphia.* Philadelphia: Academy of Natural Sciences, xiv + 1419 pp.

Raphael, S. 1970. The publication dates of the *Transactions of the Linnean Society of London*, Series I, 1791-1875. *Biol. J. Linn. Soc.* **2**:61-76.

Ride, W. D. L., Sabrosky, C. W., Bernardi, G., Melville, R. V. (eds.) 1985. *International code of zoological nomenclature.* Third edition. London: International Trust for Zoological Nomenclature, xx + 338 pp.

Sherborn, C. D. 1922. *Index animalium sive index nominum quae ab A.D. MDCCLVIII generibus et speciebus animalium imposita sunt. Sectio secunda a kalendis ianuariis, MDCCCI usque ad finem decembris MDCCCL.* Part I. London: British Museum (Natural History), cxxxi + 128 pp.

Sherborn, C. D. 1932. *Index animalium sive index nominum quae ab A.D. MDCCLVIII generibus et speciebus animalium imposita sunt. Sectio secunda a kalendis ianuariis, MDCCCI usque ad finem decembris, MDCCCL.* Epilogue, additions to bibliography, additions and corrections, and index to trivialia. London: British Museum (Natural History), vii + cxxxiii-cxlvii + 1098 pp.

Smith, H. M. 1942. The publication dates of "La Naturaleza". *Lloydia* **5**:95-96.

Stafleu, F. A., Cowan, R. S. 1976. *Taxonomic literature. A selective guide to botanical publications and collections, with dates, commentaries and types.* Volume I. A-G. Utrecht: Bohn, Scheltema & Holkema, xl + 1136 pp.

Stafleu, F. A., Cowan, R. S. 1979. *Taxonomic literature. A selective guide to botanical publications and collections, with dates, commentaries and types.* Volume II. H-Le. Utrecht: Bohn, Scheltema & Holkema, xviii + 991 pp.

Stafleu, F. A., Cowan, R. S. 1981. *Taxonomic literature. A selective guide to botanical publications and collections, with dates, commentaries and types.* Volume III. Lh-O. Utrecht: Bohn, Scheltema & Holkema, xii + 980 pp.

Stafleu, F. A., Cowan, R. S. 1983. *Taxonomic literature. A selective guide to botanical publications and collections, with dates, commentaries and types.* Volume IV. P-Sak. Utrecht: Bohn, Scheltema & Holkema, ix + 1214 pp.

Stafleu, F. A., Cowan, R. S. 1985. *Taxonomic literature. A selective guide to botanical publications and collections, with dates, commentaries and types.* Volume V. Sal-Ste. Utrecht: Bohn, Scheltema & Holkema, 1066 pp.

Stearn, W. T., Townsend, A. C. 1953. A catalogue of papers concerning the dates of publication of natural history books. Second supplement. *J. Soc. Bibliogr. Nat. Hist.* **3**:5-12.

Waterhouse, F. H. 1937. List of the dates of delivery of the sheets of the 'Proceedings' of the Zoological Society of London, from commencement in 1830 to 1859, inclusive. *Proc. Zool. Soc. Lond. Ser. A* **107**:78-83.

Watson, E. M. 1945. The dates of publication of the Journal of the West Australian Natural History Society, the Journal of the Natural History and Science Society of Western Australia and the Journal of the Royal Society of Western Australia. *J. R. Soc. West. Aust.* **29**:174-175.

Wheeler, G. 1912. On the dates of the publications of the Entomological Society of London. *Trans. Entomol. Soc. Lond.* **1911**:750-767.

List of Serials

INTRODUCTION

The list below gives the abbreviations and full titles of all 1,349 serial publications cited in the bibliography. The great majority of these serials are documented in one or both of the following serial sources, which should be consulted for more information: *Serial publications in the British Museum (Natural History)* (British Museum [Natural History], 1980) and *Zoological Record Serial Sources*™ (BIOSIS, 1992, 1994a). Most of the remaining serials can be found in at least one of the following reference works: *World List of scientific periodicals* (Brown and Stratton, 1963-1965); *Union List of serials in libraries of the United States and Canada* (Titus, 1965); the periodical list in part 1 of *Quarto catálogo dos insetos que vivem nas plantas do Brasil* (Silva *et al.*, 1967); the periodical list in volume 5 of *Index litteraturae entomologicae, Serie II* (Gaedike, 1975); *International Union List of agricultural serials* (CAB, 1990); *Chemical Abstracts Service Source Index*® (CAS, 1990); *Serial Sources for the BIOSIS Previews® Database* (BIOSIS, 1994b); and *Directory of Japanese scientific periodicals* (Anonymous, 1993). Another very useful source of information about many of these serials is MELVYL®, the University of California's online library catalog.

A serial is listed more than once if changes occur in the title, with the exception of very slight variations (e.g., *Tr. Ukr.* [or *Vseukr.*] *Akad. Nauk Fiz.-Mat. Vidd.*). Serials are cited in their original language, and are transliterated if in a script other than the Roman alphabet. For periodicals with multiple titles, in more than one language, we generally cite the title which is predominant on the journal cover.

We use a uniform set of abbreviations for all serials, conforming closely to the standards adopted by the ISSN International Centre (1994) and by BIOSIS for its *Zoological Record Serial Sources* (BIOSIS, 1992, 1994a). Where there is conflicting use of abbreviations by these two sources, we have, in general, followed BIOSIS usage, except that, unlike BIOSIS (but in conformity with ISSN), we include diacritical marks in the serial abbreviations. Serials with identical names are distinguished by adding the place of publication in parentheses following the title, e.g., *Natura* (Amst.) and *Natura* (Milan). In such instances, regardless of the language of the periodical, we cite the English word for the place of publication since it is not part of the title proper. This also corresponds to the convention adopted by BIOSIS.

LITERATURE CITED

Anonymous. 1993. *Directory of Japanese scientific periodicals.* 1992 edition. Tokyo: National Diet Library, 2704 pp.

BIOSIS. 1992. *Zoological Record Serial Sources.* 4th edition. Philadelphia: BIOSIS, xi + 400 pp.

BIOSIS. 1994a. *Zoological Record Serial Sources.* 5th edition. (1992/1994). Philadelphia: BIOSIS, x + 472 pp.

BIOSIS. 1994b. *Serial Sources for the BIOSIS Previews Database.* Volume 1994. Philadelphia: BIOSIS, vii + 450 pp.

British Museum (Natural History). 1980. *Serial publications in the British Museum (Natural History).* 3rd edition. Volumes 1-3. London: British Museum (Natural History), 1436 pp.

Brown, P., Stratton, G. B. (eds.) 1963-1965. *World List of scientific periodicals published in the years 1900-1960.* 4th edition. Volumes 1-3. London: Butterworths, xxv + xx + xxii + 1824 pp.

CAB. 1990. *International Union List of agricultural serials.* Oxford: CAB International, xii + 767 pp.

CAS. 1990. *Chemical Abstracts Service Source Index, 1907-1989 cumulative.* Volumes 1-3. Columbus, Ohio: American Chemical Society, 43 + 3551 pp.

Gaedike, R. 1975. *Index litteraturae entomologicae. Serie II: Die Welt-Literatur über die gesamte Entomologie von 1864 bis 1900.* Band V. Register. Eberswalde: Akademie der Landwirtschaftswissenschaften der Deutschen Demokratischen Republik, 238 pp. [Periodical list pp. 5-66.]

ISSN International Centre. 1994. *Liste d'abréviations de mots des titres de publications en série. List of serial title word abbreviations.* Diskette edition. Paris: ISSN International Centre.

Silva, A. G. d'A. e, Gonçalves, C. R., Galvão, D. M., Gonçalves, A. J. L., Gomes, J., Nascimento Silva, M. do, Simoni, L. de. 1967. *Quarto catálogo dos insetos que vivem nas plantas do Brasil. Parte I. Bibliografia entomológica brasileira.* Rio de Janeiro: Ministério de Agricultura, xiii + 906 pp. [Periodical list pp. 1-19.]

Titus, E. B. (ed.) 1965. *Union List of serials in libraries of the United States and Canada.* Third edition. Volumes 1-5. New York: H. W. Wilson, 4649 pp.

SERIAL ABBREVIATIONS

Abh. Ber. K. Zool. Anthropol.-Ethnogr. Mus. Dres.
Abhandlungen und Berichte des Königlichen Zoologischen und Anthropologische-Ethnographischen Museums zu Dresden

Abh. Ber. Naturkundemus. Görlitz
Abhandlungen und Berichte des Naturkundemuseums Görlitz

Abh. Ber. Staatl. Mus. Tierkd. Dres.
Abhandlungen und Berichte aus dem Staatlichen Museum für Tierkunde in Dresden

Abh. Geol. Spezialkt. Elsass-Lothringen
Abhandlungen zur Geologischen Spezialkarte von Elsass-Lothringen

Abh. Naturwiss. Ver. Bremen
Abhandlungen herausgegeben vom Naturwissenschaftlichen Verein zu Bremen

Abh. Naturwiss. Ver. Würzbg.
Abhandlungen des Naturwissenschaftlichen Vereins Würzburg

Abh. Senckenb. Naturforsch. Ges.
Abhandlungen der Senckenbergischen Naturforschenden Gesellschaft

Abiks Loodusevaatl.
Abiks Loodusevaatlejale

Abstr. Dr. Diss. Ohio State Univ.
Abstracts of Doctors' Dissertations. Ohio State University

Acta Amazonica
Acta Amazonica

Acta Arct.
Acta Arctica

Acta Biol. Exp. (Warsaw)
Acta Biologiae Experimentalis (Warsaw)

Acta Biol. Iugosl. Ser. G Biosist.
Acta Biologica Iugoslavica. Serija G. Biosistematika
Acta Cient. (P. R.)
Acta Científica (Puerto Rico)
Acta Entomol. Bohemoslov.
Acta Entomologica Bohemoslovaca
Acta Entomol. Fenn.
Acta Entomologica Fennica
Acta Entomol. Lituan.
Acta Entomologica Lituanica
Acta Fac. Rerum Nat. Univ. Comenianae Zool.
Acta Facultatis Rerum Naturalium Universitatis Comenianae. Zoologia.
Acta Faun. Entomol. Mus. Natl. Pragae
Acta Faunistica Entomologica Musei Nationalis Pragae
Acta Geol. Sin.
Acta Geologica Sinica
Acta Hymenopterol.
Acta Hymenopterologica
Acta Soc. Fauna Flora Fenn.
Acta Societatis pro Fauna et Flora Fennica
Acta Soc. Sci. Fenn.
Acta Societatis Scientiarum Fennicae
Acta Soc. Sci. Nat. Moravo-Sil.
Acta Societatis Scientiarum Naturalium Moravo-Silesiacae
Acta Terramaris
Acta Terramaris
Acta Univ. Asiae Mediae Ser. 8A. Zool.
Acta Universitatis Asiae Mediae. Series VIIIa Zoologia
Acta Zool. Cracov.
Acta Zoologica Cracoviensia
Acta Zool. Fenn.
Acta Zoologica Fennica
Acta Zool. Lilloana
Acta Zoologica Lilloana
Acta Zool. Pathol. Antverp.
Acta Zoologica et Pathologica Antverpiensia
Actes Colloq. Insectes Soc.
Actes des Colloques Insectes Sociaux
Actes Soc. Linn. Bordx.
Actes de la Société Linnéene de Bordeaux
Actes Soc. Sci. Chili
Actes de la Société Scientifique du Chili
Adv. Ecol. Res.
Advances in Ecological Research
Adv. Insect Physiol.
Advances in Insect Physiology
Afr. Entomol.
African Entomology
Aggressive Behav.
Aggressive Behavior
Agra Univ. J. Res.
Agra University Journal of Research
Agric. Gaz. N. S. W.
Agricultural Gazette of New South Wales
Agrociencia
Agrociencia
Akitu
Akitu
Allatt. Közl.
Allattani Közlemények
Allg. Forst- Jagdztg.
Allgemeine Forst- und Jagdzeitung
Allg. Z. Entomol.
Allgemeine Zeitschrift für Entomologie
Alpe
L'Alpe
Am. J. Bot.
American Journal of Botany
Am. J. Sci.
American Journal of Science
Am. Midl. Nat.
American Midland Naturalist
Am. Mus. J.
American Museum Journal
Am. Mus. Novit.
American Museum Novitates
Am. Nat.
American Naturalist
Am. Sci.
American Scientist
Am. Zool.
American Zoologist
Ameghiniana
Ameghiniana
Amtl. Ber. Versamml. Dtsch. Naturforsch. Aerzte
Amtlicher Bericht der Versammlung Deutscher Naturforscher und Aerzte
An. Acad. Bras. Cienc.
Anais da Academia Brasileira de Ciencias
An. Acad. Cienc. Méd. Fis. Nat. La Habana
Anales de la Academia de Ciencias Médicas, Fisicas y Naturales de La Habana
An. Biol.
Anales de Biología

An. Mus. Nac. Costa Rica
Anales del Museo Nacional de Costa Rica
An. Mus. Nac. Hist. Nat. B. Aires
Anales del Museo Nacional de Historia Natural de Buenos Aires
An. Mus. Nahuel Huapí
Anales del Museo Nahuel Huapí
An. Soc. Cient. Argent.
Anales de la Sociedad Cientifica Argentina
An. Soc. Entomol. Bras.
Anais da Sociedade Entomológica do Brasil
An. Soc. Esp. Hist. Nat.
Anales de la Sociedad Española de Historia Natural
An. Soc. Nordestina Zool.
Anais da Sociedade Nordestina de Zoologia
An. Stiint. Univ. "Al. I. Cuza" Iasi Sect. IIa Biol.
Analele Stiintifice de Universitatii "Al. I. Cuza" din Iasi. Sectiunea IIa Biologie
Andrias
Andrias
Angew. Parasitol.
Angewandte Parasitologie
Anim. Behav.
Animal Behavior
Ann. Accad. Aspir. Nat. Secunda Era
Annali dell'Accademia degli Aspiranti Naturalisti. Secunda Era
Ann. Cape Prov. Mus.
Annals of the Cape Provincial Museums
Ann. Entomol. Fenn.
Annales Entomologici Fennici
Ann. Entomol. Soc. Am.
Annals of the Entomological Society of America
Ann. Epiphyt.
Annales des Epiphyties
Ann. Fac. Sci. Univ. Féd. Cameroun
Annales de la Faculté des Sciences, Université Fédérale du Cameroun
Ann. Fac. Sci. Yaoundé
Annales de la Faculté des Sciences de Yaoundé
Ann. Hist.-Nat. Mus. Natl. Hung.
Annales Historico-Naturales Musei Nationalis Hungarici
Ann. K-K. Naturhist. Mus. Wien
Annalen des Kaiserlich-Königlichen Naturhistorischen Museums in Wien
Ann. Mag. Nat. Hist.
Annals and Magazine of Natural History
Ann. Mus. Civ. Stor. Nat.
Annali del Museo Civico di Storia Naturale
Ann. Mus. Civ. Stor. Nat. "Giacomo Doria"
Annali del Museo Civico di Storia Naturale "Giacomo Doria"
Ann. Mus. Hist. Nat. Nice
Annales du Muséum d'Histoire Naturelle de Nice
Ann. Mus. R. Afr. Cent. Sér. Octavo Sci. Zool.
Annales. Musée Royal de l'Afrique Centrale. Série in Octavo. Sciences Zoologiques
Ann. Mus. R. Congo Belge Nouv. Sér. Quarto Sci. Zool.
Annales du Musée Royal du Congo Belge. Nouvelle Série in Quarto. Sciences Zoologiques
Ann. N. Y. Acad. Sci.
Annals of the New York Academy of Sciences
Ann. Nat. Hist.
Annals of Natural History
Ann. Natal Mus.
Annals of the Natal Museum
Ann. Naturhist. Mus. Wien
Annalen des Naturhistorischen Museums in Wien
Ann. S. Afr. Mus.
Annals of the South African Museum
Ann. Sci. Nat.
Annales des Sciences Naturelles
Ann. Sci. Nat. Bot.
Annales des Sciences Naturelles, Botanique
Ann. Sci. Nat. Zool.
Annales des Sciences Naturelles, Zoologie
Ann. Sci. Nat. Zool. Biol. Anim.
Annales des Sciences Naturelles, Zoologie et Biologie Animale
Ann. Soc. Entomol. Belg.
Annales de la Société Entomologique de Belgique
Ann. Soc. Entomol. Fr.
Annales de la Société Entomologique de France
Ann. Soc. Entomol. Qué.
Annales de la Société Entomologique du Québec

Ann. Soc. Hortic. Hist. Nat. Hérault
Annales de la Société d'Horticulture et d'Histoire Naturelle de l'Hérault
Ann. Soc. R. Zool. Belg.
Annales de la Société Royale Zoologique de Belgique
Ann. Soc. Sci. Brux. Sér. B Sci. Phys. Nat.
Annales de la Société Scientifique de Bruxelles. Série B. Sciences Physiques et Naturelles
Ann. Stn. Biol. Besse-en-Chandesse
Annales de la Station Biologique de Besse-en-Chandesse
Ann. Univ. Abidjan Sér. E (Écol.)
Annales de l'Université d'Abidjan. Série E (Écologie)
Ann. Univ. Abidjan Sér. Sci.
Annales de l'Université d'Abidjan. Série Sciences
Ann. Univ. Mariae Curie-Sklodowska Sect. C Biol.
Annales Universitatis Mariae Curie-Sklodowska. Sectio C. Biologia
Ann. Zool. (Warsaw)
Annales Zoologici (Warsaw)
Ann. Zool. Fenn.
Annales Zoologici Fennici
Année Biol.
Année Biologique
Annot. Zool. Jpn.
Annotationes Zoologicae Japonenses
Annu. Ist. Mus. Zool. Univ. Napoli
Annuario dell'Istituto e Museo di Zoologia dell'Università di Napoli
Annu. Mus. Zool. R. Univ. Napoli
Annuario del Museo Zoologico della Reale Università de Napoli
Annu. Rep. Cocoa Res. Inst. Ghana
Annual Report. Cocoa Research Institute, Ghana
Annu. Rep. Natl. Inst. Genet. Jpn.
Annual Report. National Institute of Genetics, Japan.
Annu. Rep. Proc. Lancs. Chesh. Entomol. Soc.
Annual Report and Proceedings of the Lancashire and Cheshire Entomological Society
Annu. Rep. Smithson. Inst.
Annual Report of the Board of Regents of the Smithsonian Institution
Annu. Rev. Ecol. Syst.
Annual Review of Ecology and Systematics
Annu. Rev. Entomol.
Annual Review of Entomology
Annu. Soc. Nat. Mat. Modena
Annuario della Società dei Naturalisti e Matematici, Modena
Anz. Akad. Wiss. Wien Math.-Naturwiss. Kl.
Anzeiger der Akademie der Wissenschaften in Wien. Mathematisch-Naturwissenschaftliche Klasse
Anz. Schädlingskd.
Anzeiger für Schädlingskunde
Anz. Schädlingskd. Pflanzenschutz Umweltschutz
Anzeiger für Schädlingskunde Pflanzenschutz und Umweltschutz
Appl. Entomol. Zool.
Applied Entomology and Zoology
Arb. Morphol. Taxon. Entomol. Berl.-Dahl.
Arbeiten über Morphologische und Taxonomische Entomologie aus Berlin-Dahlem
Arb. Physiol. Angew. Entomol. Berl.-Dahl.
Arbeiten über Physiologische und Angewandte Entomologie aus Berlin-Dahlem
Arch. Biol. Sci. (Belgrade)
Archives of Biological Sciences (Belgrade)
Arch. Bot. Biogeogr. Ital.
Archivio Botanico e Biogeografico Italiano
Arch. Esc. Super. Agric. Med. Vet. Nictheroy
Archivos da Escola Superior da Agricultura e Medicina Veterinaria. Nictheroy
Arch. Forstwes.
Archiv für Forstwesen
Arch. Hydrobiol.
Archiv für Hydrobiologie
Arch. Insektenkd. Oberrheingeb. Angrenz. Länder
Archiv für Insektenkunde des Oberrheingebietes und der Angrenzenden Länder
Arch. Inst. Biol. (São Paulo)
Archivos do Instituto Biológico (São Paulo)

Arch. Inst. Biol. Veg. (Rio J.)
Archivos do Instituto de Biologia Vegetal (Rio de Janeiro)
Arch. Inst. Grand-Ducal Luxemb.
Archives. Institut Grand-Ducal de Luxembourg
Arch. Mus. Nac. (Rio J.)
Archivos do Museu Nacional (Rio de Janeiro)
Arch. Nat. Hist.
Archives of Natural History
Arch. Naturgesch.
Archiv für Naturgeschichte
Arch. Néerl. Zool.
Archives Néerlandaises de Zoologie
Arch. Sci. Phys. Nat.
Archives des Sciences Physiques et Naturelles
Arch. Svazu Ochr. Prír. Domov. Moravé
Archiv Svazu na Ochranu Prírody a Domoviny na Moravé
Arch. Zool. (Budapest)
Archivum Zoologicum (Budapest)
Arch. Zool. Exp. Gén.
Archives de Zoologie Expérimentale et Générale
Arch. Zool. Ital.
Archivio Zoologico Italiano
Arctic
Arctic
Arh. Biol. Nauka
Arhiv Bioloskih Nauka
Ari
Ari
Ark. Zool.
Arkiv för Zoologi
Arq. Biol. Tecnol. (Curitiba)
Arquivos de Biologia e Tecnologia (Curitiba)
Arq. Inst. Biol. (São Paulo)
Arquivos do Instituto Biológico (São Paulo)
Arq. Inst. Pesqui. Agron. (Recife)
Arquivos do Instituto de Pesquisas Agronômicas (Recife)
Arq. Mus. Nac. (Rio J.)
Arquivos do Museu Nacional (Rio de Janeiro)
Arq. Univ. Fed. Rural Rio J.
Arquivos da Universidade Federal Rural do Rio de Janeiro
Arq. Zool. (São Paulo)
Arquivos de Zoologia (São Paulo)
Atti Accad. Gioenia Sci. Nat.
Atti dell'Accademia Gioenia di Scienze Naturali
Atti Mus. Civ. Stor. Nat. Trieste
Atti del Museo Civico di Storia Naturale di Trieste
Atti R. Accad. Lincei Mem. Cl. Sci. Fis. Mat. Nat.
Atti della Reale Accademia dei Lincei. Memorie. Classe di Scienze Fisiche, Matematiche e Naturali
Atti R. Accad. Naz. Lincei Rend. Cl. Sci. Fis. Mat. Nat.
Atti della Reale Accademia Nazionale dei Lincei. Rendiconti. Classe di Scienze Fisiche, Matematiche e Naturali
Atti R. Accad. Sci. Fis. Mat. Napoli
Atti della Reale Accademia delle Scienze Fisiche e Matematiche. Napoli
Atti Soc. Ital. Sci. Nat. Mus. Civ. Stor. Nat. Milano
Atti della Società Italiana di Scienze Naturali e del Museo Civico di Storia Naturale di Milano
Atti Soc. Nat. Mat. Modena
Atti della Società dei Naturalisti e Matematici di Modena
Atti Soc. Toscana Sci. Nat. Mem. Ser. B
Atti della Società Toscana di Scienze Naturali. Memorie, Serie B
Attini
Attini
Audubon
Audubon
Aust. Entomol.
Australian Entomologist
Aust. Entomol. Mag.
Australian Entomological Magazine
Aust. J. Bot.
Australian Journal of Botany
Aust. J. Chem.
Australian Journal of Chemistry
Aust. J. Ecol.
Australian Journal of Ecology
Aust. J. Zool.
Australian Journal of Zoology
Aust. Zool.
Australian Zoologist

Avh. Nor. Vidensk.-Akad. Oslo I Mat.-Naturvidensk. Kl.
Avhandlingar utgitt av det Norske Videnskaps-Akademi i Oslo. I. Matematisk-Naturvidenskapelig Klasse
Avicennia
Avicennia
Bayer. Tierwelt
Bayerische Tierwelt
Behav. Ecol. Sociobiol.
Behavioral Ecology and Sociobiology
Beih. Veröff. Naturschutz Landschaftspfl. Baden-Württ.
Beihefte zu den Veröffentlichungen für Naturschutz und Landschaftspflege in Baden-Württemberg
Beitr. Entomol.
Beiträge zur Entomologie
Beitr. Kennt. Land- Süsswasserfauna Dtsch.-Südwestafr.
Beiträge zur Kenntnis der Land- und Süsswasserfauna Deutsch-Südwestafrikas
Beitr. Kunde Estlands
Beiträge zur Kunde Estlands
Beitr. Landespfl. Rheinl.-Pfalz. Beih.
Beiträge zur Landespflege in Rheinland-Pfalz. Beiheft
Beitr. Naturkd. Forsch. Südwestdtschl.
Beiträge zur Naturkundlichen Forschung in Südwestdeutschland
Beitr. Naturkd. Preuss.
Beiträge zur Naturkunde Preussens
Belg. J. Zool.
Belgian Journal of Zoology
Ber. Naturwiss.-Med. Ver. Innsb.
Bericht des Naturwissenschaftlich-Medizinischen Vereins in Innsbruck
Ber. Verh. Naturforsch. Ges. Basel
Bericht über die Verhandlungen der Naturforschenden Gesellschaft in Basel
Ber. Versamml. Bot. Zool. Ver. Rheinl.-Westfal.
Berichte über die Versammlungen des Botanischen und des Zoologischen Vereins für Rheinland-Westfalen
Berl. Entomol. Z.
Berliner Entomologische Zeitschrift
Beschäft. Berl. Ges. Naturforsch. Freunde
Beschäftigungen der Berlinischen Gesellschaft Naturforschender Freunde
Beuthen. Abh. Oberschles. Heimatforsch.
Beuthener Abhandlungen zur Oberschlesischen Heimatforschung
Biene
Die Biene
Bih. K. Sven. Vetensk.-Akad. Handl.
Bihang till Kongliga [Kungliga] Svenska Vetenskaps-Akademiens Handlingar
Biochem. Syst. Ecol.
Biochemical Systematics and Ecology
Biodivers. Conserv.
Biodiversity and Conservation
Biogeographica
Biogeographica
Biogeographica (Tokyo)
Biogeographica (Tokyo)
Biogr. Mem. Natl. Acad. Sci.
Biographical Memoirs. National Academy of Sciences
Biol. Bull. (Woods Hole)
Biological Bulletin (Woods Hole)
Biol. Cent.-Am.
Biologia Centrali-Americana
Biol. Centralbl.
Biologisches Centralblatt
Biol. Conserv.
Biological Conservation
Biol. Gabon.
Biologia Gabonica
Biol. Gallo-Hell.
Biologia Gallo-Hellenica
Biol. J. Linn. Soc.
Biological Journal of the Linnean Society
Biol. Rev. (City Coll. N. Y.)
Biological Review (City College of New York)
Biol. Rundsch.
Biologische Rundschau
Biol. Zentralbl.
Biologisches Zentralblatt
Biol. Zh. Arm.
Biologicheskii Zhurnal Armenii
Biológia (Bratisl.)
Biológia (Bratislava)
Biotemas
Biotemas
Biotropica
Biotropica
Bishop Mus. Spec. Publ.
Bishop Museum Special Publication
Bocagiana
Bocagiana

Bol. Agric. Ganad. Ind. Prov. B. Aires
Boletín de Agricultura, Ganadería e Industrias de la Provincia de Buenos Aires
Bol. Asoc. Esp. Entomol.
Boletín de la Asociación Española de Entomologia
Bol. Biol. Lab. Parasitol. Fac. Med. São Paulo
Boletim Biológico. Laboratório de Parasitologia. Faculdade de Medicina de São Paulo
Bol. Biol. Rio J.
Boletim Biológico. Rio de Janeiro
Bol. Direcc. Agric. Ganad. (Lima)
Boletín de la Dirección de Agricultura y Ganaderia (Lima)
Bol. Entomol. Venez.
Boletín de Entomologia Venezolana
Bol. Estac. Cent. Ecol.
Boletín de la Estación Central de Ecología
Bol. Fitossanit.
Boletim Fitossanitário
Bol. Minist. Educ. Entre Ríos
Boletín del Ministerio de Educación. Entre Ríos
Bol. Mus. Munic. Funchal
Boletim do Museu Municipal do Funchal
Bol. Mus. Nac. Rio J.
Boletim do Museu Nacional de Rio de Janeiro
Bol. Mus. Para. Emílio Goeldi Sér. Zool.
Boletim do Museu Paraense Emílio Goeldi. Série Zoologia
Bol. Mus. Para. Hist. Nat. Ethnogr.
Boletim do Museu Paraense de Historia Natural e Ethnographia
Bol. Mus. Rocha
Boletim do Museu Rocha
Bol. R. Soc. Esp. Hist. Nat.
Boletín de la Real Sociedad Española de Historia Natural
Bol. R. Soc. Esp. Hist. Nat. Secc. Biol.
Boletín de la Real Sociedad Española de Historia Natural. Sección Biológica
Bol. Serv. Nac. Pesqui. Agron.
Boletim. Serviço Nacional de Pesquisas Agronômicas
Bol. Soc. Biol. Concepción
Boletín de la Sociedad de Biología de Concepción
Bol. Soc. Bras. Agron.
Boletim da Sociedade Brasileira de Agronomia
Bol. Soc. Entomol. Esp.
Boletín de la Sociedad Entomológica de España
Bol. Soc. Port. Entomol.
Boletim da Sociedade Portuguesa de Entomologia
Bol. Soc. Port. Entomol. Supl.
Boletim da Sociedade Portuguesa de Entomologia. Suplemento
Bol. Univ. Nac. La Plata
Boletín de la Universidad Nacional de La Plata
Boll. Assoc. Rom. Entomol.
Bollettino dell'Associazione Romana di Entomologia
Boll. Lab. Entomol. R. Ist. Super. Agrar. Bologna
Bollettino del Laboratorio di Entomologia del Reale Istituto Superiore Agrario di Bologna
Boll. Lab. Zool. Gen. Agrar. Fac. Agrar. Portici
Bollettino del Laboratorio di Zoologia Generale e Agraria della Facoltà Agraria in Portici
Boll. Lab. Zool. Gen. Agrar. R. Sc. Super. Agric.
Bollettino del Laboratorio di Zoologia Generale e Agraria della Reale Scuola Superiore d'Agricoltura. Portici
Boll. Mus. Civ. Stor. Nat. Verona
Bollettino del Museo Civico di Storia Naturale di Verona
Boll. Mus. Reg. Sci. Nat. Torino
Bollettino del Museo Regionale di Scienze Naturali. Torino
Boll. Mus. Zool. Anat. Comp. R. Univ. Torino
Bollettino dei Musei di Zoologia ed Anatomia Comparata della Reale Università di Torino
Boll. [*R.*] *Ist. Entomol. Univ. Studi Bologna*
Bollettino [del Reale Istituto] dell'Istituto di Entomologia della Università degli Studi di Bologna
Boll. Soc. Adriat. Sci. Nat. Trieste
Bollettino della Società Adriatica di Scienze Naturali in Trieste

Boll. Soc. Entomol. Ital.
Bollettino della Società Entomologica Italiana
Boll. Zool.
Bollettino di Zoologia
Bolleti Soc. Hist. Nat. Balears
Bolleti de la Societat d'Historia Natural de les Balears
Bonn. Zool. Beitr.
Bonner Zoologische Beiträge
Boston J. Nat. Hist.
Boston Journal of Natural History
Bot. Gart. Bot. Mus. Univ. Zür.
Der Botanische Garten und das Botanische Museum der Universität Zürich
Bot. Jahrb. Syst. Pflanzengesch. Pflanzengeogr.
Botanische Jahrbücher für Systematik, Pflanzengeschichte und Pflanzengeographie
Br. Nat.
British Naturalist
Brain Behav. Evol.
Brain Behavior and Evolution
Braunschw. Naturkd. Schr.
Braunschweiger Naturkundliche Schriften
Breviora
Breviora
Brigham Young Univ. Sci. Bull. Biol. Ser.
Brigham Young University Science Bulletin. Biological Series
Brotéria Ser. Cienc. Nat.
Brotéria. Serie Ciencias Naturais
Bul. Shkencave Nat. Tiranë
Buletin i Shkencave të Natyres. Tiranë
Bul. Soc. Nat. Rom.
Buletinul Societatii Naturalistilor din România
Bul. Soc. Rom. Stiinte
Buletinul Societatii Române de Stiinte
Bul. Stiint. Acad. Repub. Pop. Rom. Sect. Biol. Stiint. Agric.
Buletin Stiintific. Academia Republicii Populare Romîne. Sectia de Biologie si Stiinte Agricole
Bull. Acad. Pol. Sci. Sér. Biol.
Bulletin de l'Académie Polonaise des Sciences. Série des Sciences Biologiques.
Bull. Acad. R. Sci. B.-Lett. Brux.
Bulletin de l'Académie Royale des Sciences et Belles-Lettres de Bruxelles
Bull. Am. Mus. Nat. Hist.
Bulletin of the American Museum of Natural History
Bull. Ann. Soc. Entomol. Belg.
Bulletin et Annales de la Société Entomologique de Belgique
Bull. Ann. Soc. R. Entomol. Belg.
Bulletin et Annales de la Société Royale Entomologique de Belgique
Bull. Assoc. Nat. Sci. Senshu Univ.
Bulletin of the Association of Natural Science, Senshu University
Bull. Bernice P. Bishop Mus.
Bulletin of the Bernice Pauahi Bishop Museum
Bull. Bimens. Soc. Entomol. Fr.
Bulletin Bimensuel de la Société Entomologique de France
Bull. Biogeogr. Soc. Jpn.
Bulletin of the Biogeographical Society of Japan
Bull. Biol. Fr. Belg.
Bulletin Biologique de la France et de la Belgique
Bull. Br. Mus. (Nat. Hist.) Entomol.
Bulletin of the British Museum (Natural History). Entomology
Bull. Brooklyn Entomol. Soc.
Bulletin of the Brooklyn Entomological Society
Bull. Colo. Biol. Assoc.
Bulletin of the Colorado Biological Association
Bull. Entomol. Res.
Bulletin of Entomological Research
Bull. Entomol. Soc. Am.
Bulletin of the Entomological Society of America
Bull. Entomol. Soc. Niger.
Bulletin of the Entomological Society of Nigeria
Bull. Inst. Agron. Stn. Rech. Gembloux
Bulletin de l'Institut Agronomique et des Stations de Recherches de Gembloux
Bull. Inst. Fondam. Afr. Noire Sér. A Sci. Nat.
Bulletin de l'Institut Fondamental d'Afrique Noire. Série A. Sciences Naturelles
Bull. Inst. Fr. Afr. Noire Sér. A Sci. Nat.
Bulletin de l'Institut Français d'Afrique Noire. Série A. Sciences Naturelles.

Bull. Inst. R. Sci. Nat. Belg.
Bulletin de l'Institut Royal des Sciences Naturelles de Belgique
Bull. Inst. Trop. Agric. Kyushu Univ.
Bulletin of the Institute of Tropical Agriculture Kyushu University
Bull. Inst. Zool. Acad. Sin. (Taipei)
Bulletin of the Institute of Zoology Academia Sinica. (Taipei)
Bull. Intér. Sect. Fr. UIEIS
Bulletin Intérieur. Section Française de l'UIEIS (Union Internationale pour l'Étude des Insectes Sociaux)
Bull. Mens. Soc. Nat. Luxemb.
Bulletin Mensuel. Société des Naturalistes Luxembourgeois
Bull. Mens. Soc. Sci. Nancy
Bulletin Mensuel de la Société des Sciences de Nancy
Bull. Misc. Inf. R. Bot. Gard. Kew
Bulletin of Miscellaneous Information. Royal Botanic Gardens, Kew
Bull. Misc. Inf. R. Bot. Gard. Kew Addit. Ser.
Bulletin of Miscellaneous Information. Royal Botanic Gardens, Kew. Additional Series
Bull. Mizunami Fossil Mus.
Bulletin of the Mizunami Fossil Museum
Bull. Mus. Comp. Zool.
Bulletin of the Museum of Comparative Zoology
Bull. Mus. Hist. Nat.
Bulletin du Muséum d'Histoire Naturelle
Bull. Mus. Natl. Hist. Nat.
Bulletin du Muséum National d'Histoire Naturelle
Bull. Mus. Natl. Hist. Nat. Écol. Gén.
Bulletin du Muséum National d'Histoire Naturelle. Écologie Générale
Bull. Mus. Natl. Hist. Nat. Sect. A. Zool. Biol. Écol. Anim.
Bulletin du Muséum National d'Histoire Naturelle. Section A. Zoologie, Biologie et Écologie Animales
Bull. Mus. R. Hist. Nat. Belg.
Bulletin du Musée Royal d'Histoire Naturelle de Belgique
Bull. Nat. Hist. Res. Cent. Univ. Baghdad
Bulletin of the Natural History Research Centre. University of Baghdad
Bull. Rech. Agron. Gembloux
Bulletin des Recherches Agronomiques de Gembloux
Bull. Romand Entomol.
Bulletin Romand d'Entomologie
Bull. SROP
Bulletin SROP [Section Regional Ouest Paléarctique, Organisation Internationale de Lutte Biologique]
Bull. Sci. Bourgogne
Bulletin Scientifique de Bourgogne
Bull. Sci. Soc. Philomath. Paris
Bulletin des Sciences par la Société Philomathique de Paris
Bull. Soc. Amis Sci. Lett. Poznan Sér. D. Sci. Biol.
Bulletin de la Société des Amis des Sciences et des Lettres de Poznan. Série D. Sciences Biologiques
Bull. Soc. Bot. Genève
Bulletin de la Société Botanique de Genève
Bull. Soc. Entomol. Belg.
Bulletin de la Société Entomologique de Belgique
Bull. Soc. Entomol. Egypte
Bulletin. Société Entomologique d'Egypte.
Bull. Soc. Entomol. Fr.
Bulletin de la Société Entomologique de France
Bull. Soc. Entomol. Ital.
Bullettino della Società Entomologica Italiana
Bull. Soc. Entomol. Ital. Resoc. Adun.
Bullettino della Società Entomologica Italiana. Resocónti di Adunanze
Bull. Soc. Entomol. Mulhouse
Bulletin de la Société Entomologique de Mulhouse
Bull. Soc. Entomol. Taiwan Prov. Chung-Hsing Univ.
Bulletin of the Society of Entomology. Taiwan Provincial Chung-Hsing University
Bull. Soc. Étude Sci. Nat. Elbeuf
Bulletin de la Société d'Étude des Sciences Naturelles d'Elbeuf
Bull. Soc. Fouad Ier Entomol.
Bulletin de la Société Fouad Ier d'Entomologie

Bull. Soc. Hist. Nat. Afr. Nord
Bulletin de la Société d'Histoire Naturelle de l'Afrique du Nord
Bull. Soc. Hist. Nat. Dep. Moselle Metz
Bulletin de la Société d'Histoire Naturelle du Département de la Moselle. Metz
Bull. Soc. Hist. Nat. Toulouse
Bulletin de la Société d'Histoire Naturelle de Toulouse
Bull. Soc. Imp. Nat. Mosc.
Bulletin de la Société Impériale des Naturalistes de Moscou
Bull. Soc. Nat. Luxemb.
Bulletin de la Société des Naturalistes Luxembourgeois
Bull. Soc. Neuchâtel. Sci. Nat.
Bulletin de la Société Neuchâteloise des Sciences Naturelles
Bull. Soc. Philomath. Paris
Bulletin de la Société Philomathique de Paris
Bull. Soc. Sci. Nat. Maroc
Bulletin de la Société des Sciences Naturelles du Maroc
Bull. Soc. Sci. Nat. Phys. Maroc
Bulletin de la Société des Sciences Naturelles et Physiques du Maroc
Bull. Soc. Sci. Nat. Tunis.
Bulletin de la Société des Sciences Naturelles de Tunisie
Bull. Soc. Vaudoise Sci. Nat.
Bulletin de la Société Vaudoise des Sciences Naturelles
Bull. Soc. Zool. Fr.
Bulletin de la Société Zoologique de France
Bull. Soc. Zool. Fr. Evol. Zool.
Bulletin de la Société Zoologique de France. Évolution et Zoologie
Bull. South. Calif. Acad. Sci.
Bulletin of the Southern California Academy of Sciences
Bull. Tianjin Inst. Geol. Miner. Resour.
Bulletin of the Tianjin Institute of Geology and Mineral Resources
Bull. Toho Gakuen
Bulletin of the Toho Gakuen
Bull. Toyama Sci. Mus.
Bulletin of the Toyama Science Museum
Bull. U. S. Geol. Geogr. Surv. Territ.
Bulletin of the United States Geological and Geographical Survey of the Territories
Bull. U. S. Geol. Surv.
Bulletin of the United States Geological Survey
Bull. Univ. Asie Cent.
Bulletin de l'Université de l'Asie Centrale
Bull. Univ. Utah Biol. Ser.
Bulletin of the University of Utah. Biological Series
Bull. Wis. Nat. Hist. Soc.
Bulletin of the Wisconsin Natural History Society
Bull. Zool. Nomencl.
Bulletin of Zoological Nomenclature
Bull. Zool. Surv. India
Bulletin of the Zoological Survey of India
Butll. Inst. Catalana Hist. Nat.
Butlleti de la Institucio Catalana d'Historia Natural
Byull. Vseross. Entomo-Fitopatol. S'ezda
Byulleten' Vserossiiskogo Entomo-Fitopatologicheskogo S'ezda
C. R. Séances Acad. Sci.
Comptes Rendus (Hebdomadaires) des Séances de l'Academie des Sciences
C. R. Séances Acad. Sci. Ser. D. Sci. Nat.
Comptes Rendus (Hebdomadaires) des Séances de l'Academie des Sciences. Série D. Sciences Naturelles
C. R. Séances Soc. Biogéogr.
Compte Rendu des Séances de la Société de Biogéographie
C. R. Somm. Séances Soc. Biogéogr.
Compte Rendu Sommaire des Séances de la Société de Biogéographie
CSIRO Div. Entomol. Rep.
CSIRO (Commonwealth Scientific and Industrial Research Organization) Division of Entomology Report
Café Cacao Thé
Café Cacao Thé
Cah. Nat. (Bull. Nat. Paris.)
Cahiers des Naturalistes (Bulletin des Naturalistes Parisiens)
Caldasia
Caldasia
Calif. Dep. Agric. Bur. Entomol. Occas. Pap.
California Department of Agriculture. Bureau of Entomology. Occasional Papers

Calif. Dep. Food Agric. Lab. Serv. Entomol. Occas. Pap.
California Department of Food and Agriculture. Laboratory Services, Entomology. Occasional Papers
Can. Entomol.
Canadian Entomologist
Can. For. Branch Bimon. Res. Notes
Canada Forestry Branch. Bimonthly Research Notes
Can. J. Genet. Cytol.
Canadian Journal of Genetics and Cytology
Can. J. Zool.
Canadian Journal of Zoology
Caribb. J. Sci.
Caribbean Journal of Science
Carinthia II
Carinthia II
Carnegie Inst. Wash. Publ.
Carnegie Institution of Washington Publication
Carolinea
Carolinea
Caryologia
Caryologia
Cas. Cesk. Spol. Entomol.
Casopis Ceskoslovenské Spolecnosti Entomologické
Cas. Ceské Spol. Entomol.
Casopis Ceské Spolecnosti Entomologické
Cas. Morav. Mus. Brne
Casopis Moravského Musea v Brne
Cas. Morav. Zemského Mus. Brne
Casopis Moravského Zemského Musea v Brne
Cas. Nár. Mus. (Prague)
Casopis Národního Musea (Prague)
Cas. Slezského Muz. Ser. A Vedy Prír.
Casopis Slezského Muzea. Serie A. Vedy Prírodni
Cat. Faunae Austriae
Catalogus Faunae Austriae
Cell Tissue Res.
Cell and Tissue Research
Checklist Specie Fauna Ital.
Checklist delle Specie della Fauna Italiana
Chem. Soc. Rev.
Chemical Society Reviews
Chin. J. Entomol.
Chinese Journal of Entomology
Chongju Sabom Taehak Nonmunjip (*J. Chongju Natl. Teach. Coll.*)
Chongju Sabom Taehak Nonmunjip (Journal of Chongju National Teachers' College)
Chongju Sabom Taehakkyo Nonmunjip (*J. Chongju Natl. Univ. Educ.*)
Chongju Sabom Taehakkyo Nonmunjip (Journal of Chongju National University of Education)
Chromosoma (Berl.)
Chromosoma (Berlin)
Cienc. Biol. Acad. Cienc. Cuba
Ciencias Biologicas. Academia de Ciencias de Cuba
Cienc. Invest.
Ciencia e Investigación
Ciênc. Cult. (São Paulo)
Ciência e Cultura (São Paulo)
Científica (Jaboticabal)
Científica (Jaboticabal)
Circ. Board Agric. Trinidad
Circular. Board of Agriculture, Trinidad
Clin. Exp. Allergy
Clinical and Experimental Allergy
Cocuyo (Havana)
Cocuyo (Havana)
Colecc. Estud. Altoaragon.
Colección de Estudios Altoaragoneses
Colecc. Monogr. (*Secr. Publ. Univ. La Laguna*)
Colección Monografías (Secretariado de Publicaciones de la Universidad de La Laguna)
Colemania
Colemania
Coleopt. Bull.
Coleopterists' Bulletin
Collana Verde
Collana Verde
Collect. Nord.
Collection Nordicana
Colloq. Int. Cent. Natl. Rech. Sci.
Colloques Internationaux du Centre National de la Recherche Scientifique
Comment. Biol.
Commentationes Biologicae
Comp. Biochem. Physiol.
Comparative Biochemistry and Physiology
Comp. Biochem. Physiol. B. Comp. Biochem.
Comparative Biochemistry and Physiology. B. Comparative Biochemistry

Comun. Acad. Repub. Pop. Rom.
Comunicarile Academiei Republicii Populare Române
Comun. Mus. Nac. Hist. Nat. "Bernardino Rivadavia"
Comunicaciones del Museo Nacional de Historia Natural "Bernardino Rivadavia"
Comun. Zool. Mus. Hist. Nat. Montev.
Comunicaciones Zoologicas del Museo de Historia Natural de Montevideo
Condor
Condor
Conn. State Geol. Nat. Hist. Surv. Bull.
Connecticut State Geological and Natural History Survey. Bulletin
Conserv. Biol.
Conservation Biology
Contr. Lab. Vertebr. Biol. Univ. Mich.
Contributions from the Laboratory of Vertebrate Biology of the University of Michigan
Contr. Sci. (Los Angel.)
Contributions in Science (Los Angeles)
Cour. Forschungsinst. Senckenb.
Courier Forschungsinstitut Senckenberg
Creat. Res. Soc. Q.
Creation Research Society Quarterly
Curr. Genet.
Current Genetics
Cytobios
Cytobios
Dan. Fauna
Danmarks Fauna
Denkschr. Akad. Wiss. Wien Math.-Naturwiss. Kl.
Denkschriften der Akademie der Wissenschaften in Wien. Mathematisch-Naturwissenschaftliche Klasse
Denkschr. Med.-Naturwiss. Ges. Jena
Denkschriften der Medizinisch-Naturwissenschaftlichen Gesellschaft zu Jena
Diss. Abstr. B. Sci. Eng.
Dissertation Abstracts. B. Sciences and Engineering
Diss. Abstr. Int. B. Sci. Eng.
Dissertation Abstracts International. B. Sciences and Engineering
Divulg. Agron.
Divulgação Agronômica
Dnev. Zool. Otd. Imp. Obshch. Lyubit. Estestvozn. Antropol. Etnogr.
Dnevnik' Zoologicheskago Otdeleniya Imperatorskago Obshchestva Lyubitelei Estestvoznaniya, Antropologii i Etnografii
Dobutsugaku Zasshi (*Zool. Mag.*)
Dobutsugaku Zasshi (Zoological Magazine)
Dokl. Akad. Nauk Arm.
Doklady Akademii Nauk Armenii
Dokl. Akad. Nauk SSSR
Doklady Akademii Nauk SSSR
Doriana
Doriana
Drosera
Drosera
Dtsch. Entomol. Natl.-Bibl.
Deutsche Entomologische National-Bibliothek
Dtsch. Entomol. Z.
Deutsche Entomologische Zeitschrift
Dusenia
Dusenia
EOS. Rev. Esp. Entomol.
EOS. Revista Española de Entomología
East Afr. Wildl. J.
East African Wildlife Journal
Ecol. Appl.
Ecological Applications
Ecol. Entomol.
Ecological Entomology
Ecol. Mediterr.
Ecologia Mediterranea
Ecol. Monogr.
Ecological Monographs
Ecol. Rev.
Ecological Review
Ecología (*Asoc. Argent. Ecol.*)
Ecología (Asociación Argentina de Ecología)
Ecología (Madr.)
Ecología (Madrid)
Ecology
Ecology
Ecotrópicos
Ecotrópicos
Edaphologia
Edaphologia
Ekol. Pol.
Ekologia Polska
Ekologiya (Mosc.)
Ekologiya (Moscow)

Entomo Shirogane
Entomo Shirogane
Entomofauna
Entomofauna
Entomol. (Trends Agric. Sci.)
Entomology (Trends in Agricultural Science)
Entomol. Abh. Staatl. Mus. Tierkd. Dres.
Entomologische Abhandlungen. Staatliches Museum für Tierkunde Dresden
Entomol. Am.
Entomologica Americana
Entomol. Annu.
Entomologist's Annual
Entomol. Anz.
Entomologischer Anzeiger
Entomol. Basil.
Entomologica Basiliensia
Entomol. Beih. Berl.-Dahl.
Entomologische Beihefte aus Berlin-Dahlem
Entomol. Ber. (Amst.)
Entomologische Berichten (Amsterdam)
Entomol. Ber. Luzern
Entomologische Berichte Luzern
Entomol. Bl. Biol. Syst. Käfer
Entomologische Blätter für Biologie und Systematik der Käfer
Entomol. Exp. Appl.
Entomologia Experimentalis et Applicata
Entomol. Fenn.
Entomologica Fennica
Entomol. Gaz.
Entomologist's Gazette
Entomol. Gen.
Entomologia Generalis
Entomol. Ger.
Entomologica Germanica
Entomol. Hell.
Entomologia Hellenica
Entomol. Issled. Kirg.
Entomologicheskie Issledovaniya v Kirgizii
Entomol. Jahrb.
Entomologisches Jahrbuch
Entomol. Listy
Entomologické Listy
Entomol. Mag. Kyoto
Entomological Magazine. Kyoto
Entomol. Medd.
Entomologiske Meddelelser
Entomol. Mitt.
Entomologische Mitteilungen
Entomol. Mon. Mag.
Entomologist's Monthly Magazine
Entomol. Nachr. Ber.
Entomologische Nachrichten und Berichte
Entomol. Nachr. Dres.
Entomologische Nachrichten. Dresden
Entomol. Nachrichtenbl. Troppau
Entomologisches Nachrichtenblatt. Troppau
Entomol. News
Entomological News
Entomol. Obozr.
Entomologicheskoe Obozrenie
Entomol. Rec. J. Var.
Entomologist's Record and Journal of Variation
Entomol. Res. Bull. (Seoul)
Entomological Research Bulletin (Seoul)
Entomol. Rev. (Wash.)
Entomological Review (Washington)
Entomol. Rev. Jpn.
Entomological Review of Japan
Entomol. Rundsch.
Entomologische Rundschau
Entomol. Scand.
Entomologica Scandinavica
Entomol. Scand. Suppl.
Entomologica Scandinavica. Supplementum
Entomol. Tidskr.
Entomologisk Tidskrift
Entomol. World Tokyo
Entomological World. Tokyo
Entomol. Z.
Entomologische Zeitschrift
Entomologen
Entomologen
Entomologist
Entomologist
Entomologiste (Paris)
Entomologiste (Paris)
Entomon
Entomon
Entomophaga
Entomophaga
Entomops
Entomops
Entomotaxonomia
Entomotaxonomia

Environ. Entomol.
Environmental Entomology
Ergeb. Hambg. Magalhaens. Sammelreise
Ergebnisse der Hamburger Magalhaensischen Sammelreise
Ergeb. Wiss. Unters. Schweiz. Nationalpark
Ergebnisse der Wissenschaftlichen Untersuchungen im Schweizerischen Nationalpark
Ergeb. Zweiten Dtsch. Zent.-Afr. Exped.
Ergebnisse der Zweiten Deutschen Zentral-Afrika Expedition
Esakia
Esakia
Esakia Spec. Issue
Esakia Special Issue
Essex Nat. (*Lond.*)
Essex Naturalist (London)
Estudi Gen. (*Coll. Univ. Girona*)
Estudi General (Collegi Universitari de Girona)
Ethol. Ecol. Evol.
Ethology Ecology and Evolution
Ethol. Ecol. Evol. Spec. Issue
Ethology Ecology and Evolution Special Issue
Ethology
Ethology
Eur. J. Entomol.
European Journal of Entomology
Evolution
Evolution
Experientia (Basel)
Experientia (Basel)
Experientiae (Viçosa)
Experientiae (Viçosa)
Ezheg. Zool. Muz.
Ezhegodnik Zoologicheskago Muzeya
Fabreries
Fabreries
Fasc. Malay. Zool.
Fasciculi Malayenses. Zoology
Faun. Abh. (Dres.)
Faunistische Abhandlungen (Dresden)
Fauna Bohemiae Septentr.
Fauna Bohemiae Septentrionalis
Fauna Bûlg.
Fauna na Bûlgariya
Fauna Entomol. Scand.
Fauna Entomologica Scandinavica
Fauna Flora Rheinl.-Pfalz
Fauna und Flora in Rheinland-Pfalz
Fauna Hawaii.
Fauna Hawaiiensis
Fauna Norv. Ser. B
Fauna Norvegica. Series B
Fauna Saudi Arab.
Fauna of Saudi Arabia
Faune Entomol. Indoch. Fr.
Faune Entomologique de l'Indochine Française
Faunus
Faunus
Feuille Jeunes Nat.
Feuille des Jeunes Naturalistes
Field Lab.
Field and Laboratory
Fieldiana Zool.
Fieldiana Zoology
Fitofilo
Fitofilo
Fla. Dep. Agric. Consum. Serv. Div. Plant Ind. Entomol. Circ.
Florida Department of Agriculture and Consumer Services. Division of Plant Industry. Entomology Circular
Fla. Entomol.
Florida Entomologist
Flora Fauna
Flora og Fauna
Folia Entomol. Hung.
Folia Entomologica Hungarica
Folia Entomol. Mex.
Folia Entomologica Mexicana
Folia Myrmecol. Termit.
Folia Myrmecologica et Termitologica
Folia Univ. Cochabamba
Folia Universitaria. Cochabamba
Folia Zool. Hydrobiol.
Folia Zoologica et Hydrobiologica
For. Res.
Forest Research
Fortschr. Zool.
Fortschritte der Zoologie
Fragm. Entomol.
Fragmenta Entomologica
Fragm. Faun. (Warsaw)
Fragmenta Faunistica (Warsaw)
Funct. Ecol.
Functional Ecology
G. Sci. Nat. Econ. Palermo
Giornale di Scienze Naturali ed Economiche di Palermo

Gen. Appl. Entomol.
General and Applied Entomology
Genera Insectorum
Genera Insectorum
Genetica
Genetica
Genetics
Genetics
Genome
Genome
Gensei
Gensei
Geol. Jahrb. Hessen
Geologisches Jahrbuch Hessen
Geol. Mag.
Geological Magazine
Glas. Prir. Muz. Beogradu Ser. B Biol. Nauke
Glasnik Prirodnjackog Muzeja u Beogradu. Serija B. Bioloske Nauke
God. Sofii. Univ. Biol. Fak. Kn. 1 Zool. Fiziol. Biohim. Zhivotn.
Godishnik na Sofiiskiya Universitet. Biologicheski Fakultet. Kniga 1. Zoologiya, Fiziologiya i Biohimiya na Zhivotnite
God. Sofii. Univ. "Kliment Okhridski" Biol. Fak. Kn. 1 Zool.
Godishnik na Sofiiskiya Universitet "Kliment Okhridski". Biologicheski Fakultet. Kniga 1. Zoologiya
Gorskostop. Nauka
Gorskostopanska Nauka
Göteb. K. Vetensk. Vitterh. Samh. Handl.
Göteborgs Kungliga Vetenskaps och Vitterhets Samhälles Handlingar
Grad. Stud. Tex. Tech Univ.
Graduate Studies. Texas Tech University
Graellsia
Graellsia
Great Basin Nat.
Great Basin Naturalist
Great Lakes Entomol.
Great Lakes Entomologist
Hakone Hakubutsu
Hakone Hakubutsu
Hakubutsu Tomo
Hakubutsu no Tomo
Handb. Exp. Pharmakol.
Handbuch der Experimentellen Pharmakologie
Handb. Identif. Br. Insects
Handbooks for the Identification of British Insects
Heilige Land
Das Heilige Land
Helv. Chim. Acta
Helvetica Chimica Acta
Hereditas (Lund)
Hereditas (Lund)
Heredity
Heredity
Hexapoda (Insecta Indica)
Hexapoda (Insecta Indica)
Hilgardia
Hilgardia
Himmel Erde
Himmel und Erde
Holarct. Ecol.
Holarctic Ecology
Humboldt
Humboldt
Hyogo Biol.
Hyogo Biology
Idea
Idea
Idia
Idia
Igaku Seibutsugaku
Igaku to Seibutsugaku
Iheringia Sér. Zool.
Iheringia. Série Zoologia
Ill. Nat. Hist. Surv. Biol. Notes
Illinois Natural History Survey. Biological Notes
Indian For. Rec. Entomol. Ser.
Indian Forest Records. Entomology Series
Indian J. Entomol.
Indian Journal of Entomology
Inf. Cient. (*Univ. Nac. Asunción*)
Informes Científicos (Universidad Nacional de Asunción)
Insect Biochem.
Insect Biochemistry
Insect Cond. La.
Insect Conditions in Louisiana
Insect World
The Insect World
Insecta Helv. Fauna
Insecta Helvetica. Fauna
Insecta Matsumurana
Insecta Matsumurana

Insecta Mundi
Insecta Mundi
Insecta Norv.
Insecta Norvegiae
Insectarium
Insectarium
Insectes
Insectes
Insectes Soc.
Insectes Sociaux
Insects Samoa
Insects of Samoa
Int. Entomol. Z.
Internationale Entomologische Zeitschrift
Int. J. Insect Morphol. Embryol.
International Journal of Insect Morphology and Embryology
Int. J. Microsc. Nat. Sci.
International Journal of Microscopy and Natural Science
Int. Sci. Rev.
Internacia Scienca Revuo
Intelligenzbl. Litt.-Ztg. Erlangen
Intelligenzblatt der Litteratur-Zeitung Erlangen
Invertebr. Taxon.
Invertebrate Taxonomy
Iowa State Coll. J. Sci.
Iowa State College Journal of Science
Ir. Nat.
Irish Naturalist
Ir. Nat. J.
Irish Naturalist's Journal
Isis Oken
Isis von Oken
Isozyme Bull.
Isozyme Bulletin
Isr. J. Entomol.
Israel Journal of Entomology
Isr. J. Zool.
Israel Journal of Zoology
Istanbul Univ. Fen Fak. Mecm. Seri B
Istanbul Universitesi Fen Fakultesi Mecmuasi. Seri B
Izv. Akad. Nauk Kaz. SSR Ser. Biol.
Izvestiya Akademii Nauk Kazakhskoi SSR. Seriya Biologicheskaya
Izv. Akad. Nauk Turkm. SSR Ser. Biol. Nauk
Izvestiya Akademii Nauk Turkmenskoi SSR. Seriya Biologicheskikh Nauk
Izv. Biol. Nauchno-Issled. Inst. Permsk. Gos. Univ.
Izvestiya Biologicheskogo Nauchno-Issledovatel'skogo Instituta pri Permskom Gosudarstvennom Universitete
Izv. Bulg. Entomol. Druzh.
Izvestiya na Bulgarskoto Entomologichno Druzhestvo
Izv. Imp. Obshch. Lyubit. Estestvozn. Antropol. Etnogr. Imp. Mosk. Univ.
Izvestiya Imperatorskago Obshchestva Lyubitelei Estestvoznaniya Antropologii i Etnografii pri Imperatorskom Moskovskom Universitete
Izv. Imp. Rus. Geogr. Obshch.
Izvestiya Imperatorskago Russkago Geograficheskago Obshchestva
Izv. Imp. Tomsk. Univ.
Izvestiya Imperatorskago Tomskago Universiteta
Izv. Imp. Varsh. Univ.
Izvestiya Imperatorskago Varshavskago Universiteta
Izv. Inst. Issled. Sib.
Izvestiya Instituta Issledovaniya Sibiri
Izv. Tomsk. Gos. Univ.
Izvestiya Tomskogo Gosudarstvennogo Universiteta
Izv. Turkest. Otd. Imp. Rus. Geogr. Obshch.
Izvestiya Turkestanskago Otdela Imperatorskago Russkago Geograficheskago Obshchestva
Izv. Zool. Inst. Muz. Bulg. Akad. Nauk.
Izvestiya na Zoologicheskiya Institut s Muzej. Bulgarska Akademiya na Naukite
J. Adv. Zool.
Journal of Advanced Zoology
J. Afr. Zool.
Journal of African Zoology
J. Agric. Univ. P. R.
Journal of Agriculture of the University of Puerto Rico
J. Allergy Clin. Immunol.
Journal of Allergy and Clinical Immunology
J. Anim. Behav.
Journal of Animal Behavior
J. Anim. Ecol.
Journal of Animal Ecology
J. Anim. Morphol. Physiol.
Journal of Animal Morphology and Physiology

J. Appl. Ecol.
Journal of Applied Ecology
J. Appl. Entomol.
Journal of Applied Entomology
J. Arid Environ.
Journal of Arid Environments
J. Ariz. Acad. Sci.
Journal of the Arizona Academy of Science
J. Asiat. Soc. Bengal Part II Nat. Sci.
Journal of the Asiatic Society of Bengal. Part II. Natural Science
J. Asiat. Soc. Bengal Part II Phys. Sci.
Journal of the Asiatic Society of Bengal. Part II. Physical Science
J. Aust. Entomol. Soc.
Journal of the Australian Entomological Society
J. Beijing For. Univ. (Engl. Ed.)
Journal of Beijing Forestry University (English Edition)
J. Biogeogr.
Journal of Biogeography
J. Bombay Nat. Hist. Soc.
Journal of the Bombay Natural History Society
J. Bot.
Journal of Botany
J. Chem. Ecol.
Journal of Chemical Ecology
J. Chromatogr.
Journal of Chromatography
J. Coll. Agric. Tohoku Imp. Univ.
Journal of the College of Agriculture, Tohoku Imperial University
J. Comp. Neurol.
Journal of Comparative Neurology
J. Comp. Physiol. A. Sens. Neural Behav. Physiol.
Journal of Comparative Physiology. A. Sensory, Neural and Behavioral Physiology
J. Dep. Sci. Calcutta Univ.
Journal of the Department of Science of Calcutta University
J. Econ. Entomol.
Journal of Economic Entomology
J. Elisha Mitchell Sci. Soc.
Journal of the Elisha Mitchell Scientific Society
J. Entomol.
Journal of Entomology
J. Entomol. Res. (New Delhi)
Journal of Entomological Research (New Delhi)
J. Entomol. Sci.
Jourmal of Entomological Science
J. Entomol. Ser. A
Journal of Entomology. Series A
J. Entomol. Ser. B
Journal of Entomology. Series B
J. Entomol. Soc. Aust. (N. S. W.)
Journal of the Entomological Society of Australia (New South Wales)
J. Entomol. Soc. B. C.
Journal of the Entomological Society of British Columbia
J. Entomol. Soc. South. Afr.
Journal of the Entomological Society of Southern Africa
J. Entomol. Zool.
Journal of Entomology and Zoology
J. Ethol.
Journal of Ethology
J. Evol. Biol.
Journal of Evolutionary Biology
J. Exp. Zool.
Journal of Experimental Zoology
J. Fac. Agric. Kyushu Univ.
Journal of the Faculty of Agriculture, Kyushu University
J. Fac. Sci. Hokkaido Univ. Ser. VI. Zool.
Journal of the Faculty of Science, Hokkaido University. Series VI. Zoology
J. Ga. Entomol. Soc.
Journal of the Georgia Entomological Society
J. Hered.
Journal of Heredity
J. Hym. Res.
Journal of Hymenoptera Research
J. Insect Behav.
Journal of Insect Behavior
J. Insect Physiol.
Journal of Insect Physiology
J. Inst. Sci. Educ. (Chongju Teach. Coll.)
Journal of the Institute of Science Education (Chongju Teachers' College)
J. Kansas Entomol. Soc.
Journal of the Kansas Entomological Society
J. Linn. Soc. Lond. Zool.
Journal of the Linnean Society of London. Zoology

J. Microsc. (Paris)
Journal de Microscopie (Paris)
J. Mol. Evol.
Journal of Molecular Evolution
J. Morphol.
Journal of Morphology
J. Mus. Godeffroy
Journal des Museum Godeffroy
J. N. Y. Entomol. Soc.
Journal of the New York Entomological Society
J. Nat. Hist.
Journal of Natural History
J. Nat. Prod. (Lloydia)
Journal of Natural Products (Lloydia)
J. Nat. Sci. (Res. Inst. Basic Sci. Won-Kwang Univ.)
Journal of Natural Science (Research Institute for Basic Science, Won-Kwang University)
J. Natl. Acad. Sci. Repub. Korea Nat. Sci. Ser.
Journal of the National Academy of Sciences. Republic of Korea. Natural Sciences Series
J. Org. Chem.
Journal of Organic Chemistry
J. Proc. Asiat. Soc. Bengal
Journal and Proceedings of the Asiatic Society of Bengal
J. Proc. Entomol. Soc. Lond.
Journal and Proceedings of the Entomological Society of London
J. Proc. Linn. Soc. Lond. Zool.
Journal and Proceedings of the Linnean Society of London. Zoology
J. Proc. R. Soc. West. Aust.
Journal and Proceedings of the Royal Society of Western Australia
J. Psychol. Neurol.
Journal für Psychologie und Neurologie
J. R. Microsc. Soc. Lond.
Journal of the Royal Microscopical Society. London
J. R. Soc. West. Aust.
Journal of the Royal Society of Western Australia
J. Sapporo Otani Jr. Coll.
Journal of the Sapporo Otani Junior College
J. Sci. Math. Phys. Nat.
Jornal de Sciencias Mathematicas, Physicas e Naturaes
J. Soc. Bibliogr. Nat. Hist.
Journal of the Society for the Bibliography of Natural History
J. Soc. Br. Entomol.
Journal of the Society for British Entomology
J. Soc. Psychol.
Journal of Social Psychology
J. Southwest For. Coll.
Journal of Southwest Forestry College
J. Tenn. Acad. Sci.
Journal of the Tennessee Academy of Science
J. Theor. Biol.
Journal of Theoretical Biology
J. Toxicol. Toxin Rev.
Journal of Toxicology. Toxin Reviews
J. Trinidad Field Nat. Club
Journal of the Trinidad Field Naturalists' Club
J. UOEH (Univ. Occup. Environ. Health)
Journal of UOEH (University of Occupational and Environmental Health)
J. Wash. Acad. Sci.
Journal of the Washington Academy of Sciences
J. West China Bord. Res. Soc. Ser. B
Journal of the West China Border Research Society. Series B
J. Zhejiang Agric. Univ.
Journal of Zhejiang Agricultural University
J. Zool. (Lond.)
Journal of Zoology (London)
Jahrb. Akad. Wiss. Lit. Mainz
Jahrbuch. Akademie der Wissenschaften und der Literatur. Mainz
Jahrb. Hambg. Wiss. Anst.
Jahrbuch der Hamburgischen Wissenschaftlichen Anstalten
Jahrb. K-K. Geol. Reichsanst. Wien
Jahrbuch der Kaiserlich-Königlichen Geologischen Reichsanstalt. Wien
Jahrb. Nassau. Ver. Naturkd.
Jahrbuch des Nassauischen Vereins für Naturkunde

Jahrb. Ver. Naturkd. Herzogthum Nassau Wiesb.
Jahrbuch des Vereins für Naturkunde im Herzogthum Nassau. Wiesbaden
Jahrb. Wiss. Bot.
Jahrbuch für Wissenschaftliche Botanik
Jorn. Cienc. Nat. Resúm.
Jornadas de Ciencias Naturales. Resúmenes
Jorn. Cienc. Nat. Resúm. Comun.
Jornadas de Ciencias Naturales. Resúmenes, Comunicaciones
Jpn. J. Ecol.
Japanese Journal of Ecology
Jpn. J. Entomol.
Japanese Journal of Entomology
Jpn. J. Genet.
Japanese Journal of Genetics
K. Dan. Vidensk. Selsk. Skr.
Kongelige Danske Videnskabernes Selskabs Skrifter
K. Sven. Vetensk.-Akad. Handl.
Kongliga Svenska Vetenskaps-Akademiens Handlingar
Kagaku (Tokyo)
Kagaku (Tokyo)
Kagaku Noggo
Kagaku no Noggo
Kans. Sch. Nat.
The Kansas School Naturalist
Kansai Konchu Zasshi
Kansai Konchu Zasshi
Kat. Fauny Pol.
Katalog Fauny Polski
Khumbu Himal Ergeb. Forschungsunternehm. Nepal Himal.
Khumbu Himal. Ergebnisse des Forschungsunternehmens Nepal Himalaya
Koedoe
Koedoe
Konowia
Konowia
Kontyû
Kontyû
Kontyû Kagaku
Kontyû Kagaku
Kontyû Kenkyu
Kontyû Kenkyu
Korean J. Appl. Entomol.
Korean Journal of Applied Entomology
Korean J. Entomol.
Korean Journal of Entomology
Korean J. Plant Prot.
Korean Journal of Plant Protection
Korean J. Syst. Zool.
Korean Journal of Systematic Zoology
Korrespondenzbl. Beil. Dtsch. Entomol. Z. "Iris"
Korrespondenzblatt. Beilage zur Deutschen Entomologischen Zeitschrift "Iris"
Kosmos (Warsaw)
Kosmos (Warsaw)
Ksiegi Pamiatkowej (Lecia Gimn. IV Jana Dlugosza Lwowie)
Ksiegi Pamiatkowej (Lecia Gimnazjum IV im. Jana Dlugosza we Lwowie)
Kyodo Shizen
Kyodo no Shizen
Kyongsang Taehak Nonmunjip (J. Kyongsang Natl. Univ.)
Kyongsang Taehak Nonmunjip (Journal of Kyongsang National University)
Lammergeyer
Lammergeyer
Leafl. Univ. Colo. Mus.
Leaflet. University of Colorado Museum
Lesn. Khoz.
Lesnoe Khoziaistvo
Levende Nat.
Levende Natuur
Life Sci.
Life Sciences
Lloydia
Lloydia
MNHN (Mus. Nac. Hist. Nat.) Not. Mens.
MNHN (Museo Nacional de Historia Natural) Noticiario Mensual
Madoqua
Madoqua
Madras J. Lit. Sci.
Madras Journal of Literature and Science
Mag. Insektenkd. (Illiger)
Magazin für Insektenkunde (Illiger)
Mag. Nat. Hist.
Magazine of Natural History
Mag. Thierreichs
Magazin des Thierreichs
Magyarorsz. Allatvilága
Magyarország Allatvilága
Mainz. Naturwiss. Arch.
Mainzer Naturwissenschaftliches Archiv
Malay. Nat. J.
Malayan Nature Journal

Malpighia
Malpighia
Märk. Tierwelt
Märkische Tierwelt
Mater. Faune Gruz.
Materialy k Faune Gruzii
Mater. Kollok. Sekts. Obshchestv. Nasek. Vses. Entomol. Obshch.
Materialy Kollokviumov Sektsii Obshchestvennykh Nasekomykh Vsesoyuznogo Entomologicheskogo Obshchestva
Mauritius Inst. Bull.
Mauritius Institute Bulletin
Md. Nat.
The Maryland Naturalist
Medd. Soc. Fauna Flora Fenn.
Meddelanden af Societas pro Fauna et Flora Fennica
Melanderia
Melanderia
Mém. Acad. Sci. Inst. Fr.
Mémoires de l'Académie des Sciences de l'Institut de France
Mem. Am. Entomol. Inst.
Memoirs of the American Entomological Institute
Mem. Am. Entomol. Soc.
Memoirs of the American Entomological Society
Mem. Assoc. Australas. Paleontol.
Memoir of the Association of Australasian Paleontologists
Mem. Biogeogr. Adriat.
Memorie de Biogeografia Adriatica
Mem. Cornell Univ. Agric. Exp. Stn.
Memoirs of the Cornell University Agricultural Experiment Station
Mem. Entomol. Soc. Can.
Memoirs of the Entomological Society of Canada
Mem. Entomol. Soc. Wash.
Memoirs of the Entomological Society of Washington
Mem. Estud. Mus. Zool. Univ. Coimbra Sér. 1. Zool. Sist.
Memórias e Estudos do Museu Zoológico da Universidade de Coimbra. Série 1. Zoologia Sistemática
Mém. Inst. Egypte
Mémoires de l'Institut d'Egypte
Mém. Inst. Fr. Afr. Noire
Mémoires de l'Institut Français d'Afrique Noire
Mem. Inst. Oswaldo Cruz Rio J.
Memórias do Instituto Oswaldo Cruz. Rio de Janeiro
Mem. Kagoshima Univ. Res. Cent. S. Pac.
Memoirs of the Kagoshima University Research Center for the South Pacific
Mem. Mus. Civ. Stor. Nat. Verona
Memorie del Museo Civico di Storia Naturale di Verona
Mem. Mus. Civ. Stor. Nat. Verona (II Ser.) Sez. Sci. Vita (A Biol.)
Memorie del Museo Civico di Storia Naturale di Verona (II Serie). Sezione Scienze della Vita (A: Biologica).
Mem. Mus. Entre Ríos
Memorias del Museo de Entre Ríos
Mém. Mus. R. Hist. Nat. Belg.
Mémoires du Musée Royal d'Histoire Naturelle de Belgique
Mem. Natl. Mus. Vic.
Memoirs of the National Museum of Victoria
Mem. Natl. Sci. Mus. (Tokyo)
Memoirs of the National Science Museum (Tokyo)
Mem. Proc. Manch. Lit. Philos. Soc.
Memoirs and Proceedings of the Manchester Literary and Philosophical Society
Mem. Qld. Mus.
Memoirs of the Queensland Museum
Mem. R. Accad. Sci. Ist. Bologna
Memorie della Reale Accademia delle Scienze dell'Istituto di Bologna
Mem. R. Accad. Sci. Torino
Memorie della Reale Accademia delle Scienze di Torino
Mem. R. Soc. Esp. Hist. Nat.
Memorias de la Real Sociedad Española de Historia Natural
Mem. Rev. Soc. Cient. "Antonio Alzate"
Memorias y Revista de la Sociedad Científica "Antonio Alzate"
Mém. Soc. Acad. Archéol. Sci. Arts Dép. Oise
Mémoires de la Société Académique d'Archéologie, Sciences et Arts du Département de l'Oise

Mem. Soc. Cuba. Hist. Nat. "Felipe Poey"
Memorias de la Sociedad Cubana de Historia Natural "Felipe Poey"
Mém. Soc. Entomol. Belg.
Mémoires de la Société Entomologique de Belgique
Mem. Soc. Entomol. Ital.
Memorie della Società Entomologica Italiana
Mém. Soc. Entomol. Qué.
Mémoires de la Société Entomologique du Québec
Mém. Soc. Hist. Nat. Afr. Nord
Mémoires de la Société d'Histoire Naturelle de l'Afrique du Nord
Mém. Soc. Linn. Nord Fr.
Mémoires de la Société Linnéenne de Nord de la France
Mém. Soc. Neuchâtel. Sci. Nat.
Mémoires de la Société Neuchâteloise des Sciences Naturelles
Mém. Soc. Port. Sci. Nat. Sér. Zool.
Mémoires publiés par la Société Portugaise des Sciences Naturelles. Série Zoologique
Mém. Soc. R. Sci. Liège
Mémoires de la Société Royale des Sciences de Liège
Mém. Soc. Vaudoise Sci. Nat.
Mémoires de la Société Vaudoise des Sciences Naturelles
Mém. Soc. Zool. Fr.
Mémoires de la Société Zoologique de France
Memo. Soc. Fauna Flora Fenn.
Memoranda Societatis pro Fauna et Flora Fennica
Memorabilia Zool.
Memorabilia Zoologica
Merkbl. Waldhyg.
Merkblätter zur Waldhygiene
Misc. Faun. Helv.
Miscellanea Faunistica Helvetiae
Misc. Inst. Miguel Lillo
Miscelánea. Instituto Miguel Lillo
Misc. Publ. Dep. Agric. N. S. W.
Miscellaneous Publications of the Department of Agriculture, New South Wales
Misc. Publ. Okla. Agric. Exp. Stn.
Miscellaneous Publication. Oklahoma Agricultural Experimental Station
Misc. Zool. (Barc.)
Miscellania Zoologica (Barcelona)
Misc. Zool. (Havana)
Miscelánea Zoologica (Havana)
Misc. Zool. Sumatr.
Miscellanea Zoologica Sumatrana
Mission Rohan-Chabot
Mission Rohan-Chabot
Mission Sci. Omo
Mission Scientifique de l'Omo
Missione Biol. Paese Borana
Missione Biologica nel Paese dei Borana
Mitt. Arbeitsgem. Naturwiss. Sibiu-Hermannstadt
Mitteilungen der Arbeitsgemeinschaft für Naturwissenschaften. Sibiu-Hermannstadt
Mitt. Bad. Landesver. Naturkd. Freibg. Breisgau
Mitteilungen des Badischen Landesvereins für Naturkunde e. V. Freiburg im Breisgau
Mitt. Bad. Landesver. Naturkd. Naturschutz Freibg. Breisgau
Mitteilungen des Badischen Landesvereins für Naturkunde und Naturschutz e. V. Freiburg im Breisgau
Mitt. Beuthen. Gesch.-Museumsver.
Mitteilungen des Beuthener Geschichts- und Museumsverein
Mitt. Dtsch. Entomol. Ges.
Mitteilungen der Deutschen Entomologischen Gesellschaft
Mitt. Dtsch. Ges. Allg. Angew. Entomol.
Mitteilungen der Deutschen Gesellschaft für Allgemeine und Angewandte Entomologie
Mitt. Entomol. Ges. Basel
Mitteilungen aus der Entomologischen Gesellschaft, Basel
Mitt. Ges. Morphol. Physiol. Münch.
Mitteilungen der Gesellschaft für Morphologie und Physiologie in München
Mitt. Hambg. Zool. Mus. Inst.
Mitteilungen aus dem Hamburgischen Zoologischen Museum und Institut
Mitt. Hermann Göring Akad. Dtsch. Forstwiss.
Mitteilungen der Hermann Göring Akademie der Deutschen Forstwissenschaften
Mitt. Münch. Entomol. Ver.
Mitteilungen der Münchener Entomologischen Verein

Mitt. Naturforsch. Ges. Soloth.
Mitteilungen der Naturforschenden Gesellschaft in Solothurn
Mitt. Naturhist. Mus. Hambg.
Mitteilungen aus dem Naturhistorischen Museum in Hamburg
Mitt. Philomath. Ges. Elsass-Lothringen
Mitteilungen der Philomathischen Gesellschaft in Elsass-Lothringen
Mitt. Schweiz. Entomol. Ges.
Mitteilungen der Schweizerischen Entomologischen Gesellschaft
Mitt. Zool. Mus. Berl.
Mitteilungen aus dem Zoologischen Museum in Berlin
Mol. Ecol.
Molecular Ecology
Mol. Phylogenet. Evol.
Molecular Phylogenetics and Evolution
Mon. Microsc. J.
Monthly Microscopical Journal
Monatsber. K. Preuss. Akad. Wiss. Berl.
Monatsberichte der Königlichen Preussischen Akademie der Wissenschaften zu Berlin
Monist
The Monist
Monit. Zool. Ital.
Monitore Zoologico Italiano
Monogr. Biol.
Monographiae Biologicae
Monogr. Inst. Sci. Technol. Manila
Monographs of the Institute of Science and Technology. Manila
Monografies (Diput. Barc.)
Monografies (Diputació de Barcelona)
Mushi
Mushi
Muz. Soucas. Rada Prírodoved.
Muzeum a Soucasnost. Rada Prírodovedná
Myrmekol. Listy
Myrmekologické Listy
N. D. Hist. Q.
North Dakota Historical Quarterly
N. Z. Entomol.
New Zealand Entomologist
N. Z. J. Sci.
New Zealand Journal of Science
N. Z. Nat. Herit.
New Zealand's Nature Heritage
Nat. Belg.
Naturalistes Belges
Nat. Can. (Qué.)
Naturaliste Canadien (Québec)
Nat. Hist. Mus. Los Angel. Cty. Sci. Bull.
Natural History Museum of Los Angeles County. Science Bulletin
Nat. Hist. Rennell Isl. Br. Solomon Isl.
Natural History of Rennell Island, British Solomon Islands
Nat. Hist. Res. Spec. Issue
Natural History Research Special Issue
Nat. Insects (Konchu Shizen)
Nature and Insects (Konchu to Shizen)
Nat. Mus. (Århus)
Natur og Museum (Århus)
Nat. Norr
Natur i Norr
Nat. Offenbarung
Natur und Offenbarung
Nat. Sicil.
Il Naturalista Siciliano
Natl. Geogr. Mag.
National Geographic Magazine
Natl. Mus. N. Z. Misc. Ser.
National Museum of New Zealand Miscellaneous Series
Natur- Landschaftsschutzgeb. Baden-Württ.
Natur- und Landschaftsschutzgebiete in Baden-Württemberg
Natura (Amst.)
Natura (Amsterdam)
Natura (Milan)
Natura (Milan)
Naturaleza (Méx.)
Naturaleza (México)
Naturalia (São Paulo)
Naturalia (São Paulo)
Naturalist (Leeds)
The Naturalist (Leeds)
Naturaliste
Naturaliste
Nature (Lond.)
Nature (London)
Nature (Paris)
Nature (Paris)
Naturkd. Jahrb. Stadt Linz
Naturkundliches Jahrbuch der Stadt Linz
Naturschutz
Naturschutz
Naturwiss. Rundsch.
Naturwissenschaftliche Rundschau

Naturwiss. Ver. Darmstadt Ber.
Naturwissenschaftlicher Verein Darmstadt. Bericht
Naturwiss. Wochenschr.
Naturwissenschaftliche Wochenschrift
Naturwissenschaften
Naturwissenschaften
Natuurhist. Maandbl.
Natuurhistorisch Maandblad
Natuurwet. Studiekring Suriname Ned. Antillen
Natuurwetenschappelijke Studiekring voor Suriname en de Nederlandse Antillen
Nauchn. Dokl. Vyssh. Shk. Biol. Nauki
Nauchnye Doklady Vysshei Shkoly. Biologicheskie Nauki
Ned. Bosb. Tijdschr.
Nederlands Bosbouw Tijdschrift
Neotropica (La Plata)
Neotropica (La Plata)
Neth. J. Zool.
Netherlands Journal of Zoology
Neue Beitr. Syst. Insektenkd.
Neue Beiträge zur Systematischen Insektenkunde
Neue Denkschr. Allg. Schweiz. Ges. Gesammten Naturwiss.
Neue Denkschriften der Allgemeinen Schweizerischen Gesellschaft für die Gesammten Naturwissenschaften
Neues Jahrb. Miner. Geol. Paläontol.
Neues Jahrbuch für Mineralogie, Geologie und Paläontologie
Neujahrsbl. Naturforsch. Ges. Zür.
Neujahrsblatt. Naturforschende Gesellschaft in Zürich
News Bull. Aust. Entomol. Soc.
News Bulletin. Australian Entomological Society
Newsl. Zool. Surv. India
Newsletter. Zoological Survey of India
Niger. Entomol. Mag.
Nigerian Entomologist's Magazine
Nor. Entomol. Tidsskr.
Norsk Entomologisk Tidsskrift
North Qld. Nat.
North Queensland Naturalist
North West. Nat.
North Western Naturalist
Norw. J. Entomol.
Norwegian Journal of Entomology
Not. Biol. Bucur.
Notationes Biologicae. Bucuresti
Not. Entomol.
Notulae Entomologicae
Not. For. Mont.
Notiziario Forestale e Montano
Notas Mus. La Plata (Zool.)
Notas del Museo de La Plata (Zoología)
Notes Leyden Mus.
Notes from the Leyden Museum
Notes R. Bot. Gard. Edinb.
Notes from the Royal Botanic Garden, Edinburgh
Notizen Geb. Natur- Heilkd.
Notizen aus dem Gebiete der Natur- und Heilkunde
Nouv. Mém. Soc. Imp. Nat. Mosc.
Nouveaux Mémoires de la Société Impériale des Naturalistes de Moscou
Nouv. Rev. Entomol.
Nouvelle Revue d'Entomologie
Nova Acta Acad. Caesareae Leopold.-Carol. Ger. Nat. Curiosorum
Nova Acta Academiae Caesareae Leopoldino-Carolinae Germanicae Naturae Curiosorum
Nova Acta Leopold.
Nova Acta Leopoldina
Nova Caled. A Zool.
Nova Caledonia. A. Zoologie
Nova Guin.
Nova Guinea
Nucleus (Calcutta)
Nucleus (Calcutta)
Nunquam Otiosus (Dres.)
Nunquam Otiosus (Dresden)
Nuova Antol. Sci. Lett. Arti
Nuova Antologia di Scienze, Lettere ed Arti
Nymphaea
Nymphaea
Nyt Mag. Naturvidensk.
Nyt Magazin for Naturvidenskaberne
Nytt Mag. Naturvidensk.
Nytt Magasin for Naturvidenskapene
Occas. Pap. Bernice P. Bishop Mus.
Occasional Papers of the Bernice Pauahi Bishop Museum
Occas. Pap. Biol. Soc. Nev.
Occasional Papers. Biological Society of Nevada

Occas. Pap. Boston Soc. Nat. Hist.
Occasional Papers of the Boston Society of Natural History
Occas. Pap. Calif. Acad. Sci.
Occasional Papers of the California Academy of Sciences
Occas. Pap. Mus. Zool. Univ. Mich.
Occasional Papers of the Museum of Zoology, University of Michigan
Occas. Pap. Natl. Mus. South. Rhod.
Occasional Papers of the National Museum of Southern Rhodesia
Occas. Pap. Natl. Mus. South. Rhod. B. Nat. Sci.
Occasional Papers of the National Museum of Southern Rhodesia. B. Natural Sciences
Oecologia (Berl.)
Oecologia (Berlin)
Oesterr. Rev.
Oesterreichische Revue
Öfvers. Fin. Vetensk.-Soc. Förh.
Öfversigt af Finska Vetenskaps-Societetens Förhandlingar
Öfvers. K. Vetensk.-Akad. Förh. Stockh.
Öfversigt af Kongliga Vetenskaps-Akademiens Förhandlingar. Stockholm
Ohio J. Sci.
Ohio Journal of Science
Oikos
Oikos
Opin. Declar. Int. Comm. Zool. Nomencl.
Opinions and Declarations rendered by the International Commission on Zoological Nomenclature
Opredeliteli Faune SSSR
Opredeliteli po Faune SSSR
Opusc. Entomol.
Opuscula Entomologica
Opusc. Inst. Sci. Indoch.
Opuscules de l'Institut Scientifique de l'Indochine
Opusc. Zool. München.
Opuscula Zoologica. München
Orient. Insects
Oriental Insects
Orsis (*Org. Sist.*)
Orsis (Organismes i Sistemes)
Osaka Shokubutsu Boeki
Osaka Shokubutsu Boeki
Österr. Zool. Z.
Österreichische Zoologische Zeitschrift
Overseas Dev. Nat. Resour. Inst. Bull.
Overseas Development Natural Resources Institute Bulletin
P. N. G. J. Agric. For. Fish.
Papua New Guinea Journal of Agriculture, Forestry and Fisheries
Pac. Insects
Pacific Insects
Pac. Insects Monogr.
Pacific Insects Monograph
Pac. Sci.
Pacific Science
Palaeogeogr. Palaeoclimatol. Palaeoecol.
Palaeogeography, Palaeoclimatology, Palaeoecology
Paleobiol.
Paleobiology
Paleontol. J.
Paleontological Journal
Paleontol. Zh.
Paleontologicheskii Zhurnal
Pan-Pac. Entomol.
Pan-Pacific Entomologist
Pap. Avulsos Zool. (São Paulo)
Papeis Avulsos de Zoologia (São Paulo)
Pap. Peabody Mus. Archaeol. Ethnol. Harv. Univ.
Papers of the Peabody Museum of Archaeology and Ethnology, Harvard University
Pedobiologia
Pedobiologia
Peking Nat. Hist. Bull.
Peking Natural History Bulletin
Pest Artic. News Summ.
Pest Articles & News Summaries
Pfälz. Heim.
Pfälzer Heimat
Pflanzenschutz Ber.
Pflanzenschutz Berichte
Philipp. J. Sci.
Philippine Journal of Science
Philipp. J. Sci. Sect. D. Gen. Biol. Ethnol. Anthropol.
Philippine Journal of Science. Section D. General Biology, Ethnology and Anthropology
Philos. Trans. R. Soc. Lond.
Philosophical Transactions of the Royal Society of London

Philos. Trans. R. Soc. Lond. B Biol. Sci.
Philosophical Transactions of the Royal Society of London. B. Biological Sciences
Physiol. Entomol.
Physiological Entomology
Physiol. Zool.
Physiological Zoology
Physis (B. Aires)
Physis (Buenos Aires)
Phytophylactica
Phytophylactica
Pilot Regist. Zool.
Pilot Register of Zoology
Poeyana Inst. Ecol. Sist. Acad. Cienc. Cuba
Poeyana. Instituto de Ecología y Sistemática, Academia de Ciencias de Cuba
Poeyana Inst. Zool. Acad. Cienc. Cuba
Poeyana. Instituto de Zoología, Academia de Ciencias de Cuba
Pol. Ecol. Stud.
Polish Ecological Studies
Pol. Pismo Entomol.
Polskie Pismo Entomologiczne
Pop. Sci. Mon.
Popular Science Monthly
Pr. Kom. Nauk Roln. Kom. Nauk Lesn.
Prace Komisji Nauk Rolniczych i Komisji Nauk Lesnych
Pr. Tow. Przyj. Nauk Wilnie Wydz. Nauk Mat. Przyr.
Prace Towarzystwa Przyjaciól Nauk w Wilnie. Wydzial Nauk Matematycznych i Przyrodniczych
Prairie Nat.
Prairie Naturalist
Prakt. Schädlingsbekämpfer
Der Praktische Schädlingsbekämpfer
Príroda (Brno)
Príroda (Brno)
Priroda (Mosc.)
Priroda (Moscow)
Probl. Attuali Sci. Cult.
Problemi Attuali di Scienza e di Cultura
Proc. Acad. Nat. Sci. Phila.
Proceedings of the Academy of Natural Sciences of Philadelphia
Proc. Acadian Entomol. Soc.
Proceedings of the Acadian Entomological Society
Proc. Am. Acad. Arts Sci.
Proceedings of the American Academy of Arts and Sciences
Proc. Am. Philos. Soc.
Proceedings of the American Philosophical Society
Proc. Biol. Soc. Wash.
Proceedings of the Biological Society of Washington
Proc. Calif. Acad. Sci.
Proceedings of the California Academy of Sciences
Proc. Ecol. Soc. Aust.
Proceedings of the Ecological Society of Australia
Proc. Entomol. Soc. B. C.
Proceedings of the Entomological Society of British Columbia
Proc. Entomol. Soc. Lond.
Proceedings of the Entomological Society of London
Proc. Entomol. Soc. Manit.
Proceedings of the Entomological Society of Manitoba
Proc. Entomol. Soc. Phila.
Proceedings of the Entomological Society of Philadelphia
Proc. Entomol. Soc. Wash.
Proceedings of the Entomological Society of Washington
Proc. Essex Inst. (Commun.)
Proceedings of the Essex Institute (Communications)
Proc. Hawaii. Entomol. Soc.
Proceedings of the Hawaiian Entomological Society
Proc. Indian Sci. Congr.
Proceedings of the Indian Science Congress
Proc. Indiana Acad. Sci.
Proceedings of the Indiana Academy of Science
Proc. Iowa Acad. Sci.
Proceedings of the Iowa Academy of Science
Proc. Isle Wight Nat. Hist. Archaeol. Soc.
Proceedings of the Isle of Wight Natural History and Archaeological Society
Proc. Jpn. Soc. Syst. Zool.
Proceedings of the Japanese Society of Systematic Zoology

Proc. Linn. Soc. N. S. W.
Proceedings of the Linnean Society of New South Wales
Proc. Malacol. Soc.
Proceedings of the Malacological Society
Proc. Manch. Lit. Philos. Soc.
Proceedings of the Manchester Literary and Philosophical Society
Proc. N. D. Acad. Sci.
Proceedings of the North Dakota Academy of Science
Proc. N. Engl. Zool. Club
Proceedings of the New England Zoological Club
Proc. Natl. Acad. Sci. U. S. A.
Proceedings of the National Academy of Sciences of the United States of America
Proc. North Cent. Branch Entomol. Soc. Am.
Proceedings of the North Central Branch. Entomological Society of America
Proc. R. Entomol. Soc. Lond. Ser. A
Proceedings of the Royal Entomological Society of London. Series A
Proc. R. Entomol. Soc. Lond. Ser. B
Proceedings of the Royal Entomological Society of London. Series B
Proc. R. Ir. Acad. Sect. B
Proceedings of the Royal Irish Academy. Section B
Proc. R. Soc. Qld.
Proceedings of the Royal Society of Queensland
Proc. R. Soc. Vic.
Proceedings of the Royal Society of Victoria
Proc. Sect. Exp. Appl. Entomol. Neth. Entomol. Soc.
Proceedings of the Section Experimental and Applied Entomology of the Netherlands Entomological Society
Proc. Sect. Sci. K. Akad. Wet. Amst.
Proceedings of the Section of Sciences. Koninklijke Akademie van Wetenschappen te Amsterdam
Proc. Somerset. Archaeol. Nat. Hist. Soc.
Proceedings of the Somersetshire Archaeological and Natural History Society
Proc. Trans. Rhod. Sci. Assoc.
Proceedings and Transactions of the Rhodesia Scientific Association
Proc. Trans. South Lond. Entomol. Nat. Hist. Soc.
Proceedings and Transactions of the South London Entomological and Natural History Society
Proc. U. S. Natl. Mus.
Proceedings of the United States National Museum
Proc. Utah Acad. Sci. Arts Lett.
Proceedings of the Utah Academy of Sciences, Arts and Letters
Proc. Wash. Acad. Sci.
Proceedings of the Washington Academy of Sciences
Proc. Zool. Soc. Lond.
Proceedings of the Zoological Society of London
Proc. Zool. Soc. Lond. Ser. A
Proceedings of the Zoological Society of London. Series A.
Programm Städt. Oberrealsch. Pesth
Programm der Städtischen Oberrealschule in Pesth
Protok. Obshch. Estestvoispyt. Imp. Kazan. Univ.
Protokoly Obshchestva Estestvoispytatelei pri Imperatorskom Kazanskom Universitete
Przegl. Zool.
Przeglad Zoologiczny
Psyche (Camb.)
Psyche (Cambridge)
Pubbl. Ist. Entomol. Agrar. Univ. Pavia
Pubblicazioni dell'Istituto di Entomologia Agraria dell'Università di Pavia
Pubbl. Stn. Zool. Napoli
Pubblicazioni della Stazione Zoologica di Napoli
Publ. Avulsas Mus. Nac. Rio J.
Publicações Avulsas. Museu Nacional. Rio de Janeiro
Publ. Cult. Cia. Diam. Angola
Publicações Culturais da Companhia de Diamantes de Angola
Publ. Dep. Zool. (Barc.)
Publicaciones del Departamento de Zoología (Barcelona)
Publ. Inst. Estud. Almer. Bol. (*Cienc.*)
Publicaciones del Instituto de Estudios Almerienses. Boletín (Ciencias)

Publ. Natuurhist. Genoot. Limburg
Publicatiës van het Natuurhistorisch Genootschap in Limburg
Publ. Univ. Liban. Sect. Sci. Nat.
Publications de l'Université Libanaise. Section des Sciences Naturelles
Purdue Univ. Agric. Exp. Stn. Res. Bull.
Purdue University Agricultural Experiment Station. Research Bulletin
Q. J. Geol. Soc. Lond.
Quarterly Journal of the Geological Society of London
Q. J. Microsc. Sci.
Quarterly Journal of Microscopical Science
Q. Rev. Biol.
Quarterly Review of Biology
Qld. Agric. J.
Queensland Agricultural Journal
Qld. J. Agric. Sci.
Queensland Journal of Agricultural Science
Quaest. Entomol.
Quaestiones Entomologicae
Quat. Res. (N. Y.)
Quaternary Research (New York)
Quid (Teresina)
Quid (Teresina)
Rapp. P.-V. Réun. Comm. Int. Explor. Sci. Mer Méditerr. Monaco
Rappports et Procès-Verbaux des Réunions de la Commission Internationale pour l'Exploration Scientifique de la Mer Méditerranée. Monaco
Rass. Soc. Alp. Giulie
Rassegna della Società Alpina delle Giulie
Rec. Auckl. Inst. Mus.
Records of the Auckland Institute and Museum
Rec. Canterbury Mus.
Records of the Canterbury Museum
Rec. Indian Mus.
Records of the Indian Museum
Rec. Zool. Surv. India
Records of the Zoological Survey of India
Rech. Agron.
Recherches Agronomiques
Recl. Zool. Suisse
Recueil Zoologique Suisse
Reclam. Reveg. Res.
Reclamation and Revegetation Research
Redia
Redia
Regist. Trimest. Colecc. Mem. Hist. Lit. Cienc. Artes
Registro Trimestre; o Colección de Memorias de Historia, Literatura, Ciencias y Artes
Reichenbachia
Reichenbachia
Rend. Semin. Fac. Sci. Univ. Cagliari
Rendiconti del Seminario della Facoltà di Scienze dell'Università di Cagliari
Rend. Sess. R. Accad. Sci. Ist. Bologna
Rendiconti delle Sessioni della Reale Accademia delle Scienze dell'Istituto di Bologna
Rend. Sess. R. Accad. Sci. Ist. Bologna Cl. Sci. Fis.
Rendiconti delle Sessioni della Reale Accademia delle Scienze dell'Istituto di Bologna. Classe di Scienze Fisiche
Rend. Unione Zool. Ital.
Rendiconti. Unione Zoologica Italiana
Rep. Entomol. U. S. Dep. Agric.
Report of the Entomologist. United States Department of Agriculture
Rep. Fac. Sci. Kagoshima Univ. (*Earth Sci. Biol.*)
Reports of the Faculty of Science, Kagoshima University (Earth Sciences and Biology)
Rep. First Sci. Exped. Manchoukuo
Report of the First Scientific Expedition to Manchoukuo
Rep. Invest. Inst. Ecol. Sist. (*Acad. Cienc. Cuba*)
Reporte de Investigación del Instituto de Ecología y Sistemática (Academia de Ciencias de Cuba)
Rep. Prog. Geol. Surv. Can.
Report on Progress. Geological Survey of Canada
Rep. Ser. Univ. Arkansas Agric. Exp. Stn.
Report Series. University of Arkansas Agricultural Experiment Station
Rep. Trans. Devon. Assoc. Adv. Sci. Lit. Art
Report and Transactions of the Devonshire Association for the Advancement of Science, Literature and Art
Res. Bull. Coll. Exp. For. Hokkaido Univ.
Research Bulletin of the College Experiment Forests, Hokkaido University

Res. J. Univ. Wyo. Agric. Exp. Stn.
Research Journal. University of Wyoming Agricultural Experiment Station
Restor. Ecol.
Restoration Ecology
Rev. Agric. (Piracicaba)
Revista de Agricultura (Piracicaba)
Rev. Agric. Cochabamba
Revista de Agricultura. Cochabamba
Rev. Asoc. Ing. Agrón. Montev.
Revista de la Asociación de Ingenieros Agrónomos. Montevideo
Rev. Biol. Acad. Répub. Pop. Roum.
Revue Biologie. Académie de la République Populaire Roumaine
Rev. Biol. Trop.
Revista de Biología Tropical
Rev. Biol. Urug.
Revista de Biología del Uruguay
Rev. Bras. Biol.
Revista Brasileira de Biologia
Rev. Bras. Entomol.
Revista Brasileira de Entomologia
Rev. Ceres
Revista Ceres
Rev. Chil. Entomol.
Revista Chilena de Entomología
Rev. Chil. Hist. Nat.
Revista Chilena de Historia Natural
Rev. Colomb. Entomol.
Revista Colombiana de Entomología
Rev. Ecuat. Hig. Med. Trop.
Revista Ecuatoriana de Higiene y Medicina Tropical
Rev. Écol. Biol. Sol
Revue d'Écologie et de Biologie du Sol
Rev. Écol. Terre Vie
Revue d'Écologie. La Terre et la Vie
Rev. Entomol. (Caen)
Revue d'Entomologie (Caen)
Rev. Entomol. (Rio J.)
Revista de Entomologia (Rio de Janeiro)
Rev. Entomol. Qué.
Revue d'Entomologie du Québec
Rev. Fac. Agron. Univ. Nac. La Pampa
Revista de la Facultad de Agronomía. Universidad Nacional de La Pampa
Rev. Fac. Humanid. Cienc. Ser. Cienc. Biol.
Revista de la Facultad de Humanidades y Ciencias. Serie Ciencias Biologicas
Rev. Fac. Sci. Tunis
Revue de la Faculté des Sciences de Tunis
Rev. Fr. Entomol.
Revue Française d'Entomologie
Rev. Fr. Entomol. (Nouv. Sér.)
Revue Française d'Entomologie (Nouvelle Série)
Rev. Gén. Sci. Pures Appl.
Revue Générale des Sciences Pures et Appliquées
Rev. Ibér. Parasitol.
Revista Ibérica de Parasitología
Rev. Jard. Bot. Mus. Hist. Nat. Parag.
Revista del Jardin Botánico y Museo de Historia Natural del Paraguay
Rev. Mag. Zool. Pure Appl.
Revue et Magasin de Zoologie Pure et Appliquée
Rev. Mus. La Plata
Revista del Museo de La Plata
Rev. Mus. Paul.
Revista do Museu Paulista
Rev. Mus. Pop. Paraná
Revista del Museo Popular de Paraná
Rev. Nicar. Entomol.
Revista Nicaraguense de Entomología
Rev. Path. Vég. Entomol. Agric. Fr.
Revue de Pathologie Végétale et d'Entomologie Agricole de France
Rev. Peru. Entomol.
Revista Peruana de Entomología
Rev. Sci. (Paris)
Revue Scientifique (Paris)
Rev. Sci. Nat. Auvergne
Revue des Sciences Naturelles d'Auvergne
Rev. Soc. Entomol. Argent.
Revista de la Sociedad Entomológica Argentina
Rev. Soc. Urug. Entomol.
Revista de la Sociedad Uruguaya de Entomología
Rev. Suisse Zool.
Revue Suisse de Zoologie
Rev. Zool. Afr. (Bruss.)
Revue Zoologique Africaine (Brussels)
Rev. Zool. Afr. (Tervuren)
Revue de Zoologie Africaine (Tervuren)
Rev. Zool. Bot. Afr.
Revue de Zoologie et de Botanique Africaines
Rigakukai
Rigakukai

Rikusuigaku Zasshi (*Jpn. J. Limnol.*)
Rikusuigaku Zasshi (Japanese Journal of Limnology)
Riv. Biol. (Rome)
Rivista di Biologia (Rome)
Riv. Biol. Colon.
Rivista di Biologia Coloniale
Riv. Biol. Gen.
Rivista de Biologia Generale
Riv. Sci. Biol.
Rivista di Scienze Biologiche
Riviera Sci.
Riviera Scientifique
Roc. Cesk. Spol. Entomol.
Rocenka Ceskoslovenské Spolecnosti Entomologické
Rocz. Nauk Lesn.
Roczniki Nauk Lesnych
Rovartani Lapok
Rovartani Lapok
Rozpr. Wiad. Muz. Dziedusz. Lwów
Rozprawy i Wiadomosci z Muzeum im. Dzieduszyckich Lwów
Rus. Entomol. Obozr.
Russkoe Entomologicheskoe Obozrenie
Rus. Zool. Zh.
Russkii Zoologicheskii Zhurnal
S. Afr. Dep. Agric. Tech. Serv. Tech. Commun.
South Africa Department of Agricultural Technical Services. Technical Communications
S. Afr. J. Agric. Sci.
South African Journal of Agricultural Science
S. Afr. J. Sci.
South African Journal of Science
S. Afr. J. Zool.
South African Journal of Zoology
S. Aust. Nat.
The South Australian Naturalist
Saishû Shiiku
Saishû to Shiiku
Sb. Entomol. Oddel. Nár. Mus. Praze
Sborník Entomologického Oddeleni Národního Musea v Praze
Sb. Klubu Prírodoved. Brne
Sborník Klubu Prírodovedeckého v Brne
Sb. Prírodoved. Klubu Trebici
Sborník Prírodovedeckého Klubu v Trebíci
Sb. Slov. Nár. Múz. Prír. Vedy
Sborník Slovenského Národného Múzea. Prírodné Vedy
Sb. Tr. Zool. Muz. (Kiev)
Sbornik Trudov Zoologicheskogo Muzeya (Kiev)
Schr. Naturforsch. Ges. Danzig
Schriften der Naturforschenden Gesellschaft in Danzig
Schr. Phys.-Ökon. Ges. Königsb.
Schriften der Physikalisch-Ökonomischen Gesellschaft zu Königsberg
Schriftenr. Bayer. Landesamt Umweltschutz
Schriftenreihe Bayerisches Landesamt für Umweltschutz
Schweiz. Entomol. Anz.
Schweizer Entomologischer Anzeiger
Schweiz. Z. Forstwes.
Schweizerische Zeitschrift für Forstwesen
Sci. Am.
Scientific American
Sci. Cult.
Science and Culture
Sci. Gerundensis
Scientia Gerundensis
Sci. Mon.
The Scientific Monthly
Sci. Rep. Fac. Educ. Gifu Univ. (*Nat. Sci.*)
Science Report of the Faculty of Education, Gifu University (Natural Science)
Sci. Rep. Tokyo Kyoiku Daigaku Sect. B
Science Reports of the Tokyo Kyoiku Daigaku. Section B
Sci. Silvae Sin.
Scientia Silvae Sinicae
Sci. Treatise Syst. Evol. Zool.
Scientific Treatise on Systematic and Evolutionary Zoology
Science (N. Y.)
Science (New York)
Science (Wash. D. C.)
Science (Washington, D. C.)
Scott. Nat.
Scottish Naturalist
Search (Sydney)
Search (Sydney)
Search Agric. (Ithaca N. Y.)
Search. Agriculture (Ithaca, New York)
Seikô
Seikô

Senckenb. Biol.
Senckenbergiana Biologica
Senckenb. Lethaea
Senckenbergiana Lethaea
Senckenbergiana
Senckenbergiana
Seoul Natl. Univ. Coll. Agric. Bull.
Seoul National University College of Agriculture Bulletin
Ser. Biol. Acad. Cienc. Cuba
Serie Biológica. Academia de Ciencias de Cuba
Ser. Entomol. (Dordr.)
Series Entomologica (Dordrecht)
Sesoko Mar. Sci. Lab. Tech. Rep.
Sesoko Marine Science Laboratory Technical Report
Sess. Entomol. ICHN-SCL
Sessió d'Entomologia ICHN-SCL
Severocesk. Prír. Príl.
Severoceskou Prírodou. Príloha
Shire Nat. Hist. Ser.
Shire Natural History Series
Shiseki Meisho Tennen Kinenbutsu
Shiseki Meisho Tennen Kinenbutsu
Sinozoologia
Sinozoologia
Sitzungsber. Abh. Naturwiss. Ges. "Isis" Dres.
Sitzungsberichte und Abhandlungen der Naturwissenschaftlichen Gesellschaft "Isis" in Dresden
Sitzungsber. Akad. Wiss. Wien Math.-Naturwiss. Kl. Abt. I
Sitzungsberichte der Akademie der Wissenschaften in Wien. Mathematisch-Naturwissenschaftliche Klasse. Abteilung I
Sitzungsber. Ges. Morphol. Physiol. Münch.
Sitzungsberichte der Gesellschaft für Morphologie und Physiologie. München
Sitzungsber. Ges. Naturforsch. Freunde Berl.
Sitzungsberichte der Gesellschaft Naturforschender Freunde zu Berlin
Sitzungsber. Kais. Akad. Wiss. Wien Math.-Naturwiss. Cl. [Kl.] Abt. I
Sitzungsberichte der Kaiserlichen Akademie der Wissenschaften in Wien. Mathematisch-Naturwissenschaftliche Classe [Klasse]. Abteilung I
Sitzungsber. Math.-Phys. Kl. K. Bayer. Akad. Wiss. Münch.
Sitzungsberichte der Mathematischen-Physikalischen Klasse der Königlich Bayerischen Akademie der Wissenschaften zu München
Sitzungsber. Naturforsch. Ges. Leipz.
Sitzungsberichte der Naturforschenden Gesellschaft zu Leipzig
Sitzungsber. Österr. Akad. Wiss. Math.-Naturwiss. Kl. Abt. I
Sitzungsberichte der Österreichischen Akademie der Wissenschaften. Mathematisch-Naturwissenschaftliche Klasse. Abteilung I
Smithson. Contrib. Bot.
Smithsonian Contributions to Botany
Smithson. Misc. Collect.
Smithsonian Miscellaneous Collections
Soc. Entomol. Stuttg.
Societas Entomologica. Stuttgart
Soc. Veneziana Sci. Nat. Lav.
Società Veneziana di Scienze Naturali. Lavori
Sociobiology
Sociobiology
Soil Biol. Biochem.
Soil Biology and Biochemistry
Soobshch. Akad. Nauk Gruz. SSR
Soobshcheniya Akademii Nauk Gruzinskoi SSR
Southwest. Entomol.
Southwestern Entomologist
Southwest. Nat.
Southwestern Naturalist
Sov. Geol.
Sovetskaya Geologiya
Sov. J. Ecol.
Soviet Journal of Ecology
Spec. Bull. Lepid. Soc. Jpn.
Special Bulletin of [the] Lepidopterological Society of Japan
Spec. Publ. Mus. Tex. Tech Univ.
Special Publications, the Museum. Texas Tech University
Speleon
Speleon
Sphecos
Sphecos
Spixiana
Spixiana

Spraw. Kom. Fizjogr. Mater. Fizjogr. Kraju
Sprawozdania Komisji Fizjograficznej oraz Materjaly do Fizjografji Kraju
Stain Technol.
Stain Technology
State Biol. Surv. Kans. Tech. Publ.
State Biological Survey of Kansas. Technical Publications
Stett. Entomol. Ztg.
Stettiner Entomologische Zeitung
Stud. Cercet. Biol. Ser. Biol. Anim.
Studii si Cercetari de Biologie. Seria Biologie Animala
Stud. Cercet. Biol. Ser. Zool.
Studii si Cercetari de Biologie. Seria Zoologie
Stud. Comun. Muz. Stiint. Nat. Bacau
Studii si Comunicari. Muzeul de Stiintele Naturi. Bacau
Stud. Entomol.
Studia Entomologica
Stud. Nat. Hist. Iowa Univ.
Studies in Natural History. Iowa University
Stud. Neotrop. Fauna Environ.
Studies on the Neotropical Fauna and Environment
Stud. Sassar. Sez. III
Studi Sassaresi. Sezione III
Stuttg. Beitr. Naturkd. Ser. B (Geol. Paläontol.)
Stuttgarter Beiträge zur Naturkunde. Serie B (Geologie und Paläontologie)
Stylops
Stylops
Suppl. Entomol.
Supplementa Entomologica
Sylwan
Sylwan
Symp. Genet. Biol. Ital.
Symposia Genetica et Biologica Italica
Syst. Entomol.
Systematic Entomology
Syst. Zool.
Systematic Zoology
Tacaya
Tacaya
Természetr. Füz.
Természetrajzi Füzetek
Terre Vie
La Terre et la Vie
Tetrahedron
Tetrahedron
Tetrahedron Lett.
Tetrahedron Letters
Tex. Agric. Exp. Stn. Bull.
Texas Agricultural Experiment Station. Bulletin
Tex. J. Agric. Nat. Resour.
Texas Journal of Agriculture and Natural Resources
Tex. J. Sci.
Texas Journal of Science
Theor. Popul. Biol.
Theoretical Population Biology
Thesis Abstr. Haryana Agric. Univ.
Thesis Abstracts. Haryana Agricultural University
Tijdschr. Entomol.
Tijdschrift voor Entomologie
Tijdschr. Plantenziekten
Tijdschrift over Plantenziekten
Tiscia (Szeged)
Tiscia (Szeged)
Tissue Cell
Tissue & Cell
Toxicon
Toxicon
Tr. Biol. Inst. (Novosibirsk)
Trudy Biologicheskogo Instituta (Novosibirsk)
Tr. Biol. Nauchn.-Issled. Inst. Tomsk. Gos. Univ.
Trudy Biologicheskogo Nauchno-Issledovatel'skogo Instituta Tomskogo Gosudarstvennogo Universiteta
Tr. Inst. Morfol. Zhivotn. Akad. Nauk SSSR
Trudy Instituta Morfologii Zhivotnykh. Akademiya Nauk SSSR
Tr. Inst. Zool. Akad. Nauk Kaz. SSR
Trudy Instituta Zoologii. Akademiya Nauk Kazakhskoi SSR
Tr. Inst. Zool. Biol. Vseukr. [or *Ukr.*] *Akad. Nauk Ser. 1 Pr. Syst. Faun.*
Trudy Instytutu Zoolohii ta Biolohii Vseukrains'koi [*or* Ukrains'koi] Akademii Nauk. Seriya 1. Pratsi z Systematyky ta Faunistyky
Tr. Kosinsk. Biol. Stn.
Trudy Kosinskoi Biologicheskoi Stantsii
Tr. Obshch. Estestvoispyt. Imp. Kazan. Univ.
Trudy Obshchestva Estestvoispytatelei pri Imperatorskom Kazanskom Universitete

Tr. Paleontol. Inst. Akad. Nauk SSSR
Trudy Paleontologicheskogo Instituta. Akademiya Nauk SSSR
Tr. Paleozool. Inst. Akad. Nauk SSSR
Trudy Paleozoologicheskogo Instituta. Akademiya Nauk SSSR
Tr. Rus. Entomol. Obshch.
Trudy Russkago Entomologicheskago Obshchestva
Tr. Rus. Estestvoispyt.
Trudy Russkikh Estestvoispytatelei
Tr. Stud. Kruzh. Izsled. Rus. Prir. Imp. Mosk. Univ.
Trudy Studencheskago Kruzhka dlya Izsledovaniya Russkoi Prirody pri Imperatorskom Moskovskom Universitete
Tr. Tomsk. Gos. Univ.
Trudy Tomskogo Gosudarstvennogo Universiteta
Tr. Turkest. Nauchn. Obshch.
Trudy Turkestanskogo Nauchnogo Obshchestva
Tr. Ukr. [or *Vseukr.*] *Akad. Nauk Fiz.-Mat. Vidd.*
Trudy. Ukrains'ka [*or* Vseukrains'ka] Akademiya Nauk. Fizichno-Matematichnoho Viddilu
Tr. Vses. Entomol. Obshch.
Trudy Vsesoyuznogo Entomologicheskogo Obshchestva
Tr. Vseukr. Akad. Nauk Pryr.-Tech. Vidd.
Trudy. Vseukrains'ka Akademiya Nauk. Pryrodnycho-Technichnoho Viddilu
Tr. Zapov. "Daurskii"
Trudy Zapovednika "Daurskii"
Tr. Zool. Inst. Akad. Nauk SSSR
Trudy Zoologicheskogo Instituta. Akademiya Nauk SSSR
Trans. Am. Entomol. Soc.
Transactions of the American Entomological Society
Trans. Am. Microsc. Soc.
Transactions of the American Microscopical Society
Trans. Am. Philos. Soc.
Transactions of the American Philosophical Society
Trans. Cardiff Nat. Soc.
Transactions of the Cardiff Naturalists' Society
Trans. Entomol. Soc. Lond.
Transactions of the Entomological Society of London
Trans. Entomol. Soc. N. S. W.
Transactions of the Entomological Society of New South Wales
Trans. Ill. Acad. Sci.
Transactions of the Illinois Academy of Science
Trans. Kans. Acad. Sci.
Transactions of the Kansas Academy of Science
Trans. Kansai Entomol. Soc.
Transactions of the Kansai Entomological Society
Trans. Kent Field Club
Transactions of the Kent Field Club
Trans. Leic. Lit. Philos. Soc.
Transactions of the Leicester Literary and Philosophical Society
Trans. Linn. Soc. Lond.
Transactions of the Linnean Society of London
Trans. Linn. Soc. Lond. Zool.
Transactions of the Linnean Society of London. Zoology
Trans. N. Y. State Agric. Soc.
Transactions of the New York State Agricultural Society
Trans. Nat. Hist. Soc. Formosa
Transactions of the Natural History Society of Formosa
Trans. Proc. N. Z. Inst.
Transactions and Proceedings of the New Zealand Institute
Trans. R. Entomol. Soc. Lond.
Transactions of the Royal Entomological Society of London
Trans. R. Soc. S. Aust.
Transactions of the Royal Society of South Australia
Trans. San Diego Soc. Nat. Hist.
Transactions of the San Diego Society for Natural History
Trans. Sapporo Nat. Hist. Soc.
Transactions of the Sapporo Natural History Society
Trans. Shikoku Entomol. Soc.
Transactions of the Shikoku Entomological Society

Trans. Soc. Br. Entomol.
Transactions of the Society for British Entomology
Trans. Tex. Acad. Sci.
Transactions of the Texas Academy of Sciences
Trav. Inst. Rech. Sahar.
Travaux de l'Institut de Recherches Sahariennes
Trav. Lab. Zool. Stn. Aquic. Grimaldi Fac. Sci. Dijon
Travaux du Laboratoire de Zoologie et de la Station Aquicole Grimaldi de la Faculté des Sciences de Dijon
Trav. Mus. Hist. Nat. "Grigore Antipa"
Travaux du Muséum d'Histoire Naturelle "Grigore Antipa"
Trav. Sci. Parc Nat. Rég. Réserves Nat. Corse
Travaux Scientifiques du Parc Naturel Régional et des Réserves Naturelles de Corse
Treballs Inst. Catalana Hist. Nat.
Treballs de la Institució Catalana d'Historia Natural
Trends Ecol. Evol.
Trends in Ecology and Evolution
Treubia
Treubia
Tromso Mus. Årsh. [*Aarsh.*]
Tromso Museums Årshefter [Aarshefter]
Trop. Biodivers.
Tropical Biodiversity
Trop. Zool.
Tropical Zoology
Tsitologiya
Tsitologiya
Turrialba
Turrialba
U. S. Dep. Agric. Agric. Handb.
United States Department of Agriculture. Agriculture Handbook
U. S. Dep. Agric. Agric. Monogr.
United States Department of Agriculture. Agriculture Monograph
U. S. Dep. Agric. Bur. Entomol. Bull.
United States Department of Agriculture. Bureau of Entomology. Bulletin
U. S. Dep. Agric. Bur. Entomol. Tech. Ser.
United States Department of Agriculture. Bureau of Entomology. Technical Series
U. S. Dep. Agric. Circ.
United States Department of Agriculture. Circular
U. S. Dep. Agric. Coop. Econ. Insect Rep.
United States Department of Agriculture. Cooperative Economic Insect Report
U. S. Dep. Agric. Tech. Bull.
United States Department of Agriculture. Technical Bulletin
Uch. Zap. Kazan. Vet. Inst.
Uchenye Zapiski Kazanskago Veterinarnago Instituta
Uebers. Arb. Veränd. Schles. Ges. Vaterl. Kult.
Uebersicht der Arbeiten und Veränderungen der Schlesischen Gesellschaft für Vaterländische Kultur
Umschau Wiss. Tech.
Umschau in Wissenschaft und Technik
Univ. Arkansas Agric. Exp. Stn. Bull.
University of Arkansas Agricultural Experiment Station. Bulletin
Univ. Calif. Agric. Exp. Stn. Circ.
University of California. College of Agriculture. Agricultural Experimental Station. Circular
Univ. Calif. Publ. Entomol.
University of California Publications in Entomology
Univ. Colo. Stud.
University of Colorado Studies
Univ. Colo. Stud. Phys. Biol. Sci.
University of Colorado Studies. Physical and Biological Sciences
Univ. Colo. Stud. Ser. Biol.
University of Colorado Studies. Series in Biology
Univ. Kans. Sci. Bull.
University of Kansas Science Bulletin
Utah Agric. Exp. Stn. Mimeogr. Ser.
Utah Agricultural Experiment Station Mimeograph Series
Vakbl. Biol.
Vakblad voor Biologen
Veda Prír.
Veda Prírodní
Verh. Dtsch. Zool. Ges.
Verhandlungen der Deutschen Zoologischen Gesellschaft
Verh. Ges. Dtsch. Naturforsch. Ärzte
Verhandlungen der Gesellschaft Deutscher Naturforscher und Ärzte

Verh. K-K. Zool.-Bot. Ges. Wien
Verhandlungen der Kaiserlich-Königlichen Zoologisch-Botanischen Gesellschaft in Wien
Verh. K. Akad. Wet. Amst. Tweede Sect.
Verhandelingen der Koninklijke Akademie van Wetenschappen te Amsterdam. Tweede Sectie
Verh. K. Vlaam. Acad. Wet. Lett. Schone Kunsten Belg. Kl. Wet.
Verhandelingen van de Koninklijke Vlaamse Academie voor Wetenschappen, Letteren en Schone Kunsten van België. Klasse der Wetenschappen
Verh. Mitt. Siebenb. Ver. Naturwiss. Hermannstadt
Verhandlungen und Mitteilungen des Siebenbürgischen Vereins für Naturwissenschaften zu Hermannstadt
Verh. Naturforsch. Ges. Basel
Verhandlungen der Naturforschenden Gesellschaft in Basel
Verh. Naturhist. Ver. Preuss. Rheinl. Westfal.
Verhandlungen des Naturhistorischen Vereins der Preussischen Rheinlande und Westfalens
Verh. Naturwiss. Ver. Hambg.
Verhandlungen des Naturwissenschaftlichen Vereins in Hamburg
Verh. Schweiz. Naturforsch. Ges.
Verhandlungen der Schweizerischen Naturforschenden Gesellschaft
Verh. Zool.-Bot. Ges. Wien
Verhandlungen der Zoologisch-Botanischen Gesellschaft in Wien
Verh. Zool.-Bot. Ver. Wien
Verhandlungen der Zoologisch-Botanischen Vereins in Wien
Veröff. Überseemus. Bremen Naturwiss.
Veröffentlichungen aus dem Überseemuseum Bremen. Naturwissenschaften
Vesmir
Vesmir
Vestn. Cesk. Zool. Spol.
Vestník Ceskoslovenské Zoologické Spolecnosti
Vestn. Klubu Prírodoved. Prostejove
Vestník Klubu Prírodovedeckého v Prostejove
Vestn. Ustred. Ustavu Geol.
Vestník Ustredního Ustavu Geologického
Vestn. Zool.
Vestnik Zoologii
Vestsi Akad. Navuk BSSR Ser. Biial. Navuk
Vestsi Akademii Navuk BSSR. Seryia Biialahichnykh Navuk
Vestsi Akad. Navuk Belarusi Ser. Biial. Navuk
Vestsi Akademii Navuk Belarusi. Seryia Biialahichnykh Navuk
Veszprém Megyei Múz. Közl. Természettud.
A Veszprém Megyei Múzeumok Közleményei. Természettudomány
Vic. Nat. (Melb.)
Victorian Naturalist (Melbourne)
Vidensk. Medd. Dan. Naturhist. Foren.
Videnskabelige Meddelelser fra Dansk Naturhistorisk Forening
Vidensk. Medd. Naturhist. Foren. Kbh.
Videnskabelige Meddelelser fra den Naturhistoriske Forening i Kjøbenhavn
Vie Milieu
Vie et Milieu
Vie Milieu Sér. C Biol. Terr.
Vie et Milieu. Série C. Biologie Terrestre
Vierteljahrsschr. Naturforsch. Ges. Zür.
Vierteljahrsschrift der Naturforschenden Gesellschaft in Zürich
Vopr. Ekol. Novosib. Gos. Univ.
Voprosy Ekologii. Novosibirskii Gosudarstvennyi Universitet
Waldhygiene
Waldhygiene
Wanderversamml. Dtsch. Entomol.
Wanderversammlung Deutscher Entomologen
Warera
Warera
Wash. State Univ. Coll. Agric. Coop. Ext. Serv. Bull.
Washington State University College of Agriculture Cooperative Extension Service. Bulletin
Wasmann J. Biol.
Wasmann Journal of Biology
West. Aust. Nat.
Western Australian Naturalist
Westermanns Monatsh.
Westermanns Monatshefte

Wet. Meded. K. Ned. Natuurhist. Ver.
Wetenschappelijke Mededelingen. Koninklijke Nederlandse Natuurhistorische Vereniging
Wien. Entomol. Ztg.
Wiener Entomologische Zeitung
Wiss. Ergeb. Dtsch. Zent.-Afr.-Exped.
Wissenschaftliche Ergebnisse der Deutschen Zentral-Afrika-Expedition
Wiss. Ergeb. Niederl. Exped. Karakorum
Wissenschaftliche Ergebnissee der Niederländischen Expeditionen in den Karakorum
Wiss. Ergeb. Reise Ostafr.
Wissenschaftliche Ergebnisse. Reise in Ostafrika
Wiss. Mitt. Bosnien Herceg.
Wissenschaftliche Mitteilungen aus Bosnien und Hercegovina
Young Nat.
Young Naturalist
Z. Angew. Entomol.
Zeitschrift für Angewandte Entomologie
Z. Angew. Zool.
Zeitschrift für Angewandte Zoologie
Z. Arbeitsgem. Österr. Entomol.
Zeitschrift der Arbeitsgemeinschaft Österreichischer Entomologen
Z. Dtsch. Geol. Ges.
Zeitschrift der Deutschen Geologischen Gesellschaft
Z. Dtsch. Ver. Wiss. Kunst São Paulo
Zeitschrift des Deutschen Vereins für Wissenschaft und Kunst in São Paulo
Z. Entomol. (Breslau)
Zeitschrift für Entomologie (Breslau)
Z. Mähr. Landesmus.
Zeitschrift des Mährischen Landesmuseums
Z. Morphol. Ökol. Tiere
Zeitschrift für Morphologie und Ökologie der Tiere
Z. Morphol. Tiere
Zeitschrift für Morphologie der Tiere
Z. Naturforsch. C J. Biosci.
Zeitschrift für Naturforschung. C, Journal of Biosciences
Z. Naturforsch. Teil B
Zeitschrift für Naturforschung. Teil B
Z. Pflanzenkr. Pflanzenschutz
Zeitschrift für Pflanzenkrankheiten und Pflanzenschutz
Z. Syst. Hymenopterol. Dipterol.
Zeitschrift für Systematische Hymenopterologie und Dipterologie
Z. Tierpsychol.
Zeitschrift für Tierpsychologie
Z. Vgl. Physiol.
Zeitschrift für Vergleichende Physiologie
Z. Wiss. Insektenbiol.
Zeitschrift für Wissenschaftliche Insektenbiologie
Z. Wiss. Zool.
Zeitschrift für Wissenschaftliche Zoologie
Z. Zellforsch. Mikrosk. Anat.
Zeitschrift für Zellforschung und Mikroskopische Anatomie
Z. Zool. Syst. Evolutionsforsch.
Zeitschrift für Zoologische Systematik und Evolutionsforschung
Zap. Imp. Akad. Nauk Fiz.-Mat. Otd.
Zapiski Imperatorskoi Akademii Nauk po Fiziko-Matematicheskomu Otdeleniyu
Zap. Imp. Rus. Geogr. Obshch. Obshch. Geogr.
Zapiski Imperatorskago Russkago Geograficheskago Obshchestva po Obshchei Geografii
Zap. Vladivost. Otd. Gos. Rus. Geogr. Obshch.
Zapiski Vladivostokskogo Otdela Gosudarstvennogo Russkogo Geograficheskogo Obshchvestva
Zast. Bilja
Zastita Bilja
Zb. Prats Zool. Muz.
Zbirnyk Prats' Zoolohichnoho Muzeyu
Zb. Slov. Nár. Múz. Prír. Vedy
Zborník Slovenského Národného Múzea. Prírodné Vedy
Zesz. Nauk. Uniw. Mikolaja Kopernika Tor.
Zeszyty Naukowe Uniwersytetu Mikolaja Kopernika w Toruniu
Zesz. Nauk. Wyz. Szk. Roln. Olszt.
Zeszyty Naukowe Wyzszej Szkoly Rolniczej w Olsztynie
Zh. Ukr. Entomol. Tov.
Zhurnal Ukrains'koho Entomolohichnoho Tovarystva
Ziva
Ziva
Zool. Anal. Complex Syst.
Zoology. Analysis of Complex Systems

Zool. Anz.
Zoologischer Anzeiger
Zool. Baetica
Zoologica Baetica
Zool. Beitr.
Zoologische Beiträge
Zool. Cat. Aust.
Zoological Catalogue of Australia
Zool. J. Lond.
Zoological Journal. London
Zool. Jahrb. Abt. Allg. Zool. Physiol. Tiere
Zoologische Jahrbücher. Abteilung für Allgemeine Zoologie und Physiologie der Tiere
Zool. Jahrb. Abt. Anat. Ontog. Tiere
Zoologische Jahrbücher. Abteilung für Anatomie und Ontogenie der Tiere
Zool. Jahrb. Abt. Syst. Geogr. Biol. Tiere
Zoologische Jahrbücher. Abteilung für Systematik, Geographie und Biologie der Tiere
Zool. Jahrb. Abt. Syst. Ökol. Geogr. Tiere
Zoologische Jahrbücher. Abteilung für Systematik, Ökologie und Geographie der Tiere
Zool. Jahrb. Suppl.
Zoologische Jahrbücher. Supplement
Zool. Meded. (Leiden)
Zoologische Mededelingen (Leiden)
Zool. Middle East
Zoology in the Middle East
Zool. Res.
Zoological Research
Zool. Sci. (Tokyo)
Zoological Science (Tokyo)
Zool. Scr.
Zoologica Scripta
Zool. Zentralbl.
Zoologisches Zentralblatt
Zool. Zh.
Zoologicheskii Zhurnal
Zoologica (N. Y.)
Zoologica (New York)
Zoologica (Stuttg.)
Zoologica (Stuttgart)
Zoologist
Zoologist
Zoomorphologie
Zoomorphologie
Zoomorphology (Berl.)
Zoomorphology (Berlin)
Zpr. Cesk. Spol. Entomol.
Zprávy Ceskoslovenské Spolecnosti Entomologické
Zpr. Kom. Prírodoved. Vyzk. Moravy Slez. Odd. Zool.
Zprávy Komise na Prírodovedecky Vyzkum Moravy a Slezska. Oddelení Zoologické
Zpr. Stud. Krajského Muz. Teplicích
Zprávy a Studie. Krajského Muzea v Teplicích
Zür. Jugend Naturforsch. Ges.
Zürcherische Jugend. Naturforschende Gesellschaft

Bibliography

Abdul-Rassoul, M. S., Dawah, H. A., Othman, N. Y. 1978. Records of insect collection (Part I) in the Natural History Research Centre, Baghdad. Bull. Nat. Hist. Res. Cent. Univ. Baghdad 7(2):1-6. [1978.04] [Formicidae pp. 4-6.]

Abe, M., Smith, D. R. 1991. The genus-group names of Symphyta (Hymenoptera) and their type species. Esakia 31:1-115. [1991.07.31] [Page 53: *Myrmicium*.]

Abe, T. 1974. Notes on the fauna of ants in Iriomote Island. [In Japanese.] Pp. 105-111 in: Ikehara, S. (ed.) Ecological studies of nature conservation of the Ryukyu Islands (1). Report for the fiscal year of 1973. [In Japanese.] Naha, Okinawa: University of the Ryukyus, 278 pp. [1974.03.30]

Abe, T. 1977. A preliminary study on the ant fauna of the Tokara Islands and Amami-Oshima. Pp. 93-102 in: Ikehara, S. (ed.) Ecological studies of nature conservation of the Ryukyu Islands - (III). Naha, Okinawa: University of the Ryukyus, 202 pp. [1977.03]

Abe, T., Maeda, A. 1977. Fauna and density of ants in sugarcane fields of the southern part of Okinawa Island. Pp. 75-91 in: Ikehara, S. (ed.) Ecological studies of nature conservation of the Ryukyu Islands - (III). Naha, Okinawa: University of the Ryukyus, 202 pp. [1977.03]

Abe, T., Nakamori, M., Uezu, K. 1976. Fauna and density of ants of a coastal limestone area in Okinawa Island. Pp. 113-120 in: Ikehara, S. (ed.) Ecological studies of nature conservation of the Ryukyu Islands - (II). Naha, Okinawa: University of the Ryukyus, 141 pp. [1976.03]

Abensperg-Traun, M., Steven, D. 1995. The effects of pitfall trap diameter on ant species richness (Hymenoptera: Formicidae) and species composition of the catch in a semi-arid eucalypt woodland. Aust. J. Ecol. 20:282-287. [1995.06]

Acosta, F. J. See: Acosta Salmerón, F. J.

Acosta Salmerón, F. J. 1978 ("1977"). Notas sobre hormigas de la provincia de Jaén (Hym. Formicidae). Bol. Asoc. Esp. Entomol. 1:133-140. [1978.04]

Acosta Salmerón, F. J. 1982 ("1981"). Sobre los caracteres morfólogicos [sic] de *Goniomma*, con algunas sugerencias sobre su taxonomía (Hym., Formicidae). EOS. Rev. Esp. Entomol. 57: 7-14. [1982.04.28]

Acosta Salmerón, F. J., Martínez, M. D. 1982. Consideraciones sobre la dulosis en el género *Strongylognathus* Mayr, 1853. (Hym. Formicidae). Bol. Asoc. Esp. Entomol. 6:121-124. [1982.12]

Acosta Salmerón, F. J., Martínez, M. D. 1984. Algunas anomalías en *Leptothorax rabaudi* Bond., 1818 [sic] (Hym., Formicidae). Bol. Asoc. Esp. Entomol. 8:41-45. [1984.06]

Acosta Salmerón, F. J., Martínez, M. D., Morales, M. 1983. Contribución al conocimiento de la mirmecofauna del encinar peninsular (1). (Hym. Formicidae). Bol. Asoc. Esp. Entomol. 6:379-391. [1983.06]

Acosta Salmerón, F. J., Martínez, M. D., Serrano Talavera, J. M. 1983. Contribución al conocimiento de la mirmecofauna del encinar peninsular. II: principales pautas autoecológicas. Bol. Asoc. Esp. Entomol. 7:297-306. [1983.11]

Acosta Salmerón, F. J., Morales Bastos, M., Serrano Talavera, J. M. 1983. Capacidad de transcripción de una mirmecocenosis en un medio adverso. Bol. Asoc. Esp. Entomol. 7:151-158. [1983.11]

Acosta Salmerón, F. J., Morales, M. A., Zorrilla, J. M. 1983. Polimorfismo y desgaste en la pieza mandibular de *Messor barbarus* (Hym., Formicidae). Graellsia 38:167-173. [1983.02.15]

Adam, A. 1912. Bau und Mechanismus des Receptaculum seminis bei den Bienen, Wespen und Ameisen. Zool. Jahrb. Abt. Anat. Ontog. Tiere 35:1-74. [1912.09.06]

Adam, A., Foerster, E. 1913. Die Ameisenfauna Oberbadens. Mitt. Bad. Landesver. Naturkd. Freibg. Breisgau 1913:205-218. [1913]

Adams, E. S. 1990. Interaction between the ants *Zacryptocerus maculatus* and *Azteca trigona*: interspecific parasitism of information. Biotropica 22:200-206. [1990.06]

Adams, E. S. 1994. Territory defense by the ant *Azteca trigona*: maintenance of an arboreal ant mosaic. Oecologia (Berl.) 97:202-208. [1994.03]

Adeli, E. 1962. Zur Ökologie der Ameisen im Gebiet des Urwaldes Rotwald (Niederösterreich). Z. Angew. Entomol. 49:290-296. [1962.03]

Adis, J., Lubin, Y. D., Montgomery, G. G. 1985 ("1984"). Arthropods from the canopy of inundated and terra firme forests near Manaus, Brazil, with critical considerations on the pyrethrum-fogging technique. Stud. Neotrop. Fauna Environ. 19:223-236. [1985]

Adlerz, G. 1885 ("1884"). Myrmecologiska studier. I. *Formicoxenus nitidulus* Nyl. Öfvers. K. Vetensk.-Akad. Förh. Stockh. 41(8):43-64. [1885]

Adlerz, G. 1886. Myrmecologiska studier. II. Svenska myror och deras lefnadsförhållanden. Bih. K. Sven. Vetensk.-Akad. Handl. 11(18):1-329. [1886]

Adlerz, G. 1887a. Myrmecologiska notiser. Entomol. Tidskr. 8:41-50. [1887]

Adlerz, G. 1887b. Notices myrmécologiques. Entomol. Tidskr. 8:155-165. [1887] [Translation of Adlerz (1887a).]

Adlerz, G. 1895. Om en myrliknande svensk spindel. Entomol. Tidskr. 16:249-253. [1895.11.15]

Adlerz, G. 1896a. Myrmecologiska notiser. Entomol. Tidskr. 17:129-141. [1896.08.26]

Adlerz, G. 1896b. Myrmekologiska studier. III. *Tomognathus sublaevis* Mayr. Bih. K. Sven. Vetensk.-Akad. Handl. Afd. IV 21(4):1-76. [1896]

Adlerz, G. 1896c ("1895"). Stridulationsorgan och ljudförnimmelser hos myror. Öfvers. K. Vetensk.-Akad. Förh. Stockh. 52:769-782. [1896]

Adlerz, G. 1902. Myrmecologiska studier. IV. *Formica suecica* n. sp., Eine neue schwedische Ameise. Öfvers. K. Vetensk.-Akad. Förh. Stockh. 59:263-265. [1902.09.12]

Adlerz, G. 1908. Zwei Gynandromorphen von *Anergates atratulus* Schenck. Ark. Zool. 5(2):1-6. [1908.11.12]

Adlerz, G. 1914. *Formica fusca-picea* Nyl., en torfmossarnas myra. Ark. Zool. 8(26):1-5. [1914.02.07]

Adolph, G. E. 1880. Ueber das Flügelgeäder des *Lasius umbratus* Nyl. Verh. Naturhist. Ver. Preuss. Rheinl. Westfal. 37:35-53. [1880]

Agassiz, J. L. R. 1846. Nomenclatoris zoologici index universalis, continens nomina systematica classium, ordinum, familiarum, et generum animalium omnium, tam viventium quam fossilium. Soloduri [= Solothurn, Switzerland]: Jent & Gassmann, viii + 393 pp. [1846.12.29] [Date of publication from Evenhuis, 1994b:478. See also Bowley & Smith (1968) and Kevan (1970).]

Agassiz, J. L. R. 1850. [Untitled. Genus *Tomognathus*, new, Agassiz.] P. 376 in: Dixon, F. The geology and fossils of the Tertiary and Cretaceous formation of Sussex. London: Longman, Brown, Green, and Longmans, xvi + 422 pp. [1850.12] [Publication date from Sherborn (1908).]

Agbogba, C. 1984. Observations sur le comportement de marche en tandem chez deux espèces de fourmis ponérines: *Mesoponera caffraria* (Smith) et *Hypoponera* sp. (Hym. Formicidae) Insectes Soc. 31:264-276. [1984.12]

Agbogba, C. 1986 ("1985"). Observations sur la récolte de substances liquides et de sucs animaux chez deux espèces d'*Aphaenogaster*: *A. senilis* et *A. subterranea* (Hym. Formicidae). Insectes Soc. 32:427-434. [1986.06]

Agosti, D. 1983. Zur Insektenfauna der Umgebung der Vogelwarte Sempach, Kanton Luzern. XIII. Hymenoptera 2: Formicidae (Ameisen). Entomol. Ber. Luzern 10:91-92. [1983.12]

Agosti, D. 1990a. What makes the Formicini the Formicini? Actes Colloq. Insectes Soc. 6:295-303. [1990.06]

Agosti, D. 1990b. Review and reclassification of *Cataglyphis* (Hymenoptera, Formicidae). J. Nat. Hist. 24:1457-1505. [1990.10.19]

Agosti, D. 1991. Revision of the oriental ant genus *Cladomyrma*, with an outline of the higher classification of the Formicinae (Hymenoptera: Formicidae). Syst. Entomol. 16:293-310. [1991.07]

Agosti, D. 1992. Revision of the ant genus *Myrmoteras* of the Malay Archipelago (Hymenoptera, Formicidae). Rev. Suisse Zool. 99:405-429. [1992.06]

Agosti, D. 1994a. A new inquiline ant (Hymenoptera: Formicidae) in *Cataglyphis* and its phylogenetic relationship. J. Nat. Hist. 28:913-919. [1994.07.29]

Agosti, D. 1994b. The phylogeny of the ant tribe Formicini (Hymenoptera: Formicidae) with the description of a new genus. Syst. Entomol. 19:93-117. [1994.08.01]

Agosti, D. 1995 ("1994"). A revision of the South American species of the ant genus *Probolomyrmex* (Hymenoptera: Formicidae). J. N. Y. Entomol. Soc. 102:429-434. [1995.07.19]

Agosti, D., Bolton, B. 1990a. The identity of *Andragnathus*, a forgotten formicine ant genus (Hym., Formicidae). Entomol. Mon. Mag. 126:75-77. [1990.04.12]

Agosti, D., Bolton, B. 1990b. New characters to differentiate the ant genera *Lasius* F. and *Formica* L. (Hymenoptera: Formicidae). Entomol. Gaz. 41:149-156. [1990.09.30]

Agosti, D., Collingwood, C. A. 1987a. A provisional list of the Balkan ants (Hym. Formicidae) and a key to the worker caste. I. Synonymic list. Mitt. Schweiz. Entomol. Ges. 60:51-62. [1987.08.31]

Agosti, D., Collingwood, C. A. 1987b. A provisional list of the Balkan ants (Hym. Formicidae) with a key to the worker caste. II. Key to the worker caste, including the European species without the Iberian. Mitt. Schweiz. Entomol. Ges. 60:261-293. [1987.12.15]

Agosti, D., Hauschteck-Jungen, E. 1988 ("1987"). Polymorphism of males in *Formica exsecta* Nyl. (Hymenoptera: Formicidae). Insectes Soc. 34:280-290. [1988.06]

Agosti, D., Mohamed, M., Arthur, C. Y. C. 1994. Has the diversity of tropical ant fauna been underestimated? An indication from leaf litter studies in a west Malaysian lowland rain forest. Trop. Biodivers. 2:270-275. [1994]

Aguayo, C. G. 1931. New ants of the genus *Macromischa*. Psyche (Camb.) 38:175-183. [1931.12] [Stamp date in MCZ library: 1932.01.26.]

Aguayo, C. G. 1932. Notes on West Indian ants. Bull. Brooklyn Entomol. Soc. 27:215-227. [1932.11.29]

Aguayo, C. G. 1937. Nueva especie de hormiga cubana. Mem. Soc. Cuba. Hist. Nat. "Felipe Poey" 11:235-236. [1937.07.10]

Aguayo, J. 1946. El extraño caso de la Historia física, política y natural de la isla de Cuba. Mem. Soc. Cuba. Hist. Nat. "Felipe Poey" 18:153-184. [1946.09.24]

Akre, R. D., Antonelli, A. L. 1976. Identification and habits of key ant pests of Washington. Wash. State Univ. Coll. Agric. Coop. Ext. Serv. Bull. 671:1-5. [1976.08]

Akre, R. D., Antonelli, A. L. 1987. Identification and habits of key ant pests of Washington. Wash. State Univ. Coll. Agric. Coop. Ext. Serv. Bull. 0671:1-8. [1987.03]

Aktaç, N. 1977 ("1976"). Studies on the myrmecofauna of Turkey I. Ants of Siirt, Bodrum and Trabzon. Istanbul Univ. Fen Fak. Mecm. Seri B 41:115-135. [1977]

Alayo, D. P. 1973. Catálogo de los himenopteros de Cuba. Habana: Instituto Cubano del Libro, 218 pp. [1973.06] [Ants pp. 121-155.]

Alayo, D. P. 1974. Introducción al estudio de los Himenópteros de Cuba. Superfamilia Formicoidea. Ser. Biol. Acad. Cienc. Cuba 53:1-58. [1974]

Alayo, D. P., Zayas Montero, L. 1977. Estudios sobre los Himenópteros de Cuba. VII. Dos nuevas especies para la fauna mirmecológica Cubana. Poeyana Inst. Zool. Acad. Cienc. Cuba 174:1-5. [1977.10.04]

Albrecht, A. 1993. *Formica cunicularia* Latreille (Hymenoptera, Formicidae) new to Finland. Entomol. Fenn. 4:13. [1993.03.29]

Alfieri, A. 1931. Contribution à l'étude de la faune myrmécologique de l'Egypte. Bull. Soc. Entomol. Egypte 15:42-48. [1931]

Ali, M. F., Attygalle, A. B., Morgan, E. D., Billen, J. P. J. 1987. The Dufour gland substances of the workers of *Formica fusca* and *Formica lemani* (Hymenoptera: Formicidae). Comp. Biochem. Physiol. B. Comp. Biochem. 88:59-63. [1987]

Ali, M. F., Billen, J. P. J., Jackson, B. D., Morgan, E. D. 1988. Secretion of the Dufour glands of two African desert ants, *Camponotus aegyptiacus* and *Cataglyphis savignyi* (Hymenoptera: Formicidae). Biochem. Syst. Ecol. 16:647-654. [1988.12.14]

Ali, M. F., Billen, J. P. J., Jackson, B. D., Morgan, E. D. 1989. The Dufour gland contents of three species of Euro-African *Messor* ants and a comparison with those of North American *Pogonomyrmex* (Hymenoptera: Formicidae). Biochem. Syst. Ecol. 17:469-477. [1989.11.08]

Ali, M. F., Morgan, E. D., Attygalle, A. B., Billen, J. P. J. 1987. Comparison of Dufour gland secretions of two species of *Lepthothorax* ants (Hymenoptera: Formicidae). Z. Naturforsch. C J. Biosci. 42:955-960. [1987.08]

Allen, C. R., Phillips, S. A., Jr., Trostle, M. R. 1993. Range expansion by the ecologically disruptive red imported fire ant into the Texas Rio Grande Valley. Southwest. Entomol. 18:315-316. [1993.12]

Allen, G. E., Buren, W. F. 1974. Microsporidan and fungal diseases of *Solenopsis invicta* Buren in Brazil. J. N. Y. Entomol. Soc. 82:125-130. [1974.08.22]

Allen, G. E., Buren, W. F., Williams, R. N., de Menezes, M., Whitcomb, W. H. 1974. The red imported fire ant, *Solenopsis invicta*; distribution and habitat in Mato Grosso, Brazil. Ann. Entomol. Soc. Am. 67:43-46. [1974.01.15]

Allen, G. W. 1989. A key to the worker castes of the ants of Kent (Hymenoptera: Formicidae). Trans. Kent Field Club 11:8-23. [1989]

Allen, P. 1957. Studies of some intercastes of *Myrmica rubra* L. (Hym., Formicidae). Entomol. Mon. Mag. 93:136-139. [1957.06.27]

Allies, A. B., Bourke, A. F. G., Franks, N. R. 1986. Propaganda substances in the cuckoo ant *Leptothorax kutteri* and the slave-maker *Harpagoxenus sublaevis*. J. Chem. Ecol. 12:1285-1293. [1986.05]

Alloway, T. M. 1979. Raiding behavior of two species of slave-making ants, *Harpagoxenus americanus* (Emery) and *Leptothorax duloticus* Wesson (Hymenoptera: Formicidae). Anim. Behav. 27:202-210. [1979.02]

Alloway, T. M. 1980. The origins of slavery in leptothoracine ants (Hymenoptera: Formicidae). Am. Nat. 115:247-261. [1980.02]

Alloway, T. M., Buschinger, A., Talbot, M., Stuart, R., Thomas, C. 1983 ("1982"). Polygyny and polydomy in three North American species of the ant genus *Leptothorax* Mayr (Hymenoptera: Formicidae). Psyche (Camb.) 89:249-274. [1983.04.27]

Allred, D. M. 1982. Ants of Utah. Great Basin Nat. 42:415-511. [1982.12.31]

Allred, D. M., Cole, A. C., Jr. 1971. Ants of the National Reactor Testing Station. Great Basin Nat. 31:237-242. [1971.12.31]

Allred, D. M., Cole, A. C., Jr. 1979. Ants from northern Arizona and southern Utah. Great Basin Nat. 39:97-102. [1979.03.31]

Almeida Filho, A. J. de. 1984. Notas sobre *Strumigenys louisianae* (Hymenoptera, Formicidae) e sua ocorrência no nordeste do Brasil. Quid (Teresina) 5:133-139.

Almeida Filho, A. J. de. 1986. Descrição de quatro machos do gênero *Ectatomma* Smith, 1858 (Hymenoptera, Formicidae, Ponerinae). Quid (Teresina) 6:24-38. [1986.12]

Almeida Filho, A. J. de. 1987a. Descrição de seis fêmeas do gênero *Ectatomma* Smith, 1858 (Hymenoptera, Formicidae, Ponerinae). An. Soc. Nordestina Zool. 1:175-183.

Almeida Filho, A. J. de. 1987b. Morfologia externa de *Ectatomma quadridens* Fabricius (Hymenoptera, Formicidae, Ponerinae). An. Soc. Nordestina Zool. 2:185-198.

Almeida Toledo, L. F. de. 1967. Histo-anatomia de glândulas de *Atta sexdens rubropilosa* Forel (Hymenoptera). Arq. Inst. Biol. (São Paulo) 34:321-329. [1967.12]

Alonso de Medina, E., Espadaler, X. 1981. Nota sobre la entomofauna de la Sierra de Prades (Tarragona) (Formícidos). Publ. Dep. Zool. (Barc.) 7:67-71. [1981]

Alpatov, V. V. 1922. The lower systematic categories and variation of ants. [In Russian.] Byull. Vseross. Entomo-Fitopatol. S'ezda 3:10. [Not seen. Cited in Zoological Record for 1922.]

Alpatov, V. V. 1924a. Die Definition der untersten systematischen Kategorien von Standpunkte des Studiums der Variabilität der Ameisen und der Crustaceen. Zool. Anz. 60:161-168. [1924.07.20]

Alpatov, V. V. 1924b. Variability and the lowest systematic categories in ant systematics. [In Russian.] Rus. Zool. Zh. 4(1/2):227-244. [1924]

Alpatov, V. V. 1924c. The ant fauna of Svyatoy Lake peatbogs near Kosine, Moscow district. [In Russian.] Tr. Kosinsk. Biol. Stn. 1:28-32. [1924]

Alpatov, V. V. 1928. Zur Systematik der Ameisen. Bemerkungen zu dem Artikel Dr. Arnoldis "Studien über die Systematik der Ameisen, Teil I". Zool. Anz. 75:138-140. [1928.02.01]

Alpatov, V. V., Palenitschko, Z. G. 1925. On the relative variability of castes and species of ants. [In Russian.] Rus. Zool. Zh. 5(4):109-116. [1925]

Alpert, G. D. 1992. Observations on the genus *Terataner* in Madagascar (Hymenoptera: Formicidae). Psyche (Camb.) 99:117-127. [1992.12.04]

Alpert, G. D., Akre, R. D. 1973. Distribution, abundance, and behavior of the inquiline ant *Leptothorax diversipilosus*. Ann. Entomol. Soc. Am. 66:753-760. [1973.07.16]

Ambach, J. 1994. Die Ameisenfauna der "Pleschinger Sandgrube" bei Linz. Naturkd. Jahrb. Stadt Linz 37-39:259-269. [1994.12]

Amstutz, M. E. 1943. The ants of the Kildeer Plain area of Ohio (Hymenoptera, Formicidae). Ohio J. Sci. 43:165-173. [1943.08.20]

Andersen, A. N. 1983a. Species diversity and temporal distribution of ants in the semi-arid mallee region of northwestern Victoria. Aust. J. Ecol. 8:127-137. [1983.06]

Andersen, A. N. 1983b. A brief survey of ants in Glenaladale National Park, with particular reference to seed-harvesting. Vic. Nat. (Melb.) 100:233-237. [1983.12]

Andersen, A. N. 1986a. Diversity, seasonality and community organization of ants at adjacent heath and woodland sites in south-eastern Australia. Aust. J. Zool. 34:53-64. [1986.02.06]

Andersen, A. N. 1986b. Patterns of ant community organization in mesic southeastern Australia. Aust. J. Ecol. 11:87-97. [1986.03]

Andersen, A. N. 1990. The use of ant communities to evaluate change in Australian terrestrial ecosystems: a review and a recipe. Proc. Ecol. Soc. Aust. 16:347-357. [1990]

Andersen, A. N. 1991a. Sampling communities of ground-foraging ants: pitfall catches compared with quadrat counts in an Australian tropical savanna. Aust. J. Ecol. 16:273-279. [1991.09]

Andersen, A. N. 1991b. Responses of ground-foraging ant communities to three experimental fire regimes in a savanna forest of tropical Australia. Biotropica 23:575-585. [1991.12]

Andersen, A. N. 1991c. The ants of southern Australia. A guide to the Bassian fauna. Melbourne: CSIRO Publications, vii + 70 pp. [1991]

Andersen, A. N. 1991d. Seed harvesting by ants in Australia. Pp. 493-503 in: Huxley, C. R., Cutler, D. F. (eds.) Ant-plant interactions. Oxford: Oxford University Press, xviii + 601 pp. [1991]

Andersen, A. N. 1992a. The rainforest ant fauna of the northern Kimberley region of western Australia (Hymenoptera: Formicidae). J. Aust. Entomol. Soc. 31:187-192. [1992.05.29]

Andersen, A. N. 1992b. Regulation of "momentary" diversity by dominant species in exceptionally rich ant communities of the Australian seasonal tropics. Am. Nat. 140:401-420. [1992.09]

Andersen, A. N. 1993a. Ants as indicators of restoration success at a uranium mine in tropical Australia. Restor. Ecol. 1:156-167. [1993.09]

Andersen, A. N. 1993b. Ant communities in the Gulf Region of Australia's semi-arid tropics: species composition, patterns of organisation, and biogeography. Aust. J. Zool. 41:399-414. [1993.11.18]

Andersen, A. N. 1995a. A classification of Australian ant communities, based on functional groups which parallel plant life-forms in relation to stress and disturbance. J. Biogeogr. 22:15-29. [1995.01]

Andersen, A. N. 1995b. [Review of: Bolton, B. 1994. Identification guide to the ant genera of the world. Cambridge, Mass.: Harvard University Press, 222 pp.] J. Aust. Entomol. Soc. 34:28. [1995.02.28]

Andersen, A. N. 1995c. Measuring more of diversity: genus richness as a surrogate for species richness in Australian ant faunas. Biol. Conserv. 73:39-43. [1995]

Andersen, A. N., Burbidge, A. H. 1991. The ants of a vine thicket near Broome: a comparison with the northwest Kimberley. J. R. Soc. West. Aust. 73:79-82. [1991.02]

Andersen, A. N., Burbidge, A. H. 1992. An overview of the ant fauna of Cape Arid National Park, Western Australia. J. R. Soc. West. Aust. 75:41-46. [1992.02]

Andersen, A. N., Majer, J. D. 1991. The structure and biogeography of rainforest ant communities in the Kimberley region of northwestern Australia. Pp. 333-346 in: McKenzie, N. L., Johnston, R. B., Kendrick, P. G. (eds.) Kimberley rainforests. Chipping Norton, N.S.W.: Surrey Beatty & Sons, xviii + 490 pp. [1991]

Andersen, A. N., McKaige, M. E. 1987. Ant communities at Rotamah Island, Victoria, with particular reference to disturbance and *Rhytidoponera tasmaniensis*. Proc. R. Soc. Vic. 99:141-146. [1987.12]

Andersen, A. N., Myers, B. A., Buckingham, K. M. 1991. The ant fauna of a mallee outlier near Melton, Victoria. Proc. R. Soc. Vic. 103:1-6. [1991.06.30]

Andersen, A. N., Patel, A. D. 1994. Meat ants as dominant members of Australian ant communities: an experimental test of their influence on the foraging success and forager abundance of other species. Oecologia (Berl.) 98:15-24. [1994.06]

Andersen, A. N., Reichel, H. 1994. The ant (Hymenoptera: Formicidae) fauna of Holmes Jungle, a rainforest patch in the seasonal tropics of Australia's Northern Territory. J. Aust. Entomol. Soc. 33:153-158. [1994.05.31]

Andersen, A. N., Yen, A. L. 1985. Immediate effects of fire on ants in the semi-arid mallee region of north-western Victoria. Aust. J. Ecol. 10:25-30. [1985.03]

Andersen, A. N., Yen, A. Y. 1992. Canopy ant communities in the semi-arid mallee region of north-western Victoria. Aust. J. Zool. 40:205-214. [1992.07.15]

Andersson, B., Douwes, P. 1989. Myran *Leptothorax affinis* funnen på Öland - ny för Nordeuropa. Entomol. Tidskr. 110:173. [1989.10.31]

Andoni, V. 1975. Të dhëna paraprake mbi inventarizimin e himenopterëve të nënrendit Aculeata. Bul. Shkencave Nat. Tiranë 29(4):99-108. [1975] [Formicidae pp. 100-102.]

Andoni, V. 1977. Kontribut mbi himenopterët e familjes Formicidae të vendit tonë. Bul. Shkencave Nat. Tiranë 31(2):93-101. [1977]

Andoni, V. 1989. Një lloj i ri milingone, *Formica rufa* L. (Formicidae- Hymenoptera) në vendin tonë. Bul. Shkencave Nat. Tiranë 43(4):60-66. [1989]

Andoni, V. 1993. La composition des espèces, la repartition géographique et quelques données bioécologiques pour les fourmis du genre *Formica* (Hymenoptera - Formicidae) en Albanie. Biol. Gallo-Hell. 20:199-208.

Andrade, M. L. de See: De Andrade, M. L.

André, E. Note: All entries here refer to André, Ernest.

André, E. 1874a. Description des fourmis d'Europe pour servir à l'étude des insectes myrmécophiles. [part] Rev. Mag. Zool. Pure Appl. (3)2:152-160. [1874.05]

André, E. 1874b. Description des fourmis d'Europe pour servir à l'étude des insectes myrmécophiles. [part] Rev. Mag. Zool. Pure Appl. (3)2:161-192. [1874.06]

André, E. 1874c. Description des fourmis d'Europe pour servir à l'étude des insectes myrmécophiles. [part] Rev. Mag. Zool. Pure Appl. (3)2:193-224. [1874.07]

André, E. 1874d. Description des fourmis d'Europe pour servir à l'étude des insectes myrmécophiles. [concl.] Rev. Mag. Zool. Pure Appl. (3)2:225-235. [1874.08]

André, E. 1881a. [Untitled. *Monomorium Abeillei*, n. sp.] P. 531 in: Emery, C. Viaggio ad Assab nei Mar Rosso dei Signori G. Doria ed O. Beccari con il R. Avviso "Esploratore" dal 16 Novembre 1879 al 26 Febbraio 1880. I. Formiche. Ann. Mus. Civ. Stor. Nat. 16:525-535. [1881.03.09]

André, E. 1881b. [Untitled. Introduced by: "M. Ernest André, de Gray, adresse les descriptions de trois nouvelles espèces de Fourmis".] Bull. Bimens. Soc. Entomol. Fr. 1881:60-62. [1881.04.23] [Reissued as André (1881d, 1881e).]

André, E. 1881c. Catalogue raisonné des Formicides provenant du voyage en Orient de M. Abeille de Perrin et description des espèces nouvelles. Ann. Soc. Entomol. Fr. (6)1:53-78. [1881.07.22]

André, E. 1881d. [Untitled. Introduced by: "M. Ernest André, de Gray, adresse les descriptions de trois nouvelles espèces de Fourmis".] [part] Ann. Soc. Entomol. Fr. (Bull.) (6)1:xlviii. [1881.07.22] [See André (1881b).]

André, E. 1881e. [Untitled. Introduced by: "M. Ernest André, de Gray, adresse les descriptions de trois nouvelles espèces de Fourmis".] [concl.] Ann. Soc. Entomol. Fr. (Bull.) (6)1:xlix-xl. [1881.10.12] [See André (1881b).]

André, E. 1881f. Les fourmis. [part] Pp. 1-48 in: André, Edm. 1881-1886. Species des Hyménoptères d'Europe et d'Algérie. Tome Deuxième. Beaune: Edmond André, 919 + 48 pp. [1881.10] [Printed by Francis Bouffait at Gray. Published by Edmond André ("chez l'auteur à Beaune (Côte-d'Or)").]

André, E. 1882a. Les fourmis. [part] Pp. 49-80 in: André, Edm. 1881-1886. Species des Hyménoptères d'Europe et d'Algérie. Tome Deuxième. Beaune: Edmond André, 919 + 48 pp. [1882.01]

André, E. 1882b. Les fourmis. [part] Pp. 81-152 in: André, Edm. 1881-1886. Species des Hyménoptères d'Europe et d'Algérie. Tome Deuxième. Beaune: Edmond André, 919 + 48 pp. [1882.04]

André, E. 1882c. Les fourmis. [part] Pp. 153-232 in: André, Edm. 1881-1886. Species des Hyménoptères d'Europe et d'Algérie. Tome Deuxième. Beaune: Edmond André, 919 + 48 pp. [1882.07]

André, E. 1882d. Les fourmis. [part] Pp. 233-280 in: André, Edm. 1881-1886. Species des Hyménoptères d'Europe et d'Algérie. Tome Deuxième. Beaune: Edmond André, 919 + 48 pp. [1882.10]

André, E. 1883a. Les fourmis. [part] Pp. 281-344 in: André, Edm. 1881-1886. Species des Hyménoptères d'Europe et d'Algérie. Tome Deuxième. Beaune: Edmond André, 919 + 48 pp. [1883.01]

André, E. 1883b. Les fourmis. [concl.] Pp. 345-404 in: André, Edm. 1881-1886. Species des Hyménoptères d'Europe et d'Algérie. Tome Deuxième. Beaune: Edmond André, 919 + 48 pp. [1883.04]

André, E. 1884a ("1883"). [Untitled. *Monomorium afrum* André n. sp.] P. 244 in: Magretti, P. Raccolte imenotterologiche nell'Africa Orientale. Bull. Soc. Entomol. Ital. 15:241-253. [1884.04.15]

André, E. 1884b ("1883"). [Untitled. *Meranoplus Magrettii* Andr. n. sp.] P. 245 in: Magretti, P. Raccolte imenotterologiche nell'Africa Orientale. Bull. Soc. Entomol. Ital. 15:241-253. [1884.04.15]

André, E. 1884c. [Untitled. Introduced by: "M. Ernest André, ayant examiné les Formicides recueillis à Aguilas par M. Weyers."] Ann. Soc. Entomol. Belg. 28:cclxxxii. [> 1884.09.06]

André, E. 1884d. Le monde des fourmis. Feuille Jeunes Nat. 15:7-9. [1884.11.01]

André, E. 1884e. [Untitled. *Monomorium afrum*, n. sp.] Pp. 540-541 in: Magretti, P. Risultati di raccolte imenotterologiche nell'Africa Orientale. Ann. Mus. Civ. Stor. Nat. 21:523-636. [1884.11.15]

André, E. 1884f. [Untitled. *Meranoplus Magrettii*, n. sp.] P. 543 in: Magretti, P. Risultati di raccolte imenotterologiche nell'Africa Orientale. Ann. Mus. Civ. Stor. Nat. 21:523-636. [1884.11.15]
André, E. 1884g. Le monde des fourmis (Suite). Feuille Jeunes Nat. 15:19-21. [1884.12.01]
André, E. 1885a. Le monde des fourmis (Suite). Feuille Jeunes Nat. 15:33-37. [1885.01.01]
André, E. 1885b. Le monde des fourmis (Suite). Feuille Jeunes Nat. 15:41-45. [1885.02.01]
André, E. 1885c. Le monde des fourmis (Fin). Feuille Jeunes Nat. 15:64-67. [1885.03.01]
André, E. 1885d. Supplément aux fourmis. Pp. 833-854 in: André, Edm. 1881-1886. Species des Hyménoptères d'Europe et d'Algérie. Tome Deuxième. Beaune: Edmond André, 919 + 48 pp. [1885.11]
André, E. 1885e. Les fourmis. Bibliothèque des Merveilles. Paris: Hachette & Cie, 347 pp. [1885]
André, E. 1886a. Appendice aux fourmis. Pp. 854-859 in: André, Edm. 1881-1886. Species des Hyménoptères d'Europe et d'Algérie. Tome Deuxième. Beaune: Edmond André, 919 + 48 pp. [1886.01]
André, E. 1886b. Errata, addenda et corrigendia. I. Fourmis. Pp. 893-894 in: André, Edm. 1881-1886. Species des Hyménoptères d'Europe et d'Algérie. Tome Deuxième. Beaune: Edmond André, 919 + 48 pp. [1886.06]
André, E. 1887. Description de quelques fourmis nouvelles ou imparfaitement connues. Rev. Entomol. (Caen) 6:280-298. [1887.11]
André, E. 1889. Hyménoptères nouveaux appartenant au groupe des Formicides. Rev. Entomol. (Caen) 8:217-231. [1889.08]
André, E. 1890. Matériaux pour servir à la faune myrmécologique de Sierra-Leone (Afrique occidentale). Rev. Entomol. (Caen) 9:311-327. [1890.11]
André, E. 1891. [Untitled. Introduced by: "M. Ch. Janet présent une note de M. Ern. André relative à une collection de Fourmis rapportée de Bornéo par M. Chaper".] Bull. Soc. Zool. Fr. 16:238-239. [1891]
André, E. 1892a. Matériaux myrmécologiques. Rev. Entomol. (Caen) 11:45-56. [1892.02]
André, E. 1892b. [Untitled. Introduced by: "M. E. André a examiné une petite collection à Hyménoptères faite à Mahé (Inde) par M. Emile Deschamps".] Bull. Soc. Zool. Fr. 17:207. [1892]
André, E. 1892c. Voyage de M. Chaper à Bornéo. Catalogue des fourmis et description des espèces nouvelles. Mém. Soc. Zool. Fr. 5:46-55. [1892]
André, E. 1893a. [Untitled. Introduced by: "M. E. André (de Gray) envoie la description d'une nouvelle espèce de Fourmi de Tunisie".] Bull. Bimens. Soc. Entomol. Fr. 1893:cxci-cxcii. [1893.05.06] [Reissued as André (1893c).]
André, E. 1893b. Description de quatre espèces nouvelles de fourmis d'Amérique. Rev. Entomol. (Caen) 12:148-152. [1893.07]
André, E. 1893c. [Untitled. Introduced by: "M. E. André (de Gray) envoie la description d'une nouvelle espèce de Fourmi de Tunisie".] Ann. Soc. Entomol. Fr. (Bull.) 62:cxci-cxcii. [1893.12.30] [First published as André (1893a).]
André, E. 1895a. Formicides de l'Ogooué (Congo français). Rev. Entomol. (Caen) 14:1-5. [1895.01]
André, E. 1895b. Notice sur les fourmis fossiles de l'ambre de la Baltique et description de deux espèces nouvelles. Bull. Soc. Zool. Fr. 20:80-84. [1895.03]
André, E. 1896a. Fourmis recueillies dans les serres du Muséum. Bull. Mus. Hist. Nat. 2:24. [1896.02.22]
André, E. 1896b. Liste des Hyménoptères appartenant aux familles des Formicides et des Mutillides recueillis au Siam et au Cambodge et offerts au Muséum par M. Pavie. Bull. Mus. Hist. Nat. 2:261-262. [1896.07.31]
André, E. 1896c. Description d'une nouvelle fourmi de France (Hymén.). Bull. Soc. Entomol. Fr. 1896:367-368. [1896.11.07]

André, E. 1896d. Fourmis nouvelles d'Asie et d'Australie. Rev. Entomol. (Caen) 15:251-265. [1896.11]

André, E. 1896e. Hyménoptères recueillis pendant les campagnes scientifiques de S. A. S. le Prince de Monaco. Bull. Soc. Zool. Fr. 21:210-211. [1896]

André, E. 1898. Description de deux nouvelles fourmis du Mexique (Hymén.). Bull. Soc. Entomol. Fr. 1898:244-247. [1898.07.09]

André, E. 1900. Sur la femelle probable de l'*Anomma nigricans* Ill. (Hyménoptère). Bull. Mus. Hist. Nat. 6:364-368. [1900.12.18]

André, E. 1902. Description de deux nouvelles fourmis du Pérou (Hymen.). Bull. Soc. Entomol. Fr. 1902:14-17. [1902.02.01]

André, E. 1903a. Hyménoptères Formicides, récoltés au Japon par M. J. Harmand. Bull. Mus. Hist. Nat. 9:128. [1903.04.28]

André, E. 1903b. Description d'une nouvelle espèce de *Dorymyrmex* et tableau dichotomique des ouvrières de ce genre (Hym.). Z. Syst. Hymenopterol. Dipterol. 3:364-365. [1903.11.01]

André, E. 1905. Description d'un genre nouveau et de deux espèces nouvelles de fourmis d'Australie. Rev. Entomol. (Caen) 24:205-208. [1905.08]

Andrewes, H. E. 1930. Catalogue of Indian insects. Part 18 - Carabidae. Calcutta: Government of India Central Publication Branch, xxii + 389 pp. [1930.12.12] [Page vii: year of publication of Macleay (1873) affirmed.]

Andrews, H. 1916a. A new ant of the genus *Messor* from Colorado. Psyche (Camb.) 23:81-83. [1916.06] [Stamp date in MCZ library: 1916.07.05.]

Andrews, H. 1916b. Ants caught on a trip to California (Hym.). Entomol. News 27:421-423. [1916.10.31]

Angus, C. J., Jones, M. K., Beattie, A. J. 1993. A possible explanation for size differences in the metapleural glands of ants (Hymenoptera: Formicidae). J. Aust. Entomol. Soc. 32:73-77. [1993.02.26]

Anonymous. 1969. Dictionary catalog of the National Agricultural Library 1862-1965. Volume 60. Taip - Thoq. New York: Rowman and Littlefield, iv + 2 + 786 pp. [1969] [Contains bibliographical details on Teranishi (1940).]

Anonymous. 1984. A list of ants collected at Unzen (Nagasaki-ken) by the members of the Myrmecologists Society (Japan) in 1983. [In Japanese.] Ari 12:5-6. [1984.05.01]

Anonymous. 1986. A list of ants collected at Akiyoshi-dai (Yamaguchi-ken) by members of the Myrmecologists Society (Japan) in 1985. Ari 14:5-6. [1986.01.31]

Anonymous. 1994. A list of ants collected at Bandai, Fukushima Prefecture by the members of the Myrmecological Society of Japan in 1992. [In Japanese.] Ari 18:31. [1994.05.30]

Antsiferov, V. M. 1980. Influence of some elements of relief and environment on the distribution of ants in mountain conditions. [In Russian.] Tr. Inst. Zool. Akad. Nauk Kaz. SSR 39:77-79. [> 1980.05.27] [Date signed for printing ("podpisano k pechati").]

Apostolov, L. G., Likhovidov, V. E. 1973. Contribution to the fauna and ecology of ants (Hymenoptera, Formicidae) from southeastern Ukraine. [In Russian.] Vestn. Zool. 1973(6):60-66. [1973.12] [November-December issue. Date signed for printing ("podpisano k pechati"): 22 November 1973.]

Arakelian, G. R. 1989. Previously unknown sexual individuals of the ant *Myrmica ravasinii* Finzi (Hymenoptera, Formicidae) from Armenia. [In Russian.] Biol. Zh. Arm. 42:733-736. [> 1989.10.27] [Date signed for printing.]

Arakelian, G. R. 1990. On the ant fauna of Khosrovsk Reserve. [In Russian.] Pp. 9-10 in: Tobias, V. I., L'vovskii, A. L. (eds.) Successes of entomology in the USSR: hymenopterous and lepidopterous insects. Materials of the 10th Congress of the All-Union Entomological Society, 11-15 September 1989. [In Russian.] Leningrad: Akademiya Nauk SSSR, 231 pp. [> 1990.04.27] [Date signed for printing ("podpisano k pechati").]

Arakelian, G. R. 1991. A new ant species of the genus *Diplorhoptrum* Mayr (Hymenoptera, Formicidae) from Armenia. [In Russian.] Dokl. Akad. Nauk Arm. 92:93-96. [> 1991.10.25] [Date signed for printing.]

Arakelian, G. R. 1994. Fauna of the Republic of Armenia. Hymenopterous insects. Ants (Formicidae). [In Russian.] Erevan: Gitutium, 153 pp. [> 1994.06.20] [Date signed for printing.]

Arakelian, G. R., Dlussky, G. M. 1991. Dacetine ants (Hymenoptera: Formicidae) of the USSR. [In Russian.] Zool. Zh. 70(2):149-152. [1991.02] [February issue. Date signed for printing ("podpisano k pechati"): 12 January 1991.]

Arakelyan, G. R. See: Arakelian, G. R.

Arnaud, P. H., Jr. 1978. A new northwestern record for the tribe Attini and northern record for the genus *Cyphomyrmex* (Hymenoptera: Formicidae). Pan-Pac. Entomol. 54:76. [1978.04.26]

Arnaud, P. H., Jr., Wale, M. M. 1973. Thomas Wrentmore Cook (1884-1962). Pan-Pac. Entomol. 49:177-181. [1973.07.11]

Arnett, R. H., Jr., Samuelson, G. A. 1986. The insect and spider collections of the world. Gainesville: E. J. Brill/Flora & Fauna Publications, 220 pp. [1986]

Arnett, R. H., Jr., Samuelson, G. A., Nishida, G. M. 1993. The insect and spider collections of the world. Second edition. Gainesville: Sandhill Crane Press, vi + 310 pp. [1993]

Arnold, G. 1905. *Formica fusca*, race *gagates*, in the New Forest. Entomol. Mon. Mag. 41:211-212. [1905.09]

Arnold, G. 1909. A large colony of *Formicoxenus nitidulus*, Nyl., and microgynes of *Formica fusca* in the New Forest. Entomol. Mon. Mag. 45:278-279. [1909.12]

Arnold, G. 1914. Nest-changing migrations of two species of ants. Proc. Trans. Rhod. Sci. Assoc. 13:25-32. [1914]

Arnold, G. 1915. A monograph of the Formicidae of South Africa. Part I. Ponerinae, Dorylinae. Ann. S. Afr. Mus. 14:1-159. [1915.02.11]

Arnold, G. 1916. A monograph of the Formicidae of South Africa. Part II. Ponerinae, Dorylinae. Ann. S. Afr. Mus. 14:159-270. [1916.05.20]

Arnold, G. 1917. A monograph of the Formicidae of South Africa. Part III. Myrmicinae. Ann. S. Afr. Mus. 14:271-402. [1917.08.09]

Arnold, G. 1920a. A monograph of the Formicidae of South Africa. Part IV. Myrmicinae. Ann. S. Afr. Mus. 14:403-578. [1920.04.16]

Arnold, G. 1920b. [Untitled. *Monomorium* (*Xeromyrmex*) *Salomonis* L., stirps *ocellatum* (Arnold).] P. 377 in: Santschi, F. Formicides africains et américains nouveaux. Ann. Soc. Entomol Fr. 88:361-390. [1920.06.30]

Arnold, G. 1922. A monograph of the Formicidae of South Africa. Part V. Myrmicinae. Ann. S. Afr. Mus. 14:579-674. [1922.10]

Arnold, G. 1924. A monograph of the Formicidae of South Africa. Part VI. Camponotinae. Ann. S. Afr. Mus. 14:675-766. [1924.04]

Arnold, G. 1926. A monograph of the Formicidae of South Africa. Appendix. Ann. S. Afr. Mus. 23:191-295. [1926.02]

Arnold, G. 1944. New species of African Hymenoptera. No. 5. Occas. Pap. Natl. Mus. South. Rhod. 2:1-38. [1944.07.17]

Arnold, G. 1946. New species of African Hymenoptera. No. 6. Occas. Pap. Natl. Mus. South. Rhod. 2:49-97. [1946.05.17]

Arnold, G. 1947. New species of African Hymenoptera. No. 7. Occas. Pap. Natl. Mus. South. Rhod. 2:131-167. [1947.11.05]

Arnold, G. 1948. New species of African Hymenoptera. No. 8. Occas. Pap. Natl. Mus. South. Rhod. 2:213-250. [1948.06.15]

Arnold, G. 1949. New species of African Hymenoptera. No. 9. Occas. Pap. Natl. Mus. South. Rhod. 2:261-275. [1949.12.29]

Arnold, G. 1951. The genus *Hagensia* Forel (Formicidae). J. Entomol. Soc. South. Afr. 14:53-56. [1951.05.31]

Arnold, G. 1952a. New species of African Hymenoptera. No. 10. Occas. Pap. Natl. Mus. South. Rhod. 2:460-493. [1952.07.02]

Arnold, G. 1952b. The genus *Terataner* Emery (Formicidae). J. Entomol. Soc. South. Afr. 15:129-131. [1952.11.30]

Arnold, G. 1953. Notes on a female *Dorylus* (*Anomma*) *Nigricans* Ill. taken with workers. J. Entomol. Soc. South. Afr. 16:141-142. [1953.11.30]

Arnold, G. 1954. New Formicidae from Kenya and Uganda. Ann. Mus. R. Congo Belge Nouv. Sér. Quarto Sci. Zool. 1:291-295. [1954.06]

Arnold, G. 1955. New species of African Hymenoptera. No. 11. Occas. Pap. Natl. Mus. South. Rhod. 2:733-762. [1955.11.02]

Arnold, G. 1956. New species of African Hymenoptera. No. 12. Occas. Pap. Natl. Mus. South. Rhod. B. Nat. Sci. 3:52-77. [1956.12]

Arnold, G. 1958. New species of African Hymenoptera. No. 13. Occas. Pap. Natl. Mus. South. Rhod. B. Nat. Sci. 3:119-143. [1958.07.16]

Arnold, G. 1959. New species of African Hymenoptera. No. 14. Occas. Pap. Natl. Mus. South. Rhod. B. Nat. Sci. 3:316-339. [1959.07.27]

Arnold, G. 1960a. Aculeate Hymenoptera from the Drakensberg Mountains, Natal. Ann. Natal Mus. 15:79-87. [1960.12.08]

Arnold, G. 1960b. New species of African Hymenoptera. No. 15. Occas. Pap. Natl. Mus. South. Rhod. B. Nat. Sci. 3:452-488. [1960.12.30]

Arnold, G. 1962. New species of African Hymenoptera. No. 16. Occas. Pap. Natl. Mus. South. Rhod. B. Nat. Sci. 3:844-855. [1962.12.31]

Arnoldi, K. V. See: Arnol'di, K. V.

Arnol'di, K. V. 1926. Studien über die Variabilität der Ameisen. I. Die ökologische und die Familienvariabilität von *Cardiocondyla stambulowi* For. Z. Morphol. Ökol. Tiere 7:254-278. [1926.12.23]

Arnol'di, K. V. 1928a. Studien über die Systematik der Ameisen. I. Allgemeiner Teil. Zool. Anz. 75:123-137. [1928.02.01]

Arnol'di, K. V. 1928b. Studien über die Systematik der Ameisen. II. *Stenamma* Westw. Zool. Anz. 75:199-215. [1928.02.15]

Arnol'di, K. V. 1928c. Studien über die Systematik der Ameisen. III. *Rossomyrmex*. Neue Gattung der Ameisen und ihre Beziehungen zu den anderen Gattungen der Formicidae. Zool. Anz. 75:299-310. [1928.03.01]

Arnol'di, K. V. 1930a. Studien über die Systematik der Ameisen. IV. *Aulacopone*, eine neue Ponerinengattung (Formicidae) in Russland. Zool. Anz. 89:139-144. [1930.06.10]

Arnol'di, K. V. 1930b. Studien über die Systematik der Ameisen. V. Der erste Vertreter der Tribe Proceratiini (Formicidae) in USSR. Zool. Anz. 91:143-146. [1930.10.05]

Arnol'di, K. V. 1930c. Studien über die Systematik der Ameisen. VI. Eine neue parasitische Ameise, mit Bezugnahme auf die Frage nach der Entstehung der Gattungsmerkmale bei den parasitären Ameisen. Zool. Anz. 91:267-283. [1930.10.20]

Arnol'di, K. V. 1930d. On the representatives of two tribes of ponerine ants new for the U.S.S.R. [In Russian.] Rus. Entomol. Obozr. 24:156-161. [1930]

Arnol'di, K. V. 1932a. Biologische Beobachtungen an der neuen paläarktischen Sklavenhalterameise *Rossomyrmex proformicarum* K. Arn., nebst einigen Bemerkungen über die Beförderungsweise der Ameisen. Z. Morphol. Ökol. Tiere 24:319-326. [1932.01.11]

Arnol'di, K. V. 1932b. Studien über die Systematik der Ameisen. VII. Die russischen Poneriden meiner Sammlung, teilweise biometrisch bearbeitet. Zool. Anz. 98:49-68. [1932.03.15]

Arnol'di, K. V. 1933a. Formicidae - ants. [In Russian.] Pp. 594-605 in: Filip'ev, I. N., Ogloblin, D. A. (eds.) Keys to insects. [In Russian.] Moskva: Ogiz, viii + 820 pp. [> 1933.12.03] [Date signed for printing ("podpisano k pechati").]

Arnol'di, K. V. 1933b. On a new genus of ant, in connection with the origin of generic traits of parasitic ants. [In Russian.] Rus. Entomol. Obozr. 25:40-51. [1933]

Arnol'di, K. V. 1934. Studien über die Systematik der Ameisen. VIII. Vorläufige Ergebnisse einer biometrischen Untersuchung einiger *Myrmica*-Arten aus dem europäischen Teile der USSR. Folia Zool. Hydrobiol. 6:151-174. [1934.10.10]

Arnol'di, K. V. 1937. Lebensformen der Ameisen. Dokl. Akad. Nauk SSSR (n.s.)16:335-338. [1937.09.02]

Arnol'di, K. V. 1948a. Ants of Talysh and the Diabar depression. Their importance for the characterization of communities of terrestrial invertebrates and for historical analysis of the fauna. [In Russian.] Tr. Zool. Inst. Akad. Nauk SSSR 7(2):206-262. [> 1948.01.27] [Date signed for printing ("podpisano k pechati").]

Arnol'di, K. V. 1948b. Fam. Formicidae - ants. [In Russian.] Pp. 769-780 in: Tarbinskii, S. P., Plavil'shchikov, N. N. (eds.) Keys to the insects of the European part of the USSR. [In Russian.] Moskva: Sel'khozgiz, 1128 pp. [1948] [Title page is dated "1948". Work was signed for printing "23/III 1946" (p. 1128). Assuming this is not a misprint, this signifies an unusual delay in publication.]

Arnol'di, K. V. 1957. On the theory of the areal in connection with the ecology and origin of populations. [In Russian.] Zool. Zh. 36:1609-1629. [1957.11] [November issue. Date signed for printing ("podpisano k pechati"): 10 November 1957.]

Arnol'di, K. V. 1964. Higher and specialized representatives of the ant genus *Cataglyphis* (Hymenoptera, Formicidae) in the fauna of the USSR. [In Russian.] Zool. Zh. 43:1800-1815. [1964.12] [December issue. Date signed for printing ("podpisano k pechati"): 26 November 1964.]

Arnol'di, K. V. 1967. New data on the ant genus *Camponotus* (Hymenoptera, Formicidae) from the USSR fauna. 1. *Camponotus* (s. str.). [In Russian.] Zool. Zh. 46:1815-1830. [1967.12] [December issue. Date signed for printing ("podpisano k pechati"): 6 December 1967.]

Arnol'di, K. V. 1968a. Zonal zoogeographic and ecologic features of myrmecofauna and the ant populations of the Russian Plain. [In Russian.] Zool. Zh. 47:1155-1178. [1968.08] [August issue. Date signed for printing ("podpisano k pechati"): 5 August 1968.]

Arnol'di, K. V. 1968b. Important additions to the myrmecofauna (Hymenoptera, Formicidae) of the USSR, and descriptions of new forms. [In Russian.] Zool. Zh. 47:1800-1822. [1968.12] [December issue. Date signed for printing ("podpisano k pechati"): 9 December 1968.]

Arnol'di, K. V. 1969. Die zonalen ökologischen und zoogeographischen Besonderheiten der Myrmekofauna der Russischen Ebene. Pedobiologia 9:215-222. [1969.05]

Arnol'di, K. V. 1970a. New species and races of the ant genus *Messor* (Hymenoptera, Formicidae). [In Russian.] Zool. Zh. 49:72-88. [1970.01] [January 1970 issue. Date signed for printing ("podpisano k pechati"): 22 December 1969.]

Arnol'di, K. V. 1970b. Review of the ant genus *Myrmica* (Hymenoptera, Formicidae) in the European part of the USSR. [In Russian.] Zool. Zh. 49:1829-1844. [1970.12] [December issue. Date signed for printing ("podpisano k pechati"): 27 November 1970.]

Arnol'di, K. V. 1971. New species and a review of the ant genus *Leptothorax* (Hymenoptera, Formicidae) of the Kazakstan plain. [In Russian.] Zool. Zh. 50:1818-1826. [1971.12] [December issue. Date signed for printing ("podpisano k pechati"): 2 December 1971.]

Arnol'di, K. V. 1972. Die zonalen ökologischen und zoogeographischen Besonderheiten der Myrmekofauna der Russischen Ebene. [Abstract.] P. 335 in: Rafes, P. M. (ed.) XIII International Congress of Entomology, Moscow, 2-9 August, 1968. Proceedings, Volume III. Leningrad: Nauka, 494 pp. [> 1972.06.08] [Date signed for printing ("podpisano k pechati").]

Arnol'di, K. V. 1975. A review of the species of the genus *Stenamma* (Hymenoptera, Formicidae) of the USSR and description of new species. [In Russian.] Zool. Zh. 54:1819-1829. [1975.12] [December issue. Date signed for printing ("podpisano k pechati"): 8 December 1975.]

Arnol'di, K. V. 1976a. Ants of the genus *Myrmica* Latr. from Central Asia and the southern Kazakstan. [In Russian.] Zool. Zh. 55:547-558. [1976.04] [April issue. Date signed for printing ("podpisano k pechati"): 18 March 1976.]

Arnol'di, K. V. 1976b. Review of the genus *Aphaenogaster* (Hymenoptera, Formicidae) in the USSR. [In Russian.] Zool. Zh. 55:1019-1026. [1976.07] [July issue. Date signed for printing ("podpisano k pechati"): 25 June 1976.]

Arnol'di, K. V. 1977a. New and little known ant species of the genus *Leptothorax* Mayr (Hymenoptera, Formicidae) in the European USSR and the Caucasus. [In Russian.] Entomol. Obozr. 56:198-204. [> 1977.02.28] [Date signed for printing ("podpisano k pechati"). Translated in Entomol. Rev. (Wash.) 56:148-153. See Arnol'di (1977c).]

Arnol'di, K. V. 1977b. Review of the harvester ants of the genus *Messor* (Hymenoptera, Formicidae) in the fauna of the USSR. [In Russian.] Zool. Zh. 56:1637-1648. [1977.11] [November issue. Date signed for printing ("podpisano k pechati"): 7 November 1977.]

Arnol'di, K. V. 1977c. New and little known species of the genus *Leptothorax* Mayr (Hymenoptera, Formicidae) in the European regions of the USSR and the Caucasus. Entomol. Rev. (Wash.) 56:148-153. [1977] [English translation of Arnol'di (1977a).]

Arnol'di, K. V., Dlussky, G. M. 1978. Superfam. Formicoidea. 1. Fam. Formicidae - ants. [In Russian.] Pp. 519-556 in: Medvedev, G. S. (ed.) Keys to the insects of the European part of the USSR. Vol. 3. Hymenoptera. Part 1. [In Russian.] Opredeliteli Faune SSSR 119:3-584. [> 1978.11.02] [Date signed for printing ("podpisano k pechati").]

Aron, S., Passera, L., Keller, L. 1994. Queen-worker conflict over sex ratio: a comparison of primary and secondary sex ratios in the Argentine ant, *Iridomyrmex humilis*. J. Evol. Biol. 7:403-418. [1994.07]

Aron, S., Vargo, E. L., Passera, L. 1995. Primary and secondary sex ratios in monogyne colonies of the fire ant. Anim. Behav. 49:749-757. [1995.03]

Ashmead, W. H. 1888. A proposed natural arrangement of the Hymenopterous families. Proc. Entomol. Soc. Wash. 1:96-99. [1888.03.01]

Ashmead, W. H. 1890. On the Hymenoptera of Colorado: descriptions of new species, notes and a list of the species found in the state. Bull. Colo. Biol. Assoc. 1:1-47. [1890] [Ants listed on pp. 36-37. No new taxa.]

Ashmead, W. H. 1892. The insect collections in the Berlin Museum. Proc. Entomol. Soc. Wash. 2:205-209. [1892.06.30]

Ashmead, W. H. 1896. The phylogeny of the Hymenoptera. Proc. Entomol. Soc. Wash. 3:323-336. [1896.10]

Ashmead, W. H. 1900. Report upon the Aculeate Hymenoptera of the Islands of St. Vincent and Grenada, with additions to the Parasitic Hymenoptera and a list of the described Hymenoptera of the West Indies. Trans. Entomol. Soc. Lond. 1900:207-367. [1900.07.14] [Ants pp. 316-323.]

Ashmead, W. H. 1902. Papers of the Harriman Alaska Expedition. XXVIII. Hymenoptera. Proc. Wash. Acad. Sci. 4:117-274. [1902.05.29] [List of ant species pp. 135-136. No new ant taxa.]

Ashmead, W. H. 1904a. A list of the Hymenoptera of the Philippine Islands, with descriptions of new species. J. N. Y. Entomol. Soc. 12:1-22. [1904.03] [List of ant species pp. 9-11. No new ant taxa.]

Ashmead, W. H. 1904b. Descriptions of new genera and species of Hymenoptera from the Philippine Islands. Proc. U. S. Natl. Mus. 28:127-158. [1904.11.05] [No new ant taxa. Repeats the list of ant species in Ashmead (1904a).]

Ashmead, W. H. 1905a. Additions to the recorded Hymenopterous fauna of the Philippine Islands, with descriptions of new species. Proc. U. S. Natl. Mus. 28:957-971. [1905.07.08] [No new ant taxa.]

Ashmead, W. H. 1905b. New Hymenoptera from the Philippines. Proc. U. S. Natl. Mus. 29:107-119. [1905.09.30] [Two new ant species described pp. 110-112. *Colobopsis albocincta*, *Aphomyrmex emeryi*.]

Ashmead, W. H. 1905c. A skeleton of a new arrangement of the families, subfamilies, tribes and genera of the ants, or the superfamily Formicoidea. Can. Entomol. 37:381-384. [1905.11.01]
Ashmead, W. H. 1905d. Doctor Ashmead's new classification of the Poneridae. Pp. 38-41 in: Cook, O. F. The social organization and breeding habits of the cotton-protecting kelep of Guatemala. U. S. Dep. Agric. Bur. Entomol. Tech. Ser. 10:1-55. [1905]
Ashmead, W. H. 1906. Classification of the foraging and driver ants, or Family Dorylidae, with a description of the genus *Ctenopyga* Ashm. Proc. Entomol. Soc. Wash. 8:21-31. [1906.07.13]
Assing, V. 1984. Investigations on the ant fauna (Hym., Formicidae) of *Calluna* heathlands in northwestern Germany. [Abstract.] P. 498 in: XVII International Congress of Entomology. Hamburg, Federal Republic of Germany, August 20-26, 1984. Abstract Volume. Hamburg: XVII International Congress of Entomology, 960 pp. [1984.08.10]
Assing, V. 1986. Distribution, densities and activity patterns of the ants (Hymenoptera: Formicidae) of *Calluna* heathlands in northwestern Germany. Entomol. Gen. 11:183-190. [1986.05.30]
Assing, V. 1989. Die Ameisenfauna (Hym.: Formicidae) nordwestdeutscher *Calluna*-Heiden. Drosera 89:49-62. [1989.11]
Assing, V. 1994. *Myrmica hellenica* Forel, 1913, in Kärnten (Hymenoptera: Formicidae). Carinthia II 104:298.
Assmann, A. 1870. Beiträge zur Insekten-Fauna der Vorwelt. I. Beitrag. Die fossilen Insekten des tertiären (miocenen) Thonlagers von Schossnitz bei Kanth. Z. Entomol. (Breslau) (N.F.)1: 33-58. [1870.03]
Atanasov, N. See: Atanassov, N.
Atanassov, N. 1934. Contribution to a study of the ant fauna (Formicidae) of Bulgaria. [In Bulgarian.] Izv. Bulg. Entomol. Druzh. 8:159-173. [1934.07.01]
Atanassov, N. 1936. Second contribution to a study of the ant fauna (Formicidae) of Bulgaria. [In Bulgarian.] Izv. Bulg. Entomol. Druzh. 9:211-236. [1936.05.01]
Atanassov, N. 1952. Regularities in the distribution and biological observations on ants in the Vitosha Mountains. [In Bulgarian.] Sofia: Bulgarska Akademiya na Naukite, 214 pp. [1952]
Atanassov, N. 1964. Studies on the systematics and ecology of ants (Formicidae, Hym.) from the Petrich region (south-west Bulgaria). [In Bulgarian.] Izv. Zool. Inst. Muz. Bulg. Akad. Nauk. 15:77-104. [> 1964.03.19] [Date signed for printing.]
Atanassov, N. 1965. Étude sur la biologie et la répartition de *Monomorium pharaonis* L. (Hym. Formicidae) en Bulgarie et dans la péninsule balkanique. Bull. Soc. Entomol. Mulhouse 1965:86-95. [1965.12]
Atanassov, N. 1974. Besonderheiten der Nahrungszusammensetzung von *Formica rufa* L. und *Formica lugubris* Zett. in Bulgarien. Waldhygiene 10:183-185. [1974]
Atanassov, N. 1982. Neue Ameisen aus den Gattungen *Messor* und *Cataglyphis* (Hymenoptera, Formicidae) für die Fauna Bulgariens. Waldhygiene 14:209-214. [1982]
Atanassov, N., Dlussky, G. M. 1992. Fauna of Bulgaria. Hymenoptera, Formicidae. [In Bulgarian.] Fauna Bûlg. 22:1-310. [1992.11] [Date signed for printing ("podpisana za pechat").]
Atanassow, N. See: Atanassov, N.
Athias-Henriot, C. 1947 ("1946.") Notes sur les caractères de la faune des fourmis aux environs de Béni-Ounif de Figuig (Sud Oranais). Bull. Soc. Hist. Nat. Afr. Nord 37:60-63. [1947.05.02]
Attygalle, A. B., Billen, J. P. J., Jackson, B. D., Morgan, E. D. 1990. Morphology and chemical contents of Dufour glands of *Pseudomyrmex* ants (Hymenoptera: Formicidae). Z. Naturforsch. C J. Biosci. 45:691-697. [1990.06]
Attygalle, A. B., Billen, J. P. J., Morgan, E. D. 1985. The postpharyngeal gland of workers of *Solenopsis geminata* (Hymenoptera: Formicidae). Actes Colloq. Insectes Soc. 2:75-86. [1985.06]
Attygalle, A. B., Evershed, R. P., Morgan, E. D., Cammaerts, M. C. 1983. Dufour gland secretions of workers of the ants *Myrmica sulcinodis* and *Myrmica lobicornis*, and comparison with six other species in *Myrmica*. Insect Biochem. 13:507-512. [1983]

Attygalle, A. B., Morgan, E. D. 1984. Chemicals from the glands of ants. Chem. Soc. Rev. 13:245-278. [1984.09]

Attygalle, A. B., Morgan, E. D. 1985. Ant trail pheromones. Adv. Insect Physiol. 18:1-30. [1985]

Attygalle, A. B., Siegel, B., Vostrowsky, O., Bestmann, H. J., Maschwitz, U. 1989. Chemical composition and function of metapleural gland secretion of the ant *Crematogaster deformis* Smith (Hymenoptera: Myrmicinae). J. Chem. Ecol. 15:317-328. [1989.01]

Ayre, G. L. 1967. The relationships between food and digestive enzymes in five species of ants (Hymenoptera: Formicidae). Can. Entomol. 99:408-411. [1967.05.05]

Azuma, M. 1938. A list of ants found in Osaka Prefecture, Japan. [In Japanese.] Entomol. World Tokyo 6:238-243. [1938.04.15]

Azuma, M. 1950a. On the myrmecological-fauna of Tomogashima, Kii Prov., with the description of new genus and new species. [In Japanese.] Hyogo Biol. 1:34-37. [1950.01.10] [See discussion of publication date in Brown & Yasumatsu (1951).]

Azuma, M. 1950b. *Camponotus* (s. str.) *herculeanus* (Linnaeus) subsp. *ligniperda* (Latreille) var. *obscuripes* Mayr 1878. [In Japanese.] Entomol. Rev. Jpn. 5:47-48. [1950.07.15]

Azuma, M. 1951. On the myrmecological-fauna of Osaka Prefecture, Japan with description of new species (Formicidae, Hymenoptera). [In Japanese and English.] Hyogo Biol. 1:86-90. [1951.01.20]

Azuma, M. 1953. On the myrmecological fauna of Mt. Rokko, Hyogo Prefecture. [In Japanese.] Warera 2:1-7. [Not seen. Cited in Onoyama (1980) and Onoyama & Terayama (1994).]

Azuma, M. 1955. A list of ants (Formicidae) from Hokkaido Is. [In Japanese.] Hyogo Biol. 3: 79-80. [1955.11.10]

Azuma, M. 1977a. On the myrmecological-fauna of Mt. Rokko, Hyogo, with description of a new species (Formicidae, Hymenoptera). [In Japanese.] Hyogo Biol. 7:112-118. [1977.02]

Azuma, M. 1977b. Formicidae. [In Japanese.] Pp. 319-324 in: Ito, S., Okutani, T., Hiura, I. (eds.) Colored illustrations of the insects of Japan. Vol. 2. [In Japanese.] Osaka: Hoikusha Co. Ltd. [Not seen. Cited in Onoyama (1980) and Onoyama & Terayama (1994).]

Baba, K. 1935. Some hymenopterous insects from Sado Island. [In Japanese.] Mushi 8:83-85. [1935.12.31] [Ants p. 85.]

Baba, K., Inoue, S. 1936. The list of insects collected in Awashima Is., Nigata. [In Japanese.] Insect World 40:401-403. [1936.11.15] [Ants p. 402.]

Baba, Y. 1987. On the ants of Shirakawa, Fukushima Prefecture. [Abstract.] [In Japanese.] Ari 15:4. [1987.10.10]

Bachmayer, F. 1960. Insektenreste aus den Congerienschichten (Pannon) von Brunn-Vösendorf (südl. von Wien) Niederösterreich. Sitzungsber. Österr. Akad. Wiss. Math.-Naturwiss. Kl. Abt. I 169:11-16. [1960]

Baena, M. L., Alberico, M. 1991. Relaciones biogeográficas de las hormigas de la Isla Gorgona. Rev. Colomb. Entomol. 17(2):24-31. [1991.12] [Possibly published later, but no precise publication date given. Issue labelled "Julio-Diciembre 1991".]

Báez, M., Ortega, G. 1978. Lista preliminar de los Himenópteros de las Islas Canarias. Bol. Asoc. Esp. Entomol. 2:185-199. [1978.12] [Ants pp. 189-190.]

Baggini, A., Pavan, M., Ronchetti, G., Valcurone, M. L. 1959. Primi cenni sui risultati del censimento in corso delle formiche del gruppo *Formica rufa* sulle Alpi italiane. Not. For. Mont. 4:1914-1916. [1959.08]

Bagnall, R. S. 1906a. *Formicoxenus nitidulus*, Nyl., in the Northumberland and Durham district. Entomol. Mon. Mag. 42:140. [1906.06]

Bagnall, R. S. 1906b. *Formicoxenus nitidulus*, Nyl., male, as British. Entomol. Mon. Mag. 42:210. [1906.09]

Bagnères, A.-G., Billen, J. P. J., Morgan, E. D. 1991. Volatile secretion of Dufour gland of workers of an army ant, *Dorylus* (*Anomma*) *molestus*. J. Chem. Ecol. 17:1633-1639. [1991.08]

Bagnères, A.-G., Morgan, E. D., Clement, J.-L. 1991. Species-specific secretions of the Dufour glands of three species of formicine ants (Hymenoptera, Formicidae). Biochem. Syst. Ecol. 19:25-33. [1991.03.07]

Bahntje, E. 1939a. Beiträge zu einer Ameisenfauna des Landes Braunschweig. [part] Entomol. Rundsch. 56:175-178. [1939.05.01]

Bahntje, E. 1939b. Beiträge zu einer Ameisenfauna des Landes Braunschweig. [part] Entomol. Rundsch. 56:199-202. [1939.05.15]

Bahntje, E. 1939c. Beiträge zu einer Ameisenfauna des Landes Braunschweig. [concl.] Entomol. Rundsch. 56:228-230. [1939.06.01]

Baiocco, L. M., Cunha, M. S. da. 1992. Histologia das glândulas exócrinas em rainhas de *Monomorium pharaonis* (Linneaus, 1758) [sic] (Hymenoptera, Formicidae). Naturalia (São Paulo) 17:139-152. [1992]

Baird, A. B. 1922 ("1921"). Some notes on the female reproductive organs in the Hymenoptera. Proc. Acadian Entomol. Soc. 7:73-88. [1922.06]

Baker, D. B. 1994. The date of the Hymenoptera section of the Exploration scientifique de l'Algérie. Arch. Nat. Hist. 21:345-350. [1994.10]

Baldridge, R. S., Rettenmeyer, C. W., Watkins, J. F., II. 1980. Seasonal, nocturnal and diurnal flight periodicities of Nearctic army ant males (Hymenoptera: Formicidae). J. Kansas Entomol. Soc. 53:189-204. [1980.01.02]

Ball, D. E., Vinson, S. B. 1984. Anatomy and histology of the male reproductive system of the fire ant, *Solenopsis invicta* Buren (Hymenoptera: Formicidae). Int. J. Insect Morphol. Embryol. 13:283-294. [1984.08]

Ball, G. E. (ed.) 1985. Taxonomy, phylogeny, and zoogeography of beetles and ants: a volume dedicated to the memory of Philip Jackson Darlington, Jr., 1904-1983. Ser. Entomol. (Dordr.) 33:1-514. [1985]

Baltazar, C. R. 1966. A catalogue of Philippine Hymenoptera (with a bibliography, 1758-1963). Pac. Insects Monogr. 8:1-488. [1966.04.30]

Banert, P., Pisarski, B. 1972. Mrówki (Formicidae) Sudetów. Fragm. Faun. (Warsaw) 18:345-359. [1972.11.25]

Barandica, J. M., López, F., Martínez, M. D., Ortuño, V. M. 1994. The larvae of *Leptanilla charonea* and *Leptanilla zaballosi* (Hymenoptera, Formicidae). Dtsch. Entomol. Z. (N.F.)41:147-153. [1994.03.04]

Barker, E. E. 1903. The bull-ants of Victoria. Vic. Nat. (Melb.) 20:104-111. [1903.12.10]

Baroni Urbani, C. 1962. Studi sulla mirmecofauna d'Italia. I. Redia 47:129-138. [1962]

Baroni Urbani, C. 1964a. Su alcune formiche raccolte in Turchia. Annu. Ist. Mus. Zool. Univ. Napoli 16:1-12. [1964.02]

Baroni Urbani, C. 1964b. Studi sulla mirmecofauna d'Italia. II. Formiche di Sicilia. Atti Accad. Gioenia Sci. Nat. (6)16:25-66. [1964.12.30]

Baroni Urbani, C. 1964c. Formiche dell'Italia appenninica (Studi sulla mirmecofauna d'Italia, III). Mem. Mus. Civ. Stor. Nat. Verona 12:149-172. [1964.12.31]

Baroni Urbani, C. 1966. Osservazioni diverse intorno al nomadismo dell'*Aphaenogaster picena* Baroni con particolare riguardo all'orientamento (Hymenoptera, Formicidae). Insectes Soc. 13:69-86. [1966.09]

Baroni Urbani, C. 1967. Le distribuzioni geografiche discontinue dei Formicidi mirmecobiotici. Arch. Bot. Biogeogr. Ital. 43[=(4)12]:355-365. [1967]

Baroni Urbani, C. 1968a. Studi sulla mirmecofauna d'Italia. V. Aspetti ecologici della Riviera del M. Cònero. Boll. Zool. 35:39-76. [1968.08.02]

Baroni Urbani, C. 1968b. Über die eigenartige Morphologie der männlichen Genitalien des Genus *Diplorhoptrum* Mayr (Hymenoptera Formicidae) und die taxonomischen Schlussfolgerungen. Z. Morphol. Tiere 63:63-74. [1968.09.17]

Baroni Urbani, C. 1968c. Ultrastruttura comparata dei muscoli dell'alitronco dei Formicidi. Boll. Zool. 35:203-217. [1968.10.02]

Baroni Urbani, C. 1968d. Monogyny in ant societies. Zool. Anz. 181:269-277. [1968.10]

Baroni Urbani, C. 1968e. Studi sulla mirmecofauna d'Italia. IV. La fauna mirmecologica delle isole Maltesi ed il suo significato ecologico e biogeografico. Ann. Mus. Civ. Stor. Nat. "Giacomo Doria" 77:408-559. [1968.12.01]

Baroni Urbani, C. 1968f. Studi sulla mirmecofaune d'Italia. VI. Il popolamento mirmecologico delle isole Maltesi. Arch. Bot. Biogeogr. Ital. 44:224-241. [1968]

Baroni Urbani, C. 1969a. Una nuova specie di *Oligomyrmex* del Sahara meridionale (Hymenoptera Formicidae). Boll. Soc. Entomol. Ital. 99-101:74-77. [1969.04.20]

Baroni Urbani, C. 1969b. Gli *Strongylognathus* del gruppo *huberi* nell'Europa occidentale: saggio di una revisione basata sulla casta operaia (Hymenoptera Formicidae). Boll. Soc. Entomol. Ital. 99-101:132-168. [1969.10.20]

Baroni Urbani, C. 1969c. Una nuova *Cataglyphis* dei Monti dell'Anatolia (Hymenoptera: Formicidae). Fragm. Entomol. 6:213-222. [1969.11.10]

Baroni Urbani, C. 1969d ("1968"). Studi sulla mirmecofauna d'Italia. VII. L'Isola di Montecristo. Atti Soc. Toscana Sci. Nat. Mem. Ser. B 75:95-107. [1969] [Volume 75 is dated "1968" but this is contradicted by receipt dates of manuscript proofs as late as 22 February 1969.]

Baroni Urbani, C. 1969e ("1968"). Studi sulla mirmecofauna d'Italia. VIII. L'Isola di Giannutri ed alcuni scogli minori dell'arcipelago Toscano. Atti Soc. Toscana Sci. Nat. Mem. Ser. B 75:325-338. [1969] [Volume 75 is dated "1968" but this is contradicted by receipt dates of manuscript proofs as late as 22 February 1969.]

Baroni Urbani, C. 1969f. Ant communities of the high-altitude Appennine grasslands. Ecology 50:488-492. [1969]

Baroni Urbani, C. 1971a. Studi sulla mirmecofauna d'Italia. IX. Una nuova specie di *Aphaenogaster* (Hymenoptera Formicidae). Boll. Soc. Entomol. Ital. 103:32-41. [1971.02.21]

Baroni Urbani, C. 1971b. Einige Homonymien in der Familie Formicidae (Hymenoptera). Mitt. Schweiz. Entomol. Ges. 44:360-362. [1971.12.20]

Baroni Urbani, C. 1971c. Catalogo delle specie di Formicidae d'Italia (Studi sulla mirmecofauna d'Italia X). Mem. Soc. Entomol. Ital. 50:5-287. [1971.12.30]

Baroni Urbani, C. 1971d. Studien zur Ameisenfauna Italiens XI. Die Ameisen des Toskanischen Archipels. Betrachtungen zur Herkunft der Inselfaunen. Rev. Suisse Zool. 78:1037-1067. [1971.12]

Baroni Urbani, C. 1972. Studi sui *Camponotus* (Hymenoptera, Formicidae). Verh. Naturforsch. Ges. Basel 82:122-135. [1972.04]

Baroni Urbani, C. 1973a. Die Gattung *Xenometra*, ein objektives Synonym (Hymenoptera, Formicidae). Mitt. Schweiz. Entomol. Ges. 46:199-201. [1973.12.28]

Baroni Urbani, C. 1973b. Katalog der Typen von Formicidae (Hymenoptera) der Sammlung des Naturhistorischen Museums Basel. Mitt. Entomol. Ges. Basel (n.s.)23:122-152. [1973.12]

Baroni Urbani, C. 1974a. Studi sulla mirmecofauna d'Italia. XII. Le Isole Pontine. Fragm. Entomol. 9:225-252. [1974.02.15]

Baroni Urbani, C. 1974b. Compétition et association dans les biocénoses de fourmis insulaires. Rev. Suisse Zool. 81:103-135. [1974.04]

Baroni Urbani, C. 1974c. Polymorphismus in der Ameisengattung *Camponotus* aus morphologischer Sicht. Pp. 543-564 in: Schmidt, G. H. (ed.) Sozialpolymorphismus bei Insekten. Probleme der Kastenbildung im Tierreich. Stuttgart: Wissenschaftliche Verlagsgesellschaft mbH, xxiv + 974 pp. [1974]

Baroni Urbani, C. 1975a. Primi reperti del genere *Calyptomyrmex* Emery nel subcontinente Indiano. Entomol. Basil. 1:395-411. [1975.10.01]

Baroni Urbani, C. 1975b. Contributo alla conoscenza dei generi *Belonopelta* Mayr e *Leiopelta* gen. n. (Hymenoptera: Formicidae). Mitt. Schweiz. Entomol. Ges. 48:295-310. [1975.12.28]

Baroni Urbani, C. 1976a. Réinterprétation du polymorphisme de la caste ouvrière chez les fourmis à l'aide de la régression polynomiale. Rev. Suisse Zool. 83:105-110. [1976.03]

Baroni Urbani, C. 1976b. Le formiche dell'arcipelago della Galita (Tunisia). Redia 59:207-223. [1976.11.10]

Baroni Urbani, C. 1977a. Ergebnisse der Bhutan-Expedition 1972 des Naturhistorischen Museums in Basel. Hymenoptera: Fam. Formicidae Genus *Mayriella*. Entomol. Basil. 2:411-414. [1977.01.07]

Baroni Urbani, C. 1977b. Ergebnisse der Bhutan-Expedition 1972 des Naturhistorischen Museums in Basel. Hymenoptera: Fam. Formicidae Genus *Stenamma*, con una nuova specie del Kashmir. Entomol. Basil. 2:415-422. [1977.01.07]

Baroni Urbani, C. 1977c. Materiali per una revisione della sottofamiglia Leptanillinae Emery (Hymenoptera: Formicidae). Entomol. Basil. 2:427-488. [1977.01.07]

Baroni Urbani, C. 1977d. Les espèces européennes du genre *Proceratium* Roger (Hymenoptera: Formicidae). Mitt. Schweiz. Entomol. Ges. 50:91-93. [1977.03.31]

Baroni Urbani, C. 1977e. Katalog der Typen von Formicidae (Hymenoptera) der Sammlung des Naturhistorischen Museums Basel (2. Teil). Mitt. Entomol. Ges. Basel (n.s.)27:61-102. [1977.09]

Baroni Urbani, C. 1977f. *Discothyrea Stumperi* n. sp. du Bhoutan, premier représentant du genre dans le Subcontinent Indien (Hymenoptera: Formicidae). Arch. Inst. Grand-Ducal Luxemb. (n.s.)37:97-101. [1977]

Baroni Urbani, C. 1978a. Contributo alla conoscenza del genere *Amblyopone* Erichson (Hymenoptera: Formicidae). Mitt. Schweiz. Entomol. Ges. 51:39-51. [1978.02.15]

Baroni Urbani, C. 1978b. Materiali per una revisione dei *Leptothorax* neotropicali appartenenti al sottogenere *Macromischa* Roger, n. comb. (Hymenoptera: Formicidae). Entomol. Basil. 3:395-618. [1978.10.18]

Baroni Urbani, C. 1978c. Analyse de quelques facteurs autécologiques influençant la microdistribution des fourmis dans les îles de l'archipel toscan. Mitt. Schweiz. Entomol. Ges. 51:367-376. [1978.12.31]

Baroni Urbani, C. 1978d. Appendix 1. Table 1. Adult populations in ant colonies. Pp. 334-335 in: Brian, M. V. (ed.) Production ecology of ants and termites. IBP vol. 13. Cambridge: Cambridge University Press, xvii + 409 pp. [1978]

Baroni Urbani, C. 1980a. Die Stellung der Gattung *Aratromyrmex* Stitz (Hymenoptera: Formicidae). Mitt. Zool. Mus. Berl. 56:95-96. [1980.04.28]

Baroni Urbani, C. 1980b. First description of fossil gardening ants. (Amber Collection Stuttgart and Natural History Museum Basel; Hymenoptera: Formicidae. I: Attini). Stuttg. Beitr. Naturkd. Ser. B (Geol. Paläontol.) 54:1-13. [1980.07.31]

Baroni Urbani, C. 1980c. *Anochetus corayi* n. sp., the first fossil Odontomachiti ant. (Amber Collection Stuttgart: Hymenoptera, Formicidae. II: Odontomachiti). Stuttg. Beitr. Naturkd. Ser. B (Geol. Paläontol.) 55:1-6. [1980.07.31]

Baroni Urbani, C. 1980d. The first fossil species of the Australian ant genus *Leptomyrmex* in amber from the Dominican Republic. (Amber Collection Stuttgart: Hymenoptera, Formicidae. III: Leptomyrmicini). Stuttg. Beitr. Naturkd. Ser. B (Geol. Paläontol.) 62:1-10. [1980.12.01]

Baroni Urbani, C. 1980e. The ant genus *Gnamptogenys* in Dominican Amber. (Amber Collection Stuttgart: Hymenoptera, Formicidae. IV: Ectatommini). Stuttg. Beitr. Naturkd. Ser. B (Geol. Paläontol.) 67:1-10. [1980.12.15]

Baroni Urbani, C. 1984 ("1983"). Clave para la determinación de los géneros de hormigas neotropicales. Graellsia 39:73-82. [1984.02.10]

Baroni Urbani, C. 1987. Comparative feeding strategies in two harvesting ants. Pp. 509-510 in: Eder, J., Rembold, H. (eds.) Chemistry and biology of social insects. München: Verlag J. Peperny, xxxv + 757 pp. [1987]

Baroni Urbani, C. 1989. Phylogeny and behavioural evolution in ants, with a discussion of the role of behaviour in evolutionary processes. Ethol. Ecol. Evol. 1:137-168. [1989.10]

Baroni Urbani, C. 1990. Comparing different hypotheses about the origins and patterns of ant diversity. Pp. 309-310 in: Veeresh, G. K., Mallik, B., Viraktamath, C. A. (eds.) Social insects and the environment. Proceedings of the 11th International Congress of IUSSI, 1990. New Delhi: Oxford & IBH Publishing Co., xxxi + 765 pp. [1990.08]

Baroni Urbani, C. 1993. The diversity and evolution of recruitment behaviour in ants, with a discussion of the usefulness of parsimony criteria in the reconstruction of evolutionary histories. Insectes Soc. 40:233-260. [1993]

Baroni Urbani, C. 1994a. The identity of the Dominican *Paraponera* (Amber Collection Stuttgart: Hymenoptera, Formicidae. V: Ponerinae, partim). Stuttg. Beitr. Naturkd. Ser. B (Geol. Paläontol.) 197:1-9. [1994.03.15]

Baroni Urbani, C. 1994b. [Untitled. *Strumigenys schleeorum* Baroni Urbani n. sp.] Pp. 35-38 in: Baroni Urbani, C., De Andrade, M. L. First description of fossil Dacetini ants with a critical analysis of the current classification of the tribe (Amber Collection Stuttgart: Hymenoptera, Formicidae. VI: Dacetini). Stuttg. Beitr. Naturkd. Ser. B (Geol. Paläontol.) 198:1-65. [1994.04.20]

Baroni Urbani, C. 1994c. [Untitled. *Acanthognathus poinari* Baroni Urbani n. sp.] Pp. 41-44 in: Baroni Urbani, C., De Andrade, M. L. First description of fossil Dacetini ants with a critical analysis of the current classification of the tribe (Amber Collection Stuttgart: Hymenoptera, Formicidae. VI: Dacetini). Stuttg. Beitr. Naturkd. Ser. B (Geol. Paläontol.) 198:1-65. [1994.04.20]

Baroni Urbani, C. 1995. Invasion and extinction in the West Indian ant fauna revisited: the example of *Pheidole* (Amber Collection Stuttgart: Hymenoptera, Formicidae. VIII: Myrmicinae, partim). Stuttg. Beitr. Naturkd. Ser. B (Geol. Paläontol.) 222:1-12. [1995.03.20]

Baroni Urbani, C., Aktaç, N. 1981. The competition for food and circadian succession in the ant fauna of a representitive Anatolian semi-steppic environment. Mitt. Schweiz. Entomol. Ges. 54:33-56. [1981.06.30]

Baroni Urbani, C., Aktaç, N., Camlitepe, Y. 1989. Disclosing the mystery of *Messor caducus* Motschulsky (Hymenoptera, Formicidae). Mitt. Schweiz. Entomol. Ges. 62:291-301. [1989.12.15]

Baroni Urbani, C., Bolton, B., Ward, P. S. 1992. The internal phylogeny of ants (Hymenoptera: Formicidae). Syst. Entomol. 17:301-329. [1992.10]

Baroni Urbani, C., Collingwood, C. A. 1976. A numerical analysis of the distribution of British Formicidae (Hymenoptera, Aculeata). Verh. Naturforsch. Ges. Basel 85:51-91. [1976.03.31]

Baroni Urbani, C., Collingwood, C. A. 1977. The zoogeography of ants (Hymenoptera, Formicidae) in northern Europe. Acta Zool. Fenn. 152:1-34. [1977.11]

Baroni Urbani, C., De Andrade, M. L. 1993. *Perissomyrmex monticola* n. sp., from Bhutan: the first natural record for a presumed Neotropical genus with a discussion on its taxonomic status. Trop. Zool. 6:89-95. [1993.05]

Baroni Urbani, C., De Andrade, M. L. 1994. First description of fossil Dacetini ants with a critical analysis of the current classification of the tribe (Amber Collection Stuttgart: Hymenoptera, Formicidae. VI: Dacetini). Stuttg. Beitr. Naturkd. Ser. B (Geol. Paläontol.) 198:1-65. [1994.04.20]

Baroni Urbani, C., Graeser, S. 1987. REM-Analysen an einer pyritisierten Ameise aus Baltischem Bernstein. Stuttg. Beitr. Naturkd. Ser. B (Geol. Paläontol.) 133:1-16. [1987.12.15]

Baroni Urbani, C., Kutter, N. 1979. Première analyse biométrique du polymorphisme de la caste ouvrière chez les fourmis du genre *Pheidologeton* (Hymenoptera: Aculeata). Mitt. Schweiz. Entomol. Ges. 52:377-389. [1979.12.31]

Baroni Urbani, C., Pisarski, B. 1977. Appendix 1. Table 2. Density of ant colonies and individuals in various regions and biotopes. Pp. 336-339 in: Brian, M. V. (ed.) Production ecology of ants and termites. IBP vol. 13. Cambridge: Cambridge University Press, xvii + 409 pp. [1977]

Baroni Urbani, C., Saunders, J. B. 1982 ("1980"). The fauna of the Dominican Republic amber: the present status of knowledge. Pp. 213-223 in: Collado, N. J. (ed.) Transactions of the 9th Caribbean Geological Conference, Santo Domingo, Dominican Republic, 1980. Volume 1. Santo Domingo: Ninth Caribbean Geological Conference, xx + 389 pp. [1982]

Baroni Urbani, C., Stemmler, O., Wittmer, W., Würmli, M. 1973. Zoologische Expedition des Naturhistorischen Museums Basel in das Königreich Bhutan. Verh. Naturforsch. Ges. Basel 83:319-336. [1973.12.31]

Baroni Urbani, C., Wilson, E. O. 1987. The fossil members of the ant tribe Leptomyrmecini (Hymenoptera: Formicidae). Psyche (Camb.) 94:1-8. [1987.10.06]

Barquín, J. 1981. Las hormigas de Canarias. Taxonomía, ecología y distribución de los Formicidae. Colecc. Monogr. (Secr. Publ. Univ. La Laguna) 3:1-584. [1981]

Barr, D., Boven, J. K. A. van, Gotwald, W. H., Jr. 1985. Phenetic studies of African army ant queens of the genus *Dorylus* (Hymenoptera: Formicidae). Syst. Entomol. 10:1-10. [1985.01]

Barr, D., Gotwald, W. H., Jr. 1982. Phenetic affinities of males of the army ant genus *Dorylus* (Hymenoptera: Formicidae: Dorylinae). Can. J. Zool. 60:2652-2658. [1982.12.10]

Barrett, C. 1927. Ant life in Central Australia. Vic. Nat. (Melb.) 44:209-212. [1927.12.07]

Barrett, K. E. J. 1963. Ants (Hym., Formicidae) from the Chobham Common Area of Surrey. Entomol. Rec. J. Var. 75:29-30. [1963.01.15]

Barrett, K. E. J. 1964. Ant records and observations for 1964. Entomol. Rec. J. Var. 76:287-289. [1964.12.15]

Barrett, K. E. J. 1965. Ant records and observations for 1965. Entomol. Rec. J. Var. 77:248-252. [1965.11.15]

Barrett, K. E. J. 1967. Ants in south Brittany. Entomol. Rec. J. Var. 79:112-116. [1967.04.15]

Barrett, K. E. J. 1968a. A survey of the distribution and present status of the wood ant, *Formica rufa* L. (Hym., Formicidae) in England and Wales. Trans. Soc. Br. Entomol. 17:217-233. [1968.03]

Barrett, K. E. J. 1968b. The distribution of ants in central southern England. Trans. Soc. Br. Entomol. 17:235-250. [1968.03]

Barrett, K. E. J. 1968c. Ants in western France. Entomologist 101:153-155. [1968.07.26]

Barrett, K. E. J. 1969 ("1968"). Some new county records for British ants (Hymenoptera: Formicidae). Entomol. Mon. Mag. 104:260. [1969.04.01]

Barrett, K. E. J. 1970. Ants in France, 1968-69. Entomologist 103:270-274. [1970.11.25]

Barrett, K. E. J. 1973 ("1972"). Some new records for British ants (Hym., Formicidae). Entomol. Mon. Mag. 108:256. [1973.09.25]

Barrett, K. E. J. 1974. Distribution maps scheme for ants in Britain. Entomol. Gaz. 25:212-216. [1974.08.05]

Barrett, K. E. J. 1977. Provisional distribution maps of ants in the British Isles. Pp. 203-216 in: Brian, M. V. Ants. London: Collins, 223 pp. [1977]

Barrett, K. E. J. 1979. Provisional atlas of the insects of the British Isles. Part 5. Hymenoptera: Formicidae, Ants. 2nd ed. Huntingdon, U.K.: Biological Records Centre, Institute of Terrestrial Ecology, 56 pp. [1979]

Barrett, K. E. J., Felton, F. 1965. The distribution of the wood ant, *Formica rufa* Linnaeus (Hymenoptera, Formicidae), in south-east England. Entomologist 98:181-191. [1965.08.19]

Bart, K. M., Farley, P. J. 1990. Ultrastructure of Dufour's gland in *Formica glacialis* (Hymenoptera: Formicidae). Trans. Am. Microsc. Soc. 109:107-110. [1990.01.24]

Barth, R. 1960. Ueber den Bewegungsmechanismus der Mandibeln von *Odontomachus chelifer* Latr. (Hymenopt., Formicidae). An. Acad. Bras. Cienc. 32:379-384. [1960.12.31]

Bartz, S. H., Hölldobler, B. 1982. Colony founding in *Myrmecocystus mimicus* Wheeler (Hymenoptera: Formicidae) and the evolution of foundress associations. Behav. Ecol. Sociobiol. 10:137-147. [1982.04]

Bass, J. A., Hays, S. B. 1976. Geographic location and identification of fire ant species in South Carolina. J. Ga. Entomol. Soc. 11:34-36. [1976.01]

Basulati, K. K., Kyzhayeva, K. Y. 1966. Study of the fauna and ecology of ants (Hymenoptera, Formicidae) of the Ukrainian Carpathians. [In Ukrainian.] Pp. 92-99 in: Pidoplichko, I. H. (ed.) Insects of the Ukrainian Carpathians and Transcarpathians. [In Ukrainian.] Kiev:

Akademiya Nauk Ukrains'koi RSR, 176 pp. [> 1966.02.01] [Date signed for printing ("pidpisano do druku").]

Baur, A., Buschinger, A., Zimmermannn, F. K. 1993. Molecular cloning and sequencing of 18S rDNA gene fragments from six different ant species. Insectes Soc. 40:325-335. [1993]

Baur, A., Chalwatzis, N., Buschinger, A., Zimmermann, F. K. 1995. Mitochondrial DNA sequences reveal close relationships between social parasitic ants and their host species. Curr. Genet. 28:242-247. [1995.08]

Bauschmann, G. 1988. Faunistisch-ökologische Untersuchungen zur Kenntnis der Ameisen des Vogelsberges (Hymenoptera, Formicidae). Entomofauna 9:69-115. [1988.02.15]

Bauschmann, G., Buschinger, A. 1992. Rote Liste gefährdeter Ameisen (Formicidae) Bayerns. Schriftenr. Bayer. Landesamt Umweltschutz 111:169-172. [1992]

Bausenwein, F. 1960. Untersuchungen über sekretorische Drüsen des Kopf- und Brustabschnittes in der *Formica rufa*-Gruppe. Cas. Cesk. Spol. Entomol. 57:31-57. [1960.01.20]

Beardsley, J. W. 1979. Notes on *Pseudomyrmex gracilis mexicanus* (Roger). Proc. Hawaii. Entomol. Soc. 23:23 [1979.05.31]

Beardsley, J. W. 1980. Note and exhibitions. *Iridomyrmex glaber* (Mayr). Proc. Hawaii. Entomol. Soc. 23:186. [1980.03.31]

Beattie, A. J. 1985. The evolutionary ecology of ant-plant mutualisms. New York: Cambridge University Press, 182 pp. [1985]

Beattie, A. J. 1989. Myrmecotrophy: plants fed by ants. Trends Ecol. Evol. 4:172-176. [1989.06]

Beattie, A. J., Turnbull, C. L., Hough, T., Knox, R. B. 1986. Antibiotic production: a possible function for the metapleural glands of ants (Hymenoptera: Formicidae). Ann. Entomol. Soc. Am. 79:448-450. [1986.05]

Beattie, A. J., Turnbull, C., Hough, T., Jobson, S., Knox, R. B. 1985. The vulnerability of pollen and fungal spores to ant secretions: evidence and some evolutionary implications. Am. J. Bot. 72:606-614. [1985.06.05]

Beattie, A. J., Turnbull, C., Knox, R. B., Williams, E. G. 1984. Ant inhibition of pollen function: a possible reason why ant pollination is rare. Am. J. Bot. 71:421-426. [1984.03.30]

Beck, D. E., Allred, D. M., Despain, W. J. 1967. Predaceous-scavenger ants in Utah. Great Basin Nat. 27:67-78. [1967.09.05]

Beck, H. 1958. Über einen weiteren Fall von lateralem Hermaphroditismus bei der Ameise *Polyergus rufescens* Latr. Zool. Anz. 161:243-249. [1958.11]

Bedziak, I. 1956. Rozmieszczenie mrówek w rezerwacie cisowym Wierzchlas. Zesz. Nauk. Uniw. Mikolaja Kopernika Tor. 1:91-103. [1956.08]

Begdon, J. 1932 ("1931"). Wymiary i wskazniki niektórych znamion mrówki *Stenamma* Westw. *westwoodi* Arn. (Westw?) *polonicum* nov. subsp., znalezionej na Pomorzu. Spraw. Kom. Fizjogr. Mater. Fizjogr. Kraju 65:113-119. [1932]

Begdon, J. 1933 ("1932"). Studia nad mrówkami Pomorza. Pol. Pismo Entomol. 11:57-97. [1933.06.15]

Begdon, J. 1954 ("1953"). Rozmieszczenie i makrotopy gatunków rodziny Formicidae na terenach nizinnych. Ann. Univ. Mariae Curie-Sklodowska Sect. C Biol. 8:435-506. [1954.06.30]

Begdon, J. 1959 ("1958"). Nowe stanowiska kilku interesujacych gatunków Formicoidea w Polsce. Ann. Univ. Mariae Curie-Sklodowska Sect. C Biol. 13:85-93. [1959.11.04]

Béique, R., Francoeur, A. 1966. Les fourmis d'une pessière à *Cladonia* (Hymenoptera: Formicidae). Nat. Can. (Qué.) 93:99-106. [1966.04]

Beláková, A. 1956. Príspevok k poznaniu mravcov (Hymenoptera - Formicidae) v hornej casti Bánovskej doliny. Acta Fac. Rerum Nat. Univ. Comenianae Zool. 1:323-329. [1956]

Beláková, A. 1961. Príspevok k poznaniu mravcov Cenkova. Biológia (Bratisl.) 16:693-696. [1961.09]

Beláková, A. 1967. Príspevok k poznaniu mravcov (Formicoidea) územia budúcej retencnej nádrze pod Vihorlatom. Sb. Slov. Nár. Múz. Prír. Vedy 13(1):97-100. [1967.04]

Belin-Depoux, M. 1990 ("1991"). Écologie et évolution des jardins de formis en Guyane Française. Rev. Écol. Terre Vie 46:1-38. [1990.12]

Bellas, T., Hölldobler, B. 1985. Constituents of mandibular and Dufour's glands of an Australian *Polyrhachis* weaver ant. J. Chem. Ecol. 11:525-538. [1985.04]

Belshaw, R., Bolton, B. 1993. The effect of forest disturbance on the leaf litter ant fauna in Ghana. Biodivers. Conserv. 2:656-666. [1993.12]

Belshaw, R., Bolton, B. 1994a. A new myrmicine ant genus from cocoa leaf litter in Ghana (Hymenoptera: Formicidae). J. Nat. Hist. 28:631-634. [1994.05.31]

Belshaw, R., Bolton, B. 1994b. A survey of the leaf litter ant fauna in Ghana, West Africa (Hymenoptera: Formicidae). J. Hym. Res. 3:5-16. [1994.10.15]

Benois, A. 1972. Principaux caractères de reconnaissance de trois espèces de fourmis envahissantes de la Côte d'Azur. Riviera Sci. 59:18-28. [1972.03.25]

Benson, R. B., Ferrière, C., Richards, O. W. 1937. The generic names of British insects. Part 5. The generic names of the British Hymenoptera Aculeata, with a check list of British species. London: Royal Entomological Society of London, pp. 81-149. [1937.08.14]

Benson, R. B., Ferrière, C., Richards, O. W. 1947a. Proposed suspension of the Règles for the names *Formica* Linnaeus, 1758, and *Camponotus* Mayr, 1861 (Class Insecta, Order Hymenoptera). Bull. Zool. Nomencl. 1:207. [1947.02.28]

Benson, R. B., Ferrière, C., Richards, O. W. 1947b. Proposed suspension of the Règles for *Ponera* Latreille, 1804 (Class Insecta, Order Hymenoptera). Bull. Zool. Nomencl. 1:216. [1947.02.28]

Benson, R. B., Nixon, G. E. J., Perkins, J. F., Richards, O. W. 1954. Support for Dr. I. H. H. Yarrow's proposal for the rephrasing of the decision taken by the International Commission regarding the name of the type species of "Formica" Linnaeus, 1758. Bull. Zool. Nomencl. 9:318. [1954.12.30]

Benson, W. W. 1985. Amazon ant-plants. Pp. 239-266 in: Prance, G. T., Lovejoy, T. E. (eds.) Key environments. Amazonia. Oxford: Pergamon Press, xiv + 442 pp. [1985]

Benson, W. W., Brandão, C. R. F. 1987. *Pheidole* diversity in the humid tropics: a survey from Serra dos Carajas, Para, Brazil. Pp. 593-594 in: Eder, J., Rembold, H. (eds.) Chemistry and biology of social insects. München: Verlag J. Peperny, xxxv + 757 pp. [1987]

Benson, W. W., Harada, A. Y. 1989 ("1988"). Local diversity of tropical and temperature [sic] ant faunas (Hymenoptera, Formicidae). Acta Amazonica 18(3/4):275-289. [1989.07]

Benson, W. W., Setz, E. Z. F. 1985. On the type localities of ants collected by James Trail in Amazonian Brazil and described by Gustav Mayr. Rev. Bras. Entomol. 29:587-590. [1985.12.30]

Bentley, B. L. 1977. Extrafloral nectaries and protection by pugnacious bodyguards. Annu. Rev. Ecol. Syst. 8:407-427. [1977]

Bequaert, J. 1913. Notes biologiques sur quelques fourmis et termites du Congo Belge. Rev. Zool. Afr. (Bruss.) 2:396-431. [1913.05.30]

Bequaert, J. 1922a. III. The predaceous enemies of ants. Bull. Am. Mus. Nat. Hist. 45:271-331. [1922.10.25]

Bequaert, J. 1922b. IV. Ants in their diverse relations to the plant world. Bull. Am. Mus. Nat. Hist. 45:333-583. [1922.10.25]

Bequaert, J. 1923. Ants accidentally introduced into New York and New Jersey; and a correction. Bull. Brooklyn Entomol. Soc. 18:165. [1923.12.11]

Bequaert, J. 1926. The date of publication of the Hymenoptera and Diptera described by Guérin in Duperrey's "Voyage de La Coquille". Entomol. Mitt. 15:186-195. [1926.03.20]

Bequaert, J. 1928. Family Formicidae. Pp. 995-1003 in: Leonard, M. D. (ed.) A list of the insects of New York. Mem. Cornell Univ. Agric. Exp. Stn. 101:1-1121. [1928.01]

Berendt, G. C. 1830. Die Insekten im Bernstein. Ein Beitrag zur Thiergeschichte der Vorwelt. Danzig: Nicolai, 38 pp. [1830.03.08]

Berg, C. 1880 ("1881"). Entomologisches aus dem Indianergebiet der Pampa. Stett. Entomol. Ztg. 42:36-72. [1880.12] [Ants pp. 71-72.]

Berg, C. 1881. Insectos. Pp. 77-115 in: Informe oficial de la comisión científica agregada al Estado mayor general de la expedición al Río Negro (Patagonia) realizada en los meses de abril, mayo y junio de 1879 bajo las órdenes del General D. Julio A. Roca. Buenos Aires: Ostwald y Martinez, xxiv + 294 pp. [1881] [Ants pp. 114-115.]

Berg, C. 1890. Enumeración sistemática y sinonimica de los formicidos argentinos, chilenos y uruguayos. An. Soc. Cient. Argent. 29:5-43. [1890.06]

Berg, R. Y. 1975. Myrmecochorous plants in Australia and their dispersal by ants. Aust. J. Bot. 23:475-508. [1975.06]

Bergroth, E. 1919. Die Erscheinungsdata zweier hemipterologischen Werke. Entomol. Mitt. 8:188-191. [1919.11.26]

Bergström, G. 1981. Chemical aspects of insect exocrine signals as a means for systematic and phylogenetic discussions in aculeate Hymenoptera. Entomol. Scand. Suppl. 15:173-184. [1981.07.01]

Bergström, G., Löfqvist, J. 1970. Chemical basis for odour communication in four species of *Lasius* ants. J. Insect Physiol. 16:2353-2375. [1970.12]

Bergström, G., Löfqvist, J. 1972a. Identification of volatile substances from Formicinae species. Pp. 355-356 in: Rafes, P. M. (ed.) XIII International Congress of Entomology, Moscow, 2-9 August, 1968. Proceedings, Volume III. Leningrad: Nauka, 494 pp. [> 1972.06.08] [Date signed for printing ("podpisano k pechati").]

Bergström, G., Löfqvist, J. 1972b. Similarities between the Dufour gland secretions of the ants *Camponotus ligniperda* (Latr.) and *Camponotus herculeanus* (L.) (Hym.). Entomol. Scand. 3:225-238. [1972.09.10]

Bergström, G., Löfqvist, J. 1973. Chemical congruence of the complex odoriferous secretions from Dufour's gland in three species of ants of the genus *Formica*. J. Insect Physiol. 19:877-907. [1973.04]

Berkelhamer, R. C. 1983. Intraspecific genetic variation and haplodiploidy, eusociality, and polygyny in the Hymenoptera. Evolution 37:540-545. [1983.05.19]

Berkelhamer, R. C. 1984. An electrophoretic analysis of queen number in three species of dolichoderine ants. Insectes Soc. 31:132-141. [1984.10]

Berland, L. 1958. Atlas des Hyménoptères de France, Belgique, Suisse. Tome II. Paris: Éditions Boubée & Cie, 184 pp. [1958.07] [Ants pp. 38-63.]

Bernard, F. 1935. Hyménoptères prédateurs des environs de Fréjus. Ann. Soc. Entomol. Fr. 104:31-72. [1935.03.31]

Bernard, F. 1936. Hyménoptères nouveaux ou peu connus en France (5e note). Remarques sur la faune des étangs méditerranéens littoraux. Bull. Soc. Entomol. Fr. 41:285-290. [1936.12.12]

Bernard, F. 1940. Fam. XXXIV. - Formicidés (Fourmis). Pp. 155-169 in: Berland, L. La faune de la France en tableaux synoptiques illustrés. Tome 7. Hyménoptères. Paris: Delagrave, 213 pp. [1940]

Bernard, F. 1945a ("1944"). Répartition des fourmis en Afrique du Nord. Bull. Soc. Hist. Nat. Afr. Nord 35:117-124. [1945.08.14]

Bernard, F. 1945b ("1944"). Notes sur l'écologie des fourmis en forêt de Mamora (Maroc). Bull. Soc. Hist. Nat. Afr. Nord 35:125-140. [1945.08.14]

Bernard, F. 1948. Les insectes sociaux du Fezzân. Comportement et biogéographie. Pp. 87-200 in: Bernard, F., Peyerimhoff, P. de. Mission scientifique du Fezzân (1944-1945). Tome V. Zoologie. Alger: Institut de Recherches Sahariennes de l'Université d'Alger, 248 pp. [1948.03.30]

Bernard, F. 1950a ("1946"). Notes sur les fourmis de France. II. Peuplement des montagnes méridionales. Ann. Soc. Entomol. Fr. 115:1-36. [1950.03]

Bernard, F. 1950b. Notes biologiques sur les cinq fourmis les plus nuisibles dans la région méditerranéenne. Rev. Path. Vég. Entomol. Agric. Fr. 29:26-42. [1950.09]

Bernard, F. 1950c. Contribution à l'étude de l'Aïr. Hyménoptères Formicidae. Mém. Inst. Fr. Afr. Noire 10:284-294. [1950]

Bernard, F. 1951a. Adaptations au milieu chez les fourmis sahariennes. Bull. Soc. Hist. Nat. Toulouse 86:88-96. [1951.05.30]

Bernard, F. 1951b. Types de répartition de la faune terrestre Nord-Africaine. C. R. Somm. Séances Soc. Biogéogr. 28:74-79. [1951.12]

Bernard, F. 1951c. Hyménoptères, super-famille des Formicoidea Ashmead 1905. Pp. 997-1104 in: Grassé, P. P. (ed.) Traité de Zoologie. Tome X. Fasc. II. Paris: Masson et Cie, pp. 976-1948. [1951]

Bernard, F. 1952. Le polymorphisme social et son déterminisme chez les fourmis. Colloq. Int. Cent. Natl. Rech. Sci. 34:123-140. [1952]

Bernard, F. 1953a. Les fourmis du Tassili des Ajjer. Pp. 121-250 in: Bernard, F. (ed.) Mission scientifique au Tassili des Ajjer (1949). Volume I. Recherches zoologiques et médicales. Paris: P. Lechevalier, 302 pp. [1953.01.31]

Bernard, F. 1953b ("1952"). La réserve naturelle intégrale du Mt Nimba. XI. Hyménoptères Formicidae. Mém. Inst. Fr. Afr. Noire 19:165-270. [1953.03]

Bernard, F. 1954 ("1953"). Une fourmi nouvelle: *Cataglyphis halophila* nichant au milieu du Chott Djerid. Bull. Soc. Sci. Nat. Tunis. 6:47-56. [1954.03.15]

Bernard, F. 1955a ("1954"). Fourmis moissonneuses nouvelles ou peu connues des montagnes d'Algérie et révision des *Messor* du groupe *structor* (Latr.). Bull. Soc. Hist. Nat. Afr. Nord 45:354-365. [< 1955.03.29] [Title page of journal is dated just "1955". Received at UC Berkeley library on 29 March 1955.]

Bernard, F. 1955b. Morphologie et comportement des fourmis lestobiotiques du genre *Epixenus* Emery. Insectes Soc. 2:273-283. [1955.12]

Bernard, F. 1955c. Liste des fourmis récoltées de 1.800 à 2.000 mètres. Pp. 28-29 in: Délye, G., Arlès, C. Promenades entomologiques au Djebel Babor. Bull. Soc. Hist. Nat. Afr. Nord 46:16-29. [1955]

Bernard, F. 1956a. Révision des *Leptothorax* (Hyménoptères Formicidae) d'Europe occidentale, basée sur la biométrie et les genitalia mâles. Bull. Soc. Zool. Fr. 81:151-165. [1956.10.15]

Bernard, F. 1956b. Remarques sur le peuplement des Baléares en fourmis. Bull. Soc. Hist. Nat. Afr. Nord 47:254-266. [1956]

Bernard, F. 1956c. Révision des fourmis paléarctiques du genre *Cardiocondyla* Emery. Bull. Soc. Hist. Nat. Afr. Nord 47:299-306. [1956]

Bernard, F. 1957a. Note sur quelques *Leptothorax* d'Europe centrale avec description de *L. carinthiacus* n. sp. (Hym. Formicidae). Bull. Soc. Entomol. Fr. 62:46-53. [1957.05.28]

Bernard, F. 1957b. *Xenometra* Emery, genre de fourmis parasite nouveau pour l'Ancien Monde (Hym. Formicidae). Bull. Soc. Entomol. Fr. 62:100-103. [1957.07.26]

Bernard, F. 1959a ("1958"). Les fourmis de l'île de Port-Cros. Contribution à l'écologie des anciennes forêts méditerranéennes. Vie Milieu 9:340-360. [1959.01]

Bernard, F. 1959b ("1958"). Notes écologiques et biologiques sur une fourmi parasite nouvelle pour la France: *Bothriomyrmex gibbus* (Soudek). Bull. Soc. Zool. Fr. 83:401-409. [1959.04.07]

Bernard, F. 1959c ("1956-58"). Missione Zavattari per l'esplorazione biogeografica delle Isole Pelagie. Les fourmis des Îles Pelagie. Comparaison avec d'autres faunes insulaires. Riv. Biol. Colon. 16:67-79. [1959.04.20]

Bernard, F. 1959d ("1958"). Fourmis des villes et fourmis du bled entre Rabat et Tanger. Bull. Soc. Sci. Nat. Phys. Maroc 38:131-142. [1959.04.30]

Bernard, F. 1959e ("1958"). Peuplement par les fourmis de sept îles du sud méditerranéen (archipels des Habibas, de la Galite et des Pelagie). C. R. Somm. Séances Soc. Biogéogr. 35:78-81. [1959.04]

Bernard, F. 1959f ("1958"). Résultats de la concurrence naturelle chez les fourmis terricoles de France et d'Afrique du Nord: évaluation numérique des sociétés dominantes. Bull. Soc. Hist. Nat. Afr. Nord 49:302-356. [1959]

Bernard, F. 1960a ("1959"). Fourmis récoltées en Corse par J. Bonfils (1957). C. R. Somm. Séances Soc. Biogéogr. 36:108-114. [1960.09]

Bernard, F. 1960b. Notes écologiques sur diverses fourmis sahariennes. Trav. Inst. Rech. Sahar. 19:51-63. [1960]

Bernard, F. 1961a. Fourmis de Majorque, de Corse et de sept petites îles du sud Méditerranéen. Colloq. Int. Cent. Natl. Rech. Sci. 94:139-157. [1961.09]

Bernard, F. 1961b. Biotopes habituels des fourmis sahariennes de plaine, d'après l'abondance de leurs nids en 60 stations très diverses. Bull. Soc. Hist. Nat. Afr. Nord 52:21-40. [1961]

Bernard, F. 1962. Peuplement des terrains rocheux par les fourmis sahariennes (Saoura, Mzab, Tassili n'Ajjer et Ahaggar). Trav. Inst. Rech. Sahar. 21:81-98. [1962.06]

Bernard, F. 1965a. Fourmis récoltées par J. Mateu dans l'Ennedi. Bull. Inst. Fr. Afr. Noire Sér. A Sci. Nat. 27:307-311. [1965.01]

Bernard, F. 1965b ("1964"). Recherches écologiques sur les fourmis des sables sahariens. Rev. Écol. Biol. Sol 1:615-638. [1965.06]

Bernard, F. 1967a ("1968"). Faune de l'Europe et du Bassin Méditerranéen. 3. Les fourmis (Hymenoptera Formicidae) d'Europe occidentale et septentrionale. Paris: Masson, 411 pp. [1967.11]

Bernard, F. 1967b. Recherches sur les fourmis des Monts-Dore. Ann. Stn. Biol. Besse-en-Chandesse 2:1-11. [1967.12]

Bernard, F. 1970 ("1969"). Les fourmis de la forêt de Mâmora (Maroc). Rev. Écol. Biol. Sol 6:483-513. [1970.03]

Bernard, F. 1971. Les fourmis de l'Île de Djerba (Tunisie). Bull. Soc. Hist. Nat. Afr. Nord 62: 3-14. [1971]

Bernard, F. 1973a. Études écologiques sur les fourmis de Breuil-Cervinia (Val d'Aoste). Rev. Écol. Biol. Sol 10:237-269. [1973.12]

Bernard, F. 1973b. Tendances calcicoles ou silicicoles chez les fourmis méditerranéennes. Pp. 16-21 in: Proceedings IUSSI VIIth International Congress, London, 10-15 September, 1973. Southampton: University of Southampton, vi + 418 pp. [1973]

Bernard, F. 1974 ("1973"). Evolution et biogéographie des *Messor* et *Cratomyrmex*, fourmis moissonneuses de l'ancien monde. C. R. Séances Soc. Biogéogr. 50:19-32. [1974.02]

Bernard, F. 1975a ("1974"). Les fourmis des rues de Kenitra (Maroc) (Hym.). Bull. Soc. Entomol. Fr. 79:178-183. [1975.03.07]

Bernard, F. 1975b. Données nouvelles sur l'écologie de la fourmi *Proformica ferreri* Bondroit., avec références particulières aux ouvrières nourrices. Insectes Soc. 22:151-168. [1975.12]

Bernard, F. 1975c ("1974"). Rapports entre fourmis et végétation près des Gorges du Verdon. Ann. Mus. Hist. Nat. Nice 2:57-79. [1975]

Bernard, F. 1976 ("1975"). Écologie des fourmis des grès d'Annot, comparées à celles de la Provence calcaire. Ann. Mus. Hist. Nat. Nice 3:33-54. [1976]

Bernard, F. 1977a. Écologie des fourmis du Parc national de Port-Cros. Bull. Mus. Natl. Hist. Nat. Écol. Gén. (3)36:53-82. [1977.04.30]

Bernard, F. 1977b. Trois fourmis nouvelles du Sahara (Hym. Formicidae). Bull. Soc. Entomol. Fr. 82:29-32. [1977.09.24]

Bernard, F. 1977c ("1976"). Trente ans de recherches sur les fourmis du Maghreb. Bull. Soc. Hist. Nat. Afr. Nord 67(1/2):81-118. [1977]

Bernard, F. 1978a ("1977"). Révision des *Diplorhoptrum* de France, fourmis plus différenciées par l'écologie que par leurs formes (Hym. Formicidae). Ann. Soc. Entomol. Fr. (n.s.)13:543-577. [1978.02.28]

Bernard, F. 1978b. *Orthocrema esterelana*, espèce nouvelle commune dans l'Estérel (Hym. Formicidae). Bull. Soc. Entomol. Fr. 83:43-46. [1978.07.18]

Bernard, F. 1978c. Prépondérance de la massue antennaire chez les cinq fourmis dominantes de la France méditerranéenne (Hym.). Bull. Soc. Entomol. Fr. 83:122-125. [1978.11.16]

Bernard, F. 1978d ("1977"). Fourmis et milieu dans le massif des Maures. Vie Milieu Sér. C Biol. Terr. 27:83-118. [1978.12]

Bernard, F. 1978e ("1976"). Contribution à la connaissance du *Tapinoma simrothi* Krausse, fourmi la plus nuisibles aux cultures du Maghreb. Bull. Soc. Hist. Nat. Afr. Nord 67(3/4):87-101. [1978]

Bernard, F. 1980a. Influence des densités végétales sur les fourmis méditerranéennes. Pp. 21-29 in: Cherix, D. (ed.) Écologie des insectes sociaux. C. R. UIEIS Section française, Lausanne, 7-8 sept. 1979. Nyon: UIEIS Section Française, xv + 160 pp. [1980.03.05]

Bernard, F. 1980b ("1979"). *Messor carthaginensis* n. sp., de Tunis, et révision des *Messor* du groupe *barbara* (Hym. Formicidae). Bull. Soc. Entomol. Fr. 84:265-269. [1980.04.25]

Bernard, F. 1980c. Evaluations quantitatives des influences humaines sur les fourmis. Ecol. Mediterr. 5:25-38. [1980.09]

Bernard, F. 1981a. Variabilité des proportions biométriques chez les *Plagiolepis*, avec description de *Pl. hoggarensis* n. sp. (Hym. Formicidae). Bull. Soc. Entomol. Fr. 86:169-172. [1981.11.06]

Bernard, F. 1981b. Revision of the genus *Messor* (harvesting ants) on a biometrical basis. Pp. 141-145 in: Howse, P. E., Clement, J.-L. (eds.) Biosystematics of social insects. Systematics Association Special Volume No. 19. London: Academic Press, 346 pp. [1981]

Bernard, F. 1982 ("1981"). Les fourmis des palmeraies: nombres, agilité, rôle, pratique. Bull. Soc. Hist. Nat. Afr. Nord 69(3/4):95-103. [1982]

Bernard, F. 1983. Encyclopédie entomologique - XLV. Les fourmis et leur milieu en France méditerranéenne. Paris: Éditions Lechevalier, 149 pp. [1983]

Bernard, F. 1984. Recherches sur les vitesses des fourmis. Actes Colloq. Insectes Soc. 1:151-160. [1984.09]

Bernard, F. 1985. Recherches sur l'évolution des fourmis moissonneuses (Hym. Formicidae). Actes Colloq. Insectes Soc. 2:45-55. [1985.06]

Bernard, F., Cagniant, H. 1963 ("1962"). Capture au Hoggar de trois *Acantholepis* nouveaux pour ce massif avec observations sur leurs modes de vie (Hym. Formicidae). Bull. Soc. Entomol. Fr. 67:161-164. [1963.02.28]

Berndt, K.-P., Fischer, E. 1983. Zur Mikromorphologie der Chitinstrukturen bei der Pharaoameise *Monomorium pharaonis* (L.) (Hymenoptera, Formicidae). Biol. Rundsch. 21:277-291. [1983]

Berndt, K.-P., Kremer, G. 1982. Heat shock-induced gynandromorphism in the pharaoh's ant, *Monomorium pharaonis* (L.). Experientia (Basel) 38:798-799. [1982.07.15]

Berndt, K.-P., Kremer, G. 1984 ("1983"). New categories in the gynandromorphism of ants. Insectes Soc. 30:461-465. [1984.04]

Berndt, K.-P., Kremer, G. 1985. A gynandromorph of Pharaoh's ant *Monomorium pharaonis* (Hymenoptera, Formicidae) with a prothoracic wing. Acta Entomol. Bohemoslov. 82:150-151. [1985.03.30]

Berndt, K.-P., Kremer, G. 1986. Die Larvenmorphologie der Pharaoameise *Monomorium pharaonis* (L.) (Hymenoptera, Formicidae). Zool. Anz. 216:305-320. [1986]

Berndt, K.-P., Wisniewski, J. 1984. Zur Teratologie der Körperanhänge bei der Pharaoameise *Monomorium pharaonis* (L.) (Hymenoptera, Formicidae). Zool. Anz. 213:313-321. [1984]

Bernstein, A. 1861. [Untitled. Introduced by: "Von Herrn Dr. Agath. Berstein in Gadok auf Java ist folgendes Schreiben eingegangen:".] Verh. K-K. Zool.-Bot. Ges. Wien 11:7-8. [1861]

Bernstein, R. A. 1979a. Evolution of niche breadth in populations of ants. Am. Nat. 114:533-544. [1979.10]

Bernstein, R. A. 1979b. Relations between species diversity and diet in communities of ants. Insectes Soc. 26:313-321. [1979.12]

Bernstein, R. A. 1980. Morphological variability of ants on and around Devils' Tower. Am. Midl. Nat. 104:185-188. [1980.08.28]

Bertkau, P. 1879a. Bericht über die wissenschaftlichen Leistungen im Gebiete der Arthropoden während der Jahre 1877-78. Arch. Naturgesch. 45(2):1-222. [1879] [Formicidae pp. 12-18. Also published as Bertkau (1879b).]

Bertkau, P. 1879b. Bericht über die wissenschaftlichen Leistungen im Gebiete der Entomologie während der Jahre 1877 und 1878. Berlin: Nicolaische Verlags-Buchhandlung, 568 pp. [1879] [Formicidae pp. 356-362.]

Beshers, S. N., Traniello, J. F. A. 1994. The adaptiveness of worker demography in the attine ant *Trachymyrmex septentrionalis*. Ecology 75:763-775. [1994.04]

Bestmann, H. J., Haak, U., Kern, F., Hölldobler, B. 1995. 2,4-dimethyl-5-hexanolide, a trail pheromone component of the carpenter ant *Camponotus herculaneus*. Naturwissenschaften 82:142-144. [1995.03]

Betrem, J. G. 1926. De mierenfauna van Meijendel. Levende Nat. 31:211-220. [1926.11.01]

Betrem, J. G. 1952. Remarks concerning the frontal parts of the head of some Hymenoptera. Pp. 97-100 in: de Wilde, J. et al. (eds.) Transactions of the IXth International Congress of Entomology. Amsterdam, August 17-24, 1951. Volume 1. Amsterdam: Ninth International Congress of Entomology, lxi + 1115 pp. [1952.12]

Betrem, J. G. 1953. Enkele opmerkingen omtrent de soorten van de *Formica rufa*-groep (Hym.). Entomol. Ber. (Amst.) 14:322-326. [1953.08.01]

Betrem, J. G. 1954. De satermier (*Formica exsecta* Nyl., 1846) en enkele van haar problemen (Hym. Form.). Entomol. Ber. (Amst.) 15:224-230. [1954.10.01]

Betrem, J. G. 1955a. On some ant types of Fabricius. Entomol. Ber. (Amst.) 15:291-293. [1955.01.01]

Betrem, J. G. 1955b. De systematische plaats van *Formica congerens* ab. *thijssei* Stärcke (Hym., Form.). Entomol. Ber. (Amst.) 15:391-393. [1955.05.01]

Betrem, J. G. 1960a. *Formica truncorum* F. niet inheems. Entomol. Ber. (Amst.) 20:130-134. [1960.07.01]

Betrem, J. G. 1960b. Ueber die Systematik der *Formica rufa*-gruppe. Tijdschr. Entomol. 103: 51-81. [1960.08.05]

Betrem, J. G. 1962a. Quelques remarques sur l'identité de la *Formica nigricans*. Entomol. Ber. (Amst.) 22:38. [1962.02.01]

Betrem, J. G. 1962b. Die neue Systematik der *Formica rufa*-Gruppe. Pp. 298-300 in: Strouhal, H., Beier, M. (eds.) XI. Internationaler Kongress für Entomologie. Wien, 17. bis 25. August 1960. Verhandlungen. Band III. Wien: Organisationskomitee des XI. Internationalen Kongresses für Entomologie, 348 pp. [1962]

Betrem, J. G. 1964. Einige Bemerkungen über *Formica*-Material aus Nordostkarelien. Waldhygiene 5:215-216. [1964]

Beyer, O. W. 1890. Der Giftapparat von *Formica rufa*, ein reducirtes Organ. Jena: G. Fischer, 50 pp. [1890]

Bezdecka, P., Skoupy, V. 1989. Mravenci (Hymenoptera, Formicidae) Vinarické hory. Muz. Soucas. Rada Prírodoved. 3:35-41. [1989]

Bhatkar, A. P. 1983. Interspecific trophallaxis in ants, its ecological and evolutionary significance. Pp. 105-123 in: Jaisson, P. (ed.) Social insects in the tropics. Volume 2. Paris: Université Paris-Nord, 252 pp. [1983.07]

Bickford, E. E. 1895a. Ueber die Morphologie und Physiologie der Ovarien der Ameisen-Arbeiterinnen. Zool. Jahrb. Abt. Syst. Geogr. Biol. Tiere 9:1-26. [1895.12.20]

Bickford, E. E. 1895b. Ueber die Morphologie und Physiologie der Ovarien der Ameisen-Arbeiterinnen. Jena: G. Fischer, 31 pp. [1895]

Bigot, Y., Hamelin, M.-H., Capy, P., Periquet, G. 1994. Mariner-like elements in hymenopteran species: insertion site and distribution. Proc. Natl. Acad. Sci. U. S. A. 91:3408-3412. [1994.04.12]

Billberg, G. J. 1820. Enumeratio insectorum in Museo Gust. Joh. Billberg. Stockholm: Gadel, 138 pp. [1820]

Billen, J. See: Billen, J. P. J.
Billen, J. P. J. 1982a. The Dufour gland closing apparatus in *Formica sanguinea* Latreille (Hymenoptera, Formicidae). Zoomorphology (Berl.) 99:235-244. [1982.03]
Billen, J. P. J. 1982b. Ultrastructure of the Dufour gland in ants (Hymenoptera, Formicidae). [Abstract.] P. 329 in: Breed, M. D., Michener, C. D., Evans, H. E. (eds.) The biology of social insects. Proceedings of the Ninth Congress of the IUSSI, Boulder, Colorado, August 1982. Boulder: Westview Press, xii + 420 pp. [1982.08]
Billen, J. P. J. 1984a. Morphology of the tibial gland in the ant *Crematogaster scutellaris*. Naturwissenschaften 71:324-325. [1984.06]
Billen, J. P. J. 1984b. Dufour gland morphology in ants (Hymenoptera, Formicidae). [Abstract.] P. 503 in: XVII International Congress of Entomology. Hamburg, Federal Republic of Germany, August 20-26, 1984. Abstract Volume. Hamburg: XVII International Congress of Entomology, 960 pp. [1984.08.10]
Billen, J. P. J. 1985a. Ultrastructure of the worker ovarioles in *Formica* ants (Hymenoptera: Formicidae). Int. J. Insect Morphol. Embryol. 14:21-32. [1985.02]
Billen, J. P. J. 1985b. Comparative ultrastructure of the poison and Dufour glands in Old and New World army ants (Hymenoptera, Formicidae). Actes Colloq. Insectes Soc. 2:17-26. [1985.06]
Billen, J. P. J. 1985c. Ultrastructure de la glande de Pavan chez *Dolichoderus quadripunctatus* (L.) (Hymenoptera, Formicidae). Actes Colloq. Insectes Soc. 2:87-95. [1985.06]
Billen, J. P. J. 1986a. Morphology and ultrastructure of the Dufour's and venom gland in the ant, *Myrmica rubra* (L.) (Hymenoptera: Formicidae). Int. J. Insect Morphol. Embryol. 15:13-25. [1986.04]
Billen, J. P. J. 1986b. Comparative morphology and ultrastructure of the Dufour gland in ants (Hymenoptera: Formicidae). Entomol. Gen. 11:165-181. [1986.05.30]
Billen, J. P. J. 1987a ("1986"). Morphology and ultrastructure of the abdominal glands in dolichoderine ants (Hymenoptera, Formicidae). Insectes Soc. 33:278-295. [1987.03]
Billen, J. P. J. 1987b. New structural aspects of the Dufour's and venom glands in social insects. Naturwissenschaften 74:340-341. [1987.07]
Billen, J. P. J. 1987c. Morphology and ultrastructure of the exocrine glands in social Hymenoptera. Pp. 81-84 in: Eder, J., Rembold, H. (eds.) Chemistry and biology of social insects. München: Verlag J. Peperny, xxxv + 757 pp. [1987]
Billen, J. P. J. 1988. Comparaison entre les fourmis australiennes des genres *Myrmecia* et *Nothomyrmecia*. Actes Colloq. Insectes Soc. 4:27-33. [1988.06]
Billen, J. P. J. 1989. Morphology of the cloacal gland in the ant *Cataglyphis savignyi*. Actes Colloq. Insectes Soc. 5:301-306. [1989.06]
Billen, J. P. J. 1990a. Phylogenetic aspects of exocrine gland development in the Formicidae. Pp. 317-318 in: Veeresh, G. K., Mallik, B., Viraktamath, C. A. (eds.) Social insects and the environment. Proceedings of the 11th International Congress of IUSSI, 1990. New Delhi: Oxford & IBH Publishing Co., xxxi + 765 pp. [1990.08]
Billen, J. P. J. 1990b. Morphology and ultrastructure of the Dufour's and venom gland in the ant *Myrmecia gulosa* (Fabr.) (Hymenoptera: Formicidae). Aust. J. Zool. 38:305-315. [1990.09.24]
Billen, J. P. J. 1990c. The sting bulb gland in *Myrmecia* and *Nothomyrmecia* (Hymenoptera: Formicidae): a new exocrine gland in ants. Int. J. Insect Morphol. Embryol. 19:133-139. [1990]
Billen, J. P. J. 1990d. A survey of the glandular systems of fire ants. Pp. 85-94 in: Vander Meer, R. K., Jaffe, K., Cedeno, A. (eds.) Applied myrmecology: a world perspective. Boulder: Westview Press, xv + 741 pp. [1990]
Billen, J. P. J. 1991. Ultrastructural organization of the exocrine glands in ants. Ethol. Ecol. Evol. Spec. Issue 1:67-73. [1991.03]
Billen, J. P. J. 1992. Origin of the trail pheromone in Ecitoninae: a behavioural and morphological examination. Pp. 203-209 in: Billen, J. P. J. (ed.) Biology and evolution of social insects. Leuven: Leuven University Press, ix + 390 pp. [1992]

Billen, J. P. J. 1993. Morphology of the exocrine system in ants. Mater. Kollok. Sekts. Obshchestv. Nasek. Vses. Entomol. Obshch. 2:1-15.

Billen, J. P. J. 1994a. Morphology and classification of the exocrine glands in ants. Ari 18:1-4. [1994.05.30]

Billen, J. P. J. 1994b. Morphology of exocrine glands in social insects: an up-date 100 years after Ch. Janet. P. 214 in: Lenoir, A., Arnold, G., Lepage, M. (eds.) Les insectes sociaux. 12ème congrès de l'Union Internationale pour l'Étude des Insectes Sociaux. Paris, Sorbonne, 21-27 août 1994. Paris: Université Paris Nord, xxiv + 583 pp. [1994.09]

Billen, J. P. J., Attygalle, A. B., Morgan, E. D., Ollett, D. G. 1987. The contents of the Dufour gland of the ant *Pogonomyrmex occidentalis*. Pp. 426-427 in: Eder, J., Rembold, H. (eds.) Chemistry and biology of social insects. München: Verlag J. Peperny, xxxv + 757 pp. [1987]

Billen, J. P. J., Boven, J. K. A. van, Evershed, R. P., Morgan, E. D. 1983. The chemical composition of the Dufour gland contents of workers of the ant *Formica cunicularia*. A test for recognition of the species. Ann. Soc. R. Zool. Belg. 113(suppl.1):283-289. [1983.12]

Billen, J. P. J., Boven, J. K. A. van. 1987. The metapleural gland in Old World army ants: a morphological and ultrastructural description (Hymenoptera, Formicidae). Rev. Zool. Afr. (Tervuren) 101:31-41. [1987.04.16]

Billen, J. P. J., Brandão, C. R. F., Paiva, R. V. S. 1995. Morphology and ultrastructure of the pygidial gland of the ant *Dinoponera australis* (Hymenoptera, Formicidae). Pap. Avulsos Zool. (São Paulo) 39:209-216. [1995.07.20]

Billen, J. P. J., Evershed, R. P., Attygalle, A. B., Morgan, E. D., Ollett, D. G. 1986. Contents of Dufour glands of workers of three species of *Tetramorium* (Hymenoptera: Formicidae). J. Chem. Ecol. 12:669-685. [1986.03]

Billen, J. P. J., Evershed, R. P., Morgan, E. D. 1984. Morphological comparison of Dufour glands in workers of *Acromyrmex octospinosus* and *Myrmica rubra*. Entomol. Exp. Appl. 35:205-207. [1984]

Billen, J. P. J., Gotwald, W. H., Jr. 1988. The crenellate lining of the Dufour gland in the genus *Aenictus*: a new character for interpreting the phylogeny of Old World army ants (Hymenoptera, Formicidae, Dorylinae). Zool. Scr. 17:293-295. [1988.06.30]

Billen, J. P. J., Jackson, B. D., Morgan, E. D. 1988a. Secretion of the Dufour gland of the ant *Nothomyrmecia macrops* (Hymenoptera: Formicidae). Experientia (Basel) 44:715-719. [1988.08.15]

Billen, J. P. J., Jackson, B. D., Morgan, E. D. 1988b. The contents of the pygidial gland of the primitive ant *Nothomyrmecia macrops* (Hymenoptera: Formicidae). Experientia (Basel) 44:794-797. [1988.09.15]

Billen, J. P. J., Kerkhofs, H. L. P. 1985. Ultrastructure of the ovarioles in *Formica* ants (Hymenoptera, Formicidae). Mitt. Dtsch. Ges. Allg. Angew. Entomol. 4:332-334. [1985.08]

Billen, J. P. J., Peeters, C. 1991. Fine structure of the gemma gland in the ant *Diacamma australe* (Hymenoptera, Formicidae). Belg. J. Zool. 121:203-210. [1991.12]

Billen, J. P. J., Peusens, G. 1984. Ultrastructure de la glande propharyngienne chez les fourmis Formicines (Hymenoptera, Formicidae). Actes Colloq. Insectes Soc. 1:121-129. [1984.09]

Billen, J. P. J., Schoeters, E. 1994. Morphology and ultrastructure of the mandibular gland in *Formica* L. ants (Hymenoptera, Formicidae). Memorabilia Zool. 48:9-16. [1994]

Billen, J. P. J., Taylor, R. W. 1993. Notes on the aberant venom gland morphology of some Australian dolichoderine and myrmicine ants (Hymenoptera, Formicidae). Belg. J. Zool. 123:159-163. [1993.12]

Bingham, C. T. 1896. A contribution to the knowledge of the Hymenopterous fauna of Ceylon. Proc. Zool. Soc. Lond. 1896:401-459. [1896.08]

Bingham, C. T. 1899. Note on *Diacamma*, a ponerine genus of ants, and of the finding of a female of *D. vagans*, Smith. J. Bombay Nat. Hist. Soc. 12:756-757. [1899.11.08]

Bingham, C. T. 1903. The fauna of British India, including Ceylon and Burma. Hymenoptera, Vol. II. Ants and Cuckoo-wasps. London: Taylor and Francis, 506 pp. [1903]

Bingham, C. T. 1905. Report on the Aculeate Hymenoptera. Fasc. Malay. Zool. 3:19-60. [1905.05.03]

Bingham, C. T. 1906a. The wild fauna and flora of the Royal Botanic Gardens, Kew. Hymenoptera. Formicidae. (Ants). Bull. Misc. Inf. R. Bot. Gard. Kew Addit. Ser. 5:27-29. [1906.03]

Bingham, C. T. 1906b. A plague of ants in the Observatory District, Cape Town, South Africa. Trans. Entomol. Soc. Lond. 1906:xxiii-xxvi. [1906.05.29]

Birket-Smith, S. J. R. 1977. Fossil insects from Spitsbergen. Acta Arct. 19:1-42. [1977] [*Myrmicium boreale* Heer referred to Ichneumonidae (p. 9).]

Birket-Smith, S. J. R. 1981. The male genitalia of Hymenoptera - a review based on morphology in Dorylidae (Formicoidea). Entomol. Scand. Suppl. 15:377-397. [1981.07.01]

Bisgaard, C. 1944. Meddelelse om nogle nye Myrer for Danmarks Fauna. Entomol. Tidskr. 24:115-126. [1944.09.01]

Bitsch, J., Peeters, C. 1992 ("1991"). Moignons alaires et morphologie thoracique chez l'ouvrière de la fourmi *Diacamma australe* (Fabricius) (Hym. Formicidae Ponerinae). Bull. Soc. Entomol. Fr. 96:213-221. [1992.01.31]

Blackburn, T., Kirby, W. F. 1880. Notes on species of aculeate Hymenoptera occurring in the Hawaiian Islands. Entomol. Mon. Mag. 17:85-89. [1880.09]

Blacker, N. C. 1989. The ants (Hymenoptera, Formicidae) of the Gower Peninsula, West Glamorgan, South Wales. Entomol. Rec. J. Var. 101:261-266. [1989.11.15]

Blacker, N. C. 1992. Some ants (Hymenoptera: Formicidae) from southern Vancouver Island, British Columbia. J. Entomol. Soc. B. C. 89:3-12. [1992.12]

Blacker, N. C. 1994. A proposed ecological relationship between the main species-groups of the ant genus *Formica* L. (Hymenoptera: Formicidae) found in northern Europe, with comments on the closely allied genus *Polyergus* Latreille. Entomol. Gaz. 45:127-140. [1994.05.06]

Blackwelder, R. E. 1947. The dates and editions of Curtis' British Entomology. Smithson. Misc. Collect. 107(5):1-27. [1947.06.12]

Blackwelder, R. E. 1949a. Studies on the dates of books on Coleoptera. I. Coleopt. Bull. 3:42-46. [1949.07.11]

Blackwelder, R. E. 1949b. Studies on the dates of books on Coleoptera. III. Coleopt. Bull. 3:92-94. [1949.12.14]

Blanchard, E. 1846. Les insectes. 2. Atlas; livraison 234 (Ins. 44). Plate 118. In: Le règne animal distribué d'après son organisation, pour servir de base à l'histoire naturelle des animaux, et d'introduction à l'anatomie comparée, par Georges Cuvier. Paris: Fortin, Masson et Cie, Libraires. [1846.11] [Publication date from Cowan (1976:41).]

Blinov, V. V. 1984. New for the fauna of Byelorussia ant species. [In Byelorussian.] Vestsi Akad. Navuk BSSR Ser. Biial. Navuk 1984(5):113-115. [> 1984.10.12] [Date signed for printing.]

Blinov, V. V. 1990. Ants of pine biocenosis of Beresinsky biosphere reserve. [In Russian.] Mater. Kollok. Sekts. Obshchestv. Nasek. Vses. Entomol. Obshch. 1:36-42. [> 1990.12.25] [Date signed for printing ("podpisano k pechati").]

Bloch, D. 1776. Beitrag zur Naturgeschichte des Kopals. Beschäft. Berl. Ges. Naturforsch. Freunde 2:91-196. [1776] [Ants pp. 168, 172, 178, 183.]

Bloch, J. A. 1940 ("1939"). Schweizerische und exotische Formicidae (Ameisen) im Solothurner Museum. Mitt. Naturforsch. Ges. Soloth. 13:19-27. [1940]

Blum, M. S. 1966. The source and specificity of trail pheromones in *Termitopone*, *Monomorium* and *Huberia*, and their relation to those of some other ants. Proc. R. Entomol. Soc. Lond. Ser. A 41:155-160. [1966.12.30]

Blum, M. S. 1973. Comparative exocrinology of the Formicidae. Pp. 23-40 in: Proceedings IUSSI VIIth International Congress, London, 10-15 September, 1973. Southampton: University of Southampton, vi + 418 pp. [1973]

Blum, M. S. 1974. Myrmicine trail pheromones: specificity, source and significance. J. N. Y. Entomol. Soc. 82:141-147. [1974.08.22]

Blum, M. S. 1980. Arthropods and economes: better fitness through ecological chemistry. Pp. 207-222 in: Gilles, R. (ed.) Animal and environmental fitness. Vol. 1. Oxford: Pergamon Press, xviii + 619 pp. [1980]
Blum, M. S. 1981. Sex pheromones in social insects: chemotaxonomic potential. Pp. 163-174 in: Howse, P. E., Clement, J.-L. (eds.) Biosystematics of social insects. Systematics Association Special Volume No. 19. London: Academic Press, 346 pp. [1981]
Blum, M. S. 1984. Poisonous ants and their venoms. Pp. 225-242 in: Tu, A. T. (ed.) Handbook of natural toxins. Volume 2. Insect poisons, allergens, and other invertebrate venoms. New York: M. Dekker, xv + 732 pp. [1984]
Blum, M. S. 1985. Alkaloidal ant venoms: chemistry and biological activities. Pp. 393-408 in: Hedin, P. A. (ed.) Bioregulators for pest control. ACS Symposium Series No. 276. Washington: American Chemical Society, xxi + 540 pp. [1985]
Blum, M. S. 1986. Ant venoms: structural and functional diversity. [Abstract.] J. Toxicol. Toxin Rev. 5:173. [1986]
Blum, M. S. 1992. Ant venoms: chemical and pharmacological properties. J. Toxicol. Toxin Rev. 11:115-164. [1992]
Blum, M. S., Ali, T. M. M., Jones, T. H., Snelling, R. R. 1994. Identification of a chemical releaser of alarm behavior for workers of *Harpegnathos saltator* Jerd. (Hymenoptera, Formicidae). Memorabilia Zool. 48:17-22. [1994]
Blum, M. S., Brand, J. M., Amante, E. 1981. o-Aminoacetophenone: identification in a primitive fungus-growing ant (*Mycocepurus goeldii*). Experientia (Basel) 37:816-817. [1981.08.15]
Blum, M. S., Brand, J. M., Duffield, R. M., Snelling, R. R. 1973. Chemistry of the venom of *Solenopsis aurea* (Hymenoptera: Formicidae). Ann. Entomol. Soc. Am. 66:702. [1973.05.15]
Blum, M. S., Callahan, P. S. 1963. The venom and poison glands of *Pseudomyrmex pallidus* (F. Smith). Psyche (Camb.) 70:69-74. [1963.07.09]
Blum, M. S., Hermann, H. R. 1969. The hymenopterous poison gland: probable functions of the main glandular elements. J. Ga. Entomol. Soc. 4:23-28. [1969.01]
Blum, M. S., Hermann, H. R. 1978a. Venoms and venom apparatuses of the Formicidae: Myrmeciinae, Ponerinae, Doylinae, Pseudomyrmecinae, Myrmicinae, and Formicinae. Handb. Exp. Pharmakol. 48:801-869. [1978]
Blum, M. S., Hermann, H. R. 1978b. Venoms and venom apparatuses of the Formicidae: Dolichoderinae and Aneuretinae. Handb. Exp. Pharmakol. 48:871-894. [1978]
Blum, M. S., Jones, T. H., Fales, H. M. 1990. Chemotaxonomic fingerprinting with ant natural products. Pp. 393-394 in: Veeresh, G. K., Mallik, B., Viraktamath, C. A. (eds.) Social insects and the environment. Proceedings of the 11th International Congress of IUSSI, 1990. New Delhi: Oxford & IBH Publishing Co., xxxi + 765 pp. [1990.08]
Blum, M. S., Jones, T. H., Lloyd, H. A., Fales, H. M., Snelling, R. R., Lubin, Y., Torres, J. 1985. Poison gland products of *Solenopsis* and *Monomorium* species. J. Entomol. Sci. 20:254-257. [1985.04]
Blum, M. S., Jones, T. H., Overal, W. L., Fales, H. M., Schmidt, J. O., Blum, N. A. 1983. Exocrine chemistry of the monotypic ant genus *Gigantiops*. Comp. Biochem. Physiol. B. Comp. Biochem. 75:15-16. [1983.05.15]
Blum, M. S., Jones, T. H., Snelling, R. R., Overal, W. L., Fales, H. M., Highet, R. J. 1982. Systematic implications of the exocrine chemistry of some *Hypoclinea* species. Biochem. Syst. Ecol. 10:91-94. [1982.05.20]
Blum, M. S., Morel, L., Fales, H. M. 1987. Chemistry of the mandibular gland secretion of the ant *Camponotus vagus*. Comp. Biochem. Physiol. B. Comp. Biochem. 86:251-252. [1987]
Blum, M. S., Moser, J. C., Cordero, A. D. 1964. Chemical releasers of social behavior. II. Source and specificity of the odor trail substances in four attine genera (Hymenoptera: Formicidae). Psyche (Camb.) 71:1-7. [1964.05.05]
Blum, M. S., Padovani, F., Amante, E. 1968. Alkanones and terpenes in the mandibular glands of *Atta* species (Hymenoptera: Formicidae). Comp. Biochem. Physiol. 26:291-299. [1968.07]

Blum, M. S., Portocarrero, C. A. 1964. Chemical releasers of social behavior - IV. The hindgut as the source of the odor trail pheromone in the Neotropical army ant genus *Eciton*. Ann. Entomol. Soc. Am. 57:793-794. [1964.11.15]

Blum, M. S., Snelling, R. R., Duffield, R. M., Hermann, H. R., Lloyd, H. A. 1988. Mandibular gland chemistry of *Camponotus* (*Myrmothrix*) *abdominalis*: chemistry and chemosystematic implications (Hymenoptera: Formicidae). Pp. 481-490 in: Trager, J. C. (ed.) Advances in myrmecology. Leiden: E. J. Brill, xxvii + 551 pp. [1988]

Blum, M. S., Wilson, E. O. 1964. The anatomical source of trail substances in formicine ants. Psyche (Camb.) 71:28-31. [1964.05.05]

Blüthgen, P. 1951. Neues oder Wissenswertes über mitteleuropäische Aculeaten und Goldwespen II. (Hym.) Bonn. Zool. Beitr. 2:229-234. [1951.12.15]

Bode, E. 1975. Ein Beitrag zur Ameisenbesiedlung (Formicidae (Hymenoptera, Hexapoda)) forstlicher Rekultivierungsflächen des Braunkohlentagebaues. Waldhygiene 11:13-20. [1975]

Boer, P., Boting, P., Dijkstra, P., Vallenduuk, H. 1995. *Formicoxenus nitidulus* in Nederland als gast in *Formica*-nesten (Hymenoptera: Formicidae: Myrmicinae). Entomol. Ber. (Amst.) 55:1-3. [1995.01]

Bogoescu, C. 1941. Contributiuni la studiul furnicilor din România. Bul. Soc. Nat. Rom. 15:12-17. [1941]

Bolton, B. 1969. Male of *Paedalgus termitolestes* W. M. Wheeler (Hymenoptera: Formicidae). Niger. Entomol. Mag. 2:14-16. [1969.12]

Bolton, B. 1971. Two new subarboreal species of the ant genus *Strumigenys* (Hym., Formicidae) from West Africa. Entomol. Mon. Mag. 107:59-64. [1971.08.27]

Bolton, B. 1972. Two new species of the ant genus *Epitritus* from Ghana, with a key to the world species (Hym., Formicidae). Entomol. Mon. Mag. 107:205-208. [1972.06.09]

Bolton, B. 1973a. The ant genera of West Africa: a synonymic synopsis with keys (Hymenoptera: Formicidae). Bull. Br. Mus. (Nat. Hist.) Entomol. 27:317-368. [1973.01.18]

Bolton, B. 1973b. The ant genus *Polyrhachis* F. Smith in the Ethiopian region (Hymenoptera: Formicidae). Bull. Br. Mus. (Nat. Hist.) Entomol. 28:283-369. [1973.07.25]

Bolton, B. 1973c. A remarkable new arboreal ant genus (Hym. Formicidae) from West Africa. Entomol. Mon. Mag. 108:234-237. [1973.09.25]

Bolton, B. 1974a. A revision of the Palaeotropical arboreal ant genus *Cataulacus* F. Smith (Hymenoptera: Formicidae). Bull. Br. Mus. (Nat. Hist.) Entomol. 30:1-105. [1974.02.15]

Bolton, B. 1974b ("1973"). New synonymy and a new name in the ant genus *Polyrhachis* F. Smith (Hym., Formicidae). Entomol. Mon. Mag. 109:172-180. [1974.05.03]

Bolton, B. 1974c. A revision of the ponerine ant genus *Plectroctena* F. Smith (Hymenoptera: Formicidae). Bull. Br. Mus. (Nat. Hist.) Entomol. 30:309-338. [1974.06.20]

Bolton, B. 1975a. A revision of the ant genus *Leptogenys* Roger (Hymenoptera: Formicidae) in the Ethiopian region with a review of the Malagasy species. Bull. Br. Mus. (Nat. Hist.) Entomol. 31:235-305. [1975.02.05]

Bolton, B. 1975b. A revision of the African ponerine ant genus *Psalidomyrmex* André (Hymenoptera: Formicidae). Bull. Br. Mus. (Nat. Hist.) Entomol. 32:1-16. [1975.04.10]

Bolton, B. 1975c. The *sexspinosa*-group of the ant genus *Polyrhachis* F. Smith (Hym. Formicidae). J. Entomol. Ser. B 44:1-14. [1975.05.27]

Bolton, B. 1976. The ant tribe Tetramoriini (Hymenoptera: Formicidae). Constituent genera, review of smaller genera and revision of *Triglyphothrix* Forel. Bull. Br. Mus. (Nat. Hist.) Entomol. 34:281-379. [1976.10.28]

Bolton, B. 1977. The ant tribe Tetramoriini (Hymenoptera: Formicidae). The genus *Tetramorium* Mayr in the Oriental and Indo-Australian regions, and in Australia. Bull. Br. Mus. (Nat. Hist.) Entomol. 36:67-151. [1977.09.29]

Bolton, B. 1979. The ant tribe Tetramoriini (Hymenoptera: Formicidae). The genus *Tetramorium* Mayr in the Malagasy region and in the New World. Bull. Br. Mus. (Nat. Hist.) Entomol. 38:129-181. [1979.03.29]

Bolton, B. 1980. The ant tribe Tetramoriini (Hymenoptera: Formicidae). The genus *Tetramorium* Mayr in the Ethiopian zoogeographical region. Bull. Br. Mus. (Nat. Hist.) Entomol. 40:193-384. [1980.07.31]

Bolton, B. 1981a. A revision of the ant genera *Meranoplus* F. Smith, *Dicroaspis* Emery and *Calyptomyrmex* Emery (Hymenoptera: Formicidae) in the Ethiopian zoogeographical region. Bull. Br. Mus. (Nat. Hist.) Entomol. 42:43-81. [1981.02.26]

Bolton, B. 1981b. A revision of six minor genera of Myrmicinae (Hymenoptera: Formicidae) in the Ethiopian zoogeographical region. Bull. Br. Mus. (Nat. Hist.) Entomol. 43:245-307. [1981.11.26]

Bolton, B. 1982. Afrotropical species of the myrmicine ant genera *Cardiocondyla*, *Leptothorax*, *Melissotarsus*, *Messor* and *Cataulacus* (Formicidae). Bull. Br. Mus. (Nat. Hist.) Entomol. 45:307-370. [1982.09.30]

Bolton, B. 1983. The Afrotropical dacetine ants (Formicidae). Bull. Br. Mus. (Nat. Hist.) Entomol. 46:267-416. [1983.08.25]

Bolton, B. 1984. Diagnosis and relationships of the myrmicine ant genus *Ishakidris* gen. n. (Hymenoptera: Formicidae). Syst. Entomol. 9:373-382. [1984.10]

Bolton, B. 1985. The ant genus *Triglyphothrix* Forel a synonym of *Tetramorium* Mayr. (Hymenoptera: Formicidae). J. Nat. Hist. 19:243-248. [1985.05.23]

Bolton, B. 1986a. A taxonomic and biological review of the tetramoriine ant genus *Rhoptromyrmex* (Hymenoptera: Formicidae). Syst. Entomol. 11:1-17. [1986.01]

Bolton, B. 1986b. Apterous females and shift of dispersal strategy in the *Monomorium salomonis*-group (Hymenoptera: Formicidae). J. Nat. Hist. 20:267-272. [1986.03.03]

Bolton, B. 1987. A review of the *Solenopsis* genus-group and revision of Afrotropical *Monomorium* Mayr (Hymenoptera: Formicidae). Bull. Br. Mus. (Nat. Hist.) Entomol. 54:263-452. [1987.06.25]

Bolton, B. 1988a. A new socially parasitic *Myrmica*, with a reassessment of the genus (Hymenoptera: Formicidae). Syst. Entomol. 13:1-11. [1988.01]

Bolton, B. 1988b. A review of *Paratopula* Wheeler, a forgotten genus of myrmicine ants (Hym., Formicidae). Entomol. Mon. Mag. 124:125-143. [1988.07.28]

Bolton, B. 1988c. *Secostruma*, a new subterranean tetramoriine ant genus (Hymenoptera: Formicidae). Syst. Entomol. 13:263-270. [1988.07]

Bolton, B. 1990a. Abdominal characters and status of the cerapachyine ants (Hymenoptera, Formicidae). J. Nat. Hist. 24:53-68. [1990.02.22]

Bolton, B. 1990b. [Untitled. A key to the living subfamilies of ants, based on the worker caste.] Pp. 33-34 in: Hölldobler, B., Wilson, E. O. The ants. Cambridge, Mass.: Harvard University Press, xii + 732 pp. [1990.03.28]

Bolton, B. 1990c. [Untitled. Keys to the ant genera of the Palearctic, Ethiopian (Afrotropical) and Malagasy, and Oriental regions. Based on the worker caste.] Pp. 34-55 in: Hölldobler, B., Wilson, E. O. The ants. Cambridge, Mass.: Harvard University Press, xii + 732 pp. [1990.03.28]

Bolton, B. 1990d. The higher classification of the ant subfamily Leptanillinae (Hymenoptera: Formicidae). Syst. Entomol. 15:267-282. [1990.07]

Bolton, B. 1990e. Army ants reassessed: the phylogeny and classification of the doryline section (Hymenoptera, Formicidae). J. Nat. Hist. 24:1339-1364. [1990.10.19]

Bolton, B. 1991. New myrmicine genera from the Oriental Region (Hymenoptera: Formicidae). Syst. Entomol. 16:1-13. [1991.01]

Bolton, B. 1992a. [Untitled. Subfamily Leptanilloidinae Bolton, new subfamily.] P. 317 in: Baroni Urbani, C., Bolton, B., Ward, P. S. The internal phylogeny of ants (Hymenoptera: Formicidae). Syst. Entomol. 17:301-329. [1992.10]

Bolton, B. 1992b. A review of the ant genus *Recurvidris* (Hym.: Formicidae), a new name for *Trigonogaster* Forel. Psyche (Camb.) 99:35-48. [1992.12.04]

Bolton, B. 1994. Identification guide to the ant genera of the world. Cambridge, Mass.: Harvard University Press, 222 pp. [1994.07.25] [Date from publisher.]

Bolton, B. 1995a. A taxonomic and zoogeographical census of the extant ant taxa (Hymenoptera: Formicidae). J. Nat. Hist. 29:1037-1056. [1995.08]

Bolton, B. 1995b. A new general catalogue of the ants of the world. Cambridge, Mass.: Harvard University Press, 504 pp. [1995.10]

Bolton, B., Belshaw, R. 1993. Taxonomy and biology of the supposedly lestobiotic ant genus *Paedalgus* (Hym.: Formicidae). Syst. Entomol. 18:181-189. [1993.07]

Bolton, B., Collingwood, C. A. 1975. Handbooks for the identification of British Insects. Hymenoptera, Formicidae. Handb. Identif. Br. Insects 6(3(c)):1-34. [1975.06]

Bolton, B., Gotwald, W. H., Jr., Leroux, J. M. 1979 ("1976"). A new West African ant of the genus *Plectroctena* with ecological notes (Hymenoptera: Formicidae). Ann. Univ. Abidjan Sér. E (Écol.) 9:371-381. [1979]

Bolton, B., Marsh, A. C. 1989. The Afrotropical thermophilic ant genus *Ocymyrmex* (Hymenoptera: Formicidae). J. Nat. Hist. 23:1267-1308. [1989.11.09]

Bonaric, J.-C. 1971a. Étude systématique et écologique des fourmis de l'Hérault. Ann. Soc. Hortic. Hist. Nat. Hérault 111:81-87. [1971.09.15]

Bonaric, J.-C. 1971b. Étude systématique et écologique des fourmis de l'Hérault (suite). Ann. Soc. Hortic. Hist. Nat. Hérault 111:119-126. [1971.11.15]

Bonaric, J.-C. 1972 ("1971"). Étude systématique et écologique des fourmis de l'Hérault (fin). Ann. Soc. Hortic. Hist. Nat. Hérault 111:169-176. [1972.01.15]

Bonavita-Cougourdan, A., Poveda, A. 1972. Étude préliminaire d'un organe mettant en rapport intestin moyen et intestin postérieur chez les larves de fourmis (Hyménoptères Formicidae). C. R. Séances Acad. Sci. Ser. D. Sci. Nat. 275:775-778. [1972.08.14]

Bond, W., Slingsby, P. 1984. Collapse of an ant-plant mutualism: the Argentine ant (*Iridomyrmex humilis*) and myrmecochorous Proteaceae. Ecology 65:1031-1037. [1984.08]

Bondroit, J. 1910 ("1909"). Les fourmis de Belgique. Ann. Soc. Entomol. Belg. 53:479-500. [1910.01.05]

Bondroit, J. 1911a. Contribution à la faune de Belgique. Notes diverses. Ann. Soc. Entomol. Belg. 55:8-13. [1911.02.03]

Bondroit, J. 1911b. Fourmis exotiques importées au jardin botanique de Bruxelles. Ann. Soc. Entomol. Belg. 55:14. [1911.02.03]

Bondroit, J. 1912. Fourmis de Hautes-Fagnes. Ann. Soc. Entomol. Belg. 56:351-352. [1912.10.31]

Bondroit, J. 1916. Un nouveau *Ponera* de France (Hym. Formicidae). Bull. Soc. Entomol. Fr. 1916:211-212. [1916.08.10]

Bondroit, J. 1917a. Notes sur quelques Formicidae de France (Hym.). Bull. Soc. Entomol. Fr. 1917:174-177. [1917.06.27]

Bondroit, J. 1917b. Diagnoses de trois nouveaux *Formica* d'Europe (Hym.). Bull. Soc. Entomol. Fr. 1917:186-187. [1917.07.17]

Bondroit, J. 1918. Les fourmis de France et de Belgique. Ann. Soc. Entomol. Fr. 87:1-174. [1918.12.24]

Bondroit, J. 1919. [Untitled. Introduced by: "M. Bondroit cite de notre pays les Fourmis suivantes."] Ann. Soc. Entomol. Belg. 59:69. [1919.07.04]

Bondroit, J. 1920a ("1919"). Notes diverses sur des fourmis d'Europe. Ann. Soc. Entomol. Belg. 59:143-158. [1920.02.25]

Bondroit, J. 1920b. Supplément aux fourmis de France et de Belgique. Ann. Soc. Entomol. Fr. 88:299-305. [1920.06.30]

Bondroit, J. 1932a ("1931"). Origine de l'ouvrière des fourmis. Ann. Soc. R. Zool. Belg. 62:13-24. [1932.07]

Bondroit, J. 1932b ("1931"). Notes sur les hyménoptères principalement les Sphégides des environs de Bruxelles. Ann. Soc. R. Zool. Belg. 62:31-44. [1932.07]

Bonetto, A. A. 1959. Las hormigas "cortadoras" de la Provincia de Santa Fé (generos: *Atta* y *Acromyrmex*). Santa Fé, Argentina: Ministerio de Agricultura y Ganadería (Dirección General de Recurzos Naturales), 79 pp. [1959]

Boomsma, J. J. 1988. Empirical analysis of sex allocation in ants: from descriptive surveys to population genetics. Pp. 42-51 in: Jong, G. de (ed.) Population genetics and evolution. Berlin: Springer Verlag, xi + 282 pp. [1988]

Boomsma, J. J. 1989. Sex-investment ratios in ants: has female bias been systematically overestimated? Am. Nat. 133:517-532. [1989.04]

Boomsma, J. J., Brouwer, A. H., Van Loon, A. J. 1990. A new polygynous *Lasius* species (Hymenoptera: Formicidae) from central Europe. II. Allozymatic confirmation of species status and social structure. Insectes Soc. 37:363-375. [1990.12]

Boomsma, J. J., Grafen, A. 1990. Intraspecific variation in ant sex ratios and the Trivers-Hare hypothesis. Evolution 44:1026-1034. [1990.08.17]

Boomsma, J. J., Grafen, A. 1991. Colony-level sex ratio selection in the eusocial Hymenoptera. J. Evol. Biol. 4:383-407. [1991.07]

Boomsma, J. J., Keller, L., Nielsen, M. G. 1995. A comparative analysis of sex ratio investment parameters in ants. Funct. Ecol. 9:743-753. [1995.10]

Boomsma, J. J., Mabelis, A. A., Verbeek, M. G. M., Los, E. C. 1987. Insular biogeography and distribution ecology of ants on the Frisian islands. J. Biogeogr. 14:21-37. [1987.01]

Boomsma, J. J., Van Loon, A. J. 1982. Structure and diversity of ant communities in successive coastal dune valleys. J. Anim. Ecol. 51:957-974. [1982.10]

Borchert, H. F., Anderson, N. L. 1973. The ants of the Bearpaw Mountains of Montana (Hymenoptera: Formicidae). J. Kansas Entomol. Soc. 46:200-224. [1973.04.30]

Borgmeier, T. 1920. Zur Lebensweise von *Odontomachus affinis* Guérin. Z. Dtsch. Ver. Wiss. Kunst São Paulo 1:31-38. [1920]

Borgmeier, T. 1923. Catalogo systematico e synonymico das formigas do Brasil. 1 parte. Subfam. Dorylinae, Cerapachyinae, Ponerinae, Dolichoderinae. Arch. Mus. Nac. (Rio J.) 24:33-103. [1923]

Borgmeier, T. 1927a. Um caso de trophobiose entre uma formiga e um parasita do caféeiro. Bol. Mus. Nac. Rio J. 3:285-289. [1927.12]

Borgmeier, T. 1927b. Algumas novas formigas brasileiras. Arch. Mus. Nac. (Rio J.) 29:57-65. [1927]

Borgmeier, T. 1927c. Catalogo systematico e synonymico das formigas do Brasil. 2 parte. Subf. Pseudomyrminae, Myrmicinae, Formicidae. Arch. Mus. Nac. (Rio J.) 29:69-164. [1927]

Borgmeier, T. 1928a. Einige neue Ameisen aus Brasilien. Zool. Anz. 75:32-39. [1928.01.02]

Borgmeier, T. 1928b. Algumas formigas do Museo Paulista. Bol. Biol. Lab. Parasitol. Fac. Med. São Paulo 12:55-70. [1928.07.10]

Borgmeier, T. 1929. Zur Kenntnis der brasilianischen Ameisen. EOS. Rev. Esp. Entomol. 5:195-214. [1929.09.15]

Borgmeier, T. 1930. Duas rainhas de *Eciton* e algumas outras formigas brasileiras. Arch. Inst. Biol. (São Paulo) 3:21-40. [1930.08]

Borgmeier, T. 1931. *Acropyga pickeli* Borgm. 1927 (Hym., Formicidae). Rev. Entomol. (Rio J.) 1:105-106. [1931.04.25]

Borgmeier, T. 1932a. A proposito de *Acropyga pickeli* Borgm. (1927) (Hym. Formicidae). Rev. Entomol. (Rio J.) 2:238-239. [1932.03.31]

Borgmeier, T. 1932b. *Leptogenys crudelis* Fr. Smith, 1858 (Hym. Formicidae). Rev. Entomol. (Rio J.) 2:485. [1932.12.28]

Borgmeier, T. 1933a. A rainha de *Eciton rogeri* Dalla Torre (Hym. Formicidae). Rev. Entomol. (Rio J.) 3:92-96. [1933.03.22]

Borgmeier, T. 1933b. Uma nova especie do genero *Leptogenys* (Hym. Formicidae). Rev. Entomol. (Rio J.) 3:226-227. [1933.07.20]

Borgmeier, T. 1933c. Nota prévia sobre *Acropyga paramaribensis* n. sp. (Hym. Formicidae). Rev. Entomol. (Rio J.) 3:263. [1933.07.20]

Borgmeier, T. 1933d. Sobre algumas especies de formigas do genero *Eciton* Latreille. Arch. Esc. Super. Agric. Med. Vet. Nictheroy 10:161-168. [1933.12]

Borgmeier, T. 1934. Contribuição para o conhecimento da fauna mirmecológica dos cafezais de Paramaribo, Guiana Holandesa (Hym. Formicidae). Arch. Inst. Biol. Veg. (Rio J.) 1:93-111. [1934.11]

Borgmeier, T. 1936a. A proposito de *Acropyga decedens* Mayr, *goeldii* For. e *pickeli* Borgm. (Hym. Formicidae). Rev. Entomol. (Rio J.) 6:304-306. [1936.07.15]

Borgmeier, T. 1936b. Sobre algumas formigas dos generos *Eciton* e *Cheliomyrmex* (Hym. Formicidae). Arch. Inst. Biol. Veg. (Rio J.) 3:51-68. [1936.12]

Borgmeier, T. 1937a. *Cardiocondyla emeryi* Forel no Brasil, e a descoberta do macho ergatoide desta especie (Hym. Formicidae). Rev. Entomol. (Rio J.) 7:129-134. [1937.07.27]

Borgmeier, T. 1937b. Formigas novas ou pouco conhecidas da América do Sul e Central, principalmente do Brasil (Hym. Formicidae). Arch. Inst. Biol. Veg. (Rio J.) 3:217-255. [1937.08]

Borgmeier, T. 1939. Nova contribuição para o conhecimento das formigas neotropicas (Hym. Formicidae). Rev. Entomol. (Rio J.) 10:403-428. [1939.09.04]

Borgmeier, T. 1940. Duas notas myrmecologicas. Rev. Entomol. (Rio J.) 11:606. [1940.06.28]

Borgmeier, T. 1948a. Die Geschlechtstiere zweier *Eciton*-Arten und einige andere Ameisen aus Mittel- und Suedamerika (Hym. Formicidae). Rev. Entomol. (Rio J.) 19:191-206. [1948.06.01]

Borgmeier, T. 1948b. Einige Ameisen aus Argentinien (Hym. Formicidae). Rev. Entomol. (Rio J.) 19:459-471. [1948.12.31]

Borgmeier, T. 1949. Formigas novas ou pouco conhecidas de Costa Rica e da Argentina (Hymenoptera, Formicidae). Rev. Bras. Biol. 9:201-210. [1949.06]

Borgmeier, T. 1950a. A fêmea dichthadiiforme e os estádios evolutivos de *Simopelta pergandei* (Forel), e a descrição de *S. bicolor*, n. sp. (Hym. Formicidae). Rev. Entomol. (Rio J.) 21:369-380. [1950.08.10]

Borgmeier, T. 1950b. Bemerkungen zu Dr. Creighton's Werk "The ants of North America". Rev. Entomol. (Rio J.) 21:381-386. [1950.08.10]

Borgmeier, T. 1950c. Uma nova espécie do gênero *Neivamyrmex* Borgmeier (Hym. Formicidae). Rev. Entomol. (Rio J.) 21:623-624. [1950.12.30]

Borgmeier, T. 1950d. Estudos sôbre *Atta* (Hym. Formicidae). Mem. Inst. Oswaldo Cruz Rio J. 48:239-263. [1950]

Borgmeier, T. 1950e. *Atta*-Studien (Hym. Formicidae). Mem. Inst. Oswaldo Cruz Rio J. 48:265-292. [1950]

Borgmeier, T. 1952 ("1951"). Algumas formigas do gênero *Macromischa* Roger (Hym. Formicidae). Arq. Mus. Nac. (Rio J.) 42:107-111. [1952.04.07]

Borgmeier, T. 1953. Vorarbeiten zu einer Revision der neotropischen Wanderameisen. Stud. Entomol. 2:1-51. [1953.07.10]

Borgmeier, T. 1954a. Uma nova *Discothyrea* com seis articulos antenais (Hymenoptera, Formicidae). Rev. Bras. Entomol. 1:191-194. [1954.03]

Borgmeier, T. 1954b. Two interesting dacetine ants from Brazil (Hym., Formicidae). Rev. Bras. Biol. 14:279-284. [1954.08]

Borgmeier, T. 1954c. Aenictini n. trib. und die Tribus-Einteilung der Dorylinen (Hym. Formicidae). Zool. Anz. 153:211-214. [1954.11]

Borgmeier, T. 1955. Die Wanderameisen der neotropischen Region. Stud. Entomol. 3:1-720. [1955]

Borgmeier, T. 1956. Ueber Rassen bei *Eciton* (Hym. Formicidae). Rev. Bras. Entomol. 4:209-212. [1956.03.05]

Borgmeier, T. 1957a. Myrmecologische Studien, I. An. Acad. Bras. Cienc. 29:103-128. [1957.03.31]

Borgmeier, T. 1957b. Die Maxillar- und Labialtaster der neotropischen Dorylinen (Hym. Formicidae). Rev. Bras. Biol. 17:387-394. [1957.09]

Borgmeier, T. 1957c. Basic questions of systematics. Syst. Zool. 6:53-69. [1957.10.20]

Borgmeier, T. 1958a. Nachtraege zu meiner Monographie der neotropischen Wanderameisen (Hym. Formicidae). Stud. Entomol. (n.s.)1:197-208. [1958.01.31]

Borgmeier, T. 1958b. [Review of: Raignier, A., Boven, J. van. 1955. Étude taxonomique, biologiques et biométrique des *Dorylus* du sous-genre *Anomma* (Hymenoptera Formicidae). Ann. Mus. R. Congo Belge Nouv. Ser. Quarto Sci. Zool. 2:1-359.] Stud. Entomol. (n.s.)1:301-303. [1958.01.31]

Borgmeier, T. 1959a. Myrmecologische Studien. II. An. Acad. Bras. Cienc. 31:309-319. [1959.06.30]

Borgmeier, T. 1959b. Revision der Gattung *Atta* Fabricius (Hymenoptera, Formicidae). Stud. Entomol. (n.s.)2:321-390. [1959.09.29]

Borgmeier, T. 1959c. A teoria da evolução. Stud. Entomol. (n.s.)2:456-460. [1959.09.29]

Borgmeier, T. 1963. Basic questions of systematics. Stud. Entomol. 6:537-562. [1963.12.10]

Borgmeier, T. 1971. List of scientific papers published by Thomas Borgmeier between the years 1920 and 1971. Stud. Entomol. 14:349-368. [1971.11.29]

Bos, H. 1887. Hets over de Nederlandsche mierenfauna. Tijdschr. Entomol. 30:181-198. [1887]

Bourke, A. F. G. 1988. Worker reproduction in the higher eusocial Hymenoptera. Q. Rev. Biol. 63:291-311. [1988.12]

Bourke, A. F. G. 1989. Comparative analysis of sex-investment ratios in slave-making ants. Evolution 43:913-918. [1989.06.22]

Bourke, A. F. G., Chan, G. L. 1994. Split sex ratios in ants with multiple mating. Trends Ecol. Evol. 9:120-122. [1994.04]

Bourke, A. F. G., Franks, N. R. 1987. Evolution of social parasites in leptothoracine ants. Pp. 37-38 in: Eder, J., Rembold, H. (eds.) Chemistry and biology of social insects. München: Verlag J. Peperny, xxxv + 757 pp. [1987]

Bourke, A. F. G., Franks, N. R. 1991. Alternative adaptations, sympatric speciation and the evolution of parasitic, inquiline ants. Biol. J. Linn. Soc. 43:157-178. [1991.07.23]

Bourke, A. F. G., Franks, N. R. 1995. Social evolution in ants. Princeton: Princeton University Press, xiii + 529 pp. [1995.10]

Bourke, A. F. G., Heinze, J. 1994. The ecology of communal breeding: the case of multiple-queen leptothoracine ants. Philos. Trans. R. Soc. Lond. B Biol. Sci. 345:359-372. [1994.09.29]

Bourne, R. A. 1973. A taxonomic study of the ant genus *Lasius* Fabricius in the British Isles (Hymenoptera: Formicidae). J. Entomol. Ser. B 42:17-27. [1973.06.29]

Boven, J. K. A. van. 1943a. Nieuwe vindplaatsen van merkwaardige mierensoorten. Natuurhist. Maandbl. 32:15-19. [1943.02.26]

Boven, J. K. A. van. 1943b. Nieuwe vindplaatsen van merkwaardige mierensoorten. (Slot). Natuurhist. Maandbl. 32:29-30. [1943.03.31]

Boven, J. K. A. van. 1944a. Nieuwe vindplaatsen van merkwaardige mierensoorten. IV. Natuurhist. Maandbl. 33:27-28. [1944.04.28]

Boven, J. K. A. van. 1944b. Voorloopige mededeeling over de mierenfauna van de Belgische Maasvallei. I. Natuurhist. Maandbl. 33:67-68. [1944.11.30]

Boven, J. K. A. van. 1944c. Voorloopige mededeeling over de mierenfauna van de Belgische Maasvallei. II. Natuurhist. Maandbl. 33:69-71. [1944.12.29]

Boven, J. K. A. van. 1945a. Voorloopige mededeeling over de mierenfauna van de Belgische Maasvallei. IV. Natuurhist. Maandbl. 34:16. [1945.03.30]

Boven, J. K. A. van. 1945b. Voorloopige mededeeling over de mierenfauna van de Belgische Maasvallei. (Slot). Natuurhist. Maandbl. 34:36-38. [1945.10.31]

Boven, J. K. A. van. 1946. Le mâle de *Plagiolepis vindobonensis* Lomn. (Hymenopt. Formicidae). Natuurhist. Maandbl. 35:9-10. [1946.02.28]

Boven, J. K. A. van. 1947a. Nieuwe vindplaatsen van merkwaardige mierensoorten. (V). Natuurhist. Maandbl. 36:5-9. [1947.02.27]

Boven, J. K. A. van. 1947b. Liste de détermination des principales espèces de fourmis belges (Hymenoptera Formicidae). Bull. Ann. Soc. Entomol. Belg. 83:163-190. [1947.05.03]

Boven, J. K. A. van. 1948. Beschrijving van een ergatoïde macropseudogyn van *Formica sanguinea*. Natuurhist. Maandbl. 37:8-10. [1948.02.27]

Boven, J. K. A. van. 1949a. Notes sur la faune des Hautes-Fagnes en Belgique. Bull. Ann. Soc. Entomol. Belg. 85:135-143. [1949.07.10]

Boven, J. K. A. van. 1949b. Varia myrmecologica 1945-1946. Natuurhist. Maandbl. 38:88-91. [1949.09.30]

Boven, J. K. A. van. 1951. Biometrische beschouwingen over het aantal oogfacetten bij de groep *Lasius flavus* De Geer (Hymenoptera Formicidae). Natuurhist. Maandbl. 40:73-76. [1951.06.30]

Boven, J. K. A. van. 1955. *Lasius* (*Chthonolasius*) *affinis* Schenck. fauna Neerl. nov. spec. (Hymenoptera Formicidae). Natuurhist. Maandbl. 44:6-10. [1955.02.25]

Boven, J. K. A. van. 1957 ("1956"). Synopsis der von P. Dr. Erich Wasmann S. J. (1859-1931) als neu beschriebenen tierformen. Publ. Natuurhist. Genoot. Limburg 9:113-141. [1957.02]

Boven, J. K. A. van. 1958. Allometrische en biometrische beschouwingen over het polymorfisme bij enkele mierensoorten (Hym. Formicidae). Verh. K. Vlaam. Acad. Wet. Lett. Schone Kunsten Belg. Kl. Wet. 56:1-134. [1958]

Boven, J. K. A. van. 1959. Vliesvleugelige insekten - Hymenoptera VI. Angeldragers (Aculeata). Mieren (Formicidae). Wet. Meded. K. Ned. Natuurhist. Ver. 30:1-32. [1959.01]

Boven, J. K. A. van. 1961a. Le polymorphisme dans la caste d'ouvrières de la fourmi voyageuse: *Dorylus* (*Anomma*) *wilverthi* Emery (Hym. Formicidae). Publ. Natuurhist. Genoot. Limburg 12:36-45. [1961.03]

Boven, J. K. A. van. 1961b. In memoriam Pater Hermann Schmitz S.J. 1878-1960. Natuurhist. Maandbl. 50:112-117. [1961.12.30]

Boven, J. K. A. van. 1967a. Formicides du Musée de Dundo (Angola) (Hymenoptera, Formicidae). Publ. Cult. Cia. Diam. Angola 76:63-75. [1967.03.14]

Boven, J. K. A. van. 1967b. La femelle de *Dorylus fimbriatus* et *termitarius* (Hymenoptera: Formicidae). Natuurhist. Maandbl. 56:55-60. [1967.04.26]

Boven, J. K. A. van. 1968. La reine de *Dorylus* (*Anomma*) *kohli* Wasmann (Hymenoptera: Formicidae). Nat. Can. (Qué.) 95:731-739. [1968.06]

Boven, J. K. A. van. 1970a. *Myrmica faniensis*, une nouvelle espèce parasite (Hymenoptera, Formicidae). Bull. Ann. Soc. R. Entomol. Belg. 106:127-132. [1970.06.15]

Boven, J. K. A. van. 1970b. Le polymorphisme des ouvrières de *Megaponera foetens* Mayr (Hymenoptera: Formicidae). Publ. Natuurhist. Genoot. Limburg 20(3/4):5-9. [1970]

Boven, J. K. A. van. 1972. Description de deux reines d'*Anomma* (Hymenoptera: Formicidae). Bull. Ann. Soc. R. Entomol. Belg. 108:133-146. [1972.09.30]

Boven, J. K. A. van. 1975. Deux nouvelles reines du genre *Dorylus* Fabricius (Hymenoptera, Formicidae). Ann. Zool. (Warsaw) 33:189-199. [1975.12.15]

Boven, J. K. A. van. 1977. De mierenfauna van België (Hymenoptera: Formicidae). Acta Zool. Pathol. Antverp. 67:1-191. [1977.06]

Boven, J. K. A. van. 1986. De mierenfauna van de Benelux (Hymenoptera: Formicidae). Wet. Meded. K. Ned. Natuurhist. Ver. 173:1-64. [1986.04]

Boven, J. K. A. van, Lévieux, J. 1970 ("1968"). Les Dorylinae de la savane de Lamto (Hymenoptera: Formicidae). Ann. Univ. Abidjan Sér. E (Écol.) 1(2):351-358. [1970.05]

Boven, J. van See: Boven, J. K. A. van

Bowley, D. R., Smith, H. M. 1968. The dates of publication of Louis Agassiz's Nomenclator Zoologicus. J. Soc. Bibliogr. Nat. Hist. 5:35-36. [1968.09]

Bown, T. M. 1982. Ichnofossils and rhizoliths of the nearshore fluvial Jebel Qatrani Formation (Oligocene), Fayum Province, Egypt. Palaeogeogr. Palaeoclimatol. Palaeoecol. 40:255-309. [1982.12]

Bradley, J. C. 1919. The synonymy and types of certain genera of Hymenoptera, especially of those discussed by the Rev. F. D. Morice and Mr. Jno. Hartley Durrant in connection with the long-forgotten "Erlangen List" of Panzer and Jurine. Trans. Entomol. Soc. Lond. 1919:50-75. [1919.08.15]

Bradley, J. C. 1975. Dates of publication of Westwood's Arcana entomologica. Proc. Biol. Soc. Wash. 88:91-94. [1975.04.23]

Brand, J. M. 1978. Fire ant venom alkaloids: their contribution to chemosystematics and biochemical evolution. Biochem. Syst. Ecol. 6:337-340. [1978.12.11]

Brand, J. M. 1985. Enantiomeric composition of an alarm pheromone component of the ants *Crematogaster castanea* and *C. liengmei*. J. Chem. Ecol. 11:177-180. [1985.02]

Brand, J. M., Blum, M. S., Barlin, M. R. 1973. Fire ant venoms: intraspecific and interspecific variation among castes and individuals. Toxicon 11:325-331. [1973.07]

Brand, J. M., Blum, M. S., Ross, H. H. 1973. Biochemical evolution in fire ant venoms. Insect Biochem. 3:45-51. [1973.03]

Brand, J. M., Duffield, R. M., MacConnell, J. G., Blum, M. S., Fales, H. M. 1973. Caste-specific compounds in male carpenter ants. Science (Wash. D. C.) 179:388-389. [1973.01.26]

Brandão, C. R. F. 1982. The Formicidae collection of the Museu de Zoologia da Universidade de São Paulo, Brasil. A list of the Attini type material. Attini 13:4-5. [1982.12]

Brandão, C. R. F. 1983. Sequential ethograms along colony development of *Odontomachus affinis* Guérin (Hymenoptera, Formicidae, Ponerinae). Insectes Soc. 30:193-203. [1983.09]

Brandão, C. R. F. 1987. Queenlessness in *Megalomyrmex* (Formicidae: Myrmicinae), with a discussion on the effects of the loss of true queens in ants. Pp. 111-112 in: Eder, J., Rembold, H. (eds.) Chemistry and biology of social insects. München: Verlag J. Peperny, xxxv + 757 pp. [1987]

Brandão, C. R. F. 1989. *Belonopelta minima* new species (Hymenoptera, Formicidae, Ponerinae) from eastern Brazil. Rev. Bras. Entomol. 33:135-138. [1989.03.31]

Brandão, C. R. F. 1990a. Phylogenetic, biogeographic, and evolutionary inferences from the description of an early Cretaceous South American Myrmeciinae. Pp. 313-314 in: Veeresh, G. K., Mallik, B., Viraktamath, C. A. (eds.) Social insects and the environment. Proceedings of the 11th International Congress of IUSSI, 1990. New Delhi: Oxford & IBH Publishing Co., xxxi + 765 pp. [1990.08]

Brandão, C. R. F. 1990b. Systematic revision of the Neotropical ant genus *Megalomyrmex* Forel (Hymenoptera: Formicidae: Myrmicinae), with the description of thirteen new species. Arq. Zool. (São Paulo) 31:411-481. [1990.09.21]

Brandão, C. R. F. 1991. Adendos ao catálogo abreviado das formigas da região Neotropical (Hymenoptera: Formicidae). Rev. Bras. Entomol. 35:319-412. [1991.08.31]

Brandão, C. R. F. 1994. Ants as ecological indicators: the case of semi-arid northeastern Brazil. P. 278 in: Lenoir, A., Arnold, G., Lepage, M. (eds.) Les insectes sociaux. 12ème congrès de l'Union Internationale pour l'Étude des Insectes Sociaux. Paris, Sorbonne, 21-27 août 1994. Paris: Université Paris Nord, xxiv + 583 pp. [1994.09]

Brandão, C. R. F., Caetano, F. H., Almeida, S. S. B. 1987. Intestinal symbionts, microorganisms and abdominal trophallaxis in the Neotropical myrmicine tribe Cephalotini (Hymenoptera: Formicidae). P. 631 in: Eder, J., Rembold, H. (eds.) Chemistry and biology of social insects. München: Verlag J. Peperny, xxxv + 757 pp. [1987]

Brandão, C. R. F., Diniz, J. L. M., Tomotake, E. M. 1991. *Thaumatomyrmex* strips millipedes for prey: a novel predatory behaviour in ants, and the first case of sympatry in the genus (Hymenoptera: Formicidae). Insectes Soc. 38:335-344. [1991]

Brandão, C. R. F., Lattke, J. E. 1990. Description of a new Ecuadorean *Gnamptogenys* species (Hymenoptera: Formicidae), with a discussion on the status of the *Alfaria* group. J. N. Y. Entomol. Soc. 98:489-494. [1990.11.14]

Brandão, C. R. F., Martins-Neto, R. G. 1990 ("1989"). [Untitled. *Cariridris* Brandão & Martins-Neto, new genus. *Cariridris bipetiolata* Brandão & Martins-Neto, new species.] Pp. 201-202 in: Brandão, C. R. F., Martins-Neto, R. G., Vulcano, M. A. The earliest known fossil ant (first southern hemisphere Mesozoic record) (Hymenoptera: Formicidae: Myrmeciinae). Psyche (Camb.) 96:195-208. [1990.04.20]

Brandão, C. R. F., Martins-Neto, R. G., Vulcano, M. A. 1990 ("1989"). The earliest known fossil ant (first southern hemisphere Mesozoic record) (Hymenoptera: Formicidae: Myrmeciinae). Psyche (Camb.) 96:195-208. [1990.04.20]

Brandão, C. R. F., Paiva, R. V. S. 1994. The Galapagos ant fauna and the attributes of colonizing ant species. Pp. 1-10 in: Williams, D. F. (ed.) Exotic ants. Biology, impact, and control of introduced species. Boulder: Westview Press, xvii + 332 pp. [1994]

Brangham, A. N. 1938. Additions to the wild fauna and flora of the Royal Botanic Gardens, Kew: XVIII. The ants of the Royal Botanic Gardens, Kew. Bull. Misc. Inf. R. Bot. Gard. Kew 1938:390-396. [1938.12.01]

Brangham, A. N. 1961 ("1960"). Some ants of the Greek island of Santorin. Entomol. Mon. Mag. 96:232-233. [1961.05.08]

Braune, M. 1972. Notizen zur Ameisenfauna der Tschechoslowakei (Hym., Formicidae). Zpr. Cesk. Spol. Entomol. 8:89-91. [1972]

Braune, M. 1975. *Sifolinia karawajewi* Arnoldi - eine für die DDR neue sozialparasitische Ameise (Hymenoptera, Formicidae). Faun. Abh. (Dres.) 5:211-214. [1975.12.19]

Brauns, J. 1901. Ueber die Lebensweise von *Dorylus* und *Aenietus* [sic]. (Hym.) Z. Syst. Hymenopterol. Dipterol. 1:14-17. [1901.01.01]

Brauns, J. 1903. Ueber das Weibchen von *Dorylus* (*Rhogmus*) *fimbriatus* Shuck. (Hym.) Z. Syst. Hymenopterol. Dipterol. 3:294-298. [1903.09.01]

Breed, M. D., Page, R. E. (eds.) 1989. The genetics of social evolution. Boulder: Westview Press, 213 pp. [1989]

Breen, J. 1977. The distribution of *Formica lugubris* Zetterstedt (Hymenoptera: Formicidae) in Ireland, with a discussion of its possible introduction. Ir. Nat. J. 19:123-127. [1977.10]

Brener, A. G. F., Ruggiero, A. 1994. Leaf-cutting ants (*Atta* and *Acromyrmex*) inhabiting Argentina: patterns in species richness and geographic range sizes. J. Biogeogr. 21:391-399. [1994.07]

Brèthes, J. 1914a. Note sur quelques Dolichodérines argentines. An. Mus. Nac. Hist. Nat. B. Aires 26:93-96. [1914.05.12]

Brèthes, J. 1914b. Sur les formes sexuelles de deux Dolichodérines. An. Mus. Nac. Hist. Nat. B. Aires 26:231-234. [1914.11.16]

Brian, M. V. 1964. Ant distribution in a southern English heath. J. Anim. Ecol. 33:451-461. [1964.10]

Brian, M. V. 1965. Social insect populations. London: Academic Press, vii + 135 pp. [1965]

Brian, M. V. 1977. Ants. London: Collins, 223 pp. [1977]

Brian, M. V. (ed.) 1978. Production ecology of ants and termites. Cambridge: Cambridge University Press, xvii + 409 pp. [1978]

Brian, M. V. 1983. Social insects. Ecology and behavioural biology. London: Chapman & Hall, x + 377 pp. [1983]

Brian, M. V., Brian, A. D. 1949. Observations on the taxonomy of the ants *Myrmica rubra* L. and *M. laevinodis* Nylander. (Hymenoptera: Formicidae.) Trans. R. Entomol. Soc. Lond. 100:393-409. [1949.12.30]

Brian, M. V., Brian, A. D. 1955. On the two forms *macrogyna* and *microgyna* of the ant *Myrmica rubra* L. Evolution 9:280-290. [1955.09.23]

Bridwell, J. C. 1899. A list of Kansas Hymenoptera. Trans. Kans. Acad. Sci. 16:203-211. [1899.06]

Briese, D. T. 1983. Different modes of reproductive behaviour (including a description of colony fission) in a species of *Chelaner* (Hymenoptera: Formicidae). Insectes Soc. 30:308-316. [1983.12]

Briese, D. T., Macauley, B. J. 1977. Physical structure of an ant community in semi-arid Australia. Aust. J. Ecol. 2:107-120. [1977.03]

Briese, D. T., Macauley, B. J. 1980. Temporal structure of an ant community in semi-arid Australia. Aust. J. Ecol. 5:121-134. [1980.03]

Briese, D. T., Macauley, B. J. 1981. Food collection within an ant community in semi-arid Australia, with special reference to seed harvesters. Aust. J. Ecol. 6:1-19. [1981.03]

Brill, J. H., Mayfield, H. T., Mar, T., Bertsch, W. 1985. Use of computerized pattern recognition in the study of the cuticular hydrocarbons of imported fire ants. I. Introduction and characterization of the cuticular hydrocarbon patterns of *Solenopsis invicta* and *S. richteri*. J. Chromatogr. 349:31-38. [1985.12.06]

Brimley, C. S. 1938. The insects of North Carolina. Raleigh: North Carolina Department of Agriculture, 560 pp. [1938.09] [Publication date from Brimley (1942:4). Ants species pp. 425-430; identifications attributed largely to M. R. Smith.]

Brimley, C. S. 1942. Supplement to insects of North Carolina. Raleigh: North Carolina Department of Agriculture, 39 pp. [1942] [Ants p. 34.]

Brischke, C. G. A. 1888. Hymenoptera aculeata der Provinzen West- und Ostpreussen. Schr. Naturforsch. Ges. Danzig N.F.7(1):85-107. [1888] [Ants (Formicariae) pp. 104-105.]

Brodie, P. B. 1895. Tertiary fossil ants in the Isle of Wight. Nature (Lond.) 52:570. [1895.10.10]

Brophy, J. J., Cavill, G. W. K., Davies, N. W., Gilbert, T. D., Philp, R. P., Plant, W. D. 1983. Hydrocarbon constituents of three species of dolichoderine ants. Insect Biochem. 13:381-390. [1983]

Brophy, J. J., Cavill, G. W. K., McDonald, J. A., Nelson, D., Plant, W. D. 1982. Volatile constituents of two species of Australian formicine ants of the genera *Notoncus* and *Polyrhachis*. Insect Biochem. 12:215-219. [1982]

Brothers, D. J. 1975. Phylogeny and classification of the aculeate Hymenoptera, with special reference to Mutillidae. Univ. Kans. Sci. Bull. 50:483-648. [1975.08.15]

Brothers, D. J., Carpenter, J. M. 1993. Phylogeny of Aculeata: Chrysidoidea and Vespoidea (Hymenoptera). J. Hym. Res. 2:227-304. [1993.09.30]

Brothers, D. J., Finnamore, A. T. 1993. Superfamily Vespoidea. Pp. 161-278 in: Goulet, H., Huber, J. T. (eds.) Hymenoptera of the world: an identification guide to families. Research Branch, Agriculture Canada Publication 1894/E. Ottawa: Canada Communications Group, vii + 668 pp. [1993]

Brough, E. J. 1977. The morphology and histology of the mandibular gland of an Australian species of *Calomyrmex* (Hymenoptera: Formicidae). Zoomorphologie 87:73-86. [1977.04.12]

Brown, C. A., Watkins, J. F., II, Eldridge, D. W. 1979. Repression of bacteria and fungi by the army ant secretion: skatole. J. Kansas Entomol. Soc. 52:119-122. [1979.02.09]

Brown, F. M. 1964. Dates of publication of the various parts of the Proceedings of the Entomological Society of Philadelphia. Trans. Am. Entomol. Soc. 89:305-308. [1964.01.31]

Brown, J. H., Davidson, D. W., Reichman, O. J. 1979. An experimental study of competition between seed-eating desert rodents and ants. Am. Zool. 19:1129-1143. [1979]

Brown, R. D. 1973. Funnel for extraction of leaf litter organisms. Ann. Entomol. Soc. Am. 66:485-486. [1973.03.15]

Brown, S. C. S. 1972. *Ponera punctatissima* (Hymenoptera: Formicidae) in Dorset. Entomol. Mon. Mag. 108:61. [1972.09.27]

Brown, W. L., Jr. 1943. A new metallic ant from the pine barrens of New Jersey. Entomol. News 54:243-248. [1943.12.29]

Brown, W. L., Jr. 1945. An unusual behavior pattern observed in a Szechuanese ant. J. West China Bord. Res. Soc. Ser. B 15:185-186. [1945]
Brown, W. L., Jr. 1947. A note upon two neglected species of *Formica* Linn. (Hym.: Formicidae). Entomol. News 58:6-9. [1947.04.17]
Brown, W. L., Jr. 1948a ("1947"). A new *Stictoponera*, with notes on the genus (Hymenoptera: Formicidae). Psyche (Camb.) 54:263-264. [1948.02.17]
Brown, W. L., Jr. 1948b. *Strumigenys karawajewi*, new name for a Sumatran ant. Entomol. News 59:44. [1948.04.22]
Brown, W. L., Jr. 1948c. A new *Discothyrea* from New Caledonia (Hymenoptera: Formicidae). Psyche (Camb.) 55:38-40. [1948.06.30]
Brown, W. L., Jr. 1948d. The status of the genus *Hercynia* J. Enzmann (Hymenoptera: Formicidae). Entomol. News 59:102. [1948.07.02]
Brown, W. L., Jr. 1948e. A preliminary generic revision of the higher Dacetini (Hymenoptera: Formicidae). Trans. Am. Entomol. Soc. 74:101-129. [1948.07.27]
Brown, W. L., Jr. 1949a. Synonymic and other notes on Formicidae (Hymenoptera). Psyche (Camb.) 56:41-49. [1949.05.17]
Brown, W. L., Jr. 1949b. A few ants from the Mackenzie River Delta. Entomol. News 60:99. [1949.06.29]
Brown, W. L., Jr. 1949c. Revision of the ant tribe Dacetini: III. *Epitritus* Emery and *Quadristruma* new genus (Hymenoptera: Formicidae). Trans. Am. Entomol. Soc. 75:43-51. [1949.07.06]
Brown, W. L., Jr. 1949d. A correction. Psyche (Camb.) 56:69. [1949.08.09] [Errata for Brown (1949a.]
Brown, W. L., Jr. 1949e. A new American *Amblyopone*, with notes on the genus (Hymenoptera: Formicidae). Psyche (Camb.) 56:81-88. [1949.08.09]
Brown, W. L., Jr. 1949f. Revision of the ant tribe Dacetini. I. Fauna of Japan, China and Taiwan. Mushi 20:1-25. [1949.09.20]
Brown, W. L., Jr. 1949g. Notes on Chinese ants: I. *Crematogaster* Lund. Mushi 20:37-38. [1949.09.20]
Brown, W. L., Jr. 1949h. Revision of the ant tribe Dacetini: IV. Some genera properly excluded from the Dacetini, with the establishment of the Basicerotini new tribe. Trans. Am. Entomol. Soc. 75:83-96. [1949.12.07]
Brown, W. L., Jr. 1950a. [Review of: Creighton, W. S. 1950. The ants of North America. Bull. Mus. Comp. Zool. 104:1-585.] Psyche (Camb.) 57:31-32. [1950.06.16]
Brown, W. L., Jr. 1950b. Supplementary notes on the feeding of dacetine ants. Bull. Brooklyn Entomol. Soc. 45:87-89. [1950.07.07]
Brown, W. L., Jr. 1950c. Revision of the ant tribe Dacetini: II. *Glamyromyrmex* Wheeler and closely related small genera. Trans. Am. Entomol. Soc. 76:27-36. [1950.08.09]
Brown, W. L., Jr. 1950d. Preliminary descriptions of seven new species of the dacetine ant genus *Smithistruma* Brown. Trans. Am. Entomol. Soc. 76:37-45. [1950.08.25]
Brown, W. L., Jr. 1950e. Revision of the ant tribe Dacetini: V. The delimitation of *Arnoldidris* new genus (Hymenoptera: Formicidae). Trans. Am. Entomol. Soc. 76:143-145. [1950.09.08]
Brown, W. L., Jr. 1950f. The status of two common North American Carpenter ants. Entomol. News 61:157-161. [1950.10.31]
Brown, W. L., Jr. 1950g. Morphological, taxonomic, and other notes on ants. Wasmann J. Biol. 8:241-250. [1950.11.05]
Brown, W. L., Jr. 1951. New synonymy of a few genera and species of ants. Bull. Brooklyn Entomol. Soc. 46:101-106. [1951.11.01]
Brown, W. L., Jr. 1952a ("1951"). *Adlerzia froggatti* Forel and some new synonymy (Hymenoptera: Formicidae). Psyche (Camb.) 58:110. [1952.04.07]
Brown, W. L., Jr. 1952b ("1951"). New synonymy in the army ant genus *Aenictus* Schuckard. Psyche (Camb.) 58:123. [1952.04.07]
Brown, W. L., Jr. 1952c ("1951"). Synonymous ant names. Psyche (Camb.) 58:124. [1952.04.07]

Brown, W. L., Jr. 1952d. Interesting northern records for eastern Hymenoptera (Formicidae and Embolemidae). Psyche (Camb.) 59:12. [1952.08.19]
Brown, W. L., Jr. 1952e. Correction to the synonymy of the ant *Camponotus formosensis* Wheeler. Psyche (Camb.) 59:19. [1952.08.19]
Brown, W. L., Jr. 1952f. *Mystrium* in Australia (Hymenoptera: Formicidae). Psyche (Camb.) 59:25. [1952.08.19]
Brown, W. L., Jr. 1952g. Contributions toward a reclassification of the Formicidae. I. Tribe Platythyreini (Hymenoptera). Breviora 6:1-6. [1952.08.29]
Brown, W. L., Jr. 1952h. Revision of the ant genus *Serrastruma*. Bull. Mus. Comp. Zool. 107: 67-86. [1952.08]
Brown, W. L., Jr. 1952i. Notes on two well-known Australian ant species. West. Aust. Nat. 3:137-138. [1952.09.15]
Brown, W. L., Jr. 1952j. The dacetine ant genus *Mesostruma* Brown. Trans. R. Soc. S. Aust. 75: 9-13. [1952.09]
Brown, W. L., Jr. 1952k. *Heteroponera* Mayr reinstated (Hymenoptera: Formicidae). Psyche (Camb.) 59:70. [1952.10.16]
Brown, W. L., Jr. 1952l. The status of some Australian *Amblyopone* species (Hym.: Formicidae). Entomol. News 63:265-267. [1952.12.15]
Brown, W. L., Jr. 1952m. On the identity of *Adlerzia* Forel (Hymenoptera: Formicidae). Pan-Pac. Entomol. 28:173-177. [1952.12.23]
Brown, W. L., Jr. 1953a ("1952"). Composition of the ant tribe Typhlomyrmicini. Psyche (Camb.) 59:104. [1953.01.30]
Brown, W. L., Jr. 1953b. Notes on Australian *Podomyrma* (Hymenoptera: Formicidae). North Qld. Nat. 21:3. [1953.03.01]
Brown, W. L., Jr. 1953c. Characters and synonymies among the genera of ants. Part I. Breviora 11:1-13. [1953.03.20]
Brown, W. L., Jr. 1953d. The neotropical species of the ant genus *Strumigenys* Fr. Smith: group of *mandibularis* Fr. Smith. J. N. Y. Entomol. Soc. 61:53-59. [1953.06.26]
Brown, W. L., Jr. 1953e. Three new ants related to *Strumigenys louisianae* Roger. Psyche (Camb.) 60:1-5. [1953.06.26]
Brown, W. L., Jr. 1953f. The neotropical species of the ant genus *Strumigenys* Fr. Smith: group of *smithii* Forel. J. N. Y. Entomol. Soc. 61:101-110. [1953.07.30]
Brown, W. L., Jr. 1953g. Revisionary studies in the ant tribe Dacetini. Am. Midl. Nat. 50:1-137. [1953.09.10]
Brown, W. L., Jr. 1953h. Characters and synonymies among the genera of ants. Part II. Breviora 18:1-8. [1953.09.23]
Brown, W. L., Jr. 1953i. An Australian *Trapeziopelta* (Hymenoptera: Formicidae). Psyche (Camb.) 60:51. [1953.09.28]
Brown, W. L., Jr. 1953j. Revisionary notes on the ant genus *Myrmecia* of Australia. Bull. Mus. Comp. Zool. 111:1-35. [1953.11]
Brown, W. L., Jr. 1953k. A revision of the dacetine ant genus *Orectognathus*. Mem. Qld. Mus. 13:84-104. [1953.12.14]
Brown, W. L., Jr. 1954a ("1953"). The Indo-Australian species of the ant genus *Strumigenys* Fr. Smith: *S. wallacei* Emery and relatives. Psyche (Camb.) 60:85-89. [1954.01.08]
Brown, W. L., Jr. 1954b. A preliminary report on dacetine ant studies in Australia. Ann. Entomol. Soc. Am. 46:465-471. [1954.02.08]
Brown, W. L., Jr. 1954c ("1953"). The neotropical species of the ant genus *Strumigenys* Fr. Smith: group of *elongata* Roger. J. N. Y. Entomol. Soc. 61:189-200. [1954.02.20]
Brown, W. L., Jr. 1954d. The synonymy of the ant *Aphaenogaster lepida* Wheeler. Pan-Pac. Entomol. 30:10. [1954.03.30]
Brown, W. L., Jr. 1954e. Remarks on the internal phylogeny and subfamily classification of the family Formicidae. Insectes Soc. 1:21-31. [1954.03]

Brown, W. L., Jr. 1954f ("1953"). The Indo-Australian species of the ant genus *Strumigenys* Fr. Smith: group of *doriae* Emery. Psyche (Camb.) 60:160-166. [1954.04.27]

Brown, W. L., Jr. 1954g. Systematic and other notes on some of the smaller species of the ant genus *Rhytidoponera* Mayr. Breviora 33:1-11. [1954.05.14]

Brown, W. L., Jr. 1954h. A review of the *coxalis* group of the ant genus *Stictoponera* Mayr. Breviora 34:1-10. [1954.07.20]

Brown, W. L., Jr. 1954i. New synonymy of an Australian *Iridomyrmex* (Hymenoptera: Formicidae). Psyche (Camb.) 61:67. [1954.08.26]

Brown, W. L., Jr. 1954j. The Indo-Australian species of the ant genus *Strumigenys* Fr. Smith: *S. chapmani* new species. Psyche (Camb.) 61:68-73. [1954.08.26]

Brown, W. L., Jr. 1954k. The ant genus *Strumigenys* Fred. Smith in the Ethiopian and Malagasy regions. Bull. Mus. Comp. Zool. 112:1-34. [1954.08]

Brown, W. L., Jr. 1954l. The neotropical species of the ant genus *Strumigenys* Fr. Smith: group of *saliens* Mayr. J. N. Y. Entomol. Soc. 62:55-62. [1954.10.29]

Brown, W. L., Jr. 1955a. The ant *Leptothorax muscorum* (Nylander) in North America. Entomol. News 66:43-50. [1955.01.28]

Brown, W. L., Jr. 1955b. *Forelifidis* M. R. Smith a synonym (Hymenoptera: Formicidae). Entomol. News 66:68. [1955.03.05]

Brown, W. L., Jr. 1955c. The identity of the British *Strongylognathus* (Hymenoptera: Formicidae). J. Soc. Br. Entomol. 5:113-114. [1955.03.22]

Brown, W. L., Jr. 1955d. A revision of the Australian ant genus *Notoncus* Emery, with notes on the other genera of Melophorini. Bull. Mus. Comp. Zool. 113:471-494. [1955.06]

Brown, W. L., Jr. 1955e. The ant *Cerapachys rufithorax* and its synonyms. Psyche (Camb.) 62:52. [1955.07.19]

Brown, W. L., Jr. 1955f. The ant *Centromyrmex donisthorpei* Menozzi, a synonym. Psyche (Camb.) 62:103. [1955.11.08]

Brown, W. L., Jr. 1955g. *Nylanderia myops* (Mann), new combination (Hymenoptera: Formicidae). Psyche (Camb.) 62:135-136. [1955.11.08]

Brown, W. L., Jr. 1955h. The first social parasite in the ant tribe Dacetini. Insectes Soc. 2:181-186. [1955.12]

Brown, W. L., Jr. 1955i. Ant taxonomy. Pp. 569-572 in: Kessel, E. W. (ed.) A century of progress in the natural sciences, 1853-1953. San Francisco: California Academy of Science, x + 807 pp. [1955]

Brown, W. L., Jr. 1956a. [Review of: Borgmeier, T. 1955. Die Wanderameisen der neotropischen Region. Stud. Entomol. 3:1-720.] Entomol. News 67:165-167. [1956.06.28]

Brown, W. L., Jr. 1956b. John Clark. Entomol. News 67:197-199. [1956.09.24]

Brown, W. L., Jr. 1956c. Some synonymies in the ant genus *Camponotus*. Psyche (Camb.) 63: 38-40. [1956.10.11]

Brown, W. L., Jr. 1956d. Notes on the ant genus *Holcoponera* Mayr, with descriptions of two new species. Insectes Soc. 3:489-497. [1956.12]

Brown, W. L., Jr. 1957a ("1956"). The identity of *Lordomyrma rugosa* Clark. Psyche (Camb.) 63:49. [1957.01.23]

Brown, W. L., Jr. 1957b ("1956"). The relationship of two African *Tetramorium* species (Hymenoptera: Formicidae). Psyche (Camb.) 63:75. [1957.01.23]

Brown, W. L., Jr. 1957c ("1955"). The neotropical species of the ant genus *Strumigenys* Fr. Smith: group of *cultriger* Mayr and *S. tococae* Wheeler. J. N. Y. Entomol. Soc. 63:97-102. [1957.03.08]

Brown, W. L., Jr. 1957d. Is the ant genus *Tetramorium* native in North America? Breviora 72:1-8. [1957.03.29]

Brown, W. L., Jr. 1957e. Biological investigations in the Selva Lacandona, Chiapas. 4. Ants from Laguna Ocotal (Hymenoptera: Formicidae). Bull. Mus. Comp. Zool. 116:228-237. [1957.04]

Brown, W. L., Jr. 1957f. Distribution and variation of the ant *Formica dakotensis* Emery. Entomol. News 68:165-167. [1957.06.10]
Brown, W. L., Jr. 1957g. Centrifugal speciation. Q. Rev. Biol. 32:247-277. [1957.09]
Brown, W. L., Jr. 1957h ("1956"). The Indo-Australian species of the ant genus *Strumigenys* Fr. Smith: Three new Philippine species. Psyche (Camb.) 63:113-118. [1957.11.18]
Brown, W. L., Jr. 1957i ("1956"). The synonymy and relationships of the ant *Pseudolasius bayoni* Menozzi. Psyche (Camb.) 63:146. [1957.11.18]
Brown, W. L., Jr. 1957j. The ending for subtribal names in zoology. Syst. Zool. 6:193-194. [1957.12]
Brown, W. L., Jr. 1958a ("1957"). The army ant *Aenictus exiguus* Clark a synonym. Psyche (Camb.) 64:5. [1958.01.10]
Brown, W. L., Jr. 1958b ("1957"). A supplement to the revisions of the dacetine ant genera *Orectognathus* and *Arnoldidris*, with keys to the species. Psyche (Camb.) 64:17-29. [1958.01.10]
Brown, W. L., Jr. 1958c. The neotropical species of the ant genus *Strumigenys* Fr. Smith: group of *cordovensis* Mayr. Stud. Entomol. (n.s.)1:217-224. [1958.01.31]
Brown, W. L., Jr. 1958d. A new Japanese species of the dacetine ant genus *Epitritus*. Mushi 31:69-72. [1958.03.15]
Brown, W. L., Jr. 1958e ("1957"). The neotropical species of the ant genus *Strumigenys* Fr. Smith: group of *marginiventris* Santschi. J. N. Y. Entomol. Soc. 65:123-128. [1958.06.27]
Brown, W. L., Jr. 1958f ("1957"). The neotropical species of the ant genus *Strumigenys* Fr. Smith: group of *ogloblini* Santschi. J. N. Y. Entomol. Soc. 65:133-137. [1958.06.27]
Brown, W. L., Jr. 1958g. Contributions toward a reclassification of the Formicidae. II. Tribe Ectatommini (Hymenoptera). Bull. Mus. Comp. Zool. 118:173-362. [1958.06]
Brown, W. L., Jr. 1958h. A review of the ants of New Zealand. Acta Hymenopterol. 1:1-50. [1958.07.30]
Brown, W. L., Jr. 1958i ("1957"). The Indo-Australian species of the ant genus *Strumigenys* Fr. Smith: *S. decollata* Mann and *S. ecliptacoca* new species. Psyche (Camb.) 64:109-114. [1958.08.14]
Brown, W. L., Jr. 1958j ("1957"). Predation of arthropod eggs by the ant genera *Proceratium* and *Discothyrea*. Psyche (Camb.) 64:115. [1958.08.14]
Brown, W. L., Jr. 1959a. The neotropical species of the ant genus *Strumigenys* Fr. Smith: group of *emeryi* Mann. Entomol. News 70:97-104. [1959.04.03]
Brown, W. L., Jr. 1959b. A revision of the dacetine ant genus *Neostruma*. Breviora 107:1-13. [1959.05.06]
Brown, W. L., Jr. 1959c. Some new species of dacetine ants. Breviora 108:1-11. [1959.05.07]
Brown, W. L., Jr. 1959d. Variation in the ant *Polyrhachis thrinax* (Hymenoptera). Entomol. News 70:164. [1959.06.15]
Brown, W. L., Jr. 1959e. Synonymy in the ant genus *Macromischa* Roger. Fla. Entomol. 42:73-74. [1959.06]
Brown, W. L., Jr. 1959f. The neotropical species of the ant genus *Strumigenys* Fr. Smith: group of *silvestrii* Emery. Stud. Entomol. (n.s.)2:25-30. [1959.09.29]
Brown, W. L., Jr. 1959g ("1958"). The Indo-Australian species of the ant genus *Strumigenys* Fr. Smith: group of *S. godeffroyi* in Borneo. Psyche (Camb.) 65:81-89. [1959.12.03]
Brown, W. L., Jr. 1959h. Appendix G. Insecta collected by the expedition. Pp. 229-230 in: Field, H. An anthropological reconnaissance in West Pakistan, 1955, with appendixes on the archaeology and natural history of Baluchistan and Bahawalpur. Pap. Peabody Mus. Archaeol. Ethnol. Harv. Univ. 52:i-xii,1-332. [1959]
Brown, W. L., Jr. 1960a. Contributions toward a reclassification of the Formicidae. III. Tribe Amblyoponini (Hymenoptera). Bull. Mus. Comp. Zool. 122:143-230. [1960.03]
Brown, W. L., Jr. 1960b ("1959"). The release of alarm and attack behavior in some New World army ants. Psyche (Camb.) 66:25-27. [1960.09.08]

Brown, W. L., Jr. 1960c ("1959"). The neotropical species of the ant genus *Strumigenys* Fr. Smith: group of *gundlachi* (Roger). Psyche (Camb.) 66:37-52. [1960.09.08]

Brown, W. L., Jr. 1960d. A new African ant of the genus *Strumigenys*, tribe Dacetini. Entomol. News 71:206. [1960.09.26]

Brown, W. L., Jr. 1961a. The neotropical species of the ant genus *Strumigenys* Fr. Smith: miscellaneous concluding studies. Psyche (Camb.) 68:58-69. [1961.12.29]

Brown, W. L., Jr. 1961b. A note on the ant *Gnamptogenys hartmani* Wheeler. Psyche (Camb.) 68:69. [1961.12.29]

Brown, W. L., Jr. 1961c. Ant. Pp. 40-41 in: Gray, P. (ed.) The encyclopedia of the biological sciences. New York: Van Nostrand Reinhold Co., xxi + 1119 pp. [1961]

Brown, W. L., Jr. 1962a. A new ant of the genus *Amblyopone* from Panama. Psyche (Camb.) 69:73-76. [1962.07.01]

Brown, W. L., Jr. 1962b. A new ant of the genus *Epitritus* from south of the Sahara. Psyche (Camb.) 69:77-80. [1962.07.01]

Brown, W. L., Jr. 1962c. The neotropical species of the ant genus *Strumigenys* Fr. Smith: synopsis and keys to the species. Psyche (Camb.) 69:238-267. [1962.12.29]

Brown, W. L., Jr. 1963a. Characters and synonymies among the genera of ants. Part III. Some members of the tribe Ponerini (Ponerinae, Formicidae). Breviora 190:1-10. [1963.09.30]

Brown, W. L., Jr. 1963b. [Review of: Gregg, R. E. 1963. The ants of Colorado, with reference to their ecology, taxonomy, and geographic distribution. Boulder: University of Colorado Press, xvi + 792 pp.] Q. Rev. Biol. 38:269. [1963.09]

Brown, W. L., Jr. 1964a. Some tramp ants of Old World origin collected in tropical Brazil. Entomol. News 75:14-15. [1964.01.07]

Brown, W. L., Jr. 1964b. The ant genus *Smithistruma*: a first supplement to the World revision (Hymenoptera: Formicidae). Trans. Am. Entomol. Soc. 89:183-200. [1964.01.31]

Brown, W. L., Jr. 1964c. [Review of: Wheeler, G. C., Wheeler, J. 1963. The ants of North Dakota. Grand Forks, North Dakota: University of North Dakota Press, viii + 326 pp.] J. Kansas Entomol. Soc. 37:87-88. [1964.03.26]

Brown, W. L., Jr. 1964d. Solution to the problem of *Tetramorium lucayanum* (Hymenoptera: Formicidae). Entomol. News 75:130-132. [1964.05.12]

Brown, W. L., Jr. 1964e. Genus *Rhoptromyrmex*, revision of, and key to species. Pilot Regist. Zool. Card No. 11. [1964.05.20]

Brown, W. L., Jr. 1964f. *Macromischoides africanus*, new synonymy of. Pilot Regist. Zool. Card No. 12. [1964.05.20]

Brown, W. L., Jr. 1964g. *Rhoptromyrmex melleus*, brief characterization of. Pilot Regist. Zool. Card No. 13. [1964.05.20]

Brown, W. L., Jr. 1964h. *Rhoptromyrmex wroughtonii*, new synonymy of, and brief characterization. Pilot Regist. Zool. Card No. 14. [1964.05.20]

Brown, W. L., Jr. 1964i. *Rhoptromyrmex opacus*, new synonymy of, and brief characterization. Pilot Regist. Zool. Card No. 15. [1964.05.20]

Brown, W. L., Jr. 1964j. *Rhoptromyrmex transversinodis*, new synonymy of, and brief characterization. Pilot Regist. Zool. Card No. 16. [1964.05.20]

Brown, W. L., Jr. 1964k. *Rhoptromyrmex globulinodis*, new synonymy of, and brief characterization. Pilot Regist. Zool. Card No. 17. [1964.05.20]

Brown, W. L., Jr. 1964l. *Hagioxenus mayri* comb. nov. Pilot Regist. Zool. Card No. 19. [1964.05.20]

Brown, W. L., Jr. 1964m. Synonymy and variation of some species of the ant genus *Anochetus*. J. Kansas Entomol. Soc. 37:212-215. [1964.08.06]

Brown, W. L., Jr. 1965a. *Colobostruma papulata* species nov. Pilot Regist. Zool. Card No. 21. [1965.04.05]

Brown, W. L., Jr. 1965b. *Colobostruma nancyae* species nov. Pilot Regist. Zool. Card No. 22. [1965.04.05]

Brown, W. L., Jr. 1965c. Contributions to a reclassification of the Formicidae. IV. Tribe Typhlomyrmecini (Hymenoptera). Psyche (Camb.) 72:65-78. [1965.06.25]

Brown, W. L., Jr. 1965d. Studies on North American ants. I. The *Formica integra* subgroup. Entomol. News 76:181-186. [1965.06.25]

Brown, W. L., Jr. 1965e. Numerical taxonomy, convergence and evolutionary reduction. Syst. Zool. 14:101-109. [1965.07.19]

Brown, W. L., Jr. 1966a. [Untitled. Synonymy of *Monomorium floreanum* Stitz under *Monomorium floricola* (Jerdon).] P. 175 in: Linsley, E. G., Usinger, R. L. Insects of the Galápagos Islands. Proc. Calif. Acad. Sci. (4)33:113-196. [1966.01.31]

Brown, W. L., Jr. 1966b. *Strumigenys rectidens* species nov. Pilot Regist. Zool. Card No. 23. [1966.04.10]

Brown, W. L., Jr. 1966c. [Review of: Brian, M. V. 1965. Social insect populations. London: Academic Press, vii + 135 pp.] Q. Rev. Biol. 41:325-326. [1966.09]

Brown, W. L., Jr. 1967a ("1966"). The ant *Cataglyphis birmana* a synonym. Psyche (Camb.) 73:277. [1967.05.24]

Brown, W. L., Jr. 1967b ("1966"). The ant *Aphaenogaster gatesi* transferred to *Pheidole*. Psyche (Camb.) 73:283. [1967.05.24]

Brown, W. L., Jr. 1967c. Studies on North American ants. II. *Myrmecina*. Entomol. News 78:233-240. [1967.11.09]

Brown, W. L., Jr. 1968a. An hypothesis concerning the function of the metapleural glands in ants. Am. Nat. 102:188-191. [1968.04.29]

Brown, W. L., Jr. 1968b ("1967"). A new *Pheidole* with reversed phragmosis (Hymenoptera: Formicidae). Psyche (Camb.) 74:331-339. [1968.06.12]

Brown, W. L., Jr. 1968c. *Strumigenys sisyrata* species nov. Pilot Regist. Zool. Card No. 24. [1968.10.15]

Brown, W. L., Jr. 1969a. The lore of the ant. [A review of four books on ants.] Audubon 71(1): 86-93. [1969.01]

Brown, W. L., Jr. 1969b. *Strumigenys lopotyle* species nov. Pilot Regist. Zool. Card No. 27. [1969.10.12]

Brown, W. L., Jr. 1969c. *Strumigenys wilsoni* species nov. Pilot Regist. Zool. Card No. 28. [1969.11.03]

Brown, W. L., Jr. 1969d. [Review of: Cole, A. C., Jr. 1968. *Pogonomyrmex* harvester ants. A study of the genus in North America. Knoxville: University of Tennessee Press, x + 222 pp.] Am. Sci. 57:65A-66A. [1969]

Brown, W. L., Jr. 1969e. Ant, paleontology of. Pp. 101-103 in: McGraw-Hill yearbook of science and technology. New York: McGraw-Hill Book Co., 424 pp. [1969]

Brown, W. L., Jr. 1970. Ant. Pp. 38-39 in: Gray, P. (ed.) The encyclopedia of the biological sciences, 2nd edition. New York: Van Nostrand Reinhold Co., xxxv + 1027 pp. [1970]

Brown, W. L., Jr. 1971a. Characters and synonymies among the genera of ants. Part IV. Some genera of subfamily Myrmicinae (Hymenoptera: Formicidae). Breviora 365:1-5. [1971.01.15]

Brown, W. L., Jr. 1971b. The identity and synonymy of *Pheidole vigilans*, a common ant of southeastern Australia (Hymenoptera: Formicidae). J. Aust. Entomol. Soc. 10:13-14. [1971.03.31]

Brown, W. L., Jr. 1971c. The Indo-Australian species of the ant genus *Strumigenys*: group of *szalayi* (Hymenoptera: Formicidae). Pp. 73-86 in: Asahina, S., et al. (eds.) Entomological essays to commemorate the retirement of Professor K. Yasumatsu. Tokyo: Hokuryukan Publishing Co., vi + 389 pp. [1971]

Brown, W. L., Jr. 1972a. [Review of: Schneirla, T. C. 1971. Army ants. A study in social organization. (Edited by H. R. Topoff.) San Francisco: W.H. Freeman & Co., xx + 349 pp.] Am. Sci. 60:507. [1972.08]

Brown, W. L., Jr. 1972b. *Asketogenys acubecca*, a new genus and species of dacetine ants from Malaya (Hymenoptera: Formicidae). Psyche (Camb.) 79:23-26. [1972.11.15]

Brown, W. L., Jr. 1972c. The geographical distribution of ants, past and present. [Abstract.] P. 94 in: 14th International Congress of Entomology, 22-30 August 1972. Abstracts. Canberra: Australian Academy of Science and Australian Entomological Society, 356 pp. [1972]

Brown, W. L., Jr. 1973a. A new species of *Miccostruma* (Hymenoptera: Formicidae) from West Africa, with notes on the genus. J. Kansas Entomol. Soc. 46:32-35. [1973.01.03]

Brown, W. L., Jr. 1973b. A comparison of the Hylean and Congo-West African rain forest ant faunas. Pp. 161-185 in: Meggers, B. J., Ayensu, E. S., Duckworth, W. D. (eds.) Tropical forest ecosystems in Africa and South America: a comparative review. Washington, D.C.: Smithsonian Institution Press, viii + 350 pp. [1973.01]

Brown, W. L., Jr. 1973c. The Indo-Australian species of the ant genus *Strumigenys*: groups of *horvathi*, *mayri* and *wallacei*. Pac. Insects 15:259-269. [1973.07.20]

Brown, W. L., Jr. 1974a. A remarkable new island isolate in the genus *Proceratium* (Hymenoptera: Formicidae). Psyche (Camb.) 81:70-83. [1974.05.21]

Brown, W. L., Jr. 1974b. *Novomessor manni* a synonym of *Aphaenogaster ensifera* (Hymenoptera: Formicidae). Entomol. News 85:45-47. [1974.08.09]

Brown, W. L., Jr. 1974c. A supplement to the revision of the ant genus *Basiceros* (Hymenoptera: Formicidae). J. N. Y. Entomol. Soc. 82:131-140. [1974.08.22]

Brown, W. L., Jr. 1974d. *Concoctio* genus nov. Pilot Regist. Zool. Card No. 29. [1974.08.31]

Brown, W. L., Jr. 1974e. *Concoctio concenta* species nov. Pilot Regist. Zool. Card No. 30. [1974.08.31]

Brown, W. L., Jr. 1974f. *Dolioponera* genus nov. Pilot Regist. Zool. Card No. 31. [1974.11.20]

Brown, W. L., Jr. 1974g. *Dolioponera fustigera* species nov. Pilot Regist. Zool. Card No. 32. [1974.11.20]

Brown, W. L., Jr. 1975. Contributions toward a reclassification of the Formicidae. V. Ponerinae, tribes Platythyreini, Cerapachyini, Cylindromyrmecini, Acanthostichini, and Aenictogitini. Search Agric. (Ithaca N. Y.) 5(1):1-115. [1975.06]

Brown, W. L., Jr. 1976a. *Cladarogenys* genus nov. Pilot Regist. Zool. Card No. 33. [1976.12.30]

Brown, W. L., Jr. 1976b. *Cladarogenys lasia* species nov. Pilot Regist. Zool. Card No. 34. [1976.12.30]

Brown, W. L., Jr. 1976c. Contributions toward a reclassification of the Formicidae. Part VI. Ponerinae, tribe Ponerini, subtribe Odontomachiti. Section A. Introduction, subtribal characters. Genus *Odontomachus*. Stud. Entomol. 19:67-171. [1976.12.30]

Brown, W. L., Jr. 1977 ("1976"). *Ctenobethylus* (Bethylidae) a new synonym of *Iridomyrmex* (Formicidae, Hymenoptera). Psyche (Camb.) 83:213-215. [1977.02.24]

Brown, W. L., Jr. 1978a ("1977"). An aberrant new genus of myrmicine ant from Madagascar. Psyche (Camb.) 84:218-224. [1978.07.19]

Brown, W. L., Jr. 1978b ("1977"). A supplement to the world revision of *Odontomachus* (Hymenoptera: Formicidae). Psyche (Camb.) 84:281-285. [1978.07.19]

Brown, W. L., Jr. 1978c. Contributions toward a reclassification of the Formicidae. Part VI. Ponerinae, tribe Ponerini, subtribe Odontomachiti. Section B. Genus *Anochetus* and bibliography. Stud. Entomol. 20:549-638. [1978.08.30]

Brown, W. L., Jr. 1980a. *Protalaridris* genus nov. Pilot Regist. Zool. Card No. 36. [1980.03.31]

Brown, W. L., Jr. 1980b. *Protalaridris armata* species nov. Pilot Regist. Zool. Card No. 37. [1980.03.31]

Brown, W. L., Jr. 1980c ("1979"). A remarkable new species of *Proceratium*, with dietary and other notes on the genus (Hymenoptera: Formicidae). Psyche (Camb.) 86:337-346. [1980.12.19]

Brown, W. L., Jr. 1981. Preliminary contributions toward a revision of the ant genus *Pheidole* (Hymenoptera: Formicidae). Part I. J. Kansas Entomol. Soc. 54:523-530. [1981.07.17]

Brown, W. L., Jr. 1982. Hymenoptera. Pp. 652-680 in: Parker, S. P. (ed.) Synopsis and classification of living organisms. Volume 2. New York: McGraw-Hill, 1232 pp. [1982]

Brown, W. L., Jr. 1983. [Review of: Hermann, H. R. (ed.) 1982. Social insects. Volume 4. New York: Academic Press, xiii + 385 pp.] Q. Rev. Biol. 58:444-445. [1983.09]

Brown, W. L., Jr. 1986 ("1985"). *Indomyrma dasypyx*, new genus and species, a myrmicine ant from peninsular India (Hymenoptera: Formicidae). Isr. J. Entomol. 19:37-49. [1986.02]

Brown, W. L., Jr. 1988a ("1987"). *Neoclystopsenella* (Bethylidae) a synonym of *Tapinoma* (Formicidae). Psyche (Camb.) 94:337. [1988.07.25]

Brown, W. L., Jr. 1988b. *Strumigenys philiporum* species nov. Pilot Regist. Zool. Card No. 38. [1988.11.15]

Brown, W. L., Jr. 1988c. *Strumigenys buleru* species nov. Pilot Regist. Zool. Card No. 39. [1988.11.15]

Brown, W. L., Jr. 1988d. *Strumigenys cochlearis* species nov. Pilot Regist. Zool. Card No. 40. [1988.11.15]

Brown, W. L., Jr. 1988e. *Strumigenys yaleopleura* species nov. Pilot Regist. Zool. Card No. 41. [1988.12.10]

Brown, W. L., Jr. 1988f. *Strumigenys anetes* species nov. Pilot Regist. Zool. Card No. 42. [1988.12.10]

Brown, W. L., Jr. 1988g. *Strumigenys paranetes* species nov. Pilot Regist. Zool. Card No. 43. [1988.12.10]

Brown, W. L., Jr. 1988h. Data on Malpighian tubule numbers in ants (Hymenoptera: Formicidae). Pp. 17-27 in: Trager, J. C. (ed.) Advances in myrmecology. Leiden: E. J. Brill, xxvii + 551 pp. [1988]

Brown, W. L., Jr. 1993 ("1992"). Two new species of *Gnamptogenys*, and an account of millipede predation by one of them. Psyche (Camb.) 99:275-289. [1993.06.02]

Brown, W. L., Jr. 1995a. [Untitled. Taxonomic changes in *Pachycondyla* attributed to Brown.] Pp. 302-311 in: Bolton, B. A new general catalogue of the ants of the world. Cambridge, Mass.: Harvard University Press, 504 pp. [1995.10]

Brown, W. L., Jr. 1995b. ("1994"). *Trichoscapa karawajewi* and its synonyms (Hymenoptera: Formicidae). Psyche (Camb.) 101:219-220. [1995.11.20]

Brown, W. L., Jr., Boisvert, R. G. 1979 ("1978"). The dacetine ant genus *Pentastruma* (Hymenoptera: Formicidae). Psyche (Camb.) 85:201-207. [1979.04.24]

Brown, W. L., Jr., Carpenter, F. M. 1979 ("1978"). A restudy of two ants from the Sicilian amber. Psyche (Camb.) 85:417-423. [1979.09.18]

Brown, W. L., Jr., Gotwald, W. H., Jr., Lévieux, J. 1971 ("1970"). A new genus of ponerine ants from West Africa (Hymenoptera: Formicidae) with ecological notes. Psyche (Camb.) 77:259-275. [1971.05.31]

Brown, W. L., Jr., Kempf, W. W. 1960. A world revision of the ant tribe Basicerotini. Stud. Entomol. (n.s.)3:161-250. [1960.12.20]

Brown, W. L., Jr., Kempf, W. W. 1961 ("1960"). The type species of the ant genus *Eurhopalothrix*. Psyche (Camb.) 67:44. [1961.02.16]

Brown, W. L., Jr., Kempf, W. W. 1968 ("1967"). *Tatuidris*, a remarkable new genus of Formicidae (Hymenoptera). Psyche (Camb.) 74:183-190. [1968.02.20]

Brown, W. L., Jr., Kempf, W. W. 1969. A revision of the neotropical dacetine ant genus *Acanthognathus* (Hymenoptera: Formicidae). Psyche (Camb.) 76:87-109. [1969.12.12]

Brown, W. L., Jr., Nutting, W. L. 1950. Wing venation and the phylogeny of the Formicidae. Trans. Am. Entomol. Soc. 75:113-132. [1950.01.04]

Brown, W. L., Jr., Taylor, R. W. 1970. Superfamily Formicoidea. Pp. 951-959 in: CSIRO (ed.) The insects of Australia. Carlton, Victoria: Melbourne University Press, xiii + 1029 pp. [1970]

Brown, W. L., Jr., Wilson, E. O. 1953. [Review of: Cook, T. W. 1953. The ants of California. Palo Alto: Pacific Books, 462 pp.] Entomol. News 64:163-164. [1953.06.19]

Brown, W. L., Jr., Wilson, E. O. 1954. The case against the trinomen. Syst. Zool. 3:174-176. [1954.12]

Brown, W. L., Jr., Wilson, E. O. 1956. Character displacement. Syst. Zool. 5:49-64. [1956.06]

Brown, W. L., Jr., Wilson, E. O. 1957a. *Dacetinops*, a new ant genus from New Guinea. Breviora 77:1-7. [1957.06.21]

Brown, W. L., Jr., Wilson, E. O. 1957b. A new parasitic ant of the genus *Monomorium* from Alabama, with a consideration of the status of genus *Epixenus* Emery. Entomol. News 68:239-246. [1957.10.22]

Brown, W. L., Jr., Wilson, E. O. 1959a. The search for *Nothomyrmecia*. West. Aust. Nat. 7:25-30. [1959.08.25]

Brown, W. L., Jr., Wilson, E. O. 1959b. The evolution of the dacetine ants. Q. Rev. Biol. 34:278-294. [1959.12]

Brown, W. L., Jr., Yasumatsu, K. 1951. On the publication date of *Polyhomoa itoi* Azuma (Hymenoptera, Formicidae). Mushi 22:93-95. [1951.04.15]

Brown, W. V., Jaisson, P., Taylor, R. W., Lacey, M. J. 1990. Novel internally branched, internal alkenes as major components of the cuticular hydrocarbons of the primitive Australian ant *Nothomyrmecia macrops* Clark (Hymenoptera: Formicidae). J. Chem. Ecol. 16:2623-2635. [1990.09]

Browne, J. T., Gregg, R. E. 1969. A study of the ecological distribution of ants in Gregory Canyon, Boulder, Colorado. Univ. Colo. Stud. Ser. Biol. 30:1-48. [1969.05]

Bruch, C. 1914. Catálogo sistemático de los formícidos argentinos. Rev. Mus. La Plata 19:211-234. [1914.03.20]

Bruch, C. 1915. Suplemento al catálogo de los formícidos argentinos. I. (Addenda et corrigenda). Rev. Mus. La Plata 19:527-537. [1915.12.24]

Bruch, C. 1916. Contribución al estudio de las hormigas de la provincia de San Luis. Rev. Mus. La Plata 23:291-357. [1916.09.22]

Bruch, C. 1917a. La forma sexuada femenina de *Cryptocerus ridiculus* Santschi. Physis (B. Aires) 3:269-270. [1917.07.30]

Bruch, C. 1917b. Costumbres y nidos de hormigas. An. Soc. Cient. Argent. 83:302-316. [1917.07] [Cover date is May-June 1917 but an article by another author in the same issue has a sign-off date of 3 July 1917 (page 279). Hence this issue cannot have been released before July 1917.]

Bruch, C. 1917c. Costumbres y nidos de hormigas. II. An. Soc. Cient. Argent. 84:154-168. [1917.10]

Bruch, C. 1917d. Hormigas de Catamarca. Physis (B. Aires) 3:430-433. [1917.12.31]

Bruch, C. 1919. Descripción de una curiosa ponerina de Córdoba *Discothyrea neotropica* n. sp. Physis (B. Aires) 4:400-402. [1919.12.31]

Bruch, C. 1921. Estudios mirmecológicos. Rev. Mus. La Plata 26:175-211. [1921.12.31]

Bruch, C. 1923. Estudios mirmecológicos con la descripción de nuevas especies de dípteros (Phoridae) por los Rr. Pp. H. Schmitz y Th. Borgmeier y de una araña (Gonyleptidae) por el Doctor Mello-Leitão. Rev. Mus. La Plata 27:172-220. [1923]

Bruch, C. 1924a. Descripción de la reina *Eciton* (*Acamatus*) *Hetschkoi* Mayr. Physis (B. Aires) 7:232-235. [1924.03.25]

Bruch, C. 1924b. Una nueva ponerina *Acanthostichus Ramosmexiae* n. spec. Physis (B. Aires) 7:260-261. [1924.03.25]

Bruch, C. 1925a. Biologia de *Pseudoatta argentina*. Physis (B. Aires) 8:106. [1925.05.23]

Bruch, C. 1925b. Macho, larva y ninfa de *Acanthostichus ramosmexiae* Bruch. Physis (B. Aires) 8:110-114. [1925.05.23]

Bruch, C. 1925c. Rectificación. Physis (B. Aires) 8:125. [1925.05.23]

Bruch, C. 1928. Estudios mirmecológicos. An. Mus. Nac. Hist. Nat. B. Aires 34:341-360. [1928.02.17]

Bruch, C. 1930. Notas preliminares acerca de *Labauchena Daguerrei* Santschi. Rev. Soc. Entomol. Argent. 3:73-80. [1930.12.26]

Bruch, C. 1931. Notas biológicas y sistemáticas acerca de *Bruchomyrma acutidens* Santschi. Rev. Mus. La Plata 33:31-55. [1931.04.17]

Bruch, C. 1932. Descripción de un género y especie nueva de una hormiga parásita (Formicidae). Rev. Mus. La Plata 33:271-275. [1932.02.29]
Bruch, C. 1934a. La hembra dictadiforme de *Acanthostichus laticornis* For. v. *obscuridens* Sants. (Hymenoptera Formicidae). Rev. Soc. Entomol. Argent. 6:3-8. [1934.05.30]
Bruch, C. 1934b. Las formas femeninas de *Eciton*. Descripción y redescripción de algunas especies de la Argentina. An. Soc. Cient. Argent. 118:113-135. [1934.09]
Bruch, C. 1934c. La obra entomológica del Doctor Angel Gallardo. Rev. Soc. Entomol. Argent. 6:235-242. [1934.12.30]
Bruch, C. 1934d. Breves observaciones sobre costumbres de *Eciton* (*Holopone*) *dulcius* For. var. *jujuyensis* For. y decripción del macho (Hym. Formic.). Rev. Soc. Entomol. Argent. 6:261-265. [1934.12.30]
Bruch, C. 1937. William Morton Wheeler. Rev. Soc. Entomol. Argent. 9:27-29. [1937.12.31]
Bruder, K. W., Gupta, A. P. 1972. Habitats and distribution of the pavement ant in New Jersey. J. Econ. Entomol. 65:1180-1181. [1972.08.15]
Brues, C. T. 1910. [Review of: Wheeler, W. M. 1910. Ants: their structure, development and behavior. New York: Columbia University Press, xxv + 663 pp.] Psyche (Camb.) 17:122-123. [1910.06] [Stamp date in MCZ library: 1910.06.27.]
Brues, C. T. 1925. *Scyphodon*, an anomalous genus of Hymenoptera of doubtful affinities. Treubia 6:93-96. [1925.02]
Brues, C. T. 1937. Professor William Morton Wheeler, with a list of his published writings. Psyche (Camb.) 44:61-96. [1937.09] [Stamp date in MCZ library: 1937.12.02.]
Brues, C. T. 1939. New Oligocene Braconidae and Bethylidae from Baltic amber. Ann. Entomol. Soc. Am. 32:251-263. [1939.06.30]
Brullé, G. A. 1833 ("1832"). Expédition scientifique de Morée. Section des sciences physiques. Tome III. Partie 1. Zoologie. Deuxième section - Des animaux articulés. [part] Paris: Levrault, pp. 289-336. [1833.03.20] [Date of publication from Sherborn & Woodward (1901c:335) and Evenhuis (1994b:485).]
Brullé, G. A. 1840. Insectes. Pp. 53-95 in: Webb, P. B., Berthelot, S. Histoire naturelle des Îles Canaries. Vol. 2. (deuxième partie - Entomologie). Paris: Mellier, 119 pp. [1840]
Bruneau de Miré, P. 1966. Une affection du caféier Robusta dans l'est du Cameroun: la "défoliation en mannequin d'osier". Café Cacao Thé 10:237-242. [1966.09] [Description of *Sphaerocrema dewasi* (= *Crematogaster wellmani*), page 240.]
Bruneau de Miré, P. 1969. Note taxonomique à propos de *Sphaerocrema dewasi* agent causal de la "défoliation en mannequin d'osier" du caféier Robusta au Cameroun. Café Cacao Thé 13: 55-56. [1969.03]
Brunnert, A., Wehner, R. 1973. Fine structure of light- and dark-adapted eyes of desert ants, *Cataglyphis bicolor* (Formicidae, Hymenoptera). J. Morphol. 140:15-29. [1973.05]
Bucher, E. H., Montenegro, R. 1974. Hábitos forrajeros de cuatro hormigas simpátridas del género *Acromyrmex* (Hymenoptera, Formicidae). Ecología (Asoc. Argent. Ecol.) 2:47-53. [1974.04]
Buckell, E. R. 1928 ("1927"). An annotated list of the ants of British Columbia. Proc. Entomol. Soc. B. C. 24:30-34. [1928.02] [Publication date from Leech (1942:35.]
Buckley, R. C. (ed.) 1982a. Ant-plant interactions in Australia. The Hague: W. Junk, x + 162 pp. [1982]
Buckley, R. C. 1982b. Ant-plant interactions: a world review. Pp. 111-141 in: Buckley, R. C. (ed.) Ant-plant interactions in Australia. The Hague: W. Junk, x + 162 pp. [1982]
Buckley, R. C. 1982c. A world bibliography of ant-plant interactions. Pp. 143-162 in: Buckley, R. C. (ed.) Ant-plant interactions in Australia. The Hague: W. Junk, x + 162 pp. [1982]
Buckley, R. C. 1987. Ant-plant-homopteran interactions. Adv. Ecol. Res. 16:53-85. [1987]
Buckley, S. B. 1860a. The cutting ant of Texas. Proc. Acad. Nat. Sci. Phila. 12:233-236. [1860.06]
Buckley, S. B. 1860b. *Myrmica* (*Atta*) *molefaciens*, "Stinging Ant" or "Mound-Making Ant," of Texas. Proc. Acad. Nat. Sci. Phila. 12:445-447. [1860.10]

Buckley, S. B. 1860c. The cutting ant of Texas (*Oecodoma Mexicana*, Sm.). Ann. Mag. Nat. Hist. (3)6:386-389. [1860.11]
Buckley, S. B. 1861. Note on ants in Texas. Proc. Acad. Nat. Sci. Phila. 13:9-10. [1861.02] [Pagebase date is January 1861, but this article was issued in a signature that included part of "February 1861" (Nolan, 1913).]
Buckley, S. B. 1866. Descriptions of new species of North American Formicidae. Proc. Entomol. Soc. Phila. 6:152-172. [1866.07] [Pagebase date is July 1866. Possibly published later, at least as a whole issue. See F. M. Brown (1964).]
Buckley, S. B. 1867. Descriptions of new species of North American Formicidae (continued from page 172.). Proc. Entomol. Soc. Phila. 6:335-350. [1867.02] [Pagebase date is February 1867. For discussion of problems with dating the parts of this serial see F. M. Brown (1964).]
Bugnion, E. 1926. Les pièces buccales, le sac infrabuccal et le pharynx des fourmis. [part] Folia Myrmecol. Termit. 1:33-44. [1926.12]
Bugnion, E. 1927a. Les pièces buccales, le sac infrabuccal et le pharynx des fourmis. [part] Folia Myrmecol. Termit. 1:59-71. [1927.02]
Bugnion, E. 1927b. Les pièces buccales, le sac infrabuccal et le pharynx des fourmis. [part] Folia Myrmecol. Termit. 1:73-88. [1927.03]
Bugnion, E. 1927c. Les pièces buccales, le sac infrabuccal et le pharynx des fourmis. [concl.] Folia Myrmecol. Termit. 1:105-134. [1927.06]
Bugnion, E. 1929. Les organes bucco-pharyngés de la fourmi coupe-feuilles du Brésil, *Atta sexdens* Lin. Zool. Anz. 82:55-78. [1929]
Bugnion, E. 1930. Les pièces buccales, le sac infrabuccal et le pharynx des fourmis. Bull. Soc. Entomol. Egypte 14:85-210. [1930]
Bünzli, G. H. 1935. Untersuchungen über coccidophile Ameisen aus den Kaffeefeldern von Surinam. Mitt. Schweiz. Entomol. Ges. 16:453-593. [1935.08.14]
Buren, W. F. 1941a. A preliminary list of Iowa ants. Iowa State Coll. J. Sci. 15:111-117. [1941.01]
Buren, W. F. 1941b. *Lasius* (*Acanthomyops*) *plumopilosus*, a new ant with plumose hairs from Iowa. Iowa State Coll. J. Sci. 15:231-235. [1941.04]
Buren, W. F. 1942. New ants from Minnesota, Iowa, and Wisconsin. Iowa State Coll. J. Sci. 16:399-408. [1942.04]
Buren, W. F. 1944a. A list of Iowa ants. Iowa State Coll. J. Sci. 18:277-312. [1944.04]
Buren, W. F. 1944b. A new fungus growing ant from Mexico. Psyche (Camb.) 51:5-7. [1944.06] [Stamp date in MCZ library: 1944.08.07.]
Buren, W. F. 1945. *Leptothorax longispinosus* subsp. *iowensis*, nom. n. (Hymenoptera, Formicidae). Proc. Entomol. Soc. Wash. 47:288. [1945.12.28]
Buren, W. F. 1950. A new *Lasius* (*Acanthomyops*) with a key to North American females. Proc. Entomol. Soc. Wash. 52:184-190. [1950.08.18]
Buren, W. F. 1959 ("1958"). A review of the species of *Crematogaster*, sensu stricto, in North America (Hymenoptera: Formicidae). Part I. J. N. Y. Entomol. Soc. 66:119-134. [1959.03.24]
Buren, W. F. 1968a. Some fundamental taxonomic problems in *Formica* (Hymenoptera: Formicidae). J. Ga. Entomol. Soc. 3:25-40. [1968.04]
Buren, W. F. 1968b. A review of the species of *Crematogaster*, sensu stricto, in North America (Hymenoptera, Formicidae). Part II. Descriptions of new species. J. Ga. Entomol. Soc. 3: 91-121. [1968.07]
Buren, W. F. 1972. Revisionary studies on the taxonomy of the imported fire ants. J. Ga. Entomol. Soc. 7:1-26. [1972.01]
Buren, W. F. 1982. Red imported fire ant now in Puerto Rico. Fla. Entomol. 65:188-189. [1982.03.30]
Buren, W. F. 1983. Artificial faunal replacement for imported fire ant control. Fla. Entomol. 66:93-100. [1983.05.15]
Buren, W. F., Allen, G. E., Whitcomb, W. H., Lennartz, F. E., Williams, R. N. 1974. Zoogeography of the imported fire ants. J. N. Y. Entomol. Soc. 82:113-124. [1974.08.22]

Buren, W. F., Allen, G. E., Williams, R. N. 1978. Approaches toward possible pest management of the imported fire ants. Bull. Entomol. Soc. Am. 24:418-421. [1978.12.15]

Buren, W. F., Hermann, H. R., Blum, M. S. 1970. The widespread occurrence of mandibular grooves in aculeate Hymenoptera. J. Ga. Entomol. Soc. 5:185-196. [1970.10]

Buren, W. F., Naves, M. A., Carlysle, T. C. 1977. False phragmosis and apparent specialization for subterranean warfare in *Pheidole lamia* Wheeler (Hymenoptera: Formicidae). J. Ga. Entomol. Soc. 12:100-108. [1977.04]

Buren, W. F., Nickerson, J. C., Thompson, C. R. 1976 ("1975"). Mixed nests of *Conomyrma insana* and *C. flavopecta* - Evidence of parasitism (Hymenoptera: Formicidae). Psyche (Camb.) 82:306-314. [1976.04.20]

Burgman, M. A., Crozier, R. H., Taylor, R. W. 1980. Comparisons of different methods of determining affinities of nine ant species of the genus *Camponotus*. Aust. J. Zool. 28:151-160. [1980.05.07]

Burmeister, C. H. C. 1831. Derselbe über Bernsteininsecten. Isis Oken 1831:1100. [1831]

Burnham, L. 1979 ("1978"). Survey of social insects in the fossil record. Psyche (Camb.) 85: 85-133. [1979.01.26]

Burrill, A. C., Smith, M. R. 1918. A preliminary list of the ants of Wisconsin. Ohio J. Sci. 18:229-232. [1918.05.06]

Burrill, A. C., Smith, M. R. 1919. A key to the species of Wisconsin ants, with notes on their habits. Ohio J. Sci. 19:279-292. [1919.03.29]

Bursakov, S. S. 1984. Two new species of ants of the genus *Tetramorium* (Hymenoptera, Formicidae) from southeast Kazakhstan. [In Russian.] Zool. Zh. 63:399-405. [1984.03] [March issue. Date signed for printing ("podpisano k pechati"): 21 February 1984.]

Burton, J. L., Franks, N. R. 1985. The foraging ecology of the army ant *Eciton rapax*: an ergonomic enigma? Ecol. Entomol. 10:131-141. [1985.05]

Buschinger, A. 1966a ("1965"). *Leptothorax* (*Mychothorax*) *kutteri* n. sp., eine sozialparasitische Ameise (Hymenoptera, Formicidae). Insectes Soc. 12:327-334. [1966.03]

Buschinger, A. 1966b. Untersuchungen an *Harpagoxenus sublaevis* Nyl. (Hym. Formicidae). I. Freilandbeobachtungen zu Verbreitung und Lebensweise. Insectes Soc. 13:5-16. [1966.06]

Buschinger, A. 1966c. *Leptothorax* (*Mychothorax*) *muscorum* Nylander und *Leptothorax* (*M.*) *gredleri* Mayr zwei gute Arten. Insectes Soc. 13:165-172. [1966.12]

Buschinger, A. 1968a. Mono- und Polygynie bei Arten der Gattung *Leptothorax* Mayr (Hymenoptera Formicidae). Insectes Soc. 15:217-225. [1968.12]

Buschinger, A. 1968b. Zur Verbreitung und Lebenweise des Tribus Leptothoracini (Hymenoptera, Formicidae) in Nordbayern. Bayer. Tierwelt 1:115-128. [1968]

Buschinger, A. 1970. Neue Vorstellungen zur Evolution des Sozialparasitismus und der Dulosis bei Ameisen (Hym., Formicidae). Biol. Zentralbl. 89:273-299. [1970.06]

Buschinger, A. 1971a. Zur Verbreitung und Lebensweise sozialparasitischer Ameisen des Schweizer Wallis (Hym., Formicidae). Zool. Anz. 186:47-59. [1971.02]

Buschinger, A. 1971b. "Locksterzeln" und Kopula der sozialparasitischen Ameise *Leptothorax kutteri* Buschinger (Hym., Form.). Zool. Anz. 186:242-248. [1971.04]

Buschinger, A. 1971c. Zur Verbreitung der Sozialparasiten von *Leptothorax acervorum* (Fabr.) (Hym., Formicidae). Bonn. Zool. Beitr. 22:322-331. [1971.09]

Buschinger, A. 1971d. Weitere Untersuchungen zum Begattungsverhalten sozialparasitischer Ameisen (*Harpagoxenus sublaevis* Nyl. und *Doronomyrmex pacis* Kutter, Hym., Formicidae). Zool. Anz. 187:184-198. [1971.10]

Buschinger, A. 1972a. Kreuzung zweier sozialparasitischer Ameisenarten, *Doronymyrmex pacis* Kutter und *Leptothorax kutteri* Buschinger (Hym., Formicidae). Zool. Anz. 189:169-179. [1972.10]

Buschinger, A. 1972b. Begattungsverhalten sozialparasitischer Ameisen aus dem Tribus Leptothoracini (*Harpagoxenus sublaevis*, *Doronomyrmex pacis*, *Leptothorax kutteri*). Insectes Soc. 19:416. [1972]

Buschinger, A. 1973. Ameisen des Tribus Leptothoracini (Hym., Formicidae) als Zwischenwirte von Cestoden. Zool. Anz. 191:369-380. [1973.12]

Buschinger, A. 1974a. Zur Biologie der Sozialparasitischen Ameise *Leptothorax goesswaldi* Kutter (Hym., Formicidae). Insectes Soc. 21:133-144. [1974.12]

Buschinger, A. 1974b. Polymorphismus und Kastendetermination im Ameisentribus Leptothoracini. Pp. 604-623 in: Schmidt, G. H. (ed.) Sozialpolymorphismus bei Insekten. Stuttgart: Wissenschaftliche Verlagsgesellschaft mbH, xxiv + 974 pp. [1974]

Buschinger, A. 1974c. Monogynie und Polygynie in Insektensozietäten. Pp. 862-896 in: Schmidt, G. H. (ed.) Sozialpolymorphismus bei Insekten. Stuttgart: Wissenschaftliche Verlagsgesellschaft mbH, xxiv + 974 pp. [1974]

Buschinger, A. 1974d. Polymorphismus und Polyethismus sozialparasitischer Hymenopteren. Pp. 897-934 in: Schmidt, G. H. (ed.) Sozialpolymorphismus bei Insekten. Stuttgart: Wissenschaftliche Verlagsgesellschaft mbH, xxiv + 974 pp. [1974]

Buschinger, A. 1975a. Eine genetische Komponente im Polymorphismus der dulotischen Ameise *Harpagoxenus sublaevis*. Naturwissenschaften 62:239-240. [1975.05]

Buschinger, A. 1975b. Sexual pheromones in ants. Pp. 225-233 in: Noirot, C., Howse, P. E., Le Masne, G. (eds.) Pheromones and defensive secretions in social insects. Dijon: French Section of the IUSSI, vi + 248 pp. [1975.12]

Buschinger, A. 1975c. Die Ameisenfauna des Bausenberges, der nordöstlichen Eifel und Voreifel (Hym., Formicidae) mit einer quantitativen Auswertung von Fallenfängen. Beitr. Landespfl. Rheinl.-Pfalz. Beih. 4:251-273. [1975]

Buschinger, A. 1979a. Functional monogyny in the American guest ant *Formicoxenus hirticornis* (Emery) (= *Leptothorax hirticornis*), (Hym., Form.). Insectes Soc. 26:61-68. [1979.06]

Buschinger, A. 1979b. *Doronomyrmex pocahontas* n. sp., a parasitic ant from Alberta, Canada (Hym. Formicidae). Insectes Soc. 26:216-222. [1979.09]

Buschinger, A. 1979c. Zur Ameisenfauna von Südhessen unter besonderer Berücksichtigung von geschützten und schutzwürdigen Gebieten. Naturwiss. Ver. Darmstadt Ber. (N.F.)3:7-32.

Buschinger, A. 1981. Biological and systematic relationships of social parasitic Leptothoracini from Europe and North America. Pp. 211-222 in: Howse, P. E., Clement, J.-L. (eds.) Biosystematics of social insects. Systematics Association Special Volume No. 19. London: Academic Press, 346 pp. [1981]

Buschinger, A. 1982. *Epimyrma goesswaldi* Menozzi 1931 = *Epimyrma ravouxi* (André 1896). Morphologischer und biologischer Nachweis der Synonymie (Hym., Formicidae). Zool. Anz. 208:352-358. [1982]

Buschinger, A. 1983 ("1982"). *Leptothorax faberi* n. sp., an apparently parasitic ant from Jasper National Park, Canada (Hymenoptera: Formicidae). Psyche (Camb.) 89:197-209. [1983.04.27]

Buschinger, A. 1985a. The *Epimyrma* species of Corsica (Hymenoptera, Formicidae). Spixiana 8:277-280. [1985.12.30]

Buschinger, A. 1985b. New records of rare parasitic ants (Hym., Form.) in the French Alps. Insectes Soc. 32:321-324. [1985.12]

Buschinger, A. 1986. Evolution of social parasitism in ants. Trends Ecol. Evol. 1:155-160. [1986.12]

Buschinger, A. 1987a. *Teleutomyrmex schneideri* Kutter 1950 and other parasitic ants found in the Pyrenees. Spixiana 10:81-83. [1987.03.01]

Buschinger, A. 1987b. Synonymy of *Leonomyrma* Arnoldi 1968 with *Chalepoxenus* Menozzi 1922 (Hymenoptera: Formicidae). Psyche (Camb.) 94:117-126. [1987.10.06]

Buschinger, A. 1987c. Symposium: biosystematics of the ant tribe Leptothoracini. Introduction. Pp. 27-28 in: Eder, J., Rembold, H. (eds.) Chemistry and biology of social insects. München: Verlag J. Peperny, xxxv + 757 pp. [1987]

Buschinger, A. 1987d. Biological arguments for a systematic rearrangement of the ant tribe Leptothoracini. P. 43 in: Eder, J., Rembold, H. (eds.) Chemistry and biology of social insects. München: Verlag J. Peperny, xxxv + 757 pp. [1987]

Buschinger, A. 1987e. Polymorphism and reproductive division of labor in advanced ants. Pp. 257-258 in: Eder, J., Rembold, H. (eds.) Chemistry and biology of social insects. München: Verlag J. Peperny, xxxv + 757 pp. [1987]

Buschinger, A. 1989a. Evolution, speciation, and inbreeding in the parasitic ant genus *Epimyrma* (Hymenoptera, Formicidae). J. Evol. Biol. 2:265-283. [1989.06]

Buschinger, A. 1989b. Workerless *Epimyrma kraussei* Emery 1915, the first parasitic ant of Crete. Psyche (Camb.) 96:69-74. [1989.12.18]

Buschinger, A. 1990a. Evolutionary transitions between types of social parasitism in ants, hypotheses and evidence. Pp. 145-146 in: Veeresh, G. K., Mallik, B., Viraktamath, C. A. (eds.) Social insects and the environment. Proceedings of the 11th International Congress of IUSSI, 1990. New Delhi: Oxford & IBH Publishing Co., xxxi + 765 pp. [1990.08]

Buschinger, A. 1990b. Sympatric speciation and radiative evolution of socially parasitic ants - Heretic hypotheses and their factual background. Z. Zool. Syst. Evolutionsforsch. 28:241-260. [1990.12]

Buschinger, A. 1990c. Regulation of worker and queen formation in ants with special reference to reproduction and colony development. Pp. 37-57 in: Engels, W. (ed.) Social insects: an evolutionary approach to castes and reproduction. Berlin: Springer-Verlag, 265 pp. [1990]

Buschinger, A. 1992. Genetik der Kastenbildung bei Ameisen. Naturwiss. Rundsch. 45:85-92. [1992.03]

Buschinger, A. 1995. Life history of the parasitic ant *Epimyrma bernardi* Espadaler, 1982 (Insecta, Hymenoptera, Formicidae). Spixiana 18:75-81. [1995.03.01]

Buschinger, A., Alloway, T. M. 1978. Caste polymorphism in *Harpagoxenus canadensis* M. R. Smith (Hym., Formicidae). Insectes Soc. 25:339-350. [1978.12]

Buschinger, A., Alloway, T. M. 1979. Sexual behaviour in the slave-making ant, *Harpagoxenus canadensis* M. R. Smith, and sexual pheromone experiments with *H. canadensis*, *H. americanus* (Emery), and *H. sublaevis* (Nylander) (Hymenoptera; Formicidae). Z. Tierpsychol. 49:113-119. [1979.09]

Buschinger, A., Cagniant, H., Ehrhardt, W., Heinze, J. 1989 ("1988"). *Chalepoxenus brunneus*, a workerless "degenerate slave-maker" ant (Hymenoptera, Formicidae). Psyche (Camb.) 95:253-363. [1989.06.13]

Buschinger, A., Douwes, P. 1993. Socially parasitic ants of Greece. Biol. Gallo-Hell. 20:183-189.

Buschinger, A., Ehrhardt, W., Fischer, K. 1981. *Doronomyrmex pacis*, *Epimyrma stumperi* und *E. goesswaldi* (Hym., Formicidae) neu für Frankreich. Insectes Soc. 28:67-70. [1981.03]

Buschinger, A., Ehrhardt, W., Fischer, K., Ofer, J. 1988. The slave-making ant genus *Chalepoxenus* (Hymenoptera, Formicidae). I. Review of literature, range, slave species. Zool. Jahrb. Abt. Syst. Ökol. Geogr. Tiere 115:383-401. [1988]

Buschinger, A., Ehrhardt, W., Winter, U. 1980. The organization of slave raids in dulotic ants - a comparative study (Hymenoptera; Formicidae). Z. Tierpsychol. 53:245-264. [1980]

Buschinger, A., Fischer, K. 1991. Hybridization of chromosome-polymorphic populations of the inquiline ant, *Doronomyrmex kutteri* (Hym., Formicidae). Insectes Soc. 38:95-103. [1991]

Buschinger, A., Fischer, K., Guthy, H. P., Jessen, K., Winter, U. 1987 ("1986"). Biosystematic revision of *Epimyrma kraussei*, *E. vandeli*, and *E. foreli* (Hymenoptera, Formicidae). Psyche (Camb.) 93:253-276. [1987.03.17]

Buschinger, A., Francoeur, A. 1983. The guest ant, *Symmyrmica chamberlini*, rediscovered near Salt Lake City, Utah (Hymenoptera, Formicidae). Psyche (Camb.) 90:297-305. [1983.12.09]

Buschinger, A., Francoeur, A. 1992 ("1991"). Queen polymorphism and functional monogyny in the ant, *Leptothorax sphagnicolus* Francoeur. Psyche (Camb.) 98:119-133. [1992.02.07]

Buschinger, A., Francoeur, A., Fischer, K. 1981 ("1980"). Functional monogyny, sexual behavior, and karyotype of the guest ant, *Leptothorax provancheri* Emery (Hymenoptera, Formicidae). Psyche (Camb.) 87:1-12. [1981.04.27]

Buschinger, A., Heinze, J. 1992. Polymorphism of female reproductives in ants. Pp. 11-23 in: Billen, J. P. J. (ed.) Biology and evolution of social insects. Leuven: Leuven University Press, ix + 390 pp. [1992]

Buschinger, A., Heinze, J. 1993. *Doronomyrmex pocahontas*: not a workerless parasite but still an enigmatic taxon (Hymenoptera, Formicidae). Insectes Soc. 40:423-432. [1993]

Buschinger, A., Jessen, K., Cagniant, H. 1990. The life history of *Epimyrma algeriana*, a slave-making ant with facultative polygyny (Hymenoptera, Formicidae). Zool. Beitr. (N.F.)33: 23-49. [1990.08.30]

Buschinger, A., Klump, B. 1988. Novel strategy of host-colony exploitation in a permanently parasitic ant, *Doronomyrmex goesswaldi*. Naturwissenschaften 75:577-578. [1988.11]

Buschinger, A., Maschwitz, U. 1984. Defensive behavior and defensive mechanisms in ants. Pp. 95-150 in: Hermann, H. R. (ed.) Defensive mechanisms in social insects. New York: Praeger, xii + 259 pp. [1984]

Buschinger, A., Peeters, C., Crozier, R. H. 1990 ("1989"). Life-pattern studies of an Australian *Sphinctomyrmex* (Formicidae: Ponerinae; Cerapachyini): functional polygyny, brood periodicity and raiding behavior. Psyche (Camb.) 96:287-300. [1990.04.20]

Buschinger, A., Schumann, R. D. 1994. New records of *Leptothorax wilsoni* from western North America. Psyche (Camb.) 101:13-18. [1994.06.28]

Buschinger, A., Schumann, R. D., Heinze, J. 1994. First records of the guest ant *Formicoxenus quebecensis* Francoeur from western Canada (Hymenoptera, Formicidae). Psyche (Camb.) 101:53-57. [1994.06.28]

Buschinger, A., Stoewesand, H. 1971. Teratologische Untersuchungen an Ameisen (Hymenoptera: Formicidae). Beitr. Entomol. 21:211-241. [1971]

Buschinger, A., Winter, U. 1982. Evolutionary trends in the parasitic ant genus *Epimyrma*. Pp. 266-269 in: Breed, M. D., Michener, C. D., Evans, H. E. (eds.) The biology of social insects. Proceedings of the Ninth Congress of the IUSSI, Boulder, Colorado, August 1982. Boulder: Westview Press, xii + 420 pp. [1982.08]

Buschinger, A., Winter, U. 1983. Population studies of the dulotic ant, *Epimyrma ravouxi*, and the degenerate slavemaker, *E. kraussei* (Hymenoptera: Formicidae). Entomol. Gen. 8:251-266. [1983.09.20]

Buschinger, A., Winter, U. 1984. Biosystematics of the social parasitic ant genus *Epimyrma* (Hym., Formicidae). [Abstract.] P. 497 in: XVII International Congress of Entomology. Hamburg, Federal Republic of Germany, August 20-26, 1984. Abstract Volume. Hamburg: XVII International Congress of Entomology, 960 pp. [1984.08.10]

Buschinger, A., Winter, U. 1985. Life history and male morphology of the workerless parasitic ant *Epimyrma corsica* (Hymenoptera: Formicidae). Entomol. Gen. 10:65-75. [1985.05.31]

Buschinger, A., Winter, U., Faber, W. 1984 ("1983"). The biology of *Myrmoxenus gordiagini* Ruzsky, a slave-making ant (Hymenoptera, Formicidae). Psyche (Camb.) 90:335-342. [1984.04.06]

Butler, A. G. 1874. Insects. In: Richardson, J., Gray, J. E. (eds.) The zoology of the voyage of H.M.S. Erebus & Terror, under the command of Captain Sir James Clark Ross, R.N., F.R.S., during the years 1839 to 1843. Vol. 2 [part]. London: E. W. Janson, pp. 25-51, pl. 7-10. [1874]

Byars, L. F. 1949. The Mexican leaf-cutting ant in the United States. J. Econ. Entomol. 42:545. [1949.06]

Byars, L. F. 1951. A new fungus-growing ant from Arizona. Proc. Entomol. Soc. Wash. 53:109-111. [1951.04.12]

Bytinski-Salz, H. 1953. The zoogeography of the ants in the Near East. Istanbul Univ. Fen Fak. Mecm. Seri B 18:67-74. [1953.01]

Bytinski-Salz, H. 1965. Vein variations in *Plagiolepis* (Hym. Form.) and their possible phylogenetic significance. Pp. 71-73 in: Freeman, P. (ed.) XIIth International Congress of Entomology. London, 8-16 July, 1964. Proceedings. London: XIIth International Congress of Entomology, 842 pp. [1965]

Caesar, C. J. 1913. Die Stirnaugen der Ameisen. Zool. Jahrb. Abt. Anat. Ontog. Tiere 35:161-242. [1913.01.13]

Caetano, F. H. 1980. Dados sobre a genitália interna de içás de *Atta capiguara* (Gonçalves, 1944) (Hymenoptera - Formicidae). Pap. Avulsos Zool. (São Paulo) 33:247-252. [1980.10.15]

Caetano, F. H. 1984. Morfologia comparado do trato digestivo de formigas da subfamília Myrmicinae (Hymenoptera: Formicidae). Pap. Avulsos Zool. (São Paulo) 35:257-305. [1984.12.28]

Caetano, F. H. 1988. Anatomia, histologia e histoquímica do sistema digestivo e excretor de operárias de formigas (Hymenoptera, Formicidae). Naturalia (São Paulo) 13:129-174. [1988]

Caetano, F. H. 1990a. Can we use the digestive tract for phylogenetic studies in ants? Pp. 321-322 in: Veeresh, G. K., Mallik, B., Viraktamath, C. A. (eds.) Social insects and the environment. Proceedings of the 11th International Congress of IUSSI, 1990. New Delhi: Oxford & IBH Publishing Co., xxxi + 765 pp. [1990.08]

Caetano, F. H. 1990b. Morphology of the digestive tract and associated excretory organs of ants. Pp. 119-129 in: Vander Meer, R. K., Jaffe, K., Cedeno, A. (eds.) Applied myrmecology: a world perspective. Boulder: Westview Press, xv + 741 pp. [1990]

Caetano, F. H., Cruz Landim, C. da. 1984 ("1983"). Ultra-estrutura das células colunares do ventrículo de *Camponotus arboreus* (Hymenoptera, Formicidae) e suas implicações funcionais. Naturalia (São Paulo) 8:91-100. [1984]

Caetano, F. H., Cruz Landim, C. da. 1987. Microorganisms in the gut of ants of the tribe Cephalotini: location and relationship with intestinal structures. [Abstract.] Pp. 629-630 in: Eder, J., Rembold, H. (eds.) Chemistry and biology of social insects. München: Verlag J. Peperny, xxxv + 757 pp. [1987]

Caetano, F. H., Jaffe, K., Crewe, R. W. 1994. The digestive tract of the *Cataulacus* ants: presence of the microorganisms in the ileum. P. 391 in: Lenoir, A., Arnold, G., Lepage, M. (eds.) Les insectes sociaux. 12ème congrès de l'Union Internationale pour l'Étude des Insectes Sociaux. Paris, Sorbonne, 21-27 août 1994. Paris: Université Paris Nord, xxiv + 583 pp. [1994.09]

Caetano, F. H., Lage Filho, A. L. 1983 ("1982"). Anatomia e histologia do trato digestivo de formigas do gênero *Odontomachus* (Hymenoptera, Ponerinae). Naturalia (São Paulo) 7:125-134. [1983]

Caetano, F. H., Pimentel, M. A. L., Mathias, M. I. C., Tomotake, M. E. M. 1989. Morfologia interna de operárias de *Dolichoderus decollatus* (Fr. Smith, 1858) (Formicidae: Dolichoderinae). III. Genitália interna. An. Soc. Entomol. Bras. 18(suppl.):101-108. [1989.12.20]

Caetano, F. H., Tomotake, M. E. M., Mathias, M. I. C., Pimentel, M. A. L. 1990. Morfologia interna de operárias de *Dolichoderus attelaboides* (Fabricius, 1775) (Hymenoptera, Formicidae, Dolichoderinae). II. Glândula mandibular e mandíbulas. Rev. Bras. Entomol. 34:615-618. [1990.11.15]

Caetano, F. H., Tomotake, M. E. M., Pimentel, M. A. L., Mathias, M. I. C. 1990. Morfologia interna de operárias de *Dolichoderus attelaboides* (Fabricius, 1775) (Formicidae: Dolichoderinae). I. Trato digestivo e sistema excretor anexo. Naturalia (São Paulo) 15:57-65. [1990]

Caetano, F. H., Torres, Jr., A. H., Mathias, M. I. C., Tomotake, M. E. M. 1995 ("1994"). Apocrine secretion in the ant, *Pachycondyla striata*, ventriculus (Formicidae: Ponerinae). Cytobios 80:235-242. [1995]

Cagniant, H. 1961. Étude de fourmis récoltées par le Professeur H. Janetschek dans la Sierra Nevada. Bull. Soc. Hist. Nat. Afr. Nord 52:104-117. [1961]

Cagniant, H. 1964 ("1962"). Étude de quelques fourmis marocaines. Statistique provisoire des Formicidae du Maroc. Bull. Soc. Hist. Nat. Afr. Nord 53:83-118. [1964]

Cagniant, H. 1966a. Description des trois castes de *Leptothorax tyndalei* (Forel) (Hym. Formicidae). Bull. Soc. Entomol. Fr. 71:17-21. [1966.06.18]

Cagniant, H. 1966b. Note sur le peuplement en fourmis d'une montagne de la région d'Alger, l'Atlas de Blida. Bull. Soc. Hist. Nat. Toulouse 102:278-284. [1966.06.30]

Cagniant, H. 1966c. Nouvelle description d'*Aphaenogaster* (*Attomyrma*) *crocea* (André) Hyménoptère Formicidae. Représentation des trois castes. Notes biologiques. Bull. Soc. Zool. Fr. 91:61-69. [1966.10.27]

Cagniant, H. 1966d. Description des males de *Leptothorax annibalis* et *Camponotus atlantis* (Hym. Formicidae). Représentation des trois castes chez ces deux espèces. Ann. Soc. Entomol. Fr. (n.s.)2:967-974. [1966.12]

Cagniant, H. 1966e ("1965"). Clé dichotomique des fourmis de l'Atlas blidéen. Bull. Soc. Hist. Nat. Afr. Nord 56:26-40. [1966]

Cagniant, H. 1967a. Description de *Messor bernardi* n. sp. (Hym. Formicidae). Représentation des trois castes. Bull. Soc. Entomol. Fr. 72:131-139. [1967.06.27]

Cagniant, H. 1967b. *Leptothorax barryi* n. sp. Hyménoptère Formicidae Myrmicinae d'Algérie. Bull. Soc. Entomol. Fr. 72:272-275. [1967.12.29]

Cagniant, H. 1968a. Liste préliminaire de fourmis forestières d'Algérie. Résultats obtenus de 1963 à 1966. Bull. Soc. Hist. Nat. Toulouse 104:138-147. [1968.05.30]

Cagniant, H. 1968b. Description de *Leptothorax monjauzei* n. sp. d'Algérie (Hym. Formicidae Myrmicinae). Représentation de trois castes et notes biologiques. Bull. Soc. Entomol. Fr. 73:83-90. [1968.07.11]

Cagniant, H. 1968c. Description d'*Epimyrma algeriana* (nov. sp.) (Hyménoptères Formicidae, Myrmicinae), fourmi parasite. Représentation des trois castes. Quelques observations biologiques, écologiques et éthologiques. Insectes Soc. 15:157-170. [1968.12]

Cagniant, H. 1968d ("1966"). Description des génitalia de mâles de fourmis. Bull. Soc. Hist. Nat. Afr. Nord 57:77-85. [1968]

Cagniant, H. 1969a ("1968"). Description d'*Aphaenogaster crocea faureli* n. subsp. d'Algérie (Hym. Formicidae, Myrmicinae). Bull. Soc. Entomol. Fr. 73:232-235. [1969.02.26]

Cagniant, H. 1969b ("1968"). Du nouveau sur la répartition des *Epimyrma* d'Algérie (Hyménoptères - Formicidae - Myrmicinae). Bull. Soc. Hist. Nat. Toulouse 104:427-429. [1969.02.28]

Cagniant, H. 1969c. Sur deux *Aphaenogaster* rares d'Algérie (Hyménoptères Formicidae, Myrmicinae). Insectes Soc. 16:103-114. [1969.12]

Cagniant, H. 1970a ("1969"). Deuxième liste de fourmis d'Algérie récoltées principalement en forêt. (1re partie). Bull. Soc. Hist. Nat. Toulouse 105:405-430. [1970.02.05]

Cagniant, H. 1970b ("1969"). Nouvelle description de *Leptothorax spinosus* (Forel) d'Algérie. Représentation des trois castes et notes biologiques (Hym. Formicidae Myrmicinae). Bull. Soc. Entomol. Fr. 74:201-208. [1970.02.26]

Cagniant, H. 1970c. Deuxième liste de fourmis d'Algérie, récoltées principalement en forêt (Deuxième partie). Bull. Soc. Hist. Nat. Toulouse 106:28-40. [1970.06.30]

Cagniant, H. 1970d. Une nouvelle fourmi parasite d'Algérie: *Sifolinia kabylica* (nov. sp.), Hyménoptères. Formicidae. Myrmicinae. Insectes Soc. 17:39-47. [1970.06]

Cagniant, H. 1971. Description du mâle de *Stenamma africanum* Santschi (Hym. Formicidae Myrmicinae). Bull. Soc. Entomol. Fr. 76:98-101. [1971.07.28]

Cagniant, H. 1973a ("1972"). Note sur les peuplements de fourmis en forêt d'Algérie. Bull. Soc. Hist. Nat. Toulouse 108:386-390. [1973.02.28]

Cagniant, H. 1973b. Description et représentation des trois castes de *Chalepoxenus kutteri* (nov. sp.) (Hyménoptères Formicidae - Myrmicidae). Insectes Soc. 20:145-156. [1973.12]

Cagniant, H. 1977 ("1976"). Distribution, écologie et nid de la fourmi *Cataglyphis cursor* Fonscolombe. Hyménoptères Formicidae. Vie Milieu Sér. C Biol. Terr. 26:265-276. [1977.11]

Cagniant, H. 1982. Contribution à la connaissance des fourmis marocaines. *Aphaenogaster dejeani*, n. sp. (Hyménoptères, Formicoïdea, Myrmicidae). Nouv. Rev. Entomol. 12:281-286. [1982.10.30]

Cagniant, H. 1983. Contribution à la connaissance des fourmis marocaines. *Chalepoxenus tramieri*, nov. sp. Nouv. Rev. Entomol. 13:319-322. [1983.10.30]
Cagniant, H. 1984. Contribution à la connaissance des fourmis marocaines. *Aphaenogaster espadaleri* n. sp. (Hymenoptera, Myrmicidae). Nouv. Rev. Entomol. (n.s.)1:387-395. [1984.12.30]
Cagniant, H. 1985a. Contribution à la connaissance des fourmis marocaines: *Chalepoxenus brunneus* n. sp. (Hymenoptera, Myrmicidae). Nouv. Rev. Entomol. (n.s.)2:141-146. [1985.06.25]
Cagniant, H. 1985b. Contribution à la connaissance des fourmis marocaines. Étude des *Camponotus* du groupe *cruentatus* au Maroc (Hyménoptères - Formicidae). Bull. Soc. Hist. Nat. Toulouse 121:77-84. [1985.06]
Cagniant, H. 1986. Contribution à la connaissance des fourmis marocaines: description des sexués et compléments à la définition de l'espèce *Aphaenogaster theryi* Santschi 1923 (Hyménoptères, Formicoidea, Myrmicidae). Bull. Soc. Hist. Nat. Toulouse 122:139-144. [1986.10]
Cagniant, H. 1987a ("1986"). *Aphaenogaster nadigi* Santschi, bona species (n. status). EOS. Rev. Esp. Entomol. 62:31-43. [1987.02.27]
Cagniant, H. 1987b ("1986"). Contribution à la connaissance des fourmis marocaines. *Camponotus vagus ifranensis* n. ssp. (Hym. Formicidae). Bull. Soc. Entomol. Fr. 91:117-123. [1987.03.30]
Cagniant, H. 1987c ("1986"). Contribution à la connaissance des fourmis marocaines. *Leptothorax personatus* n. sp. (Hym. Formicoidea). Bull. Soc. Entomol. Fr. 91:243-250. [1987.06.22]
Cagniant, H. 1987d. Contribution à la connaissance des fourmis marocaines. Nouvelle description et compléments à la définition de l'espèce *Aphaenogaster praedo* Emery (Hyménoptères, Formicoïdea, Myrmicidae). Problèmes posés par le groupe "*praedo*". Bull. Soc. Hist. Nat. Toulouse 123:159-165. [1987.10]
Cagniant, H. 1988a. Contribution à la connaissance des fourmis marocaines. Description des trois castes d'*Aphaenogaster torossiani* n. sp. et notes biologiques (Hym. Formicoidea Myrmicidae). Bull. Soc. Entomol. Fr. 92:241-250. [1988.03.07]
Cagniant, H. 1988b. Contribution à la connaissance des fourmis marocaines. Description des trois castes d'*Aphaenogaster wilsoni* n. sp. (Hymenoptera, Myrmicidae). Nouv. Rev. Entomol. (n.s.)5:49-55. [1988.04.20]
Cagniant, H. 1988c. Contribution à la connaissance des fourmis marocaines: *Aphaenogaster baronii* n. sp. (Hyménoptères, Formicoidea, Myrmicidae). Bull. Soc. Hist. Nat. Toulouse 124:43-50. [1988.07]
Cagniant, H. 1989a. Contribution à la connaissance des fourmis marocaines. Description des trois castes d'*Aphaenogaster weulersseae* n. sp.; notes biologiques et écologiques; étude comparée de trois populations (Hym. Formicoidea Myrmicidae). Bull. Soc. Entomol. Fr. 94:113-125. [1989.08.30]
Cagniant, H. 1989b. Essai d'application de quelques indices et modèles de distributions d'abondances à trois peuplements de fourmis terricoles. Orsis (Org. Sist.) 4:113-124. [1989]
Cagniant, H. 1990a ("1989"). Contribution à la connaissance des fourmis marocaines. *Aphaenogaster gemella* au Maroc: nouvelle description d'*Aphaenogaster gemella* ssp. *marocana* Forel (n. status) de la région de Tanger. Problèmes biogéographiques soulevés par l'espèce *Aphaenogaster gemella* (Roger) (Hyménoptères, Formicoïdea, Myrmicidae). Bull. Soc. Hist. Nat. Toulouse 125:47-54. [1990.01]
Cagniant, H. 1990b. Contribution à la connaissance des fourmis marocaines: *Aphaenogaster atlantis* Santschi, 1929 bona species (n. status) (Hymenoptera, Formicidae, Myrmicinae). Bull. Mus. Natl. Hist. Nat. Sect. A Zool. Biol. Écol. Anim. (4)12:123-133. [1990.08.02]
Cagniant, H. 1990c. Contribution à la connaissance des fourmis marocaines. Révision de l'espèce *Aphaenogaster curiosa* Santschi (n. status) (Hymenoptera, Formicidae). J. Afr. Zool. 104:457-475. [1990.10.30]

Cagniant, H. 1990d. Contribution à la connaissance des fourmis marocaines. *Aphaenogaster sicardi* n. sp. (Hymenoptera, Formicidae, Myrmicinae). Bull. Mus. Natl. Hist. Nat. Sect. A Zool. Biol. Écol. Anim. (4)12:443-453. [1990.11.24]

Cagniant, H. 1990e. Contribution à la connaissance des fourmis marocaines: *Aphaenogaster miniata* n. sp. Description des trois castes, populations et justification de son statut de bonne espèce (Hymenoptera, Formicidae). Nouv. Rev. Entomol. (n.s.)7:143-154. [1990.11.30]

Cagniant, H. 1991a. Contribution à la connaissance des fourmis marocaines. - Répartition et variations géographiques d'*Aphaenogaster wilsoni* Cagniant, 1988 (Hymenoptera, Formicidae, Myrmicinae). J. Afr. Zool. 105:49-61. [1991.02.28]

Cagniant, H. 1991b. Étude des populations marocaines de la superespèce *Camponotus cruentatus* Latreille. Mise en evidence de *Camponotus obscuriventris* sp. n. (Hymenoptera, Formicidae). EOS. Rev. Esp. Entomol. 67:35-54. [1991.11.15]

Cagniant, H. 1991c. Contribution à la connaissance des fourmis marocaines. *Camponotus hölldobleri* n. sp. (Hymenoptera, Formicidae). Bull. Soc. Zool. Fr. Evol. Zool. 116:37-48. [1991]

Cagniant, H. 1992a. Contribution à la connaissance des fourmis marocaines. *Aphaenogaster fallax* sp. n. (Hymenoptera, Formicidae, Myrmicinae). J. Afr. Zool. 106:197-210. [1992.07.10]

Cagniant, H. 1992b. Étude des populations marocaines d'*Aphaenogaster* (suprasp.) *senilis* (Hymenoptera, Formicidae, Myrmicinae). Bull. Mus. Natl. Hist. Nat. Sect. A Zool. Biol. Écol. Anim. (4)14:179-199. [1992.07.24]

Cagniant, H. 1992c. Contribution à la connaissance des fourmis marocaines. Description des trois castes d'*Aphaenogaster dejeani* Cagniant, 1982 (Hymenoptera, Formicidae). Bull. Soc. Zool. Fr. Evol. Zool. 117:65-73. [1992]

Cagniant, H. 1992d. Contribution à la connaissance des fourmis marocaines. *Aphaenogaster tinauti* n. sp. (Hymenoptera, Formicidae). Bull. Soc. Zool. Fr. Evol. Zool. 117:365-374. [1992]

Cagniant, H. 1994. Contribution à la connaissance des fourmis marocaines. Description d'*Aphaenogaster rifensis* sp. n. Révision de la superespèce *Aphaenogaster* (supersp.) *gemella* Rog. (n. taxon) (Hymenoptera, Formicidae). Bull. Soc. Zool. Fr. Evol. Zool. 119:15-29. [1994]

Cagniant, H., Espadaler, X. 1991. État actuel des connaissances sur les fourmis du Maroc: projet d'une faune. Actes Colloq. Insectes Soc. 7:133-136. [1991.06]

Cagniant, H., Espadaler, X. 1993a. *Camponotus guanchus* Santschi, 1908, stat. nov. et études des populations de *Camponotus sicheli* Mayr, 1866 (Hymenoptera, Formicidae). J. Afr. Zool. 107:419-438. [1993.10.29]

Cagniant, H., Espadaler, X. 1993b. Liste des espèces de fourmis du Maroc. Actes Colloq. Insectes Soc. 8:89-93.

Cagniant, H., Espadaler, X., Colombel, P. 1991. Biométrie et répartition de quelques populations d'*Aphaenogaster* (suprasp.) *senilis* (Hymenopteres Formicidae) du Bassin Méditerranéen Occidental et du Maroc. Vie Milieu 41:61-71. [1991.03]

Cagniant, H., Ledoux, A. 1974. Nouvelle description d'*Aphaenogaster senilis* sur des exemplaires de la région de Banyuls-sur-Mer (P.-O.), France. Vie Milieu Sér. C Biol. Terr. 24:97-110. [1974.12]

Caldas, A., Moutinho, P. R. S. 1993. Composição e diversidade da fauna de formigas (Hymenoptera, Formicidae) em áreas sob remoção experimental de árvores na reserva florestal de Linhares, ES, Brasil. Rev. Bras. Entomol. 37:299-304. [1993.06.30]

Calvert, P. P. 1919. Gundlach's work on the Odonata of Cuba: a critical study. Trans. Am. Entomol. Soc. 45:335-396. [1919.12.23] [See Gundlach (1888).]

Cameron, H. D. 1988. Geoffroy in Fourcroy, 1785. Sphecos 17:6-7. [1988.10]

Cameron, P. 1886. On a new species of *Strumigenys* (*S. lewisi*) from Japan. Proc. Manch. Lit. Philos. Soc. 25:229-232. [1886]

Cameron, P. 1891. Appendix. Hym. Formicidae. Pp. 89-95 in: Whymper, E. Travels amongst the Great Andes. London: J. Murray, xxii + 147 pp. [1891]

Cameron, P. 1892. Synonymical notes on Cynipidae and Formicidae. Entomol. Mon. Mag. 28:67. [1892.03]
Cameron, P. 1898. Notes on a collection of Hymenoptera from Greymouth, New Zealand, with descriptions of new species. Mem. Proc. Manch. Lit. Philos. Soc. 42(1):1-53. [1898.02.04]
Camilo, G. R., Phillips, S. A., Jr. 1990. Evolution of ant communities in response to invasion by the fire ant *Solenopsis invicta*. Pp. 190-198 in: Vander Meer, R. K., Jaffe, K., Cedeno, A. (eds.) Applied myrmecology: a world perspective. Boulder: Westview Press, xv + 741 pp. [1990]
Cammaerts, M.-C., Cammaerts, R., Bruge, H. 1986. Contribution à l'étude de la forme microgyne de *Myrmica rubra* L. (Hymenoptera, Formicidae). Actes Colloq. Insectes Soc. 3:211-217. [1986]
Cammaerts, M.-C., Evershed, R. P., Morgan, E. D. 1981a. Comparative study of the Dufour gland secretions of workers of four species of *Myrmica* ants. J. Insect Physiol. 27:59-65. [1981]
Cammaerts, M.-C., Evershed, R. P., Morgan, E. D. 1981b. Comparative study of the mandibular gland secretion of four species of *Myrmica* ants. J. Insect Physiol. 27:225-231. [1981]
Cammaerts, M.-C., Evershed, R. P., Morgan, E. D. 1981c. Comparative study of pheromones emitted by different species of *Myrmica*. Pp. 185-192 in: Howse, P. E., Clement, J.-L. (eds.) Biosystematics of social insects. Systematics Association Special Volume No. 19. London: Academic Press, 346 pp. [1981]
Cammaerts, M.-C., Evershed, R. P., Morgan, E. D. 1982. Mandibular gland secretions of workers of *Myrmica rugulosa* and *M. schencki*: comparison with four other *Myrmica* species. Physiol. Entomol. 7:119-125. [1982.06]
Cammaerts, M.-C., Evershed, R. P., Morgan, E. D. 1983. The volatile components of the mandibular gland secretion of workers of the ants *Myrmica lobicornis* and *Myrmica sulcinodis*. J. Insect Physiol. 29:659-664. [1983]
Cammaerts, R., Cammaerts, M.-C. 1985. Un système primitif d'approvisionnement chez *Manica rubida* (Hymenoptera, Formicidae). Actes Colloq. Insectes Soc. 2:151-157. [1985.06]
Cammaerts, R., Cammaerts, M.-C. 1988. Four ants (Hym.: Formicidae) new to the Belgian fauna. Entomol. Rec. J. Var. 100:37-38. [1988.01.25]
Cammaerts, R., Pasteels, J. M., Roisin, Y. 1985. Identification et distribution de *Tetramorium caespitum* (L.) et *T. impurum* (Foerster) en Belgique (Hymenoptera Formicidae). Actes Colloq. Insectes Soc. 2:109-118. [1985.06]
Campione, B. M., Novak, J. A., Gotwald, W. H., Jr. 1983. Taxonomy and morphology of the West African army ant, *Aenictus asantei* n. sp. (Hymenoptera: Formicidae). Ann. Entomol. Soc. Am. 76:873-883. [1983.09]
Campos, F. 1960. Contribución preliminar al estudio de las hormigas (Brachycera) [sic] del Ecuador. Rev. Ecuat. Hig. Med. Trop. 17:67-69. [1960.06]
Carbonell Mas, C. S. 1943. Las hormigas cortadoras del Uruguay. Rev. Asoc. Ing. Agrón. Montev. 15(3):30-39. [1943.09]
Carlin, N. F. 1988. Species, kin, and other forms of recognition in the brood discrimination of ants. Pp. 267-295 in: Trager, J. C. (ed.) Advances in myrmecology. Leiden: E. J. Brill, xxvii + 551 pp. [1988]
Carlin, N. F. 1989. Discrimination within and between colonies of social insects: two null hypotheses. Neth. J. Zool. 39:86-100. [1989.11]
Carlin, N. F., Halpern, R., Hölldobler, B., Schwartz, P. 1987. Early learning and the recognition of conspecific cocoons by carpenter ants (*Camponotus* spp.). Ethology 75:306-316. [1987.08]
Carlin, N. F., Hölldobler, B. 1983. Nestmate and kin recognition in interspecific mixed colonies of ants. Science (Wash. D. C.) 222:1027-1029. [1983.12.02]
Carlin, N. F., Hölldobler, B. 1986. The kin recognition system of carpenter ants (*Camponotus* spp.). I. Hierarchical cues in small colonies. Behav. Ecol. Sociobiol. 19:123-134. [1986.07]
Carlton, C. E. 1987. Identification of Arkansas fire ant (Hymenoptera: Formicidae: *Solenopsis* spp.) workers. Rep. Ser. Univ. Arkansas Agric. Exp. Stn. 301:1-8. [1987.11]

Carney, W. P. 1969. Behavioral and morphological changes in carpenter ants harboring dicrocoeliid metacercariae. Am. Midl. Nat. 82:605-611. [1969.11.10]

Carniel, A. 1988 ("1987"). Contributo alla conoscenza della mirmecofauna del Cansiglio (Prealpi Carniche) (Hymenoptera). Boll. Soc. Entomol. Ital. 119:179-190. [1988.02.29]

Carniel, A. 1991. Formicidi raccolti nella foresta di Tarvisio (Alpi Carniche). Boll. Soc. Entomol. Ital. 123:141-148. [1991.09.20]

Carniel, A. 1992. Segnalazioni faunistiche italiane. 216. *Leptothorax carinthiacus* Bernard, 1957 (Hymenoptera Formicidae). Boll. Soc. Entomol. Ital. 124:80. [1992.05.31]

Carniel, A., Battisti, A. 1994. Formicidi della chioma de *Pinus nigra* in popolamenti appenninici (Hymenoptera). Boll. Soc. Entomol. Ital. 126:31-39. [1994.07.31]

Carpenter, F. M. 1927. Notes on a collection of amber ants. Psyche (Camb.) 34:30-32. [1927.02] [Stamp date in MCZ library: 1927.04.04.]

Carpenter, F. M. 1929. A fossil ant from the Lower Eocene (Wilcox) of Tennessee. J. Wash. Acad. Sci. 19:300-301. [1929.08.19]

Carpenter, F. M. 1930. The fossil ants of North America. Bull. Mus. Comp. Zool. 70:1-66. [1930.01]

Carpenter, F. M. 1931. Insects from the Miocene (Latah) of Washington. Introduction. Ann. Entomol. Soc. Am. 24:307-309. [1931.06.30]

Carpenter, F. M. 1935. A new name for *Lithomyrmex* Carp. (Hymenoptera). Psyche (Camb.) 42:91. [1935.06] [Stamp date in MCZ library: 1935.08.13.]

Carpenter, F. M. 1992. Treatise on invertebrate paleontology. Part R. Arthropoda 4. Volume 4. Superclass Hexapoda. Boulder: Geological Society of America, pp. i-ii, 279-655. [1992.12.09] [Series editor is R. L. Kaesler. Pagination continuous with that of volume 3 which otherwise has the same title, year and author. Formicidae treated on pages 490-495. Date of publication from Evenhuis (1994b:485).]

Carroll, C. R. 1979. A comparative study of two ant faunas: the stem-nesting ant communities of Liberia, West Africa and Costa Rica, Central America. Am. Nat. 113:551-562. [1979.04]

Carroll, C. R., Janzen, D. H. 1973. Ecology of foraging by ants. Annu. Rev. Ecol. Syst. 4:231-257. [1973]

Carroll, J. F. 1975. Biology and ecology of ants of the genus *Aphaenogaster* in Florida. Diss. Abstr. Int. B. Sci. Eng. 36:2051-2052. [1975.11]

Carroll, J. F., Kimbrough, J. W., Whitcomb, W. H. 1981. Mycophagy by *Aphaenogaster* spp. (Hymenoptera: Formicidae). Proc. Entomol. Soc. Wash. 83:326-331. [1981.04.30]

Carter, W. G. 1962a. Ants of the North Carolina Piedmont. J. Elisha Mitchell Sci. Soc. 78:1-18. [1962.05]

Carter, W. G. 1962b. Ant distribution in North Carolina. J. Elisha Mitchell Sci. Soc. 78:150-204. [1962.11]

Carvalho, M. B. de, Carvalho, R. F. de. 1939. Primeira contribução para um catálogo de insetos de Pernambuco. Arq. Inst. Pesqui. Agron. (Recife) 2:27-60. [1939] [Ants pp. 58-59.]

Carvalho, M. B. de, Oliveira Freitas, A. de. 1960. Terceira contribução para um catálogo de insetos de Pernambuco. Arq. Inst. Pesqui. Agron. (Recife) 5:95-114. [1960] [Ants pp. 107-108.]

Casevitz-Weulersse, J. 1974. Fourmis récoltées en Corse et en Sardaigne. Ann. Soc. Entomol. Fr. (n.s.)10:611-621. [1974.09]

Casevitz-Weulersse, J. 1983. Les larves de *Crematogaster* (*Acrocoelia*) *scutellaris* (Olivier) (Hym. Formicidae). Bull. Soc. Entomol. Fr. 88:258-267. [1982.12.23]

Casevitz-Weulersse, J. 1984. Les larves à expansions latérales de *Crematogaster* (*Acrocoelia*) *scutellaris* (Olivier) (Hym. Formicidae). Actes Colloq. Insectes Soc. 1:131-138. [1984.09]

Casevitz-Weulersse, J. 1986. A propos de fourmis de la Corse. Actes Colloq. Insectes Soc. 3:261-272. [1986]

Casevitz-Weulersse, J. 1990a. Étude des peuplements de fourmis de la Corse (Hymenoptera, Formicidae). Rev. Écol. Biol. Sol 27:29-59. [1990.01]

Casevitz-Weulersse, J. 1990b. Étude systématique de la myrmécofaune corse (Hymenoptera, Formicidae) (Première partie). Bull. Mus. Natl. Hist. Nat. Sect. A Zool. Biol. Écol. Anim. (4)12:135-163. [1990.08.02]
Casevitz-Weulersse, J. 1990c. Étude systématique de la myrmécofaune corse (Hymenoptera, Formicidae) (Deuxième partie). Bull. Mus. Natl. Hist. Nat. Sect. A Zool. Biol. Écol. Anim. (4)12:415-442. [1990.11.24]
Casevitz-Weulersse, J. 1990d. Données écologiques sur la myrmécofaune corse. Actes Colloq. Insectes Soc. 6:35-42. [1990.06]
Casevitz-Weulersse, J. 1991. Les fourmis de Corse. Insectes 81:2-4. [1991.06]
Casevitz-Weulersse, J. 1992a. Analyse biogéographique de la myrmécofaune corse et comparaison avec celle des régions voisines. C. R. Séances Soc. Biogéogr. 68:105-129. [1992.09]
Casevitz-Weulersse, J. 1992b. La myrmécofaune de la Réserve Naturelle de Scandola. Inventaire spécifique (1984/1986-1991). Trav. Sci. Parc Nat. Rég. Réserves Nat. Corse 36:85-95. [1992.12]
Casevitz-Weulersse, J., Cherix, D. 1991. Francis Bernard: 1908-1990. Actes Colloq. Insectes Soc. 7:1-11. [1991.06]
Casevitz-Weulersse, J., Prost, M. 1991. Fourmis de la Côte-d'Or présentes dans les collections du Muséum d'Histoire Naturelle de Dijon. Bull. Sci. Bourgogne 44:53-72. [1991.09.09]
Casolari, C., Casolari Moreno, R. 1980. Cataloghi I- Collezione imenotterologica di Massimiliano Spinola. Torino: Museo Regionale di Scienze Naturali, 166 pp. [1980.03.31]
Cavill, G. W. K., Ford, D. L., Locksley, H. D. 1956. The chemistry of ants. I. Terpenoid constituents of some Australian *Iridomyrmex* species. Aust. J. Chem. 9:288-293. [1956.05]
Cavill, G. W. K., Hinterberger, H. 1960a. The chemistry of ants. IV. Terpenoid constituents of some *Dolichoderus* and *Iridomyrmex* species. Aust. J. Chem. 13:514-519. [1960.11]
Cavill, G. W. K., Hinterberger, H. 1960b. Dolichoderine ant extractives. Pp. 53-59 in: Pavan, M., Eisner, T. (eds.) XI. Internationaler Kongress für Entomologie. Wien 1960. Verhandlungen. Band III. Symposium 3, Symposium 4. Pavia: Istituto di Entomologia Agraria dell'Università di Pavia, 293 pp. [1960]
Cavill, G. W. K., Robertson, P. L. 1965. Ant venoms, attractants, and repellents. Science (Wash. D. C.) 149:1337-1345. [1965.09.17]
Cavill, G. W. K., Robertson, P. L., Brophy, J. J., Clark, D. V., Duke, R., Orton, C.J., Plant, W. D. 1982. Defensive and other secretions of the Australian cocktail ant, *Iridomyrmex nitidiceps*. Tetrahedron 38:1931-1938. [1982]
Cavill, G. W. K., Robertson, P. L., Brophy, J. J., Duke, R. K., McDonald, J., Plant, W. D. 1984. Chemical ecology of the meat ant, *Iridomyrmex purpureus* sens. strict. Insect Biochem. 14:505-513. [1984]
Cavill, G. W. K., Robertson, P. L., Whitfield, F. B. 1964. Venom and venom apparatus of the bull ant, *Myrmecia gulosa* (Fabr.). Science (Wash. D. C.) 146:79-80. [1964.10.09]
Ceballos, G. 1943. Las tribus de los himenópteros de España. [part] Madrid: Instituto Español de Entomología, pp. 289-420. [1943.12.30] [The entire work was published in seven parts between 1941 and 1943. This, the last, part was published concurrently with volume 19, number 4 of EOS (Revista Española de Entomología). Ants pp. 311-338.]
Ceballos, G. 1956. Catálogo de los Himenópteros de España. Madrid: Instituto Español de Entomologia, 554 pp. [1956] [Ants pp. 295-321.]
Ceballos, G. 1959. Primer suplemento al Catálogo de los Himenópteros de España. EOS. Rev. Esp. Entomol. 35:215-242. [1959.09.30] [Ants pp. 224-225.]
Ceballos, G. 1964. Segundo suplemento al Catálogo de los Himenópteros de España. EOS. Rev. Esp. Entomol. 40:43-97. [1964.08.31] [Ants p. 81.]
Ceballos, P., Ronchetti, G. 1966. Le formiche del gruppo *Formica rufa* sui Pirenei orientali spagnoli, nelle province di Lerida e Gerona. Mem. Soc. Entomol. Ital. 45:153-168. [1966.12.20]

Cecconi, G. 1908. Contributo alla fauna delle Isole Tremiti. Boll. Mus. Zool. Anat. Comp. R. Univ. Torino 23(583):1-53. [1908.05.20] [Ants pp. 23-24, including (p. 24) descriptions of new taxa by Emery.]

Cerdá, X., Retana, J. 1988. Descripción de la comunidad de hormigas de un prado sabanoide en Canet de Mar (Barcelona). Ecología (Madr.) 2:333-341. [1988]

Cerdá, X., Retana, J., Bosch, J. 1991. Hormigas de Port-Bou (Gerona): una aproximación a su estudio ecológico. Ecología (Madr.) 5:413-425. [1991]

Cerdá, X., Retana, J., Carpintero, S., Cros, S. 1992. Petals as the main resource collected by the ant *Cataglyphis floricola*. Sociobiology 20:315-319. [1992] [*Cataglyphis floricola*: nomen nudum.]

Chahine-Hanna, N. H. 1981. Nouvelle description de *Tapinoma simrothi* var. *phoenicium* (Emery) (Hym - Formicoïdae - Dolichoderidae) sur les exemplaires de la région côtière de Hadath-Liban. Ecol. Mediterr. 7(1):155-162. [1981]

Chapela, I. H., Rehner, S. A., Schultz, T. R., Mueller, U. G. 1994. Evolutionary history of the symbiosis between fungus-growing ants and their fungi. Science (Wash. D. C.) 266:1691-1694. [1994.12.09]

Chapman, J. A. 1954. Swarming of ants on western United States mountain summits. Pan-Pac. Entomol. 30:93-102. [1954.06.24]

Chapman, J. A. 1957. A further consideration of summit ant swarms. Can. Entomol. 89:389-395. [1957.10.25]

Chapman, J. W. 1963. Some new and interesting Philippine ants (Hymenoptera: Formicidae). Philipp. J. Sci. 92:247-263. [1963.10.03]

Chapman, J. W. 1965 ("1964"). Studies on the ecology of the army ants of the Philippines genus *Aenictus* Schuckard (Hymenoptera: Formicidae). Philipp. J. Sci. 93:551-595. [1965.03.12]

Chapman, J. W., Capco, S. R. 1951. Check list of the ants (Hymenoptera: Formicidae) of Asia. Monogr. Inst. Sci. Technol. Manila 1:1-327. [1951]

Chappell, J. 1886. The insects of Llangollen and its vicinity. Young Nat. 7:57-61. [1886.04] [Nomen nudum (*Myrmica nodicornis*, p. 58).]

Chapuisat, M. 1994. Microsatellites for measuring relatedness in *Formica* ants. P. 45 in: Lenoir, A., Arnold, G., Lepage, M. (eds.) Les insectes sociaux. 12ème congrès de l'Union Internationale pour l'Étude des Insectes Sociaux. Paris, Sorbonne, 21-27 août 1994. Paris: Université Paris Nord, xxiv + 583 pp. [1994.09]

Charsley, R. S. 1877a. New British ant, *Ponera ochracea*? Entomol. Mon. Mag. 14:69-70. [1877.08]

Charsley, R. S. 1877b. A new species of ant found in Britain. Entomol. Mon. Mag. 14:162-163. [1877.12]

Cheesman, L. E., Crawley, W. C. 1928. A contribution towards the insect fauna of French Oceania. - Part III. Formicidae. Ann. Mag. Nat. Hist. (10)2:514-525. [1928.11]

Cheng, L. 1982. Notes on Formicidae - New to China. Entomotaxonomia 4:252. [1982.12]

Cheng, L., Ye, Q., Yang, Y. 1992. *Atopomyrmex srilankensis* - a new record from China. Entomotaxonomia 14:244. [1992.12]

Cherix, D. (ed.) 1980. Écologie des insectes sociaux. C. R. UIEIS Section française - Lausanne, 7-8 septembre 1979. Nyon: UIEIS Section Française, xv + 160 pp. [1980.03.05]

Cherix, D. 1983. Pseudogynes (= sécrétergates) et répartition des individus à l'intérieur d'une fourmilière de *Formica lugubris* Zett (Hymenoptera, Formicidae). Insectes Soc. 30:184-192. [1983.09]

Cherix, D., Burgat, M. 1979. A propos de la distribution verticale des fourmis du groupe *rufa* dans les parties centrale et occidentale du Jura suisse. Bull. SROP 1979(II-3):37-48. [1979]

Cherix, D., Burgat, M. 1981. La distribuzione altimetrica delle formiche del gruppo *Formica rufa* nella parte centrale e occidentale del Giura svizzero. Collana Verde 59:51-62. [1981]

Cherix, D., Higashi, S. 1979. Distribution verticale des fourmis dans le Jura vaudois et recensement préliminaire des bourdons (Hymenoptera, Formicidae et Apidae). Bull. Soc. Vaudoise Sci. Nat. 74:315-324. [1979.12.21]

Cherrett, J. M. 1989. Leaf-cutting ants. Pp. 473-488 in: Lieth, H., Werger, M. J. A. (eds.) Tropical rain forest ecosystems: biogeographical and ecological studies. Ecosystems of the world 14B. Amsterdam: Elsevier, xvii + 713 pp. [1989]

Cherrett, J. M., Cherrett, F. J. 1989. A bibliography of the leaf-cutting ants, *Atta* spp. and *Acromyrmex* spp., up to 1975. Overseas Dev. Nat. Resour. Inst. Bull. 14:1-58. [1989]

Chew, R. M. 1977. Some ecological characteristics of the ants of a desert-shrub community in southeastern Arizona. Am. Midl. Nat. 98:33-49. [1977.07.23]

Chew, R. M., De Vita, J. 1980. Foraging characteristics of a desert ant assemblage: functional morphology and species separation. J. Arid Environ. 3:75-83. [1980.03]

Chhotani, D., Maiti, P. 1970. Contribution to the knowledge of the Formicidae of the Andaman Islands. [Abstract.] Proc. Indian Sci. Congr. 57(3):438. [1970]

Chhotani, O. B., Ray, K. K. 1976 ("1975"). Fauna of Rajasthan, India, Hymenoptera. Rec. Zool. Surv. India 71:13-49. [1976] [Ants pp. 35-43.]

Chilson, L. M. 1953. Insect records from Johnston Island. Proc. Hawaii. Entomol. Soc. 15:81-84. [1953.03.27]

Chilson, L. M. 1959. New ant records for Palmyra Island. Proc. Hawaii. Entomol. Soc. 17:18. [1959.08.31]

Choe, J. C. 1988. Worker reproduction and social evolution in ants (Hymenoptera: Formicidae). Pp. 163-187 in: Trager, J. C. (ed.) Advances in myrmecology. Leiden: E. J. Brill, xxvii + 551 pp. [1988]

Choi, B.-M. 1985. Study on distribution of ants (Formicidae) from Korea (1). Ant fauna in Mt. Songni. [In Korean.] Chongju Sabom Taehak Nonmunjip (J. Chongju Natl. Teach. Coll.) 22:401-437. [1985.02]

Choi, B.-M. 1986. Studies on the distribution of ants (Formicidae) in Korea (3). [In Korean.] Chongju Sabom Taehak Nonmunjip (J. Chongju Natl. Teach. Coll.) 23:317-386. [1986.02] [Not seen. Cited in Choi & Kim (1987:369) and Terayama & Choi (1991:64).]

Choi, B.-M. 1987. Taxonomic study on ants (Formicidae) in Korea (1). On the genus *Monomorium*. [In Korean.] J. Inst. Sci. Educ. (Chongju Natl. Teach. Coll.) 11:17-30. [1987.02]

Choi, B.-M. 1988. Studies on the distribution of ants (Formicidae) in Korea (5). Ant fauna in Is. Kanghwado. [In Korean.] Chongju Sabom Taehak Nonmunjip (J. Chongju Natl. Teach. Coll.) 25:217-231. [1988.05]

Choi, B.-M. 1995. Taxonomic study on ants (tribe: Dacetini) in Korea. [In Korean.] Korean J. Entomol. 25:189-196. [1995.07.25]

Choi, B.-M., Bang, J.-R. 1992a. Studies on the distribution of ants (Formicidae) in Korea (9). Ant fauna in Mt. Togyusan. [In Korean.] Korean J. Appl. Entomol. 31:101-112. [1992.06]

Choi, B.-M., Bang, J.-R. 1992b. Studies on the distribution of ants (Formicidae) in Korea (10). Ant distribution in Gangweon Do. [In Korean.] J. Inst. Sci. Educ. (Chongju Natl. Teach. Coll.) 14:12-30. [1992.12]

Choi, B.-M., Bang, J.-R. 1992c. Studies on the distribution of ants (Formicidae) in Korea (11). Ant distribution in Gyeongsang Buk Do. [In Korean.] J. Inst. Sci. Educ. (Chongju Natl. Teach. Coll.) 14:31-49. [1992.12]

Choi, B.-M., Bang, J.-R. 1993. Studies on the distribution of ants (Formicidae) in Korea (12). The analysis of ant communities in 23 islands. [In Korean.] Chongju Sabom Taehakkyo Nonmunjip (J. Chongju Natl. Univ. Educ.) 30:317-330. [1993.03]

Choi, B.-M., Kim, C.-H. 1987. Studies on the distribution of ants (Formicidae) in Korea (4). Ant fauna in Is. Hongdo and Is. Taehukusando. Chongju Sabom Taehak Nonmunjip (J. Chongju Natl. Teach. Coll.) 24:357-370. [1987]

Choi, B.-M., Kim, C.-H., Bang, J.-R. 1993. Studies on the distribution of ants (Formicidae) in Korea (13). A checklist of ants from each province (Do), with taxonomic notes. [In Korean.] Chongju Sabom Taehakkyo Nonmunjip (J. Chongju Natl. Univ. Educ.) 30:331-380. [1993.03]
Choi, B.-M., Kondoh, M., Choi, M.-K. 1985. Study on the distribution of ants (Formicidae) from Korea (2). Myrmecofauna in Mt. Halla. [In Korean.] Chongju Sabom Taehak Nonmunjip (J. Chongju Natl. Teach. Coll.) 22:439-462. [1985]
Choi, B.-M., Lee, I.-H. 1995. Studies on the distribution of ants (Formicidae) in Korea (14). Ant fauna in Island Sohuksando. [In Korean.] Korean J. Appl. Entomol. 34:191-197. [1995.09.30]
Choi, B.-M., Ogata, K., Terayama, M. 1993. Comparative studies of ant faunas of Korea and Japan. I. Faunal comparison among islands of southern Korea and northern Kyushu, Japan. Bull. Biogeogr. Soc. Jpn. 48:37-49. [1993.07.31]
Choi, B.-M., Park, K.-S. 1991a. Studies on the distribution of ants (Formicidae) in Korea (6). The vegetation, the species composition and the colony density of ants in Mt. Namsam, Seoul. [In Korean.] Korean J. Appl. Entomol. 30:65-79. [1991.03.30]
Choi, B.-M., Park, K.-S. 1991b. Studies on the distribution of ants (Formicidae) in Korea (7). Ant fauna in Mt. Kyeryongsan. [In Korean.] Korean J. Appl. Entomol. 30:80-85. [1991.03.30]
Chopard, L. 1921. La fourmi d'Argentine *Iridomyrmex humilis* var. *arrogans* Santschi dans le midi de la France. Ann. Epiphyt. 7:237-265. [1921]
Chou, L.-Y., Terayama, M. 1991. Name lists of insects in Taiwan - Hymenoptera: Apocrita: Formicidae. [In Chinese.] Chin. J. Entomol. 11:75-84. [1991.03]
Christ, J. L. 1791. Naturgeschichte, Klassification und Nomenclatur der Insekten vom Bienen, Wespen und Ameisengeschlecht. Frankfurt: Hermann, 535 pp. [1791]
Chujo, M. 1939. Ants of Sozan in Taihoku-Prefecture, Formosa. Trans. Kansai Entomol. Soc. 8: 3-5. [1939.04.20]
Cîrdei, F., Bulimar, F. 1965. Noi contributii asupra studiului formicidelor (Fam. Formicidae) din Moldova. An. Stiint. Univ. "Al. I. Cuza" Iasi Sect. IIa Biol. 11:311-319. [1965]
Cîrdei, F., Bulimar, F., Boisteanu, T. 1970 ("1969"). Contributii la studiul formicidelor (Fam. Formicidae) din Carpatii Orientali. Stud. Comun. Muz. Stiint. Nat. Bacau 1969:151-156. [> 1970.04.30] [Date approved for printing ("bun de tipar").]
Cîrdei, F., Bulimar, F., Varvara, M. 1964. Date cu privire la raspîndirea speciilor genului *Lasius* Fabr. (Fam. Formicidae) în Moldova. An. Stiint. Univ. "Al. I. Cuza" Iasi Sect. IIa Biol. 10:109-112. [1964]
Clark, D. B., Guayasamín, C., Pazmiño, O., Donoso, C., Páez de Villacís, Y. 1982. The tramp ant *Wasmannia auropunctata*: autecology and effects on ant diversity and distribution on Santa Cruz Island, Galapagos. Biotropica 14:196-207. [1982.09]
Clark, J. 1924a ("1923"). Australian Formicidae. J. Proc. R. Soc. West. Aust. 9:72-89. [1924.01.04] [Publication date from Watson (1945:175).]
Clark, J. 1924b. Australian Formicidae. J. Proc. R. Soc. West. Aust. 10:75-89. [1924.04.30]
Clark, J. 1925a. The ants of Victoria. Part I. Vic. Nat. (Melb.) 42:58-64. [1925.07.10]
Clark, J. 1925b. The ants of Victoria. Part II. Vic. Nat. (Melb.) 42:135-144. [1925.10.09]
Clark, J. 1926. Australian Formicidae. J. R. Soc. West. Aust. 12:43-51. [1926.01.25]
Clark, J. 1927. The ants of Victoria. Part III. Vic. Nat. (Melb.) 44:33-40. [1927.06.07]
Clark, J. 1928a. Australian Formicidae. J. R. Soc. West. Aust. 14:29-41. [1928.04.24]
Clark, J. 1928b. Ants from North Queensland. Vic. Nat. (Melb.) 45:169-171. [1928.10.04]
Clark, J. 1928c. Excursion through Western District of Victoria. Entomological Reports. Formicidae. Vic. Nat. (Melb.) 45(Suppl.):39-44. [1928.10]
Clark, J. 1929a. Contributions to the fauna of Rottnest Island. No. III. The ants. J. R. Soc. West. Aust. 15:55-56. [1929.04.22]
Clark, J. 1929b. Results of a collecting trip to the Cann River, East Gippsland. Vic. Nat. (Melb.) 46:115-123. [1929.10.10]
Clark, J. 1930a. Some new Australian Formicidae. Proc. R. Soc. Vic. (n.s.)42:116-128. [1930.03.10]

Clark, J. 1930b. The Australian ants of the genus *Dolichoderus* (Formicidae). Sugenus *Hypoclinea* Mayr. Aust. Zool. 6:252-268. [1930.08.20]
Clark, J. 1930c. New Formicidae, with notes on some little-known species. Proc. R. Soc. Vic. (n.s.)43:2-25. [1930.08.30]
Clark, J. 1934a. Notes on Australian ants, with descriptions of new species and a new genus. Mem. Natl. Mus. Vic. 8:5-20. [1934.09]
Clark, J. 1934b. New Australian ants. Mem. Natl. Mus. Vic. 8:21-47. [1934.09]
Clark, J. 1934c. Ants from the Otway Ranges. Mem. Natl. Mus. Vic. 8:48-73. [1934.09]
Clark, J. 1936. A revision of Australian species of *Rhytidoponera* Mayr (Formicidae). Mem. Natl. Mus. Vic. 9:14-89. [1936.11]
Clark, J. 1938. The Sir Joseph Banks Islands. Reports of the McCoy Society for Field Investigation and Research. Part 10. Formicidae (Hymenoptera). Proc. R. Soc. Vic. (n.s.)50:356-382. [1938.05.23]
Clark, J. 1941a. Notes on the Argentine ant and other exotic ants introduced into Australia. Mem. Natl. Mus. Vic. 12:59-70. [1941.03]
Clark, J. 1941b. Australian Formicidae. Notes and new species. Mem. Natl. Mus. Vic. 12:71-94. [1941.03]
Clark, J. 1943. A revision of the genus *Promyrmecia* Emery (Formicidae). Mem. Natl. Mus. Vic. 13:83-149. [1943.09]
Clark, J. 1951. The Formicidae of Australia. Vol. 1. Subfamily Myrmeciinae. Melbourne: CSIRO, 230 pp. [1951]
Clausen, R. 1938. Untersuchungen über den männlichen Copulationsapparat der Ameisen, speziell der Formicinae. Mitt. Schweiz. Entomol. Ges. 17:233-346. [1938.06.15]
Claver, S., Fowler, H. G. 1993. The ant fauna (Hymenoptera, Formicidae) of the Ñacuñan Biosphere Reserve. Naturalia (São Paulo) 18:189-193. [1993]
Cobelli, R. 1906. Le formiche del promontorio di Sezza (Istria). Verh. K.-K. Zool.-Bot. Ges. Wien 56:477-480. [1906.10.19]
Cockerell, T. D. A. 1898. Miscellaneous notes. Proc. Entomol. Soc. Wash. 4:64-65. [1898.03.21]
Cockerell, T. D. A. 1906. A new fossil ant. Entomol. News 17:27-28. [1906.01]
Cockerell, T. D. A. 1915. British fossil insects. Proc. U. S. Natl. Mus. 49:469-499. [1915.12.11]
Cockerell, T. D. A. 1920. Fossil arthropods in the British Museum. - I. Ann. Mag. Nat. Hist. (9)5:273-279. [1920.03]
Cockerell, T. D. A. 1921. Some Eocene insects from Colorado and Wyoming. Proc. U. S. Natl. Mus. 59:29-39. [1921.06.27]
Cockerell, T. D. A. 1923a. The earliest known ponerine ant. Entomologist 56:51-52. [1923.03]
Cockerell, T. D. A. 1923b. Fossil insects from the Eocene of Texas. Am. J. Sci. 5:397-410. [1923.05]
Cockerell, T. D. A. 1927. Fossil insects from the Miocene of Colorado. Ann. Mag. Nat. Hist. (9)19:161-166. [1927.02]
Cohic, F. 1948. Observations morphologiques et écologiques sur *Dorylus* (*Anomma*) *nigricans* Illiger (Hymenoptera, Dorylidae). Rev. Fr. Entomol. 14:229-276. [1948.01.31]
Cokendolpher, J. C. 1990. The ants (Hymenoptera, Formicidae) of western Texas. Part III. Additions and corrections. Spec. Publ. Mus. Tex. Tech Univ. 31:1-19. [1990.02.23]
Cokendolpher, J. C., Brown, J. D. 1985. Air-dry method for studying chromosomes of insects and arachnids. Entomol. News 96:114-118. [1985.05.15]
Cokendolpher, J. C., Francke, O. F. 1983. Gynandromorphic desert fire ant, *Solenopsis aurea* Wheeler (Hymenoptera: Formicidae). J. N. Y. Entomol. Soc. 91:242-245. [1983.09.27]
Cokendolpher, J. C., Francke, O. F. 1985a ("1984"). Karyotype of *Conomyrma flava* (McCook) (Hymenoptera: Formicidae). J. N. Y. Entomol. Soc. 92:349-351. [1985.02.05]
Cokendolpher, J. C., Francke, O. F. 1985b. Temperature preferences of four species of fire ants (Hymenoptera: Formicidae: *Solenopsis*). Psyche (Camb.) 92:91-104. [1985.06.28]

Cokendolpher, J. C., Francke, O. F. 1990. The ants (Hymenoptera, Formicidae) of western Texas. Part II. Subfamilies Ecitoninae, Ponerinae, Pseudomyrmecinae, Dolichoderinae, and Formicinae. Spec. Publ. Mus. Tex. Tech Univ. 30:1-76. [1990.02.23]

Cokendolpher, J. C., Phillips, S. A., Jr. 1989. Rate of spread of the red imported fire ant, *Solenopsis invicta* (Hymenoptera: Formicidae), in Texas. Southwest. Nat. 34:443-449. [1989.09]

Cole, A. C., Jr. 1932a. The relation of the ant, *Pogonomyrmex occidentalis* Cr., to its habitat. Ohio J. Sci. 32:133-146. [1932.03]

Cole, A. C., Jr. 1932b. Notes on the ant *Pogonomyrmex californicus*, Buckley (Hym.: Formicidae). Entomol. News 43:113-115. [1932.05.12]

Cole, A. C., Jr. 1932c. Nests of the ant, *Formica subpolita* Mayr, in the western United States. Ohio J. Sci. 32:247-248. [1932.05]

Cole, A. C., Jr. 1933a. Ant communities of a section of the sagebrush semi-desert in Idaho, with special reference to the vegetation (Hymenop.: Formicidae). Entomol. News 44:16-19. [1933.01.12]

Cole, A. C., Jr. 1933b. Descriptions of two new ants of the genus *Pheidole* (Hymenoptera: Formicidae). Ann. Entomol. Soc. Am. 26:616-618. [1933.12.28]

Cole, A. C., Jr. 1933c. An ecological study of the ant fauna (Hymenoptera: Formicidae) of the southern and northern desert shrub regions of the United States. Abstr. Dr. Diss. Ohio State Univ. 11:97-105. [1933]

Cole, A. C., Jr. 1934a. Studies of semidesert ants (Hymenoptera: Formicidae). [part] Entomol. News 45:96-101. [1934.04.10]

Cole, A. C., Jr. 1934b. An ecological study of the ants of the southern desert shrub region of the United States. Ann. Entomol. Soc. Am. 27:388-405. [1934.10.10]

Cole, A. C., Jr. 1934c. An annotated list of the ants of the Snake River Plains, Idaho (Hymenoptera: Formicidae). Psyche (Camb.) 41:221-227. [1934.12] [Stamp date in MCZ library: 1935.03.07.]

Cole, A. C., Jr. 1936a. An annotated list of the ants of Idaho (Hymenoptera: Formicidae). Can. Entomol. 68:34-39. [1936.02.26]

Cole, A. C., Jr. 1936b. Descriptions of seven new western ants. (Hymenop.: Formicidae). Entomol. News 47:118-121. [1936.05.08]

Cole, A. C., Jr. 1937a. An annotated list of the ants of Arizona (Hymen.: Formicidae). [part] Entomol. News 48:97-101. [1937.04.13]

Cole, A. C., Jr. 1937b. An annotated list of the ants of Arizona (Hym.: Formicidae). [concl.] Entomol. News 48:134-140. [1937.05.13]

Cole, A. C., Jr. 1938a. Suggestions concerning taxonomic nomenclature of the hymenopterous family Formicidae, and descriptions of three new ants. Am. Midl. Nat. 19:236-241. [1938.01.15]

Cole, A. C., Jr. 1938b. A new ant from Idaho's semidesert. Am. Midl. Nat. 19:678-681. [1938.05.25]

Cole, A. C., Jr. 1938c. Descriptions of new ants from the western United States. Am. Midl. Nat. 20:368-373. [1938.09.29]

Cole, A. C., Jr. 1939a. A new ant from the Great Smoky Mountains, Tennessee. Am. Midl. Nat. 22:413-417. [1939.09.27]

Cole, A. C., Jr. 1939b. Pseudogynes of *Formica neogagates* Emery. Am. Midl. Nat. 22:418-419. [1939.09.27]

Cole, A. C., Jr. 1940a. A new ant from Indiana. Am. Midl. Nat. 23:224-226. [1940.02.06]

Cole, A. C., Jr. 1940b. A guide to the ants of the Great Smoky Mountains National Park, Tennessee. Am. Midl. Nat. 24:1-88. [1940.07.31]

Cole, A. C., Jr. 1942. The ants of Utah. Am. Midl. Nat. 28:358-388. [1942.10.13]

Cole, A. C., Jr. 1943a ("1942"). Synonyms of *Formica difficilis* Emery (Hymenoptera: Formicidae). Ann. Entomol. Soc. Am. 35:389. [1943.01.07]

Cole, A. C., Jr. 1943b. A new subspecies of *Formica moki* Wheeler (Hymenoptera: Formicidae). Am. Midl. Nat. 29:183-184. [1943.02.27]

Cole, A. C., Jr. 1947 ("1946"). A description of *Formica parcipappa*, a new ant from Idaho (Hymenoptera: Formicidae). Ann. Entomol. Soc. Am. 39:616-618. [1947.01.23]

Cole, A. C., Jr. 1948. A synonym of *Pheidole dentata* var. *commutata* Mayr (Hymenoptera, Formicidae). Proc. Entomol. Soc. Wash. 50:82. [1948.04.30]

Cole, A. C., Jr. 1949a. A study of the genus *Gesomyrmex* Mayr, and a description of a species new to the genus (Hymenoptera: Formicidae). Ann. Entomol. Soc. Am. 42:71-76. [1949.05.25]

Cole, A. C., Jr. 1949b. The ants of Mountain Lake, Virginia. J. Tenn. Acad. Sci. 24:155-156. [1949.07]

Cole, A. C., Jr. 1949c. Notes on *Gesomyrmex* (Hymenoptera: Formicidae). Entomol. News 60:181. [1949.10.24]

Cole, A. C., Jr. 1949d. The ants of Bikini Atoll, Marshall Islands (Hymenoptera). Pan-Pac. Entomol. 25:172-174. [1949.11.25]

Cole, A. C., Jr. 1952a. A new subspecies of *Trachymyrmex smithi* (Hymenoptera: Formicidae) from New Mexico. J. Tenn. Acad. Sci. 27:159-162. [1952.04]

Cole, A. C., Jr. 1952b. A new *Pheidole* (Hymenoptera: Formicidae) from Florida. Ann. Entomol. Soc. Am. 45:443-444. [1952.10.25]

Cole, A. C., Jr. 1952c. Notes on the *Pheidole pilifera* (Roger) complex and a description of a new subspecies (Hymenoptera: Formicidae). J. Tenn. Acad. Sci. 27:278-280. [1952.10]

Cole, A. C., Jr. 1953a. A checklist of the ants (Hymenoptera: Formicidae) of the Great Smoky Mountains National Park, Tennessee. J. Tenn. Acad. Sci. 28:34-35. [1953.01]

Cole, A. C., Jr. 1953b. Studies of New Mexico ants (Hymenoptera: Formicidae). I. A new species of *Myrmica*. J. Tenn. Acad. Sci. 28:81-82. [1953.01]

Cole, A. C., Jr. 1953c. Studies of New Mexico ants. II. A description of a new subspecies of *Aphaenogaster huachucana* (Hymenoptera: Formicidae). J. Tenn. Acad. Sci. 28:82-84. [1953.01]

Cole, A. C., Jr. 1953d. Studies of New Mexico ants. III. The ponerines and dorylines (Hymenoptera: Formicidae). J. Tenn. Acad. Sci. 28:84-85. [1953.01]

Cole, A. C., Jr. 1953e. Notes on the genus *Leptothorax* in New Mexico and a description of a new species. Proc. Entomol. Soc. Wash. 55:27-30. [1953.02.16]

Cole, A. C., Jr. 1953f. Studies of New Mexico ants. IV. The genera *Myrmica*, *Manica*, *Aphaenogaster*, and *Novomessor* (Hymenoptera: Formicidae). J. Tenn. Acad. Sci. 28:242-244. [1953.07]

Cole, A. C., Jr. 1953g. Studies of New Mexico ants. V. The genus *Pheidole* with synonymy (Hymenoptera: Formicidae). J. Tenn. Acad. Sci. 28:297-299. [1953.10]

Cole, A. C., Jr. 1953h. Studies of New Mexico ants. VI. The genera *Monomorium*, *Solenopsis*, *Myrmecina*, and *Trachymyrmex* (Hymenoptera: Formicidae). [part] J. Tenn. Acad. Sci. 28:299-300. [1953.10]

Cole, A. C., Jr. 1953i. Studies of New Mexico ants. VI. The genera *Monomorium*, *Solenopsis*, *Myrmecina*, and *Trachymyrmex* (Hymenoptera: Formicidae). [concl.] J. Tenn. Acad. Sci. 28:316. [1953.10]

Cole, A. C., Jr. 1953j. *Brachymyrmex depilis* subsp. *flavescens* Grundmann a synonym of *Brachymyrmex depilis* Emery (Hymenoptera: Formicidae). Entomol. News 64:266. [1953.12.08]

Cole, A. C., Jr. 1954a. Studies of New Mexico ants. VIII. A solution to the *Formica densiventris* Viereck problem (Hymenoptera: Formicidae). J. Tenn. Acad. Sci. 29:89-90. [1954.01]

Cole, A. C., Jr. 1954b. Studies of New Mexico ants. VII. The genus *Pogonomyrmex* with synonymy and a description of a new species (Hymenoptera: Formicidae). J. Tenn. Acad. Sci. 29:115-121. [1954.04]

Cole, A. C., Jr. 1954c. Studies of New Mexico ants. XI. The genus *Formica* with a description of a new species (Hymenoptera: Formicidae). J. Tenn. Acad. Sci. 29:163-167. [1954.04]

Cole, A. C., Jr. 1954d. Studies of New Mexico ants. X. The genus *Leptothorax* (Hymenoptera: Formicidae). J. Tenn. Acad. Sci. 29:240-241. [1954.07]

Cole, A. C., Jr. 1954e. Studies of New Mexico ants. IX. *Pogonomyrmex apache* Wheeler a synonym of *Pogonomyrmex sancti-hyacinthi* Wheeler (Hymenoptera: Formicidae). J. Tenn. Acad. Sci. 29:266-271. [1954.10]

Cole, A. C., Jr. 1954f. Studies of New Mexico ants. XII. The genera *Brachymyrmex*, *Camponotus*, and *Prenolepis* (Hymenoptera: Formicidae). J. Tenn. Acad. Sci. 29:271-272. [1954.10]

Cole, A. C., Jr. 1954g. Studies of New Mexico ants. XIII. The genera *Acanthomyops*, *Myrmecocystus*, and *Polyergus* (Hymenoptera: Formicidae). J. Tenn. Acad. Sci. 29:284-285. [1954.10]

Cole, A. C., Jr. 1955a. Studies of New Mexico ants. XIV. A description of a new species of *Pheidole* Westwood (Hymenoptera: Formicidae). J. Tenn. Acad. Sci. 30:47-49. [1955.01]

Cole, A. C., Jr. 1955b. Studies of New Mexico ants. XV. Additions, corrections, and new synonymy. J. Tenn. Acad. Sci. 30:49-50. [1955.01]

Cole, A. C., Jr. 1955c. Studies of Nevada ants. I. Notes on *Veromessor lariversi* M. R. Smith and a description of the queen (Hymenoptera: Formicidae). J. Tenn. Acad. Sci. 30:51-52. [1955.01]

Cole, A. C., Jr. 1956a. Studies of Nevada ants. II. A new species of *Lasius* (*Chthonolasius*) (Hymenoptera: Formicidae). J. Tenn. Acad. Sci. 31:26-27. [1956.01]

Cole, A. C., Jr. 1956b. A new species of *Leptothorax* from Arizona (Hymenoptera: Formicidae). J. Tenn. Acad. Sci. 31:28-31. [1956.01]

Cole, A. C., Jr. 1956c. Observations of some members of the genus *Pheidole* in the southwestern United States with synonymy (Hymenoptera: Formicidae). J. Tenn. Acad. Sci. 31:112-118. [1956.04]

Cole, A. C., Jr. 1956d. In defense of the integrity of an ant. J. Tenn. Acad. Sci. 31:212-214. [1956.07]

Cole, A. C., Jr. 1956e. *Leptothorax stenotyle* (n. nov.) for *Leptothorax angustinodus* Cole. J. Tenn. Acad. Sci. 31:214. [1956.07]

Cole, A. C., Jr. 1956f. Studies of Nevada ants. III. The status of *Formica nevadensis* Wheeler (Hymenoptera: Formicidae). J. Tenn. Acad. Sci. 31:256-257. [1956.10]

Cole, A. C., Jr. 1956g. Studies of Nevada ants. IV. Descriptions of sexual castes of three members of the *rufa* group of the genus *Formica* L. J. Tenn. Acad. Sci. 31:257-260. [1956.10]

Cole, A. C., Jr. 1956h. New synonymy in the genus *Manica* Jurine (Hymenoptera: Formicidae). J. Tenn. Acad. Sci. 31:260-262. [1956.10]

Cole, A. C., Jr. 1957a. *Paramyrmica*, a new North American genus of ants allied to *Myrmica* Latreille. (Hymenoptera: Formicidae). J. Tenn. Acad. Sci. 32:37-42. [1957.01]

Cole, A. C., Jr. 1957b. A new *Leptothorax* from Texas (Hymenoptera: Formicidae). J. Tenn. Acad. Sci. 32:42-45. [1957.01]

Cole, A. C., Jr. 1957c. Descriptions of sexual castes of some ants in the genera *Myrmica*, *Manica* and *Xiphomyrmex* from the western United States (Hymenoptera: Formicidae). J. Tenn. Acad. Sci. 32:208-213. [1957.07]

Cole, A. C., Jr. 1957d. Another new *Leptothorax* from Texas (Hymenoptera: Formicidae). J. Tenn. Acad. Sci. 32:213-215. [1957.07]

Cole, A. C., Jr. 1958a. A remarkable new species of *Lasius* (*Chthonolasius*) from California (Hymenoptera: Formicidae). J. Tenn. Acad. Sci. 38:75-77. [1958.01]

Cole, A. C., Jr. 1958b ("1957"). Notes on western ants (Hymenoptera: Formicidae). J. N. Y. Entomol. Soc. 65:129-131. [1958.06.27]

Cole, A. C., Jr. 1958c. North American *Leptothorax* of the *nitens-carinatus* complex (Hymenoptera: Formicidae). Ann. Entomol. Soc. Am. 51:535-538. [1958.11.29]

Cole, A. C., Jr. 1963a. A new species of *Veromessor* from the Nevada Test Site and notes on related species (Hymenoptera: Formicidae). Ann. Entomol. Soc. Am. 56:678-682. [1963.09.15]

Cole, A. C., Jr. 1963b. A preliminary synopsis of the subgenera and complexes of the ant genus *Pogonomyrmex* Mayr in North America. Symp. Genet. Biol. Ital. 12:51-59. [1963]
Cole, A. C., Jr. 1965. Discovery of the worker caste of *Pheidole* (*P.*) *inquilina*, new combination (Hymenoptera: Formicidae). Ann. Entomol. Soc. Am. 58:173-175. [1965.03.15]
Cole, A. C., Jr. 1966a. Keys to the subgenera, complexes, and species of the genus *Pogonomyrmex* (Hymenoptera: Formicidae) in North America, for identification of the workers. Ann. Entomol. Soc. Am. 59:528-530. [1966.05.15]
Cole, A. C., Jr. 1966b. Ants of the Nevada Test Site. Brigham Young Univ. Sci. Bull. Biol. Ser. 7(3):1-27. [1966.06]
Cole, A. C., Jr. 1968. *Pogonomyrmex* harvester ants. A study of the genus in North America. Knoxville, Tenn.: University of Tennessee Press, x + 222 pp. [1968]
Cole, A. C., Jr., Jones, J. W., Jr. 1948. A study of the weaver ant, *Oecophylla smaragdina* (Fab.). Am. Midl. Nat. 39:641-651. [1948.07.28]
Cole, B. J. 1980. Repertoire convergence in two mangrove ants, *Zacryptocerus varians* and *Camponotus* (*Colobopsis*) sp. Insectes Soc. 27:265-275. [1980.09]
Cole, B. J. 1983a ("1982"). The guild of sawgrass-inhabiting ants in the Florida Keys. Psyche (Camb.) 89:351-356. [1983.04.27]
Cole, B. J. 1983b. Assembly of mangrove ant communities: patterns of geographical distribution. J. Anim. Ecol. 52:339-347. [1983.06]
Cole, B. J. 1983c. Assembly of mangrove ant communities: colonization abilities. J. Anim. Ecol. 52:349-355. [1983.06]
Cole, B. J. 1983d. Multiple mating and the evolution of social behavior in the Hymenoptera. Behav. Ecol. Sociobiol. 12:191-201. [1983.06]
Cole, B. J. 1985. Size and behavior in ants: constraints on complexity. Proc. Natl. Acad. Sci. U. S. A. 82:8548-8551. [1985.12]
Cole, B. J. 1986. The social behavior of *Leptothorax allardycei* (Hymenoptera, Formicidae): time budgets and the evolution of worker reproduction. Behav. Ecol. Sociobiol. 18:165-173. [1986.01]
Colina, G. O. 1981. Hormigas (Hymenoptera: Formicidae) de la Reserva de la Biosfera del Bolsón de Mapimí y distribución de recursos dentro del "guild" de hormigas granívoras: datos preliminares. [Abstract]. Folia Entomol. Mex. 48:19-20. [1981.09]
Collingwood, C. See: Collingwood, C. A.
Collingwood, C. A. 1950. Ants in N. Scotland. Entomol. Rec. J. Var. 62:41-42. [1950.04.15]
Collingwood, C. A. 1951a. Ants in Scotland. Entomol. Rec. J. Var. 63:306-307. [1951.12.15]
Collingwood, C. A. 1951b. The distribution of ants in north-west Scotland. Scott. Nat. 63:45-49. [1951]
Collingwood, C. A. 1953. Ants in Galloway. Entomol. Rec. J. Var. 65:297-298. [1953.10.15]
Collingwood, C. A. 1954. Rare ants (Hym., Formicidae) in Dorset. Entomol. Mon. Mag. 90:43-44. [1954.03.06]
Collingwood, C. A. 1955a. Ants in S.W. Scotland. Entomol. Rec. J. Var. 67:11-12. [1955.01.15]
Collingwood, C. A. 1955b. Ants in the South Midlands. Entomol. Gaz. 6:143-149. [1955.07.13]
Collingwood, C. A. 1956a. Ant hunting in France. Entomologist 89:106-108. [1956.05.29]
Collingwood, C. A. 1956b. A rare parasitic ant (Hym., Formicidae) in France. Entomol. Mon. Mag. 92:197. [1956.08.06]
Collingwood, C. A. 1956c. Aberrations in British ants of the genus *Formica*. J. Soc. Br. Entomol. 5:193-196. [1956.11.23]
Collingwood, C. A. 1956d. Distribution of ants allied to *Formica fusca* L. and *F. rufa* L. in Britain. Entomologist 89:291-294. [1956.12.31]
Collingwood, C. A. 1957a. A collection of ants (Hym. Formicidae) in the Leicester City Museum. [part] Entomol. Rec. J. Var. 69:167-170. [1957.08.15]
Collingwood, C. A. 1957b. A collection of ants (Hym. Formicidae) in the Leicester City Museum. [concl.] Entomol. Rec. J. Var. 69:183-187. [1957.09.15]

Collingwood, C. A. 1957c. The species of ants of the genus *Lasius* in Britain. J. Soc. Br. Entomol. 5:204-214. [1957.11.29]
Collingwood, C. A. 1958a. Summit ant swarms. Entomol. Rec. J. Var. 70:65-67. [1958.03.15]
Collingwood, C. A. 1958b. A key to the species of ants (Hymenoptera, Formicidae) found in Britain. Trans. Soc. Br. Entomol. 13:69-96. [1958.04.30]
Collingwood, C. A. 1958c. A survey of Irish Formicidae. Proc. R. Ir. Acad. Sect. B 59:213-219. [1958.05.16]
Collingwood, C. A. 1958d. The ants of the genus *Myrmica* in Britain. Proc. R. Entomol. Soc. Lond. Ser. A 33:65-75. [1958.06.30]
Collingwood, C. A. 1959. Scandinavian ants. Entomol. Rec. J. Var. 71:77-83. [1959.03.15]
Collingwood, C. A. 1961a ("1960"). The third Danish Expedition to Central Asia. Zoological Results 27. Formicidae (Insecta) from Afghanistan. Vidensk. Medd. Dan. Naturhist. Foren. 123:51-79. [1961.01.15]
Collingwood, C. A. 1961b ("1960"). *Myrmica schencki* Em. (Hym., Formicidae) and other ants in Somerset. Entomol. Mon. Mag. 96:130. [1961.03.02]
Collingwood, C. A. 1961c. Ergebnisse der Deutschen Afghanistan-Expedition 1956 der Landessammlungen für Naturkunde Karlsruhe. Formicidae (Hymenoptera Aculeata). Beitr. Naturkd. Forsch. Südwestdtschl. 19:289-290. [1961.04.01]
Collingwood, C. A. 1961d. New vice-county records for British ants. Entomol. Rec. J. Var. 73: 90-93. [1961.04.15]
Collingwood, C. A. 1961e. Ants in Finland. Entomol. Rec. J. Var. 73:190-195. [1961.09.15]
Collingwood, C. A. 1961f. Ants in the Scottish Highlands. Scott. Nat. 70:12-21. [1961.10]
Collingwood, C. A. 1962a. *Myrmica puerilis* Staercke, 1942, an ant new to Britain. Entomol. Mon. Mag. 98:18-20. [> 1962.01] [Precise publication dates not available for this volume of Entomol. Mon. Mag.]
Collingwood, C. A. 1962b. New records for British ants, 1961-1962. Entomol. Rec. J. Var. 74:234-236. [1962.11.15]
Collingwood, C. A. 1962c. Some ants (Hym. Formicidae) from north-east Asia. Entomol. Tidskr. 83:215-230. [1962.12.31]
Collingwood, C. A. 1963a. Notes on some South European and Mediterranean ants. Entomol. Rec. J. Var. 75:114-119. [1963.04.15]
Collingwood, C. A. 1963b. The *Lasius* (*Chthonolasius*) *umbratus* (Hym., Formicidae) species complex in north Europe. Entomologist 96:145-158. [1963.07.26]
Collingwood, C. A. 1963c. Three ant species new to Norway. Entomol. Rec. J. Var. 75:225-228. [1963.09.15]
Collingwood, C. A. 1964a. New locality records for British ants, 1963. Entomol. Rec. J. Var. 76:58-59. [1964.02.15]
Collingwood, C. A. 1964b. The identification and distribution of British ants. 1. A revised key to the species found in Britain. Trans. Soc. Br. Entomol. 16:93-114,121. [1964.12]
Collingwood, C. A. 1965. The distribution of ants (Hymenoptera) of the *Formica fusca* species group in Europe. P. 446 in: Freeman, P. (ed.) XIIth International Congress of Entomology. London, 8-16 July, 1964. Proceedings. London: XIIth International Congress of Entomology, 842 pp. [1965]
Collingwood, C. A. 1966. Notes on British ants, 1964-1965. Entomol. Rec. J. Var. 78:23-25. [1966.01.15]
Collingwood, C. A. 1969. The identity of *Crematogaster "depressa"*. Annu. Rep. Cocoa Res. Inst. Ghana 1967-1968:70-71. [1969]
Collingwood, C. A. 1970a. The first European Regional Congress of Myrmecology. Entomologist 103:25-26. [1970.02.05]
Collingwood, C. A. 1970b. Formicidae (Hymenoptera: Aculeata) from Nepal. Khumbu Himal Ergeb. Forschungsunternehm. Nepal Himal. 3:371-387. [1970.10.01]

Collingwood, C. A. 1971. A synopsis of the Formicidae of north Europe. Entomologist 104:150-176. [1971.08.10]

Collingwood, C. A. 1974. A revised list of Norwegian ants (Hymenoptera: Formicidae). Nor. Entomol. Tidsskr. 21:31-35. [1974]

Collingwood, C. A. 1976a. Ants (Hymenoptera: Formicidae) from North Korea. Ann. Hist.-Nat. Mus. Natl. Hung. 68:295-309. [1976]

Collingwood, C. A. 1976b. Mire invertebrate fauna at Eidskog, Norway. III. Formicidae (Hymenoptera Aculeata). Norw. J. Entomol. 23:185-187. [1976]

Collingwood, C. A. 1978. A provisional list of Iberian Formicidae with a key to the worker caste (Hym. Aculeata). EOS. Rev. Esp. Entomol. 52:65-95. [1978.05.10]

Collingwood, C. A. 1979. The Formicidae (Hymenoptera) of Fennoscandia and Denmark. Fauna Entomol. Scand. 8:1-174. [1979]

Collingwood, C. A. 1981. Ants (Hymenoptera: Formicidae) from Korea, 2. Folia Entomol. Hung. 42[34]:25-30. [1981.06.30]

Collingwood, C. A. 1982. Himalayan ants of the genus *Lasius* (Hymenoptera: Formicidae). Syst. Entomol. 7:283-296. [1982.07]

Collingwood, C. A. 1985. Hymenoptera: Fam. Formicidae of Saudi Arabia. Fauna Saudi Arab. 7:230-302. [1985]

Collingwood, C. A. 1987. Taxonomy and zoogeography of the *Formica rufa* L. species group. Pp. 65-67 in: Eder, J., Rembold, H. (eds.) Chemistry and biology of social insects. München: Verlag J. Peperny, xxxv + 757 pp. [1987]

Collingwood, C. A. 1991a. New records for British ants. Entomol. Rec. J. Var. 103:92-94. [1991.03.25]

Collingwood, C. A. 1991b. Especies raras de hormigas del género *Lasius* en España (Hymenoptera, Formicidae). Bol. Asoc. Esp. Entomol. 15:215-219. [1991.12.30]

Collingwood, C. A. 1993a. Description of the host ant of *Cossyphodinus bremeri* Ferrer sp. nov. (Hymenoptera, Formicidae). Pp. 229-231 in: Ferrer, J., Collingwood, C. A. New species of *Cossyphodinus* from Africa (Coleoptera, Tenebrionidae) and description of the host ant *Messor ferreri* sp. nov. (Hymenoptera, Formicidae). Entomofauna 14:221-232. [1993.05.10]

Collingwood, C. A. 1993b. A comparative study of the ant fauna of five Greek islands. Biol. Gallo-Hell. 20:191-197.

Collingwood, C. A., Agosti, D. 1987. Symposium: taxonomy and zoogeography of *Formica rufa* species. General discussion. Pp. 71-72 in: Eder, J., Rembold, H. (eds.) Chemistry and biology of social insects. München: Verlag J. Peperny, xxxv + 757 pp. [1987]

Collingwood, C. A., Barrett, K. E. J. 1964. The identification and distribution of British ants. 2. The vice-county distribution of indigenous ants in the British Isles. Trans. Soc. Br. Entomol. 16:114-121. [1964.12]

Collingwood, C. A., Barrett, K. E. J. 1966. Additions to the vice-county distribution of British ants, 1966. Entomologist 99:254-256. [1966.10.18]

Collingwood, C. A., Hughes, J. 1987. Ant species in Yorkshire, England. Naturalist (Leeds) 112:95-101. [1987.09]

Collingwood, C. A., Kugler, J. 1994. *Solenopsis dentata* (Hymenoptera, Formicidae): a new species from Israel. Isr. J. Entomol. 28:119-122. [1994.12.15]

Collingwood, C. A., Satchell, J. E. 1956. The ants of the South Lake District. J. Soc. Br. Entomol. 5:159-164. [1956.05.25]

Collingwood, C. A., Van Harten, A. 1993. The ants (Hymenoptera: Formicidae) of the Cape Verde Islands. Cour. Forschungsinst. Senckenb. 159:411-414. [1993.07.01]

Collingwood, C. A., Yarrow, I. H. H. 1969. A survey of Iberian Formicidae (Hymenoptera). EOS. Rev. Esp. Entomol. 44:53-101. [1969.10.01]

Colombel, P. 1971. Étude biométrique du couvain et des adultes d'*Odontomachus haematodes* (Hym. Form. Poneridae). Ann. Fac. Sci. Univ. Féd. Cameroun 6:53-71. [1971.04]

Comín, P., De Haro, A. 1980. Datos iniciales para un estudio ecológico de las hormigas de Menorca (Hym. Formicidae). Bolleti Soc. Hist. Nat. Balears 24:23-48. [1980]
Comín, P., Espadaler, X. 1984. Ants of the Pityusic Islands (Hym. Formicidae). Pp. 287-301 in: Kuhbier, H., Alcover, J. A., Guerau d'Arellano Tur, C. (eds.) Biogeography and ecology of the Pityusic Islands (Monogr. Biol., Vol. 52). The Hague: W. Junk, xvi + 704 pp. [1984]
Comín, P., Furió, V. 1986. Distribución biogeográfica de las hormigas (Hymenoptera, Formicidae) en las islas del mediterráneo occidental. Bolleti Soc. Hist. Nat. Balears 30:67-79. [1986]
Comín del Río, P. See: Comín, P.
Comstock, J. H. 1881. Dates of publication of entomological reports. Rep. Entomol. U. S. Dep. Agric. 1880:275. [1881]
Conconi, J. R. D. de See: Ramos-Elorduy de Conconi, J.
Consani, M. 1947. Reperti corologici sulle formiche italiane. Redia 32:179-182. [1947]
Consani, M. 1948 ("1947"). Primo contributo alla conoscenza della fauna entomologica del Matese. Imenotteri (Formicidae). Boll. Assoc. Rom. Entomol. 2:28-29. [1948.05.10]
Consani, M. 1949. Formiche raccolte nell'Appennino Abruzzese dal Sig. Pio Bisleti. Boll. Assoc. Rom. Entomol. 4:11-12. [1949.07.28]
Consani, M. 1951. Formiche dell'Africa Orientale I. Boll. Ist. Entomol. Univ. Studi Bologna 18:167-172. [1951.05.15]
Consani, M. 1952 ("1951"). [Untitled. *Crematogaster* (*Acrocoelia*) *nigriceps* ssp. *saganensis* Cons. n. ssp.] P. 65 in: Menozzi, C., Consani, M. Missione biologica Sagan-Omo diretta dal Prof. E. Zavattari. Hymenoptera Formicidae. Riv. Biol. Colon. 11:57-71. [1952.12]
Consani, M. 1953a ("1952"). Un caso di omonimia nel genere *Crematogaster* Lund (Hymenoptera, Formicidae). Boll. Soc. Entomol. Ital. 82:100. [1953.01.31]
Consani, M. 1953b. Formiche di Puglia e delle isole Tremiti. Mem. Biogeogr. Adriatica 2:25-31. [1953]
Consani, M., Zangheri, P. 1952. Fauna di Romagna. Imenotteri - Formicidi. Mem. Soc. Entomol. Ital. 31:38-48. [1952.11.05]
Coody, C. J., Watkins, J. F., II. 1986. The correlation of eye size with circadian flight periodicity of Nearctic army ant males of the genus *Neivamyrmex* (Hymenoptera; Formicidae; Ecitoninae). Tex. J. Sci. 38:3-7. [1986.02]
Cook, J. L., Martin, J. B., Gold, R. E. 1994. First record of *Tapinoma melanocephalum* (Hymenoptera: Formicidae) in Texas. Southwest. Entomol. 19:409-410. [1994.12]
Cook, O. F. 1905. The social organization and breeding habits of the cotton-protecting kelep of Guatemala. U. S. Dep. Agric. Bur. Entomol. Tech. Ser. 10:1-55. [1905]
Cook, T. W. 1953. The ants of California. Palo Alto, California: Pacific Books, 462 pp. [1953]
Cori, K., Finzi, B. 1931. Aufzählung der von Karl Cori 1914 auf süddalmatinischen Inseln gesammelten Ameisen. Anz. Akad. Wiss. Wien Math.-Naturwiss. Kl. 68:237-240. [1931]
Coronado Padilla, R., Morales, A., Espinosa M., E. 1972. Distribución geográfica de las especies de hormigas arrieras existentes en la República Mexicana. Folia Entomol. Mex. 23-24:95-96. [1972.01]
Costa, A. 1883. Notizie ed osservazioni sulla geo-fauna Sarda. Memoria seconda. Risultamento di ricerche fatte in Sardegna nella primavera del 1882. Atti R. Accad. Sci. Fis. Mat. Napoli (2)1(2):1-109. [1883.06.30]
Costa, A. 1884. Notizie ed osservazioni sulla geo-fauna Sarda. Memoria terza. Risultamento delle ricerche fatte in Sardegna nella estate del 1883. Atti R. Accad. Sci. Fis. Mat. Napoli (2)1(9): 1-64. [1884.07.04]
Costa Lima, A. da. 1931. A proposito da *Acropyga pickeli* Borgm., 1927 (Hymenoptera: Formicoidea). Bol. Biol. Rio J. 17:2-9. [1931.08.10]
Cotti, G. 1963. Bibliografia ragionata 1930-1961 del gruppo *Formica rufa* in Italiano, Deutsch, English. Collana Verde 8:1-413. [1963.06.07]
Coulon, L. 1924 ("1923"). Les Formicides du Musée d'Elbeuf (collection européenne). Bull. Soc. Étude Sci. Nat. Elbeuf 42:121-134. [1924]

Covelo de Zolessi, L. See: Zolessi, L. C. de
Cover, S. P. and collaborators. 1990. [Untitled. Keys to the ant genera of the Nearctic and Neotropical regions. Based on the worker caste.] Pp. 62-81 in: Hölldobler, B., Wilson, E. O. The ants. Cambridge, Mass.: Harvard University Press, xii + 732 pp. [1990.03.28]
Cover, S. P., Tobin, J. E., Wilson, E. O. 1990. The ant community of a tropical lowland rainforest site in Peruvian Amazonia. Pp. 699-700 in: Veeresh, G. K., Mallik, B., Viraktamath, C. A. (eds.) Social insects and the environment. Proceedings of the 11th International Congress of IUSSI, 1990. New Delhi: Oxford & IBH Publishing Co., xxxi + 765 pp. [1990.08]
Cowan, C. F. 1969. Notes on Griffith's Animal Kingdom of Cuvier (1824-1835). J. Soc. Bibliogr. Nat. Hist. 5:137-140. [1969.04]
Cowan, C. F. 1970. The insects of the Coquille Voyage. J. Soc. Bibliogr. Nat. Hist. 5:358-360. [1970.10]
Cowan, C. F. 1971. On Guérin's Iconographie: particularly the insects. J. Soc. Bibliogr. Nat. Hist. 6:18-29. [1971.10]
Cowan, C. F. 1976. On the Disciples' Edition of Cuvier's Règne Animal. J. Soc. Bibliogr. Nat. Hist. 8:32-64. [1976.11]
Coyle, F. A. 1966. Defensive behavior and associated morphological features in three species of the ant genus *Paracryptocerus*. Insectes Soc. 13:93-104. [1966.09]
Craig, R., Crozier, R. H. 1978a. Caste-specific locus expression in ants. Isozyme Bull. 11:64-65. [1978.02]
Craig, R., Crozier, R. H. 1978b. No evidence for role of heterozygosity in ant caste determination. Isozyme Bull. 11:66-67. [1978.02]
Craig, R., Crozier, R. H. 1979. Relatedness in the polygynous ant *Myrmecia pilosula*. Evolution 33:335-341. [1979.05.15]
Crawley, W. C. 1911. *Formica fusca*, L., var. *glebaria*, Nyl., a form new to Britain. Entomol. Rec. J. Var. 23:96. [1911.04.15]
Crawley, W. C. 1912a. *Leptothorax tuberum*, Fab., subsp. *corticalis*, Schenk, an ant new to Britain. Entomol. Rec. J. Var. 24:63-65. [1912.03.15]
Crawley, W. C. 1912b. *Anergates atratulus*, Schenk., a British ant, and the acceptance of a queen by *Tetramorium caespitum*, L. Entomol. Rec. J. Var. 24:218-219. [1912.09.15]
Crawley, W. C. 1914a. A revision of the genus *Leptothorax*, Mayr, in the British Isles. [part] Entomol. Rec. J. Var. 26:89-96. [1914.04.08]
Crawley, W. C. 1914b. A revision of the genus *Leptothorax*, Mayr, in the British Isles. [concl.] Entomol. Rec. J. Var. 26:106-109. [1914.05.15]
Crawley, W. C. 1915a. Ants from north and central Australia, collected by G. F. Hill. - Part I. Ann. Mag. Nat. Hist. (8)15:130-136. [1915.01]
Crawley, W. C. 1915b. Ants from north and south-west Australia (G. F. Hill, Rowland Turner) and Christmas Island, Straits Settlements. - Part II. Ann. Mag. Nat. Hist. (8)15:232-239. [1915.02]
Crawley, W. C. 1916a. A new species of ponerine ant captured by an *Asilus*. Entomologist 49: 30-31. [1916.02]
Crawley, W. C. 1916b. Ants from British Guiana. Ann. Mag. Nat. Hist. (8)17:366-378. [1916.05]
Crawley, W. C. 1917. Ants and aphides in west Somerset. Proc. Somerset. Archaeol. Nat. Hist. Soc. 62:148-163. [1917]
Crawley, W. C. 1918. Some new Australian ants. Entomol. Rec. J. Var. 30:86-92. [1918.05.15]
Crawley, W. C. 1920a. Ants from Mesopotamia and north-west Persia. Entomol. Rec. J. Var. 32:162-166. [1920.09.15]
Crawley, W. C. 1920b. Ants from Mesopotamia and north-west Persia (concluded). Entomol. Rec. J. Var. 32:177-179. [1920.10.15]
Crawley, W. C. 1920c. A new species of ant imported into England. Entomol. Rec. J. Var. 32:180-181. [1920.10.15]
Crawley, W. C. 1920d. A gynandromorph of *Monomorium floricola*, Jerd. Entomol. Rec. J. Var. 32:217-218. [1920.12.15]

Crawley, W. C. 1921. New and little-known species of ants from various localities. Ann. Mag. Nat. Hist. (9)7:87-97. [1921.01]
Crawley, W. C. 1922a. New ants from Australia. Ann. Mag. Nat. Hist. (9)9:427-448. [1922.04]
Crawley, W. C. 1922b. Formicidae. - A new species and variety. Entomol. Rec. J. Var. 34:85-86. [1922.05.15]
Crawley, W. C. 1922c. Notes on some Australian ants. Biological notes by E. B. Poulton, D.Sc., M.A., F.R.S., and notes and descriptions of new forms by W. C. Crawley, B.A., F.E.S., F.R.M.S. [part] Entomol. Mon. Mag. 58:118-120. [1922.05]
Crawley, W. C. 1922d. Notes on some Australian ants. Biological notes by E. B. Poulton, D.Sc., M.A., F.R.S., and notes and descriptions of new forms by W. C. Crawley, B.A., F.E.S., F.R.M.S. [concl.] Entomol. Mon. Mag. 58:121-126. [1922.06]
Crawley, W. C. 1922e. New ants from Australia (concluded from vol. ix. p. 449). Ann. Mag. Nat. Hist. (9)10:16-36. [1922.07]
Crawley, W. C. 1923a. Myrmecological notes. Entomol. Rec. J. Var. 35:29-32. [1923.01.15]
Crawley, W. C. 1923b. Myrmecological notes. - New Australian Formicidae. Entomol. Rec. J. Var. 35:177-179. [1923.12.15]
Crawley, W. C. 1924a. Ants from Sumatra, with biological notes by Edward Jacobson. Ann. Mag. Nat. Hist. (9)13:380-409. [1924.04]
Crawley, W. C. 1924b. A Nigerian ant imported into England. Entomol. Rec. J. Var. 36:91-92. [1924.06.15]
Crawley, W. C. 1925a. Formicidae. A new genus. Entomol. Rec. J. Var. 37:40-41. [1925.03.15]
Crawley, W. C. 1925b. New ants from Australia. - II. Ann. Mag. Nat. Hist. (9)16:577-598. [1925.11]
Crawley, W. C. 1925c. Two myrmecological notes. Entomol. Rec. J. Var. 37:170-171. [1925.12.15]
Crawley, W. C. 1926. A revision of some old types of Formicidae. Trans. Entomol. Soc. Lond. 1925:373-393. [1926.02.05]
Crawley, W. C., Baylis, H. A. 1921. *Mermis* parasitic on ants of the genus *Lasius*. J. R. Microsc. Soc. Lond. 1921:353-372. [1921.12] [Article divided into two parts, by Crawley (pp. 353-364) and Baylis (pp. 365-372), respectively.]
Crawley, W. C., Donisthorpe, H. 1913. The founding of colonies by queen ants. Pp. 11-77 in: Jordan, K., Eltringham, H. (eds.) 2nd International Congress of Entomology, Oxford, August 1912. Volume II, Transactions. London: Hazell, Watson & Viney Ltd., 489 pp. [1913.10.06]
Creighton, W. S. 1927. The slave-raids of *Harpagoxenus americanus*. Psyche (Camb.) 34:11-29. [1927.02] [Stamp date in MCZ library: 1927.04.04.]
Creighton, W. S. 1928a. Notes on three abnormal ants. Psyche (Camb.) 35:51-55. [1928.03] [Stamp date in MCZ library: 1928.09.14.]
Creighton, W. S. 1928b. A new species of *Thaumatomyrmex* from Cuba. Psyche (Camb.) 35:162-166. [1928.09] [Stamp date in MCZ library: 1929.03.15.]
Creighton, W. S. 1929. New forms of *Odontoponera transversa*. Psyche (Camb.) 36:150-154. [1929.06] [Stamp date in MCZ library: 1929.07.09.]
Creighton, W. S. 1930a. A review of the genus *Myrmoteras* (Hymenoptera, Formicidae). J. N. Y. Entomol. Soc. 38:177-192. [1930.07.12]
Creighton, W. S. 1930b. The New World species of the genus *Solenopsis* (Hymenop. Formicidae). Proc. Am. Acad. Arts Sci. 66:39-151. [1930.12]
Creighton, W. S. 1932. A new female of *Acamatus* from Texas. Psyche (Camb.) 39:73-78. [1932.09] [Stamp date in MCZ library: 1932.12.09.]
Creighton, W. S. 1933. *Cyathomyrmex*, a new name for the subgenus *Cyathocephalus* Emery. Psyche (Camb.) 40:98-100. [1933.09] [Stamp date in MCZ library: 1933.11.13.]
Creighton, W. S. 1934. Descriptions of three new North American ants with certain ecological observations on previously described forms. Psyche (Camb.) 41:185-200. [1934.12] [Stamp date in MCZ library: 1935.03.07.]

Creighton, W. S. 1935. Two new species of *Formica* from western United States. Am. Mus. Novit. 773:1-8. [1935.01.28]
Creighton, W. S. 1937. Notes on the habits of *Strumigenys*. Psyche (Camb.) 44:97-109. [1937.12] [Stamp date in MCZ library: 1938.03.01.]
Creighton, W. S. 1938. On formicid nomenclature. J. N. Y. Entomol. Soc. 46:1-9. [1938.04.01]
Creighton, W. S. 1939a. A generic reallocation for *Myrmoteras kuroiwae*. J. N. Y. Entomol. Soc. 47:39-40. [1939.04.07]
Creighton, W. S. 1939b. A new subspecies of *Crematogaster minutissima* with revisionary notes concerning that species (Hymenoptera: Formicidae). Psyche (Camb.) 46:137-140. [1939.12] [Stamp date in MCZ library: 1940.01.10.]
Creighton, W. S. 1940a. A revision of the North American variants of the ant *Formica rufa*. Am. Mus. Novit. 1055:1-10. [1940.04.15]
Creighton, W. S. 1940b. A revision of the forms of *Stigmatomma pallipes*. Am. Mus. Novit. 1079:1-8. [1940.07.17]
Creighton, W. S. 1945. Observations on the subgenus *Rhachiocrema* (Hymenoptera: Formicidae) with the description of a new species from Borneo. Psyche (Camb.) 52:109-118. [1945.10.26]
Creighton, W. S. 1950a. The ants of North America. Bull. Mus. Comp. Zool. 104:1-585. [1950.04]
Creighton, W. S. 1950b. *Polyhomoa* Azuma, a synonym of *Kyidris* Brown (Hymenoptera: Formicidae). Psyche (Camb.) 57:93-94. [1950.12.29]
Creighton, W. S. 1951. Studies on Arizona ants. 1. The habits of *Camponotus ulcerosus* Wheeler and its identity with *C. bruesi* Wheeler. Psyche (Camb.) 58:47-64. [1951.11.19]
Creighton, W. S. 1952a ("1951"). Studies on Arizona ants. 2. New data on the ecology of *Aphaenogaster huachucana* and a description of the sexual forms. Psyche (Camb.) 58:89-99. [1952.04.07]
Creighton, W. S. 1952b. Studies on Arizona ants (3). The habits of *Pogonomyrmex huachucanus* Wheeler and a description of the sexual castes. Psyche (Camb.) 59:71-81. [1952.10.16]
Creighton, W. S. 1953a. New data on the habits of the ants of the genus *Veromessor*. Am. Mus. Novit. 1612:1-18. [1953.03.20]
Creighton, W. S. 1953b ("1952"). *Pseudomyrmex apache*, a new species from the southwestern United States (Hymenoptera: Formicidae). Psyche (Camb.) 59:131-142. [1953.04.27]
Creighton, W. S. 1953c ("1952"). Studies on Arizona ants (4). *Camponotus* (*Colobopsis*) *papago*, a new species from southern Arizona. Psyche (Camb.) 59:148-162. [1953.04.27]
Creighton, W. S. 1953d. New data on the habits of *Camponotus* (*Myrmaphaenus*) *ulcerosus* Wheeler. Psyche (Camb.) 60:82-84. [1953.09.28]
Creighton, W. S. 1953e. A new subspecies of *Xenomyrmex stolli* from northeastern Mexico (Hymenoptera, Formicidae). Am. Mus. Novit. 1634:1-5. [1953.12.14]
Creighton, W. S. 1953f. The rediscovery of *Leptothorax silvestrii* (Santschi) (Hymenoptera, Formicidae). Am. Mus. Novit. 1635:1-7. [1953.12.15]
Creighton, W. S. 1954. Additional studies on *Pseudomyrmex apache* (Hymenoptera: Formicidae). Psyche (Camb.) 61:9-15. [1954.06.30]
Creighton, W. S. 1955. Studies on the distribution of the genus *Novomessor* (Hymenoptera: Formicidae). Psyche (Camb.) 62:89-97. [1955.11.08]
Creighton, W. S. 1956. Notes on *Myrmecocystus lugubris* Wheeler and its synonym, *Myrmecocystus yuma* Wheeler (Hymenoptera, Formicidae). Am. Mus. Novit. 1807:1-4. [1956.12.07]
Creighton, W. S. 1957a ("1956"). Studies on the North American representatives of *Ephebomyrmex* (Hymenoptera: Formicidae). Psyche (Camb.) 63:54-66. [1957.01.23]
Creighton, W. S. 1957b ("1955"). Observations on *Pseudomyrmex elongata* Mayr (Hymenoptera: Formicidae). J. N. Y. Entomol. Soc. 63:17-20. [1957.03.08]
Creighton, W. S. 1957c. A study of the genus *Xenomyrmex* (Hymenoptera, Formicidae). Am. Mus. Novit. 1843:1-14. [1957.09.12]

Creighton, W. S. 1958 ("1957"). A revisionary study of *Pheidole vasliti* Pergande (Hymenoptera: Formicidae). J. N. Y. Entomol. Soc. 65:203-212. [1958.06.27]
Creighton, W. S. 1965a ("1964"). The habits of *Pheidole* (*Ceratopheidole*) *clydei* Gregg (Hymenoptera: Formicidae). Psyche (Camb.) 71:169-173. [1965.03.06]
Creighton, W. S. 1965b. Studies on southwestern ants belonging to *Camponotus*, subgenus *Myrmobrachys* (Hymenoptera, Formicidae). Am. Mus. Novit. 2239:1-9. [1965.12.17]
Creighton, W. S. 1966a ("1965"). The habits and distribution of *Macromischa subditiva* Wheeler (Hymenoptera: Formicidae). Psyche (Camb.) 72:282-286. [1966.05.10]
Creighton, W. S. 1966b. The habits of *Pheidole ridicula* Wheeler with remarks on habit patterns in the genus *Pheidole* (Hymenoptera: Formicidae). Psyche (Camb.) 73:1-7. [1966.08.31]
Creighton, W. S. 1969. Studies on *Camponotus* (*Myrmaphaenus*) *andrei* Forel. Am. Mus. Novit. 2393:1-6. [1969.10.10]
Creighton, W. S. 1971. New data on the distribution and habits of *Leptothorax* (*Nesomyrmex*) *wilda* (Hymenoptera: Formicidae). J. Ga. Entomol. Soc. 6:207-210. [1971.10]
Creighton, W. S., Crandall, R. H. 1954. New data on the habits of *Myrmecocystus melliger* Forel. Biol. Rev. (City Coll. N. Y.) 16:2-6. [1954.03]
Creighton, W. S., Gregg, R. E. 1954. Studies on the habits and distribution of *Cryptocerus texanus* Santschi (Hymenoptera: Formicidae). Psyche (Camb.) 61:41-57. [1954.08.26]
Creighton, W. S., Gregg, R. E. 1955. New and little-known species of *Pheidole* (Hymenoptera: Formicidae) from the southwestern United States and northern Mexico. Univ. Colo. Stud. Ser. Biol. 3:1-46. [1955.09]
Creighton, W. S., Nutting, W. L. 1965. The habits and distribution of *Cryptocerus rohweri* Wheeler (Hymenoptera, Formicidae). Psyche (Camb.) 72:59-64. [1965.06.25]
Creighton, W. S., Snelling, R. R. 1967 ("1966"). The rediscovery of *Camponotus* (*Myrmaphaenus*) *yogi* Wheeler (Hymenoptera: Formicidae). Psyche (Camb.) 73:187-195. [1967.01.25]
Creighton, W. S., Snelling, R. R. 1974. Notes on the behavior of three species of *Cardiocondyla* in the United States (Hymenoptera: Formicidae). J. N. Y. Entomol. Soc. 82:82-92. [1974.08.22]
Creighton, W. S., Tulloch, G. S. 1930. Notes on *Euponera gilva* (Roger) (Hymenoptera, Formicidae). Psyche (Camb.) 37:71-79. [1930.03] [Stamp date in MCZ library: 1930.06.10.]
Cresson, E. T. 1865a. Catalogue of Hymenoptera in the collection of the Entomological Society of Philadelphia, from Colorado Territory. [part] Proc. Entomol. Soc. Phila. 4:242-313. [1865.05] [Publication date from information supplied in F. M. Brown (1964:307).]
Cresson, E. T. 1865b. Catalogue of Hymenoptera in the collection of the Entomological Society of Philadelphia, from Colorado Territory. [concl.] Proc. Entomol. Soc. Phila. 4:426-488. [1865.08] [Publication date from information supplied in F. M. Brown (1964:307). Ants pp. 426-428.]
Cresson, E. T. 1872. Hymenoptera Texana. Trans. Am. Entomol. Soc. 4:153-292. [1872.11]
Cresson, E. T. 1887. Synopsis of the families and genera of the Hymenoptera of America, north of Mexico, together with a catalogue of the described species, and bibliography. Trans. Am. Entomol. Soc., Suppl. Vol. 1887:1-351. [1887]
Cresson, E. T. 1916. The Cresson types of Hymenoptera. Mem. Am. Entomol. Soc. 1:1-141. [1916.06.24]
Cresson, E. T. 1928. The types of Hymenoptera in the Academy of Natural Sciences of Philadelphia other than those of Ezra T. Cresson. Mem. Am. Entomol. Soc. 5:1-90. [1928.09.14]
Crewe, R. M. 1972. Alarm pheromones: their chemical composition and raison d'etre in myrmicine ants. Diss. Abstr. Int. B. Sci. Eng. 32:5846. [1972.04]
Crewe, R. M., Blum, M. S. 1970. Alarm pheromones in the genus *Myrmica* (Hymenoptera: Formicidae): their composition and species specificity. Z. Vgl. Physiol. 70:363-373. [1970.12.21]

Crewe, R. M., Blum, M. S. 1971. 6-methyl-5-hepten-2-one. Chemotaxonomic significance in an *Iridomyrmex* species (Hymenoptera: Formicidae). Ann. Entomol. Soc. Am. 64:1007-1010. [1971.09.15]

Crewe, R. M., Blum, M. S. 1972. Alarm pheromones of the Attini: their phylogenetic significance. J. Insect Physiol. 18:31-42. [1972.01]

Crewe, R. M., Blum, M. S., Collingwood, C. A. 1972. Comparative analysis of alarm pheromones in the ant genus *Crematogaster*. Comp. Biochem. Physiol. B. Comp. Biochem. 43:703-716. [1972.11.15]

Crewe, R. M., Brand, J. M., Fletcher, D. J. C., Eggers, S. H. 1970. The mandibular gland chemistry of some South African species of *Crematogaster* (Hymenoptera: Formicidae). J. Ga. Entomol. Soc. 5:42-47. [1970.01]

Crewe, R. M., Peeters, C. P., Villet, M. 1984. Frequency distribution of worker sizes in *Megaponera foetens* (Fabricius). S. Afr. J. Zool. 19:247-248. [1984.07]

Crosland, M. W. J. 1988a. Effect of a gregarine parasite on the color of *Myrmecia pilosula* (Hymenoptera: Formicidae). Ann. Entomol. Soc. Am. 81:481-484. [1988.05]

Crosland, M. W. J. 1988b. Inability to discriminate between related and unrelated larvae in the ant *Rhytidoponera confusa* (Hymenoptera: Formicidae). Ann. Entomol. Soc. Am. 81:844-850. [1988.09]

Crosland, M. W. J. 1989. Intraspecific aggression in the primitive ant genus *Myrmecia*. Insectes Soc. 36:161-172. [1989.09]

Crosland, M. W. J., Crozier, R. H. 1986. *Myrmecia pilosula*, an ant with only one pair of chromosomes. Science (Wash. D. C.) 231:1278. [1986.03.14]

Crosland, M. W. J., Crozier, R. H., Imai, H. T. 1988. Evidence for several sibling biological species centred on *Myrmecia pilosula* (F. Smith) (Hymenoptera: Formicidae). J. Aust. Entomol. Soc. 27:13-14. [1988.02.29]

Crosland, M. W. J., Crozier, R. H., Jefferson, E. 1988. Aspects of the biology of the primitive ant genus *Myrmecia* F. (Hymenoptera: Formicidae). J. Aust. Entomol. Soc. 27:305-309. [1988.11.29]

Crowson, R. A. 1965. Some thoughts concerning the insects of the Baltic amber. P. 133 in: Freeman, P. (ed.) XIIth International Congress of Entomology, London, 8-16 July, 1964. Proceedings. London: XIIth International Congress of Entomology, 842 pp. [1965]

Crozier, R. H. 1968a. An acetic acid dissociation, air drying technique for insect chromosomes, with aceto-lactic orcein staining. Stain Technol. 43:171-173. [1968.05]

Crozier, R. H. 1968b. Interpopulation karyotype differences in Australian *Iridomyrmex* of the "*detectus*" group (Hymenoptera: Formicidae: Dolichoderinae). J. Aust. Entomol. Soc. 7:25-27. [1968.06.30]

Crozier, R. H. 1968c. The chromosomes of three Australian dacetine ant species (Hymenoptera: Formicidae). Psyche (Camb.) 75:87-90. [1968.08.06]

Crozier, R. H. 1969a ("1968"). Cytotaxonomic studies on some Australian dolichoderine ants (Hymenoptera: Formicidae). Caryologia 21:241-259. [1969.01.10]

Crozier, R. H. 1969b. Chromosome number polymorphism in an Australian ponerine ant. Can. J. Genet. Cytol. 11:333-339. [1969.06]

Crozier, R. H. 1970a. Karyotypes of twenty-one ant species (Hymenoptera: Formicidae), with reviews of the known ant karyotypes. Can. J. Genet. Cytol. 12:109-128. [1970.03]

Crozier, R. H. 1970b. Pericentric rearrangement polymorphism in a North American dolichoderine ant (Hymenoptera: Formicidae). Can. J. Genet. Cytol. 12:541-546. [1970.09]

Crozier, R. H. 1970c. On the potential for genetic variability in haplo-diploidy. Genetica 41:551-556. [1970]

Crozier, R. H. 1973. Apparent differential selection at an isozyme locus between queens and workers of the ant *Aphaenogaster rudis*. Genetics 73:313-318. [1973.03.23]

Crozier, R. H. 1975. Animal cytogenetics. 3. Insecta. (7) Hymenoptera. Berlin: Gebrüder Borntraeger, 95 pp. [1975]

Crozier, R. H. 1976. Genetic boundaries in the ant *Aphaenogaster rudis*. Isozyme Bull. 9:58. [1976.02]

Crozier, R. H. 1977a. Genetic differentiation between populations of the ant *Aphaenogaster "rudis"* in the southeastern United States. Genetica 47:17-36. [1977.04.20]

Crozier, R. H. 1977b. Evolutionary genetics of the Hymenoptera. Annu. Rev. Entomol. 22:263-288. [1977]

Crozier, R. H. 1979. Genetics of sociality. Pp. 223-286 in: Hermann, H. R. (ed.) Social insects. Volume 1. New York: Academic Press, xv + 437 pp. [1979]

Crozier, R. H. 1981. Genetic aspects of ant evolution. Pp. 356-370 in: Atchley, W. R., Woodruff, D. C. (eds.) Essays in evolution and speciation in honor of M. J. D. White. Cambridge: Cambridge University Press, ix + 436 pp. [1981]

Crozier, R. H. 1983. Genetics and insect systematics: retrospect and prospect. Pp. 80-92 in: Highley, E., Taylor, R. W. (eds.) Australian systematic entomology: A bicentenary perspect. Melbourne: CSIRO, vii + 147 pp. [1983]

Crozier, R. H., Consul, P. C. 1976. Conditions for genetic polymorphism in social Hymenoptera under selection at the colony level. Theor. Popul. Biol. 10:1-9. [1976.08]

Crozier, R. H., Dobric, N., Imai, H. T., Graur, D., Cornuet, J.-M., Taylor, R. W. 1995. Mitochondrial-DNA sequence evidence on the phylogeny of Australian jack-jumper ants of the *Myrmecia pilosula* complex. Mol. Phylogenet. Evol. 4:20-30. [1995.03]

Crozier, R. H., Pamilo, P. 1986. Relatedness within and between colonies of a queenless ant species of the genus *Rhytidoponera* (Hymenoptera: Formicidae). Entomol. Gen. 11:113-117. [1986.05.30]

Crozier, R. H., Pamilo, P., Taylor, R. W., Crozier, Y. C. 1986. Evolutionary patterns in some putative Australian species in the ant genus *Rhytidoponera*. Aust. J. Zool. 34:535-560. [1986.09.02]

Cruz Landim, C. da. 1990. Cephalic exocrine glands of ants: a morphological view. Pp. 102-118 in: Vander Meer, R. K., Jaffe, K., Cedeno, A. (eds.) Applied myrmecology: a world perspective. Boulder: Westview Press, xv + 741 pp. [1990]

Cruz Landim, C. da, Caetano, F. H. 1981. The histochemistry and fine structure of the vitellarium in *Atta* (Formicidae, Myrmicinae). Rev. Bras. Biol. 41:363-370. [1981.05]

Culver, D. C. 1972. A niche analysis of Colorado ants. Ecology 53:126-131. [1972]

Culver, D. C. 1974. Species packing in Caribbean and North Temperate ant communities. Ecology 55:974-988. [1974]

Cumber, R. A. 1959. Distributional and biological notes on sixteen North Island species of Formicidae (Hymenoptera). N. Z. Entomol. 2(4):10-14. [1959.12]

Cumber, R. A. 1968. Notes on a colony of *Orectognathus antennatus* Fr. Smith (Hymenoptera: Myrmicidae) from the Coromandel peninsula. N. Z. Entomol. 4(1):43-44. [1968.03]

Curtis, J. 1829. *Myrmecina Latreillii*. Plate 265 [plus 2 unnumbered pages of text] in: Curtis, J. British entomology; being illustrations and descriptions of the genera of insects found in Great Britain and Ireland. Volume 6. London: published by the author, plates 242-289. [1829.06.01] [For publication details on the entire work see Blackwelder (1947, 1949a) and Evenhuis et al. (1989).]

Curtis, J. 1839. *Formica rufa*. The red, hill, or horse Ant, or Pismire. Plate 752 [plus 2 unnumbered pages of text] in: Curtis, J. British entomology; being illustrations and descriptions of the genera of insects found in Great Britain and Ireland. Volume 16. London: published by the author, plates 722-769. [1839.08.01]

Curtis, J. 1854. On the genus *Myrmica* and other indigenous ants. Trans. Linn. Soc. Lond. 21:211-220. [1854.11.21] [Publication date from Raphael (1970).]

Cushman, J. H., Lawton, J. H., Manly, B. F. J. 1993. Latitudinal patterns in European ant assemblages: variation in species richness and body size. Oecologia (Berl.) 95:30-37. [1993.08]

Czechowska, W. 1976. Myrmekofauna Pieninskiego Parku Narodowego (Hymenoptera, Formicoidea). Fragm. Faun. (Warsaw) 21:115-144. [1976.12.30]

Czechowski, W. 1991. Comparison of the myrmecofaunas (Hymenoptera, Formicoidea) of tree stands and lawns in Warsaw parks. Fragm. Faun. (Warsaw) 35:179-183. [1991.12.15]

Czechowski, W. 1993. Hybrids in red wood ants (Hymenoptera, Formicidae). Ann. Zool. (Warsaw) 44:43-54. [1993.02.15]

Czechowski, W. 1994. Publications of Professor Bohdan Pisarski. Memorabilia Zool. 48:293-296. [1994]

Czechowski, W., Czechowska, W., Palmowska, J. 1990. Arboreal myrmecofauna of Warsaw parks. Fragm. Faun. (Warsaw) 34:37-45. [1990.11.15]

Czechowski, W., Pisarski, B. 1990a. Ants (Hymenoptera, Formicoidea) of the Vistula escarpment in Warsaw. Fragm. Faun. (Warsaw) 33:109-128. [1990.05.31]

Czechowski, W., Pisarski, B. 1990b. Ants (Hymenoptera, Formicoidea) of linden-oak-hornbeam forests and thermophilous oak forests of the Mazovian Lowland. 1. Nest density. Fragm. Faun. (Warsaw) 34:133-141. [1990.11.15]

Czechowski, W., Pisarski, B., Czechowska, W. 1990. Ants (Hymenoptera, Formicoidea) of moist meadows on the Mazovian Lowland. Fragm. Faun. (Warsaw) 34:47-60. [1990.11.15]

Czechowski, W., Pisarski, B., Yamauchi, K. 1995. Succession of ant communities (Hymenoptera, Formicidae) moist pine forests. Fragm. Faun. (Warsaw) 38:447-488. [Not seen. Cited in BIOSIS.]

Daguerre, J. B. 1945. Hormigas del género *Atta* Fabricius de la Argentina (Hymenop. Formicidae). Rev. Soc. Entomol. Argent. 12:438-460. [1945.12.30]

Dahl, F. 1901. Das Leben der Ameisen im Bismarck-Archipel, nach eigenen Beobachtungen vergleichend dargestellt. Mitt. Zool. Mus. Berl. 2:1-70. [1901.04.03]

Dalla Torre, C. G. de See: Dalla Torre, K. W. von.

Dalla Torre, K. W. von. 1889. Die Hymenopteren von Helgoland. Wien. Entomol. Ztg. 8:46-48. [1889.02.28]

Dalla Torre, K. W. von. 1892. Hymenopterologische Notizen. Wien. Entomol. Ztg. 11:89-93. [1892.03.18]

Dalla Torre, K. W. von. 1893. Catalogus Hymenopterorum hucusque descriptorum systematicus et synonymicus. Vol. 7. Formicidae (Heterogyna). Leipzig: W. Engelmann, 289 pp. [1893]

Dalla Torre, K. W. von. 1908. Die Ameisen von Tirol und Voralberg. Entomol. Jahrb. 17:170-171. [1908]

Dalla Torre, K. W. von, Friese, H. 1898. Die hermaphroditen und gynandromorphen Hymenopteren. Ber. Naturwiss.-Med. Ver. Innsb. 24:1-96. [1898]

Dallas, W. S. 1867. Insecta. Pp. 268-576 in: Günther, A. (ed.) The record of zoological literature. 1866. Volume third. London: John van Voorst, x + 649 pp. [1867] [See Taschenberg (1866).]

Daloze, D., Braekman, J.-C., Vanhecke, P., Merlin, P., Pasteels, J., Boevé, J.-L., Francke, W. 1987. The defensive chemistry of ants of the genera *Tetraponera* and *Crematogaster*. [Abstract.] Pp. 421-422 in: Eder, J., Rembold, H. (eds.) Chemistry and biology of social insects. München: Verlag J. Peperny, xxxv + 757 pp. [1987]

Datta, S. K., Raychaudhuri, D. 1983. Taxonomy of the aphidocolous ants (Hymenoptera: Formicidae) of Sikkim and hilly areas of West Bengal (1) Subfamily: Myrmicinae. Akitu (n.s.)56:1-14. [1983.12.05]

Datta, S. K., Raychaudhuri, D. 1985. A new species of ant (Hymenoptera: Formicidae) from Nagaland, north-east India. Sci. Cult. 51:271-273. [1985.08]

Datta, S. K., Raychaudhuri, D., Agarwala, B. K. 1982. Study on aphid tending ants in India. I. New records of aphid and ant species in their association. Entomon 7:327-328. [1982.09]

Datta, S. K., Raychaudhuri, D., Agarwala, B. K. 1983. Study on aphid tending ants in India - II. New records of aphid and ant species in their association. Entomon 8:23-25. [1983.03]

Davidson, D. W. 1977a. Species diversity and community organization in desert seed-eating ants. Ecology 58:711-724. [1977]

Davidson, D. W. 1977b. Foraging ecology and community organization in desert seed-eating ants. Ecology 58:725-737. [1977]

Davidson, D. W. 1978. Size variability in the worker caste of a social insect (*Veromessor pergandei* Mayr) as a function of the competitive environment. Am. Nat. 112:523-532. [1978.06]

Davidson, D. W., Foster, R. B., Snelling, R. R., Lozada, P. W. 1991. Variable composition of some tropical ant-plant symbioses. Pp. 145-162 in: Price, P. W., Lewinsohn, T. M., Fernandes, G. W., Benson, W. W. (eds.) Plant-animal interactions. Evolutionary ecology in tropical and temperate regions. New York: J. Wiley, xiv + 639 pp. [1991]

Davidson, D. W., McKey, D. 1993a. The evolutionary ecology of symbiotic ant-plant relationships. J. Hym. Res. 2:13-83. [1993.09.30]

Davidson, D. W., McKey, D. 1993b. Ant-plant symbioses: stalking the Chuyachaqui. Trends Ecol. Evol. 8:326-332. [1993.09]

Davies, N. B., Bourke, A. F. G., Brooke, M. de L. 1989. Cuckoos and parasitic ants: interspecific brood parasitism as an evolutionary arms race. Trends Ecol. Evol. 4:274-278. [1989.09]

Davies, S. J., Villet, M. H., Blomefield, T. M., Crewe, R. M. 1994. Reproduction and division of labour in *Leptogenys schwabi* Forel (Hymenoptera Formicidae), a polygynous, queenless ponerine ant. Ethol. Ecol. Evol. 6:507-517. [1994.12]

Davis, W. T. 1914. The fungus-growing ant on Long Island, New York. J. N. Y. Entomol. Soc. 22:64-65. [1914.03]

Davis, W. T., Bequaert, J. 1922. An annotated list of the ants of Staten Island and Long Island, New York. Bull. Brooklyn Entomol. Soc. 17:1-25. [1922.06.15]

Dazzini Valcurone, M., Fanfani, A. 1982. Nouve formazioni glandolari del gastro in *Dolichoderus* (*Hypoclinea*) *doriae* Em. (Formicidae, Dolichoderinae). Pubbl. Ist. Entomol. Agrar. Univ. Pavia 19:1-18. [1982]

Dazzini Valcurone, M., Fanfani, A. 1985. Investigations on Formicidae: Pavan's gland and other glands of the gaster in Dolichoderinae. Pubbl. Ist. Entomol. Agrar. Univ. Pavia 31:1-20. [1985]

De Andrade, M. L. 1992. First fossil "true *Macromischa*" in amber from the Dominican Republic (Hymenoptera, Formicidae). Mitt. Schweiz. Entomol. Ges. 65:341-351. [1992.12]

De Andrade, M. L. 1993. [Untitled. *Perissomyrmex monticola* de Andrade n. sp.] Pp. 90-93 in: Baroni Urbani, C., De Andrade, M. L. *Perissomyrmex monticola* n. sp., from Bhutan: the first natural record for a presumed Neotropical genus with a discussion on its taxonomic status. Trop. Zool. 6:89-95. [1993.05]

De Andrade, M. L. 1994a. [Untitled. *Strumigenys electrina* de Andrade n. sp.] Pp. 38-41 in: Baroni Urbani, C., De Andrade, M. L. First description of fossil Dacetini ants with a critical analysis of the current classification of the tribe (Amber Collection Stuttgart: Hymenoptera, Formicidae. VI: Dacetini). Stuttg. Beitr. Naturkd. Ser. B (Geol. Paläontol.) 198:1-65. [1994.04.20]

De Andrade, M. L. 1994b. [Untitled. Descriptions of new taxa: *Rhopalothrix inopinata* de Andrade n. sp.; *Strumigenys nepalensis* de Andrade n. sp.; *Strumigenys assamensis* de Andrade n. sp.] Pp. 54-64 in: Baroni Urbani, C., De Andrade, M. L. First description of fossil Dacetini ants with a critical analysis of the current classification of the tribe (Amber Collection Stuttgart: Hymenoptera, Formicidae. VI: Dacetini). Stuttg. Beitr. Naturkd. Ser. B (Geol. Paläontol.) 198:1-65. [1994.04.20]

De Andrade, M. L. 1994c. Fossil Odontomachiti ants from the Dominican Republic (Amber Collection Stuttgart: Hymenoptera, Formicidae. VII: Odontomachiti). Stuttg. Beitr. Naturkd. Ser. B (Geol. Paläontol.) 199:1-28. [1994.04.29]

De Andrade, M. L. 1995. The ant genus *Aphaenogaster* in Dominican and Mexican amber (Amber Collection Stuttgart: Hymenoptera, Formicidae. IX: Pheidolini). Stuttg. Beitr. Naturkd. Ser. B (Geol. Paläontol.) 223:1-11. [1995.03.20]

Debaisieux, P. 1934. Les organes scolopidiaux des insectes; l'organe subgénual des fourmis. Ann. Soc. Sci. Brux. Sér. B Sci. Phys. Nat. 54:338-345. [1934.12.17]

Debouge, M.H., Caspar, C. 1983. Contribution à la faunistique des fourmis de la Corse (Hymenoptera, Formicidae). Bull. Ann. Soc. R. Entomol. Belg. 119:202-221. [1983.11.30]

De Geer, C. 1773. Mémoires pour servir à l'histoire des insectes. Tome troisième. Stockholm: Pierre Hesselberg, 696 pp. [1773]

De Geer, C. 1778. Mémoires pour servir à l'histoire des insectes. Tome septième. Stockholm: Pierre Hesselberg, 950 pp. [1778]

De Haro, A. 1971 ("1970"). Los formícidos, grupo de gran interés zoológico (Hym. Formicidae). Graellsia 26:59-98. [1971.03.15]

De Haro, A. 1974 ("1973"). Formícidos (Hymenoptera, Formicidae) del Valle de Batuecas y zona occidental de la Cordillera Central (Salamanca). [Abstract.] Bol. R. Soc. Esp. Hist. Nat. Secc. Biol. 71:372. [1974.10.15]

De Haro, A. 1976 ("1974"). Formícidos del Valle de las Batuecas y parte occidental de la Cordillera Central (Salamanca). Bol. R. Soc. Esp. Hist. Nat. Secc. Biol. 72:229-235. [1976.06.30]

De Haro, A. 1981. Particularitats de la mirmecofauna del Cap de Gata (Almería). Butll. Inst. Catalana Hist. Nat. 47:139-142. [1981]

De Haro, A., Collingwood, C. A. 1977. Prospección mirmecológica por Andalucia. Bol. Estac. Cent. Ecol. 6(12):85-90. [1977]

De Haro, A., Collingwood, C. A. 1981. Formícidos de las Sierras de Prades-Montsant, Sierras de Cavalls-Alfara-Montes Blancos (Tarragona). Bol. Estac. Cent. Ecol. 10:55-58. [1981]

De Haro, A., Collingwood, C. A. 1988. Prospección mirmecológica por las sierras de Aitana-Alfaro y los cabos de la Nao-San Antonio (Alicante) y su comparación con la fauna balear y de Córcega-Cerdeña. Orsis (Org. Sist.) 3:165-172. [1988]

De Haro, A., Collingwood, C. A. 1991. Prospección mirmecológica en la Cordillera Ibérica. Orsis (Org. Sist.) 6:109-126. [1991]

De Haro, A., Collingwood, C. A. 1992. Prospección mirmecológica por Extremadura (España) y Sao Brás-Almodovar, Alcácer do Sal, Serra da Estrela (Portugal). Bol. Soc. Port. Entomol. Supl. 3(1):95-104. [1992]

De Haro, A., Collingwood, C. A. 1994. Prospección mirmecológica por el littoral mediterráneo de Marruecos (Cabo Negro, Martil, Oued Lau) y su comparación con la zona meridional ibérica. Orsis (Org. Sist.) 9:97-104. [1994]

De Haro, A., Collingwood, C. A., Comín, P. 1986. Prospección mirmecológica por Ibiza y Formentera (Baleares). Orsis (Org. Sist.) 2:115-120. [1986]

De Haro, A., Collingwood, C. A., Douwes, P. 1995. Nota preliminar sobre sistemática molecular gen-aloenzimática de algunas formas españolas y marroquíes del grupo albicans del género *Cataglyphis* (Hym., Formicidae). Orsis (Org. Sist.) 10:73-81. [1995]

De Haro Vera, A. See: De Haro, A.

Deichmüller, J. V. 1881. Fossile Insecten aus dem Diatomeenschiefer von Kutschlin bei Bilin, Böhmen. Nova Acta Acad. Caesareae Leopold.-Carol. Ger. Nat. Curiosorum 42:293-331. [1881]

Dejean, A. 1985a. Microévolution du comportement de captures des proies chez les dacetines de la sous-tribu des Strumigeniti (Hymenoptera, Formicidae, Myrmicinae). Actes Colloq. Insectes Soc. 2:239-247. [1985.06]

Dejean, A. 1985b. Étude éco-éthologique de la prédation chez les fourmis du genre *Smithistruma* (Formicidae - Myrmicinae - Dacetini). II. Attraction des proies principales (collemboles). Insectes Soc. 32:158-172. [1985.10]

Dejean, A. 1987 ("1986"). Étude du comportement de prédation dans le genre *Strumigenys* (Formicidae - Myrmicinae). Insectes Soc. 33:388-405. [1987.06]

Dejean, A. 1988 ("1987"). New cases of archaic foundation of societies in Myrmicinae (Formicidae): study of prey capture by queens of Dacetini. Insectes Soc. 34:211-221. [1988.06]

Dejean, A., Akoa, A., Djieto Lordon, C., Lenoir, A. 1994. Mosaic ant territories in an African secondary rain forest (Hymenoptera: Formicidae). Sociobiology 23:275-292. [1994]

De Kock, A. E., Giliomee, J. H. 1989. A survey of the Argentine ant, *Iridomyrmex humilis* (Mayr), (Hymenoptera: Formicidae) in South African fynbos. J. Entomol. Soc. South. Afr. 52:157-164. [1989.03]

Delabie, J. H. C. 1994. Primeiro registro de *Tetramorium lucayanum* Wheeler na América continental (Hymenoptera: Formicidae).] An. Soc. Entomol. Bras. 23:141-142. [1994.04]

Delabie, J. H. C. 1995. Formigas associadas aos nectários extraflorais de *Epidendrum cinnabarum* Salzm. (Orchidaceae) numa área de restinga na Bahia. An. Soc. Entomol. Bras. 24:479-487. [1995.12]

Delabie, J. H. C., do Nascimento, I. C., Cazorla, I. M., Casimiro, A. B. 1994. Application of a biogeographic model to evaluate the ant diversity in different unities of territory in the South-American tropics. P. 336 in: Lenoir, A., Arnold, G., Lepage, M. (eds.) Les insectes sociaux. 12ème congrès de l'Union Internationale pour l'Étude des Insectes Sociaux. Paris, Sorbonne, 21-27 août 1994. Paris: Université Paris Nord, xxiv + 583 pp. [1994.09]

Delabie, J. H. C., do Nascimento, I. C., Pacheco, P., Casimiro, A. B. 1995. Community structure of house-infesting ants (Hymenoptera: Formicidae) in southern Bahia, Brazil. Fla. Entomol. 78:264-270. [1995.06.23]

Delage, B. 1966. Sur une fonction particulière des glandes pharyngiennes des fourmis. C. R. Séances Acad. Sci. Ser. D. Sci. Nat. 263:1743-1744. [1966.12.05]

Delage-Darchen, B. 1971. Contribution à l'étude écologique d'une savane de Côte d'Ivoire (Lamto). Les fourmis des strates herbacée et arborée. Biol. Gabon. 7:461-496. [1971.12]

Delage-Darchen, B. 1972a. Une fourmi de Côte d'Ivoire: *Melissotarsus titubans* Del., n. sp. Insectes Soc. 19:213-226. [1972.12]

Delage-Darchen, B. 1972b. Le polymorphisme larvaire chez les fourmis *Nematocrema* d'Afrique. Insectes Soc. 19:259-277. [1972.12]

Delage-Darchen, B. 1973. Evolution de l'aile chez les fourmis *Crematogaster* (Myrmicinae) d'Afrique. Insectes Soc. 20:221-242. [1973.12]

Delage-Darchen, B. 1976. Les glandes post-pharyngiennes des fourmis: connaissances actuelles sur leur structure, leur fonctionnement, leur rôle. Année Biol. (4)15:63-76. [1976.02]

Delage-Darchen, B. 1978. Les stades larvaires de *Crematogaster* (*Sphaerocrema*) *striatula*, fourmi forestière d'Afrique [Hym. Formicidae]. Ann. Soc. Entomol. Fr. (n.s.)14:293-299. [1978.12.05]

Della Lucia, T. M. C. (ed.) 1993. As formigas cortadeiras. Viçosa: Editora Folha de Viçosa, 262 pp. [1993.11.21]

Della Lucia, T. M. C., Fowler, H. G., Moreira, D. D. O. 1993. Espécies de formigas cortadeiras no Brasil. Pp. 26-31 in: Della Lucia, T. M. C. (ed.) As formigas cortadeiras. Viçosa: Editora Folha de Viçosa, 262 pp. [1993.11.21]

Della Lucia, T. M. C., Loureiro, M. C., Chandler, L., Freire, J. A. H., Galvão, J. D., Fernandes, B. 1982. Ordenação de comunidades de Formicidae em quatro agroecossistemas em Viçosa, Minas Gerais. Experientiae (Viçosa) 28:67-94. [1982.06]

Della Santa, E. 1988a. *Stenamma petiolatum* Emery (Hymenoptera: Formicidae) en Suisse. Mitt. Schweiz. Entomol. Ges. 61:361-364. [1988.12.15]

Della Santa, E. 1988b. Observation d'une anomalie de la morphologie chez une de *Stenamma westwoodi* Emery (Formicidae). Bull. Romand Entomol. 6:101-103. [1988.12]

Della Santa, E. 1994a ("1993"). Une petite moissonneuse provençale aux yeux en forme de virgule: la fourmi *Goniomma blanci* (André, 1881) (Hymenoptera - Formicidae). Bull. Romand Entomol. 12:61-66. [1994.11]

Della Santa, E. 1994b. Guide pour l'identification des principales espèces de fourmis de Suisse. Misc. Faun. Helv. 3:1-124.

Délye, G. 1960. Fourmis du Tassili des Ajjer. Bull. Soc. Hist. Nat. Afr. Nord 51:259-272. [1960]

Délye, G. 1961. *Monomorium* (*Equesimessor*) *chobauti* Em. (Hyménoptères formicidae) à Beni-Abbès (Saoura). Nid. sexués (=*Holcomyrmex Faf* Forel). Bull. Soc. Hist. Nat. Afr. Nord 52:67-72. [1961]

Délye, G. 1964a ("1962"). *Cataglyphis* (*Paraformica*) *emmae* Forel (Hyménoptères Formicidae). Sexués et "soldats". Bull. Soc. Hist. Nat. Afr. Nord 53:21-27. [1964]

Délye, G. 1964b. Sur le peuplement myrmécologique de quelques ergs du Sahara nord-occidental. Trav. Inst. Rech. Sahar. 23:165-170. [1964]

Délye, G. 1965a. *Cataglyphis* (*Paraformica*) emmae Forel sexués et "soldats". Bull. Soc. Entomol. Fr. 70:52-56. [1965.07.03]

Délye, G. 1965b. Anatomie et fonctionnement des stigmates de quelques fourmis (Hyménoptères Formicidae). Insectes Soc. 12:285-290. [1965.12]

Délye, G. 1970 ("1969"). Répartition des fourmis dans les grands massifs de dunes du Sahara nord-occidental (Hym.). Bull. Soc. Entomol. Fr. 74:224-227. [1970.05.30]

Délye, G. 1971. *Oxyopomyrmex emeryi* Santschi (Hym. Formicidae) dans le Grand Erg Occidental. Description des sexués. Nouv. Rev. Entomol. 1:211-214. [1971.06.30]

Dennis, C. A. 1938. The distribution of ant species in Tennessee with reference to ecological factors. Ann. Entomol. Soc. Am. 31:267-308. [1938.06.29]

De Santis, L. 1940. Las principales hormigas dañinas de la provincia de Buenos Aires. I. La hormiga invasora. Bol. Agric. Ganad. Ind. Prov. B. Aires 20(5/6):5-16. [1940.06]

Dethier, M., Cherix, D. 1982. Note sur les Formicidae du Parc national suisse. Mitt. Schweiz. Entomol. Ges. 55:125-138. [1982.07.31]

Detrain, C., Pasteels, J. M. 1987. Morphological and biochemical differences in the abdominal glands of *Pheidole pallidula* (Myrmicinae). Pp. 447-448 in: Eder, J., Rembold, H. (eds.). Chemistry and biology of social insects. München: Verlag J. Peperny, xxxv + 757 pp. [1987]

Devi, C. M., Singh, T. K. 1987. Aphidocolous ants (Hymenoptera: Formicidae) in Manipur. Entomon 12:309-313. [1987.12]

Dewitz, H. 1877. Ueber Bau und Entwickelung des Stachels der Ameisen. Z. Wiss. Zool. 28:527-556. [1877.04.23]

Deyrup, M. 1988. First record of *Epitritus* from North America (Hymenoptera: Formicidae). Fla. Entomol. 71:217-218. [1988.06.20]

Deyrup, M. 1991. *Technomyrmex albipes*, a new exotic ant in Florida (Hymenoptera: Formicidae). Fla. Entomol. 74:147-148. [1991.03.29]

Deyrup, M., Carlin, N., Trager, J., Umphrey, G. 1988. A review of the ants of the Florida Keys. Fla. Entomol. 71:163-176. [1988.06.20]

Deyrup, M., Johnson, C., Wheeler, G. C., Wheeler, J. 1989. A preliminary list of the ants of Florida. Fla. Entomol. 72:91-101. [1989.03.31]

Deyrup, M., Trager, J. 1985 ("1984"). *Strumigenys rogeri*, an African dacetine ant new to the U.S. (Hymenoptera: Formicidae). Fla. Entomol. 67:512-516. [1985.01.11]

Deyrup, M., Trager, J. 1986. Ants of the Archbold Biological Station, Highlands County, Florida (Hymenoptera: Formicidae). Fla. Entomol. 69:206-228. [1986.05.23]

Deyrup, M., Trager, J., Carlin, N. 1985. The genus *Odontomachus* in the southeastern United States (Hymenoptera: Formicidae). Entomol. News 96:188-195. [1985.12.01]

Deyrup, M. A. See: Deyrup, M.

De Zolessi, L. C. See: Zolessi, L. C. de

Diehl-Fleig, E., de Araújo, A. M., Cavalli-Molina, S. 1994. Genetic and social structure of the leaf-cutting ants *Acromyrmex heyeri* and *A. striatus* (Hymenoptera: Formicidae). P. 410 in: Lenoir, A., Arnold, G., Lepage, M. (eds.) Les insectes sociaux. 12ème congrès de l'Union Internationale pour l'Étude des Insectes Sociaux. Paris, Sorbonne, 21-27 août 1994. Paris: Université Paris Nord, xxiv + 583 pp. [1994.09]

Dietz, B. H., Brandão, C. R. F. 1993. Comportamento de caça e dieta de *Acanthognathus rudis* Brown & Kempf, com comentários sobre a evolução da predação em Dacetini (Hymenoptera, Formicidae, Myrmicinae). Rev. Bras. Entomol. 37:683-692. [1993.12.31]

Diffie, S., Meer, R. K. V., Bass, M. H. 1988. Discovery of hybrid fire ant populations in Georgia and Alabama. J. Entomol. Sci. 23:187-191. [1988.04.28]

Dill, M., Mashchwitz, U. 1994. The migrating herdsmen ants of southeast Asia: a species-rich complex of trophobiotic symbioses of *Dolichoderus* ants and their mealybugs (Homoptera: Pseudococcidae: Allomyrmococcini). P. 199 in: Lenoir, A., Arnold, G., Lepage, M. (eds.) Les insectes sociaux. 12ème congrès de l'Union Internationale pour l'Étude des Insectes Sociaux. Paris, Sorbonne, 21-27 août 1994. Paris: Université Paris Nord, xxiv + 583 pp. [1994.09]

Diller, E. 1990. Die von Spix und Martius 1817-1820 in Brasilien gesammelten und von J. A. M. Perty 1833 bearbeiteten Hymenopteren in der Zoologischen Staatssammlung München (Insecta, Hymenoptera). Spixiana 13:61-81. [1990.03.31] [Page 69: designation of lectotype of *Formica cuneata* Perty, 1833.]

Diniz, J. L. M. 1975. *Leptothorax* (*Nesomyrmex*) *mirassolis*, nova espécie de formiga da região norte-ocidental do estado de São Paulo, Brasil (Hymenoptera, Formicidae). Rev. Bras. Entomol. 19:79-83. [1975.06.30]

Diniz, J. L. M. 1990. Revisão sistemática da tribo Stegomyrmicini, com a descripção de uma nova espécie (Hymenoptera, Formicidae). Rev. Bras. Entomol. 34:277-295. [1990.09.30]

Diniz, J. L. M., Brandão, C. R. F. 1993. Biology and myriapod predation by the Neotropical myrmicine ant *Stegomyrmex vizottoi* (Hymenoptera: Formicidae). Insectes Soc. 40:301-311. [1993]

Diniz-Filho, J. A. F., Von Zuben, C. J., Fowler, H. G., Schlindwein, M. N., Bueno, O. C. 1994. Multivariate morphometrics and allometry in a polymorphic ant. Insectes Soc. 41:153-163. [1994]

Diver, C. 1936. A new genus of ants in Britain. Nature (Lond.) 137:458. [1936.03.14]

Diver, C. 1940. The problem of closely related species living in the same area. Pp. 303-328 in: Huxley, J. S. (ed.) The new systematics. Oxford: Clarendon Press, viii + 583 pp. [1940] [Taxonomy of *Myrmica* and *Lasius* discussed (pp. 317-320).]

Dixey, F. A., Longstaff, G. B. 1907. Entomological observations and captures during the visit of the British Association to South Africa in 1905. Trans. Entomol. Soc. Lond. 1907:309-381. [1907.09.26]

Dixey, L. R., Gardiner, P. C. 1934. Heterogony in *Messor barbarus* L. var. *capitatus* Latreille. Ann. Mag. Nat. Hist. (10)13:619-627. [1934.06]

Dlusskiy, G. M. See: Dlussky, G. M.

Dlussky, G. M. 1962. Ants of the northern slopes of the Talasskii Alatau range. [In Russian.] Tr. Inst. Zool. Akad. Nauk Kaz. SSR 18:177-188. [> 1962.04.14] [Date signed for printing ("podpisano k pechati").]

Dlussky, G. M. 1963a. Two new species of ants (Hymenoptera, Formicidae) from eastern Transbaikalia. [In Russian.] Entomol. Obozr. 42:190-194. [> 1963.03.15] [Date signed for printing ("podpisano k pechati"). Translated in Entomol. Rev. (Wash.) 42:104-106. See Dlussky (1963b).]

Dlussky, G. M. 1963b. Two new species of ants (Hymenoptera, Formicidae) from eastern Transbaikalia. Entomol. Rev. (Wash.) 42:104-106. [1963] [English translation of Dlussky (1963a).]

Dlussky, G. M. 1964. The ants of the subgenus *Coptoformica* of the genus Formica (Hymenoptera, Formicidae) of the USSR. [In Russian.] Zool. Zh. 43:1026-1040. [1964.07] [July issue. Date signed for printing ("podpisano k pechati"): 26 June 1964.]

Dlussky, G. M. 1965a. Ants of the genus *Formica* L. of Mongolia and northeast Tibet (Hymenoptera, Formicidae). Ann. Zool. (Warsaw) 23:15-43. [1965.05.20]

Dlussky, G. M. 1965b. Entomological research in Kirghizia. Three new species of ants from Kirghizia. [In Russian.] Entomol. Issled. Kirg. 4:27-33. [> 1965.06.30] [Date signed for printing ("podpisano k pechati").]

Dlussky, G. M. 1967a. Ants of the genus *Formica* (Hymenoptera, Formicidae, g. *Formica*). [In Russian.] Moskva: Nauka Publishing House, 236 pp. [> 1967.03.14] [Date signed for printing ("podpisano k pechati").]

Dlussky, G. M. 1967b. Ants of the genus *Formica* from the Baltic amber. [In Russian.] Paleontol. Zh. 1967(2):80-89. [> 1967.05.19] [Date signed for printing ("podpisano k pechati"). Translated in Paleontol. J. 1(2):69-77. See Dlussky (1967c).]

Dlussky, G. M. 1967c. Ants of the genus *Formica* from the Baltic Amber. Paleontol. J. 1(2):69-77. [1967] [English translation of Dlussky (1967b).]

Dlussky, G. M. 1969a. Ants of the genus *Proformica* Ruzs. of the USSR and contiguous countries (Hymenoptera, Formicidae). [In Russian.] Zool. Zh. 48:218-232. [1969.02] [February issue. Date signed for printing ("podpisano k pechati"): 31 January 1969.]

Dlussky, G. M. 1969b. First finding of an ant from the subfamily Leptanillinae (Hymenoptera, Formicidae) in the USSR. [In Russian.] Zool. Zh. 48:1666-1671. [1969.11] [November issue. Date signed for printing ("podpisano k pechati"): 24 October 1969.]

Dlussky, G. M. 1972. The evolution of ant nest construction. Pp. 359-360 in: Rafes, P. M. (ed.) XIII International Congress of Entomology, Moscow, 2-9 August, 1968. Proceedings, Volume III. Leningrad: Nauka, 494 pp. [> 1972.06.08] [Date signed for printing ("podpisano k pechati").]

Dlussky, G. M. 1974. Ants of the bed-takyr complex of the central Karakum. [In Russian.] Nauchn. Dokl. Vyssh. Shk. Biol. Nauki 1974(9):19-23. [> 1974.09.04] [Date signed for printing ("podpisano k pechati"). September 1974 issue.]

Dlussky, G. M. 1975a. Superfamily Formicoidea Latreille, 1802. Family Formicidae Latreille, 1802. [In Russian.] Pp. 114-122 in: Rasnitsyn, A. P. (ed.) Hymenoptera Apocrita of Mesozoic. [In Russian.] Tr. Paleontol. Inst. Akad. Nauk USSR 147:1-134. [> 1975.01.07] [Date signed for printing ("podpisano k pechati").]

Dlussky, G. M. 1975b. The ants of saxaul forests from Murgab delta. [In Russian.] Pp. 159-185 in: Mawaer, B. M., Pravdin, F. N. (eds.) Insects as components of biogeocenosis of the saxaul forest. [In Russian.] Moskva: Nauka, 221 pp. [> 1975.09.05] [Date signed for printing ("podpisano k pechati").]

Dlussky, G. M. 1981a. Miocene ants (Hymenoptera, Formicidae) of the USSR. [In Russian.] Pp. 64-83 in: Vishnyakova, V. N., Dlussky, G. M., Pritykina, L. N. New fossil insects from the territories of the USSR. [In Russian.] Tr. Paleontol. Inst. Akad. Nauk SSSR 183:1-87. [> 1981.09.03] [Date signed for printing ("podpisano k pechati").]

Dlussky, G. M. 1981b. Desert ants. [In Russian.] Moskva: Nauka, 230 pp. [> 1981.10.21] [Date signed for printing ("podpisano k pechati").]

Dlussky, G. M. 1983. A new family of Upper Cretaceous Hymenoptera: an "intermediate link" between the ants and the scolioids. [In Russian.] Paleontol. Zh. 1983(3):65-78. [> 1983.08.29] [Date signed for printing ("podpisano k pechati"). Translated in Paleontol. J. 17(3):63-76. See Dlussky (1984).]

Dlussky, G. M. 1984. A new family of Upper Cretaceous Hymenoptera: an "intermediate link" between the ants and the scolioids. Paleontol. J. 17(3):63-76. [1984] [English translation of Dlussky (1983).]

Dlussky, G. M. 1987. New Formicoidea (Hymenoptera) of the Upper Cretaceous. [In Russian.] Paleontol. Zh. 1987(1):131-135. [> 1987.02.04] [Date signed for printing ("podpisano k pechati"). Translated in Paleontol. J. 21(1):146-150. See Dlussky (1988b).]

Dlussky, G. M. 1988a. Ants of Sakhalin amber (Paleocene?). [In Russian.] Paleontol. Zh. 1988(1):50-61. [> 1988.02.09] [Date signed for printing ("podpisano k pechati"). Translated in Paleontol. J. 22(1-3):50-61. See Dlussky (1988c).]

Dlussky, G. M. 1988b. New Formicoidea (Hymenoptera) of the Upper Cretaceous. Paleontol. J. 21(1):146-150. [1988] [English translation of Dlussky (1987).]

Dlussky, G. M. 1988c. Ants from (Paleocene?) Sakhalin amber. Paleontol. J. 22(1):50-61. [1988] [English translation of Dlussky (1988a).]

Dlussky, G. M. 1993a. Ants (Hymenoptera: Formicidae) of Fiji, Tonga, and Samoa, and the problem of island faunas formation. 1. Statement of the problem. [In Russian.] Zool. Zh. 72(5):66-76. [1993.05] [May issue. Date signed for printing ("podpisano k pechati"): 6 April 1993. Translated in Entomol. Rev. (Wash.) 74(1):68-78. See Dlussky (1995).]

Dlussky, G. M. 1993b. Ants (Hymenoptera, Formicidae) of Fiji, Tonga, and Samoa, and the problem of island faunas formation. 2. Tribe Dacetini. [In Russian.] Zool. Zh. 72(6):52-65. [1993.06] [June issue. Date signed for printing ("podpisano k pechati"): 19 May 1993. Translated in Entomol. Rev. (Wash.) 73(1):110-122. See Dlussky (1994).]

Dlussky, G. M. 1994a. Zoogeography of southwestern Oceania. [In Russian.] Pp. 48-93 in: Puzatchenko, Y. G., Golovatch, S. I., Dlussky, G. M., Diakonov, K. N., Zakharov, A. A., Korganova, G. A. Animal populations of the islands of southwestern Oceania (ecogeographic studies). [In Russian.] Moskva: Nauka, 254 pp. [> 1993.12.07] [Date signed for printing ("podpisano k pechati"). Copyright date 1994. English summary on page 248.]

Dlussky, G. M. 1994b. Ants (Hymenoptera, Formicidae) of Fiji, Tonga, and Samoa and the problem of formation of island fauna. 2. Tribe Dacetini. Entomol. Rev. (Wash.) 73(1):110-122. [1994.07] [English translation of Dlussky (1993b).]

Dlussky, G. M. 1995. Ants (Hymenoptera, Formicidae) of Fiji, Tonga, and Samoa and the problem of formation of island fauna. 1. Statement of the problem. Entomol. Rev. (Wash.) 74(1):68-78. [1995.05] [English translation of Dlussky (1993a).]

Dlussky, G. M., Fedoseeva, E. B. 1988. Origin and early stages of evolution in ants. [In Russian.] Pp. 70-144 in: Ponomarenko, A. G. (ed.) Cretaceous biocenotic crisis and insect evolution. [In Russian.] Moskva: Nauka, 232 pp. [> 1987.12.22] [Date signed for printing ("podpisano k pechati"). Title page dated "1988".]

Dlussky, G. M., Pisarski, B. 1970. Formicidae aus der Mongolei. Ergebnisse der Mongolisch-Deutschen Biologischen Expeditionen seit 1962, Nr. 46. Mitt. Zool. Mus. Berl. 46:85-90. [1970.04.24]

Dlussky, G. M., Pisarski, B. 1971. Rewizja polskich gatunków mrówek (Hymenoptera: Formicidae) z rodzaju *Formica* L. Fragm. Faun. (Warsaw) 16:145-224. [1971.03.25]

Dlussky, G. M., Radchenko, A. G. 1990. The ants of Vietnam. S/f Pseudomyrmicinae. S/f Myrmicinae. Tribes Calyptomyrmecini, Meranoplini and Cataulacini. [In Russian.] Pp. 119-125 in: Akimov, I. A., Emelianov, I. G., Zerova, M. D. et al. (eds.) News of faunistics and systematics. [In Russian.] Kiev: Naukova Dumka, 184 pp. [> 1990.11.06] [Not seen. Citation from Radchenko (pers. comm.). Date signed for printing ("podpisano k pechati"): 6 November 1990.]

Dlussky, G. M., Radchenko, A. G. 1994a. Ants of the genus *Diplorhoptrum* (Hymenoptera, Formicidae) from the central Palearctic. [In Russian.] Zool. Zh. 73(2):102-111. [1994.02] [February issue. Date signed for printing ("podpisano k pechati"): 11 January 1994. Translated in Entomol. Rev. (Wash.) 73(5):156-167. See Dlussky & Radchenko (1994b).]

Dlussky, G. M., Radchenko, A. G. 1994b. Ants of the genus *Diplorhoptrum* (Hymenoptera, Formicidae) from the central Palearctic region. Entomol. Rev. (Wash.) 73(5):156-167. [1994.10] [English translation of Dlussky & Radchenko (1994a).]

Dlussky, G. M., Soyunov, O. S. 1988. Ants of the genus *Temnothorax* Mayr (Hymenoptera: Formicidae) of the USSR. [In Russian.] Izv. Akad. Nauk Turkm. SSR Ser. Biol. Nauk 1988(4):29-37. [> 1988.08.26] [Date signed for printing ("podpisano k pechati").]

Dlussky, G. M., Soyunov, O. S., Zabelin, S. I. 1990 ("1989"). Ants of Turkmenistan. [In Russian.] Ashkhabad: Ylym Press, 273 pp. [> 1990.06.07] [Date signed for printing ("podpisano k pechati").]

Dlussky, G. M., Zabelin, S. I. 1985. Ant fauna (Hymenoptera, Formicidae) of the River Sumbar Basin (south-west Kopetdag). [In Russian.] Pp. 208-246 in: Nechaevaya, N. T. (ed.) The vegetation and animal world of western Kopetdag. [In Russian.] Ashkhabad: Ylym, 278 pp. [> 1985.10.22] [Date signed for printing ("podpisano k pechati").]

Dlussky, G. M., Zakharov, A. A. 1965. Distribution of ants in various forest types. [In Russian.] Lesn. Khoz. 18(8):55-57. [> 1965.07.21] [Date signed for printing ("podpisano k pechati"). August 1965 issue.]

Dmitrienko, V. K. 1972a. Intraspecific variation of ants of the genus *Formica*. [In Russian.] Pp. 78-79 in: Cherepanov, A. I., Folitarek, S. S., Maksimov, A. A., et al. (eds.) Zoological problems of Siberia (Reports of the fourth Conference of Zoologists of Siberia). [In Russian.] Novosibirsk: Nauka, 556 pp. [> 1972.02.24] [Date signed for printing ("podpisano k pechati").]

Dmitrienko, V. K. 1972b. Ecological and geographical aspects of the myrmecofauna in the light taiga. Pp. 360-361 in: Rafes, P. M. (ed.) XIII International Congress of Entomology, Moscow, 2-9 August, 1968. Proceedings, Volume III. Leningrad, Nauka, 494 pp. [> 1972.06.08] [Date signed for printing ("podpisano k pechati").]

Dmitrienko, V. K. 1974. Characteristics of landscape ecological complexes of ants in Buryatya. [In Russian.] Pp. 40-41 in: Cherepanov, A. I. (ed.) Entomology questions of Siberia. [In Russian.] Novosibirsk: Nauka, 206 pp. [> 1974.01.16] [Date signed for printing ("podpisano k pechati").]

Dmitrienko, V. K. 1979. Biocoenotic complexes of ants in Buryatya. [In Russian.] Pp. 93-108 in: Petrenko, E. S. (ed.) Fauna of the forests of the Lake Baikal basin. [In Russian.] Novosibirsk: Nauka, 157 pp. [> 1979.09.17] [Date signed for printing ("podpisano k pechati").]

Dmitrienko, V. K., Petrenko, E. S. 1965. Ant fauna of the forests of central Yakutia. [In Russian.] Pp. 73-86 in: Konikov, A. S. (ed.) Research on forest protection in Siberia. [In Russian.] Moskva: Nauka, 111 pp. [> 1965.05.11] [Date signed for printing ("podpisano k pechati").]

Dmitrienko, V. K., Petrenko, E. S. 1976. Ants of the taiga biocoenoses of Siberia. [In Russian.] Novosibirsk: Nauka, 220 pp. [> 1975.12.12] [Date signed for printing ("podpisano k pechati"). Title page dated 1976.]

Dobrzanski, J. 1965. Genesis of social parasitism among ants. Acta Biol. Exp. (Warsaw) 25:59-71. [1965.01]

Doflein, F. 1920. Mazedonische Ameisen. Beobachtungen über ihre Lebensweise. Jena: G. Fischer, 74 pp. [1920]

Don, A. W. 1974. Ants and termites. N. Z. Nat. Herit. 3(33):909-913.

Don, A. W. 1994. Ants (Hymenoptera: Formicidae) from the Three Kings Islands, New Zealand. N. Z. Entomol. 17:22-29. [1994.07]

Don, A. W., Jones, T. H. 1993. The stereochemistry of 3-butyl-5-(5-hexenyl)-pyrrolizidine from populations of *Monomorium antarcticum* (Smith) (Hymenoptera: Formicidae) and its possible role as a unique taxonomic character. N. Z. Entomol. 16:45-48. [1993.07]

Donisthorpe, H. 1908a. Additions to the wild fauna and flora of the Royal Botanic Gardens, Kew: VII. I. Fauna. Hymenoptera. Formicidae (Ants). Bull. Misc. Inf. R. Bot. Gard. Kew 1908:121-122. [1908.04]

Donisthorpe, H. 1908b. Ants found in Great Britain. Trans. Leic. Lit. Philos. Soc. 12:221-233. [1908.07]

Donisthorpe, H. 1909a. Additions to the wild fauna and flora of the Royal Botanic Gardens, Kew: IX. Fauna. Hymenoptera. Formicidae (Ants). Bull. Misc. Inf. R. Bot. Gard. Kew 1909:250-251. [1909.09]

Donisthorpe, H. 1909b. *Formica sanguinea*, Latr. at Bewdley, with an account of a slave-raid, and description of two gynandromorphs. Zoologist 1909:463-466. [1909.12]

Donisthorpe, H. 1911a. *Lasius mixtus*, Nyl., in Britain. Entomol. Rec. J. Var. 23:236-238. [1911.09]

Donisthorpe, H. 1911b. Additions to the wild fauna and flora of the Royal Botanic Gardens: XII. Hymenoptera. Formicidae (Ants). Bull. Misc. Inf. R. Bot. Gard. Kew 1911:367-369. [1911.12.02]

Donisthorpe, H. 1911c. A revised list of the British ants. Entomologist 44:389-391. [1911.12]

Donisthorpe, H. 1912a. Myrmecophilous notes for 1911. Entomol. Rec. J. Var. 24:4-10. [1912.01.15]

Donisthorpe, H. 1912b. Some races of ants new to Britain. Entomol. Rec. J. Var. 24:306. [1912.12.15]

Donisthorpe, H. 1913a. Some notes on the genus Myrmica. [part] Entomol. Rec. J. Var. 25:1-8. [1913.01.15]

Donisthorpe, H. 1913b. Some notes on the genus Myrmica, Latr. [concl.] Entomol. Rec. J. Var. 25:42-48. [1913.02.15]

Donisthorpe, H. 1913c. Myrmecophilous notes for 1912. Entomol. Rec. J. Var. 25:61-68. [1913.03.15]

Donisthorpe, H. 1913d. Ants and myrmecophiles on Lundy. Entomol. Rec. J. Var. 25:267-269. [1913.11.15]

Donisthorpe, H. 1913e. On some remarkable associations between ants of different species. Annu. Rep. Proc. Lancs. Ches. Entomol. Soc. 36:38-56. [1913]

Donisthorpe, H. 1914a. Myrmecophilous notes for 1913. Entomol. Rec. J. Var. 26:37-45. [1914.02.15]

Donisthorpe, H. 1914b. Three myrmecological notes. Entomol. Rec. J. Var. 26:136-138. [1914.06.15]

Donisthorpe, H. 1915a. Genital armature of the male ant. Trans. Entomol. Soc. Lond. 1915:l-liii. [1915.08.05]

Donisthorpe, H. 1915b. Marriage-flights of *Donisthorpea* species on August 8th, etc. Entomol. Rec. J. Var. 27:206-207. [1915.09.15]

Donisthorpe, H. 1915c. The type of *Camponotus* (*Myrmoturba*) *maculatus*, F. Entomol. Rec. J. Var. 27:221-222. [1915.10.15]

Donisthorpe, H. 1915d. Descriptions of a pterergate and two gynandromorphs of *Myrmica scabrinodis* Nyl., with a list of all the known cases of the latter. Entomol. Rec. J. Var. 27:258-260. [1915.11.15]

Donisthorpe, H. 1915e. *Myrmica schencki* Emery, an ant new to Britain. Entomol. Rec. J. Var. 27:265-266. [1915.12.15]

Donisthorpe, H. 1915f. British ants, their life-history and classification. Plymouth: Brendon & Son Ltd., xv + 379 pp. [1915]

Donisthorpe, H. 1916a. Myrmecophilous notes for 1915. Entomol. Rec. J. Var. 28:1-4. [1916.01.15]

Donisthorpe, H. 1916b. *Epitritus wheeleri*, n. sp., an ant new to science; with notes on the genus *Epitritus*, Emery. Entomol. Rec. J. Var. 28:121-122. [1916.06.15]

Donisthorpe, H. 1916c. The ants of the Netherlands and their guests. [Review of: Schmitz, H. 1915. De Nederlandsche mieren en haar gasten. Maastricht: C. Goffin, 146 + iv pp.] Entomol. Rec. J. Var. 28:228-229. [1916.10.15]

Donisthorpe, H. 1916d. Synonymy of some genera of ants. [part] Entomol. Rec. J. Var. 28:241-244. [1916.11.15]

Donisthorpe, H. 1916e. Synonymy of some genera of ants. [concl.] Entomol. Rec. J. Var. 28:275-277. [1916.12.15]

Donisthorpe, H. 1917a. Myrmecophilous notes for 1916. [part] Entomol. Rec. J. Var. 29:30-33. [1917.01.15]

Donisthorpe, H. 1917b. Myrmecophilous notes for 1916. [concl.] Entomol. Rec. J. Var. 29:48-52. [1917.03.15]

Donisthorpe, H. 1917c. [Review of: Wheeler, W. M. 1915. The ants of the Baltic amber. Schr. Phys. Ökon. Ges. Königsb. 55:1-142.] Entomol. Rec. J. Var. 29:112-116. [1917.06.15]

Donisthorpe, H. 1917d. *Dolichoderus* (*Hypoclinea*) *crawleyi* n. sp., a species of ant new to science; with a few notes on the genus. Entomol. Rec. J. Var. 29:201-202. [1917.10.15]

Donisthorpe, H. 1918a. Some notes on a paper by Dr. Leach on ants and gnats in 1825. Entomol. Rec. J. Var. 30:8-9. [1918.01.15]

Donisthorpe, H. 1918b. Myrmecophilous notes for 1917. Entomol. Rec. J. Var. 30:21-24. [1918.02.15]

Donisthorpe, H. 1918c. A list of ants from Mesopotamia; with a description of a new species and a new variety. Entomol. Rec. J. Var. 30:165-167. [1918.10.15]

Donisthorpe, H. 1919a. Myrmecophilous notes for 1918. [part] Entomol. Rec. J. Var. 31:1-5. [1919.01.17]

Donisthorpe, H. 1919b. Myrmecophilous notes for 1918. [concl.] Entomol. Rec. J. Var. 31:21-26. [1919.02.15]

Donisthorpe, H. 1920a. [Review of: Bondroit, J. 1918. Les fourmis de France et de Belgique. Ann. Soc. Entomol. Fr. 87:1-174.] Entomol. Rec. J. Var. 32:71-75. [1920.04.15]

Donisthorpe, H. 1920b. British Oligocene ants. Ann. Mag. Nat. Hist. (9)6:81-94. [1920.07]

Donisthorpe, H. 1921. [Review of: Forel, A. 1921. Le monde social des fourmis du globe comparé à celui de l'homme. Tome 1. Genève: Libraire Kundig, xiv + 192 pp.] Entomol. Rec. J. Var. 33:59-60. [1921.03.15]

Donisthorpe, H. 1922a. Myrmecophilous notes for 1921. Entomol. Rec. J. Var. 34:1-5. [1922.01.15]

Donisthorpe, H. 1922b ("1921"). The subfamilies of Formicidae. Trans. Entomol. Soc. Lond. 1921:xl-xlvi. [1922.04.13]

Donisthorpe, H. 1922c. On some abnormalities in ants. Entomol. Rec. J. Var. 34:81-85. [1922.05.15]

Donisthorpe, H. 1922d. Some notes on *Ponera punctatissima* Roger. Entomol. Mon. Mag. 58[=(3)8]:134-137. [1922.06]

Donisthorpe, H. 1923a. Myrmecophilous notes for 1922. Entomol. Rec. J. Var. 35:1-9. [1923.01.15]

Donisthorpe, H. 1923b. *Acanthomyops* (*Donisthorpea*) *brunneus*, Latr., a species of Formicidae new to Britain. Entomol. Rec. J. Var. 35:21-23. [1923.02.15]

Donisthorpe, H. 1923c. [Review of: Forel, A. 1921. Le monde social des fourmis du globe comparé à celui de l'homme. Tome 2. Genève: Libraire Kundig, iii + 184 pp.] Entomol. Rec. J. Var. 35:38-40. [1923.02.15]

Donisthorpe, H. 1924a. [Review (part) of: Wheeler, W. M. 1923. Social life among the insects. New York: Harcourt, Brace & Co., vii + 375 pp.] Entomol. Rec. J. Var. 36:30-32. [1924.02.15]

Donisthorpe, H. 1924b. [Review (concl.) of: Wheeler, W. M. 1923. Social life among the insects. New York: Harcourt, Brace & Co., vii + 375 pp.] Entomol. Rec. J. Var. 36:42-47. [1924.03.15]

Donisthorpe, H. 1924c. [Review of: Forel, A. 1922. Le monde social des fourmis du globe comparé à celui de l'homme. Tome 3. Genève: Libraire Kundig, vii + 227 pp.] Entomol. Mon. Mag. 60:89-93. [1924.04]

Donisthorpe, H. 1924d. [Review of: Forel, A. 1923. Le monde social des fourmis du globe comparé à celui de l'homme. Tome 4. Genève: Libraire Kundig, vii + 172 pp.] Entomol. Mon. Mag. 60:140-142. [1924.06]

Donisthorpe, H. 1924e. [Review of: Forel, A. 1923. Le monde social des fourmis du globe comparé à celui de l'homme. Tome 5. Genève: Libraire Kundig, vi + 174 pp.] Entomol. Rec. J. Var. 36:173-177. [1924.12.15]

Donisthorpe, H. 1926a. Ants and myrmecophiles at Bordighera. [part] Entomol. Rec. J. Var. 38: 5-8. [1926.01.15]

Donisthorpe, H. 1926b. Ants and myrmecophiles at Bordighera. [concl.] Entomol. Rec. J. Var. 38:17-18. [1926.02.15]

Donisthorpe, H. 1926c. The ants (Formicidae), and some myrmecophiles, of Sicily. [part] Entomol. Rec. J. Var. 38:161-165. [1926.12.15]
Donisthorpe, H. 1927a. The ants (Formicidae), and some myrmecophiles, of Sicily. [concl.] Entomol. Rec. J. Var. 39:6-9. [1927.01.15]
Donisthorpe, H. 1927b. Gynandromorphism in ants. Proc. Entomol. Soc. Lond. 1:92-93. [1927.05.05]
Donisthorpe, H. 1927c. British ants, their life history and classification (2nd edn.). London: G. Routledge and Sons, xvi + 436 pp. [1927]
Donisthorpe, H. 1927d. The guests of British ants, their habits and life-histories. London: G. Routledge and Sons, xxii + 244 pp. [1927]
Donisthorpe, H. 1929a. The Formicidae (Hymenoptera) taken by Major R. W. G. Hingston, M.C., I.M.S. (ret.), on the Mount Everest Expedition, 1924. Ann. Mag. Nat. Hist. (10)4:444-449. [1929.11]
Donisthorpe, H. 1929b. Gynandromorphism in ants. Zool. Anz. 82:92-96. [1929]
Donisthorpe, H. 1930a. [Review of: Krausse, A. 1929. Ameisenkunde. Eine Einführung in die Systematik und Biologie der Ameisen. Stuttgart: A. Kernen, 172 pp.] Entomol. Rec. J. Var. 42:63-64. [1930.04.15]
Donisthorpe, H. 1930b. The ants (Formicidae) and guests (myrmecophiles) of Windsor Forest and District. [part] Entomol. Rec. J. Var. 42(suppl.):1-4. [1930.05.15]
Donisthorpe, H. 1930c. The ants (Formicidae) and guests (myrmecophiles) of Windsor Forest and District. [part] Entomol. Rec. J. Var. 42(suppl.):5-8. [1930.07.15]
Donisthorpe, H. 1930d. A new subspecies of *Acanthomyops* (Hymenoptera, Formicidae) from Kashmir. Ann. Mag. Nat. Hist. (10)6:225-226. [1930.08]
Donisthorpe, H. 1930e. The ants (Formicidae) and guests (myrmecophiles) of Windsor Forest and District. [part] Entomol. Rec. J. Var. 42(suppl.):9-12. [1930.09.15]
Donisthorpe, H. 1930f. *Cardiocondyla bicolor*, sp. n. (Hymenoptera, Formicidae), a species of myrmecine ant new to science. Ann. Mag. Nat. Hist. (10)6:366. [1930.09]
Donisthorpe, H. 1930g. The ants (Formicidae) and guests (myrmecophiles) of Windsor Forest and District. [part] Entomol. Rec. J. Var. 42(suppl.):13-16. [1930.10.15]
Donisthorpe, H. 1930h. The ants (Formicidae) and guests (myrmecophiles) of Windsor Forest and District. [concl.] Entomol. Rec. J. Var. 42(suppl.):17-18. [1930.11.15]
Donisthorpe, H. 1930i. Formicidae. P. 21 in: Lawson, A. K. (ed.) A check list of the fauna of Lancashire and Chesire. Part 1. Arbroath, Scotland: The Lancashire and Chesire Fauna Committee, xii + 115 pp. [1930]
Donisthorpe, H. 1931a. Erich Wasmann, S. J., Hon. F. E. S. Entomol. Rec. J. Var. 43:76. [1931.04.15]
Donisthorpe, H. 1931b. *Camponotus* (*Tanaemyrmex*) *britteni*, sp. n. (Hymenoptera, Formicidae), a formicine ant new to science. Ann. Mag. Nat. Hist. (10)8:129-131. [1931.07]
Donisthorpe, H. 1931c. Auguste Forel. Entomol. Rec. J. Var. 43:176. [1931.11.15]
Donisthorpe, H. 1931d. Descriptions of some new species of ants. Ann. Mag. Nat. Hist. (10)8:494-501. [1931.11]
Donisthorpe, H. 1932a. A new species of *Camponotus* (Hym., Formicidae) from Colombia. Stylops 1:88-89. [1932.04.15]
Donisthorpe, H. 1932b. On the identity of some ants from Ceylon described by F. Walker. Ann. Mag. Nat. Hist. (10)9:574-576. [1932.06]
Donisthorpe, H. 1932c. On the identity of Smith's types of Formicidae (Hymenoptera) collected by Alfred Russell Wallace in the Malay Archipelago, with descriptions of two new species. Ann. Mag. Nat. Hist. (10)10:441-476. [1932.11]
Donisthorpe, H. 1933a. A new species of *Aphaenogaster* (Hym. Formicidae) from India. Stylops 2:24. [1933.01.31]
Donisthorpe, H. 1933b. Descriptions of three new species of Formicidae, and a synonymical note. Ann. Mag. Nat. Hist. (10)11:194-198. [1933.02]

Donisthorpe, H. 1933c. Additional records of ants and myrmecophiles in Britain since the 2nd edition of "British Ants" (1927) and my paper on the ants and myrmecophiles of Windsor Forest (1930). Entomol. Rec. J. Var. 45:132-136. [1933.10.15]
Donisthorpe, H. 1933d. On a small collection of ants made by Dr. F. W. Edwards in Argentina. Ann. Mag. Nat. Hist. (10)12:532-538. [1933.11]
Donisthorpe, H. 1934. Report on the Insecta collected by Colonel R. Meinertzhagen in the Ahaggar Mountains. VIII.- Formicidae. Ann. Mag. Nat. Hist. (10)13:190-192. [1934.02]
Donisthorpe, H. 1935. The ants of Christmas Island. Ann. Mag. Nat. Hist. (10)15:629-635. [1935.06]
Donisthorpe, H. 1936a. *Rhopalomastix janeti* (Hym. Formicidae) a species of ant new to science. Entomol. Rec. J. Var. 48:55-56. [1936.05.15]
Donisthorpe, H. 1936b. *Acropyga* (*Rhizomyrma*) *robae* sp. nov. (Hym. Formicidae), a new S. American ant, with remarks on the genus, etc. Entomologist 69:108-111. [1936.05]
Donisthorpe, H. 1936c. *Strongylognathus diveri* sp. n. (Hym., Formicidae), a genus and species new to the British list, with notes on the genus. Entomol. Mon. Mag. 72:111-116. [1936.05]
Donisthorpe, H. 1936d. A new study of *Mesoxena mistura* Smith. Proc. R. Entomol. Soc. Lond. Ser. B 5:119-120. [1936.06.15]
Donisthorpe, H. 1936e. The ants of the Azores. Entomol. Mon. Mag. 72:130-133. [1936.06]
Donisthorpe, H. 1936f. Five new species of ant (Formicidae) from various localities. Ann. Mag. Nat. Hist. (10)18:524-530. [1936.11]
Donisthorpe, H. 1937a. Some new forms of Formicidae and a correction. Ann. Mag. Nat. Hist. (10)19:619-628. [1937.06]
Donisthorpe, H. 1937b. A new species of *Harpegnathos* Jerd., with some remarks on the genus, and other known species (Hym. Formicidae). Entomol. Mon. Mag. 73:196-201. [1937.09]
Donisthorpe, H. 1937c. Generic names, &c., of the British Formicidae. [part] Entomol. Rec. J. Var. 49:131-132. [1937.11.15]
Donisthorpe, H. 1937d. Generic names, &c., of the British Formicidae. [concl.] Entomol. Rec. J. Var. 49:143-145. [1937.12.15]
Donisthorpe, H. 1937e. A new subgenus and three new species of *Polyrhachis* Smith. Entomologist 70:273-275. [1937.12]
Donisthorpe, H. 1938a. Five new species of ant, chiefly from New Guinea. Ann. Mag. Nat. Hist. (11)1:140-148. [1938.01]
Donisthorpe, H. 1938b. Some ants of the subgenus *Planimyrma* Viehmeyer of the genus *Aphaenogaster* Mayr. Entomol. Mon. Mag. 74:30-32. [1938.02]
Donisthorpe, H. 1938c. The subgenus *Cyrtomyrma* Forel of *Polyrhachis* Smith, and descriptions of new species, etc. Ann. Mag. Nat. Hist. (11)1:246-267. [1938.03]
Donisthorpe, H. 1938d. New species and varieties of ants from New Guinea. Ann. Mag. Nat. Hist. (11)1:593-599. [1938.06]
Donisthorpe, H. 1938e. A pterergate of *Acanthomyops* (*Chtonolasius*) *flavus* F. (Hym. Formicidae). Entomol. Mon. Mag. 74:178-180. [1938.08]
Donisthorpe, H. 1938f. Progress in our knowledge of British Coleoptera, ants, and myrmecophiles. Entomol. Rec. J. Var. 50:126-129. [1938.10.15]
Donisthorpe, H. 1938g. An ergatandromorph of *Myrmica laevinodis* Nyl., and the list of gynandromorphs, etc., brought up to date (Hym., Formicidae). Entomologist 71:251-252. [1938.11]
Donisthorpe, H. 1938h. New species of ants and a new subgenus of *Dolichoderus* from various localities. Ann. Mag. Nat. Hist. (11)2:498-504. [1938.11]
Donisthorpe, H. 1939a. The genus *Lioponera* Mayr (Formicidae, Cerapachyinae), with descriptions of two new species and an ergatandromorph. Ann. Mag. Nat. Hist. (11)3:252-257. [1939.03]
Donisthorpe, H. 1939b. On the occurrence of dealated males in the genus *Dorylus* Fab. (Hym. Formicidae). Proc. R. Entomol. Soc. Lond. Ser. A 14:79-81. [1939.06.15]

Donisthorpe, H. 1939c. *Typhlomyrmex richardsi* (Hym., Formicidae), a new species of ponerine ant from British Guiana. Entomol. Mon. Mag. 75:161-162. [1939.07]

Donisthorpe, H. 1939d. Descriptions of several species of ants (Hymenopt.) taken by Dr. O. W. Richards in British Guiana. Proc. R. Entomol. Soc. Lond. Ser. B 8:152-154. [1939.08.15]

Donisthorpe, H. 1940a. Descriptions of new species of ants (Hym., Formicidae) from various localities. Ann. Mag. Nat. Hist. (11)5:39-48. [1940.01]

Donisthorpe, H. 1940b. *Lordomyrma infundibuli* (Hym., Formicidae), a new species of ant from Dutch New Guinea. Entomol. Mon. Mag. 76:45-47. [1940.02.01]

Donisthorpe, H. 1940c. Some new forms of *Odontomachus* (Hym., Formicidae). Entomologist 73:106-109. [1940.05]

Donisthorpe, H. 1940d. Mimicry in ants. Entomol. Mon. Mag. 76:254. [1940.11.04]

Donisthorpe, H. 1941a. *Lordomyrma niger* sp. n. (Hym., Formicidae), with a key and notes on the genus. Entomol. Mon. Mag. 77:36-38. [1941.02.03]

Donisthorpe, H. 1941b. New ants from Waigeu Island, New Guinea, and the Solomons. Entomologist 74:36-42. [1941.02.17]

Donisthorpe, H. 1941c. Descriptions of new species of ants from New Guinea. Ann. Mag. Nat. Hist. (11)7:129-144. [1941.02]

Donisthorpe, H. 1941d. Glanures myrmecologiques. [part] Entomol. Rec. J. Var. 53:27-28. [1941.03.15]

Donisthorpe, H. 1941e. The ants of Norfolk Island. Entomol. Mon. Mag. 77:90-93. [1941.04.02]

Donisthorpe, H. 1941f. Glanures myrmecologiques. [concl.] Entomol. Rec. J. Var. 53:36-38. [1941.04.15]

Donisthorpe, H. 1941g. A new species of *Echinopla* (Hym., Formicidae), with some notes on the genus. Entomologist 74:115-116. [1941.05.09]

Donisthorpe, H. 1941h. The ants of Japen Island, Dutch New Guinea (Hym., Formicidae). Trans. R. Entomol. Soc. Lond. 91:51-64. [1941.07.15]

Donisthorpe, H. 1941i. [Review of: Snodgrass, R. E. 1941. The male genitalia of Hymenoptera. Smithson. Misc. Collect. 99(14):1-86.] Entomol. Rec. J. Var. 53:80-84. [1941.07.15]

Donisthorpe, H. 1941j. A new genus and species of Formicidae (Hym.) from Papua. Entomol. Mon. Mag. 77:175. [1941.07.31]

Donisthorpe, H. 1941k. Descriptions of new ants (Hym., Formicidae) from various localities. Ann. Mag. Nat. Hist. (11)8:199-210. [1941.09]

Donisthorpe, H. 1941l. Dr Felix Santschi. Entomol. Rec. J. Var. 53:99-100. [1941.09.15]

Donisthorpe, H. 1941m. Synonymical notes, etc., on Formicidae (Hym.). Entomol. Mon. Mag. 77:237-240. [1941.10.01]

Donisthorpe, H. 1941n. A new vice-county record for two common British ants. Entomol. Rec. J. Var. 53:109. [1941.10.15]

Donisthorpe, H. 1941o. Description of a new species of *Crematogaster* Lund, subgenus *Physocrema* Forel, with a list of, and a key to, the known species of the subgenus. Entomologist 74:225-227. [1941.10.17]

Donisthorpe, H. 1942a. The Formicidae (Hym.) of the Armstrong College Expedition to the Siwa Oasis. Ann. Mag. Nat. Hist. (11)9:26-33. [1942.01]

Donisthorpe, H. 1942b. Descriptions of a few ants from the Philippine Islands, and a male of *Polyrhachis bihamata* Drury from India. Ann. Mag. Nat. Hist. (11)9:64-72. [1942.01]

Donisthorpe, H. 1942c. *Myopopone wollastoni* sp. n., with notes on other forms in the genus and descriptions of the males of two species (Hym., Formicidae). Entomol. Mon. Mag. 78:29-31. [1942.02.05]

Donisthorpe, H. 1942d. Ants from the Colombo Museum Expedition to Southern India, September-October 1938. Ann. Mag. Nat. Hist. (11)9:449-461. [1942.06]

Donisthorpe, H. 1942e. Laboulbeniaceae and ants. Entomol. Mon. Mag. 78:193-199. [1942.09.02]

Donisthorpe, H. 1942f. New species of ants (Hym., Formicidae) from the Gold Coast, Borneo, Celebes, New Guinea and New Hebrides. Ann. Mag. Nat. Hist. (11)9:701-709. [1942.09]

Donisthorpe, H. 1942g. A new species of *Diplomorium* (Hym. Formicidae), with some notes on the genus. Entomologist 75:217-218. [1942.10.09]

Donisthorpe, H. 1942h. Notes on the subgenus *Orthonotomyrmex* Ashmead of *Camponotus* Mayr, and description of a new species (Hym., Formicidae). Entomol. Mon. Mag. 78:248-251. [1942.11.02]

Donisthorpe, H. 1942i. Notes on the genus *Pseudolasius* Emery with the description of *Pseudolasius karawajewi* sp. n. (Hym., Formicidae). Proc. R. Entomol. Soc. Lond. Ser. B 11:166-169. [1942.11.14]

Donisthorpe, H. 1943a. Descriptions of new ants, chiefly from Waigeu Island, N. Dutch New Guinea. Ann. Mag. Nat. Hist. (11)10:167-176. [1943.03]

Donisthorpe, H. 1943b. Ants from the Colombo Museum Expedition to Southern India, September-October, 1938. Ann. Mag. Nat. Hist. (11)10:196-208. [1943.03]

Donisthorpe, H. 1943c. *Ireneidris myops* gen. et sp. n. (Hym., Formicidae) from Waigeu Island, Dutch New Guinea. Entomol. Mon. Mag. 79:81-82. [1943.04.06]

Donisthorpe, H. 1943d. A myrmecophilous woodlouse. Nature (Lond.) 151:675-676. [1943.06.12]

Donisthorpe, H. 1943e. The ants (Hym., Formicidae) of Waigeu Island, North Dutch New Guinea. Ann. Mag. Nat. Hist. (11)10:433-475. [1943.07]

Donisthorpe, H. 1943f. Myrmecological gleanings. Proc. R. Entomol. Soc. Lond. Ser. B 12:115-116. [1943.08.24]

Donisthorpe, H. 1943g. A list of the type-species of the genera and subgenera of the Formicidae. [part] Ann. Mag. Nat. Hist. (11)10:617-688. [1943.09]

Donisthorpe, H. 1943h. A list of the type-species of the genera and subgenera of the Formicidae. [concl.] Ann. Mag. Nat. Hist. (11)10:721-737. [1943.11]

Donisthorpe, H. 1944a. *Messor aegyptiacus* Emery subspecies *canaliculatus* subsp. n. (*striaticeps* Santschi? (1923) nec Er. André (1883)) (Hym. Formicidae). Proc. R. Entomol. Soc. Lond. Ser. B 13:1-3. [1944.02.25]

Donisthorpe, H. 1944b. Notes arising out of Mr. W. D. Hincks's review of Mr. Donisthorpe's "A list of the type-species of the genera and subgenera of the Formicidae". Entomol. Mon. Mag. 80:59. [1944.03.29]

Donisthorpe, H. 1944c. A new subgenus and three new species of *Polyrhachis* F. Smith (Hym., Formicidae). Entomol. Mon. Mag. 80:64-66. [1944.03.29]

Donisthorpe, H. 1944d. Two new species of *Pristomyrmex* Mayr (Hym. Formicidae), with some notes on the genus. Proc. R. Entomol. Soc. Lond. Ser. B 13:81-84. [1944.08.15]

Donisthorpe, H. 1944e. A new species of *Bothriomyrmex* Emery (Hym. Formicidae), and some notes on the genus. Proc. R. Entomol. Soc. Lond. Ser. B 13:100-103. [1944.10.25]

Donisthorpe, H. 1945a. New and rare species of *Crematogaster* Lund (Hym. Formicidae) from the Gold Coast, etc. Entomologist 78:10-11. [1945.01.16]

Donisthorpe, H. 1945b. A new species and a new variety of *Technomyrmex* Mayr (Hym., Formicidae) from New Guinea. Entomol. Mon. Mag. 81:57-58. [1945.03.15]

Donisthorpe, H. 1945c. A new species of *Triglyphothrix* Forel (Hym., Formicidae) from Uganda with some notes on the genus. Entomol. Mon. Mag. 81:76-77. [1945.04.14]

Donisthorpe, H. 1945d. On a small collection of ants (Hym., Formicidae) from West Africa, associated with Coccidae. Ann. Mag. Nat. Hist. (11)12:265-273. [1945.09.18]

Donisthorpe, H. 1946a. British ants as at present known, with some remarks, and a list of the same. Entomol. Rec. J. Var. 58:64-65. [1946.05.15]

Donisthorpe, H. 1946b. Formicidae, Stephens (1829). Entomol. Rec. J. Var. 58:89-91. [1946.06.15]

Donisthorpe, H. 1946c. A new subspecies of *Messor* Forel, and a new variety of *Aphaenogaster* Mayr (Hym. Formicidae) from Turkey. Proc. R. Entomol. Soc. Lond. Ser. B 15:53-54. [1946.06.15]

Donisthorpe, H. 1946d. Undescribed forms of *Camponotus* (*Colobopsis*) *vitiensis* from the Fiji Islands (Hymenoptera, Formicidae). Proc. R. Entomol. Soc. Lond. Ser. B 15:69-70. [1946.06.15]

Donisthorpe, H. 1946e. Fifty gynandromorphous ants taken in a single colony of *Myrmica sabuleti* Meinert in Ireland. Entomologist 79:121-131. [1946.06.20]

Donisthorpe, H. 1946f. *Ponera punctatissima* Roger (Hym., Formicidae) in Ireland, with some notes on the species. Entomol. Mon. Mag. 82:230. [1946.09.02]

Donisthorpe, H. 1946g ("1945"). New species of ants (Hym. Formicidae) from the Island of Mauritius. Ann. Mag. Nat. Hist. (11)12:776-782. [1946.09.19]

Donisthorpe, H. 1946h. *Ireneopone gibber* (Hym., Formicidae), a new genus and species of myrmicine ant from Mauritius. Entomol. Mon. Mag. 82:242-243. [1946.10.01]

Donisthorpe, H. 1946i. The ants (Hym. Formicidae) of Mauritius. Ann. Mag. Nat. Hist. (11)13: 25-35. [1946.10.31]

Donisthorpe, H. 1946j. A new genus and species of Formicidae (Hym.) from Mauritius. Proc. R. Entomol. Soc. Lond. Ser. B 15:145-147. [1946.12.23]

Donisthorpe, H. 1947a ("1946"). New species of ants from China and Mauritius. Ann. Mag. Nat. Hist. (11)13:283-286. [1947.01.22]

Donisthorpe, H. 1947b. *Pheidole* (*Pheidolacanthinus*) *striatus* sp. n. from New Guinea, with notes on the subgenus *Pheidolacanthinus* F. Smith (Hym., Formicidae). Entomol. Mon. Mag. 83:172-174. [1947.07.31]

Donisthorpe, H. 1947c ("1946"). Ants from New Guinea, including new species and a new genus. Ann. Mag. Nat. Hist. (11)13:577-595. [1947.08.15]

Donisthorpe, H. 1947d. The Formicidae of the Channel Islands. Entomologist 80:180-182. [1947.08.20]

Donisthorpe, H. 1947e. Some new ants from New Guinea. Ann. Mag. Nat. Hist. (11)14:183-197. [1947.12.23]

Donisthorpe, H. 1947f. Results of the Armstrong College Expedition to Siwa Oasis (Libyan Desert), 1935: A second list of the Formicidae (Hymenoptera). Bull. Soc. Fouad Ier Entomol. 31:109-111. [1947.12.31]

Donisthorpe, H. 1947g. Some gynandromorph ants and a possible pterergate from Ireland. Entomologist 80:277-279. [1947.12.31]

Donisthorpe, H. 1948a ("1947"). *Liomyrmex reneae* sp. n. (Hym., Formicidae) with a list of the species and some notes on the genus *Liomyrmex* Mayr. Entomol. Mon. Mag. 83:293-294. [1948.01.09]

Donisthorpe, H. 1948b ("1947"). A second instalment of the Ross Collection of ants from New Guinea. Ann. Mag. Nat. Hist. (11)14:297-317. [1948.02.04]

Donisthorpe, H. 1948c. A new genus and species of ant from New Britain. Entomol. Rec. J. Var. 60:65-66. [1948.05.15]

Donisthorpe, H. 1948d. A third instalment of the Ross Collection of ants from New Guinea. Ann. Mag. Nat. Hist. (11)14:589-604. [1948.06.07]

Donisthorpe, H. 1948e. A new species of *Colobopsis* (Hym., Formicidae) from New Guinea with a few notes on the subgenus. Entomol. Mon. Mag. 84:121-122. [1948.06.10]

Donisthorpe, H. 1948f. *Microbolbos testaceus*, a new genus and species of ponerine ant. Entomologist 81:170-171. [1948.07.13]

Donisthorpe, H. 1948g. A fourth instalment of the Ross Collection of ants from New Guinea. Ann. Mag. Nat. Hist. (12)1:131-143. [1948.09.17]

Donisthorpe, H. 1948h. A redescription of the types of *Strumigenys mandibularis* F. Smith, and *Cephaloxys capitata* F. Smith. Psyche (Camb.) 55:78-81. [1948.10.23]

Donisthorpe, H. 1949a ("1948"). A new genus and species of dacetine ant (Hym., Formicidae) from New Guinea. Entomol. Mon. Mag. 84:281. [1949.01.11]

Donisthorpe, H. 1949b ("1948"). A fifth instalment of the Ross Collection of ants from New Guinea. Ann. Mag. Nat. Hist. (12)1:487-506. [1949.01.14]

Donisthorpe, H. 1949c ("1948"). A sixth instalment of the Ross Collection of ants from New Guinea. Ann. Mag. Nat. Hist. (12)1:744-759. [1949.02.04]

Donisthorpe, H. 1949d. Further gynandromorph ants from Ireland. Entomologist 82:38-40. [1949.02.15]

Donisthorpe, H. 1949e. *Lordomyrma bensoni* (Hym., Formicidae), a species of ant new to science from New Guinea. Entomol. Mon. Mag. 85:94. [1949.04.25]

Donisthorpe, H. 1949f. A new *Camponotus* from Madagascar and a small collection of ants from Mauritius. Ann. Mag. Nat. Hist. (12)2:271-275. [1949.07.15]

Donisthorpe, H. 1949g. A species of *Calyptomyrmex* Emery (Hym., Formicidae) from New Guinea. Entomol. Mon. Mag. 85:186. [1949.07.22]

Donisthorpe, H. 1949h. A seventh instalment of the Ross Collection of ants from New Guinea. Ann. Mag. Nat. Hist. (12)2:401-422. [1949.08.29]

Donisthorpe, H. 1950a. An eighth instalment of the Ross Collection of ants from New Guinea. Ann. Mag. Nat. Hist. (12)3:338-341. [1950.04]

Donisthorpe, H. 1950b. Two new species of ants from Turkey. Entomol. Rec. J. Var. 62:60-61. [1950.06.15]

Donisthorpe, H. 1950c. Two more new ants from Turkey. Entomol. Rec. J. Var. 62:68-69. [1950.07.15]

Donisthorpe, H. 1950d. Two new species of ants, and a few others from Turkey. Ann. Mag. Nat. Hist. (12)3:638-640. [1950.07]

Donisthorpe, H. 1950e. A first instalment of the ants of Turkey. Ann. Mag. Nat. Hist. (12)3:1057-1067. [1950.12]

Donisthorpe, H., Crawley, W. C. 1914. Polymorphism in ants. Trans. Entomol. Soc. Lond. 1914: x-xiv. [1914.06.25]

Donisthorpe, H., Morley, D. B. W. 1945. A list of scientific terms used in myrmecology. Proc. R. Entomol. Soc. Lond. Ser. A 20:43-49. [1945.06.30]

Donisthorpe, H. S. J. K. See: Donisthorpe, H.

Donnelly, D., Giliomee, J. H. 1985a. Community structure of epigaeic ants (Hymenoptera: Formicidae) in fynbos vegetation in the Jonkershoek Valley. J. Entomol. Soc. South. Afr. 48:247-257. [1985.10]

Donnelly, D., Giliomee, J. H. 1985b. Community structure of epigaeic ants in a pine plantation and in newly burnt fynbos. J. Entomol. Soc. South. Afr. 48:259-265. [1985.10]

Dorow, W. H. O. 1995. Revision of the ant genus *Polyrhachis* Smith, 1857 (Hymenoptera: Formicidae: Formicinae) on subgenus level with keys, checklist of species and bibliography. Cour. Forschungsinst. Senckenb. 185:1-113. [1995.10.20]

Dorow, W. H. O., Kohout, R. J. 1995. A review of the subgenus *Hemioptica* Roger of the genus *Polyrhachis* Fr. Smith with description of a new species (Hymenoptera: Formicidae: Formicinae). Zool. Meded. (Leiden) 69:93-104. [1995.07.31]

Dorow, W. H. O., Maschwitz, U. 1990. The *arachne*-group of *Polyrhachis* (Formicidae: Formicinae): weaver ants cultivating Homoptera on bamboo. Insectes Soc. 37:73-89. [1990.03]

Douglas, A. 1956. Supposedly parasitic Bulldog ant. West. Aust. Nat. 5:120. [1956.07.11]

Douglas, A., Brown, W. L., Jr. 1959. *Myrmecia inquilina* new species: the first parasite among the lower ants. Insectes Soc. 6:13-19. [1959.09]

Dours, J. A. 1873. Catalogue synonymique des Hyménoptères de France. Mém. Soc. Linn. Nord Fr. 3:1-230. [1873] [Ants pp. 164-170.]

Douwes, P. 1976a. Intressanta fynd av myror (Hym., Formicidae).] Entomologen 5:25. [1976.06.30]

Douwes, P. 1976b. Sveriges myror - illustrerade bestämningstabeller över arbetarna. Entomologen 5:37-54. [1976.12.23]

Douwes, P. 1977. *Sifolinia karavajevi*, en för Sverige ny myra (Hym., Formicidae). Entomol. Tidskr. 98:147-148. [1977.12.10]

Douwes, P. 1979. *Formica rufa*-gruppens systematik. Entomol. Tidskr. 100:187-191. [1979.12.10]
Douwes, P. 1981a. Hur man känner igen de olika arterna av den vanliga stackmyran. Entomol. Tidskr. 102:80-82. [1981.06.01]
Douwes, P. 1981b. Intraspecific and interspecific variation in workers of the *Formica rufa* group (Hymenoptera: Formicidae) in Sweden. Entomol. Scand. Suppl. 15:213-223. [1981.07.01]
Douwes, P. 1983. Fynd av myror i Sverige. Entomol. Tidskr. 104:37-38. [1983.02.15]
Douwes, P. 1990. Morphology of the parasitic myrmicine ant. Pp. 147-148 in: Veeresh, G. K., Mallik, B., Viraktamath, C. A. (eds.) Social insects and the environment. Proceedings of the 11th International Congress of IUSSI, 1990. New Delhi: Oxford & IBH Publishing Co., xxxi + 765 pp. [1990.08]
Douwes, P. 1995. Sveriges myror. Entomol. Tidskr. 116:83-99. [1995.11]
Douwes, P., Buschinger, A. 1983. Två för Nordeuropa nya myror. Entomol. Tidskr. 104:1-4. [1983.02.15]
Douwes, P., Jessen, K., Buschinger, A. 1988. *Epimyrma adlerzi* sp. n. (Hymenoptera: Formicidae) from Greece: morphology and life history. Entomol. Scand. 19:239-249. [1988.09.28]
Douwes, P., Stille, B. 1987. The use of enzyme electrophoresis in *Leptothorax* classification. Pp. 29-30 in: Eder, J., Rembold, H. (eds.) Chemistry and biology of social insects. München: Verlag J. Peperny, xxxv + 757 pp. [1987]
Douwes, P., Stille, B. 1991. Hybridization and variation in the *Leptothorax tuberum* group (Hymenoptera: Formicidae). Z. Zool. Syst. Evolutionsforsch. 29:165-175. [1991.09]
Dowton, M., Austin, A. D. 1994. Molecular phylogeny of the insect order Hymenoptera: Apocritan relationships. Proc. Natl. Acad. Sci. U. S. A. 91:9911-9915. [1994.10]
Drury, D. 1773. Illustrations of natural history. Wherein are exhibited upwards of two hundred and twenty figures of exotic insects, according to their different genera. Vol. 2. London: B. White, vii + 90 pp. [1773]
Drury, D. 1782. Illustrations of natural history. Wherein are exhibited upwards of two hundred figures of exotic insects. Vol. 3. London: B. White, xxvi + 76 pp. [1782]
DuBois, M. B. 1979. New records of ants in Kansas. State Biol. Surv. Kans. Tech. Publ. 8:47-55. [1979.05.09]
DuBois, M. B. 1980a. New records of ants in Kansas, II. State Biol. Surv. Kans. Tech. Publ. 9: 35-51. [1980.05.13]
DuBois, M. B. 1980b. Notes on the inquilinous *Monomorium* of North America (Hymenoptera: Formicidae). J. Kansas Entomol. Soc. 53:626. [1980.07.31]
DuBois, M. B. 1981a. New records of ants in Kansas, III. State Biol. Surv. Kans. Tech. Publ. 10:32-44. [1981.04.29]
DuBois, M. B. 1981b. Two new species of inquilinous *Monomorium* from North America (Hymenoptera: Formicidae). Univ. Kans. Sci. Bull. 52:31-37. [1981.06.12]
DuBois, M. B. 1985. Distribution of ants in Kansas: subfamilies Ponerinae, Ecitoninae, and Myrmicinae (Hymenoptera: Formicidae). Sociobiology 11:153-187. [1985]
DuBois, M. B. 1986. A revision of the native New World species of the ant genus *Monomorium* (*minimum* group) (Hymenoptera: Formicidae). Univ. Kans. Sci. Bull. 53:65-119. [1986.03.24]
DuBois, M. B. 1988. Distribution of army ants (Hymenoptera: Formicidae) in Illinois. Entomol. News 99:157-160. [1988.07.07]
DuBois, M. B. 1993. What's in a name? A clarification of *Stenamma westwoodi*, *S. debile*, and *S. lippulum* (Hymenoptera: Formicidae: Myrmicinae). Sociobiology 21:299-334. [1993]
DuBois, M. B. 1994. Checklist of Kansas ants. Kans. Sch. Nat. 40(2):3-16. [1994.04]
DuBois, M. B. 1995. Biodiversity of ants in Kansas (Hymenoptera: Formicidae). Sociobiology 26:305-320. [1995]
DuBois, M. B., Danoff-Burg, J. 1994. Distribution of ants in Kansas: subfamilies Dolichoderinae and Formicinae (Hymenoptera: Formicidae). Sociobiology 24:147-178. [1994]

DuBois, M. B., LaBerge, W. E. 1988. Annotated list of ants in Illinois (Hymenoptera: Formicidae). Pp. 133-156 in: Trager, J. C. (ed.) Advances in myrmecology. Leiden: E. J. Brill, xxvii + 551 pp. [1988]

Duelli, P. 1977. Das soziale Trageverhalten bei neotropischen Ameisen der Gattung *Pseudomyrmex* (Hym., Formicidae): eine Verhaltensnorm als Hinweis für Phylogenie und Taxonomie? Insectes Soc. 24:359-365. [1977.12]

Duelli, P., Duelli-Klein, R. 1976. Freilandversuche zum Heimfindevermögen südamerikanischer Ameisen (Formicidae: Ponerinae, Dolichoderinae, Formicinae). Stud. Entomol. 19:409-419. [1976.12.30]

Duelli, P., Näf, W., Baroni Urbani, C. 1989. Flughöhen verschiedener Ameisenarten in der Hochrheinebene. Mitt. Schweiz. Entomol. Ges. 62:29-35. [1989.07.14]

Duffield, R. M., Blum, M. S. 1975. Identification, role and systematic significance of 3-octanone in the carpenter ant, *Camponotus schaefferi* Whr. Comp. Biochem. Physiol. B. Comp. Biochem. 51:281-282. [1975.07.15]

Duffield, R. M., Brand, J. M., Blum, M. S. 1977. 6-methyl-5-hepten-2-one in *Formica* species: identification and function as an alarm pheromone (Hymenoptera: Formicidae). Ann. Entomol. Soc. Am. 70:309-310. [1977.05.16]

Duffield, R. M., Wheeler, J. W., Blum, M. S. 1980. Methyl anthranilate: identification and possible function in *Aphaenogaster fulva* and *Xenomyrmex floridanus*. Fla. Entomol. 63:203-206. [1980.06.20]

Duffield, R. M., Wheeler, J. W., Snelling, R. R. 1988. Mellein in the mandibular glands of worker *Camponotus ferrugineus* (Fabr.): an anomoly [sic] in the subgenus *Camponotus*. Pp. 475-480 in: Trager, J. C. (ed.) Advances in myrmecology. Leiden: E. J. Brill, xxvii + 551 pp. [1988]

Dufour, L. 1856. Note sur la *Formica barbara*. Ann. Soc. Entomol. Fr. (3)4:341-343. [1856.11.12]

Dufour, L. 1857. Mélanges entomologiques. Ann. Soc. Entomol. Fr. (3)5:39-70. [1857.06.24]

Dufour, L. 1862. Notices entomologiques. Ann. Soc. Entomol. Fr. (4)2:131-148. [1862.06.05]

Dufour, L. 1864. Note sur une nouvelle espèce de fourmi (*Formica Vinsonnella*). Ann. Soc. Entomol. Fr. (4)4:210. [1864.10.12]

Dufour, L., Perris, E. 1840. Sur les insectes Hyménoptères qui nichent dans l'intérieur des tiges sèches de la Ronce. Ann. Soc. Entomol. Fr. 9:5-53. [1840.03]

Duméril, C. 1860. Entomologie analytique. [part] Mém. Acad. Sci. Inst. Fr. 31:665-1340. [1860] [Ants pp. 897-931.]

Du Merle, P. 1978. Les peuplements de fourmis et les peuplements d'acridiens du Mont Ventoux II. - Les peuplements de fourmis. Terre Vie 32(suppl. 1):161-218. [1978.03]

Du Merle, P. 1982. Fréquentation des strates arbustive et arborescente par les fourmis en montagne méditerranéenne française. Insectes Soc. 29:422-444. [1982.10]

Du Merle, P., Barbero, M., Bartes, J. P., Callot, G., Mazet, R. 1984. Observations préliminaires sur l'écologie d'espèces des genres *Leptothorax*, *Lasius*, et *Plagiolepis*, et sur la position taxonomique des *Lasius niger* du sud de la France (Hym., Formicidae). Actes Colloq. Insectes Soc. 1:171-179. [1984.09]

Dumpert, K. 1972a. Alarmstoffrezeptoren auf der Antenne von *Lasius fuliginosus* (Latr.) (Hymenoptera, Formicidae). Z. Vgl. Physiol. 76:403-425. [1972.03.02]

Dumpert, K. 1972b. Bau und Verteilung der Sensillen auf der Antennengeissel von *Lasius fuliginosus* (Latr.) (Hymenoptera, Formicidae). Z. Morphol. Tiere 73:95-116. [1972.09.05]

Dumpert, K. 1978. Das Sozialleben der Ameisen. Berlin: Paul Parey, 253 pp. [1978]

Dumpert, K. 1981. The social biology of ants. [Translated by C. Johnson.] Boston: Pitman, vi + 298 pp. [1981]

Dumpert, K. 1986 ("1985"). *Camponotus* (*Karavaievia*) *texens* sp. n. and *C.* (*K.*) *gombaki* sp. n. from Malaysia in comparison with other *Karavaievia* species (Formicidae: Formicinae). Psyche (Camb.) 92:557-573. [1986.04.27]

Dumpert, K. 1989. Taxonomy. Pp. 217-224 in: Dumpert, K., Maschwitz, U., Nässig, W., Dorow, W. *Camponotus* (*Karavaievia*) *asli* sp. n. and *C.* (*K.*) *montanus* sp. n., two weaver ant species from Malaysia (Formicidae: Formicinae). Zool. Beitr. (N.F.)32:217-231. [1989.09.12]

Dumpert, K. 1995. [Untitled. *Camponotus* (*Karavaievia*) *orinus* Dumpert, new name.] P. 115 in: Bolton, B. A new general catalogue of the ants of the world. Cambridge, Mass.: Harvard University Press, 504 pp. [1995.10]

Dumpert, K., Maschwitz, U., Nässig, W., Dorow, W. 1989. *Camponotus* (*Karavaievia*) *asli* sp. n. and *C.* (*K.*) *montanus* sp. n., two weaver ant species from Malaysia (Formicidae: Formicinae). Zool. Beitr. (N.F.)32:217-231. [1989.09.12]

Duncan, F. D., Crewe, R. M. 1993. A comparison of the energetics of foraging of three species of *Leptogenys* (Hymenoptera, Formicidae). Physiol. Entomol. 18:372-378. [1993.12]

Duncan, F. M. 1937. On the dates of publication of the Society's 'Proceedings', 1859-1926. Proc. Zool. Soc. Lond. Ser. A 107:71-77. [1937.04.15] [Duncan's contribution is followed by two appendices, by F. H. Waterhouse and H. Peavot, respectively, providing dates for the 'Proceedings', 1830-1858, and the 'Transactions', 1833-1869.]

Dunton, R. F., Vinson, S. B., Johnston, J. S. 1991. Unique isozyme electromorphs in polygynous red imported fire ant populations. Biochem. Syst. Ecol. 19:453-460. [1991.09.18]

Dupuis, C. 1986. Dates de publication de l'"Histoire naturelle générale et particulière des Crustacés et des Insectes" (1802-1805) par Latreille dans le "Buffon de Sonnini". Ann. Soc. Entomol. Fr. (n.s.)22:205-210. [1986.06.30]

Eady, R. D. 1968. Some illustrations of microsculpture in the Hymenoptera. Proc. R. Entomol. Soc. Lond. Ser. A 43:66-72. [1968.07.12]

Eastlake Chew, A., Chew, R. M. 1980. Body size as a determinant of small-scale distributions of ants in evergreen woodland southeastern Arizona. Insectes Soc. 27:189-202. [1980.09]

Echols, H. W. 1964. *Gnamptogenys hartmani* (Hymenoptera: Formicidae) discovered in Louisiana. Ann. Entomol. Soc. Am. 57:137. [1964.01.15]

Eckert, J. E., Mallis, A. 1937. Ants and their control in California. Univ. Calif. Agric. Exp. Stn. Circ. 342:1-37. [1937.07] [Revised edition appeared September, 1941.]

Eder, J., Rembold, H. (eds.) 1987. Chemistry and biology of social insects. München: Verlag J. Peperny, xxxv + 757 pp. [1987]

Ehrhardt, S. 1931. Über Arbeitsteilung bei *Myrmica*- und *Messor*-Arten. Z. Morphol. Ökol. Tiere 20:755-812. [1931.03.14]

Ehrhardt, W. 1987. Biosystematics of the slavemaking ant genus *Chalepoxenus*. Pp. 39-40 in: Eder, J., Rembold, H. (eds.) Chemistry and biology of social insects. München: Verlag J. Peperny, xxxv + 757 pp. [1987]

Eichhorn, O. 1964. Zur Verbreitung und Ökologie der hügelbauenden Waldameisen in den Ostalpen. Z. Angew. Entomol. 54:253-289. [1964.08]

Eichhorn, O. 1971a. Zur Verbreitung und Ökologie der Ameisen der Hauptwaldtypen mitteleuropäischer Gebirgswälder. Z. Angew. Entomol. 67:170-179. [1971.01]

Eichhorn, O. 1971b. Zur Verbreitung und Ökologie von *Formica fusca* L. und *F. lemani* Bondroit in den Hauptwaldtypen der mitteleuropäischen Gebirgswälder (zugleich ein Beitrag zum "Weisstannenproblem" im Thüringer Wald). Z. Angew. Entomol. 68:337-344. [1971.08]

Eichhorn, O. 1972. Beborstungsunterschiede bei Arbeiterinnen der *Serviformica*-Gruppe (Hym., Formicidae) und Hinweise auf ihre Ökologie. Waldhygiene 9:261-264. [1972]

Eichler, W. 1972. Pharaoameisen-Fundorte in der DDR. Angew. Parasitol. 13:245-246. [1972.12]

Eichler, W. 1974. Die Pharaoameise in der UdSSR. Dtsch. Entomol. Z. (N.F.)21:151-157. [1974.05.20]

Eichwald, E. 1841. Fauna Caspio-Caucasia nonnullis observationibus novis. Nouv. Mém. Soc. Imp. Nat. Mosc. 7:1-290. [1841]

Eidmann, H. 1925. Die Koloniegründung der Ameisen. Sitzungsber. Ges. Morph. Physiol. Münch. 36:78-82. [1925]

Eidmann, H. 1926a. Die Koloniegründung der einheimischen Ameisen. Z. Vgl. Physiol. 3:776-826. [1926.03.29]

Eidmann, H. 1926b. Koloniegründung bei Ameisen. Pp. 70-77 in: Jordan, K., Horn, W. (eds.) Verhandlungen des III. Internationalen Entomologen-Kongresses, Zürich, 19.-25. Juli 1925. Band II. Vorträge. Weimar: G. Uschmann, 646 pp. [1926.08]

Eidmann, H. 1926c. Die Ameisenfauna der Balearen. Z. Morphol. Ökol. Tiere 6:694-742. [1926.11.06]

Eidmann, H. 1927. Zur Kenntnis der Insektenfauna der balearischen Inseln. Entomol. Mitt. 16: 24-37. [1927.01.05] [Ants pp. 33-34.]

Eidmann, H. 1928. Weitere Beobachtungen über die Koloniegründung einheimischer Ameisen. Z. Vgl. Physiol. 7:39-55. [1928.02.14]

Eidmann, H. 1930. Entomologische Ergebnisse einer Reise nach Ostasien. Verh. Zool.-Bot. Ges. Wien 79:308-335. [1930.10.31]

Eidmann, H. 1933. Zur Kenntnis der Ameisenfauna von Südlabrador. Zool. Anz. 101:201-221. [1933.01.15]

Eidmann, H. 1936a. Ökologisch-faunistische Studien an südbrasilianischen Ameisen. [part] Arb. Physiol. Angew. Entomol. Berl.-Dahl. 3:26-48. [1936.03.18] [This and the succeeding paper include new taxa attributed to Menozzi, but Eidmann is responsible for any descriptions and should be credited with authorship of the names (Article 50, ICZN).]

Eidmann, H. 1936b. Ökologisch-faunistische Studien an südbrasilianischen Ameisen. [concl.] Arb. Physiol. Angew. Entomol. Berl.-Dahl. 3:81-114. [1936.05.04]

Eidmann, H. 1937. Die Gäste und Gastverhältnisse der Blattschneiderameise *Atta sexdens* L. Z. Morphol. Ökol. Tiere 32:391-462. [1937.02.27]

Eidmann, H. 1938. Zur Kenntnis der Lebensweise der Blattschneiderameise *Acromyrmex subterraneus* For. var. *eidmanni* Santschi und ihrer Gäste. Rev. Entomol. (Rio J.) 8:291-314. [1938.06.25]

Eidmann, H. 1941a. Zur Ökologie und Zoogeographie der Ameisenfauna von Westchina und Tibet. Wissenschaftliche Ergebnisse der 2. Brooke Dolan-Expedition 1934-1935. Z. Morphol. Ökol. Tiere 38:1-43. [1941.11.18] [Includes nomina nuda, attributed to Menozzi.]

Eidmann, H. 1941b. Zur Kenntnis der *Crematogaster impressa* Em. (Hym. Formicidae) und ihrer Gäste. 12. Beitrag zu den wissenschaftlichen Ergebnissen der Forschungsreise H. Eidmann nach Spanisch-Guinea 1939/40. Zool. Anz. 136:207-220. [1941.12.31]

Eidmann, H. 1942a. Zur Kenntnis der Ameisenfauna des Nanga Parbat. Zool. Jahrb. Abt. Syst. Ökol. Geogr. Tiere 75:239-266. [1942.07.01]

Eidmann, H. 1942b. Die Überwinterung der Ameisen. Z. Morphol. Ökol. Tiere 39:217-275. [1942.12.31]

Eidmann, H. 1943. Successionen westafrikanischer Holzinsekten 25. Beitrag zu den Ergebnissen der Westafrika-Expedition Eidmann 1939/40. Mitt. Hermann Göring Akad. Dtsch. Forstwiss. 3(1):240-271. [1943]

Eidmann, H. 1944. Die Ameisenfauna von Fernando Poo. 27. Beitrag zu den Ergebnissen der Westafrika-Expedition. Zool. Jahrb. Abt. Syst. Ökol. Geogr. Tiere 76:413-490. [1944.01.18] [Includes "new species", attributed to Menozzi, but described previously (see Menozzi, 1942a).]

Eisner, T. 1957. A comparative morphological study of the proventriculus of ants (Hymenoptera: Formicidae). Bull. Mus. Comp. Zool. 116:439-490. [1957.07]

Eisner, T., Brown, W. L., Jr. 1958. The evolution and social significance of the ant proventriculus. Pp. 503-508 in: Becker, E. C. et al. (eds.) Proceedings of the Tenth International Congress of Entomology, Montreal, August 17-25, 1956. Volume 2. Ottawa: Mortimer Ltd., 1055 pp. [1958.12]

Eisner, T., Happ, G. M. 1962. The infrabuccal pocket of a formicine ant: a social filtration device. Psyche (Camb.) 69:107-116. [1962.10.31]

Eisner, T., Wilson, E. O. 1952. The morphology of the proventriculus of a formicine ant. Psyche (Camb.) 59:47-60. [1952.10.16]

Elias, S. A. 1982. Holocene insect fossils from two sites at Ennadai Lake, Keewatin, Northwest Territories, Canada. Quat. Res. (N. Y.) 17:371-390. [1982.05]

Elmes, G. W. 1971. An experimental study on the distribution of heathland ants. J. Anim. Ecol. 40:495-499. [1971.06]

Elmes, G. W. 1973. Miniature queens of the ant *Myrmica rubra* L. (Hymenoptera, Formicidae). Entomologist 106:133-136. [1973.07.26]

Elmes, G. W. 1976. Some observations on the microgyne form of *Myrmica rubra* L. (Hymenoptera, Formicidae). Insectes Soc. 23:3-22. [1976.03]

Elmes, G. W. 1978. A morphometric comparison of three closely related species of *Myrmica* (Formicidae), including a new species from England. Syst. Entomol. 3:131-145. [1978.04]

Elmes, G. W. 1980. Queen number in colonies of ants of the genus *Myrmica.* Insectes Soc. 27: 43-60. [1980.03]

Elmes, G. W. 1981. An aberrant form of *Myrmica scabrinodis* Nylander (Hym. Formicidae). Insectes Soc. 28:27-31. [1981.03]

Elmes, G. W. 1982. The phenology of five species of *Myrmica* (Hymenoptera Formicidae) from South Dorset, England. Insectes Soc. 29:548-560. [1982.12]

Elmes, G. W. 1991a. The social biology of *Myrmica* ants. Actes Colloq. Insectes Soc. 7:17-34. [1991.06]

Elmes, G. W. 1991b. Mating strategy and isolation between the two forms, *macrogyna* and *microgyna*, of *Myrmica ruginodis* (Hym. Formicidae). Ecol. Entomol. 16:411-423. [1991.11]

Elmes, G. W. 1994. A population of the social parasite *Myrmica hirsuta* Elmes (Hymenoptera, Formicidae) recorded from Jutland, Denmark, with a first description of the worker caste. Insectes Soc. 41:437-442. [1994]

Elmes, G. W., Clarke, R. T. 1981. A biometric investigation of variation of workers of *Myrmica ruginodis* Nylander (Formicidae). Pp. 121-140 in: Howse, P. E., Clement, J.-L. (eds.) Biosystematics of social insects. Systematics Association Special Volume No. 19. London: Academic Press, 346 pp. [1981]

Elmes, G. W., Thomas, J. A. 1985. Morphometrics as a tool in identification: a case study of *Myrmica* from France (Hymenoptera, Formicidae). Actes Colloq. Insectes Soc. 2:97-108. [1985.06]

Elmes, G. W., Wardlaw, J. C. 1981. The quantity and quality of overwintered larvae in five species of *Myrmica* (Hymenoptera: Formicidae). J. Zool. (Lond.) 193:429-446. [1981.03.09]

Elmes, G. W., Wardlaw, J. C. 1983a. A comparison of the effect of temperature on the development of large hibernated larvae of four species of *Myrmica* (Hymenoptera:Formicidae). Insectes Soc. 30:106-118. [1983.07]

Elmes, G. W., Wardlaw, J. C. 1983b. A comparison of the effect of a queen upon the development of large hibernated larvae of six species of the genus *Myrmica* (Hym. Formicidae). Insectes Soc. 30:134-148. [1983.09]

Elton, E. T. G. 1976 ("1975"). Females of *Formica rufa* L. (Hymenoptera, Formicidae) with enlarged labial glands. Insectes Soc. 22:405-414. [1976.03]

Emerson, A. E. 1971. [Review of: Evans, M. A., Evans, H. E. 1970. William Morton Wheeler, biologist. Cambridge, Mass.: Harvard University Press, 363 pp.] Science (Wash. D. C.) 172:679. [1971.05.14]

Emery, C. 1869a. Formicidarum italicorum species duae novae. Bull. Soc. Entomol. Ital. 1:135-137. [1869.04]

Emery, C. 1869b. Enumerazione dei formicidi che rinvengonsi nei contorni di Napoli con descrizioni di specie nuove o meno conosciute. Ann. Accad. Aspir. Nat. Secunda Era 2:1-26. [1869]

Emery, C. 1869c. Descrizione di una nuova formica italiana. Annu. Mus. Zool. R. Univ. Napoli 5:117-118. [1869]

Emery, C. 1870. Studi mirmecologici. Bull. Soc. Entomol. Ital. 2:193-201. [1870.06]
Emery, C. 1875a ("1876"). Ueber hypogaeische Ameisen. Stett. Entomol. Ztg. 37:71-76. [1875.12]
Emery, C. 1875b. Le formiche ipogee con descrizioni di specie nuove o poco note. Ann. Mus. Civ. Stor. Nat. 7:465-474. [1875]
Emery, C. 1875c. Aggiunta alla nota sulle formiche ipogee. Ann. Mus. Civ. Stor. Nat. 7:895. [1875]
Emery, C. 1877a. Catalogo delle formiche esistenti nelle collezioni del Museo Civico di Genova. Parte prima. Formiche provenienti dal Viaggio dei signori Antinori, Beccari e Issel nel Mar Rosso e nel paese dei Bogos. [part] Ann. Mus. Civ. Stor. Nat. 9:363-368. [1877.03.27]
Emery, C. 1877b. Saggio di un ordinamento naturale dei Mirmicidei, e considerazioni sulla filogenesi delle formiche. Bull. Soc. Entomol. Ital. 9:67-83. [1877.03]
Emery, C. 1877c. Catalogo delle formiche esistenti nelle collezioni del Museo Civico di Genova. Parte prima. Formiche provenienti dal Viaggio dei signori Antinori, Beccari e Issel nel Mar Rosso e nel paese dei Bogos. [concl.] Ann. Mus. Civ. Stor. Nat. 9:369-381. [1877.04.03]
Emery, C. 1878a. Liste des fourmis de la collection de feu Camille van Volxem, avec la description d'une espèce nouvelle. Ann. Soc. Entomol. Belg. 21:viii-x. [> 1878.01.05]
Emery, C. 1878b. Catalogo delle formiche esistenti nelle collezioni del Museo Civico di Genova. Parte seconda. Formiche dell'Europa e delle regioni limitrofe in Africa e in Asia. [part] Ann. Mus. Civ. Stor. Nat. 12:43-48. [1878.02.12]
Emery, C. 1878c. Catalogo delle formiche esistenti nelle collezioni del Museo Civico di Genova. Parte seconda. Formiche dell'Europa e delle regioni limitrofe in Africa e in Asia. [concl.] Ann. Mus. Civ. Stor. Nat. 12:49-59. [1878.03.19]
Emery, C. 1880. Crociera del Violante, comandato dal capitano armatore Enrico d'Albertis, durante l'anno 1877. Formiche. Ann. Mus. Civ. Stor. Nat. 15:389-398. [1880.04.09]
Emery, C. 1881a. Spedizione italiana nell'Africa equatoriale. Risultati zoologici. Formiche. Ann. Mus. Civ. Stor. Nat. 16:270-276. [1881.01.13]
Emery, C. 1881b. Viaggio ad Assab nel Mar Rosso dei Signori G. Doria ed O. Beccari con il R. Avviso "Esploratore" dal 16 novembre 1879 al 26 febbraio 1880. I. Formiche. Ann. Mus. Civ. Stor. Nat. 16:525-535. [1881.03.09]
Emery, C. 1882a ("1881"). [Untitled. 3e Genre. - *Parasyscia*, Emery, nov. gen.] Pp. 235-236 in: André, Ern. Les fourmis [part]. Pp. 233-280 in: André, Edm. 1881-1886. Species des Hyménoptères d'Europe et d'Algérie. Tome Deuxième. Beaune: Edmond André, 919 + 48 pp. [1882.10]
Emery, C. 1882b. Le crociere dell'yacht "Corsaro" del capitano armatore Enrico d'Albertis. II. Formiche. Ann. Mus. Civ. Stor. Nat. 18:448-452. [1882.12.28]
Emery, C. 1883. Alcune formiche della Nuova Caledonia. Bull. Soc. Entomol. Ital. 15:145-151. [1883.09.25]
Emery, C. 1884a. Materiali per lo studio della fauna Tunisina raccolti da G. e L. Doria. III. Rassegna delle formiche della Tunisia. [part] Ann. Mus. Civ. Stor. Nat. 21[=(2)1]:373-384. [1884.10.06]
Emery, C. 1884b. Materiali per lo studio della fauna Tunisina raccolti da G. e L. Doria. III. Rassegna delle formiche della Tunisia. [concl.] Ann. Mus. Civ. Stor. Nat. 21[=(2)1]:384-386. [1884.10.09]
Emery, C. 1884c. [Untitled. *Colobopsis Clerodendri* Emery.] P. 51 in: Beccari, O. 1884-1886. Malesia. Raccolta di osservazioni botaniche intorno alla piante dell'Arcipelago Indo-malese e Papuano. Vol. 2. Genova: Tipografia del R. Istituto Sordo-Muti, 340 pp. [1884.12]
Emery, C. 1884d. [Untitled. *Iridomyrmex hospes* Emery.] Pp. 63-64 in: Beccari, O. 1884-1886. Malesia. Raccolta di osservazioni botaniche intorno alla piante dell'Arcipelago Indo-malese e Papuano. Vol. 2. Genova: Tipografia del R. Istituto Sordo-Muti, 340 pp. [1884.12] [Publication date from p. 340.]
Emery, C. 1885. [Untitled. Introduced by: "Prof. C. Emery mi ha comunicato il seguente prospetto dei gruppi principali di formiche, cui appartengono le specie trovate finora in queste piante".]

Pp. 211-212 in: Beccari, O. 1884-1886. Malesia. Raccolta di osservazioni botaniche intorno alla piante dell'Arcipelago Indo-malese e Papuano. Vol. 2. Genova: Tipografia del R. Istituto Sordo-Muti, 340 pp. [1885.09.28] [Publication date from p. 340.]

Emery, C. 1886a. Ueber dimorphe und flügellose Männchen bei Hymenopteren. Biol. Zentralbl. 5:686-689. [1886.01.15]

Emery, C. 1886b. Alcune formiche africane. Bull. Soc. Entomol. Ital. 18:355-366. [1886.06.29]

Emery, C. 1887a ("1886"). Catalogo delle formiche esistenti nelle collezioni del Museo Civico di Genova. Parte terza. Formiche della regione Indo-Malese e dell'Australia. [part] Ann. Mus. Civ. Stor. Nat. 24[=(2)4]:209-240. [1887.01.20]

Emery, C. 1887b ("1886"). Catalogo delle formiche esistenti nelle collezioni del Museo Civico di Genova. Parte terza. Formiche della regione Indo-Malese e dell'Australia. [part] Ann. Mus. Civ. Stor. Nat. 24[=(2)4]:241-256. [1887.01.24]

Emery, C. 1887c ("1886"). Catalogo delle formiche esistenti nelle collezioni del Museo Civico di Genova. Parte terza. Formiche della regione Indo-Malese e dell'Australia. [part] Ann. Mus. Civ. Stor. Nat. 24[=(2)4]:256-258. [1887.01.26]

Emery, C. 1887d ("1886"). Mimetismo e costumi parassitari dei *Camponotus lateralis* Ol. Bull. Soc. Entomol. Ital. 18:412-413. [1887.03.20]

Emery, C. 1887e. Catalogo delle formiche esistenti nelle collezioni del Museo Civico di Genova. Parte terza. Formiche della regione Indo-Malese e dell'Australia (continuazione e fine). [part] Ann. Mus. Civ. Stor. Nat. 25[=(2)5]:427-432. [1887.10.25]

Emery, C. 1887f. Catalogo delle formiche esistenti nelle collezioni del Museo Civico di Genova. Parte terza. Formiche della regione Indo-Malese e dell'Australia (continuazione e fine). [part] Ann. Mus. Civ. Stor. Nat. 25[=(2)5]:433-448. [1887.11.14]

Emery, C. 1887g. Catalogo delle formiche esistenti nelle collezioni del Museo Civico di Genova. Parte terza. Formiche della regione Indo-Malese e dell'Australia (continuazione e fine). [part] Ann. Mus. Civ. Stor. Nat. 25[=(2)5]:449-464. [1887.11.25]

Emery, C. 1887h. Catalogo delle formiche esistenti nelle collezioni del Museo Civico di Genova. Parte terza. Formiche della regione Indo-Malese e dell'Australia (continuazione e fine). [concl.] Ann. Mus. Civ. Stor. Nat. 25[=(2)5]:465-473. [1887.11.29]

Emery, C. 1887i. Catalogo delle formiche esistenti nelle collezioni del Museo Civico di Genova. Parte terza (Supplemento). Formiche raccolte dal sig. Elio Modigliani in Sumatra e nell'isola Nias. [part] Ann. Mus. Civ. Stor. Nat. 25[=(2)5]:528. [1887.12.12]

Emery, C. 1888a. Catalogo delle formiche esistenti nelle collezioni del Museo Civico di Genova. Parte terza (Supplemento). Formiche raccolte dal sig. Elio Modigliani in Sumatra e nell'isola Nias. [concl.] Ann. Mus. Civ. Stor. Nat. 25[=(2)5]:529-534. [1888.01.02]

Emery, C. 1888b ("1887"). Le tre forme sessuali del *Dorylus helvolus* L. e degli altri Dorilidi. Bull. Soc. Entomol. Ital. 19:344-351. [1888.01.20]

Emery, C. 1888c ("1887"). Formiche della provincia di Rio Grande do Sûl nel Brasile, raccolte dal dott. Hermann von Ihering. Bull. Soc. Entomol. Ital. 19:352-366. [1888.01.20]

Emery, C. 1888d. Über den sogenannten Kaumagen einiger Ameisen. Z. Wiss. Zool. 46:378-412. [1888.05.23]

Emery, C. 1888e. Alcune formiche della Repubblica Argentina raccolte dal Dott. C. Spegazzini. Ann. Mus. Civ. Stor. Nat. 26[=(2)6]:690-694. [1888.12.20]

Emery, C. 1889a. Alleanze difensive tra piante e formiche. Nuova Antol. Sci. Lett. Arti 103[=(3)19]:578-591. [1889.02.01]

Emery, C. 1889b. Intorno ad alcune formiche della fauna paleartica. Ann. Mus. Civ. Stor. Nat. 27[=(2)7]:439-443. [1889.08.14]

Emery, C. 1889c. Formiche di Birmania e del Tenasserim raccolte da Leonardo Fea (1885-87). [part] Ann. Mus. Civ. Stor. Nat. 27[=(2)7]:485-512. [1889.09.04]

Emery, C. 1889d. Formiche di Birmania e del Tenasserim raccolte da Leonardo Fea (1885-87). [concl.] Ann. Mus. Civ. Stor. Nat. 27[=(2)7]:513-520. [1889.09.09]

Emery, C. 1890a ("1889"). Alcune considerazioni sulla fauna mirmecologica dell'Africa. Bull. Soc. Entomol. Ital. 21:69-75. [1890.06.30]
Emery, C. 1890b. Voyage de M. E. Simon au Venezuela (Décembre 1887 - Avril 1888). Formicides. Ann. Soc. Entomol. Fr. (6)10:55-76. [1890.08.08]
Emery, C. 1890c. Studii sulle formiche della fauna neotropica. Bull. Soc. Entomol. Ital. 22:38-80. [1890.09.15]
Emery, C. 1891a. Zur Biologie der Ameisen. Biol. Centralbl. 11:165-180. [1891.04.01]
Emery, C. 1891b. Le formiche dell'ambra Siciliana nel Museo Mineralogico dell'Università di Bologna. Mem. R. Accad. Sci. Ist. Bologna (5)1:141-165 [pagination of separate: 567-591]. [1891] [Session of 12 April 1891.]
Emery, C. 1891c. Exploration scientifique de la Tunisie. Zoologie. - Hyménoptères. Révision critique des fourmis de la Tunisie. Paris: Imprimerie Nationale, iii + 21 pp. [1891]
Emery, C. 1892a. Sopra alcune formiche raccolte dall'Ingegnere L. Bricchetti Robecchi nel paese dei Somali. [part] Ann. Mus. Civ. Stor. Nat. 32[=(2)12]:110-112. [1892.01.27]
Emery, C. 1892b. Sopra alcune formiche raccolte dall'Ingegnere L. Bricchetti Robecchi nel paese dei Somali. [concl.] Ann. Mus. Civ. Stor. Nat. 32[=(2)12]:113-122. [1892.01.28]
Emery, C. 1892c ("1891"). Note sinonimiche sulle formiche. Bull. Soc. Entomol. Ital. 23:159-167. [1892.02.27]
Emery, C. 1892d. Origines de la faune actuelle des fourmis de l'Europe. Bull. Soc. Vaudoise Sci. Nat. 27:258-260. [1892.02]
Emery, C. 1892e. [Untitled. From table of contents: "Sur une fourmi nouvelle d'Assinie et remarques sur les Dorylides d'Afrique."] Bull. Bimens. Soc. Entomol. Fr. 1892:liii-lv. [1892.03.05] [Reissued as Emery (1892h).]
Emery, C. 1892f ("1891"). Voyage de M. Ch. Alluaud dans le territoire d'Assinie (Afrique occidentale) en juillet et août 1886. Formicides. Ann. Soc. Entomol. Fr. 60:553-574. [1892.04.13]
Emery, C. 1892g. Älteres über Ameisen in Dornen afrikanischer Akazien. Zool. Anz. 15:237. [1892.06.27]
Emery, C. 1892h. [Untitled. From table of contents: "Sur une fourmi nouvelle d'Assinie et remarques sur les Dorylides d'Afrique."] Ann. Soc. Entomol. Fr. (Bull.) 61:liii-lv. [1892.10.12] [First published as Emery (1892e).]
Emery, C. 1893a ("1892"). [Untitled. Introduced by: "M. C. Emery, de Bologne, envoie les diagnoses de cinq nouveaux genres de Formicides".] Bull. Bimens. Soc. Entomol. Fr. 1892:cclxxv-cclxxvii. [1893.01.07] [Reissued as Emery (1893c).]
Emery, C. 1893b. Zirpende und springende Ameisen. Biol. Centralbl. 13:189-190. [1893.04.01]
Emery, C. 1893c ("1892"). [Untitled. Introduced by: "M. C. Emery, de Bologne, envoie les diagnoses de cinq nouveaux genres de Formicides".] Ann. Soc. Entomol. Fr. (Bull.) 61:cclxxv-cclxxvii. [1893.04.28] [First published as Emery (1893a).]
Emery, C. 1893d. Ueber die Herkunft der Pharao-Ameise. Biol. Centralbl. 13:435-436. [1893.07.15]
Emery, C. 1893e. Voyage de M. Ch. Alluaud aux îles Canaries. Formicides. Ann. Soc. Entomol. Fr. 62:81-88. [1893.07.31]
Emery, C. 1893f. Notice sur quelques fourmis des îles Galapagos. Ann. Soc. Entomol. Fr. 62: 89-92. [1893.07.31]
Emery, C. 1893g. Formicides de l'Archipel Malais. Rev. Suisse Zool. 1:187-229. [1893.09.28]
Emery, C. 1893h. Voyage de M. E. Simon à l'île de Ceylan (janvier-février 1892). Formicides. Ann. Soc. Entomol. Fr. 62:239-258. [1893.10.25] [Includes an appendix: Formicides de l'île d'Aden (pp. 256-258).]
Emery, C. 1893i. Voyage de M. E. Simon aux îles Philippines (mars et avril 1890). Formicides. Ann. Soc. Entomol. Fr. 62:259-270. [1893.10.25]
Emery, C. 1893j. Intorno ad alcune formiche della collezione Spinola. Boll. Mus. Zool. Anat. Comp. R. Univ. Torino 8(163):1-3. [1893.12.01]

Emery, C. 1893k. Beiträge zur Kenntniss der nordamerikanischen Ameisenfauna. Zool. Jahrb. Abt. Syst. Geogr. Biol. Tiere 7:633-682. [1893.12.23]

Emery, C. 1893l. Studio monografico sul genere *Azteca* Forel. Mem. R. Accad. Sci. Ist. Bologna (5)3:119-152 [pagination of separate: 319-352]. [1893] [Session of 27 March 1893.]

Emery, C. 1894a. Die Entstehung und Ausbildung des Arbeiterstandes bei den Ameisen. Biol. Centralbl. 14:53-59. [1894.01.15]

Emery, C. 1894b. Mission scientifique de M. Ch. Alluaud aux îles Séchelles (mars, avril, mai 1892). 2e mémoire. Formicides. Ann. Soc. Entomol. Fr. 63:67-72. [1894.04.10]

Emery, C. 1894c. Descriptions de deux fourmis nouvelles. Ann. Soc. Entomol. Fr. 63:72-74. [1894.04.10]

Emery, C. 1894d. Studi sulle formiche della fauna neotropica. VI-XVI. Bull. Soc. Entomol. Ital. 26:137-241. [1894.06.27]

Emery, C. 1894e. Viaggio del Dott. E. Festa in Palestina, nel Libano e regioni vicine. XI. Descrizione di un nuovo *Camponotus*. Boll. Mus. Zool. Anat. Comp. R. Univ. Torino 9(185):1-2. [1894.10.25]

Emery, C. 1894f. Viaggio del dottor Alfredo Borelli nella Repubblica Argentina e nel Paraguay. VIII. Formiche. Boll. Mus. Zool. Anat. Comp. R. Univ. Torino 9(186):1-4. [1894.10.25]

Emery, C. 1894g. *Camponotus sexguttatus* Fab. e *C. sexguttatus* Sm. et auct. Boll. Mus. Zool. Anat. Comp. R. Univ. Torino 9(187):1-4. [1894.10.25]

Emery, C. 1894h. [Untitled.] Pp. 373-401 in: Ihering, H. von. Die Ameisen von Rio Grande do Sul. Berl. Entomol. Z. 39:321-447. [1894.10] [Emery's contribution is in the form of species descriptions and comments on pp. 373-374, 376, 377, 378-379, 384, 384-385, 387, 388, 389, 390, 393-394, 395 and 401.]

Emery, C. 1894i. Descrizione di una nuova Formica di Sicilia. Nat. Sicil. 14:28. [1894.12]

Emery, C. 1894j. Rassegna degl'Imenotteri raccolti nel Mozambico dal Cav. Fornasini esistenti nel Museo Zoologico della R. Università di Bologna. I. Formicidi. Mem. R. Accad. Sci. Ist. Bologna (5)4:46-47 [pagination of separate: 112-113]. [1894] [Session of 10 December 1893.]

Emery, C. 1894k. Alcune formiche dell'isola di Creta. Bull. Soc. Entomol. Ital. Resoc. Adun. 26:7-10. [1894]

Emery, C. 1894l. Estudios sobre las hormigas de Costa Rica. An. Mus. Nac. Costa Rica 1888-1889:45-64. [1894]

Emery, C. 1894m. Estudio contributivo á la biologia de las hormigas. An. Mus. Nac. Costa Rica 1888-1889:65-67. [1894]

Emery, C. 1895a. Le problème des Doryles (Hyménoptères). Bull. Soc. Entomol. Fr. 1895:lxxi-lxxiv. [1895.03.09]

Emery, C. 1895b. Esplorazione del Giuba e dei suoi affluenti compiuta dal Cap. V. Bottego durante gli anni 1892-93 sotto gli auspicii della Società Geografica Italiana. Risultati zoologici. X. Formiche. Ann. Mus. Civ. Stor. Nat. 35[=(2)15]:175-184. [1895.04.09]

Emery, C. 1895c. [Untitled. *Pachycondyla Fauveli* n. sp.] P. 175 in: Wasmann, E. Die Ameisen- und Termitengäste von Brasilien. Verh. K-K. Zool.-Bot. Ges. Wien 45:137-179. [1895.04]

Emery, C. 1895d. Beiträge zur Kenntniss der nordamerikanischen Ameisenfauna. (Schluss). Zool. Jahrb. Abt. Syst. Geogr. Biol. Tiere 8:257-360. [1895.05.11]

Emery, C. 1895e ("1894"). Note sur les fourmis du Chili avec descriptions de deux espèces nouvelles. Actes Soc. Sci. Chili 4:213-216. [1895.05.29]

Emery, C. 1895f ("1894"). On a new species of ant from New Zealand. Trans. Proc. N. Z. Inst. 27:635-636. [1895.05]

Emery, C. 1895g. Mission scientifique de M. Ch. Alluaud dans le territoire de Diego-Suarez (Madagascar-nord) (Avril-août 1893). Formicides. Ann. Soc. Entomol. Belg. 39:336-345. [> 1895.07.06]

Emery, C. 1895h. Descriptions de quelques fourmis nouvelles d'Australie. Ann. Soc. Entomol. Belg. 39:345-358. [> 1895.07.06]

Emery, C. 1895i. Voyage de M. E. Simon dans l'Afrique australe (janvier-avril 1893). 3e mémoire. Formicides. Ann. Soc. Entomol. Fr. 64:15-56. [1895.08.16]

Emery, C. 1895j. On the origin of European and North American ants. Nature (Lond.) 52:399-400. [1895.08.22]

Emery, C. 1895k. Deuxième note sur les fourmis du Chili. Actes Soc. Sci. Chili 5:10-18. [1895.08.31]

Emery, C. 1895l. Die Gattung *Dorylus* Fab. und die systematische Eintheilung der Formiciden. Zool. Jahrb. Abt. Syst. Geogr. Biol. Tiere 8:685-778. [1895.10.18]

Emery, C. 1895m. Viaggio di Leonardo Fea in Birmania e regioni vicine. LXIII. Formiche di Birmania del Tenasserim e dei Monti Carin raccolte da L. Fea. Parte II. Ann. Mus. Civ. Stor. Nat. 34[=(2)14]:450-483. [1895]

Emery, C. 1895n. Sopra alcune formiche della fauna mediterranea. Mem. R. Accad. Sci. Ist. Bologna (5)5:59-75 [pagination of separate: 291-307]. [1895] [Session of 21 April 1895.]

Emery, C. 1895o. [Untitled. *Crematogaster constructor* n. sp.] Pp. 135-136 in: Mayr, G. Afrikanische Formiciden. Ann. K-K. Naturhist. Mus. Wien 10:124-154. [1895] [This apparently postdates Emery's description in Ann. Soc. Entomol. Fr. 64:29 (16 August 1895).]

Emery, C. 1895p. Ueber die Ameisenfauna von Nordamerika. Verh. Ges. Dtsch. Naturforsch. Ärzte 66(2,1.Hälfte):141-142. [1895]

Emery, C. 1896a. Formiche raccolte dal dott. E. Festa nei pressi del golfo di Darien. Boll. Mus. Zool. Anat. Comp. R. Univ. Torino 11(229):1-4. [1896.03.01]

Emery, C. 1896b. Alcune forme nuove del genere *Azteca* For. e note biologiche. Boll. Mus. Zool. Anat. Comp. R. Univ. Torino 11(230):1-7. [1896.03.01]

Emery, C. 1896c. Sur les fourmis des genres *Sysphincta* et *Proceratium* (Hym.). Bull. Soc. Entomol. Fr. 1896:101-102. [1896.03.07]

Emery, C. 1896d. Sur les fourmis du genre *Macromischa* Rog. (Hym.). Bull. Soc. Entomol. Fr. 1896:102-103. [1896.03.07]

Emery, C. 1896e. Clef analytique des genres de la famille des Formicides, pour la détermination des neutres. Ann. Soc. Entomol. Belg. 40:172-189. [> 1896.05.02]

Emery, C. 1896f. Formicides récoltés à Buitenzorg (Java), par M. Massart. Ann. Soc. Entomol. Belg. 40:245-249. [> 1896.06.06]

Emery, C. 1896g. Studi sulle formiche della fauna neotropica. XVII-XXV. Bull. Soc. Entomol. Ital. 28:33-107. [1896.10.31]

Emery, C. 1896h. Formiciden, gesammelt in Paraguay von Dr. J. Bohls. Zool. Jahrb. Abt. Syst. Geogr. Biol. Tiere 9:625-638. [1896.11.15]

Emery, C. 1896i ("1897"). Formiche raccolte dal Cap. V. Bottego nella regione dei Somali. Ann. Mus. Civ. Stor. Nat. 37[=(2)17]:153-160. [1896.12.15]

Emery, C. 1896j. Saggio di un catalogo sistematico dei generi *Camponotus*, *Polyrhachis* e affini. Mem. R. Accad. Sci. Ist. Bologna (5)5:363-382 [pagination of separate: 761-780]. [1896] [Session of 8 March 1896.]

Emery, C. 1896k. Le polymorphisme des fourmis et la castration alimentaire. Pp. 395-410 in: Hoek, P. P. C. (ed.) Compte-rendu des séances du Troisième Congrès International de Zoologie. Leyde, 16-21 Septembre 1895. Leyden: E. J. Brill, 543 pp. [1896]

Emery, C. 1897a ("1896"). Description d'une fourmi nouvelle d'Algérie (Hymén.). Bull. Soc. Entomol. Fr. 1896:418-419. [1897.01.09]

Emery, C. 1897b. Descriptions de deux fourmis (Hymén.). Bull. Soc. Entomol. Fr. 1897:12-14. [1897.01.23]

Emery, C. 1897c. Formicidarum species novae vel minus cognitae in collectione Musaei Nationalis Hungarici quas in Nova-Guinea, colonia germanica, collegit L. Biró. Természetr. Füz. 20:571-599. [1897.11.01]

Emery, C. 1897d. Viaggio di Lamberto Loria nella Papuasia orientale. XVIII. Formiche raccolte nella Nuova Guinea dal Dott. Lamberto Loria. [part] Ann. Mus. Civ. Stor. Nat. 38[=(2)18]:546-576. [1897.11.16]

Emery, C. 1897e. Viaggio di Lamberto Loria nella Papuasia orientale. XVIII. Formiche raccolte nella Nuova Guinea dal Dott. Lamberto Loria. [concl.] Ann. Mus. Civ. Stor. Nat. 38[=(2)18]:577-594. [1897.11.22]

Emery, C. 1897f. Formiche raccolte da Don Eugenio dei Principi Ruspoli, durante l'ultimo suo viaggio nelle regioni dei Somali e dei Galla. Ann. Mus. Civ. Stor. Nat. 38[=(2)18]:595-605. [1897.11.22]

Emery, C. 1897g. Anhang. Verzeichniss der auf der zweiten Reise nach Kleinasien (1897) gesammelten Ameisen, mit einer Neubeschreibung. P. 239 in: Escherich, K. Zur Kenntniss der Myrmecophilen Kleinasiens. I. Coleoptera. Wien. Entomol. Ztg. 16:229-239. [1897.11.30]

Emery, C. 1897h. Revisione del genere *Diacamma* Mayr. Rend. Sess. R. Accad. Sci. Ist. Bologna (n.s.)1:147-167. [1897] [Session of 25 April 1897.]

Emery, C. 1898a. Aggiunte e correzioni al saggio di un catalogo sistematico dei generi *Camponotus*, *Polyrachis* [sic] e affini. Rend. Sess. R. Accad. Sci. Ist. Bologna (n.s.)2:225-231. [1898] [Session of 22 May 1898.]

Emery, C. 1898b. Descrizioni di formiche nuove malesi e australiane. Note sinonimiche. Rend. Sess. R. Accad. Sci. Ist. Bologna (n.s.)2:231-245. [1898] [Session of 22 May 1898.]

Emery, C. 1898c. Beiträge zur Kenntniss der palaearktischen Ameisen. Öfvers. Fin. Vetensk.-Soc. Förh. 20:124-151. [1898]

Emery, C. 1899a. Formiche dell'ultima spedizione Bottego. Ann. Mus. Civ. Stor. Nat. 39[=(2)19]:499-501. [1899.01.10]

Emery, C. 1899b. Glanures myrmécologiques (Hymén.). Bull. Soc. Entomol. Fr. 1899:17-20. [1899.02.04]

Emery, C. 1899c. Ergebnisse einer Reise nach dem Pacific (Schauinsland 1896-1897). Formiciden. Zool. Jahrb. Abt. Syst. Geogr. Biol. Tiere 12:438-440. [1899.09.28]

Emery, C. 1899d. Fourmis d'Afrique. Ann. Soc. Entomol. Belg. 43:459-504. [1899.11.06]

Emery, C. 1899e. Formiche di Madagascar raccolte dal Sig. A. Mocquerys nei pressi della Baia di Antongil (1897-1898). Bull. Soc. Entomol. Ital. 31:263-290. [1899.12.31]

Emery, C. 1899f. Intorno alle larve di alcune formiche. Rend. Sess. R. Accad. Sci. Ist. Bologna (n.s.)3:93. [1899] [Session of 7 May 1899.]

Emery, C. 1899g. Intorno alle larve di alcune formiche. Mem. R. Accad. Sci. Ist. Bologna (5)8: 3-10. [1899] [Session of 7 May 1899.]

Emery, C. 1900a. Intorno al torace delle formiche e particolarmente dei neutri. Bull. Soc. Entomol. Ital. 32:103-119. [1900.05.20]

Emery, C. 1900b. Formicidarum species novae vel minus cognitae in collectione Musaei Nationalis Hungarici quas in Nova-Guinea, colonia germanica, collegit L. Biró. Publicatio secunda. Természetr. Füz. 23:310-338. [1900.08.01] [Publication date is that of separate. Whole volume dated 15 September 1900.]

Emery, C. 1900c. Formiche raccolte da Elio Modigliani in Sumatra, Engano e Mentawei. [part] Ann. Mus. Civ. Stor. Nat. 40[=(2)20]:661-688. [1900.12.20]

Emery, C. 1900d. Formiche raccolte da Elio Modigliani in Sumatra, Engano e Mentawei. [part] Ann. Mus. Civ. Stor. Nat. 40[=(2)20]:689-720. [1900.12.21]

Emery, C. 1900e. Nuovi studi sul genere *Eciton*. Mem. R. Accad. Sci. Ist. Bologna (5)8:173-188 [pagination of separate: 511-526]. [1900] [Session of 25 March 1900.]

Emery, C. 1900f ("1899"). Ueber Ameisenlarven. [Including discussion pp. 234-235 by Forel, Emery, Brunn and Hertwig]. Verh. Ges. Dtsch. Naturforsch. Ärzte 71(2,1.Hälfte):233-235. [1900]

Emery, C. 1901a. Formiche raccolte da Elio Modigliani in Sumatra, Engano e Mentawei. [concl.] Ann. Mus. Civ. Stor. Nat. 40[=(2)20]:721-722. [1901.02.04]

Emery, C. 1901b. Notes sur les sous-familles des Dorylines et Ponérines (Famille des Formicides). Ann. Soc. Entomol. Belg. 45:32-54. [1901.03.07]

Emery, C. 1901c. Remarques sur un petit groupe de *Pheidole* (Hymén. Formic.) de la région sonorienne. Bull. Soc. Entomol. Fr. 1901:119-121. [1901.03.24]

Emery, C. 1901d. Note sulle Doriline. Bull. Soc. Entomol. Ital. 33:43-56. [1901.05.31]
Emery, C. 1901e. Spicilegio mirmecologico. Bull. Soc. Entomol. Ital. 33:57-63. [1901.05.31]
Emery, C. 1901f. A propos de la classification des Formicides. Réponse à l'article publié sous le même titre par M. le Prof. Forel. Ann. Soc. Entomol. Belg. 45:197-198. [1901.07.03]
Emery, C. 1901g. Le formiche in rapporto alla fauna di Selebes. Monit. Zool. Ital. 12:178. [1901.07.05]
Emery, C. 1901h. Ameisen gesammelt in Ceylon von Dr. W. Horn 1899. Dtsch. Entomol. Z. 1901:113-122. [1901.07]
Emery, C. 1901i. Formiciden von Celebes. Zool. Jahrb. Abt. Syst. Geogr. Biol. Tiere 14:565-580. [1901.08.06]
Emery, C. 1901j. Der Geschlechtspolymorphismus der Treiberameisen und die flügellose Urform der Ameisenweibchen. Naturwiss. Wochenschr. 17:54-55. [1901.11.03]
Emery, C. 1901k. Formicidarum species novae vel minus cognitae in collectione Musaei Nationalis Hungarici, quas in Nova-Guinea, colonia germanica, collegit L. Biró. Publicatio tertia. Természetr. Füz. 25:152-160. [1901.12.15] [Publication date is that of separate. Whole volume dated 20 April 1902.]
Emery, C. 1901l. Studi sul polimorfismo e la metamorfosi nel genere *Dorylus*. Mem. R. Accad. Sci. Ist. Bologna (5)9:183-201 [pagination of separate: 415-433]. [1901] [Session of 12 May 1901.]
Emery, C. 1901m. [Untitled. Descriptions of new taxa: *Messor barbarus* Linn. var. *lobulifera* Emery n. var.; *Formica nasuta* Nyl. subspec. *mongolica* Emery n. subspec.] P. 159 in: Mocsáry, S., Szépligeti, G. Hymenopterák (Hymenopteren). Pp. 121-169 in: Horváth, G. Zichy Jenó Gróf harmadik ázsiai utazásának állattani eredményei. Vol. 2. Budapest: V. Hornyánsky, xli + 470 pp. [1901]
Emery, C. 1902a. Description d'une nouvelle espèce de fourmi du Brésil. Bull. Soc. Entomol. Fr. 1902:181. [1902.06.06]
Emery, C. 1902b. [Untitled. *Camponotus punctulatus* Mayr, subsp. *termitarius* Em. n. subsp.] P. 297 in: Wasmann, E. Neues über die zusammengesetzten Nester und gemischten Kolonien der Ameisen. [part]. Allg. Z. Entomol. 7:293-298. [1902.08.15]
Emery, C. 1902c. Note mirmecologiche. Rend. Sess. R. Accad. Sci. Ist. Bologna (n.s.)6:22-34. [1902] [Session of 17 November 1901.]
Emery, C. 1903. Intorno ad alcune specie di *Camponotus* dell'America Meridionale. Rend. Sess. R. Accad. Sci. Ist. Bologna (n.s.)7:62-81. [1903] [Session of 8 February 1903.]
Emery, C. 1904a. Le affinità del genere *Leptanilla* e i limiti delle Dorylinae. Arch. Zool. Ital. 2:107-116. [1904.08.24]
Emery, C. 1904b. Zur Kenntniss des Polymorphismus der Ameisen. Zool. Jahrb. Suppl. 7:587-610. [1904]
Emery, C. 1905a. [Untitled. *Leptothorax tuberum* F. subsp. *exilis* Emery var. *dichroa* n. var.] P. 452 in: Mantero, G. Materiali per una fauna dell'Archipelago Toscano. Isola del Giglio. II. Ann. Mus. Civ. Stor. Nat. 41[=(3)1]:449-454. [1905.01.11]
Emery, C. 1905b. Éthologie, phylogénie et classification. Pp. 160-174 in: Bedot, M. (ed.) Compte-rendu des séances du Sixième Congrès International de Zoologie tenu à Berne du 14 au 16 août 1904. Genève: W. Kündig & Fils, xii + 733 pp. [1905.05.25]
Emery, C. 1905c. [Untitled. Discussion of paper by A. Forel.] P. 456 in: Bedot, M. (ed.) Compte-rendu des séances du Sixième Congrès International de Zoologie tenu à Berne du 14 au 16 août 1904. Genève: W. Kündig & Fils, xii + 733 pp. [1905.05.25]
Emery, C. 1905d. Sur l'origine des fourmilières. Pp. 459-462 in: Bedot, M. (ed.) Compte-rendu des séances du Sixième Congrès International de Zoologie tenu à Berne du 14 au 16 août 1904. Genève: W. Kündig & Fils, xii + 733 pp. [1905.05.25]
Emery, C. 1905e. Deux fourmis de l'ambre de la Baltique (Hym.). Bull. Soc. Entomol. Fr. 1905:187-189. [1905.09.25]

Emery, C. 1905f. Revisione delle specie del genere *Atta* appartenenti ai sottogeneri *Moellerius* e *Acromyrmex*. Mem. R. Accad. Sci. Ist. Bologna (6)2:39-54 [pagination of separate: 107-122]. [1905.11]

Emery, C. 1905g. Le forme paleartiche del *Camponotus maculatus* F. Rend. Sess. R. Accad. Sci. Ist. Bologna (n.s.)9:27-44. [1905] [Session of 18 December 1904.]

Emery, C. 1906a. Über W. H. Ashmeads neues System der Ameisen. Zool. Anz. 29:717-718. [1906.02.23]

Emery, C. 1906b. Note sur *Prenolepis vividula* Nyl. et sur la classification des espèces du genre *Prenolepis*. Ann. Soc. Entomol. Belg. 50:130-134. [1906.04.25]

Emery, C. 1906c ("1905"). Studi sulle formiche della fauna neotropica. XXVI. Bull. Soc. Entomol. Ital. 37:107-194. [1906.06.16]

Emery, C. 1906d. Résultats du voyage du S. Y. Belgica en 1897-1898-1899. Rapports scientifiques. Zoologie. Hyménoptères. Formicidae. Anvers: J.-E. Buschmann, 1 p. [1906.08.20]

Emery, C. 1906e. Zur Kenntnis des Polymorphismus der Ameisen. Biol. Centralbl. 26:624-630. [1906.09.01]

Emery, C. 1906f. Rassegna critica delle specie paleartiche del genere *Myrmecocystus*. Mem. R. Accad. Sci. Ist. Bologna (6)3:47-61 [pagination of separate: 173-187]. [1906.11]

Emery, C. 1907. Una formica nuova italiana spettante ad un nuovo genere. Rend. Sess. R. Accad. Sci. Ist. Bologna (n.s.)11:49-51. [1907] [Session of 12 May 1907.]

Emery, C. 1908a. Beiträge zur Monographie der Formiciden des paläarktischen Faunengebietes. Dtsch. Entomol. Z. 1908:165-205. [1908.03.01]

Emery, C. 1908b. Remarques sur les observations de M. de Lannoy touchant l'existence de *Lasius mixtus* dans les fourmilières de *Lasius fuliginosus*. Ann. Soc. Entomol. Belg. 52:182-183. [1908.04.30]

Emery, C. 1908c. Descriptions d'une genre nouveau et de plusieurs formes nouvelles de fourmis du Congo. Ann. Soc. Entomol. Belg. 52:184-189. [1908.04.30]

Emery, C. 1908d. Beiträge zur Monographie der Formiciden des paläarktischen Faunengebietes. (Hym.) (Fortsetzung.) III. Die mit *Aphaenogaster* verwandte Gattungengruppe. Dtsch. Entomol. Z. 1908:305-338. [1908.05.01]

Emery, C. 1908e. [Untitled. Descriptions of new taxa: *Tetramorium caespitum* var. *diomedea* Emery, nova varietas; *Strongylognathus huberi* For., subsp. *rehbinderi* For., var. *cecconii* Emery, nova varietas.] P. 24 in: Cecconi, G. Contributo alla fauna delle Isole Tremiti. Boll. Mus. Zool. Anat. Comp. R. Univ. Torino 23(583):1-53. [1908.05.20]

Emery, C. 1908f. Beiträge zur Monographie der Formiciden des paläarktischen Faunengebietes. (Hym.) Teil III. Dtsch. Entomol. Z. 1908:437-465. [1908.07.01]

Emery, C. 1908g. Beiträge zur Monographie der Formiciden des paläarktischen Faunengebietes. (Hym.) Teil IV. Parasitische und Gast-Myrmicinen mit Ausnahme von *Strongylognathus*. Dtsch. Entomol. Z. 1908:549-558. [1908.09.01]

Emery, C. 1908h. *Myrmecocystus viaticus* et formes voisines. Bull. Soc. Vaudoise Sci. Nat. 44:213-217. [1908.09]

Emery, C. 1908i. Le formiche e gli alberi in Italia. Alpe 6:152-154. [1908.10]

Emery, C. 1908j. Beiträge zur Monographie der Formiciden des paläarktischen Faunengebietes. (Hym.) Teil V. Dtsch. Entomol. Z. 1908:663-686. [1908.11.01]

Emery, C. 1908k. [Untitled. *Azteca Muelleri* Em. nov. var. *Wacketi* C. Em.] P. 391 in: Forel, A. Ameisen aus Sao Paulo (Brasilien), Paraguay etc. gesammelt von Prof. Herm. v. Ihering, Dr. Lutz, Dr. Fiebrig, etc. Verh. K-K. Zool.-Bot. Ges. Wien 58:340-418. [1908.12.18]

Emery, C. 1909a. Beiträge zur Monographie der Formiciden des paläarktischen Faunengebietes. (Hym.) Teil VI. Dtsch. Entomol. Z. 1909:19-37. [1909.01.03]

Emery, C. 1909b. Beiträge zur Monographie der Formiciden des paläarktischen Faunengebietes. (Hym.) Teil VII. Dtsch. Entomol. Z. 1909:179-204. [1909.03.01]

Emery, C. 1909c. [Untitled. Introduced by: "M. Lameere donne lecture de la lettre suivante que lui a adressée M. C. Emery:".] Ann. Soc. Entomol. Belg. 53:133-134. [1909.04.30] [Emery disclaims authorship of *Rhoptromyrmex opacus*, attributed incorrectly to him by Forel (1909a).]
Emery, C. 1909d. Beiträge zur Monographie der Formiciden des paläarktischen Faunengebietes. (Hym.) Teil VIII. Dtsch. Entomol. Z. 1909:355-376. [1909.05.01]
Emery, C. 1909e. Über den Ursprung der dulotischen, parasitischen und myrmekophilen Ameisen. Biol. Centralbl. 29:352-362. [1909.05.01]
Emery, C. 1909f. Beiträge zur Monographie der Formiciden des paläarktischen Faunengebietes. (Hym.) Teil IX. Dtsch. Entomol. Z. 1909:695-712. [1909.11.01]
Emery, C. 1909g. Intorno all'origine delle formiche dulotiche parassitiche e mirmecofile. Rend. Sess. R. Accad. Sci. Ist. Bologna Cl. Sci. Fis. (n.s.)13:36-51. [1909] [Session of 17 January 1909.]
Emery, C. 1910a. Beiträge zur Monographie der Formiciden des paläarktischen Faunengebietes. (Hym.) Teil X. Dtsch. Entomol. Z. 1910:127-132. [1910.03.05]
Emery, C. 1910b. Hymenoptera. Fam. Formicidae. Subfam. Dorylinae. Genera Insectorum 102: 1-34. [1910.06.25] [Date of publication from Evenhuis (1994a).]
Emery, C. 1910c. Antwort auf C. Schrottky's Nomenklaturfragen in No 9 dieser Zeitschrift. Dtsch. Entomol. Natl.-Bibl. 1:95. [1910.12.15]
Emery, C. 1911a. Einiges über die Ernährung der Ameisenlarven und die Entwicklung des temporären Parasitismus bei *Formica*. Dtsch. Entomol. Natl.-Bibl. 2:4-6. [1911.01.01]
Emery, C. 1911b. La fondazione di formicai da femmine fecondate di *Pheidole pallidula* e di *Tetramorium caespitum*. Sulle intolleranza o fratellanza fra le formiche di formicai differenti. Boll. Lab. Zool. Gen. Agrar. R. Sc. Super. Agric. 5:134-139. [1911.01.10]
Emery, C. 1911c. Instrument pour mesurer exactement les parties des insectes. Ann. Soc. Entomol. Belg. 55:211-212. [1911.08.04]
Emery, C. 1911d. Fragments myrmécologiques. Ann. Soc. Entomol. Belg. 55:213-225. [1911.08.04]
Emery, C. 1911e. Hymenoptera. Fam. Formicidae. Subfam. Ponerinae. Genera Insectorum 118: 1-125. [1911.09.09] [Date of publication from Evenhuis (1994a).]
Emery, C. 1911f. Formicidae. Résultats de l'expédition scientifique néerlandaise à la Nouvelle-Guinée en 1903 sous les auspices de Arthur Wichmann. Nova Guin. 5:531-539. [1911.11.03]
Emery, C. 1911g. Formicidae. Résultats de l'expédition scientifique néerlandaise à la Nouvelle-Guinée en 1907 et 1909 sous les auspices de Dr. H. A. Lorentz. Nova Guin. 9:249-259. [1911]
Emery, C. 1912a. Revision der *Rhytidoponera* (subg. *Chalcoponera*) der *metallica*-Gruppe. (Hym.-Formic.). Dtsch. Entomol. Z. 1912:77-81. [1912.01.29]
Emery, C. 1912b. Études sur les Myrmicinae. [I-IV.] Ann. Soc. Entomol. Belg. 56:94-105. [1912.05.09]
Emery, C. 1912c. Formiche raccolte durante i viaggi di S. A. R. la Duchessa Elena d'Aosta nella regione dei grandi laghi dell'Africa equatoriale. Annu. Mus. Zool. R. Univ. Napoli (n.s.)3(26):1-2. [1912.05.22]
Emery, C. 1912d. Les espèces-type des genres et sous-genres de la famille des Formicides. Ann. Soc. Entomol. Belg. 56:271-273. [1912.08.02] [Includes, at end of article: Appendice à mes "Études sur les Myrmicinae".]
Emery, C. 1912e. Beiträge zur Monographie der Formiciden des paläarktischen Faunengebietes. Teil XI. Dtsch. Entomol. Z. 1912:651-672. [1912.11.30]
Emery, C. 1912f. Der Wanderzug der Steppen- und Wüstenameisen von Zentral-Asien nach Süd-Europa und Nord-Afrika. Zool. Jahrb. Suppl. 15("Erster Band"):95-104. [1912] [Zool. Jahrb. Suppl. 15 itself consists of three volumes, each individually paginated. The preceding article appears in the first volume.]
Emery, C. 1913a ("1912"). Hymenoptera. Fam. Formicidae. Subfam. Dolichoderinae. Genera Insectorum 137:1-50. [1913.03.08] [Date of publication from Evenhuis (1994a).]

Emery, C. 1913b. Über die Abstammung der europäischen arbeiterinnenlosen Ameise *Anergates*. Biol. Centralbl. 33:258-260. [1913.05.20]

Emery, C. 1913c. Études sur les Myrmicinae. [V-VII.] Ann. Soc. Entomol. Belg. 57:250-262. [1913.09.05]

Emery, C. 1913d. La nervulation de l'aile antérieure des Formicides. Rev. Suisse Zool. 21:577-587. [1913.09]

Emery, C. 1913e. Le origini e le migrazioni della fauna mirmecologica d'Europa. Rend. Sess. R. Accad. Sci. Ist. Bologna Cl. Sci. Fis. (n.s.)17:29-46. [1913] [Session of 26 January 1913.]

Emery, C. 1914a. Formiche d'Australia e di Samoa raccolte dal Prof. Silvestri nel 1913. Boll. Lab. Zool. Gen. Agrar. R. Sc. Super. Agric. 8:179-186. [1914.01.30]

Emery, C. 1914b. *Cephalotes* et *Cryptocerus*. Le type du genre *Crematogaster*. Ann. Soc. Entomol. Belg. 58:37-39. [1914.02.06]

Emery, C. 1914c. Wissenschaftliche Ergebnisse der Bearbeitung von O. Leonhard's Sammlungen. 5. Südeuropäische Ameisen (Hym.). Entomol. Mitt. 3:156-159. [1914.05.01]

Emery, C. 1914d. Note sulle formiche della collezione sarda e della collezione dell'Italia meridionale, radunate da Achille Costa, e conservate nel Museo Zoologico della R. Università di Napoli. Annu. Mus. Zool. R. Univ. Napoli (n.s.)4(18):1-3. [1914.10.07]

Emery, C. 1914e. Intorno alla classificazione dei Myrmicinae. Rend. Sess. R. Accad. Sci. Ist. Bologna Cl. Sci. Fis. (n.s.)18:29-42. [1914] [Session of 11 January 1914.]

Emery, C. 1914f. Les fourmis de la Nouvelle-Calédonie et des îles Loyalty. Nova Caled. A Zool. 1:393-437. [1914]

Emery, C. 1915a. Contributo alla conoscenza delle formiche delle isole italiane. Descrizione di forme mediterrannee nuove o critiche. Ann. Mus. Civ. Stor. Nat. 46[=(3)6]:244-270. [1915.01.17]

Emery, C. 1915b. *Sima* oder *Tetraponera*? Zool. Anz. 45:265-266. [1915.02.23]

Emery, C. 1915c. Su due formiche della Tripolitania. Boll. Lab. Zool. Gen. Agrar. R. Sc. Super. Agric. 9:378. [1915.04.07]

Emery, C. 1915d. Sur le type de *Camponotus maculatus* (*Formica maculata* F.) (Hym. Formicidae). Bull. Soc. Entomol. Fr. 1915:79-80. [1915.04.13]

Emery, C. 1915e. Formiche raccolte nell'Eritrea dal Prof. F. Silvestri. Boll. Lab. Zool. Gen. Agrar. R. Sc. Super. Agric. 10:3-26. [1915.05.10]

Emery, C. 1915f. Escursioni zoologiche del Dr. Enrico Festa nell'Isola di Rodi. XII. Formiche. Boll. Mus. Zool. Anat. Comp. R. Univ. Torino 30(701):1-7. [1915.05.30]

Emery, C. 1915g. Noms de sous-genres et de genres proposés pour la sous-famille des Myrmicinae. Modifications à la classification de ce groupe (Hymenoptera Formicidae). Bull. Soc. Entomol. Fr. 1915:189-192. [1915.07.31]

Emery, C. 1915h. Histoire d'une société expérimentale de *Polyergus rufescens*. Rev. Suisse Zool. 23:385-400. [1915.11]

Emery, C. 1915i. Les *Pheidole* du groupe *megacephala* (Formicidae). Rev. Zool. Afr. (Bruss.) 4:223-250. [1915.12.15]

Emery, C. 1915j. Le formiche del genere *Solenopsis* abitanti l'Africa. Rend. Sess. R. Accad. Sci. Ist. Bologna Cl. Sci. Fis. (n.s.)19:57-66. [1915] [Session of 7 March 1915.]

Emery, C. 1915k. Definizione del genere *Aphaenogaster* e partizione di esso in sottogeneri. *Parapheidole* e *Novomessor* nn. gg. Rend. Sess. R. Accad. Sci. Ist. Bologna Cl. Sci. Fis. (n.s.)19:67-75. [1915] [Session of 21 March 1915.]

Emery, C. 1915l. La vita delle formiche. Torino: Fratelli Bocca, vii + 254 pp. [1915]

Emery, C. 1916a ("1915"). Fauna entomologica italiana. I. Hymenoptera.-Formicidae. Bull. Soc. Entomol. Ital. 47:79-275. [1916.11.30]

Emery, C. 1916b. Formiche d'Italia nuove o critiche. Rend. Sess. R. Accad. Sci. Ist. Bologna Cl. Sci. Fis. (n.s.)20:53-66. [1916] [Session of 12 March 1916.]

Emery, C. 1917a. Questions de nomenclature et synonymies relatives à quelques genres et espèces de Formicides (Hym.). Bull. Soc. Entomol. Fr. 1917:94-97. [1917.03.28]

Emery, C. 1917b. Formiche ibride. Rend. Sess. R. Accad. Sci. Ist. Bologna Cl. Sci. Fis. (n.s.)21:23-29. [1917] [Session of 26 November 1916.]
Emery, C. 1919a. Errata au mémoire. Les *Pheidole* du groupe *megacephala* (Formicidae). Rev. Zool. Afr. (Bruss.) 6:169-170. [1919.03.10]
Emery, C. 1919b. Sur le genre *Tranopelta* et sur le type du genre *Cremastogaster* (Hym. Formicidae). Bull. Soc. Entomol. Fr. 1919:60-62. [1919.04.07]
Emery, C. 1919c. Notes critiques de myrmécologie. [I-IV.] Ann. Soc. Entomol. Belg. 59:100-107. [1919.08.01]
Emery, C. 1919d. Formiche dell'Isola Cocos. Rend. Sess. R. Accad. Sci. Ist. Bologna Cl. Sci. Fis. (n.s.)23:36-40. [1919] [Session of 8 December 1918.]
Emery, C. 1920a. Notes critiques de myrmécologie. [VI-VIII.] Ann. Soc. Entomol. Belg. 60:59-61. [1920.05.20]
Emery, C. 1920b. Le genre *Camponotus* Mayr. Nouvel essai de la subdivision en sous-genres. Rev. Zool. Afr. (Bruss.) 8:229-260. [1920.10.01]
Emery, C. 1920c. [Review of: Forel, A. 1920. Les fourmis de la Suisse. Seconde édition. La Chaux-de-Fonds: Imprimerie Cooperative, xvi + 333 pp.] Riv. Biol. (Rome) 2:527-528. [1920.10]
Emery, C. 1920d. Studi sui *Camponotus*. Bull. Soc. Entomol. Ital. 52:3-48. [1920.12.06]
Emery, C. 1920e. La distribuzione geografica attuale delle formiche. Tentativo di spiegarne la genesi col soccorso di ipotesi filogenetiche e paleogeografiche. Atti R. Accad. Lincei Mem. Cl. Sci. Fis. Mat. Nat. (5)13:357-450. [1920]
Emery, C. 1921a. Specific names repeated in the Linnean genus *Formica*. Psyche (Camb.) 28: 24-26. [1921.02] [Stamp date in MCZ library: 1921.04.25.]
Emery, C. 1921b. Formiche raccolte a Budrum (Anatolia) da Raffaele Varriale, Cap. medico nella R. Marina. Ann. Mus. Civ. Stor. Nat. "Giacomo Doria" 49[=(3)9]:208-218. [1921.03.31]
Emery, C. 1921c. Hymenoptera. Fam. Formicidae. Subfam. Myrmicinae. [part] Genera Insectorum 174A:1-94 + 7 plates. [1921.06.01] [Date of publication from Evenhuis (1994a).]
Emery, C. 1921d. Notes critiques de myrmécologie. [IX-X.] Ann. Soc. Entomol. Belg. 61:313-319. [1921.09.02]
Emery, C. 1921e. Le genre *Polyrhachis*. Classification; espèces nouvelles ou critiques. Bull. Soc. Vaudoise Sci. Nat. 54:17-25. [1921.09.16]
Emery, C. 1921f. Quels sont les facteurs du polymorphisme du sexe féminin chez les fourmis? Rev. Gén. Sci. Pures Appl. 32:737-741. [1921.12.30]
Emery, C. 1921g. Formiche tessitrici del genere *Oecophylla* fossili e viventi. Rend. Sess. R. Accad. Sci. Ist. Bologna Cl. Sci. Fis. (n.s.)25:99-105. [1921] [Session of 22 May 1921.]
Emery, C. 1922a. L'ouverture cloacale des Formicinae ouvrières et femelles. Bull. Soc. Entomol. Belg. 4:62-65. [1922.05.26]
Emery, C. 1922b. Il genere *Lasius* (F.) Mayr e particolarmente le forme mediterranee del gruppo *umbratus* Nyl. Boll. Soc. Entomol. Ital. 54:9-15. [1922.06.20]
Emery, C. 1922c. Hymenoptera. Fam. Formicidae. Subfam. Myrmicinae. [part] Genera Insectorum 174B:95-206. [1922.09.02] [Date of publication from Evenhuis (1994a).]
Emery, C. 1922d. *Messor barbarus* (L.). Appunti de sinonimia e di sistematica. Boll. Soc. Entomol. Ital. 54:92-99. [1922.10.22]
Emery, C. 1922e. Quelques fourmis nouvelles minuscules. Ann. Hist.-Nat. Mus. Natl. Hung. 19:107-112. [1922.12.29]
Emery, C. 1922f. Aggiunte alla memoria: "La distribuzione geografica attuale delle formiche". Atti R. Accad. Naz. Lincei Rend. Cl. Sci. Fis. Mat. Nat. (5)31(1):72-75. [1922]
Emery, C. 1922g. Le specie americane del genere *Melophorus* (*Lasiophanes*). Rend. Sess. R. Accad. Sci. Ist. Bologna Cl. Sci. Fis. (n.s.)26:90-95. [1922] [Session of 12 February 1922.]
Emery, C. 1923a. Einige exotische Ameisen des Deutschen Entomologischen Institutes. Entomol. Mitt. 12:60-62. [1923.01.05]
Emery, C. 1923b. Berichtigung. Entomol. Mitt. 12:205. [1923.07.30]

Emery, C. 1924a. Formiche della Cirenaica raccolte dal Dott. Enrico Festa e dal Prof. Filippo Silvestri. Boll. Soc. Entomol. Ital. 56:6-11. [1924.01.24]
Emery, C. 1924b. Alcune formiche di Malta. Boll. Soc. Entomol. Ital. 56:11-12. [1924.01.24]
Emery, C. 1924c. Formiche di Spagna raccolte dal Prof. Filippo Silvestri. Boll. Lab. Zool. Gen. Agrar. R. Sc. Super. Agric. 17:164-171. [1924.03.29]
Emery, C. 1924d. [Review of: Wheeler, W. M. 1923. Social life among the insects. New York: Harcourt, Brace & Co., vii + 375 pp.] Riv. Biol. (Rome) 6:222-223. [1924.04]
Emery, C. 1924e. [Untitled. Taxonomic comments.] Pp. 90, 92-95 in: Schkaff, B. Formiche di Costantinopoli. Boll. Soc. Entomol. Ital. 56:90-96. [1924.06.15]
Emery, C. 1924f ("1922"). Hymenoptera. Fam. Formicidae. Subfam. Myrmicinae. [concl.] Genera Insectorum 174C:207-397. [1924.06.24] [Publication date from Evenhuis (1994a).]
Emery, C. 1924g. [Review of: Wheeler, W. M. (with the collaboration of J. Bequaert, I. W. Bailey, F. Santschi and W. M. Mann) 1922. Ants of the American Museum Congo Expedition. A contribution to the myrmecology of Africa. Bull. Am. Mus. Nat. Hist. 45:1-1139.] Riv. Biol. (Rome) 6:533-534. [1924.10]
Emery, C. 1924h. Casi di anomalia e di parassitismo nelle formiche. Rend. Sess. R. Accad. Sci. Ist. Bologna Cl. Sci. Fis. (n.s.)28:82-89. [1924] [Session of 27 April 1924.]
Emery, C. 1925a ("1924"). Notes critiques de myrmécologie. Ann. Soc. Entomol. Belg. 64:177-191. [1925.04] [Note added in press (p. 179) is dated April, 1925.]
Emery, C. 1925b. Revision des espèces paléarctiques du genre *Tapinoma*. Rev. Suisse Zool. 32:45-64. [1925.05]
Emery, C. 1925c. Les espèces européennes et orientales du genre *Bothriomyrmex*. Bull. Soc. Vaudoise Sci. Nat. 56:5-22. [1925.11.30]
Emery, C. 1925d. Hymenoptera. Fam. Formicidae. Subfam. Formicinae. Genera Insectorum 183: 1-302. [1925.12.31] [Publication date from Evenhuis (1994a).]
Emery, C. 1925e. I *Camponotus* (*Myrmentoma*) paleartici del gruppo *lateralis*. Rend. Sess. R. Accad. Sci. Ist. Bologna Cl. Sci. Fis. (n.s.)29:62-72. [1925] [Session of 11 January 1925.]
Emery, C. 1926. Ultime note mirmecologiche. Boll. Soc. Entomol. Ital. 58:1-9. [1926.01.29]
Emery, C., Cavanna, G. 1880. Escursione in Calabria (1877-78). Formicidei. Bull. Soc. Entomol. Ital. 12:123-126. [1880.08.15]
Emery, C., Forel, A. 1879. Catalogue des Formicides d'Europe. Mitt. Schweiz. Entomol. Ges. 5:441-481. [1879.02]
Enzmann, E. V. 1944. Systematic notes on the genus *Pseudomyrma*. Psyche (Camb.) 51:59-103. [1944.12] [Stamp date in MCZ library: 1945.02.20.]
Enzmann, E. V. 1952. *Woitkowskia*, a new genus of army ants. Proc. Iowa Acad. Sci. 59:442-447. [1952.12]
Enzmann, J. 1946a. A new form of *Myrmecina*. J. N. Y. Entomol. Soc. 54:13-15. [1946.04.17]
Enzmann, J. 1946b. A new house-invading ant from Massachusetts. J. N. Y. Entomol. Soc. 54:47-49. [1946.04.17]
Enzmann, J. 1946c. *Crematogaster lineolata cerasi*, the cherry ant of Asa Fitsch; (with a survey of the American forms of *Crematogaster*, subgenus *Acrocoelia*). J. N. Y. Entomol. Soc. 54: 89-97. [1946.06.18]
Enzmann, J. 1947a. *Hercynia*, a new genus of myrmicine ants. J. N. Y. Entomol. Soc. 55:43-46. [1947.03.27]
Enzmann, J. 1947b. New forms of *Aphaenogaster* and *Novomessor*. J. N. Y. Entomol. Soc. 55:147-152. [1947.05.26]
Erichson, W. F. 1839. Bericht über die Leistungen im Gebiete der Naturgeschichte während des Jahres 1838. IX. Insecten. Arch. Naturgesch. 5(2):281-375. [1839]
Erichson, W. F. 1842. Beitrag zur Insecten-Fauna von Vandiemensland, mit besonderer Berücksichtigung der geographischen Verbreitung der Insecten. Arch. Naturgesch. 8(1):83-287. [1842]

Errard, C. 1984. Evolution, en fonction de l'âge, des relations sociales dans les colonies mixtes hétérospécifiques chez les fourmis des genres *Camponotus* et *Pseudomyrmex*. Insectes Soc. 31:185-198. [1984.10]

Errard, C. 1986. Interactions biotope-phylogénèse sur la tolerance inter-spécifique chez les fourmis. Actes Colloq. Insectes Soc. 3:143-152. [1986]

Errard, C., Jaisson, P. 1985. Étude des relations sociales dans les colonies mixtes hétérospécifiques chez les fourmis (Hymenoptera, Formicidae). Folia Entomol. Mex. 61:135-146. [1985.06.28]

Esben-Petersen, P. 1922. New species of Neuroptera in the British Museum. Ann. Mag. Nat. Hist. (9)10:617-621. [1922.12] [Provides date of publication for Mulsant (1842).]

Escalante G., J. A. See: Escalante Gutiérrez, J. A.

Escalante Gutiérrez, J. A. 1976 ("1975"). Hormigas de la provincia de la Convención, Cusco. Rev. Peru. Entomol. 18:125-126. [1976.06.30]

Escalante Gutiérrez, J. A. 1977 ("1976"). Hormigas del valle de K'osñipata (Paucartambo, Cusco). Rev. Peru. Entomol. 19:107-108. [1977.08.11]

Escalante Gutiérrez, J. A. 1980 ("1979"). Notas sobre las hormigas del Cusco. Rev. Peru. Entomol. 22:111-112. [1980.08.11]

Escalante Gutiérrez, J. A. 1993 ("1991"). Especies de hormigas conocidas del Perú (Hymenoptera: Formicidae). Rev. Peru. Entomol. 34:1-13. [1993.01.31]

Escherich, K. 1906. Die Ameise. Schilderung ihrer Lebensweise. Braunschweig: Friedrich Vieweg und Sohn, xx + 232 pp. [1906.03.15]

Escherich, K. 1917. Die Ameise, Schilderung ihrer Lebensweise. Zweite verbesserte und vermehrte Auflage. Braunschweig: Friedrich Vieweg und Sohn, xvi + 348 pp. [1917]

Escherich, K. 1942. Die Forstinsekten Mitteleuropas. Ein Lehr- und Handbuch. V. Hymenoptera (Hautflügler) und Diptera (Zweiflügler). Berlin: Paul Parey, x + 746 pp. [1942] [Ants pp. 421-473.]

Escherich, K., Ludwig, A. 1907 ("1906"). Beiträge zur Kenntnis der elsässischen Ameisenfauna. Mitt. Philomath. Ges. Elsass-Lothringen 3:381-389. [1907]

Espadaler, X. 1978 ("1977"). Descripción de los sexuados de *Tapinoma pygmaeum* (Dufour, 1857) (Hymenoptera, Formicidae). Vie Milieu Sér. C Biol. Terr. 27:119-128. [1978.12]

Espadaler, X. 1979. Citas nuevas o interesantes de hormigas (Hym. Formicidae) para España. Bol. Asoc. Esp. Entomol. 3:95-101. [1979.11]

Espadaler, X. 1981a ("1980"). *Sifolinia lemasnei* (Bernard, 1968) en España (Hymenoptera, Formicidae). Bol. Asoc. Esp. Entomol. 4:121-124. [1981.05]

Espadaler, X. 1981b ("1979"). Una nueva hormiga de la Península Ibérica (Hymenoptera, Formicidae). Misc. Zool. (Barc.) 5:77-81. [1981]

Espadaler, X. 1981c. Les formigues granívores de la Mediterrània occidental. Treballs Inst. Catalana Hist. Nat. 9:39-44. [1981]

Espadaler, X. 1981d. Biometria de les *Myrmica* pirinenques (Hymenoptera, Formicidae). Els índexs cefàlics. Estudi Gen. (Coll. Univ. Girona) 1:189-196. [1981]

Espadaler, X. 1982a. *Epimyrma bernardi* n. sp., a new parasitic ant (Hymenoptera, Formicidae). Spixiana 5:1-6. [1982.03.01]

Espadaler, X. 1982b. *Xenhyboma mystes* Santschi, 1919 = *Monomorium medinae* Forel, 1892. Évidence biologique de la synonymie (Hymenoptera, Formicidae). Nouv. Rev. Entomol. 12:111-113. [1982.03.30]

Espadaler, X. 1983a ("1981"). *Camponotus universitatis* Forel, 1890, retrouvé en France. Vie Milieu 31:341-342. [1983.03]

Espadaler, X. 1983b. Sobre formigues trobades en coves (Hymenoptera, Formicidae). Speleon 26-27:53-56.

Espadaler, X. 1984. *Leptothorax nadigi* Kutter, 1925 y *Goniomma blanci* (André, 1881): descripción de los machos (Hym. Formicidae). Bol. Asoc. Esp. Entomol. 8:135-141. [1984.06]

Espadaler, X. 1986a ("1985"). *Goniomma kugleri*, a new granivorous ant from the Iberian Peninsula (Hymenoptera: Formicidae). Isr. J. Entomol. 19:61-66. [1986.02]

Espadaler, X. 1986b. *Formica decipiens* Bondr., 1918: descripción del macho y dos adiciones a la fauna ibérica (Hym., Formicidae). Bol. Asoc. Esp. Entomol. 10:45-50. [1986.05]

Espadaler, X. 1986c. VIII. Formícidos de los alrededores de La Laguna de Sariñena (Huesca). Descripción del macho de *Camponotus foreli* Emery (Hym. Formicidae). Colecc. Estud. Altoaragon. 6:109-126. [1986]

Espadaler, X. 1987 ("1986"). Formigues del Montseny. Pp. 101-103 in: Terradas, J., Miralles, J. (eds.) El patrimoni biològic del Montseny. Catàlegs de flora i fauna, 1. Barcelona: Servei de Pars Naturals, xviii + 171 pp. [1987]

Espadaler, X. 1992. Formigues del Garraf: coneixement actual i clau d'identificació. Monografies (Diput. Barc.) 19:9-13. [1992.05]

Espadaler, X., Agosti, D. 1985. *Monomorium hesperium* Emery: description de la femelle (Hymenoptera, Formicidae). Mitt. Schweiz. Entomol. Ges. 58:295-297. [1985.12.15]

Espadaler, X., Agosti, D. 1987. *Monomorium boltoni* n. sp. from São Nicolau (Cape Verde Islands) (Hymenoptera, Formicidae). Mitt. Schweiz. Entomol. Ges. 60:295-299. [1987.12.15]

Espadaler, X., Ascaso, C. 1990. Adición a las hormigas (Hymenoptera, Formicidae) del Montseny (Barcelona). Orsis (Org. Sist.) 5:141-147. [1990]

Espadaler, X., Báez, M. 1993. *Myrmecina graminicola* (Latr., 1802) (Hymenoptera, Formicidae) in Madeira. Bocagiana 167:1-3. [1993.07.01]

Espadaler, X., Cagniant, H. 1987. Contribution à la connaissance des fourmis marocaines. Description du mâle de *Proformica theryi* Santschi, 1936 (Hymenoptera, Formicidae). Nouv. Rev. Entomol. (n.s.)4:133-138. [1987.06.30]

Espadaler, X., Cagniant, H. 1992 ("1991"). *Plagiolepis xene* Stärcke, the first inquiline ant from the Balearic Islands, Spain. Psyche (Camb.) 98:351-354. [1992.09.04]

Espadaler, X., Collingwood. C. A. 1982. Notas sobre *Leptothorax* Mayr, 1855, con descripción de *L. gredosi* n. sp. (Hym. Formicidae). Bol. Asoc. Esp. Entomol. 6:41-48. [1982.12]

Espadaler, X., Du Merle, P. 1989. *Leptothorax fuentei* Santschi, 1919, en France (Hymenoptera, Formicidae). Vie Milieu 39:121-123. [1989.05]

Espadaler, X., Du Merle, P., Plateaux, L. 1983. Redescription de *Leptothorax grouvellei* Bondroit, 1918. Notes biologiques et écologiques (Hymenoptera, Formicidae). Insectes Soc. 30:274-286. [1983.12]

Espadaler, X., Franch, J. 1978 ("1977"). *Leptothorax nadigi* Kutter, 1925 (Hym. Formicidae) en España. Bol. Asoc. Esp. Entomol. 1:161-162. [1978.04]

Espadaler, X., López-Soria, L. 1991. Rareness of certain Mediterranean ant species: fact or artifact? Insectes Soc. 38:365-377. [1991]

Espadaler, X., Martí, S. 1994 La feromona de pista de *Crematogaster* Lund (Hymenoptera, Formicidae): vàlida per a tot el gènere? Sess. Entomol. ICHN-SCL 8:81-86. [1994.12]

Espadaler, X., Muñoz Batet, J. 1979. *Goniomma blanci* (André, 1881) (Hym., Formicidae): descripción de la hembra. Bol. Asoc. Esp. Entomol. 3:11-15. [1979.11]

Espadaler, X., Nieves, J. L. 1983. Hormigas (Hymenoptera, Formicidae) pobladoras de agallas abandonadas de cinípidos (Hymenoptera, Cynipidae) sobre *Quercus* sp. en la Península Ibérica. Bol. Estac. Cent. Ecol. 12:89-93. [1983]

Espadaler, X., Plateaux, L., Casevitz Weulersse, J. 1984. *Leptothorax melas*, n. sp., de Corse. Notes écologiques et biologiques (Hymenoptera, Formicidae). Rev. Fr. Entomol. (Nouv. Sér.) 6:123-132. [1984.09.15]

Espadaler, X., Restrepo, C. 1983. Els gèneres *Epimyrma* Emery i *Chalepoxenus* Menozzi, formigues paràsites socials (Hymenoptera: Formicidae), a la Península Ibèrica. Estat actual del coneixement. Butll. Inst. Catalana Hist. Nat. 49:123-126. [1983]

Espadaler, X., Retana, J., Cerdá, X. 1990. The caste system of *Camponotus foreli* Emery (Hymenoptera: Formicidae). Sociobiology 17:299-312. [1990]

Espadaler, X., Riasol Boixart, J. M. 1981. Secretergates de *Formica* sp.: una morfologia de origen patologico en hormigas. Rev. Ibér. Parasitol. 41:539-549. [1981.12]

Espadaler, X., Riasol Boixart, J. M. 1983. Cisticercoides de Cyclophyllidea en hormigas *Leptothorax* Mayr. Modificaciones morfologicas y etologicas del huesped intermediario. Rev. Ibér. Parasitol. 43:219-227. [1983.09]

Espadaler, X., Riasol, J. M. 1983. Distribución, variabilidad y sinonimias en *Aphaenogaster ibérica* [sic] Emery, 1908 y dos adiciones a la fauna ibérica (Hymenoptera, Formicidae). Pp. 219-228 in: Actas del I Congreso Ibérico de Entomología. Volume 1. León: Universidad de León, 408 pp. [1983]

Espadaler, X., Roda, F. 1984. Formigues (Hymenoptera, Formicidae) de la Meda Gran. Pp. 245-254 in: Ros, J., Olivella, I., Gili, J. M. (eds.) Els sistemes naturals de les illes Medes. Arxius de la Secció de Ciències, LXXIII. Barcelona: Institut d'Estudis Catalana, 828 pp. [1984.07.31]

Espadaler, X., Rodríguez, R. 1989. The male of *Leptothorax risi* Forel, 1892 (Hymenoptera, Formicidae). Orsis (Org. Sist.) 4:141-144. [1989]

Espadaler, X., Suñer, D. 1995. Per què hi ha formigues del Montgrí (Girona) que no es troben a l'illa Meda Gran. Orsis (Org. Sist.) 10:91-97. [1995]

Espadaler Gelabert, X. See: Espadaler, X.

Espelie, K. E., Hermann, H. R. 1988. Congruent cuticular hydrocarbons: biochemical convergence of a social wasp, an ant and a host plant. Biochem. Syst. Ecol. 16:505-508. [1988.10.21]

Essig, E. O. 1926. Insects of western North America. New York: Macmillan, xi + 1035 pp. [1926.09] [Ants pp. 853-869.]

Ettershank, G. 1965. A new species of *Megalomyrmex* from the Chilean Andes (Formicidae, Hymenoptera). Psyche (Camb.) 72:55-58. [1965.06.25]

Ettershank, G. 1966. A generic revision of the world Myrmicinae related to *Solenopsis* and *Pheidologeton* (Hymenoptera: Formicidae). Aust. J. Zool. 14:73-171. [1966.02]

Ettershank, G., Brown, W. L., Jr. 1964a. *Monomorium solleri* comb. nov. Pilot Regist. Zool. Card No. 18. [1964.05.20]

Ettershank, G., Brown, W. L., Jr. 1964b. The Malpighian tubules as meristic characters in ants (Hym., Formicidae). Entomol. Mon. Mag. 100:5-7. [1964.09.08]

Evans, J. D. 1995. Relatedness threshold for the production of female sexuals in colonies of a polygynous ant, *Myrmica tahoensis*, as revealed by microsatellite DNA analysis. Proc. Natl. Acad. Sci. U. S. A. 92:6514-6517. [1995.07]

Evans, M. A., Evans, H. E. 1970. William Morton Wheeler, biologist. Cambridge, Mass.: Harvard University Press, 363 pp. [1970.12.11] [Date from publisher.]

Evans, W. 1912. A list of the ants (Heterogyna or Formicidae) of the Forth area. Scott. Nat. 1912:104-108. [1912.05]

Evans, W. 1913. *Tetramorium caespitum* (L.) - an ant new to Scotland - in the Forth area. Scott. Nat. 1913:116-117. [1913.05]

Evenhuis, N. L. 1994a. The publication and dating of P. A. Wytsman's Genera Insectorum. Arch. Nat. Hist. 21:49-66. [1994.02]

Evenhuis, N. L. 1994b. Catalogue of the fossil flies of the world. Leiden: Bachhuys Publishers, 8 + 600 pp. [1994]

Evenhuis, N. L., Thompson, F. C., Pont, A. C., Pyle, B. L. 1989. Literature cited. Pp. 809-991 in: Evenhuis, N. L. (ed.) Catalog of the Diptera of the Australasian and Oceanian regions. Bishop Mus. Spec. Publ. 86:1-1155. [1989.08.23] [Date from Evenhuis (1994b:495).]

Evershed, R. P., Morgan, E. D. 1981. Chemical investigations of the Dufour gland contents of attine ants. Insect Biochem. 11:343-351. [1981]

Evershed, R. P., Morgan, E. D. 1983. The amounts of trail pheromone substances in the venom of workers of four species of attine ants. Insect Biochem. 13:469-474. [1983]

Evershed, R. P., Morgan, E. D., Cammaerts, M. C. 1981. Identification of the trail pheromone of the ant *Myrmica rubra* L., and related species. Naturwissenschaften 68:374-376. [1981.07]

Evershed, R. P., Morgan, E. D., Cammaerts, M. C. 1982. 3-ethyl-2,5-dimethylpyrazine, the trail pheromone from the venom gland of eight species of *Myrmica* ants. Insect Biochem. 12:383-391. [1982]

Ezhikov, T. 1923. Über den Character der Variabilität der Ameisen-Ovarien. Rus. Zool. Zh. 3:353-357. [1923]

Ezhikov, T. 1934. Individual variability and dimorphism of social insects. Am. Nat. 68:333-344. [1934.08]

Ezikov, J. See: Ezhikov, T.

Faber, W. 1967. Beiträge zur Kenntnis sozialparasitischer Ameisen. I. *Lasius* (*Austrolasius* n. sg.) *reginae* n. sp., eine neue temporär sozialparasitische Erdameise aus Österreich (Hym. Formicidae). Pflanzenschutz Ber. 36:73-107. [1967.11]

Faber, W. 1969. Beiträge zur Kenntnis sozialparasitischer Ameisen. 2. *Aporomyrmex ampeloni* nov. gen., nov. spec. (Hym. Formicidae), ein neuer permanenter Sozialparasit bei *Plagiolepis vindobonensis* Lomnicki aus Österreich. Pflanzenschutz Ber. 39:39-100. [1969.02]

Fabres, G., Brown, W. L., Jr. 1978. The recent introduction of the pest ant *Wasmannia auropunctata* into New Caledonia. J. Aust. Entomol. Soc. 17:139-142. [1978.06.01]

Fabricius, J. C. 1775. Systema entomologiae, sistens insectorum classes, ordines, genera, species adiectis synonymis, locis, descriptionibus, observationibus. Flensburgi et Lipsiae [= Flensburg and Leipzig]: Korte, 832 pp. [1775.04.30] [Date of publication from Evenhuis (1994b). Ants pp. 391-396.]

Fabricius, J. C. 1776. Genera insectorum eorumque characteres naturales secundum numerum, figuram, situm et proportionem omnium partium oris adjecta mantissa specierum nuper detectarum. Chilonii [= Kiel]: Bartsch, 310 pp. [1776] [Ants p. 130.]

Fabricius, J. C. 1782 ("1781"). Species insectorum exhibentes eorum differentias specificas, synonyma, auctorum loca natalia, metamorphosin adiectis observationibus, descriptionibus. Tome I. Hamburgi et Kilonii [= Hamburg and Kiel]: C. E. Bohn, 552 pp. [1782] [Date based on Hope's (1845) translation of Fabricius' autobiography.Ants pp. 488-494.]

Fabricius, J. C. 1787. Mantissa insectorum sistens eorum species nuper detectas adiectis characteribus, genericis, differentiis, specificis, emendationibus, observationibus. Tome I. Hafniae [= Copenhagen]: C. G. Proft, 348 pp. [1787] [Ants pp. 307-311.]

Fabricius, J. C. 1793. Entomologia systematica emendata et aucta. Secundum classes, ordines, genera, species, adjectis synonimis, locis observationibus, descriptionibus. Tome 2. Hafniae [= Copenhagen]: C. G. Proft, 519 pp. [1793] [Ants pp. 349-365.]

Fabricius, J. C. 1798. Supplementum entomologiae systematicae. Hafniae [= Copenhagen]: Proft and Storch, 572 pp. [1798.05.20] [Date of publication from Evenhuis (1994b). Ants pp. 279-281.]

Fabricius, J. C. 1804. Systema Piezatorum secundum ordines, genera, species, adjectis synonymis, locis, observationibus, descriptionibus. Brunswick: C. Reichard, xiv + 15-439 + 30 pp. [1804] [Date of publication from Hedicke (1941). Ants pp. 395-428.]

Fales, H. M., Blum, M. S., Bian, Z., Jones, T. H., Don, A. W. 1984. Volatile compounds from ponerine ants in the genus *Mesoponera*. J. Chem. Ecol. 10:651-665. [1984.04]

Fales, H. M., Blum, M. S., Southwick, E. W., Williams, D. L., Roller, P. P., Don, A. W. 1988. Structure and synthesis of tetrasubstituted pyrazines in ants in the genus *Mesoponera*. Tetrahedron 44:5045-5050. [1988]

Fales, H. M., Jones, T. H., Jaouni, T., Blum, M. S., Schmidt, J. O. 1992. Phenylalkenals in ponerine (*Leptogenys* sp.) and myrmicine (*Pogonomyrmex* sp.) ants. J. Chem. Ecol. 18:847-854. [1992.06]

Fancello, L., Leo, P. 1991 ("1990"). Le attuali conoscenze sui formicidi Dacetini di Sardegna e Sicilia (Hymenoptera, Formicidae). Boll. Assoc. Rom. Entomol. 45:125-129. [1991.12.30]

Fanfani, A., Dazzini Valcurone, M. 1984. Nuovi dati relativi alla "glandola di Pavan" in *Iridomyrmex humilis* Mayr (Formicidae Dolichoderinae). Pubbl. Ist. Entomol. Agrar. Univ. Pavia 28:1-9. [1984.10]

Fanfani, A., Dazzini Valcurone, M. 1986. Glandole delle ponerine e ricerche sulle glandole del gastro di *Megaponera foetens* (Fabr.) (Hymenoptera Formicidae). Probl. Attuali Sci. Cult. 260:115-132. [1986]

Fanfani, A., Dazzini Valcurone, M. 1990. Glandola tibiale e glandola del basitarso di *Crematogaster striatula* Emery (Formicidae, Myrmicinae). Probl. Attuali Sci. Cult. 265:121-124. [1990]

Fanfani, A., Dazzini Valcurone, M. 1991a. Metapleural glands of some Dolichoderinae ants. Ethol. Ecol. Evol. Spec. Issue 1:95-98. [1991.03]

Fanfani, A., Dazzini, M. V. 1991b. Le glandole metatoraciche di *Crematogaster striatula* Emery (Hymenoptera, Formicidae, Myrmicinae). Fragm. Entomol. 23:191-200. [1991.09.10]

Fanfani, A., Giovannotti, M. 1994. Morphology and ultrastructural organization of the metapleural glands in *Crematogaster scutellaris* (Formicidae, Myrmicinae). P. 420 in: Lenoir, A., Arnold, G., Lepage, M. (eds.) Les insectes sociaux. 12ème congrès de l'Union Internationale pour l'Étude des Insectes Sociaux. Paris, Sorbonne, 21-27 août 1994. Paris: Université Paris Nord, xxiv + 583 pp. [1994.09]

Faulds, W. 1970. A second species of *Iridomyrmex* Mayr established in New Zealand. N. Z. Entomol. 4(4):18-19. [1970.11]

Febvay, G., Kermarrec, A. 1981. Morphologie et fonctionnement du filtre infrabuccal chez une attine *Acromyrmex octospinosus* (Reich) (Hymenoptera: Formicidae): rôle de la poche infrabuccale. Int. J. Insect Morphol. Embryol. 10:441-449. [1981.12]

Feener, D. H., Jr. 1981. Notes on the biology of *Pheidole lamia* (Hymenoptera: Formicidae) at its type locality (Austin, Texas). J. Kansas Entomol. Soc. 54:269-277. [1981.04.29]

Feener, D. H., Jr. 1982. Intra- and interspecific patterns of allometry in the ant genus *Pheidole*. [Abstract.] P. 250 in: Breed, M. D., Michener, C. D., Evans, H. E. (eds.) The biology of social insects. Proceedings of the Ninth Congress of the IUSSI, Boulder, Colorado, August 1982. Boulder: Westview Press, xii + 420 pp. [1982.08]

Feener, D. H., Jr. 1987. Response of *Pheidole morrisi* to two species of enemy ants, and a general model of defense behavior in *Pheidole*. J. Kansas Entomol. Soc. 60:569-575. [1987.12.29]

Feller, C., Cherix, D. 1984. Première contribution à l'ètude de *Formica bruni* Kutter (Hymenoptera, Formicidae). Mitt. Schweiz. Entomol. Ges. 57:231-232. [1984.10.15]

Fellers, J. H. 1987. Interference and exploitation in a guild of woodland ants. Ecology 68:1466-1478. [1987.10]

Fellers, J. H., Fellers, G. M. 1976. Tool use in a social insect and its implications for competitive interactions. Science (Wash. D. C.) 192:70-72. [1976.04.02]

Fellers, J. H., Fellers, G. M. 1983 ("1982"). Status and distribution of ants in the Crater District of Haleakala National Park. Pac. Sci. 36:427-437. [1983.08.12]

Fellowes, J. R. 1994. Community structure of Hong Kong ants. P. 421 in: Lenoir, A., Arnold, G., Lepage, M. (eds.) Les insectes sociaux. 12ème congrès de l'Union Internationale pour l'Étude des Insectes Sociaux. Paris, Sorbonne, 21-27 août 1994. Paris: Université Paris Nord, xxiv + 583 pp. [1994.09]

Felton, J. C. 1962. The ants *Myrmica puerilis* Staercke and *M. scabrinodis* Nyl. in Kent. Entomol. Mon. Mag. 98:121. [> 1962.08] [Precise publication dates are not available for volume 98 of Entomol. Mon. Mag.]

Felton, J. C. 1965. Some records of ants (Hym. Formicidae) from east Kent. Entomol. Mon. Mag. 101:14-15. [1965.11.22]

Felton, J. C. 1968. *Myrmica puerilis* (Staercke) (Hym., Formicidae) from the Belgian coast. Entomol. Mon. Mag. 104:68. [1968.08.19]

Fenger, H. 1863. Anatomie und Physiologie des Giftapparates bei den Hymenopteren. Arch. Naturgesch. 29(1):137-178. [1863] [Ants pp. 175-177.]

Fenner, H. W. 1895. Arizona ants. Entomol. News 6:214-216. [1895.08.28]

Fernández, F. 1992 ("1991"). Las hormigas cazadoras del género *Ectatomma* (Formicidae: Ponerinae) en Colombia. Caldasia 16:551-564. [1992.04.28]

Fernández, F. 1993. Hormigas de Colombia III: los géneros *Acanthoponera* Mayr, *Heteroponera* Mayr y *Paraponera* Fr. Smith (Formicidae: Ponerinae: Ectatommini). Caldasia 17:249-258. [1993.12]

Fernández, F., Palacio, E. E. 1993. Las hormigas en Colombia. Perspectivas y estado actual de su conocimiento. Tacaya 1:3-4. [1993.10]

Fernández, F., Palacio, E. E. 1995. Hormigas de Colombia IV: nuevos registros de géneros y especies. Caldasia 17:587-596. [1995.03.26]

Fernández, F., Schneider, L. 1989. Reconocimiento de hormigas en la Reserva La Macarena. Rev. Colomb. Entomol. 15(1):38-44. [> 1989.06] [Probably published later, but no precise publication date given. Issue labelled "Enero-Junio de 1989".]

Fernández, I., Ballesta, M., Tinaut, A. 1994. Worker polymorphism in *Proformica longiseta* (Hymenoptera: Formicidae). Sociobiology 24:39-46. [1994]

Ferreira Brandão, C. R. See: Brandão, C. R. F.

Ferster, B., Prusak, Z. 1994. A preliminary checklist of the ants (Hymenoptera: Formicidae) of Everglades National Park. Fla. Entomol. 77:508-512. [1994.12.22]

Ferster, B., Traniello, J. F. A. 1995. Polymorphism and foraging behavior in *Pogonomyrmex badius* (Hymenoptera: Formicidae): worker size, foraging distance, and load size associations. Environ. Entomol. 24:673-678. [1995.06]

Fiala, J. 1934 ("1933"). Poznámky ke znamosti o rozsírení mravencu na Morave. Sb. Klubu Prírodoved. Brne 16:151-152. [1934]

Fiebrig, K. 1935. Apuntes zoológicos. I. Hormigas. Rev. Jard. Bot. Mus. Hist. Nat. Parag. 4:119-126. [1935]

Fielde, A. M. 1915. On certain vesicles found in the integument of ants. Proc. Acad. Nat. Sci. Phila. 67:36-40. [1915.03.24]

Finnegan, R. J. 1966. An unusual species of ant, *Formica fossaceps* Buren, in Quebec. Can. For. Branch Bimon. Res. Notes 22(6):4. [1966.12]

Finnegan, R. J. 1975. Introduction of a predacious red wood ant, *Formica lugubris* (Hymenoptera: Formicidae), from Italy to eastern Canada. Can. Entomol. 107:1271-1274. [1975.12.22]

Finnegan, R. J. 1977. Establishment of a predacious red wood ant, *Formica obscuripes* (Hymenoptera: Formicidae), from Manitoba to eastern Canada. Can. Entomol. 109:1145-1148. [1977.08.12]

Finzi, B. 1922a ("1921"). Primo contributo alla conoscenza della fauna mirmecologica della Venezia Giulia. Bull. Soc. Entomol. Ital. 53:118-120. [1922.02.15]

Finzi, B. 1922b ("1921"). Note alla fauna mirmecologica della Venezia Giulia. Rend. Unione Zool. Ital. 1921:29-32. [1922]

Finzi, B. 1923a. Risultati scientifici della spedizione Ravasini-Lona in Albania. III. Formiche. Boll. Soc. Entomol. Ital. 55:1-4. [1923.01.22]

Finzi, B. 1923b. Formiche. In: Un'escursione coleotterologica sul Monte Cavallo ed al Cansiglio. Rass. Soc. Alp. Giulie 24:26.

Finzi, B. 1924a. Formiche dell'isola d'Elba e Monte Argentario. Boll. Soc. Entomol. Ital. 56: 12-15. [1924.01.24]

Finzi, B. 1924b. Secondo contributo alla conoscenza della fauna mirmecologica della Venezia Giulia. Boll. Soc. Entomol. Ital. 56:120-123. [1924.10.29]

Finzi, B. 1926. Le forme europee del genere *Myrmica* Latr. Primo contributo. Boll. Soc. Adriat. Sci. Nat. Trieste 29:71-119. [1926]

Finzi, B. 1927a. Terzo contributo alla conoscenza della fauna mirmecologica della Venezia Giulia. Boll. Soc. Entomol. Ital. 59:7-10. [1927.01.31]

Finzi, B. 1927b. Nota sui *Camponotus* (*Myrmentoma*) *lateralis*, *piceus*, *dalmaticus*. Folia Myrmecol. Termit. 1:51-52. [1927.02]

Finzi, B. 1928a. *Formica cinerea* Mayr e varietà paleartiche. Boll. Soc. Entomol. Ital. 60:65-75. [1928.05.31]

Finzi, B. 1928b. Quarto contributo alla conoscenza della fauna mirmecologica della Venezia Giulia. Boll. Soc. Entomol. Ital. 60:128-130. [1928.10.24]

Finzi, B. 1928c. Weitere Beiträge zur Kenntnis der Fauna Griechenlands und der Inseln des Aegäischen Meeres. 1. Ameisen aus Griechenland und von den Aegäischen Inseln. Sitzungsber. Akad. Wiss. Wien Math.-Naturwiss. Kl. Abt. I 137:787-792. [1928]

Finzi, B. 1929. Le forme italiane del genere *Messor*. Boll. Soc. Entomol. Ital. 61:75-94. [1929.06.03]

Finzi, B. 1930a. Contributo allo studio degli *Aphaenogaster* paleartici (Formicidae-Myrmicinae). Boll. Soc. Entomol. Ital. 62:151-156. [1930.10.15]

Finzi, B. 1930b. Wissenschaftliche Ergebnisse einer zoologischen Forschungsreise nach Westalgerien und Marokko. III. Teil. Ameisen aus Marokko und Westalgerien. Sitzungsber. Akad. Wiss. Wien Math.-Naturwiss. Kl. Abt. I 139:14-16. [1930]

Finzi, B. 1930c. Hymenopteren aus Palästina und Syrien. (Zoologische Studienreise von R. Ebner 1928 mit Unterstützung der Akademie der Wissenschaften in Wien.) Formicidae. Sitzungsber. Akad. Wiss. Wien Math.-Naturwiss. Kl. Abt. I 139:22-24. [1930]

Finzi, B. 1930d. Zoologische Forschungsreise nach den Jonischen Inseln und dem Peloponnes. XII. Teil. Die Ameisen der Jonischen Inseln. Sitzungsber. Akad. Wiss. Wien Math.-Naturwiss. Kl. Abt. I 139:309-319. [1930]

Finzi, B. 1932a. Sopra alcune formiche dell'Isola di Rodi. Boll. Soc. Entomol. Ital. 64:24-27. [1932.02.15]

Finzi, B. 1932b. Ergebnisse einer zoologischen Forschungsreise nach Marokko. Unternommen 1930 mit Unterstützung der Akademie der Wissenschaften in Wien von Franz Werner und Richard Ebner. V. Ameisen aus Marokko. Sitzungsber. Akad. Wiss. Wien Math.-Naturwiss. Kl. Abt. I 141:243-244. [1932]

Finzi, B. 1933. Raccolte entomologiche nell'Isola di Capraia fatte da C. Mancini e F. Capra (1927-1931). II. Formicidae. Mem. Soc. Entomol. Ital. 11:162-165. [1933.01.25]

Finzi, B. 1936. Risultati scientifici della spedizione di S. A. S. il Principe Alessandro della Torre e Tasso nell'Egitto e peninsola del Sinai. XI. Formiche. Bull. Soc. Entomol. Egypte 20:155-210. [1936.12.28]

Finzi, B. 1939a. Materiali zoologici dell'Eritrea raccolti da G. Müller durante la spedizione dell'Istituto Sieroterapico Milanese e conservati al Museo di Trieste. Parte III. Hymenoptera: Formicidae. Atti Mus. Civ. Stor. Nat. Trieste 14:153-168. [1939.01.31]

Finzi, B. 1939b. Quinto contributo alla conoscenza della fauna mirmecologica della Venezia Giulia. Boll. Soc. Entomol. Ital. 71:86-90. [1939.06.05]

Finzi, B. 1939c. Ergebnisse der von Franz Werner und Otto v. Wettstein auf den Ägäischen Inseln unternommenen Sammelreisen. Ameisen. Sitzungsber. Akad. Wiss. Wien Math.-Naturwiss. Kl. Abt. I 148:153-161. [1939]

Finzi, B. 1940 ("1939"). Formiche della Libia. Mem. Soc. Entomol. Ital. 18:155-166. [1940.02.05]

Fisher, B. L. 1995. Ant diversity patterns along an elevational gradient in eastern Madagascar. [Abstract.] Pp. 19, 84-85 in: Patterson, B. D., Goodman, S. M., Sedlock, J. L. (eds.) Environmental change in Madagascar. Chicago: Field Museum of Natural History, 143 pp. [1995]

Fitch, A. 1855 ("1854"). Report [upon the noxious and other insects of the State of New-York]. Trans. N. Y. State Agric. Soc. 14:705-880. [1855] [Letter of presentation of the report to the state assembly is dated 3 April 1855, but the published version in the Transactions contains a postscript dated 7 August 1855. See Oswald & Penny (1991:74) for other bibliographical details.]

Fjellberg, A. 1975. Occurrence of *Formica uralensis* Ruzsky (Hymenoptera, Formicidae) in Pasvik, North Norway. Norw. J. Entomol. 22:83. [1975]

Flanders, S. E. 1958 ("1957"). Regulation of caste in social Hymenoptera. J. N. Y. Entomol. Soc. 65:97-105. [1958.03.25]

Fletcher, D. J. C. 1971. The glandular source and social functions of trail pheromones in two species of ants (*Leptogenys*). J. Entomol. Ser. A 46:27-37. [1971.08.10]

Foerster, A. 1850a. Hymenopterologische Studien. 1. Formicariae. Aachen: Ernst Ter Meer, 74 pp. [1850]

Foerster, A. 1850b. Eine Centurie neuer Hymenopteren. Zweite Dekade. Verh. Naturhist. Ver. Preuss. Rheinl. Westfal. 7:485-500. [1850]

Foerster, E. 1912. Vergleichend-anatomische Untersuchungen über Stechapparat der Ameisen. Zool. Jahrb. Abt. Anat. Ontog. Tiere 34:347-380. [1912.07.11]

Fonscolombe, B. de. 1846. Notes sur huit espèces nouvelles d'Hyménoptères et de Neuroptères, trouvées aux environs d'Aix. Ann. Soc. Entomol. Fr. (2)4:39-51. [1846.07.08]

Fontenla, J. L. See: Fontenla Rizo, J. L.

Fontenla Rizo, J. L. 1992a. Relaciones estructurales de dos comunidades de hormigas en el Jardín Botánico de Cienfuegos, Cuba. Rep. Invest. Inst. Ecol. Sist. (Acad. Cienc. Cuba) Ser. Zool. 16:1-23. [1992.10]

Fontenla Rizo, J. L. 1992b. Mirmecofauna de la caña de azúcar en Cuba. Análisis preliminar de su composición. Rep. Invest. Inst. Ecol. Sist. (Acad. Cienc. Cuba) 1992:1-28. [1992.12] [No volume or issue number given.]

Fontenla Rizo, J. L. 1993a. Composición y estructura de comunidades de hormigas en un sistema de formaciones vegetales costeras. Poeyana Inst. Ecol. Sist. Acad. Cienc. Cuba 441:1-19. [1993.12.24]

Fontenla Rizo, J. L. 1993b. Notas sobre el género *Macromischa* (Hymenoptera: Formicidae) en Cuba, y descripción de una nueva especie. Poeyana Inst. Ecol. Sist. Acad. Cienc. Cuba 442: 1-7. [1993.12.24]

Fontenla Rizo, J. L. 1993c. Mirmecofauna de Isla de la Juventud y de algunos cayos del archipiélago cubano. Poeyana Inst. Ecol. Sist. Acad. Cienc. Cuba 444:1-7. [1993.12.24]

Fontenla Rizo, J. L. 1993d ("1992"). Mirmecofauna de la caña de azúcar en Cuba. Aspectos biogeográficos. Cienc. Biol. Acad. Cienc. Cuba 25:61-75. [1993] [Volume 25 of this serial is dated "1992" but contains a notice of Alwyn Gentry's death on 3 August 1993.]

Fontenla Rizo, J. L. 1994a. Biogeográfia de *Macromischa* (Hymenoptera: Formicidae) en Cuba. Avicennia 1:19-29.

Fontenla Rizo, J. L. 1994b. Mirmecofauna de la Península de Hicacos, Cuba. Avicennia 1:79-85.

Fontenla Rizo, J. L. 1994c. Mirmecofauna de un hábitat-isla y del agroecosistema circundante. Cienc. Biol. Acad. Cienc. Cuba 26:40-55.

Fontenla Rizo, J. L. 1994d. Adiciones a la mirmecofauna (Hymenoptera: Formicidae) de la Isla de la Juventud, Archipiélago de los Canarreos y Archipiélago Sabana-Camagüey. Cocuyo (Havana) 1:4-5. [1994.11]

Fontenla Rizo, J. L. 1995a. Un comentario sobre las "hormigas locas" (*Paratrechina*) cubanas, con énfasis en *P. fulva*. Cocuyo (Havana) 2:6-7. [1995.03]

Fontenla Rizo, J. L. 1995b. Reflexiones sobre las hormigas "vagabundas" de Cuba. Cocuyo (Havana) 3:11-22. [1995.06]

Fontenla Rizo, J. L., Hernández Triana, L. M. 1994. Caracterización ecológica de la mirmecofauna de un cañaveral. Cienc. Biol. Acad. Cienc. Cuba 26:56-68.

Fontenla Rizo, J. L., Hernández, L. M. 1993. Relaciones de coexistencia en comunidades de hormigas en un agroecosistema de caña de azúcar. Poeyana Inst. Ecol. Sist. Acad. Cienc. Cuba 438:1-16. [1993.12.24]

Forbes, J. 1938. Anatomy and histology of the worker of *Camponotus herculeanus pennsylvanicus* DeGeer (Formicidae, Hymenoptera). Ann. Entomol. Soc. Am. 31:181-195. [1938.06.29]

Forbes, J. 1952. The genitalia and terminal segments of the male carpenter ant, *Camponotus pennsylvanicus* DeGeer (Formicidae, Hymenoptera). J. N. Y. Entomol. Soc. 60:157-171. [1952.10.21]

Forbes, J. 1954. The anatomy and histology of the male reproductive system of *Camponotus pennsylvanicus* DeGeer (Formicidae, Hymenoptera). J. Morphol. 95:523-555. [1954.11]

Forbes, J. 1956a. A comparison of the male reproductive systems of two doryline ants. [Abstract.] Bull. Entomol. Soc. Am. 2(3):14. [1956.09]

Forbes, J. 1956b. Observations on the gastral digestive tract in the male carpenter ant, *Camponotus pennsylvanicus* DeGeer (Formicidae, Hymenoptera). Insectes Soc. 3:505-511. [1956.12]

Forbes, J. 1958a. The gastral digestive organs in the male carpenter ant, *Camponotus pennsylvanicus* DeGeer, prior to and at swarming. P. 567 in: Becker, E. C. et al. (eds.) Proceedings of the Tenth International Congress of Entomology, Montreal, August 17-25, 1956. Volume 1. Ottawa: Mortimer Ltd., 941 pp. [1958.12]

Forbes, J. 1958b. The male reproductive system of the army ant, *Eciton hamatum* Fabricius. Pp. 593-596 in: Becker, E. C. et al. (eds.) Proceedings of the Tenth International Congress of Entomology, Montreal, August 17-25, 1956. Volume 1. Ottawa: Mortimer Ltd., 941 pp. [1958.12]

Forbes, J. 1967. The male genitalia and terminal gastral segments of two species of the primitive ant genus *Myrmecia* (Hymenoptera: Formicidae). J. N. Y. Entomol. Soc. 75:35-42. [1967.05.03]

Forbes, J., Brassel, R. W. 1962. The male genitalia and terminal segments of some members of the genus *Polyergus* (Hymenoptera: Formicidae). J. N. Y. Entomol. Soc. 70:79-87. [1962.06.16]

Forbes, J., Do-Van-Quy, D. 1965. The anatomy and histology of the male reproductive system of the legionary ant, *Neivamyrmex harrisi* (Haldeman) (Hymenoptera: Formicidae). J. N. Y. Entomol. Soc. 73:95-111. [1965.06.16]

Forbes, J., Hagopian, M. 1966 ("1965"). The male genitalia and terminal segments of the ponerine ant *Rhytidoponera metallica* F. Smith (Hymenoptera: Formicidae). J. N. Y. Entomol. Soc. 73:190-194. [1966.01.31]

Forbes, J., McFarlane, A. M. 1961. The comparative anatomy of digestive glands in the female castes and the male of *Camponotus pennsylvanicus* DeGeer (Formicidae, Hymenoptera). J. N. Y. Entomol. Soc. 69:92-103. [1961.06]

Ford, F. C., Forbes, J. 1980. Anatomy of the male reproductive systems of the adults and pupae of two doryline ants, *Dorylus* (*Anomma*) *wilverthi* Emery and *D.* (*A.*) *nigricans* Illiger. J. N. Y. Entomol. Soc. 88:133-142. [1980.07.16]

Ford, F. C., Forbes, J. 1984 ("1983"). Histology of the male reproductive systems in the adults and pupae of two Doryline ants, *Dorylus* (*Anomma*) *wilverthi* Emery and *D.* (*A.*) *nigricans* Illiger (Hymenoptera: Formicidae). J. N. Y. Entomol. Soc. 91:355-376. [1984.03.14]

Forder, J. C., Marsh, A. C. 1989. Social organization and reproduction in *Ocymyrmex foreli* (Formicidae: Myrmicinae). Insectes Soc. 36:106-115. [1989.09]

Forel, A. 1869. Observations sur les moeurs du *Solenopsis fugax*. Mitt. Schweiz. Entomol. Ges. 3:105-128. [1869.08]

Forel, A. 1870. Notices myrmécologiques. Mitt. Schweiz. Entomol. Ges. 3:306-312. [1870.10]

Forel, A. 1873 ("1872"). Sur les rapports que peuvent avoir entre elles les fourmis d'espèces différentes. Verh. Schweiz. Naturforsch. Ges. 55:43-45. [1873]

Forel, A. 1874. Les fourmis de la Suisse. Systématique, notices anatomiques et physiologiques, architecture, distribution géographique, nouvelles expériences et observations de moeurs. Neue Denkschr. Allg. Schweiz. Ges. Gesammten Naturwiss. 26:1-452. [1874]

Forel, A. 1876. Études myrmécologiques en 1875 avec remarques sur un point de l'anatomie des coccides. Bull. Soc. Vaudoise Sci. Nat. 14:33-62. [1876.02]

Forel, A. 1877. Ueber die systematische Bedeutung des Kaumagens und des Giftapparates der Ameisen. Amtl. Ber. Versamml. Dtsch. Naturforsch. Aertze 50:190. [1877]

Forel, A. 1878a. Der Giftapparat und die Analdrüsen der Ameisen. Z. Wiss. Zool. 30(suppl.): 28-68. [1878.04.23]

Forel, A. 1878b. Beitrag zur Kenntniss der Sinnesempfindungen der Insekten. Mitt. Münch. Entomol. Ver. 2:1-21. [1878.08]

Forel, A. 1878c. Études myrmécologiques en 1878 (première partie) avec l'anatomie du gésier des fourmis. Bull. Soc. Vaudoise Sci. Nat. 15:337-392. [1878.10]

Forel, A. 1879a. Études myrmécologiques en 1879 (deuxième partie [1re partie en 1878]). Bull. Soc. Vaudoise Sci. Nat. 16:53-128. [1879.03]

Forel, A. 1879b. *Aphaenogaster* (?) *Schaufussi* Forel n. sp. Nunquam Otiosus (Dres.) 3:465-466. [1879]

Forel, A. 1880a. An Weingeist-exemplaren der Honigameise (*Myrmecocystus melliger* Llave, *M. mexicanus* Wesmael) gemachte Beobachtungen. Mitt. Ges. Morphol. Physiol. Münch. 1880:1-2. [1880.01] [Not seen. Cited in Kutter (1931b:181.]

Forel, A. 1880b. [Untitled. Introduced by: "12. Die wissenschaftlichen Mittheilungen eröffnete Herr Prof. Forel mit einem Vortrag über einige durch ihre Lebensweise interessanten Ameisen, die er zugleich vorzeigte."] Mitt. Schweiz. Entomol. Ges. 6:20. [1880.09.10]

Forel, A. 1881. Die Ameisen der Antille St. Thomas. Mitt. Münch. Entomol. Ver. 5:1-16. [1881.04]

Forel, A. 1884. [Untitled. Introduced by: "Prof. Forel zeigt Fragmente des aus einer Art Carton bestehenden Nestes einer südwestafrikanischen *Crematogaster*-Art, welche von Herrn Dr. Max Buchner dort gesammelt wurde."] Mitt. Schweiz. Entomol. Ges. 7:3-4. [1884.05.10]

Forel, A. 1885a ("1884"). Études myrmécologiques en 1884 avec une description des organes sensoriels des antennes. Bull. Soc. Vaudoise Sci. Nat. 20:316-380. [1885.02]

Forel, A. 1885b. Indian ants of the Indian Museum, Calcutta. J. Asiat. Soc. Bengal Part II Nat. Sci. 54:176-182. [1885.12.29]

Forel, A. 1886a. Einige Ameisen aus Itajahy (Brasilien). Mitt. Schweiz. Entomol. Ges. 7:210-217. [1886.01]

Forel, A. 1886b. Espèces nouvelles de fourmis américaines. Ann. Soc. Entomol. Belg. 30:xxxviii-xlix. [> 1886.02.06]

Forel, A. 1886c. Diagnoses provisoires de quelques espèces nouvelles de fourmis de Madagascar, récoltées par M. Grandidier. Ann. Soc. Entomol. Belg. 30:ci-cvii. [> 1886.05.01]

Forel, A. 1886d. Indian ants of the Indian Museum, Calcutta, No. 2. J. Asiat. Soc. Bengal Part II Nat. Sci. 55:239-249. [1886.10.01]

Forel, A. 1886e. Nouvelles fourmis de Grèce récoltées par M. E. von Oertzen. Ann. Soc. Entomol. Belg. 30:clix-clxviii. [> 1886.10.02]

Forel, A. 1886f. Expériences et remarques critiques sur les sensations des insectes. Recl. Zool. Suisse 4:1-50. [1886.11.01]

Forel, A. 1886g. Expériences et remarques critiques sur les sensations des insectes. Deuxième partie. Nouvelles et anciennes expériences. (Suite). Recl. Zool. Suisse 4:145-240. [1886.11.01]

Forel, A. 1886h. Études myrmécologiques en 1886. Ann. Soc. Entomol. Belg. 30:131-215. [> 1886.11.06]

Forel, A. 1887. Fourmis récoltées à Madagascar par le Dr. Conrad Keller. Mitt. Schweiz. Entomol. Ges. 7:381-389. [1887.10]

Forel, A. 1888. Appendices à mon mémoire sur les sensations des insectes. Recl. Zool. Suisse 4:515-523. [1888.04.20]

Forel, A. 1889 ("1888"). Ameisen aus den Sporaden, den Cykladen und Griechenland, gesammelt 1887 von Herrn von Oertzen. Berl. Entomol. Z. 32:255-265. [1889.02]

Forel, A. 1890a. Un parasite de la *Myrmecia forficata* Fabr. Ann. Soc. Entomol. Belg. 34:viii-x. [> 1890.02.01]

Forel, A. 1890b. Fourmis de Tunisie et de l'Algérie orientale. Ann. Soc. Entomol. Belg. 34:lxi-lxxvi. [> 1890.04.05]

Forel, A. 1890c. *Aenictus-Typhlatta* découverte de M. Wroughton. Nouveaux genres de Formicides. Ann. Soc. Entomol. Belg. 34:cii-cxiv. [> 1890.06.07] [Translated in J. Bombay Nat. Hist. Soc. 5:388-397. See Forel (1890h).]

Forel, A. 1890d. Ueber neuere Beobachtungen, die Lebensweise der Ameisengäste und gewisser Ameisen betreffend. Humboldt 9:190-194. [1890.06]

Forel, A. 1890e. Une nouvelle fourmi. Naturaliste 12:217-218. [1890.09.15]

Forel, A. 1890f. Eine myrmekologische Ferienreise nach Tunesien und Ostalgerien nebst einer Beobachtung des Herrn Gleadow in Indien über *Aenictus*. Humboldt 9:296-306. [1890.09]

Forel, A. 1890g. Norwegische Ameisen und Drüsenkitt als Material zum Nestbau der Ameisen. Mitt. Schweiz. Entomol. Ges. 8:229-233. [1890.12]

Forel, A. 1890h. *Aenictus* (*Typhlatta*) and some new genera of Formicidae. [Translated by R. C. Wroughton.] J. Bombay Nat. Hist. Soc. 5:388-397. [1890] [English translation of Forel (1890c).]

Forel, A. 1891a. Un nouveau genre de Myrmicides. Ann. Soc. Entomol. Belg. 35:cccvii-cccviii. [> 1891.06.06]

Forel, A. 1891b. Ueber die Ameisensubfamilie der Doryliden. Verh. Ges. Dtsch. Naturforsch. Ärzte 63(2):162-164. [1891]

Forel, A. 1891c. Les Formicides. [part] In: Grandidier, A. Histoire physique, naturelle, et politique de Madagascar. Volume XX. Histoire naturelle des Hyménoptères. Deuxième partie (28e fascicule). Paris: Hachette et Cie, v + 237 pp. [1891]

Forel, A. 1892a. Notes myrmécologiques. Ann. Soc. Entomol. Belg. 36:38-43. [> 1892.02.06]

Forel, A. 1892b. Die Akazien-*Crematogaster* von Prof. Keller aus dem Somaliland. Zool. Anz. 15:140-143. [1892.04.11]

Forel, A. 1892c. Die Ameisen Neu-Seelands. Mitt. Schweiz. Entomol. Ges. 8:331-343. [1892.05]

Forel, A. 1892d. Attini und Cryptocerini. Zwei neue *Apterostigma*-Arten. Mitt. Schweiz. Entomol. Ges. 8:344-349. [1892.05]

Forel, A. 1892e. Liste der aus dem Somaliland von Hrn. Prof. Dr. Conr. Keller aus der Expedition des Prinzen Ruspoli im August und September 1891 zurückgebrachten Ameisen. Mitt. Schweiz. Entomol. Ges. 8:349-354. [1892.05]

Forel, A. 1892f. Critique de: Peter Cameron. Hemenoptera [sic], Formicidae; Extracted from supplementary appendix to Travels amongst the Great Andes of the Equator by Edw. Whymper. London 1891. Ann. Soc. Entomol. Belg. 36:255-256. [> 1892.07.02]

Forel, A. 1892g. Quelques fourmis de la faune méditerranéenne. Ann. Soc. Entomol. Belg. 36:452-457. [> 1892.09.03]

Forel, A. 1892h. Le mâle des *Cardicondyla* [sic] et la reproduction consanguine perpétuée. Ann. Soc. Entomol. Belg. 36:458-461. [> 1892.09.03]

Forel, A. 1892i. A propos de ma critique de M. Cameron. Ann. Soc. Entomol. Belg. 36:462. [> 1892.09.03]

Forel, A. 1892j. Die Ameisenfauna Bulgariens. (Nebst biologischen Beobachtungen.) Verh. K-K. Zool.-Bot. Ges. Wien 42:305-318. [1892.09]

Forel, A. 1892k. Les Formicides de l'Empire des Indes et de Ceylan. Part I. J. Bombay Nat. Hist. Soc. 7:219-245. [1892.10.01]

Forel, A. 1892l. Nouvelles espèces de Formicides de Madagascar (récoltées par M. Sikora). Première série. Ann. Soc. Entomol. Belg. 36:516-535. [> 1892.12.03]

Forel, A. 1892m. Rectification à ma communication sur le mâle des *Cardiocondyla* faite à la séance du 3 septembre 1892. Ann. Soc. Entomol. Belg. 36:536. [> 1892.12.03]

Forel, A. 1892n. Hérmaphrodite de l'*Azteca instabilis* Smith. Bull. Soc. Vaudoise Sci. Nat. 28:268-270. [1892.12]

Forel, A. 1892o. Les Formicides. [concl.] In: Grandidier, A. Histoire physique, naturelle, et politique de Madagascar. Volume XX. Histoire naturelle des Hyménoptères. Deuxième partie. Supplèment au 28e fascicule. Paris: Hachette et Cie, pp. 229-280. [1892]

Forel, A. 1892p. Die Nester der Ameisen. Neujahrsbl. Naturforsch. Ges. Zür. 95:1-36. [1892] [Translated in Annu. Rep. Smithson. Inst. 1894:479-505 (see Forel, 1896f) and in Int. J. Microsc. Nat. Sci. 16:347-381 (Forel, 1897e).]

Forel, A. 1893a. Observations nouvelles sur la biologie de quelques fourmis. Bull. Soc. Vaudoise Sci. Nat. 29:51-53. [1893.03]

Forel, A. 1893b. Sur la classification de la famille des Formicides, avec remarques synonymiques. Ann. Soc. Entomol. Belg. 37:161-167. [> 1893.04.01]

Forel, A. 1893c. Les Formicides de l'Empire des Indes et de Ceylan. Part II. J. Bombay Nat. Hist. Soc. 7:430-439. [1893.04.23]

Forel, A. 1893d. [Untitled. Introduced by: "M. Forel nous expédie la note rectificative qui suit au sujet de son article sur la classification des Formicides."] Ann. Soc. Entomol. Belg. 37:285-286. [> 1893.06.03]

Forel, A. 1893e. Les Formicides de l'Empire des Indes et de Ceylan. Part III. J. Bombay Nat. Hist. Soc. 8:17-36. [1893.07.01]

Forel, A. 1893f. Nouvelles fourmis d'Australie et des Canaries. Ann. Soc. Entomol. Belg. 37:454-466. [> 1893.09.02]

Forel, A. 1893g. Quelques fourmis de la faune méditerranéenne. An. Soc. Esp. Hist. Nat. (Actas) (2)2[=22]:90-94. [1893.11.01] [This is a translation of Forel (1892g), with corrected type locality information by M. Medina.]

Forel, A. 1893h. Note sur les Attini. Ann. Soc. Entomol. Belg. 37:586-607. [> 1893.12.02]

Forel, A. 1893i. Note préventive sur un nouveau genre et une nouvelle espèce de Formicide (Camponotide). Ann. Soc. Entomol. Belg. 37:607-608. [> 1893.12.02]

Forel, A. 1893j. Formicides de l'Antille St. Vincent, récoltées par Mons. H. H. Smith. Trans. Entomol. Soc. Lond. 1893:333-418. [1893.12.29] [Date from Wheeler, G. (1912).]

Forel, A. 1894a. Algunas hormigas de Canarias recogidas por el Sr. Cabrera y Diaz. An. Soc. Esp. Hist. Nat. (Actas) (2)2[=22]:159-162. [1894.01.31]

Forel, A. 1894b. Abessinische und andere afrikanische Ameisen, gesammelt von Herrn Ingenieur Alfred Ilg, von Herrn Dr. Liengme, von Herrn Pfarrer Missionar P. Berthoud, Herrn Dr. Arth. Müller etc. Mitt. Schweiz. Entomol. Ges. 9:64-100. [1894.01]

Forel, A. 1894c. Les Formicides de l'Empire des Indes et de Ceylan. Part IV. J. Bombay Nat. Hist. Soc. 8:396-420. [1894.02.01]

Forel, A. 1894d. Les Formicides de la Province d'Oran (Algérie). Bull. Soc. Vaudoise Sci. Nat. 30:1-45. [1894.03]

Forel, A. 1894e. Quelques fourmis de Madagascar (récoltées par M. le Dr. Voltzkow); de Nouvelle Zélande (récoltées par M. W. W. Smith); de Nouvelle Calédonie (récoltées par M. Sommer); de Queensland (Australie) (récoltées par M. Wiederkehr); et de Perth (Australie occidentale) (récoltées par M. Chase). Ann. Soc. Entomol. Belg. 38:226-237. [> 1894.05.05]

Forel, A. 1894f. Polymorphisme et ergatomorphisme des fourmis. Arch. Sci. Phys. Nat. (3)32:373-380. [1894.10]

Forel, A. 1895a. Nouvelles fourmis de diverses provenances, surtout d'Australie. Ann. Soc. Entomol. Belg. 39:41-49. [> 1895.01.04]

Forel, A. 1895b. A fauna das formigas do Brazil. Bol. Mus. Para. Hist. Nat. Ethnogr. 1:89-139. [1895.04]

Forel, A. 1895c. Die Ameisen- und Termitengäste von Brasilien. Anhang. Beschreibung einiger neuer brasilianischer Ameisenarten. Verh. K-K. Zool.-Bot. Ges. Wien 45:178-179. [1895.04] [See also Wasmann (1895a).]

Forel, A. 1895d. Nouvelles fourmis de l'Imerina oriental (Moramanga etc.). Ann. Soc. Entomol. Belg. 39:243-251. [> 1895.05.04]

Forel, A. 1895e. Südpalaearctische Ameisen. Mitt. Schweiz. Entomol. Ges. 9:227-234. [1895.05]

Forel, A. 1895f. Les Formicides de l'Empire des Indes et de Ceylan. Part V. J. Bombay Nat. Hist. Soc. 9:453-472. [1895.06.20]

Forel, A. 1895g. Nouvelles fourmis d'Australie, récoltées à The Ridge, Mackay, Queensland, par M. Gilbert Turner. Ann. Soc. Entomol. Belg. 39:417-428. [> 1895.09.07]

Forel, A. 1895h. Une nouvelle fourmi melligère. Ann. Soc. Entomol. Belg. 39:429-430. [> 1895.09.07]

Forel, A. 1895i. Quelques fourmis du centre de Madagascar. Ann. Soc. Entomol. Belg. 39:485-488. [> 1895.12.07]

Forel, A. 1895j. Ueber den Polymorphismus und Ergatomorphismus der Ameisen. Verh. Ges. Dtsch. Naturforsch. Ärzte 66(2,1.Hälfte):142-147. [1895]

Forel, A. 1895k. [Untitled. *Ponera Gleadowi* n. sp.] Pp. 60-61 [pagination of separate: 292-293] in: Emery, C. Sopre alcune formiche della fauna mediterranea. Mem. R. Accad. Sci. Ist. Bologna (5)5:59-75 [pagination of separate: 291-307]. [1895] [Session of 21 April 1895.]

Forel, A. 1896a. [Untitled. *Azteca chartifex* For. n. sp.] P. 4 in: Emery, C. Alcune forme nuove del genere *Azteca* Forel e note biologiche. Boll. Mus. Zool. Anat. Comp. R. Univ. Torino 11(230):1-7. [1896.03.01]

Forel, A. 1896b. Quelques particularités de l'habitat des fourmis de l'Amérique tropicale. Ann. Soc. Entomol. Belg. 40:167-171. [> 1896.05.02]

Forel, A. 1896c. Zur Fauna und Lebensweise der Ameisen im columbischen Urwald. Mitt. Schweiz. Entomol. Ges. 9:401-411. [1896.11]

Forel, A. 1896d. Fourmis dans les forêts vierges de la Colombie et des Antilles. Arch. Sci. Phys. Nat. (4)2:616-617. [1896.12]

Forel, A. 1896e. Die Fauna und die Lebensweise der Ameisen im kolumbischen Urwald und in den Antillen. Verh. Schweiz. Naturforsch. Ges. 79:148-150. [1896]

Forel, A. 1896f. Ants' nests. Annu. Rep. Smithson. Inst. 1894:479-505. [1896] [English translation of Forel (1892p).]

Forel, A. 1897a. Deux fourmis d'Espagne. Ann. Soc. Entomol. Belg. 41:132-133. [1897.05.01]

Forel, A. 1897b. Quelques Formicides de l'Antille de Grenada récoltés par M. H. H. Smith. Trans. Entomol. Soc. Lond. 1897:297-300. [1897.09.01]

Forel, A. 1897c. Communication verbale sur les moeurs des fourmis de l'Amérique tropicale. Ann. Soc. Entomol. Belg. 41:329-332. [1897.11.15]

Forel, A. 1897d. Ameisen aus Nossi-Bé, Majunga, Juan de Nova (Madagaskar), den Aldabra-Inseln und Sansibar, gesammelt von Herrn Dr. A. Voeltzkow aus Berlin. Mit einem Anhang über die von Herrn Privatdocenten Dr. A. Brauer in Marburg auf den Seychellen und von Herrn Perrot auf Ste. Marie (Madagaskar) gesammelten Ameisen. Abh. Senckenb. Naturforsch. Ges. 21:185-208. [1897]

Forel, A. 1897e. Ants' nests. Int. J. Microsc. Nat. Sci. 16:347-381. [1897] [English translation of Forel (1892p).]

Forel, A. 1898. La parabiose chez les fourmis. Bull. Soc. Vaudoise Sci. Nat. 34:380-384. [1898.12]

Forel, A. 1899a. Heterogyna (Formicidae). Fauna Hawaii. 1:116-122. [1899.03.20]

Forel, A. 1899b. Formicidae. [part]. Biol. Cent.-Am. Hym. 3:1-24. [1899.04]

Forel, A. 1899c. Trois notices myrmécologiques. Ann. Soc. Entomol. Belg. 43:303-310. [1899.06.28]

Forel, A. 1899d. Formicidae. [part]. Biol. Cent.-Am. Hym. 3:25-56. [1899.06]

Forel, A. 1899e. Formicidae. [part]. Biol. Cent.-Am. Hym. 3:57-80. [1899.08]

Forel, A. 1899f. Formicidae. [part]. Biol. Cent.-Am. Hym. 3:81-104. [1899.09]

Forel, A. 1899g. [Untitled. Letter from Faisons, North Carolina, dated 28 July 1899 and addressed to the Société Entomologique de Belgique.] Ann. Soc. Entomol. Belg. 43:438-447. [1899.10.04]

Forel, A. 1899h. Formicidae. [part]. Biol. Cent.-Am. Hym. 3:105-136. [1899.11]

Forel, A. 1899i. Formicidae. [concl.] Biol. Cent.-Am. Hym. 3:137-160. [1899.12]

Forel, A. 1899j. Von Ihrer Königl. Hoheit der Prinzessin Therese von Bayern auf einer Reise in Südamerika gesammelte Insekten. I. Hymenopteren. a. Fourmis. Berl. Entomol. Z. 44:273-277. [1899.12]

Forel, A. 1899k. A sketch of the biology of ants. Pp. 433-452 in: Story, W. E., Wilson, L. N. (eds.) Clark University 1889-1899. Decennial Celebration. Worcester, Mass.: Clark University, vi + 567 pp. [1899]

Forel, A. 1900a. Un nouveau genre et une nouvelle espèce de Myrmicide. Ann. Soc. Entomol. Belg. 44:24-26. [1900.01.30]

Forel, A. 1900b. Ponerinae et Dorylinae d'Australie récoltés par MM. Turner, Froggatt, Nugent, Chase, Rothney, J.-J. Walker, etc. Ann. Soc. Entomol. Belg. 44:54-77. [1900.03.09]

Forel, A. 1900c. Ébauche sur les moeurs des fourmis de l'Amérique du Nord. Riv. Sci. Biol. 2:180-192. [1900.03] [Translated in Psyche (Camb.) 9:231-239, 243-245. See Forel (1901h, 1901i). This is Forel's letter and Appendix to the Société Entomologique de Belgique (Forel, 1899g).]

Forel, A. 1900d. Les Formicides de l'Empire des Indes et de Ceylan. Part VI. J. Bombay Nat. Hist. Soc. 13:52-65. [1900.04.15]

Forel, A. 1900e. Formicidae. Index. Biol. Cent.-Am. Hym. 3:161-169. [1900.04]

Forel, A. 1900f. Les Formicides de l'Empire des Indes et de Ceylan. Part VII. J. Bombay Nat. Hist. Soc. 13:303-332. [1900.07.29]

Forel, A. 1900g. Expériences et remarques critiques sur les sensations des insectes. Riv. Sci. Biol. 2:561-601. [1900.08]

Forel, A. 1900h. Fourmis du Japon. Nids en toile. *Strongylognathus Huberi* et voisins. Fourmilière triple. *Cyphomymrex Wheeleri*. Fourmis importées. Mitt. Schweiz. Entomol. Ges. 10:267-287. [1900.10]

Forel, A. 1900i ("1899"). Ueber nordamerikanische Ameisen. Verh. Ges. Dtsch. Naturforsch. Ärzte 71(2,1.Hälfte):239-242. [1900]

Forel, A. 1901a. Les Formicides de l'Empire des Indes et de Ceylan. Part VIII. J. Bombay Nat. Hist. Soc. 13:462-477. [1901.01.20]

Forel, A. 1901b. Critique des expériences faites dès 1887 avec quelques nouvelles expériences. Troisième partie (1901). Riv. Biol. Gen. 3:7-62. [1901.02]

Forel, A. 1901c. Formiciden aus dem Bismarck-Archipel, auf Grundlage des von Prof. Dr. F. Dahl gesammelten Materials. Mitt. Zool. Mus. Berl. 2:4-37. [1901.04.03]

Forel, A. 1901d. I. Fourmis mexicaines récoltées par M. le professeur W.-M. Wheeler. II. A propos de la classification des fourmis. Ann. Soc. Entomol. Belg. 45:123-141. [1901.04.30]

Forel, A. 1901e. [Untitled. Introduced by: "Faccio seguire le descrizioni mandatemi dal Prof. Forel di due specie africaine della sua collezione".] Pp. 48-49 in: Emery, C. Note sulle Doriline. Bull. Soc. Entomol. Ital. 33:43-56. [1901.05.31]

Forel, A. 1901f. Sensations des insectes. Continuation de la critique des expériences faites dès 1887. Quatrième partie (1901). Riv. Biol. Gen. 3:241-282. [1901.05]

Forel, A. 1901g. Einige neue Ameisen aus Südbrasilien, Java, Natal und Mossamedes. Mitt. Schweiz. Entomol. Ges. 10:297-311. [1901.06]

Forel, A. 1901h. Sketch of the habits of North American ants. I. [Translated by A. P. Morse.] Psyche (Camb.) 9:231-239. [1901.08] [English translation of Forel (1900c).]

Forel, A. 1901i. Sketch of the habits of North American ants. II. [Translated by A. P. Morse.] Psyche (Camb.) 9:243-245. [1901.09] [English translation of Forel (1900c).]

Forel, A. 1901j. Variétés myrmécologiques. Ann. Soc. Entomol. Belg. 45:334-382. [1901.12.05]

Forel, A. 1901k. Nouvelles espèces de Ponerinae. (Avec un nouveau sous-genre et une espèce nouvelle d'*Eciton*). Rev. Suisse Zool. 9:325-353. [1901.12.18]

Forel, A. 1901l. Fourmis termitophages, Lestobiose, *Atta tardigrada*, sous-genres d'*Euponera*. Ann. Soc. Entomol. Belg. 45:389-398. [1901.12.30]

Forel, A. 1901m. Formiciden des Naturhistorischen Museums zu Hamburg. Neue *Calyptomyrmex*-, *Dacryon*-, *Podomyrma*- und *Echinopla*-Arten. Mitt. Naturhist. Mus. Hambg. 18:43-82. [1901]

Forel, A. 1901n. Die psychischen Fähigkeiten der Ameisen und einiger anderer Insekten; mit einem Anhang über die Eigentümlichkeiten des Geruchsinnes bei jenen Tieren. Vorträge gehalten den 13. August 1901 am V. Internationalen Zoologen-Kongress zu Berlin. München: Ernst Reinhardt Verlagsbuchlandlung, 57 pp. [1901]

Forel, A. 1902a. Les fourmis du Sahara algérien récoltées par M. le Professeur A. Lameere et le Dr. A. Diehl. Ann. Soc. Entomol. Belg. 46:147-158. [1902.05.01]

Forel, A. 1902b. Quatre notices myrmécologiques. Ann. Soc. Entomol. Belg. 46:170-182. [1902.05.01]

Forel, A. 1902c. Myrmicinae nouveaux de l'Inde et de Ceylan. Rev. Suisse Zool. 10:165-249. [1902.06.30]

Forel, A. 1902d. Variétés myrmécologiques. Ann. Soc. Entomol. Belg. 46:284-296. [1902.07.03]
Forel, A. 1902e. [Untitled. *Strumigenys Lujae* Forel n. sp.] Pp. 294-295 in: Wasmann, E. Neues über die zusammengesetzten Nester und gemischten Kolonien der Ameisen [part]. Allg. Z. Entomol. 7:293-298. [1902.08.15]
Forel, A. 1902f. [Untitled. *Tapinoma Heyeri* Forel n. sp.] P. 296 in: Wasmann, E. Neues über die zusammengesetzten Nester und gemischten Kolonien der Ameisen [part]. Allg. Z. Entomol. 7:293-298. [1902.08.15]
Forel, A. 1902g. Les Formicides de l'Empire des Indes et de Ceylan. Part IX. J. Bombay Nat. Hist. Soc. 14:520-546. [1902.10.18]
Forel, A. 1902h. Beispiele phylogenetischer Wirkungen und Rückwirkungen bei den Instinkten und dem Körperbau der Ameisen als Belege für die Evolutionslehre und psychophysiologische Identitätslehre. J. Psychol. Neurol. 1:99-110. [1902.10]
Forel, A. 1902i. Fourmis d'Algérie récoltées par M. le Dr. K. Escherich. Ann. Soc. Entomol. Belg. 46:462-463. [1902.12.04]
Forel, A. 1902j. Fourmis nouvelles d'Australie. Rev. Suisse Zool. 10:405-548. [1902.12.30]
Forel, A. 1902k. Descriptions of some ants from the Rocky Mountains of Canada (Alberta and British Columbia), collected by Edward Whymper. Trans. Entomol. Soc. Lond. 1902:699-700. [1902.12.30]
Forel, A. 1902l. Die psychischen Fähigkeiten der Ameisen und einiger anderer Insekten. Pp. 141-169 in: Matschie, P. (ed.) Verhandlungen des V. Internationalen Zoologen-Congresses zu Berlin 12.-16. August 1901. Jena: Gustav Fischer, xxvi + 1187 pp. [1902]
Forel, A. 1902m. [Untitled. Discussion of paper by R. F. Scharff. Introduced by: "Herr Dr. A. Forel (Chigny bei Morges) fragt den Vortragenden, ob er Vergleichungen mit den Alpen angestellt habe."] P. 361 in: Matschie, P. (ed.) Verhandlungen des V. Internationalen Zoologen-Congresses zu Berlin 12.-16. August 1901. Jena: Gustav Fischer, xxvi + 1187 pp. [1902]
Forel, A. 1902n. Die Eigentümlichkeiten des Geruchssinnes bei den Insekten. Pp. 806-815 in: Matschie, P. (ed.) Verhandlungen des V. Internationalen Zoologen-Congresses zu Berlin 12.-16. August 1901. Jena: Gustav Fischer, xxvi + 1187 pp. [1902]
Forel, A. 1903a. Les Formicides de l'Empire des Indes et de Ceylan. Part X. J. Bombay Nat. Hist. Soc. 14:679-715. [1903.02.10]
Forel, A. 1903b. Die Sitten und Nester einiger Ameisen der Sahara bei Tugurt und Biskra. Mitt. Schweiz. Entomol. Ges. 10:453-459. [1903.02]
Forel, A. 1903c. Faune myrmécologique des noyers dans le canton de Vaud. Bull. Soc. Vaudoise Sci. Nat. 39:83-94. [1903.03]
Forel, A. 1903d. Recherches biologiques récentes de Miss Adèle Fielde sur les fourmis. Bull. Soc. Vaudoise Sci. Nat. 39:95-99. [1903.03]
Forel, A. 1903e. Mélanges entomologiques, biologiques et autres. Ann. Soc. Entomol. Belg. 47:249-268. [1903.07.02]
Forel, A. 1903f. Les fourmis des îles Andamans et Nicobares. Rapports de cette faune avec ses voisines. Rev. Suisse Zool. 11:399-411. [1903.09.05]
Forel, A. 1903g. Ants and some other insects. An inquiry into the psychic powers of these animals with an appendix on the peculiarities of their olfactory sense. [Translated by W. M. Wheeler.] Monist 14:33-66. [1903.10] [See also Forel (1904a).]
Forel, A. 1903h. Einige neue Ameisen aus Süd-Angola. Pp. 559-564 in: Baum, H. Kunene-Sambesi-Expedition, 1903. Berlin: Verlag des Kolonial-Wirtschaftlichen Komitees, 593 pp. [1903]
Forel, A. 1904a. Ants and some other insects. An inquiry into the psychic powers of these animals with an appendix on the peculiarities of their olfactory sense. [concl.] [Translated by W. M. Wheeler.] Monist 14:177-193. [1904.01]
Forel, A. 1904b. Fourmis de British Columbia récoltées par M. Ed. Whymper. Ann. Soc. Entomol. Belg. 48:152-155. [1904.03.29]

Forel, A. 1904c ("1903"). Note sur les fourmis du Musée Zoologique de l'Académie Impériale des Sciences à St. Pétersbourg. Ezheg. Zool. Muz. 8:368-388. [1904.03]

Forel, A. 1904d. Miscellanea myrmécologiques. Rev. Suisse Zool. 12:1-52. [1904.04.18]

Forel, A. 1904e. Fourmis du Musée de Bruxelles. Ann. Soc. Entomol. Belg. 48:168-177. [1904.05.04]

Forel, A. 1904f. In und mit Pflanzen lebende Ameisen aus dem Amazonas-Gebiet und aus Peru, gesammelt von Herrn E. Ule. Zool. Jahrb. Abt. Syst. Geogr. Biol. Tiere 20:677-707. [1904.08.31]

Forel, A. 1904g. Dimorphisme du mâle chez les fourmis et quelques autres notices myrmécologiques. Ann. Soc. Entomol. Belg. 48:421-425. [1904.12.23]

Forel, A. 1904h. Ueber Polymorphismus und Variation bei den Ameisen. Zool. Jahrb. Suppl. 7:571-586. [1904]

Forel, A. 1904i. Formiciden. Ergeb. Hambg. Magalhaens. Sammelreise 7(8):1-7. [1904]

Forel, A. 1904j. Hymenoptera, Formicidae. Pp. 13-15 in: Wasmann, E. Termitophilen aus dem Sudan. In: Jägerskiöld, L. A. Results of the Swedish Zoological Expedition to Egypt and the White Nile, 1901. Part 1 (no. 13). Uppsala: Library of the Royal University of Uppsala, 21 pp. [1904]

Forel, A. 1904k. The psychical faculties of ants and some other insects. Annu. Rep. Smithson. Inst. 1903:587-599. [1904]

Forel, A. 1904l. Ants and some other insects. An inquiry into the psychic powers of these animals with an appendix on the peculiarities of their olfactory sense. [Translated by W. M. Wheeler.] Chicago: Open Court Publishing Company, 49 pp. [1904] [Reprint of Forel (1903g and 1904a).]

Forel, A. 1905a. Einige biologische Beobachtungen des Herrn Prof. Dr. E. Göldi an brasilianischen Ameisen. Biol. Centralbl. 25:170-181. [1905.03.15]

Forel, A. 1905b. Sklaverei, Symbiose und Schmarotzertum bei Ameisen. Mitt. Schweiz. Entomol. Ges. 11:85-90. [1905.04]

Forel, A. 1905c. Einige neue biologische Beobachtungen über Ameisen. Pp. 449-455 in: Bedot, M. (ed.) Compte-rendu des séances du Sixième Congrès International de Zoologie tenu à Berne du 14 au 16 août 1904. Genève: W. Kündig & Fils, xii + 733 pp. [1905.05.25]

Forel, A. 1905d. A revision of the species of the Formicidae (ants) of New Zealand. Trans. Proc. N. Z. Inst. 37:353-355. [1905.06]

Forel, A. 1905e. Miscellanea myrmécologiques II (1905). Ann. Soc. Entomol. Belg. 49:155-185. [1905.08.31]

Forel, A. 1905f. Ameisen aus Java. Gesammelt von Prof. Karl Kraepelin 1904. Mitt. Naturhist. Mus. Hambg. 22:1-26. [1905]

Forel, A. 1906a. Moeurs des fourmis parasites des genres *Wheeleria* et *Bothriomyrmex*. Rev. Suisse Zool. 14:51-69. [1906.03.28]

Forel, A. 1906b. Les fourmis de l'Himalaya. Bull. Soc. Vaudoise Sci. Nat. 42:79-94. [1906.03]

Forel, A. 1906c. Fourmis d'Asie mineure et de la Dobrudscha récoltées par M. le Dr. Oscar Vogt et Mme Cécile Vogt, Dr. méd. Ann. Soc. Entomol. Belg. 50:187-190. [1906.06.26]

Forel, A. 1906d. Fourmis néotropiques nouvelles ou peu connues. Ann. Soc. Entomol. Belg. 50:225-249. [1906.08.30]

Forel, A. 1907a. Nova speco kaj nova gentonomo de formikoj. Int. Sci. Rev. 4:144-145. [1907.05]

Forel, A. 1907b. [Review of: Escherich, K. 1906. Die Ameise. Schilderung ihrer Lebensweise. Braunschweig: Friedrich Vieweg und Sohn, xx + 232 pp.] Mitt. Schweiz. Entomol. Ges. 11:265-266. [1907.05]

Forel, A. 1907c. La faune malgache des fourmis et ses rapports avec les faunes de l'Afrique, de l'Inde, de l'Australie etc. Rev. Suisse Zool. 15:1-6. [1907.06.28]

Forel, A. 1907d. Formicides du Musée National Hongrois. Ann. Hist.-Nat. Mus. Natl. Hung. 5: 1-42. [1907.06.30]

Forel, A. 1907e. Fourmis nouvelles de Kairouan et d'Orient. Ann. Soc. Entomol. Belg. 51:201-208. [1907.08.02]

Forel, A. 1907f. Fourmis d'Ethiopie récoltées par M. le baron Maurice de Rothschild en 1905. Rev. Entomol. (Caen) 26:129-144. [1907.09]

Forel, A. 1907g. The Percy Sladen Trust Expedition to the Indian Ocean in 1905, under the leadership of Mr. J. Stanley Gardiner, M.A. VI. Fourmis des Seychelles, Amirantes, Farquhar et Chagos. Trans. Linn. Soc. Lond. Zool. (2)12:91-94. [1907.09]

Forel, A. 1907h. Formiciden aus dem Naturhistorischen Museum in Hamburg. II. Teil. Neueingänge seit 1900. Mitt. Naturhist. Mus. Hambg. 24:1-20. [1907.10.15]

Forel, A. 1907i. Ameisen von Madagaskar, den Comoren und Ostafrika. Wiss. Ergeb. Reise Ostafr. 2:75-92. [1907]

Forel, A. 1907j. Formicidae. In: Michaelsen, W., Hartmeyer, R. (eds.) Die Fauna Südwest-Australiens. Band I, Lieferung 7. Jena: Gustav Fischer, pp. 263-310. [1907]

Forel, A. 1908a. Catálogo systemático da collecção de formigas do Ceará. Bol. Mus. Rocha 1(1):62-69. [1908.01]

Forel, A. 1908b. Fourmis de Ceylan et d'Égypte récoltées par le Prof. E. Bugnion. *Lasius carniolicus*. Fourmis de Kerguelen. Pseudandrie? *Strongylognathus testaceus*. Bull. Soc. Vaudoise Sci. Nat. 44:1-22. [1908.03]

Forel, A. 1908c. Fourmis de Costa-Rica récoltées par M. Paul Biolley. Bull. Soc. Vaudoise Sci. Nat. 44:35-72. [1908.03]

Forel, A. 1908d. [Untitled. Introduced by: "Prof. Dr. A. Forel hatte die Güte, mir zum Zwecke der Publikation nachfolgende Diagnose einzusenden: *Pheidole Anastasii* Emery var. *cellarum* A. Forel nov. var."] Bot. Gart. Bot. Mus. Univ. Zür. 1907:6. [< 1908.04.21] [Received at the BMNH 21 April 1908.]

Forel, A. 1908e. Lettre à la Société Entomologique de Belgique. Ann. Soc. Entomol. Belg. 52:180-181. [1908.04.30]

Forel, A. 1908f. Konflikt zwischen zwei Raubameisenarten. Biol. Centralbl. 28:445-447. [1908.07.01]

Forel, A. 1908g. Remarque sur la réponse de M. le prof. Emery. Bull. Soc. Vaudoise Sci. Nat. 44:218. [1908.09]

Forel, A. 1908h. Ameisen aus Sao Paulo (Brasilien), Paraguay etc. gesammelt von Prof. Herm. v. Ihering, Dr. Lutz, Dr. Fiebrig, etc. Verh. K-K. Zool.-Bot. Ges. Wien 58:340-418. [1908.12.18]

Forel, A. 1908i. E. Formiciden. P. 105 in: Wissenschaftliche Ergebnisse der Expedition Filchner nach China und Tibet, 1903-1905. X. Band. - I. Teil. Berlin: Ernst Siegfried Mittler & Sohn, 244 pp. [1908]

Forel, A. 1908j. The senses of insects. [Translated by Macleod Yearsley.] London: Methuen, xiv + 324 pp. [1908]

Forel, A. 1909a. Ameisen aus Guatemala usw., Paraguay und Argentinien (Hym.). Dtsch. Entomol. Z. 1909:239-269. [1909.03.01]

Forel, A. 1909b. Fourmis du Musée de Bruxelles. Fourmis de Benguela récoltées par M. Creighton Wellman, et fourmis du Congo récoltées par MM. Luja, Kohl et Laurent. Ann. Soc. Entomol. Belg. 53:51-73. [1909.03.05]

Forel, A. 1909c. Fourmis d'Espagne récoltées par M. O. Vogt et Mme Cécile Vogt, Docteurs en médecine. Ann. Soc. Entomol. Belg. 53:103-106. [1909.04.02]

Forel, A. 1909d. Professor Dr. Gustav Mayr. Mitt. Schweiz. Entomol. Ges. 11:361-364. [1909.05]

Forel, A. 1909e. A propos des "fourmilières-boussoles". Bull. Soc. Vaudoise Sci. Nat. 45:341-343. [1909.06]

Forel, A. 1909f. Faune antarctique des fourmis. Mitt. Schweiz. Entomol. Ges. 11:381-382. [1909.11]

Forel, A. 1909g. La faune xérothermique des fourmis et l'angle du Valais. Arch. Sci. Phys. Nat. (4)28:506-508. [1909.11]

Forel, A. 1909h. Fondation des fourmilières de *Formica sanguinea*. Arch. Sci. Phys. Nat. (4)28:508-509. [1909.11]

Forel, A. 1909i. Ameisen aus Java und Krakatau beobachtet und gesammelt von Herrn Edward Jacobson. Notes Leyden Mus. 31:221-232. [1909.12.20]

Forel, A. 1909j. Schlussanhang. Pp. 252-253 in: Jacobson, E. Ameisen aus Java und Krakatau beobachtet und gesammelt von Herrn Edward Jacobson, bestimmt und beschrieben von Dr. A. Forel. II. Biologischer Theil. Notes Leyden Mus. 31:233-253. [1909.12.20]

Forel, A. 1909k. Études myrmécologiques en 1909. Fourmis de Barbarie et de Ceylan. Nidification des *Polyrhachis*. Bull. Soc. Vaudoise Sci. Nat. 45:369-407. [1909.12]

Forel, A. 1909l. [Forward to paper by E. Jacobson: Ein Moskito als Gast und diebischer Schmarotzer der *Crematogaster difformis* Smith und eine andere schmarotzende Fliege. Tijdschr. Entomol. 52:159-164.] Tijdschr. Entomol. 52:158-159. [1909.12.31]

Forel, A. 1910a. Glanures myrmécologiques. Ann. Soc. Entomol. Belg. 54:6-32. [1910.02.04]

Forel, A. 1910b. Formicides australiens reçus de MM. Froggatt et Rowland Turner. Rev. Suisse Zool. 18:1-94. [1910.03.22]

Forel, A. 1910c. Ameisen aus der Kolonie Erythräa. Gesammelt von Prof. Dr. K. Escherich (nebst einigen in West-Abessinien von Herrn A. Ilg gesammelten Ameisen). Zool. Jahrb. Abt. Syst. Geogr. Biol. Tiere 29:243-274. [1910.07.22]

Forel, A. 1910d. Fourmis des Philippines. Philipp. J. Sci. Sect. D. Gen. Biol. Ethnol. Anthropol. 5:121-130. [1910.07]

Forel, A. 1910e. Zoologische und anthropologische Ergebnisse einer Forschungsreise im westlichen und zentralen Südafrika ausgeführt in den Jahren 1903-1905 von Dr. Leonhard Schultze. Vierter Band. Systematik und Tiergeographie. D) Formicidae. Denkschr. Med.-Naturwiss. Ges. Jena 16:1-30. [< 1910.12.02] [No exact publication date is available but a comparison of references to species names in Forel (1910e) with those in Forel (1910f) suggests that the former was published first.]

Forel, A. 1910f. Note sur quelques fourmis d'Afrique. Ann. Soc. Entomol. Belg. 54:421-458. [1910.12.02]

Forel, A. 1910g. Das Sinnesleben der Insekten. München: Ernst Reinhardt, xv + 393 pp. [1910]

Forel, A. 1911a. Fourmis de Bornéo, Singapore, Ceylan, etc. récoltées par MM. Haviland, Green, Winkler, Will, Hose, Roepke et Waldo. Rev. Suisse Zool. 19:23-62. [1911.01]

Forel, A. 1911b. Ameisen aus Java beobachtet und gesammelt von Herrn Edward Jacobson. II. Theil. Notes Leyden Mus. 33:193-218. [1911.04.29]

Forel, A. 1911c. Ameisen aus Ceylon, gesammelt von Prof. K. Escherich (einige von Prof. E. Bugnion). Pp. 215-228 in: Escherich, K. Termitenleben auf Ceylon. Jena: Gustav Fischer, xxxii + 262 pp. [1911.04] [Apparently published by April 1911 since Forel (1911d) reviews Escherich's book in that month in Mitt. Schweiz. Entomol. Ges.]

Forel, A. 1911d ("1910"). [Review of: Escherich, K. 1911. Termitenleben auf Ceylon. Jena: Gustav Fischer, xxxii + 262 pp.] Mitt. Schweiz. Entomol. Ges. 12:54-55. [1911.04]

Forel, A. 1911e. Ameisen des Herrn Prof. v. Ihering aus Brasilien (Sao Paulo usw.) nebst einigen anderen aus Südamerika und Afrika (Hym.). Dtsch. Entomol. Z. 1911:285-312. [1911.05.04]

Forel, A. 1911f. Fourmis nouvelles ou intéressantes. Bull. Soc. Vaudoise Sci. Nat. 47:331-400. [1911.06]

Forel, A. 1911g. Die Ameisen des K. Zoologischen Museums in München. Sitzungsber. Math.-Phys. Kl. K. Bayer. Akad. Wiss. Münch. 11:249-303. [1911.07]

Forel, A. 1911h. Fourmis d'Afrique et d'Asie. I. Fourmis d'Afrique surtout du Musée du Congo Belge. Rev. Zool. Afr. (Bruss.) 1:274-283. [1911.08.31]

Forel, A. 1911i. Fourmis d'Afrique et d'Asie. Annexe. II. Quelques fourmis d'Asie. Rev. Zool. Afr. (Bruss.) 1:284-286. [1911.08.31]

Forel, A. 1911j. Aperçu sur la distribution géographique et la phylogénie des fourmis. Pp. 81-100 in: Ier Congrès International d'Entomologie, Bruxelles, août 1910. Vol. II, Mémoires. Bruxelles: Hayez, 520 pp. [1911.10.30]

Forel, A. 1911k. Une colonie polycalique de *Formica sanguinea* sans esclaves dans le canton de Vaud. Pp. 101-104 in: Ier Congrès International d'Entomologie, Bruxelles, août 1910. Vol. II, Mémoires. Bruxelles: Hayez, 520 pp. [1911.10.30]

Forel, A. 1911l. Sur le genre *Metapone* n. g. nouveau groupe des Formicides et sur quelques autres formes nouvelles. Rev. Suisse Zool. 19:445-459. [1911.12]

Forel, A. 1912a. H. Sauter's Formosa-Ausbeute. Formicidae (Hym.). Entomol. Mitt. 1:45-61. [1912.03.01]

Forel, A. 1912b. H. Sauter's Formosa-Ausbeute. Formicidae (Hym.) (Schluss). Entomol. Mitt. 1:67-81. [1912.03.01]

Forel, A. 1912c. Einige interessante Ameisen des Deutschen Entomologischen Museums zu Berlin-Dahlem. Entomol. Mitt. 1:81-83. [1912.03.01]

Forel, A. 1912d. Formicides néotropiques. Part I. Ann. Soc. Entomol. Belg. 56:28-49. [1912.03.02]

Forel, A. 1912e. Ameisen aus Java beobachtet und gesammelt von Edward Jacobson. III. Theil. Notes Leyden Mus. 34:97-112. [1912.04.01]

Forel, A. 1912f. Formicides néotropiques. Part II. 3me sous-famille Myrmicinae Lep. (Attini, Dacetii, Cryptocerini). Mém. Soc. Entomol. Belg. 19:179-209. [1912.04.15]

Forel, A. 1912g. Formicides néotropiques. Part III. 3me sous-famille Myrmicinae (suite). Genres *Cremastogaster* et *Pheidole*. Mém. Soc. Entomol. Belg. 19:211-237. [1912.04.15]

Forel, A. 1912h. Formicides néotropiques. Part IV. 3me sous-famille Myrmicinae Lep. (suite). Mém. Soc. Entomol. Belg. 20:1-32. [1912.06.05]

Forel, A. 1912i. Formicides néotropiques. Part V. 4me sous-famille Dolichoderinae Forel. Mém. Soc. Entomol. Belg. 20:33-58. [1912.06.10]

Forel, A. 1912j. Formicides néotropiques. Part VI. 5me sous-famille Camponotinae Forel. Mém. Soc. Entomol. Belg. 20:59-92. [1912.06.20]

Forel, A. 1912k. Die Weibchen der "Treiberameisen" *Anomma nigricans* Illiger und *Anomma Wilverthi* Emery, nebst einigen andern Ameisen aus Uganda. Mitt. Naturhist. Mus. Hambg. 29:173-181. [1912.08.15]

Forel, A. 1912l. The Percy Sladen Trust Expedition to the Indian Ocean in 1905, under the leadership of Mr. J. Stanley Gardiner, M.A. Volume 4. No. XI. Fourmis des Seychelles et des Aldabras, reçues de M. Hugh Scott. Trans. Linn. Soc. Lond. Zool. (2)15:159-167. [1912.09]

Forel, A. 1912m. Quelques fourmis de Tokio. Ann. Soc. Entomol. Belg. 56:339-342. [1912.10.03]

Forel, A. 1912n. Descriptions provisoires de genres, sous-genres, et espèces de Formicides des Indes orientales. Rev. Suisse Zool. 20:761-774. [1912.12]

Forel, A. 1912o. Einige neue und interessante Ameisenformen aus Sumatra etc. Zool. Jahrb. Suppl. 15("Erster Band"):51-78. [1912] [Zool. Jahrb. Suppl. 15 itself consists of three volumes, each individually paginated. The preceding article appears in the first volume.]

Forel, A. 1913a. Fourmis de Rhodesia, etc. récoltées par M. G. Arnold, le Dr. H. Brauns et K. Fikendey. Ann. Soc. Entomol. Belg. 57:108-147. [1913.05.02]

Forel, A. 1913b. Formicides du Congo Belge récoltés par MM. Bequaert, Luja, etc. Rev. Zool. Afr. (Bruss.) 2:306-351. [1913.05.30]

Forel, A. 1913c. Fourmis de Nigérie. Rev. Zool. Afr. (Bruss.) 2:352-353. [1913.05.30]

Forel, A. 1913d. Fourmis de la faune méditerranéenne récoltées par MM. U. et J. Sahlberg. Rev. Suisse Zool. 21:427-438. [1913.06]

Forel, A. 1913e. Notes sur ma collection de fourmis. Ann. Soc. Entomol. Belg. 57:202. [1913.07.04]

Forel, A. 1913f. Quelques fourmis des Indes, du Japon et d'Afrique. Rev. Suisse Zool. 21:659-673. [1913.09]

Forel, A. 1913g. H. Sauter's Formosa-Ausbeute: Formicidae II. Arch. Naturgesch. (A)79(6):183-202. [1913.09]

Forel, A. 1913h. Fourmis de Tasmanie et d'Australie récoltées par MM. Lae, Froggatt etc. Bull. Soc. Vaudoise Sci. Nat. 49:173-195. [1913.09]

Forel, A. 1913i. Quelques fourmis du Musée du Congo Belge (1). Ann. Soc. Entomol. Belg. 57:347-359. [1913.12.05]
Forel, A. 1913j. Notes sur quelques *Formica*. Ann. Soc. Entomol. Belg. 57:360-361. [1913.12.05]
Forel, A. 1913k. Ameisen aus Rhodesia, Kapland usw. (Hym.) gesammelt von Herrn G. Arnold, Dr. H. Brauns und Anderen. Dtsch. Entomol. Z. 1913(Suppl.):203-225. [1913.12.15]
Forel, A. 1913l. Wissenschaftliche Ergebnisse einer Forschungsreise nach Ostindien ausgeführt im Auftrage der Kgl. Preuss. Akademie der Wissenschaften zu Berlin von H. v. Buttel-Reepen. II. Ameisen aus Sumatra, Java, Malacca und Ceylon. Gesammelt von Herrn Prof. Dr. v. Buttel-Reepen in den Jahren 1911-1912. Zool. Jahrb. Abt. Syst. Geogr. Biol. Tiere 36:1-148. [1913.12.19]
Forel, A. 1913m. Fourmis d'Argentine, du Brésil, du Guatémala & de Cuba reçues de M. M. Bruch, Prof. v. Ihering, Mlle Báez, M. Peper et M. Rovereto. Bull. Soc. Vaudoise Sci. Nat. 49:203-250. [1913.12]
Forel, A. 1914a. Le genre *Camponotus* Mayr et les genres voisins. Rev. Suisse Zool. 22:257-276. [1914.05]
Forel, A. 1914b. Deux nouveautés myrmécologiques. Yvorne, Switzerland: published by the author, 1 p. [1914.09.01]
Forel, A. 1914c. Einige amerikanische Ameisen. Dtsch. Entomol. Z. 1914:615-620. [1914.12.10]
Forel, A. 1914d. Formicides d'Afrique et d'Amérique nouveaux ou peu connus. Bull. Soc. Vaudoise Sci. Nat. 50:211-288. [1914.12]
Forel, A. 1914e. Quelques fourmis de Colombie. Pp. 9-14 in: Fuhrmann, O., Mayor, E. Voyage d'exploration scientifique en Colombie. Mém. Soc. Neuchâtel. Sci. Nat. 5(2):1-1090. [1914]
Forel, A. 1915a. Fauna Simalurensis. Hymenoptera Aculeata, Fam. Formicidae. Tijdschr. Entomol. 58:22-43. [1915.03.15]
Forel, A. 1915b. Results of Dr. E. Mjöbergs Swedish Scientific Expeditions to Australia 1910-13. 2. Ameisen. Ark. Zool. 9(16):1-119. [1915.06.16]
Forel, A. 1915c. Formicides d'Afrique et d'Amérique nouveaux ou peu connus. IIe partie. Bull. Soc. Vaudoise Sci. Nat. 50:335-364. [1915.06]
Forel, A. 1915d. Fauna insectorum helvetiae. Hymenoptera. Formicidae. Die Ameisen der Schweiz. Mitt. Schweiz. Entomol. Ges. 12(Beilage):1-77. [1915]
Forel, A. 1916. Fourmis du Congo et d'autres provenances récoltées par MM. Hermann Kohl, Luja, Mayné, etc. Rev. Suisse Zool. 24:397-460. [1916.04]
Forel, A. 1917. Cadre synoptique actuel de la faune universelle des fourmis. Bull. Soc. Vaudoise Sci. Nat. 51:229-253. [1917.03.08]
Forel, A. 1918a. Études myrmécologiques en 1917. Bull. Soc. Vaudoise Sci. Nat. 51:717-727. [1918.04.05]
Forel, A. 1918b. Quelques fourmis de Madagascar récoltées par le Dr. Friederichs et quelques remarques sur d'autres fourmis. Bull. Soc. Vaudoise Sci. Nat. 52:151-156. [1918.10.02]
Forel, A. 1919. Geographie und Wanderungen der Ameisen, ihre Bedeutung für die Evolutionslehre. Westermanns Monatsh. 127:213-216. [1919.10]
Forel, A. 1920a ("1919"). Deux fourmis nouvelles du Congo. Bull. Soc. Vaudoise Sci. Nat. 52:479-481. [1920.02.16]
Forel, A. 1920b. Les fourmis de la Suisse. Notices anatomiques et physiologiques, architecture, distribution géographique, nouvelles expériences et observations de moeurs. Seconde édition revue et corrigée. La Chaux-de-Fonds: Imprimerie Coopérative, xvi + 333 pp. [1920.10] [Reviewed in the October issue of Riv. Biol. (Rome) (Emery, 1920c).]
Forel, A. 1921a ("1920"). Fourmis trouvées dans des galles de *Cordia* et d'*Agonandra*, etc. Bull. Soc. Bot. Genève (2)12:201-208. [1921.08.31]
Forel, A. 1921b. Quelques fourmis des environs de Quito (Ecuador) récoltées par Mlle Eléonore Naumann. Bull. Soc. Vaudoise Sci. Nat. 54:131-135. [1921.11.15]

Forel, A. 1921c. Le monde social des fourmis comparé à celui de l'homme. Tome 1. Genèse, formes, anatomie, classification. géographie, fossiles. Genève: Librairie Kundig, xiv + 192 pp. [1921]

Forel, A. 1921d. Le monde social des fourmis comparé à celui de l'homme. Tome 2. Sensations, physiologie, fourmis et plantes, hôtes, parasites, nids. Genève: Librairie Kundig, iii + 184 pp. [1921]

Forel, A. 1922a ("1921"). Remarque sur "C. Emery, Hymenoptera, Fam. Formicidae" dans Genera insectorum de P. Wytsman. Bull. Soc. Vaudoise Sci. Nat. 54:205-207. [1922.01.15]

Forel, A. 1922b. Glanures myrmécologiques en 1922. Rev. Suisse Zool. 30:87-102. [1922.09]

Forel, A. 1922c. Hyménoptères, Formicides. Pp. 941-963 in: Bouvier, E.-L. (ed.) Voyage de M. le baron Maurice de Rothschild en Éthiopie et en Afrique orientale anglaise (1904-1905). Résultats scientifiques. Animaux articulés. Deuxième partie. Paris: Imprimerie Nationale, pp. 483-1041. [1922.12.01]

Forel, A. 1922d. Le monde social des fourmis comparé à celui de l'homme. Tome 3. Appareils d'observation. Fondation des foumilières. Moeurs à l'intérieur des nids. Bétail, jardins, fourmis parasites. (Avec appendice du Dr. E. Bugnion.) Genève: Librairie Kundig, vii + 227 pp. [1922]

Forel, A. 1923a. Le monde social des fourmis comparé à celui de l'homme. Tome 4. Alliances et guerres, parabiose, lestobiose, esclavagisme. Genève: Librairie Kundig, vii + 172 pp. [1923]

Forel, A. 1923b. Le monde social des fourmis comparé à celui de l'homme. Tome 5. Moeurs specialisées. Epilogue, les fourmis. Les termites et l'homme. Genève: Librairie Kundig, vi + 174 pp. [1923]

Forel, A. 1925a. Le Professeur Dr. Carlo Emery. Bull. Ann. Soc. Entomol. Belg. 65:198-199. [1925.06.30]

Forel, A. 1925b. *Monomorium Pharaonis* in Genfer Hotels. Mitt. Schweiz. Entomol. Ges. 13:427-428. [1925.07.15]

Forel, A. 1925c. Prof. Dr. Carlo Emery. Bull. Soc. Vaudoise Sci. Nat. 56:23-24. [1925.11.30]

Forel, A. 1928a. The social world of the ants compared with that of man. Volume 1. [Translated by C. K. Ogden.] London: G. P. Putnam's Sons, Ltd., xlv + 551 pp. [1928.02]

Forel, A. 1928b. The social world of the ants compared with that of man. Volume 2. [Translated by C. K. Ogden.] London: G. P. Putman's Sons, Ltd., xx + 444 pp. [1928.02]

Forel, A. 1948. Die Welt der Ameisen. (Ausgewählt und übersetzt von Heinrich Kutter.) Zürich: Rotapfel, 275 pp. [1948]

Forskål, P. 1775. Descriptiones animalium, avium, amphibiorum, piscium, insectorum, vermium; quae in itinere orientali observavit Petrus Forskål. Post mortum auctoris edidit Carsten Niebuhr. Hauniae [= Copenhagen]: Moeller, xxxiv + 164 pp. [1775]

Forsslund, K.-H. 1947. Svenska myror. 1-10. Entomol. Tidskr. 68:67-80. [1947.04.05]

Forsslund, K.-H. 1949. Svenska myror. 11-14. Entomol. Tidskr. 70:19-32. [1949.05.05]

Forsslund, K.-H. 1950. Om *Lasius niger* var. *alienobrunnea* E. Strand 1903 (Hym. Formicidae). Nor. Entomol. Tidsskr. 8:51-52. [1950.09.01]

Forsslund, K.-H. 1957a. Catalogus insectorum sueciae. XV. Hymenoptera: Fam. Formicidae. Opusc. Entomol. 22:70-78. [1957.03.09]

Forsslund, K.-H. 1957b. Svenska myror. 15-19. Entomol. Tidskr. 78:32-40. [1957.06.15]

Förster, B. 1891. Die Insekten des "Plattigen Steinmergels" von Brunstatt. Abh. Geol. Spezialkt. Elsass-Lothringen 3:333-594. [1891.06.20] [Date of publication from Evenhuis (1994b).]

Forsyth, A. 1981. Sex ratio and parental investment in an ant population. Evolution 35:1252-1253. [1981.12.22]

Fortelius, W., Pamilo, P., Rosengren, R., Sundström, L. 1987. Male size dimorphism and alternative reproductive tactics in *Formica exsecta* ants (Hymenoptera, Formicidae). Ann. Zool. Fenn. 24:45-54. [1987.05.15]

Fowler, H. G. 1977. *Acromyrmex (Moellerius) landolti* Forel en el Paraguay: las subspecies *balzani* (Emery) y *fracticornis* (Forel) (Insecta: Hymenoptera). Neotropica (La Plata) 23: 39-44. [1977.06.30]

Fowler, H. G. 1978. New records of *Atta* and *Acromyrmex* species in Paraguay. Attini 4:2. [1978.07]

Fowler, H. G. 1979a ("1978"). Variación sexual diferencial en *Acromyrmex rugosus rugosus* (Hymenoptera - Formicidae - Attini). Neotropica (La Plata) 24:141-144. [1979.08.01]

Fowler, H. G. 1979b. Las hormigas cortadoras del Paraguay de los géneros *Atta* Fabricius y *Acromyrmex* Mayr: bionomia, distribución y sistemàtica. Inf. Cient. (Univ. Nac. Asunción) 2:30-70. [Not seen. Cited in Fowler (1983a:138) and Fowler (1985a:33).]

Fowler, H. G. 1981 ("1980"). Nuevos registros de hormigas para el Paraguay (Hymenoptera Formicidae). Neotropica (La Plata) 26:183-186. [1981.04.01]

Fowler, H. G. 1982a. A new species of *Trachymyrmex* fungus-growing ant (Hymenoptera: Myrmicinae: Attini) from Paraguay. J. N. Y. Entomol. Soc. 90:70-73. [1982.06.07]

Fowler, H. G. 1982b. Evolution of the foraging behavior of leaf-cutting ants (*Atta* and *Acromyrmex*). [Abstract.] P. 33 in: Breed, M. D., Michener, C. D., Evans, H. E. (eds.) The biology of social insects. Proceedings of the Ninth Congress of the IUSSI, Boulder, Colorado, August 1982. Boulder: Westview Press, xii + 420 pp. [1982.08]

Fowler, H. G. 1983a. Distribution patterns of Paraguayan leaf-cutting ants (*Atta* and *Acromyrmex*) (Formicidae: Attini). Stud. Neotrop. Fauna Environ. 18:121-138. [1983]

Fowler, H. G. 1983b. Glandular and structural variation with respect to worker size variation in the carpenter ant, *Camponotus pennsylvanicus* (DeGeer) (Hymenoptera: Formicidae). Sociobiology 8:199-207. [1983]

Fowler, H. G. 1984 ("1983"). Latitudinal gradients and diversity of the leaf-cutting ants (*Atta* and *Acromyrmex*) (Hymenoptera: Formicidae). Rev. Biol. Trop. 31:213-216. [1984.05]

Fowler, H. G. 1985a. Leaf-cuttings ants of the genera *Atta* and *Acromyrmex* of Paraguay (Hymenoptera Formicidae). Dtsch. Entomol. Z. (N.F.)32:19-34. [1985.03.15]

Fowler, H. G. 1985b ("1984"). Colony-level regulation of forager caste ratios in response to caste perturbations in the carpenter ant, *Camponotus pennsylvanicus* (De Geer) (Hymenoptera: Formicidae). Insectes Soc. 31:461-472. [1985.04]

Fowler, H. G. 1986. Polymorphism and colony ontogeny in North American carpenter ants (Hymenoptera: Formicidae: *Camponotus pennsylvanicus* and *Camponotus ferrugineus*). Zool. Jahrb. Abt. Allg. Zool. Physiol. Tiere 90:297-316. [1986]

Fowler, H. G. 1988a ("1987"). Worker polymorphism in field colonies of carpenter ants (Hymenoptera: Formicidae: *Camponotus*): stochastic selection? Insectes Soc. 34:204-210. [1988.06]

Fowler, H. G. 1988b. Taxa of the neotropical grass-cutting ants, *Acromyrmex* (Hymenoptera: Formicidae: Attini). Científica (Jaboticabal) 16:281-295. [1988]

Fowler, H. G. 1992. Native fauna simplification by introduction of an exotic ant following hydroelectric dam construction in northeastern Brazil. Ciênc. Cult. (São Paulo) 44:345-346. [1992.10]

Fowler, H. G. 1994 ("1993"). Relative representation of *Pheidole* (Hymenoptera: Formicidae) in local ground ant assemblages of the Americas. An. Biol. 19:29-37.

Fowler, H. G. 1995. The population status of the endangered Brazilian endemic leaf-cutting ant *Atta robusta* (Hymenoptera: Formicidae). Biol. Conserv. 74:147-150. [1995]

Fowler, H. G., Bernardi, J. V. E., Delabie, J. C., Forti, L. C., Pereira-da-Silva, V. 1990. Major ant problems of South America. Pp. 3-14 in: Vander Meer, R. K., Jaffe, K., Cedeno, A. (eds.) Applied myrmecology: a world perspective. Boulder: Westview Press, xv + 741 pp. [1990]

Fowler, H. G., Bernardi, J. V. E., di Romagnano, L. F. T. 1990. Community structure and *Solenopsis invicta* in São Paulo. Pp. 199-207 in: Vander Meer, R. K., Jaffe, K., Cedeno, A. (eds.) Applied myrmecology: a world perspective. Boulder: Westview Press, xv + 741 pp. [1990]

Fowler, H. G., Campiolo, S., Pesquero, M. A., Porter, S. D. 1995. Notes on a southern record for *Solenopsis geminata* (Hymenoptera: Formicidae). Iheringia Sér. Zool. 79:173.
Fowler, H. G., Claver, S. 1991. Leaf-cutter ant assemblies: effects of latitude, vegetation, and behaviour. Pp. 51-59 in: Huxley, C. R., Cutler, D. F. (eds.) Ant-plant interactions. Oxford: Oxford University Press, xviii + 601 pp. [1991]
Fowler, H. G., Delabie, J. H. 1995. A new record of *Cylindromyrmex striatus* and range extension of *C. brasiliensis* in Brazil (Hymenoptera, Formicidae). Rev. Biol. Trop. 43:327-328. [1995.09]
Fowler, H. G., Della Lucia, T. M. C., Moreira, D. D. O. 1993. Posição taxonômica das formigas cortadeiras. Pp. 4-25 in: Della Lucia, T. M. C. (ed.) As formigas cortadeiras. Viçosa: Editora Folha de Viçosa, 262 pp. [1993.11.21]
Fowler, H. G., Forti, L. C., Pereira-da-Silva, V., Saes, N. B. 1986. Economics of grass-cutting ants. Pp. 18-35 in: Lofgren, C. S., Vander Meer, R. K. (eds.) Fire ants and leaf cutting ants: biology and management. Boulder: Westview Press, xv + 435 pp. [1986]
Fowler, H. G., Robinson, S. W. 1979. Field identification and relative pest status of Paraguayan leaf-cutting ants. Turrialba 29:11-16. [1979.03]
Fowler, H. G., Schlindwein, M. N., Medeiros, M. A. de. 1994. Exotic ants and community simplification in Brazil: a review of the impact of exotic ants on native ant assemblages. Pp. 151-162 in: Williams, D. F. (ed.) Exotic ants. Biology, impact, and control of introduced species. Boulder: Westview Press, xvii + 332 pp. [1994]
Fowler, H. G., Schlittler, F. M., Schlindwein, M. N. 1995. ATTINE: a computerized database of leaf-cutting ant (*Atta* and *Acromyrmex*) literature. J. Appl. Entomol. 119:255. [1995.04]
Franch, J., Espadaler, X. 1988. Ants as colonizing agents of pine stumps in San Juan de la Peña (Huesca, Spain). Vie Milieu 38:149-154. [1988.10]
Franch Batlle, J. See: Franch, J.
Francke, O. F., Cokendolpher, J. C., Horton, A. H., Phillips, S. A., Jr., Potts, L. R. 1983. Distribution of fire ants in Texas. Southwest. Entomol. 8:32-41. [1983.03]
Francke, O. F., Merickel, F. W. 1982 ("1981"). Two new species of *Pogonomyrmex* harvester ants from Texas (Hymenoptera: Formicidae). Pan-Pac. Entomol. 57:371-379. [1982.03.15]
Francoeur, A. 1965. Écologie des populations de fourmis dans un bois de chênes rouges et d'érables rouges. Nat. Can. (Qué.) 92:263-276. [1965.11]
Francoeur, A. 1966a. Le genre "*Stenamma*" Westwood au Québec (Hymenoptera, Formicidae). Ann. Soc. Entomol. Qué. 11:115-119. [1966.05]
Francoeur, A. 1966b. La faune myrmécologique de l'érablière à sucre (Aceretum saccharophori, Dansereau) de la région de Québec. Nat. Can. (Qué.) 93:443-472. [1966.10]
Francoeur, A. 1968. Une nouvelle espèce du genre *Myrmica* au Québec (Formicidae, Hymenoptera). Nat. Can. (Qué.) 95:727-730. [1968.06]
Francoeur, A. 1969. [Review of: Bernard, F. 1967 ("1968"). Les fourmis (Hymenoptera Formicidae) d'Europe occidentale et septentrionale. Paris: Masson, 411 pp.] Nat. Can. (Qué.) 96:153-154. [1969.02]
Francoeur, A. 1972. Reclassification des espèces néarctiques du groupe *fusca*, genre *Formica* (Hymenoptera: Formicidae). [Abstract]. Ann. Soc. Entomol. Qué. 17:111. [1972.09]
Francoeur, A. 1973. Révision taxonomique des espèces néarctiques du groupe *fusca*, genre *Formica* (Formicidae, Hymenoptera). Mém. Soc. Entomol. Qué. 3:1-316. [1973.09]
Francoeur, A. 1974. Nouvelles données et remarques sur la répartition nordique de quelques Formicides (Hyménoptères) néarctiques. Nat. Can. (Qué.) 101:935-936. [1974.12]
Francoeur, A. 1975 ("1974"). Notes for a revision of the ant genus *Formica*. 1. New identifications and synonymies for some nearctic specimens from Emery, Forel and Mayr collections. Entomol. News 85:257-264. [1975.08.30]
Francoeur, A. 1977a. The taxonomic status and biogeographic significance of the Sumatran *Formica* (Formicidae, Hymenoptera). Psyche (Camb.) 84:11-12. [1977.11.30]

Francoeur, A. 1977b. Synopsis taxonomique et économique des fourmis du Québec (Formicidae, Hymenoptera). Ann. Soc. Entomol. Qué. 22:205-212. [1977.11]

Francoeur, A. 1979a. Les fourmis du Québec. 1. Introduction. 2. La famille des Formicidae. 3. La sous-famille des Ponerinae. Ann. Soc. Entomol. Qué. 24:12-47. [1979.01]

Francoeur, A. 1979b. Formicoidea. Pp. 502-503 in: Danks, H. V. (ed.) Canada and its insect fauna. Mem. Entomol. Soc. Can. 108:1-573. [1979.04.12]

Francoeur, A. 1981a. Un mâle sans yeux composés de *Formica subsericea* (Formicidae, Hymenoptera). Nat. Can. (Qué.) 108:107-110. [1981.03]

Francoeur, A. 1981b. Les fourmis de la presqu'île de Forillon comté de Gaspé-est, Québec (Formicidae, Hymenoptera). Fabreries 7:78-83. [1981.04]

Francoeur, A. 1981c. Le groupe néarctique *Myrmica lampra* (Formicidae, Hymenoptera). Can. Entomol. 113:755-759. [1981.10.23]

Francoeur, A. 1982. Reclassification of the Nearctic species in the ant genus *Myrmica* (Formicidae, Hymenoptera). [Abstract.] Pp. 405-406 in: Breed, M. D., Michener, C. D., Evans, H. E. (eds.) The biology of social insects. Proceedings of the Ninth Congress of the IUSSI, Boulder, Colorado, August 1982. Boulder: Westview Press, xii + 420 pp. [1982.08]

Francoeur, A. 1983. The ant fauna near the tree-line in northern Quebec (Formicidae, Hymenoptera). Collect. Nord. 47:177-180. [1983.10]

Francoeur, A. 1984. La biosystématique des fourmis. Fabreries 11:11-15. [1984]

Francoeur, A. 1985a. [Untitled. *Formicoxenus quebecensis* Francoeur sp. nov.] Pp. 378-379 in: Francoeur, A., Loiselle, R., Buschinger, A. Biosystématique de la tribu Leptothoracini (Formicidae, Hymenoptera). 1. Le genre *Formicoxenus* dans la région holarctique. Nat. Can. (Qué.) 112:343-403. [1985.09]

Francoeur, A. 1985b. Deuxième mention de *Formica argentea* au Québec (Formicidae, Hymenoptera). Fabreries 12:10-11. [1985]

Francoeur, A. 1986a. The altitudinal range of *Pogonomyrmex pima* (Formicidae: Hymenoptera). Proc. Entomol. Soc. Wash. 88:594. [1986.07.31]

Francoeur, A. 1986b. Deux nouvelles fourmis néarctiques: *Leptothorax retractus* et *L. sphagnicolus* (Formicidae, Hymenoptera). Can. Entomol. 118:1151-1164. [1986.11]

Francoeur, A. 1986c. Supplément aux fourmis de la presqu'île de Forillon, Gaspé-est, Québec (Formicides, Hyménoptères). Fabreries 12:56-58. [1986]

Francoeur, A., Béique, R. 1966a. Les Formicides (Hyménoptères) de Provancher. Can. Entomol. 98:140-145. [1966.02.09]

Francoeur, A., Béique, R. 1966b. Un genre de Formicidae (Hymenoptera) nouveau pour le Québec. Nat. Can. (Qué.) 93:439. [1966.08]

Francoeur, A., Béique, R. 1968. Additions à la faune myrmécologique du Québec. Nat. Can. (Qué.) 95:227-229. [1968.02]

Francoeur, A., Elias, S. A. 1985. *Dolichoderus taschenbergi* Mayr (Hymenoptera: Formicidae) from an early Holocene fossil insect assemblage in the Colorado Front Range. Psyche (Camb.) 92:303-307. [1985.11.28]

Francoeur, A., Loiselle, R. 1984. Description du mâle et notice sur la biologie de *Myrmica quebecensis* Francoeur (Formicidae, Hymenoptera). Rev. Entomol. Qué. 29:3-11. [1984.01]

Francoeur, A., Loiselle, R. 1988a. Évolution du strigile chez les Formicides (Hyménoptères). Nat. Can. (Qué.) 115:333-353. [1988.12]

Francoeur, A., Loiselle, R. 1988b. The male of *Leptothorax wilda* with notes on the subgenus *Nesomyrmex* (Formicidae, Hymenoptera). Pp. 43-54 in: Trager, J. C. (ed.) Advances in myrmecology. Leiden: E. J. Brill, xxvii + 551 pp. [1988]

Francoeur, A., Loiselle, R., Buschinger, A. 1984. Taxonomic revision of the ant genus *Formicoxenus* (Formicidae, Hymenoptera). [Abstract.] P. 528 in: XVII International Congress of Entomology. Hamburg, Federal Republic of Germany, August 20-26, 1984. Abstract Volume. Hamburg: XVII International Congress of Entomology, 960 pp. [1984.08.10]

Francoeur, A., Loiselle, R., Buschinger, A. 1985. Biosystématique de la tribu Leptothoracini (Formicidae, Hymenoptera). 1. Le genre *Formicoxenus* dans la région holarctique. Nat. Can. (Qué.) 112:343-403. [1985.09]

Francoeur, A., Loiselle, R., Buschinger, A. 1987. Taxonomic revision of the ant genus *Formicoxenus* (Formicidae, Hymenoptera). P. 44 in: Eder, J., Rembold, H. (eds.) Chemistry and biology of social insects. München: Verlag J. Peperny, xxxv + 757 pp. [1987]

Francoeur, A., Loiselle, R., LePrince, D. 1982. Les fourmis du camp Kéno, comté de Portneuf, Québec (Formicidae: Hymenoptera). Fabreries 9:27-35. [1982.12]

Francoeur, A., Maldague, M. 1966. Classification des micromilieux de nidification des fourmis. Nat. Can. (Qué.) 93:473-478. [1966.10]

Francoeur, A., Pépin, D. 1975. Productivité de la fourmi *Formica dakotensis* dans la pessière tourbeuse. I. Densité observée et densité estimée des colonies. Insectes Soc. 22:135-150. [1975.12]

Francoeur, A., Snelling, R. R. 1979. Notes for a revision of the ant genus *Formica*. 2. Reidentifications for some specimens from the T. W. Cook collection and new distribution data (Hymenoptera: Formicidae). Contr. Sci. (Los Angel.) 309:1-7. [1979.03.16]

François, J. 1958. Contribution à l'étude écologique des Formicides (Insectes Hyménoptères) de la région Dijonnaise. Trav. Lab. Zool. Stn. Aquic. Grimaldi Fac. Sci. Dijon 25:1-36. [1958.05.12]

Françoso, M. F. L., Brandão, C. R. F. 1993. Classificação superior dos Formicidae. Biotemas 6:121-132. [1993.04]

Frank, S. R. 1987. Variable sex ratio among colonies of ants. Behav. Ecol. Sociobiol. 20:195-201. [1987.03]

Franks, N. R. 1990. [Review of: Hölldobler, B., Wilson, E. O. 1990. The ants. Cambridge, Mass.: Harvard University Press, xii + 732 pp.] Science (Wash. D. C.) 248:897-898. [1990.05.18]

Franks, N. R., Bossert, W. H. 1983. The influence of swarm raiding army ants on the patchiness and diversity of a tropical leaf litter ant community. Pp. 151-163 in: Sutton, S. L., Whitmore, T. C., Chadwick, A. C. (eds.) Tropical rain forest: ecology and management. Oxford: Blackwell, xii + 498 pp. [1983]

Franks, N. R., Bourke, A. F. G. 1990. The evolution of inquiline ant parasites: the interspecific versus the intraspecific hypothesis. Pp. 149-150 in: Veeresh, G. K., Mallik, B., Viraktamath, C. A. (eds.) Social insects and the environment. Proceedings of the 11th International Congress of IUSSI, 1990. New Delhi: Oxford & IBH Publishing Co., xxxi + 765 pp. [1990.08]

Franks, N. R., Hölldobler, B. 1987. Sexual competition during colony reproduction in army ants. Biol. J. Linn. Soc. 30:229-243. [1987.04.03]

Franks, N. R., Ireland, B., Bourke, A. F. G. 1990. Conflicts, social economics and life history strategies in ants. Behav. Ecol. Sociobiol. 27:175-181. [1990.09.28]

Franks, N. R., Norris, P. J. 1987. Constraints on the division of labour in ants: D'Arcy Thompson's cartesian transformations applied to worker polymorphism. Pp. 253-270 in: Pasteels, J. M., Deneubourg, J.-L. (eds.) From individual to collective behavior in social insects. (Experientia: Supplementum, Volume 54). Basel: Birkhäuser Verlag, 433 pp. [1987]

Franks, N. R., Partridge, L. W. 1993. Lanchester battles and the evolution of combat in ants. Anim. Behav. 45:197-199. [1993.01]

Franks, N. R., Scovell, E. 1983. Dominance and reproductive success among slave-making worker ants. Nature (Lond.) 304:724-725. [1983.08.25]

Frauenfeld, G. See: Frauenfeld, G R. von.

Frauenfeld, G. R. von. 1856. Beitrag zur Fauna Dalmatien's. Verh. Zool.-Bot. Ver. Wien 6:431-448. [1856] [Ants pp. 433, 435.]

Frauenfeld, G. R. von. 1867. Zoologische Miscellen. XI. Verh. K-K. Zool.-Bot. Ges. Wien 17:425-502. [1867]

Freeland, J. 1958. Biological and social patterns in the Australian bulldog ants of the genus *Myrmecia*. Aust. J. Zool. 6:1-18. [1958.05]

Freeland, J., Crozier, R. H., Marc, J. 1982. On the occurrence of arolia in ant feet. J. Aust. Entomol. Soc. 21:257-262. [1982.11.30]
Freitas, A. V. L. 1995. Nest relocation and prey specialization in the ant *Leptogenys propefalcigera* Roger (Formicidae: Ponerinae) in an urban area in south-eastern Brazil. Insectes Soc. 42:453-456. [1995] [*Leptogenys propefalcigera*, nomen nudum.]
Fresneau, D., Garcia Perez, J., Jaisson, P. 1982. Evolution of polyethism in ants: observational results and theories. Pp. 129-155 in: Jaisson, P. (ed.) Social insects in the tropics. Volume 1. Paris: Université Paris-Nord, 280 pp. [1982.06]
Fritzsche, R. 1978. Pflanzenschädlinge. Band 9. Hautflügler. Leipzig: Neumann Verlag, 212 pp. [1978] [Ants pp. 172-178, including keys.]
Froggatt, W. W. 1896. Honey ants. Pp. 385-392 in: Spencer, B. (ed.) Report on the work of the Horn Scientific Expedition to Central Australia. Part II. Zoology. London: Dulau & Co., iv + 431 pp. [1896.02]
Froggatt, W. W. 1905. Domestic insects: ants. Agric. Gaz. N. S. W. 16:861-866. [1905.09.02]
Froggatt, W. W. 1906. Domestic insects: ants. With catalogue of Australasian species. Misc. Publ. Dep. Agric. N. S. W. 889:1-35. [1906] [Reprint of Froggatt (1905), with additions and emendations.]
Froggatt, W. W. 1914. Scientific notes on an expedition into the interior of Australia carried out by Capt. S. A. White, M.B.O.U., from July to October, 1913. Insecta - Hymenoptera. Trans. R. Soc. S. Aust. 38:459. [1914.12]
Fröhlich, C. 1969. Untersuchungen zur Morphologie der Gelenksinnesfelder der Wiesenameise *Formica pratensis* Retzius, 1783. Beitr. Entomol. 19:281-288. [1969]
Frohlich, K. O., Kürschner, I. 1964. Die Giftorgane der Wiesenameise (*Formica nigricans* Emery, 1909). Hymenoptera: Formicidae. Beitr. Entomol. 14:507-524. [1964.06]
Fromantin, J., Soulié, J. 1961. Notes systématiques sur le genre *Cremastogaster* avec description de trois espèces nouvelles du Cambodge (Hymenoptera Formicoidea). Bull. Soc. Hist. Nat. Toulouse 96:87-112. [1961.05.30]
Frumhoff, P. C., Ward, P. S. 1992. Individual-level selection, colony-level selection, and the association between polygyny and worker monomorphism in ants. Am. Nat. 139:559-590. [1992.03]
Fujiyama, I. 1970. Fossil insects from the Chôjabaru Formation, Iki Island, Japan. Mem. Natl. Sci. Mus. (Tokyo) 3:65-74. [1970.10.20]
Fukai, T. 1908. On the characters and habits of *Polyrhachis lamellidens* Sm. [In Japanese.] Insect World 12:271-273. [1908.07.15]
Fukai, T. 1911. On *Camponotus marginatus*. [In Japanese.] Hakubutsu Tomo 11:40-41. [Not seen. Cited in Okano (1989) and Onoyama & Terayama (1994).]
Fullaway, D. T. 1914. A list of Laysan Island insects. Proc. Hawaii. Entomol. Soc. 3:20-22. [1914.09] [Ants (4 spp.) p. 20.]
Funakubo, H. 1939. A list of Hymenoptera from Kitakomagun, Yamanashi Prefecture. [In Japanese.] Entomol. World Tokyo 7:323-335. [1939.06.15] [Formicidae p. 332.]
Gadagkar, R., Nair, P., Chandrashekara, K., Bhat, D. M. 1993. Ant species richness and diversity in some selected localities in Western Ghats, India. Hexapoda (Insecta Indica) 5:79-94.
Gadau, J., Heinze, J., Dick, B., Hölldobler, B. 1994. Multilocus DNA-fingerprinting in ants. P. 44 in: Lenoir, A., Arnold, G., Lepage, M. (eds.) Les insectes sociaux. 12ème congrès de l'Union Internationale pour l'Étude des Insectes Sociaux. Paris, Sorbonne, 21-27 août 1994. Paris: Université Paris Nord, xxiv + 583 pp. [1994.09]
Gahan, A. B., Rohwer, S. A. 1917a. Lectotypes of the species of Hymenoptera (except Apoidea) described by Abbé Provancher. [part] Can. Entomol. 49:298-308. [1917.09.01]
Gahan, A. B., Rohwer, S. A. 1917b. Lectotypes of the species of Hymenoptera (except Apoidea) described by Abbé Provancher. [part] Can. Entomol. 49:391-400. [1917.11.01]
Gahan, A. B., Rohwer, S. A. 1917c. Lectotypes of the species of Hymenoptera (except Apoidea) described by Abbé Provancher. [part] Can. Entomol. 49:427-433. [1917.12.06]

Gahan, A. B., Rohwer, S. A. 1918a. Lectotypes of the species of Hymenoptera (except Apoidea) described by Abbé Provancher. [part] Can. Entomol. 50:101-106. [1918.03.15]
Gahan, A. B., Rohwer, S. A. 1918b. Lectotypes of the species of Hymenoptera (except Apoidea) described by Abbé Provancher. [concl.] Can. Entomol. 50:196-201. [1918.06.05]
Gahl, H., Maschwitz, U. 1977. Eine Ameise aus dem Mittel-Eozän von Messel bei Darmstadt (Hessen). Geol. Jahrb. Hessen 105:69-73. [1977]
Gaige, F. M. 1914a. Description of a new subspecies of *Pogonomyrmex occidentalis* Cresson from Nevada. Proc. Biol. Soc. Wash. 27:93-96. [1914.05.11]
Gaige, F. M. 1914b. Results of the Mershon expedition to the Charity Islands, Lake Huron. The Formicidae of Charity Island. Occas. Pap. Mus. Zool. Univ. Mich. 5:1-29. [1914.12.15]
Gaige, F. M. 1916. The Formicidae of the Shiras expedition to Whitefish Point, Michigan, in 1914. Occas. Pap. Mus. Zool. Univ. Mich. 25:1-4. [1916.04.15]
Gallardo, A. 1912. Observaciones sobre una hormiga invasora *Iridomyrmex humilis* Mayr. Physis (B. Aires) 1:133-138. [1912.12.31]
Gallardo, A. 1913. Dos palabras más acerca de la hormiga invasora *Iridomyrmex humilis* Mayr. Physis (B. Aires) 1:264-265. [1913.09.30]
Gallardo, A. 1915. Observaciones sobre algunas hormigas de la República Argentina. An. Mus. Nac. Hist. Nat. B. Aires 27:1-35. [1915.06.03]
Gallardo, A. 1916a. Fauna mirmecológica de Tandil y la Ventana. Physis (B. Aires) 2:128-131. [1916.02.12]
Gallardo, A. 1916b. Las hormigas de la República Argentina. Subfamilia Dolicoderinas. An. Mus. Nac. Hist. Nat. B. Aires 28:1-130. [1916.05.18]
Gallardo, A. 1916c. Notas acerca de la hormiga *Trachymyrmex pruinosus* Emery. An. Mus. Nac. Hist. Nat. B. Aires 28:241-252. [1916.09.04]
Gallardo, A. 1916d. Notas complementarias sobre las Dolicoderinas Argentinas. An. Mus. Nac. Hist. Nat. B. Aires 28:257-261. [1916.09.04]
Gallardo, A. 1916e. Notes systématiques et éthologiques sur les fourmis attines de la République Argentine. An. Mus. Nac. Hist. Nat. B. Aires 28:317-344. [1916.09.25]
Gallardo, A. 1917. Notes critiques sur les "Formicides sud-américains nouveaux ou peu connus du Docteur Santschi." Physis (B. Aires) 3:48-51. [1917.03.17]
Gallardo, A. 1918a. Hormigas dolicoderinas de los Andes de Mendoza. Physis (B. Aires) 4:28-31. [1918.05.15]
Gallardo, A. 1918b. Hormigas argentinas. Subfam. Dorylinae. Tribu Ecitonii. Rev. Mus. Pop. Paraná 1(2):1-3. [Not seen. Cited in Bruch (1934c:241.]
Gallardo, A. 1918c. Las hormigas de la República Argentina. Subfamilia Ponerinas. An. Mus. Nac. Hist. Nat. B. Aires 30:1-112. [1918.12.15]
Gallardo, A. 1919a. Una nueva prodorilina *Acanthostichus afflictus* [male]. An. Mus. Nac. Hist. Nat. B. Aires 30:237-242. [1919.07.22]
Gallardo, A. 1919b. Hormigas del Neuquén y Río Negro. An. Mus. Nac. Hist. Nat. B. Aires 30:243-254. [1919.08.02]
Gallardo, A. 1920. Las hormigas de la República Argentina. Subfamilia Dorilinas. An. Mus. Nac. Hist. Nat. B. Aires 30:281-410. [1920.02.28]
Gallardo, A. 1929a. Note sur les moeurs de la fourmi *Pseudoatta argentina*. Rev. Soc. Entomol. Argent. 2:197-202. [1929.10.31]
Gallardo, A. 1929b. Notas sobre las Dorilinas argentinas. An. Mus. Nac. Hist. Nat. B. Aires 36: 43-48. [1929.12.26]
Gallardo, A. 1930. Sobre el género *Dorymyrmex* Mayr en la Argentina. Rev. Chil. Hist. Nat. 34:143-148. [1930.12.24]
Gallardo, A. 1931a. Deux nouvelles espèces de "*Pogonomyrmex*" de la République Argentine (Hyménoptères Formicides). Rev. Mus. La Plata 33:185-188. [1931.11.18]

Gallardo, A. 1931b. Algunas formas sexuales aún no descriptas de las hormigas del género *Crematogaster* de la República Argentina. Rev. Soc. Entomol. Argent. 3:297-304. [1931.11.25]

Gallardo, A. 1931c. Doctor Augusto Forel. Rev. Soc. Entomol. Argent. 3:337-342. [1931.11.25]

Gallardo, A. 1932a. Las hormigas de la República Argentina. Subfamilia Mirmicinas, sección Promyrmicinae. An. Mus. Nac. Hist. Nat. B. Aires 37:37-87. [1932.05.21]

Gallardo, A. 1932b. El ingeniero Carlos Janet. An. Soc. Cient. Argent. 113:228-230. [1932.05]

Gallardo, A. 1932c. Las hormigas de la República Argentina. Subfamilia Mirmicinas, segunda sección Eumyrmicinae, tribu Myrmicini (F. Smith), género *Pogonomyrmex* Mayr. An. Mus. Nac. Hist. Nat. B. Aires 37:89-170. [1932.10.30]

Gallardo, A. 1932d. El subgénero *Elasmopheidole* en la República Argentina (Himenopteros Formicidos). Rev. Chil. Hist. Nat. 36:178-182. [1932]

Gallardo, A. 1934a. Un nuevo ejemplar femenino de *Eciton rogeri* Dalla Torre (Himenóp. Formicid.). Rev. Soc. Entomol. Argent. 6:2. [1934.05.30]

Gallardo, A. 1934b. Las hormigas de la República Argentina. Subfamilia Mirmicinas, segunda sección Eumyrmicinae, tribu Crematogastrini (Forel), género *Crematogaster* Lund. An. Mus. Nac. Hist. Nat. B. Aires 38:1-84. [1934.08.31]

Gallé, L. 1966. Ecological and zoocoenological investigation of the Formicoidea fauna of the flood area of the Tisza River. Tiscia (Szeged) 2:113-118. [1966]

Gallé, L. 1972. Formicidae populations of the ecosystems in the environs of Tiszafüred. Tiscia (Szeged) 7:59-68. [1972]

Gallé, L. 1979. Adatok a Bakony hegység hangya-(Hymenoptera: Formicoidea) faunájának ismeretéhez. Veszprém Megyei Múz. Közl. Természettud. 14:239-244. [1979]

Gallé, L. 1981. The formicoid fauna of the Hortobágy. Pp. 307-311 in: Mahunka, S. (ed.) The fauna of the Hortobágy National Park. Budapest: Akadémiai Kiadó, 415 pp. [1981]

Gallé, L. 1986. Habitat and niche analysis of grassland ants (Hymenoptera: Formicidae). Entomol. Gen. 11:197-211. [1986.05.30]

Gallé, L. 1990. Assembly of sand-dune forest ant communities. Memorabilia Zool. 44:1-6. [1990]

Gallé, L. 1991. Structure and succession of ant assemblages in a north European sand dune area. Holarct. Ecol. 14:31-37. [1991.01]

Gallé, L., Kovács, E., Margóczi, K. 1995. Ants on trees: an example of metacommunities in extremely small patches. [Abstract.] P. 85 in: Demeter, A., Peregovits, L. (eds.) Ecological processes: current status and perspectives. Abstracts. 7th European Ecological Congress. Budapest: Hungarian Biological Society, 294 pp. [1995.08.20]

Galvagni, E. 1902. Beiträge zur Kenntnis der Fauna einiger dalmatinischer Inseln. Verh. K-K. Zool.-Bot. Ges. Wien 52:362-388. [1902.08.08] [Ants p. 380.]

Gama, V. 1978. Desenvolvimento pós-embrionário das glandulas componentes do sistema salivar de *Camponotus* (*Myrmothrix*) *rufipes* (Fabricius, 1775), (Hymenoptera, Formicidae). Arq. Zool. (São Paulo) 29:133-183. [1978.06.27]

Gama, V. 1985. O sistema salivar de *Camponotus* (*Myrmothrix*) *rufipes* (Fabricius, 1775) (Hymenoptera: Formicidae). Rev. Bras. Biol. 45:317-359. [1985.11.30]

Gama, V., Bueno, O. C., Cruz Landim, C. da. 1976 ("1975"). O aparelho do ferrão de *Camponotus rufipes* (Fabricius) (Hymenoptera: Formicidae) e suas estruturas anexas. Rev. Bras. Biol. 35:589-593. [1976.07.14]

Gama, V., Cruz Landim, C. da. 1983 ("1982"). Estudo comparativo das glândulas do sistema salivar de formigas (Hymenoptera, Formicidae). Naturalia (São Paulo) 7:145-165. [1983]

Gama, V., Cruz Landim, C. da. 1984. Morfologia do tubo digestivo de *Camponotus* (*Myrmothrix*) *rufipes* (Fabricius, 1775) (Hymenoptera, Formicidae) durante a metamorfose. Naturalia (São Paulo) 9:43-55. [1984]

Gams, H. 1921. Zur Ameisengeographie von Mitteleuropa. Naturwiss. Wochenschr. 36[=(n.s.)20]:414-416. [1921.07.10]

Ganglbauer, L., Heyden, L. von. 1906. Über die Entomologia parisiensis von Geoffroy und Fourcroy. Wien. Entomol. Ztg. 25:301-302. [1906.08.15]

Gantes, H. 1949. Morphologie externe et croissance de quelques larves de Formicidés. Bull. Soc. Hist. Nat. Afr. Nord 40:71-97. [1949.09.15]

García-Pérez, J. A., MacKay, W. P., González-Villarreal, D., Camacho-Trujillo, R. 1994 ("1992"). Estudio preliminar de la mirmecofauna del Parque Nacional Chipinque, Nuevo León, México y su distribución altitudinal. Folia Entomol. Mex. 86:185-190. [1994.06.30]

Gardner, W. 1892. A preliminary list of the Hymenoptera - Aculeata of Lancashire and Cheshire, with notes on the habits of the genera. [part] Br. Nat. 2:21-23. [1892.01] [Ants pp. 22-23.]

Garling, L. 1979. Origin of ant-fungus mutualism: a new hypothesis. Biotropica 11:284-291. [1979.12]

Gaspar, C. 1964a. Sur *Ponera coarctata* Latreille et ses relations thermiques (Hymenoptera, Formicidae). Bull. Inst. Agron. Stn. Rech. Gembloux 32:33-35. [1964]

Gaspar, C. 1964b. Étude myrmécologique d'une région naturelle de Belgique: la Famenne. Survey des fourmis de la région (Hymenoptera, Formicidae). Bull. Inst. Agron. Stn. Rech. Gembloux 32:427-434. [1964]

Gaspar, C. 1965a. Introduction à l'étude des fourmis. Nat. Belg. 46:64-79. [1965.02]

Gaspar, C. 1965b. Notes sur l'écologie et l'éthologie des espèces du genre *Lasius* (Hymenoptera Formicidae). Insectes Soc. 12:219-230. [1965.12]

Gaspar, C. 1966a. Étude myrmécologique des tourbières dans les Hautes-Fagnes en Belgique (Hymenoptera, Formicidae). Rev. Écol. Biol. Sol 3:301-312. [1966.06]

Gaspar, C. 1966b. Étude myrmécologique d'une région naturelle de Belgique: la Famenne. C. - Observations faites en 1965. Bull. Rech. Agron. Gembloux 1:25-29. [1966]

Gaspar, C. 1968. Les fourmis de la Drôme et des Basses-Alpes en France (Hymenoptera, Formicidae). Nat. Can. (Qué.) 95:747-766. [1968.06]

Gaspar, C. 1971a. Les fourmis de la Famenne. I. Une étude zoogéographique. Bull. Inst. R. Sci. Nat. Belg. 47(20):1-116. [1971.11.30]

Gaspar, C. 1971b. Les fourmis de la Famenne. II. Une étude zoosociologique. Rev. Écol. Biol. Sol 8:553-607. [1971.12]

Gaspar, C. 1972. Les fourmis de la Famenne. III. Une étude écologique. Rev. Écol. Biol. Sol 9: 99-125. [1972.03]

Gateva, R. 1975. Morphological, biological and ecological characteristics of ants of the genus *Formica*, found in Bulgaria. [In Bulgarian.] Gorskostop. Nauka 12(6):48-59. [> 1975.12.23] [Date signed for printing.]

Gauld, I., Bolton, B. (eds.) 1988. The Hymenoptera. Oxford: Oxford University Press, xii + 322 pp. [1988]

Gavrilov, K. 1964 ("1963"). Dr. Nicolas Kusnezov. Acta Zool. Lilloana 19:4-23. [1964.06.10]

Geer, C. de See: De Geer, C.

Gehring, W. J., Wehner, R. 1995. Heat shock protein synthesis and thermotolerance in *Cataglyphis*, an ant from the Sahara desert. Proc. Natl. Acad. Sci. U. S. A. 92:2994-2998. [1995.03.28]

Gené, G. 1841. [Untitled. Introduced by: "Il Prof. Cav. Gené legge alcune sue osservazioni sulle abitudini di tre specie d'*Imenotteri*."] Pp. 398-400 in: Atti della Terza Riunione degli Scienziati Italiani tenuta in Firenze nel settembre del 1841. Firenze: Galileiana, 791 pp. [1841]

Geoffroy, E. L. 1785. [Untitled. Descriptions of new taxa, attributable to Geoffroy.] In: Fourcroy, A. F. de. Entomologia parisiensis, sive catalogus insectorum quae in agro parisiensi reperiuntur. Pars secunda. Paris: Via et Aedibus Serpentineis, [1] + pp. 233-544. [1785] [For discussion of authorship see Ganglbauer & Heyden (1906) and Cameron (1988). Ants pp. 451-453.]

Germar, E. F. 1837. Fauna insectorum Europae. Fasciculus 19. Insectorum protogaeae specimen sistens insecta carbonum fossilium. Halle: Kümmel, pl. 1-25. [1837]

Gerstäcker, A. 1859. [Untitled. Introduced by: "Hr. Peters berichtete über sein Reisewerk, von dem die Insecten bis zum 64., die Botanik bis zum 34. Bogen gedruckt sind und theilte den Schluss der Diagnosen der von Hrn. Dr. Gerstäcker bearbeiteten Hymenopteren mit."] Monatsber. K. Preuss. Akad. Wiss. Berl. 1858:261-264. [1859]

Gerstäcker, A. 1862. Hymenoptera Hautflügler. Pp. 439-526 in: Peters, W. C. H. Naturwissenschaftliche Reise nach Mossambique auf Befehl seiner Majestät des Königs Friedrich Wilhelm IV in den Jahren 1842 bis 1848 ausgeführt. Berlin: G. Reimer, 21 + 566 pp. [1862]

Gerstäcker, A. 1863. Ueber ein merkwürdiges neues Hymenopteron aus der Abtheilung der Aculeata. Stett. Entomol. Ztg. 24:76-93. [1863.03]

Gerstäcker, A. 1871. Beitrag zur Insektenfauna von Zanzibar. Arch. Naturgesch. 37(1):345-363. [1871]

Gerstäcker, A. 1872. Hymenopterologische Beiträge. Stett. Entomol. Ztg. 33:250-308. [1872.07]

Gerstäcker, A. 1873. Die Gliederthier-Fauna des Sansibar-Gebietes. Nach dem von Dr. O. Kersten während der v. d. Decken'schen Ost-Afrikanischen Expedition im Jahre 1862 und von C. Cooke auf der Insel Sansibar im Jahre 1864 gesammelten Material. Leipzig: C. F. Winter, xii + 542 pp. [1873] [Formicid taxa described as new in this work appear earlier in Gerstäcker (1871).]

Gerstaecker, A. See: Gerstäcker, A.

Gertsch, P., Pamilo, P., Varvio, S. L. 1995. Microsatellites reveal high genetic diversity within colonies of *Camponotus* ants. Mol. Ecol. 4:257-260. [1995.04]

Gestro, R. 1926. In memoria di Carlo Emery. Ann. Mus. Civ. Stor. Nat. "Giacomo Doria" 52:149-155. [1926.06.05]

Ghigi, A. 1925. Carlo Emery. Riv. Biol. (Rome) 7:711-739. [1925.12]

Ghilarov, M. S., Arnol'di, K. V. 1969. Steppe elements in the soil arthropod fauna of north-west Caucasus Mountains. Mem. Soc. Entomol. Ital. 48:103-112. [1969.12.30]

Gibson, G. A. P. 1985. Some pro- and mesothoracic structures important for phylogenetic analysis of Hymenoptera, with a review of terms used for the structures. Can. Entomol. 117:1395-1443. [1985.11]

Giebel, C. G. 1856. Fauna der Vorwelt mit steter Berücksichtigung der lebenden Thiere. Monographisch dargestellt. Zweiter Band. Gliederthiere. Erste Abtheilung. Insecten und Spinnen. Leipzig: Brockhaus, xviii + 511 pp. [1856.10.06] [For additional bibliographical details see Evenhuis (1994b:500). Ants pp. 160-180.]

Gillespie, D. S., Cole, A. C., Jr. 1954. Measurements in the worker caste of two species of *Eciton* (*Neivamyrmex* Borgmeier) (Hymenoptera: Formicidae). Am. Midl. Nat. 44:203-204. [1950.07.17]

Giordani Soika, A. 1931. Primo contributo alla conoscenza degli imenotteri del Lido di Venezia. Boll. Soc. Entomol. Ital. 63:99-103. [1931.07.20]

Girard, M. 1879. Traité élémentaire d'entomologie. Volume 2. Paris: Librairie J.-B. Baillière et Fils, 1028 pp. [1879]

Gistel, J. 1856. Die Mysterien der europäischen Insectenwelt. Kempten: Dannheimer, 12 + 532 pp. [1856]

Gladstone, D. E. 1981. Why there are no ant slave rebellions. Am. Nat. 117:779-781. [1981.05]

Glancey, B. M., Lofgren, C. S. 1986. A naturally occurring teratology in the red imported fire ant (Hymenoptera: Formicidae). Fla. Entomol. 69:764-767. [1986.12.22]

Glancey, B. M., Nickerson, J. C., Wojcik, D., Trager, J., Banks, W. A., Adams, C. 1987. The increasing incidence of the polygynous form of the red imported fire ant, *Solenopsis invicta* (Hymenoptera: Formicidae) in Florida. Fla. Entomol. 70:400-402. [1987.09.25]

Glancey, B. M., St. Romain, M. K., Crozier, R. H. 1976. Chromosome numbers of the red and black imported fire ants, *Solenopsis invicta* and *S. richteri*. Ann. Entomol. Soc. Am. 69:469-470. [1976.05.17]

Glancey, B. M., Wojcik, D. P., Craig, C. H., Mitchell, J. A. 1976. Ants of Mobile County, AL, as monitored by bait transects. J. Ga. Entomol. Soc. 11:191-197. [1976.07]
Gleim, K.-H. 1980a. Ameisen, die unsere Honigerzeuger hegen und schützen. Biene 116:296-297. [1980.07]
Gleim, K.-H. 1980b. Ameisen, die unsere Honigerzeuger hegen und schützen. II. Biene 116:346-349. [1980.08]
Gleim, K.-H. 1981a. Die Erkennungsmerkmale der für den Waldschutz und die Honigtauproduktion wichtigen Waldameisen. Biene 117:101-104. [1981.03]
Gleim, K.-H. 1981b. Unsere Waldameisen, ihre Erkennungsmerkmale und ihre Bedeutung für die Lachnidenhege. Biene 117:150-153. [1981.04]
Glöckner, W. E. 1958. Histologische Untersuchungen an der Diebsameise *Solenopsis fugax* Latr. waehrend der Metamorphose. Stud. Entomol. (n.s.)1:529-544. [1958.11.29]
Glover, P. E. 1967. Notes on some ants in northern Somalia. East Afr. Wildl. J. 5:65-73. [1967.08]
Gmelin, J. F. 1790. Caroli a Linné, systema naturae per regna tria naturae, secundum classes, ordines, genera, species, cum characteribus, differentiis, synonymis, locis. Editio decima tertia, aucta, reformata. Tomus I. Pars V. Lipsiae [= Leipzig]: G. E. Beer, pp. 2225-3020. [1790.12.06] [Publication date from Hopkinson (1908).]
Goaga, A., Paraschivescu, D. 1993 ("1991"). Cercetari asupra formicidelor (Ins., Formicidae) din colectia muzeului de stiintele naturii din Bacau. Nymphaea 21:113-120. [1993]
Godden, C., Cosens, D. 1987. Ant species (Hym., Formicidae) native to North Knapdale, Argyllshire. Entomol. Mon. Mag. 123:209-216. [1987.11.30]
Godfrey, R. 1907. Notes on the animal life of the hothouses of the Royal Botanic Garden, Edinburgh. Notes R. Bot. Gard. Edinb. 4:99-103. [1907.04]
Goeldi, E. 1911a. Der Ameisenstaat. [part] Himmel Erde 23:289-307. [1911.04]
Goeldi, E. 1911b. Der Ameisenstaat. [part] Himmel Erde 23:349-365. [1911.05]
Goeldi, E. 1911c. Der Ameisenstaat. [concl.] Himmel Erde 23:395-406. [1911.06]
Goetsch, W. 1932. Beiträge zur Biologie südamerikanischer Ameisen. 1. Teil: Wüstenameisen. Z. Morphol. Ökol. Tiere 25:1-30. [1932.04.21]
Goetsch, W. 1933a. Chilenische Wüsten-, Steppen- und Wald-Ameisen. Sitzungsber. Ges. Morph. Physiol. Münch. 42:27-35. [1933] [Includes new taxa (*Dorymyrmex goetschi* and *Forelius eidmanni*), attributed to Menozzi, but Goetsch is responsible for the descriptions and should be credited with authorship (Article 50, ICZN).]
Goetsch, W. 1933b. Formididae [sic] Chilensis (Hormigas chilenas). Bol. Soc. Biol. Concepción 7:11-28. [1933]
Goetsch, W. 1934. Untersuchungen über die Zusammenarbeit im Ameisenstaat. Z. Morphol. Ökol. Tiere 28:319-401. [1934.05.19]
Goetsch, W. 1935. Biologie und Verbreitung chilenischer Wüsten-, Steppen- und Waldameisen. Zool. Jahrb. Abt. Syst. Ökol. Geogr. Tiere 67:235-318. [1935.11.06]
Goetsch, W. 1936. Formicidae Mediterraneae. Beiträge zur Biologie und Verbreitung der Ameisen am Golfe von Neapel. Pubbl. Stn. Zool. Napoli 15:392-422. [1936.06.15]
Goetsch, W. 1937a. Formicidae Mediterraneae. Beiträge zur Kenntnis der Ameisen am Golfe von Neapel. II. Teil. Formicinen der Insel Capri und Ischia. Pubbl. Stn. Zool. Napoli 16:273-315. [1937.08.31]
Goetsch, W. 1937b. Die Entstehung der "Soldaten" im Ameisenstaat. Naturwissenschaften 25:803-808. [1937.12.10]
Goetsch, W. 1937c. Die Staaten der Ameisen. Berlin: J. Springer, vii + 159 pp. [1937]
Goetsch, W. 1939. Die Staaten argentinischer Blattschneider-Ameisen. Zoologica (Stuttg.) 35(3/4)[=Heft 96]:1-105. [1939]
Goetsch, W. 1942. Beiträge zur Biologie spanischer Ameisen. EOS. Rev. Esp. Entomol. 18:175-241. [1942.09.29]
Goetsch, W. 1949. Beiträge zur Biologie und Verbreitung der Ameisen in Kärnten und in den Nachbargebieten. Österr. Zool. Z. 2:39-69. [1949.04.11]

Goetsch, W. 1951. Ameisen- und Termiten-Studien in Ischia, Capri und Neapel. Zool. Jahrb. Abt. Syst. Ökol. Geogr. Tiere 80:64-98. [1951.06.29]

Goetsch, W. 1953. Die Staaten der Ameisen. Zweite, ergänzte Auflage. Berlin: Springer-Verlag, viii + 152 pp. [1953]

Goetsch, W. 1957. The ants. [English translation of "Die Staaten der Ameisen", 2nd edition.] Ann Arbor: University of Michigan Press, 173 pp. [1957]

Goetsch, W., Eisner, H. E. 1930. Beiträge zur Biologie körnersammelnder Ameisen. II. Teil. Z. Morphol. Ökol. Tiere 16:371-452. [1930.01.13]

Goetsch, W., Menozzi, C. 1935. Die Ameisen Chiles. Konowia 14:94-102. [1935.03.15]

Golbach, R. 1964 ("1963"). Un caso teratologico en *Cephalotes atratus* Latr. (Hym. Formicidae). Acta Zool. Lilloana 19:503-505. [1964.06.10]

Goldstein, E. L. 1976 ("1975"). Island biogeography of ants. Evolution 29:750-762. [1976.03.12]

Goll, W. 1967. Strukturuntersuchungen am Gehirn von *Formica*. Z. Morphol. Ökol. Tiere 59:143-210. [1967.06.12]

Gonçalves, C. R. 1942. Contribuição para o conhecimento do gênero *Atta* Fabr., das formigas saúvas. Bol. Soc. Bras. Agron. 5:333-358. [1942.09]

Gonçalves, C. R. 1944. Descrição de uma nova saúva brasileira (Hym., Form.). Rev. Bras. Biol. 4:233-238. [1944.06.22] [Date from Rev. Bras. Biol. 6:145.]

Gonçalves, C. R. 1947a ("1945"). Saúvas do sul e centro do Brasil. Bol. Fitossanit. 2:183-218. [1947]

Gonçalves, C. R. 1947b ("1946"). [Untitled. Introduced by: "De volta de uma viagem aos Estados do Ceará, Piauí, Pará, Amazonas, Mato Grosso, e aos Territórios do Amapá e Guaporé, o Agr. Fitossanitarista Cincinato R. Gonçalves realizou uma palestra à 24 de janeiro de 1946, sôbre as principais observações que fêz sôbre as formigas cortadeiras e as pragas da carnaubeira ..."] Bol. Fitossanit. 3:51-54. [1947]

Gonçalves, C. R. 1954 ("1951"). Saúvas do nordeste do Brasil (Atta spp., Formicidae). Bol. Fitossanit. 5:1-42. [1954.07]

Gonçalves, C. R. 1955. Nota suplementar sôbre as saúvas do nordeste do Brasil. Bol. Fitossanit. 6:21-26. [1955]

Gonçalves, C. R. 1960. Distribuição, biologia e ecologia das saúvas. Divulg. Agron. 1:2-10. [Not seen. Cited in Loureiro & Soares (1990).]

Gonçalves, C. R. 1961. O genero *Acromyrmex* no Brasil (Hym. Formicidae). Stud. Entomol. 4:113-180. [1961.10.11]

Gonçalves, C. R. 1963. Nota sobre a sistemática de *Atta sexdens* (L., 1758) e de suas sub-espécies. (Hym., Formicidae). Bol. Fitossanit. 9:1-3. [1963]

Gonçalves, C. R. 1967a. *Acromyrmex muticinodus* (Forel, 1901), sinônimo de *Acromyrmex niger* (F. Smith, 1858) (Hym., Formicidae). Rev. Bras. Entomol. 12:17-20. [1967.02]

Gonçalves, C. R. 1967b. As formigas cortadeiras da Amazônia, dos gêneros *Atta* Fabr. e *Acromyrmex* Mayr (Hym., Formicidae). Pp. 181-202 in: Lent, H. (ed.) Atas do simpósio sôbre a biota amazônica. Vol. 5: Zoologia. Rio de Janeiro: Conselho Nacional de Pesquisas, 603 pp. [1967]

Gonçalves, C. R. 1971. As saúvas de Mato Grosso, Brasil (Hymenoptera, Formicidae). Arq. Mus. Nac. (Rio J.) 54:249-253. [1971]

Gonçalves, C. R. 1982. *Acromyrmex subterraneus* Forel 1893 y *A. coronatus* (Fabricius 1804), especies de hormigas nuevas para Venezuela. Attini 12:5. [1982.04]

Gonçalves, C. R. 1983a ("1982"). Descrição de *Acromyrmex diasi*, uma nova espécie de formiga cortadeira de folhas (Hym., Formicidae). Rev. Bras. Biol. 42:485-487. [1983.01.30]

Gonçalves, C. R. 1983b ("1982"). *Atta silvai*, nova espécie de formiga saúva (Hymenoptera, Formicidae). Arq. Univ. Fed. Rural Rio J. 5:173-178. [1983.12.16]

Gonçalves, C. R. 1986. Filogenia do gênero *Atta* Fabricius (Hymenoptera, Formicidae). Publ. Avulsas Mus. Nac. Rio J. 65:13-17. [1986.05]

Goñi, B. 1980. Contribución al conocimiento cariotípico en hormigas neotropicales, II. *Camponotus* (Hymenoptera: Formicidae). Jorn. Cienc. Nat. Resúm. 1:57-58. [1980.09.29]

Goñi, B. 1981. Cariotipo de tres especies del género *Acromyrmex* del Uruguay (Hymenoptera: Formicidae). Jorn. Cienc. Nat. Resúm. Comun. 2:15. [1981.09.20]

Goñi, B., Imai, H. T., Kubota, M., Kondo, M., Yong, H.-S., Tho, Y. P. 1982. Chromosome observations of tropical ants in western Malaysia and Singapore. Annu. Rep. Natl. Inst. Genet. Jpn. 32:71-73. [1982]

Goñi, B., Zolessi, L. C. de, Imai, H. T. 1982. Chromosome observations of some Uruguayan ants (Hymenoptera: Formicidae). Annu. Rep. Natl. Inst. Genet. Jpn. 32:69-70. [1982]

Goñi, B., Zolessi, L. C. de, Imai, H. T. 1984 ("1983"). Karyotypes of thirteen ant species from Uruguay (Hymenoptera, Formicidae). Caryologia 36:363-371. [1984.02.29]

González Pérez, J. L. 1987. Estudio eto-ecológico de algunos formicoideos cubanos. I. Cienc. Biol. Acad. Cienc. Cuba 18:53-63. [1987]

Gordon, D. M. 1984. Species-specific patterns in the social activities of harvester ant colonies (*Pogonomyrmex*). Insectes Soc. 31:74-86. [1984.07]

Gösswald, K. 1930. Die Biologie einer neuen *Epimyrma*art aus dem mittleren Maingebiet. Z. Wiss. Zool. 136:464-484. [1930.07]

Gösswald, K. 1932. Ökologische Studien über die Ameisenfauna des mittleren Maingebietes. Z. Wiss. Zool. 142:1-156. [1932.08]

Gösswald, K. 1933. Weitere Untersuchungen über die Biologie von *Epimyrma gösswaldi* Men. und Bemerkungen über andere parasitische Ameisen. Z. Wiss. Zool. 144:262-288. [1933.10]

Gösswald, K. 1934. Die Grundzüge der stammesgeschichtlichen Entwicklung des Ameisenparasitismus, neu beleuchtet durch die Entdeckung einer weiteren parasitischen Ameise. Entomol. Beih. Berl.-Dahl. 1:57-62. [1934.08.07]

Gösswald, K. 1939. Über den Sozialparasitismus der Ameisen. Pp. 1149-1155 in: Jordan, K., Hering, E. M. (eds.) VII. Internationaler Kongress für Entomologie. Verhandlungen. Band II. Weimar: Internationaler Kongress für Entomologie, pp. 618-1424. [1939.07]

Gösswald, K. 1941. Rassenstudien an der roten Waldameise *Formica rufa* L. auf systematischer, ökologischer, physiologischer und biologischer Grundlage. Z. Angew. Entomol. 28:62-124. [1941.07]

Gösswald, K. 1942. Art- und Rassenunterschiede bei der Roten Waldameise. Naturschutz 23:109-115. [1942.11]

Gösswald, K. 1944. Rassenstudien an der Roten Waldameise im Lichte der Ganzheitsforschung. Anz. Schädlingskd. 20:1-8. [1944]

Gösswald, K. 1951a. Zur Biologie, Ökologie und Morphologie einer neuen Varietät der Kleinen Roten Waldameise: *Formica minor pratensoides*. Z. Angew. Entomol. 32:433-457. [1951.05]

Gösswald, K. 1951b. Zur Ameisenfauna des Mittleren Maingebietes mit Bermerkungen über Veränderungen seit 25 Jahren. Zool. Jahrb. Abt. Syst. Ökol. Geogr. Tiere 80:507-532. [1951.12.20]

Gösswald, K. 1951c. Versuche zum Sozialparasitismus der Ameisen bei der Gattung *Formica* L. Zool. Jahrb. Abt. Syst. Ökol. Geogr. Tiere 80:533-582. [1951.12.20]

Gösswald, K. 1951d. Die rote Waldameise im Dienste der Waldhygiene. Forstwirtschaftliche Bedeutung, Nutzung, Lebensweise, Zucht, Vermehrung und Schutz. Lüneburg: Metta Kinau Verlag, 160 pp. [1951]

Gösswald, K. 1953. Histologische Untersuchungen an der arbeiterlosen Ameise *Teleutomyrmex schneideri* Kutter (Hym. Formicidae). Mitt. Schweiz. Entomol. Ges. 26:81-128. [1953.06.20]

Gösswald, K. 1954. Unsere Ameisen. I. Teil. Stuttgart: Franckh'sche Verlagshandlung, 88 pp. [1954]

Gösswald, K. 1955. Unsere Ameisen. II. Teil. Stuttgart: Franckh'sche Verlagshandlung, 80 pp. [1955]

Gösswald, K. 1959. Einig zur Zusammenarbeit für die Gesundung des Waldes! Waldhygiene 3: 2-36. [1959]

Gösswald, K. 1961a. Un viaggio in Italia fra le formiche del gruppo *Formica rufa*. Collana Verde 7:28-60. [1961.04.15]

Gösswald, K. 1961b ("1960"). Untersuchungen zum Paarungs- und Adoptionsverhalten verschiedener *Formica*-Arten. Pp. 612-617 in: Strouhal, H., Beier, M. (eds.) XI. Internationaler Kongress für Entomologie. Wien, 17. bis 25. August 1960. Verhandlungen. Band I. Wien: Organisationskomitee des XI. Internationalen Kongresses für Entomologie, xliv + 803 pp. [1961] [Publication date ("Ende 1961") from statement in Band II.]

Gösswald, K. 1981. Artunterschiede der Waldameisen in Aussehen, Lebensweise, Organisation, Verhalten, Nest- und Strassenbau, Ökologie und Verbreitung. Merkbl. Waldhyg. 1:1-32. [1981]

Gösswald, K. 1982a. Artunterschiede der Waldameisen in Aussehen, Lebensweise, Organisation, Verhalten, Nest- und Strassenbau, Ökologie und Vorbereitung. Waldhygiene 14:161-192. [1982]

Gösswald, K. 1982b. Ökologie und geographische Verbreitung der Waldameisen-Arten (Hym., Formicidae). Z. Angew. Zool. 69:29-77. [1982]

Gösswald, K. 1985. Organisation und Leben der Ameisen. Stuttgart: Wissenschaftliche Verlagsgesellschaft mbH, 355 pp. [1985]

Gösswald, K. 1987. Über System und Nomenklatur der *Formica*-Gruppe (Hym. Formicidae). Waldhygiene 17:97-112. [1987]

Gösswald, K. 1989. Die Waldameise. Band 1. Biologische Grundlagen, Ökologie und Verhalten. Wiesbaden: AULA-Verlag, xi + 660 pp. [1989]

Gösswald, K. 1990. Die Waldameise. Band 2. Die Waldameise im Ökosystem Wald, ihr Nutzen und ihre Hege. Wiesbaden: AULA-Verlag, x + 510 pp. [1990]

Gösswald, K., Halberstadt, K. 1961. Zur Ameisenfauna der Rhön. Abh. Naturwiss. Ver. Würzbg. 2:27-34. [1961.08]

Gösswald, K., Kneitz, G. 1965. Zur Verbreitung der Waldameisen im Bayerischen Wald (Gen. *Formica*, Hym., Formicidae). Collana Verde 16:145-174. [1965.09.10]

Gösswald, K., Kneitz, G., Pirnke, F.-R. 1968. Zur Verbreitung der Waldameisen (Formicidae, *Formica*) in einem Gebirgsmassiv der Steiermark. Waldhygiene 7:166-189. [1968]

Gösswald, K., Kneitz, G., Schirmer, G. 1965. Die geographische Verbreitung der hügelbauenden *Formica*-Arten (Hym., Formicidae) in Europa. Zool. Jahrb. Abt. Syst. Ökol. Geogr. Tiere 92:369-404. [1965.08.31]

Gösswald, K., Schirmer, G. 1965. Zur geographischen Verbreitung der hügelbauenden *Formica*-Arten. Collana Verde 16:133-144. [1965.09.10]

Gösswald, K., Schmidt, G. H. 1959a. Papierchromatographische Untersuchungen zur Art- und Rassendifferenzierung. Eine Studie zum Problem der Waldameisensystematik. Umschau Wiss. Tech. 59:265-269. [1959.05.01]

Gösswald, K., Schmidt, G. H. 1959b. Zur morphologischen und biochemischen Differenzierung der Waldameisen (Hym. Form., Gen. *Formica*) und ihrer waldhygienischen Bedeutung. Waldhygiene 3:37-46. [1959]

Gösswald, K., Schmidt, G. H. 1960. Neue Wege zur Unterscheidung der Waldameisenformen (Hymenoptera, Formicidae). Entomophaga 5:13-31. [1960.02.24]

Gösswald, K., Schmidt, G. H. 1961 ("1960"). Untersuchungen zum Flügelabwurf und Begattungsverhalten einiger *Formica*-Arten (Ins. Hym.) im Hinblick auf ihre systematische Differenzierung. Insectes Soc. 7:298-321. [1961]

Gösswald, K., Schmidt, G. H., Kloft, W., Baggini, A., Pavan, M., Ronchetti, G. 1961. Ricerche morfologico-biometriche sulla differenziazione del "gruppo *Formica nigricans*" e sulla sua diffusione in Italia (Hym. Formicidae). Collana Verde 7:12-27. [1961.04.15]

Gotwald, W. H., Jr. 1969a. Comparative morphological studies on the ants with particular reference to the mouthparts. [Abstract.] Diss. Abstr. B. Sci. Eng. 29:4215-4216. [1969.05]

Gotwald, W. H., Jr. 1969b. Comparative morphological studies of the ants, with particular reference to the mouthparts (Hymenoptera: Formicidae). Mem. Cornell Univ. Agric. Exp. Stn. 408:1-150. [1969.07]

Gotwald, W. H., Jr. 1970. Mouthpart morphology of the ant *Aneuretus simoni*. Ann. Entomol. Soc. Am. 63:950-952. [1970.07.15]

Gotwald, W. H., Jr. 1971. Phylogenetic affinities of the ant genus *Cheliomyrmex* (Hymenoptera: Formicidae). J. N. Y. Entomol. Soc. 79:161-173. [1971.10.14]

Gotwald, W. H., Jr. 1973. Mouthpart morphology of the African ant *Oecophylla longinoda* Latreille (Hymenoptera: Formicidae). J. N. Y. Entomol. Soc. 81:72-78. [1973.08.20]

Gotwald, W. H., Jr. 1974. Predatory behavior and food preferences of driver ants in selected African habitats. Ann. Entomol. Soc. Am. 67:877-886. [1974.11.15]

Gotwald, W. H., Jr. 1976. Behavioral observations on African army ants of the genus *Aenictus* (Hymenoptera: Formicidae). Biotropica 8:59-65. [1976.03]

Gotwald, W. H., Jr. 1977. The origins and disperal of army ants of the subfamily Dorylinae. Pp. 126-127 in: Velthuis, H. H. W., Wiebes, J. T. (eds.) Proceedings of the Eighth International Congress of the IUSSI, Wageningen, The Netherlands, September 5-10, 1977. Wageningen: Centre for Agricultural Publishing and Documentation, 325 pp. [1977]

Gotwald, W. H., Jr. 1978. Trophic ecology and adaptation in tropical Old World ants of the subfamily Dorylinae (Hymenoptera: Formicidae). Biotropica 10:161-169. [1978.09]

Gotwald, W. H., Jr. 1979. Phylogenetic implications of army ant zoogeography (Hymenoptera: Formicidae). Ann. Entomol. Soc. Am. 72:462-467. [1979.07.15]

Gotwald, W. H., Jr. 1980. Ant. Pp. 35-38 in: Academic American Encyclopedia. Volume 2. Princeton: Aretê Publishing Co., 384 pp. [1980]

Gotwald, W. H., Jr. 1982. Army ants. Pp. 157-254 in: Hermann, H. R. (ed.) Social insects. Volume 4. New York: Academic Press, 385 pp. [1982]

Gotwald, W. H., Jr. 1985. Reflections on the evolution of army ants (Hymenoptera, Formicidae). Actes Colloq. Insectes Soc. 2:7-16. [1985.06]

Gotwald, W. H., Jr. 1987. The relationship of form and function in army ant queens. Pp. 255-256 in: Eder, J., Rembold, H. (eds.) Chemistry and biology of social insects. München: Verlag J. Peperny, xxxv + 757 pp. [1987]

Gotwald, W. H., Jr. 1988. On becoming an army ant. Pp. 227-235 in: Trager, J. C. (ed.) Advances in myrmecology. Leiden: E. J. Brill, xxvii + 551 pp. [1988]

Gotwald, W. H., Jr. 1995. Army ants: the biology of social predation. Ithaca, New York: Cornell University Press, xviii + 302 pp. [1995] [Received July, 1995.]

Gotwald, W. H., Jr., Barr, D. 1980. Quantitative studies on major workers of the ant genus *Dorylus* (Hymenoptera: Formicidae: Dorylinae). Ann. Entomol. Soc. Am. 73:231-238. [1980.03.15]

Gotwald, W. H., Jr., Barr, D. 1988 ("1987"). Quantitative studies on workers of the Old World army ant genus *Aenictus* (Hymenoptera: Formicidae). Insectes Soc. 34:261-273. [1988.06]

Gotwald, W. H., Jr., Brown, W. L., Jr. 1967 ("1966"). The ant genus *Simopelta* (Hymenoptera: Formicidae). Psyche (Camb.) 73:261-277. [1967.05.24]

Gotwald, W. H., Jr., Burdette, A. W. 1981. Morphology of the male internal reproductive system in army ants: phylogenetic implications (Hymenoptera: Formicidae). Proc. Entomol. Soc. Wash. 83:72-92. [1981.02.17]

Gotwald, W. H., Jr., Cunningham-van Someren, G. R. 1976. Taxonomic and behavioral notes on the African ant, *Aenictus eugenii* Emery, with a description of the queen (Hymenoptera: Formicidae). J. N. Y. Entomol. Soc. 84:182-188. [1976.10.28]

Gotwald, W. H., Jr., Kupiec, B. M. 1975. Taxonomic implications of doryline worker ant morphology: *Cheliomyrmex morosus* (Hymenoptera: Formicidae). Ann. Entomol. Soc. Am. 68:961-971. [1975.11.17]

Gotwald, W. H., Jr., Leroux, J. M. 1980. Taxonomy of the African army ant, *Aenictus decolor* (Mayr), with a description of the queen (Hymenoptera: Formicidae). Proc. Entomol. Soc. Wash. 82:599-608. [1980.10.24]

Gotwald, W. H., Jr., Lévieux, J. 1972. Taxonomy and biology of a new West African ant belonging to the genus *Amblyopone* (Hymenoptera: Formicidae). Ann. Entomol. Soc. Am. 65:383-396. [1972.03.15]

Gotwald, W. H., Jr., Schaefer, R. F., Jr. 1982. Taxonomic implications of doryline worker ant morphology: *Dorylus* subgenus *Anomma* (Hymenoptera: Formicidae). Sociobiology 7:187-204. [1982]

Gowdey, C. C. 1917. A list of Ugandan Coccidae, their food-plants and natural enemies. Bull. Entomol. Res. 8:187-189. [1917.12]

Grandi, G. 1935. Contributi alla conoscenza degli Imenotteri Aculeati. XV. Boll. R. Ist. Entomol. Univ. Studi Bologna 8:27-121. [1935.11.10] [Ants pp. 98-104.]

Grandi, G. 1943 ("1942"). In memoria di Carlo Menozzi. Boll. R. Ist. Entomol. Univ. Studi Bologna 14:193-201. [1943.08.30]

Gravenhorst, J. L. C. 1807. Vergleichende Uebersicht des Linnéischen und einiger neueren zoologischen Systeme. Göttingen: Dieterich, 20 + 476 pp. [1807]

Gray, B. 1971. A morphometric study of the ant species, *Myrmecia dispar* (Clark) (Hymenoptera: Formicidae). Insectes Soc. 18:95-109. [1971.09]

Gray, B. 1974 ("1973"). A morphometric study of worker variation in three *Myrmecia* species (Hymenoptera: Formicidae). Insectes Soc. 20:323-331. [1974.03]

Gray, B., Lamb, K. P. 1968. Some observations on the Malpighian tubules and ovarioles in *Myrmecia dispar* (Clark) (Hymenoptera: Formicidae). J. Aust. Entomol. Soc. 7:80-81. [1968.06.30]

Gray, G. R. 1832. [Untitled. *Camptognatha testacea*, attributed to Gray.] P. 516 in: Griffith, E. and others. The animal kingdom arranged in conformity with its organization, by the Baron Cuvier, with supplementary additions to each order by Edward Griffith, F.L.S., A.S., and others. Volume the Fifteenth [= Class Insecta, Volume 2]. London: Whittaker, Treacher & Co., 796 pp. [1832] [Individual parts of Volume 15 may be more precisely dateable; see discussion in Cowan (1969).]

Gray, J. E. 1841. Contributions towards the geographical distribution of the Mammalia of Australia, with notes on some recently discovered species. Appendix C. Pp. 397-414 in: Grey, G. Journals of two expeditions of discovery in north west and western Australia during the years 1837, 38 and 39. Volume 2. London: T. & W. Boone, 482 pp. [1841.11]

Greaves, T. 1971. The distribution of the three forms of the meat ant *Iridomyrmex purpureus* (Hymenoptera: Formicidae) in Australia. J. Aust. Entomol. Soc. 10:15-21. [1971.03.31]

Gredler, M. V. 1858. Die Ameisen von Tirol. Achtes Programm des Gymnasiums in Bozen. Bozen: Ebersche Buchdruckerei, 34 pp. [1858]

Gredler, M. V. 1859. [Untitled. Introduced by: "Herr Sekretär G. Frauenfeld legt eine von Herrn Professor V. Gredler in Bozen eingegangene Notiz zur geografischen Verbreitung der Ameisen in Oesterreich vor."] Verh. K.-K. Zool.-Bot. Ges. Wien 9(Sitzungsber.):127-128. [1859]

Greenberg, L., Fletcher, D. J. C., Vinson, S. B. 1985. Differences in worker size and mound distribution in monogynous and polygynous colonies of the fire ant *Solenopsis invicta* Buren. J. Kansas Entomol. Soc. 58:9-18. [1985.04.08]

Greenberg, L., Williams, H. J., Vinson, S. B. 1990. A comparison of venom and hydrocarbon profiles from alates in Texas monogyne and polygyne fire ants. Pp. 95-101 in: Vander Meer, R. K., Jaffe, K., Cedeno, A. (eds.) Applied myrmecology: a world perspective. Boulder: Westview Press, xv + 741 pp. [1990]

Greenslade, P. J. M. 1970. Observations on the inland variety (v. *viridiaeneus* Viehmeyer) of the meat ant *Iridomyrmex purpureus* (Frederick Smith) (Hymenoptera: Formicidae). J. Aust. Entomol. Soc. 9:227-231. [1970.12.31]

Greenslade, P. J. M. 1972. Comparative ecology of four tropical ant species. Insectes Soc. 19:195-212. [1972.12]

Greenslade, P. J. M. 1974a. The identity of *Iridomyrmex purpureus* form *viridiaeneus* Viehmeyer (Hymenoptera: Formicidae). J. Aust. Entomol. Soc. 13:247-248. [1974.09.13]

Greenslade, P. J. M. 1974b. Distribution of two forms of the meat ant, *Iridomyrmex purpureus* (Hymenoptera: Formicidae), in parts of South Australia. Aust. J. Zool. 22:489-504. [1974.11]

Greenslade, P. J. M. 1976. Distribution of two forms of the meat ant *Iridomyrmex purpureus* (Hymenoptera: Formicidae) on Kangaroo Island and in the Yorke Peninsula. Aust. J. Zool. 24:557-564. [1976.12.20]

Greenslade, P. J. M. 1979. A guide to ants of South Australia. Adelaide: South Australian Museum (Special Educational Bulletin Series), 44 pp. [1979]

Greenslade, P. J. M. 1982. Competition and community organisation in an Australian ant fauna. [Abstract.] P. 65 in: Breed, M. D., Michener, C. D., Evans, H. E. (eds.) The biology of social insects. Proceedings of the Ninth Congress of the IUSSI, Boulder, Colorado, August 1982. Boulder: Westview Press, xii + 420 pp. [1982.08]

Greenslade, P. J. M. 1985a. Some effects of season and geographical aspect on ants (Hymenoptera: Formicidae) in the Mt Lofty Ranges, South Australia. Trans. R. Soc. S. Aust. 109:17-23. [1985.06.28]

Greenslade, P. J. M. 1985b. Preliminary observations on ants (Hymenoptera: Formicidae) of forests and woodlands in the Alligator Rivers region, N.T. Proc. Ecol. Soc. Aust. 13:153-160. [1985]

Greenslade, P. J. M. 1987. Environment and competition as determinants of local geographical distribution of five meat ants, *Iridomyrmex purpureus* and allied species (Hymenoptera: Formicidae). Aust. J. Zool. 35:259-273. [1987.06.12]

Greenslade, P. J. M., Greenslade, P. 1973. Ants of a site in arid southern Australia. Pp. 145-149 in: Proceedings IUSSI VIIth International Congress, London, 10-15 September, 1973. Southampton: University of Southampton, vi + 418 pp. [1973]

Greenslade, P. J. M., Greenslade, P. 1977. Some effects of vegetation cover and disturbance on a tropical ant fauna. Insectes Soc. 24:163-182. [1977.06]

Greenslade, P. J. M., Halliday, R. B. 1982. Distribution and speciation in meat ants, *Iridomyrmex purpureus* and related species (Hymenoptera: Formicidae). Pp. 249-255 in: Barker, W. R., Greenslade, P. J. M. (eds.) Evolution of the flora and fauna of arid Australia. Frewville, South Australia: Peacock Publications, 392 pp. [1982]

Greenslade, P. J. M., Halliday, R. B. 1983. Colony dispersion and relationships of meat ants *Iridomyrmex purpureus* and allies in an arid locality in South Australia. Insectes Soc. 30:82-99. [1983.07]

Greenslade, P. J. M., Thompson, C. H. 1981. Ant distribution, vegetation, and soil relationships in the Cooloola-Noosa River area, Queensland. Pp. 192-207 in: Gillison, A. N., Anderson, D. J. (eds.) Vegetation classification in Australia. Canberra: C.S.I.R.O. & Australian National University Press, xxi + 229 pp. [1981]

Gregg, E. V. 1945. A statistical study of taxonomic categories in ants (Formicidae: *Lasius neoniger* and *Lasius americanus*). Ann. Entomol. Soc. Am. 38:529-548. [1945.12.31]

Gregg, R. E. 1938. [Review of: Wheeler, W. M. 1937. Mosaics and other anomalies among ants. Cambridge, Mass.: Harvard University Press, 95 pp.] Ecology 19:312-314. [1938.04]

Gregg, R. E. 1942. The origin of castes in ants with special reference to *Pheidole morrisi* Forel. Ecology 23:295-308. [1942.07]

Gregg, R. E. 1945a ("1944"). The ants of the Chicago region. Ann. Entomol. Soc. Am. 37:447-480. [1945.01.19]

Gregg, R. E. 1945b. The worker caste of *Harpagoxenus canadensis* Smith (Formicidae). Can. Entomol. 77:74-76. [1945.08.09]

Gregg, R. E. 1945c. Two new forms of *Monomorium* (Formicidae). Psyche (Camb.) 52:62-69. [1945.10.26]

Gregg, R. E. 1946. The ants of northeastern Minnesota. Am. Midl. Nat. 35:747-755. [1946.05.30]

Gregg, R. E. 1947. Altitudinal indicators among the Formicidae. Univ. Colo. Stud. Phys. Biol. Sci. 2:385-403. [1947.04]

Gregg, R. E. 1948. An unusual nest of the prairie mound-building ant (Hymenoptera, Formicidae). Proc. Entomol. Soc. Wash. 50:183-186. [1948.10.18]

Gregg, R. E. 1949a. A note on *Pheidole* (*Macropheidole*) *rhea* Wheeler (Hymenoptera: Formicidae). Psyche (Camb.) 56:70-73. [1949.08.09]

Gregg, R. E. 1949b. A new ant from southwestern United States (Hymenoptera, Formicidae). Proc. Entomol. Soc. Wash. 51:171-174. [1949.08.17]

Gregg, R. E. 1950. A new species of *Pheidole* from the Southwest. J. N. Y. Entomol. Soc. 58: 89-93. [1950.06.23]

Gregg, R. E. 1951. Two new species of exotic ants. Psyche (Camb.) 58:77-84. [1951.11.19]

Gregg, R. E. 1952a. A new ant of the genus *Pheidole* from Colorado. Am. Mus. Novit. 1557:1-4. [1952.05.05]

Gregg, R. E. 1952b. The female of *Formica opaciventris* Emery (Formicidae). Psyche (Camb.) 59:13-19. [1952.08.19]

Gregg, R. E. 1953a. Notes on the ant, *Leptothorax obliquicanthus* Cole (Hymenoptera: Formicidae). Breviora 22:1-4. [1953.10.13]

Gregg, R. E. 1953b. The soldier caste of *Pheidole* (*Ceratopheidole*) *clydei* Gregg (Hymenoptera, Formicidae). Am. Mus. Novit. 1637:1-7. [1953.12.16]

Gregg, R. E. 1953c. Morphological considerations affecting the taxonomy of certain genera of ants (Hymenoptera, Formicidae). Proc. Entomol. Soc. Wash. 55:324-330. [1953.12.29]

Gregg, R. E. 1954a ("1953"). Taxonomic notes on the ant, *Camponotus cooperi* Gregg. Psyche (Camb.) 60:102-104. [1954.01.08]

Gregg, R. E. 1954b. Geographical distribution of the genus *Myrmoteras*, including the description of a new species (Hymenoptera: Formicidae). Psyche (Camb.) 61:20-30. [1954.06.30]

Gregg, R. E. 1955a. A new species of ant belonging to the *Pheidole pilifera* complex (Hymenoptera: Formicidae). Psyche (Camb.) 62:19-28. [1955.05.23]

Gregg, R. E. 1955b. The rediscovery of *Veromessor lobognathus* (Andrews) (Hymenoptera: Formicidae). Psyche (Camb.) 62:45-52. [1955.07.19]

Gregg, R. E. 1956. An extension of range for the ant, *Pheidole lamia* Wheeler (Hymenoptera: Formicidae). Entomol. News 67:37-39. [1956.02.03]

Gregg, R. E. 1957 ("1956"). A new species of *Myrmoteras* from Ceylon (Hymenoptera: Formicidae). Psyche (Camb.) 63:41-45. [1957.01.23]

Gregg, R. E. 1958. Two new species of *Metapone* from Madagascar (Hymenoptera: Formicidae). Proc. Entomol. Soc. Wash. 60:111-121. [1958.04.18]

Gregg, R. E. 1959 ("1958"). Key to the species of *Pheidole* (Hymenoptera: Formicidae) in the United States. J. N. Y. Entomol. Soc. 66:7-48. [1959.01.20]

Gregg, R. E. 1961. The status of certain myrmicine ants in western North America with a consideration of the genus *Paramyrmica* Cole (Hymenoptera: Formicidae). J. N. Y. Entomol. Soc. 69:209-220. [1961.12]

Gregg, R. E. 1963. The ants of Colorado, with reference to their ecology, taxonomy, and geographic distribution. Boulder: University of Colorado Press, xvi + 792 pp. [1963]

Gregg, R. E. 1964. Distribution of the ant genus *Formica* in the mountains of Colorado. Leafl. Univ. Colo. Mus. 13:59-69. [1964.08]

Gregg, R. E. 1969a. Geographic distribution of the ant genus *Formica* (Hymenoptera: Formicidae). Proc. Entomol. Soc. Wash. 71:38-49. [1969.03.25]

Gregg, R. E. 1969b. New species of *Pheidole* from Pacific Coast islands (Hymenoptera: Formicidae). Entomol. News 80:93-101. [1969.05.02]

Gregg, R. E. 1972a. A new species of *Stenamma* (Hymenoptera: Formicidae) from Utah. Great Basin Nat. 32:35-39. [1972.03.31]

Gregg, R. E. 1972b. The northward distribution of ants in North America. Can. Entomol. 104:1073-1091. [1972.08.16]

Gregg, R. E. 1973. A new species of *Camponotus* (Hymenoptera: Formicidae) from Nevada. Southwest. Nat. 18:39-43. [1973.03.30]

Gregg, R. E. 1974. William Steel Creighton - An appreciation. J. N. Y. Entomol. Soc. 82:67-75. [1974.08.22]

Gribodo, G., Emery, C. 1882. Ordo Hymenoptera. Pp. 81-85 in: Cavanna, G. Parte II.- Catalogo degli animali raccolti al Vulture, al Pollino ed in altri luoghi dell'Italia meridionale e centrale. Bull. Soc. Entomol. Ital. 14:31-87. [1882.05.20]

Griep, E. 1940. Die Ameisen von Bellinchen a. d. Oder. (Hym., Form.). Märk. Tierwelt 4:224-230. [1940.09.15]

Griffin, F. J. 1932. On the dates of publication and contents of the parts of Westwood (J. O.). Introduction to the Modern classification of Insects, 1838-1840. Proc. Entomol. Soc. Lond. 6:83-84. [1932.03.22]

Griffin, F. J. 1936. On the dates of publication of Motschulsky (V. de), 'Etudes entomologiques,' I.-XI., 1853-1862. Ann. Mag. Nat. Hist. (10)17:256-257. [1936.02]

Gronenberg, W. 1995. The fast mandible strike in the trap-jaw ant *Odontomachus*. I. Temporal properties and morphological characteristics. J. Comp. Physiol. A. Sens. Neural Behav. Physiol. 176:391-398. [1995.03]

Gronenberg, W., Peeters, C. 1993. Central projections of the sensory hairs on the gemma of the ant *Diacamma*: substrate for behavioural modulation? Cell Tissue Res. 273:401-415. [1993.09]

Gronenberg, W., Tautz, J., Hölldobler, B. 1993. Fast trap jaws and giant neurons in the ant *Odontomachus*. Science (Wash. D. C.) 262:561-563. [1993.10.22]

Grundmann, A. W. 1952. A new *Brachymyrmex* from northern Utah. J. Kansas Entomol. Soc. 25:117. [1952.07.15]

Grundmann, A. W., Peterson, B. V. 1953. House infesting ants in Salt Lake City, Utah. J. Kansas Entomol. Soc. 26:59-60. [1953.05.28]

Guérin-Méneville, F. E. 1831. Insectes, plate 8. In: Duperrey, L. I. (ed.) 1830-1831. Voyage autour du monde, exécuté par ordre du Roi, sur la corvette de sa Majesté, La Coquille, pendant les années 1822, 1823, 1824 et 1825. Zoologie. Atlas, Insectes. Paris: H. Bertrand, pls. 1-21, 14 bis. [1831.11.15] [Date of publication from Guérin-Méneville (1838:271). See also Cowan (1970).]

Guérin-Méneville, F. E. 1838. Première division. Crustacés, arachnides et insectes. In: Duperrey, L. I. (ed.) Voyage autour du monde, executé par ordre du Roi, sur la corvette de sa Majesté, La Coquille, pendant les années 1822, 1823, 1824 et 1825. Zoologie. Tome deuxième. Part 2. Paris: H. Bertrand, xii + 9-320 pp. [1838] [Date of publication from Bequaert (1926) and Cowan (1970).]

Guérin-Méneville, F. E. 1844a. Iconographie du règne animal de G. Cuvier, ou représentation d'après nature de l'une des espèces les plus remarquables, et souvent non encore figurées, de chaque genre d'animaux. Insectes. Paris: J. B. Baillière, 576 pp. [1844.09.07] [Date of publication from Cowan (1971). Ants pp. 421-428.]

Guérin-Méneville, F. E. 1844b. Observations sur un insecte qui attaque les olives, dans nos départements méridionaux, et cause une diminution très-considerable dans la récolte de l'huile. C. R. Séances Acad. Sci. 19:1147-1150. [1844.12.02]

Guérin-Méneville, F. E. 1849. Insectes. Pp. 241-398 in: Lefebvre, T. 1847-1851. Voyage en Abyssinie exécuté pendant les années 1839-1843. Pt. IV. Tom. VI. Histoire naturelle - Zoologie. Paris: Bertrand, 398 pp. [1849] [Publication date from Sherborn & Woodward (1901b:162). Famille des Hétérogynes: pp. 351-353.]

Guérin-Méneville, F. E. 1852. Notice sur une nouvelle espèce de fourmi découverte à Saint-Domingue par M. Auguste Sallé, et qui fait son nid dans les plaines marécageuses, sur les buissons. Rev. Mag. Zool. Pure Appl. (2)4:73-79. [1852.02]

Guérin-Méneville, F. E. 1857. Animaux articulés à pieds articulés. In: Sagra, R. de la. Histoire physique, politique et naturelle de l'ile de Cuba. [Tome VII.] Paris: Arthus Bertrand, lxxxvii + 868 pp. [1857] [Also published in a Spanish edition in the same year; see Aguayo (1946). Ants pp. 755-758 (four species recorded).]

Guiglia, D. 1965. Le specie di Imenotteri descritte da F. E. Guérin Méneville che si trovano a München (Zoologische Sammlung des Bayerischen Staates). Opusc. Zool. Münch. 87:1-2. [1965.12.15]

Guillou, E. J. F. le See: Le Guillou, E. J. F.

Guittini, U., Ronchetti, G. 1972. Nouvo contributo alla conoscenza della colonizzazione naturale delle formiche del gruppo *Formica rufa* in provincia di Novara. Boll. Soc. Entomol. Ital. 104:197-210. [1972.12.20]

Gulick, L. 1913. Synoptic list of ants reported from the Hawaiian Islands. Proc. Hawaii. Entomol. Soc. 2:306-311. [1913.07]

Gulmahamad, H. 1995. The genus *Liometopum* Mayr (Hymenoptera: Formicidae) in California, with notes on nest architecture and structural importance. Pan-Pac. Entomol. 71:82-86. [1995.10.09]

Gundlach, J. 1888 ("1886"). Contribución a la entomología cubana. Tomo II. Parte Segunda. Himenópteros. Habana: Impresa "La Antilla" de Cacho-Negrete, 5-187 + i-viii pp. [Also published in parts (pliegos) as supplements to An. Acad. Cienc. Med. Fis. Nat. La Habana (for details see Calvert, 1919). Ants pp. 77-105, 182-183. Pliegos 10-14 (corresponding to pp. 73-114) appeared between approximately 15 December 1886 and 15 May 1887, while pliego 23 (pp. 177-184) was issued on 15 February or 15 March 1888.]

Hacobian, B. S. 1992. New distribution records of the green tree ant *Oecophylla smaragdina* (Fabricius) (Hymenoptera: Formicidae: Formicinae) and three associated lycaenid butterflies. Aust. Entomol. Mag. 19:111-113. [1992.11.27]

Haddow, A. J., Yarrow, I. H. H., Lanchaster, G. A., Corbet, P. S. 1966. Nocturnal flight cycle in the males of African doryline ants (Hymenoptera: Formicidae). Proc. R. Entomol. Soc. Lond. Ser. A 41:103-106. [1966.09.27]

Hagan, H. R. 1954a. The reproductive system of the army-ant queen, *Eciton* (*Eciton*). Part 1. General anatomy. Am. Mus. Novit. 1663:1-12. [1954.06.10]

Hagan, H. R. 1954b. The reproductive system of the army-ant queen, *Eciton* (*Eciton*). Part 2. Histology. Am. Mus. Novit. 1664:1-17. [1954.06.10]

Hagan, H. R. 1954c. The reproductive system of the army-ant queen, *Eciton* (*Eciton*). Part 3. The oöcyte cycle. Am. Mus. Novit. 1665:1-20. [1954.06.10]

Hahn, C. W. 1832. Die Wanzenartigen Insecten. Getreu nach der Natur abgebildet und beschrieben. Band I, Heft 3. Nürnberg: Zeh, pp. 80-117. [1832.11] [Publication date from Sherborn (1914). See also Bergroth (1919).]

Halberstadt, K. 1963. Zur Pigmentierung der Formicinenkutikula. Symp. Genet. Biol. Ital. 12:214-220. [1963]

Haldeman, S. S. 1844. Descriptions of insects, presumed to be undescribed. Proc. Acad. Nat. Sci. Phila. 2:53-55. [1844.06] [Published between 30 April 1844 and 21 June 1844. See Nolan (1913).]

Haldeman, S. S. 1849a. On the identity of *Anomma* with *Dorylus*, suggested by specimens which Dr. Savage found together, and transmitted to illustrate his paper on the driver ants. Proc. Acad. Nat. Sci. Phila. 4:200-202. [1849.08]

Haldeman, S. S. 1849b. On several new Hymenoptera of the genera *Ampulex*, *Sigalphus*, *Chelonus* and *Dorylus*. Proc. Acad. Nat. Sci. Phila. 4:203-204. [1849.08]

Haldeman, S. S. 1852. Appendix C. - Insects. Pp. 366-378 in: Stansbury, H. An expedition to the Valley of the Great Salt Lake of Utah; including a description of its geography, natural history, and minerals, and an analysis of its waters. London: Sampson Low, Son & Co., 487 pp. [1852]

Haliday, A. H. 1836. Descriptions, etc. of the Hymenoptera. Pp. 316-331 in: Curtis, J., Haliday, A. H., Walker, F. Descriptions, etc. of the insects collected by Captain P. P. King R. N., F. R. S., in the survey of the Straits of Magellan. Trans. Linn. Soc. Lond. 17:315-359. [1836.06.21] [Publication date from Evenhuis et al. (1989:981). Raphael (1970) estimates a publication date between 21 June and 9 July 1936. Ants pp. 328-329.]

Hall, D. W., Smith, I. C. 1951. Studies in Pharaoh's ant, *Monomorium pharaonis* (L.). (6) External characters, size variation and cephalic ratios. Entomol. Mon. Mag. 87:217-221. [1951.08.20]

Hall, D. W., Smith, I. C. 1952. Studies in Pharaoh's ant, *Monomorium pharaonis* (L.). (7) Thoracic structures, typical and atypical. Entomol. Mon. Mag. 88:97-102. [1952.05.08]

Hall, D. W., Smith, I. C. 1953. Atypical forms of the wingless worker and the winged female in *Monomorium pharaonis* (L.) (Hymenoptera: Formicidae). Evolution 7:127-135. [1953.06.22]

Hall, D. W., Smith, I. C. 1954. Studies in Pharaoh's ant, *Monomorium pharaonis* (L.). (9) Somatic mosaics. Entomol. Mon. Mag. 90:176-182. [1954.10.06]

Hallett, H. M. 1912 ("1911"). A list of Hymenoptera Aculeata recorded for the county of Glamorgan. Trans. Cardiff Nat. Soc. 44:92-99. [1912]

Halliday, R. B. 1975. Electrophoretic variation of amylase in meat ants, *Iridomyrmex purpureus*, and its taxonomic significance. Aust. J. Zool. 23:271-276. [1975.05]

Halliday, R. B. 1979. Esterase variation at three loci in meat ants. J. Hered. 70:57-61. [1979.02]

Halliday, R. B. 1981. Heterozygosity and genetic distance in sibling species of meat ants (*Iridomyrmex purpureus* group). Evolution 35:234-242. [1981.04.03]

Halliday, R. B. 1983. Social organization of meat ants *Iridomyrmex purpureus* analyzed by gel electrophoresis of enzymes. Insectes Soc. 30:45-56. [1983.07]

Halverson, D. D., Wheeler, J., Wheeler, G. C. 1976. Natural history of the sandhill ant, *Formica bradleyi* (Hymenoptera: Formicidae). J. Kansas Entomol. Soc. 49:280-303. [1976.04.26]

Hamann, H. H. F. 1957. On a new record of *Gesomyrmex* Mayr (Formicidae). Idea 10(4):1-6. [1957.08.15]

Hamann, H. H. F., Klemm, W. 1967. Ergebnisse der zoologischen Nubien-Expedition 1962. Teil XXXIV. Hymenoptera - Formicidae. Ann. Naturhist. Mus. Wien 70:411-421. [1967.10]

Hamann, H. H. F., Klemm, W. 1976. Ergebnisse der von Dr. O. Paget und Dr. E. Kritscher auf Rhodos durchgeführten zoologischen Exkursionen. XVI. Formicidae. Ann. Naturhist. Mus. Wien 80:669-679. [1976.11]

Hamilton, W. D. 1972. Altruism and related phenomena, mainly in social insects. Annu. Rev. Ecol. Syst. 3:193-232. [1972]

Hamm, A. H. 1902. *Formica exsecta* in South Devon. Entomol. Mon. Mag. 38:266. [1902.11]

Handlirsch, A. 1906. Die fossilen Insekten und die Phylogenie der rezenten Formen. Ein Handbuch für Paläontologen und Zoologen. Lieferung 4. Leipzig: Wilhelm Engelmann, pp. 481-640. [1906.10]

Handlirsch, A. 1907. Die fossilen Insekten und die Phylogenie der rezenten Formen. Ein Handbuch für Paläontologen und Zoologen. Lieferung 6. Leipzig: Wilhelm Engelmann, pp. 801-960. [1907.06]

Handlirsch, A. 1908. Die fossilen Insekten und die Phylogenie der rezenten Formen. Ein Handbuch für Paläontologen und Zoologen. Lieferung 8. Leipzig: Wilhelm Engelmann, pp. 1121-1280. [1908.01]

Hansen, L. D., Akre, R. D. 1985. Biology of carpenter ants in Washington State (Hymenoptera: Formicidae: *Camponotus*). Melanderia 43:i-v, 1-61. [1985]

Hansen, S. R. 1978. Resource utilization and coexistence of three species of *Pogonomyrmex* ants in an Upper Sonoran grassland community. Oecologia (Berl.) 35:109-117. [1978.07.12]

Harada, A. Y. 1987. Uma nova espécie do gênero *Monacis* Roger, da Amazônia (Hymenoptera: Formicidae). Acta Amazonica 16/17:599-606. [1987.10]

Harada, A. Y. 1990. Ant pests of the Tapinomini tribe. Pp. 298-315 in: Vander Meer, R. K., Jaffe, K., Cedeno, A. (eds.) Applied myrmecology: a world perspective. Boulder: Westview Press, xv + 741 pp. [1990]

Harada, A. Y., Benson, W. W. 1988. Espécies de *Azteca* (Hymenoptera, Formicidae) especializadas em *Cecropia* spp. (Moraceae): distribuição geográfica e considerações ecológicas. Rev. Bras. Entomol. 32:423-435. [1988.12.30]

Haro, A. de See: De Haro, A.

Harrington, W. H. 1891. Note on *Amblyopone pallipes*, Hald. Can. Entomol. 23:138-139. [1891.06.04]

Harris, R. A. 1979. A glossary of surface sculpturing. Calif. Dep. Food Agric. Lab. Serv. Entomol. Occ. Pap. 28:1-31. [1979.06]

Harvey, P. R. 1995. A review of notable bees, wasps and ants recently recorded from Essex. Essex Nat. (Lond.) 12:17-32. [Not seen. Cited in BIOSIS.]

Hasegawa, E. 1995 ("1994"). Sex allocation in the ant *Colobopsis nipponicus* (Wheeler). I. Population sex ratio. Evolution 48:1121-1129. [1995.01.26]

Hasegawa, E., Yamaguchi, T. 1995. Population structure, local mate competition, and sex-allocation pattern in the ant *Messor aciculatus*. Evolution 49:260-265. [1995.08.22]

Hashimoto, Y. 1990. Unique features of sensilla on the antennae of Formicidae (Hymenoptera). Appl. Entomol. Zool. 25:491-501. [1990.11]

Hashimoto, Y. 1991a. Phylogenetic study of the family Formicidae based on the sensillum structures on the antennae and labial palpi (Hymenoptera, Aculeata). Jpn. J. Entomol. 59:125-140. [1991.03.25]

Hashimoto, Y. 1991b. Phylogenetic implications of the spur structures of the hind tibia in the Formicidae (Hymenoptera). Jpn. J. Entomol. 59:289-294. [1991.06.25]

Hashimoto, Y. 1992. The phylogeny of the Formicidae and its relationship to other Aculeata. [Abstract.] P. 52 in: Proceedings XIX International Congress of Entomology. Abstracts. Beijing, China, June 28 - July 4, 1992. Beijing: XIX International Congress of Entomology, xi + 730 pp. [1992.07.04]

Hashimoto, Y. 1995. Unique habits of stomodeal trophallaxis in the ponerine ant *Hypoponera* sp. Insectarium 32(6):16-22. [1995.06]

Hashimoto, Y., Yamauchi, K., Hasegawa, E. 1995. Unique habits of stomodeal trophallaxis in the ponerine ant *Hypoponera* sp. Insectes Soc. 42:137-144. [1995]

Hashmi, A. A. 1973a. A revision of the Neotropical ant subgenus *Myrmothrix* (genus *Camponotus*). Diss. Abstr. Int. B. Sci. Eng. 33:5891. [1973.06]

Hashmi, A. A. 1973b. A revision of the Neotropical ant subgenus *Myrmothrix* of genus *Camponotus* (Hymenoptera: Formicidae). Stud. Entomol. 16:1-140. [1973.10.31]

Haskins, C. P. 1928. Notes on the behavior and habits of *Stigmatomma pallipes* Haldeman. J. N. Y. Entomol. Soc. 36:179-184. [1928.08.02]

Haskins, C. P. 1941. Notes on the method of colony foundation of the ponerine ant *Bothroponera soror* Emery. J. N. Y. Entomol. Soc. 49:211-216. [1941.05.06]

Haskins, C. P. 1952 ("1951"). Note on a gynandromorph in *Amblyopone australis* Erichson (Hymenoptera: Formicidae). J. N. Y. Entomol. Soc. 59:221-224. [1952.01.31]

Haskins, C. P. 1970. Researches in the biology and social behavior of primitive ants. Pp. 355-388 in: Aronson, L. R., Tobach, E., Lehrman, D. S., Rosenblatt, J. S. (eds.) Development and evolution of behavior. San Francisco: W. H. Freeman, xviii + 656 pp. [1970]

Haskins, C. P. 1979 ("1978"). Sexual calling behavior in highly primitive ants. Psyche (Camb.) 85:407-415. [1979.09.18]

Haskins, C. P., Haskins, E. F. 1950. Note on the method of colony foundation of the ponerine ant *Brachyponera* (*Euponera*) *lutea* Mayr. Psyche (Camb.) 57:1-9. [1950.06.16]

Haskins, C. P., Haskins, E. F. 1951a ("1950"). Notes on the biology and social behavior of the archaic ponerine ants of the genera *Myrmecia* and *Promyrmecia*. Ann. Entomol. Soc. Am. 43:461-491. [1951.01.20]

Haskins, C. P., Haskins, E. F. 1951b. Note on the method of colony foundation of the ponerine ant *Amblyopone australis* Erichson. Am. Midl. Nat. 45:432-445. [1951.06.06]

Haskins, C. P., Haskins, E. F. 1955. The pattern of colony foundation in the archaic ant *Myrmecia regularis*. Insectes Soc. 2:115-126. [1955.06]

Haskins, C. P., Haskins, E. F. 1964. Notes on the biology and social behavior of *Myrmecia inquilina*. The only known myrmeciine social parasite. Insectes Soc. 11:267-282. [1964.12]

Haskins, C. P., Haskins, E. F. 1974. Notes on the necrophoric behaviour in the archaic ant *Myrmecia vindex*. Psyche (Camb.) 81:258-267. [1974.09.26]

Haskins, C. P., Haskins, E. F. 1981 ("1980"). Notes on female and worker survivorship in the archaic ant genus *Myrmecia*. Insectes Soc. 27:345-350. [1981.03]

Haskins, C. P., Hewitt, R. E., Haskins, E. F. 1973. Release of aggressive and capture behaviour in the ant *Myrmecia gulosa* F. by exocrine products of the ant *Camponotus*. J. Entomol. Ser. A 47:125-139. [1973.04.17]

Haskins, C. P., Whelden, R. M. 1954. Note on the exchange of ingluvial food in the genus *Myrmecia*. Insectes Soc. 1:33-37. [1954.03]

Haskins, C. P., Whelden, R. M. 1965. "Queenlessness", worker sibship, and colony versus population structure in the formicid genus *Rhytidoponera*. Psyche (Camb.) 72:87-112. [1965.06.25]

Haskins, C. P., Zahl, P. A. 1971. The reproductive pattern of *Dinoponera grandis* Roger (Hymenoptera, Ponerinae) with notes on the ethology of the species. Psyche (Camb.) 78:1-11. [1971.12.30]

Haug, G. W. 1932. Description of the male of *Strumigenys louisianae* subsp. *laticephala* M. R. Smith (Hymen.: Formicidae). Ann. Entomol. Soc. Am. 25:170-172. [1932.03.31]

Hauschteck, E. 1961. Die Chromosomen von fünf Ameisenarten. Rev. Suisse Zool. 68:218-223. [1961.07]

Hauschteck, E. 1962. Die Chromosomen einiger in der Schweiz vorkommender Ameisenarten. Vierteljahrsschr. Naturforsch. Ges. Zür. 107:213-220. [1962.12.31]

Hauschteck, E. 1963. Chromosomes of Swiss ants. [Abstract.] P. 140 in: Geerts, S. J. (ed.) Genetics today. Proceedings of the XI International Congress of Genetics, The Hague, The Netherlands, September, 1963. Volume 1. Abstracts. New York: MacMillan Co., 332 pp. [1963]

Hauschteck, E. 1965. Halbe haploide Chromosomenzahl im Hoden von *Myrmica sulcinodis* Nyl. (Formicidae). Experientia (Basel) 21:323-325. [1965.06.15]

Hauschteck-Jungen, E., Jungen, H. 1976. Ant chromosomes. I. The genus *Formica*. Insectes Soc. 23:513-524. [1976]

Hauschteck-Jungen, E., Jungen, H. 1983. Ant chromosomes. II. Karyotypes of western palearctic species. Insectes Soc. 30:149-164. [1983.09]

Hayashi, N. 1994. An approach to chemotaxonomy of ants. [Abstract.] [In Japanese.] Ari 17:1-2. [1994.04.30]

Hayashi, N., Komae, H. 1980. Components of the ant secretions. Biochem. Syst. Ecol. 8:293-295. [1980.08.15]

Hayashi, N., Komae, H. 1984. Chemical components of the ant secretions. [Abstract.] [In Japanese.] Ari 12:1-1-2. [1984.05.01]

Hayashida, K. 1957. Ecological distribution of ants in Sapporo and vicinity. (Preliminary report.) J. Fac. Sci. Hokkaido Univ. Ser. VI. Zool. 13:173-177. [1957.08]

Hayashida, K. 1959. Ecological distribution of ants in Mt. Atusanupuri, an active volcano in Akan National Park, Hokkaido. J. Fac. Sci. Hokkaido Univ. Ser. VI. Zool. 14:252-260. [1959.12]

Hayashida, K. 1961 ("1960"). Studies on the ecological distribution of ants in Sapporo and its vicinity (1 et 2). Insectes Soc. 7:125-162. [1961.06]

Hayashida, K. 1964. Studies on the ecological distribution of ants in Kutchan and its adjacent area. J. Sapporo Otani Jr. Coll. 2:107-129. [Not seen. Cited in Onoyama & Terayama (1994).]

Hayashida, K. 1966. Notes on the symbiotic ants of *Formica sanguinea* Latreille. [In Japanese.] Ari 4:2-4. [1966.10.20]

Hayashida, K. 1971. Vertical distribution of ants in the southern part of the Hidaka mountains. [In Japanese.] Mem. Natl. Sci. Mus. (Tokyo) 4:29-38. [1971.11.30]

Hayashida, K., Meada, S. 1960. Studies on the ecological distribution of ants in Akkeshi. J. Fac. Sci. Hokkaido Univ. Ser. VI. Zool. 14:305-319. [1960.12]

Hayes, W. P. 1925a. A preliminary list of the ants of Kansas (Hymenoptera, Formicidae). [part] Entomol. News 36:10-12. [1925.01.05]
Hayes, W. P. 1925b. A preliminary list of the ants of Kansas (Hymenoptera, Formicidae). [part] Entomol. News 36:39-43. [1925.02.05]
Hayes, W. P. 1925c. A preliminary list of the ants of Kansas (Hymenoptera, Formicidae). [concl.] Entomol. News 36:69-73. [1925.03.05]
Hazeltine, W. F. 1967. Female genitalia of Hymenoptera and comparative morphology of male and female genital segments of Bombinae (Hymenoptera, Apidae). Purdue Univ. Agric. Exp. Stn. Res. Bull. 833:1-36. [1967.10]
Headley, A. E. 1943. The ants of Ashtabula County, Ohio (Hymenoptera, Formicidae). Ohio J. Sci. 43:22-31. [1943.02.26]
Heatwole, H. 1991. The ant assemblage of a sand-dune desert in the United Arab Emirates. J. Arid Environ. 21:71-79. [1991.07]
Heatwole, H., Muir, R. 1991. Foraging, abundance and biomass of ants in the pre-Saharan steppe of Tunisia. J. Arid Environ. 21:337-350. [1991.11]
Hedicke, H. 1941. Über das Erscheinungsjahr von Fabricius' Systema Piezatorum. Mitt. Dtsch. Entomol. Ges. 10:82-83. [1941.10.01] [Provides evidence about the year of publication of Fabricius (1804).]
Heer, O. 1849. Die Insektenfauna der Tertiärgebilde von Oeningen und von Radoboj in Croatien. Zweiter Theil: Heuschrecken, Florfliegen, Aderflüger, Schmetterlinge und Fliegen. Leipzig: W. Engelmann, vi + 264 pp. [1849] [Also published as Heer (1850).]
Heer, O. 1850. Die Insektenfauna der Tertiärgebilde von Oeningen und von Radoboj in Croatien. Zweite Abtheilung: Heuschrecken, Florfliegen, Aderflüger, Schmetterlinge und Fliegen. Neue Denkschr. Allg. Schweiz. Ges. Gesammten Naturwiss. 11(1):1-264. [1850]
Heer, O. 1852. Ueber die Haus-Ameise Madeiras. Zür. Jugend Naturforsch. Ges. 54:1-24. [1852]
Heer, O. 1856a. On the house ant of Madeira (translated by R. T. Lowe). [part] Ann. Mag. Nat. Hist. (2)17:209-224. [1856.03]
Heer, O. 1856b. On the house ant of Madeira (translated by R. T. Lowe). [concl.] Ann. Mag. Nat. Hist. (2)17:322-333. [1856.04]
Heer, O. 1864. Die Urwelt der Schweiz. [Lieferungen 7-11.] Zürich: Friedrich Schulthess, pp. 289-496. [1864.08.18] [Publication date from Evenhuis (1994b:504). Ants discussed and illustrated pp. 386-388.]
Heer, O. 1867. Fossile Hymenopteren aus Oeningen und Radoboj. Neue Denkschr. Allg. Schweiz. Ges. Gesammten Naturwiss. 22(4):1-42. [1867]
Heer, O. 1870. Die Miocene Flora und Fauna Spitzbergens. K. Sven. Vetensk.-Akad. Handl. (n.s. 3)8(7):1-98. [1870]
Heer, O. 1879. Die Urwelt der Schweiz. Zweite, umgearbeitete und vermehrte Auflage. Zürich: Friedrich Schulthess, xix + 713 pp. [1879] [Ants pp. 412-414.]
Hefetz, A. 1993. Hymenopteran exocrine secretions as a tool for chemosystematic analysis - possibilities and constraints. Biochem. Syst. Ecol. 21:163-169. [1993.01.31]
Hefetz, A., Lenoir, A. 1992. Dufour's gland composition in the desert ant *Cataglyphis*: species specificity and population differences. Z. Naturforsch. C J. Biosci. 47:285-289. [1992.04]
Hefetz, A., Orion, T. 1982. Pheromones of ants of Israel. I. The alarm-defense system of some larger Formicinae. Isr. J. Entomol. 16:87-97. [1982.12.30]
Heinze, J. 1986. A new find of *Paratrechina longicornis* (Latreille 1802) on the Azores (Hymenoptera, Formicidae). Bocagiana 101:1-3. [1986.09.15]
Heinze, J. 1987a. Three species of social parasitic ants new to Turkey. Insectes Soc. 34:65-68. [1987.09]
Heinze, J. 1987b. The application of electrophoretical data on species differentiation in the ant tribe Leptothoracini. Pp. 31-32 in: Eder, J., Rembold, H. (eds.) Chemistry and biology of social insects. München: Verlag J. Peperny, xxxv + 757 pp. [1987]

Heinze, J. 1989a. Alternative dispersal strategies in a North American ant. Naturwissenschaften 76:477-478. [1989.10]

Heinze, J. 1989b. A biochemical approach toward the systematics of the *Leptothorax "muscorum"* group in North America (Hymenoptera: Formicidae). Biochem. Syst. Ecol. 17:595-601. [1989.12.21]

Heinze, J. 1989c. *Leptothorax wilsoni* n.sp., a new parasitic ant from eastern North America (Hymenoptera: Formicidae). Psyche (Camb.) 96:49-62. [1989.12.18]

Heinze, J. 1991. Biochemical studies on the relationship between socially parasitic ants and their hosts. Biochem. Syst. Ecol. 19:195-206. [1991.06.06]

Heinze, J. 1992. Ecological correlates of functional monogyny and queen dominance in leptothoracine ants. Pp. 25-33 in: Billen, J. P. J. (ed.) Biology and evolution of social insects. Leuven: Leuven University Press, ix + 390 pp. [1992]

Heinze, J. 1993a. Habitat structure, dispersal strategies and queen number in two boreal *Leptothorax* ants. Oecologia (Berl.) 96:32-39. [1993.10]

Heinze, J. 1993b. Life histories of subarctic ants. Arctic 46:354-358. [1993.12]

Heinze, J. 1994. Genetic colony and population structure of the ant *Leptothorax* cf. *canadensis*. Can. J. Zool. 72:1477-1480. [1994.12.02]

Heinze, J., Alloway, T. M. 1992 ("1991"). *Leptothorax paraxenus* n. sp., a workerless social parasite from North America (Hymenoptera: Formicidae). Psyche (Camb.) 98:195-206. [1992.02.07]

Heinze, J., Buschinger, A. 1987. Queen polymorphism in a non-parasitic *Leptothorax* species (Hymenoptera, Formicidae). Insectes Soc. 34:28-43. [1987.09]

Heinze, J., Buschinger, A. 1988. Electrophoretic variability of esterases in the ant tribe, Leptothoracini. Biochem. Syst. Ecol. 16:217-221. [1988.01.14]

Heinze, J., Buschinger, A. 1989a ("1988"). Polygyny and functional monogyny in *Leptothorax* ants (Hymenoptera: Formicidae). Psyche (Camb.) 95:309-325. [1989.06.13]

Heinze, J., Buschinger, A. 1989b. Queen polymorphism in *Leptothorax* spec.A: its genetic and ecological background (Hymenptera: Formicidae). Insectes Soc. 36:139-155. [1989.09]

Heinze, J., Gadau, J., Hölldobler, B., Nanda, I., Schmid, M., Scheller, K. 1994. Genetic variability in the ant *Camponotus floridanus* detected by multilocus DNA fingerprinting. Naturwissenschaften 81:34-36. [1994.01]

Heinze, J., Hölldobler, B. 1993a. Queen polymorphism in an Australian weaver ant, *Polyrhachis* cf. *doddi*. Psyche (Camb.) 100:83-92. [1993.09.27]

Heinze, J., Hölldobler, B. 1993b. Fighting for a harem of queens: physiology of reproduction in *Cardiocondyla* male ants. Proc. Natl. Acad. Sci. U. S. A. 90:8412-8414. [1993.09]

Heinze, J., Hölldobler, B. 1994. Fighting and thelytokous parthenogenesis in the ant, *Platythyrea punctata*. P. 245 in: Lenoir, A., Arnold, G., Lepage, M. (eds.) Les insectes sociaux. 12ème congrès de l'Union Internationale pour l'Étude des Insectes Sociaux. Paris, Sorbonne, 21-27 août 1994. Paris: Université Paris Nord, xxiv + 583 pp. [1994.09]

Heinze, J., Hölldobler, B. 1995. Thelytokous parthenogenesis and dominance hierarchies in the ponerine ant, *Platythyrea punctata*. Naturwissenschaften 82:40-41. [1995.01]

Heinze, J., Hölldobler, B., Cover, S. P. 1992. Queen polymorphism in the North American harvester ant, *Ephebomyrmex imberbiculus*. Insectes Soc. 39:267-273. [1992]

Heinze, J., Hölldobler, B., Peeters, C. 1994. Conflict and cooperation in ant societies. Naturwissenschaften 81:489-497. [1994.11]

Heinze, J., Hölldobler, B., Trenkle, S. 1995. Reproductive behavior of the ant *Leptothorax (Dichothorax) pergandei*. Insectes Soc. 42:309-315. [1995]

Heinze, J., Kauffmann, S. 1993. The socially parasitic ants of Turkey (Hymenoptera, Formicidae). Zool. Middle East 8:31-35. [1993]

Heinze, J., Kauffmann, S., Hülsen, B. 1993. *Doronomyrmex pacis* Kutter 1945, a socially parasitic ant new to Germany (Insecta, Hymenoptera, Formicidae). Spixiana 16:171-172. [1993.07.01]

Heinze, J., Kühnholz, S., Schilder, K., Hölldobler, B. 1993. Behavior of ergatoid males in the ant, *Cardiocondyla nuda*. Insectes Soc. 40:273-282. [1993]

Heinze, J., Lipski, N., Hölldobler, B., Bourke, A. F. G. 1995. Geographical variation in the social and genetic structure of the ant, *Leptothorax acervorum*. Zool. Anal. Complex Syst. 98:127-135. [1995]

Heinze, J., Ortius, D. 1992 ("1991"). Social organization of *Leptothorax acervorum* from Alaska (Hymenoptera: Formicidae). Psyche (Camb.) 98:227-240. [1992.02.07]

Heinze, J., Schulz, A., Radchenko, A. G. 1993. Redescription of the ant *Leptothorax* (s. str.) *scamni* Ruzsky, 1905. Psyche (Camb.) 100:177-183. [1993.12.20]

Heinze, J., Stuart, R. J., Alloway, T. M., Buschinger, A. 1992. Host specificity in the slave-making ant *Harpagoxenus canadensis* M. R. Smith. Can. J. Zool. 70:167-170. [1992.03.27]

Hemming, F. 1954. "*Formica*" Linnaeus, 1758: Report on proposed action, under the plenary powers, to give valid force to the decision taken by the Commission in Paris: Action needed because of circumstances not then known to the Commission. Bull. Zool. Nomencl. 9:309-312. [1954.12.30]

Herbers, J. M. 1984 ("1983"). Social organization in *Leptothorax* ants: within- and between-species patterns. Psyche (Camb.) 90:361-386. [1984.04.06]

Herbers, J. M. 1985. Seasonal structuring of a north temperate ant community. Insectes Soc. 32:224-240. [1985.12]

Herbers, J. M. 1988. *Manica mutica* in Gunnison County, Colorado. Pp. 157-159 in: Trager, J. C. (ed.) Advances in myrmecology. Leiden: E. J. Brill, xxvii + 551 pp. [1988]

Herbers, J. M. 1994. Structure of an Australian ant community with comparisons to North American counterparts (Hymenoptera: Formicidae). Sociobiology 24: 293-306. [1994]

Hermann, H. R. 1968a. Group raiding in *Termitopone commutata* (Roger) (Hymenoptera: Formicidae). J. Ga. Entomol. Soc. 3:23-24. [1968.01]

Hermann, H. R. 1968b. The hymenopterous poison apparatus. V. *Aneuretus simoni*. Ann. Entomol. Soc. Am. 61:1315-1317. [1968.09.16]

Hermann, H. R. 1968c. The hymenopterous poison apparatus. VII. *Simopelta oculata* (Hymenoptera: Formicidae: Ponerinae). J. Ga. Entomol. Soc. 3:163-166. [1968.10]

Hermann, H. R. 1969a. The hymenopterous poison apparatus. VIII. *Leptogenys* (*Lobopelta*) *elongata* (Hymenoptera: Formicidae). J. Kansas Entomol. Soc. 42:239-243. [1969.06.06]

Hermann, H. R. 1969b. The hymenopterous poison apparatus: evolutionary trends in three closely related subfamilies of ants (Hymenoptera: Formicidae). J. Ga. Entomol. Soc. 4:123-141. [1969.07]

Hermann, H. R. 1973. Formation of preforage aggregations in ponerine ants (Hymenoptera: Formicidae), a possible step toward group raiding. J. Ga. Entomol. Soc. 8:185-186. [1973.07]

Hermann, H. R. 1975. The ant-like venom apparatus of *Typhoctes peculiaris*, a primitive mutillid wasp. Ann. Entomol. Soc. Am. 68:882-884. [1975.09.15]

Hermann, H. R. (ed.) 1979. Social insects. Volume 1. New York: Academic Press, xv + 437 pp. [1979]

Hermann, H. R. (ed.) 1981. Social insects. Volume 2. New York: Academic Press, xiii + 491 pp. [1981]

Hermann, H. R. (ed.) 1982a. Social insects. Volume 3. New York: Academic Press, xiii + 459 pp. [1982]

Hermann, H. R. (ed.) 1982b. Social insects. Volume 4. New York: Academic Press, xiii + 385 pp. [1982]

Hermann, H. R. 1983. Directional trajectory manipulation and venom dispersal in formicine ants (Hymenoptera: Formicidae). Sociobiology 8:105-117. [1983]

Hermann, H. R. (ed.) 1984a. Defensive mechanisms in social insects. New York: Praeger, xii + 259 pp. [1984]

Hermann, H. R. 1984b. Elaboration and reduction of the venom apparatus in aculeate Hymenoptera. Pp. 201-243 in: Hermann, H. R. (ed.) Defensive mechanisms in social insects. New York: Praeger, xii + 259 pp. [1984]

Hermann, H. R., Baer, R., Barlin, M. 1975. Histology and function of the venom gland system in formicine ants. Psyche (Camb.) 82:67-73. [1975.06.08]

Hermann, H. R., Blum, M. S. 1965. Morphology and histology of the reproductive system of the imported fire ant queen, *Solenopsis saevissima richteri*. Ann. Entomol. Soc. Am. 58:81-89. [1965.01.15]

Hermann, H. R., Blum, M. S. 1966. The morphology and histology of the hymenopterous poison apparatus. I. *Paraponera clavata* (Formicidae). Ann. Entomol. Soc. Am. 59:397-409. [1966.03.15]

Hermann, H. R., Blum, M. S. 1967a. The morphology and histology of the hymenopterous poison apparatus. II. *Pogonomyrmex badius* (Formicidae). Ann. Entomol. Soc. Am. 60:661-668. [1967.05.15]

Hermann, H. R., Blum, M. S. 1967b. The morphology and histology of the hymenopterous poison apparatus. III. *Eciton hamatum* (Formicidae). Ann. Entomol. Soc. Am. 60:1282-1291. [1967.11.15]

Hermann, H. R., Blum, M. S. 1968. The hymenopterous poison apparatus. VI. *Camponotus pennsylvanicus* (Hymenoptera: Formicidae). Psyche (Camb.) 75:216-227. [1968.11.15]

Hermann, H. R., Blum, M. S. 1981. Defensive mechanisms in the social Hymenoptera. Pp. 77-197 in: Hermann, H. R. (ed.) Social insects. Volume 2. New York: Academic Press, 491 pp. [1981]

Hermann, H. R., Blum, M. S., Wheeler, J. W., Overal, W. L., Schmidt, J. O., Chao, J. 1984. Comparative anatomy and chemistry of the venom apparatus and mandibular glands in *Dinoponera grandis* (Guérin) and *Paraponera clavata* (F.) (Hymenoptera: Formicidae: Ponerinae). Ann. Entomol. Soc. Am. 77:272-279. [1984.05]

Hermann, H. R., Chao, J.-T. 1983. Furcula, a major component of the hymenopterous venom apparatus. Int. J. Insect Morphol. Embryol. 12:321-337. [1983.12]

Hermann, H. R., Douglas, M. E. 1976a. Sensory structures on the venom apparatus of a primitive ant species. Ann. Entomol. Soc. Am. 69:681-686. [1976.07.15]

Hermann, H. R., Douglas, M. E. 1976b. Comparative survey of the sensory structures on the sting and ovipositor of hymenopterous insects. J. Ga. Entomol. Soc. 11:223-239. [1976.07]

Hermann, H. R., Moser, J. C., Hunt, A. N. 1970. The hymenopterous poison apparatus. X. Morphological and behavioral changes in *Atta texana* (Hymenoptera: Formicidae). Ann. Entomol. Soc. Am. 63:1552-1558. [1970.11.16]

Hermann, H. R., Jr. See: Hermann, H. R.

Hervé, P. 1969. Les espèces françaises de la tribu des Dacetini (Hymenoptera: Formicidae). Entomops 2:155-158. [1969.01.15]

Hess, C. G. 1958. The ants of Dallas County, Texas, and their nesting sites; with particular reference to soil texture as an ecological factor. Field Lab. 26:3-72. [1958.04]

Heuer, H. G. 1979. Bemerkungen zur Ameisenfauna (Hymenoptera: Formicidae) der Nordseeinsel Borkum. Abh. Naturwiss. Ver. Bremen 39:41-45. [1979]

Higashi, M. 1939. The insect-fauna of the Tottori sand-dune. [In Japanese.] Trans. Kansai Entomol. Soc. 8:25-46. [1939.04.20] [Ants pp. 42-43.]

Higashi, S., Peeters, C. 1990. Worker polymorphism and nest structure in *Myrmecia brevinoda* Forel (Hymenoptera: Formicidae). J. Aust. Entomol. Soc. 29:327-331. [1990.11.30]

Higashi, S., Sato, H., Sugawara, H., Fukuda, H. 1985. Myrmecofaunal changes since the 1977-78 eruptions on Mt. Usu. Jpn. J. Ecol. 35:469-479. [1985.12.30]

Higashi, S., Yamauchi, K. 1979. Influence of a supercolonial ant *Formica* (*Formica*) *yessensis* Forel on the distribution of other ants in Ishikari coast. Jpn. J. Ecol. 29:257-264. [1979.09.30]

Higgins, L. G. 1963. Dates of publication of the Novara Reise. J. Soc. Bibliogr. Nat. Hist. 4:153-159. [1963.11]

Hilburn, D. J., Marsh, P. M., Schauff, M. E. 1990. Hymenoptera of Bermuda. Fla. Entomol. 73:161-176. [1990.03.30] [Ants pp. 172-173.]
Hilton, W. A. 1919. Ants from the Claremont Laguna region. J. Entomol. Zool. 11:38. [1919.06]
Hinkle, G., Wetterer, J. K., Schultz, T. R., Sogin, M. L. 1994. Phylogeny of the attine ant fungi based on analysis of small subunit ribosomal RNA gene sequences. Science (Wash. D. C.) 266:1695-1697. [1994.12.09]
Hinton, H. E. 1951. Myrmecophilous Lycaenidae and other Lepidoptera - a summary. Proc. Trans. South Lond. Entomol. Nat. Hist. Soc. 1949-1950:111-175. [1951.04]
Hirai, H., Yamamoto, M.-T., Ogura, K., Satta, Y., Yamada, M., Taylor, R. W., Imai, H. T. 1994. Multiplication of 28S rDNA and NOR activity in chromosome evolution among ants of the *Myrmecia pilosula* species complex. Chromosoma (Berl.) 103:171-178. [1994.06]
Hirano, I. 1963. The list of Japanese references of insects. No. 223. Formicoidea. [In Japanese.] Osaka Shokubutsu Boeki 87:36-64.
Hocking, B. 1970. Insect associations with the swollen thorn acacias. Trans. R. Entomol. Soc. Lond. 122:211-255. [1970.12.31]
Hoffman, D. R. 1993. Allergens in Hymenoptera venom XXIV: the amino acid sequences of imported fire ant venom allergens Sol i II, Sol i III, and Sol i IV. J. Allergy Clin. Immunol. 91:71-78. [1993.01]
Hoffman, D. R., Guralnick, M. W., Vander Meer, R. K., Smith, A. M. 1989. Comparison of venom allergens from two species of imported fire ants. [Abstract.] J. Allergy Clin. Immunol. 83:232. [1989.01]
Hoffman, D. R., Smith, A. M., Schmidt, M., Moffitt, J. E., Guralnick, M. 1990. Allergens in Hymenoptera venom. XXII. Comparison of venoms from two species of imported fire ants, *Solenopsis invicta* and *richteri*. J. Allergy Clin. Immunol. 85:988-996. [1990.06]
Högmo, O. 1995. Busksmalmyran *Leptothorax parvulus* funnen på Gotland - ny för Nordeuropa. Entomol. Tidskr. 116:127-128. [1995.11]
Hohmann, H., La Roche, F., Ortega, G., Barquín, J. 1993. Bienen, Wespen und Ameisen der Kanarischen Inseln. Veröff. Überseemus. Bremen Naturwiss. 12:14-712. [1993] [Formicidae pp. 58-60, 145-166.]
Hohorst, B. 1972. Biometrische Untersuchungen an *Formica* (*Serviformica*) *rufibarbis* Fabricius (Hymenoptera: Formicidae). Insectes Soc. 19:405-407. [1972]
Holgersen, H. 1938. Bidrag til Norges Formicidefauna. Nor. Entomol. Tidsskr. 5:74-78. [1938.06.24]
Holgersen, H. 1940. Myrmekologiske notiser I. Nor. Entomol. Tidsskr. 5:183-187. [1940.01]
Holgersen, H. 1942a. Myrmekologiske notiser II. Nor. Entomol. Tidsskr. 6:93-98. [1942.03.02]
Holgersen, H. 1942b. Ants of northern Norway (Hym., Form.). Tromso Mus. Årsh. 63(2):1-34. [1942.10.09]
Holgersen, H. 1943a. *Formica gagatoides* Ruzs. in Norway. Tromso Mus. Årsh. 64:3-17. [1943.05]
Holgersen, H. 1943b. Insecta, ex Sibiria meridionali et Mongolia, in itinere Orjan Olsen 1914 collecta. C. Hymenoptera. 1. Formicidae. D. Hemiptera. 1. Homoptera cicadina. Nor. Entomol. Tidsskr. 6:162-163. [1943.09.01]
Holgersen, H. 1943c. Bestemmelsestabell over norske maur (Hym., Formicidae). Nor. Entomol. Tidsskr. 6:164-182. [1943.09.01]
Holgersen, H. 1943d. *Ponera punctatissima* Rog. (Hym. Form.) funnet i Norge. Nor. Entomol. Tidsskr. 6:183-186. [1943.09.01]
Holgersen, H. 1943e. Ant studies in Rogaland (south-western Norway). Avh. Nor. Vidensk.-Akad. Oslo I Mat.-Naturvidensk. Kl. 1943(7):1-75. [1943.10.15]
Holgersen, H. 1944a. The ants of Norway (Hymenoptera, Formicidae). Nytt Mag. Naturvidensk. 84:165-203. [1944.09.08]
Holgersen, H. 1944b. Tre maur-arter som bor ettersokes i Sverige (Hym., Formicidae). Entomol. Tidskr. 65:199-202. [1944.12.12]

Holl, F. 1829. Handbuch der Petrefactenkunde. Bd. 2 [part] Dresden: P. O. Hilschersche Buchhandlung, pp. 117-232. [1829]

Hölldobler, B. 1960. Über die Ameisenfauna in Finnland-Lappland. Waldhygiene 3:229-238. [1960]

Hölldobler, B. 1961. Temperaturunabhaengige rhythmische Erscheinungen bei Rossameisenkolonien (*Camponotus ligniperda* Latr. und *Camponotus herculeanus* L.) (Hym. Form.). Insectes Soc. 8:13-22. [1961.12]

Hölldobler, B. 1962a. Zur Frage der Oligogynie bei *Camponotus ligniperda* Latr. und *Camponotus herculeanus* L. (Hym. Formicidae). Z. Angew. Entomol. 49:337-352. [1962.05]

Hölldobler, B. 1962b. Über die forstliche Bedeutung der Rossameisen (*Camponotus ligniperda* Latr. und *Camponotus herculaneus* L. (Hym., Form.)). Waldhygiene 4:228-250. [1962]

Hölldobler, B. 1973. Zur Ethologie der chemischen Verständigung bei Ameisen. Nova Acta Leopold. (n.s.)37(2):259-292. [1973.06.20]

Hölldobler, B. 1976a. Recruitment behavior, home range orientation and territoriality in harvester ants, *Pogonomyrmex*. Behav. Ecol. Sociobiol. 1:3-44. [1976.02.06]

Hölldobler, B. 1976b. Tournaments and slavery in a desert ant. Science (Wash. D. C.) 192:912-914. [1976.05.28]

Hölldobler, B. 1976c. The behavioral ecology of mating in harvester ants (Hymenoptera: Formicidae: *Pogonomyrmex*). Behav. Ecol. Sociobiol. 1:405-423. [1976.12.30]

Hölldobler, B. 1977. Communication in social Hymenoptera. Pp. 418-470 in: Sebeok, T. A. (ed.) How animals communicate. Bloomington, Indiana: Indiana Univ. Press, xxi + 1128 pp. [1977]

Hölldobler, B. 1978. Ethological aspects of chemical communication in ants. Pp. 75-115 in: Rosenblatt, J. S., Hinde, R.A., Beer, C., Busnel, M.-C. (eds.) Advances in the study of behavior. Volume 8. New York: Academic Press, xiv + 261 pp. [1978]

Hölldobler, B. 1980. Canopy orientation: a new kind of orientation in ants. Science (Wash. D. C.) 210:86-88. [1980.10.03]

Hölldobler, B. 1981. Zur Evolution von Rekrutierungssignalen bei Ameisen. Nova Acta Leopold. (n.s.)54:431-447. [1981.12.15]

Hölldobler, B. 1982a. The cloacal gland, a new pheromone gland in ants. Naturwissenschaften 69:186-187. [1982.04]

Hölldobler, B. 1982b ("1981"). Trail communication in the dacetine ant *Orectognathus versicolor* (Hymenoptera: Formicidae). Psyche (Camb.) 88:245-257. [1982.05.28]

Hölldobler, B. 1982c. Chemical communication in ants: new exocrine glands and their behavioral function. Pp. 312-317 in: Breed, M. D., Michener, C. D., Evans, H. E. (eds.) The biology of social insects. Proceedings of the Ninth Congress of the IUSSI, Boulder, Colorado, August 1982. Boulder: Westview Press, xii + 420 pp. [1982.08]

Hölldobler, B. 1982d. Communication, raiding behavior and prey storage in *Cerapachys* (Hymenoptera: Formicidae). Psyche (Camb.) 89:3-23. [1982.12.17]

Hölldobler, B. 1984a. Communication during foraging and nest-relocation in the African stink ant, *Paltothyreus tarsatus* Fabr. (Hymenoptera, Formicidae, Ponerinae). Z. Tierpsychol. 65:40-52. [1984.05]

Hölldobler, B. 1984b. Evolution of insect communication. Pp. 349-377 in: Lewis, T. (ed.) Insect communication. 12th Symposium of the Royal Entomological Society of London. London: Academic Press, xvii + 414 pp. [1984]

Hölldobler, B. 1985 ("1984"). A new exocrine gland in the slave raiding ant genus *Polyergus*. Psyche (Camb.) 91:225-235. [1985.04.22]

Hölldobler, B. 1986 ("1985"). Liquid food transmission and antennation signals in ponerine ants. Isr. J. Entomol. 19:89-99. [1986.02]

Hölldobler, B. 1989 ("1988"). Chemical communication in *Meranoplus* (Hymenoptera: Formicidae). Psyche (Camb.) 95:139-151. [1989.06.13]

Hölldobler, B. 1995. The chemistry of social regulation: multicomponent signals in ant societies. Proc. Natl. Acad. Sci. U. S. A. 92:19-22. [1995.01.03]

Hölldobler, B., Bartz, S. H. 1985. Sociobiology of reproduction in ants. Fortschr. Zool. 31:237-257. [1985]

Hölldobler, B., Engel, H. 1979 ("1978"). Tergal and sternal glands in ants. Psyche (Camb.) 85:285-330. [1979.09.18]

Hölldobler, B., Engel, H., Taylor, R. W. 1982. A new sternal gland in ants and its function in chemical communication. Naturwissenschaften 69:90-91. [1982.02]

Hölldobler, B., Engel-Siegel, H. 1982. Tergal and sternal glands in male ants. Psyche (Camb.) 89:113-132. [1982.12.17]

Hölldobler, B., Engel-Siegel, H. 1985 ("1984"). On the metapleural gland of ants. Psyche (Camb.) 91:201-224. [1985.04.22]

Hölldobler, B., Haskins, C. P. 1977. Sexual calling behavior in primitive ants. Science (Wash. D. C.) 195:793-794. [1977.02.25]

Hölldobler, B., Lumsden, C. J. 1980. Territorial strategies in ants. Science (Wash. D. C.) 210:732-739. [1980.11.14]

Hölldobler, B., Markl, H. 1990 ("1989"). Notes on interspecific, mixed colonies in the harvester ant genus *Pogonomyrmex*. Psyche (Camb.) 96:237-238. [1990.04.20]

Hölldobler, B., Möglich, M. 1980. The foraging system of *Pheidole militicida* (Hymenoptera: Formicidae). Insectes Soc. 27:237-264. [1980.09]

Hölldobler, B., Obermayer, M., Wilson, E. O. 1992. Communication in the primitive cryptobiotic ant *Prionopelta amabilis* (Hymenoptera: Formicidae). J. Comp. Physiol. A. Sens. Neural Behav. Physiol. 171:9-16. [1992.08]

Hölldobler, B., Oldham, N. J., Morgan, E. D., König, W. A. 1995. Recruitment pheromones in the ants *Aphaenogaster albisetosus* and *A. cockerelli* (Hymenoptera: Formicidae). J. Insect Physiol. 41:739-744. [1995.09]

Hölldobler, B., Palmer, J. M. 1989a. A new tarsal gland in ants and the possible role in chemical communication. Naturwissenschaften 76:385-386. [1989.08]

Hölldobler, B., Palmer, J. M. 1989b. Footprint glands in *Amblyopone australis* (Formicidae, Ponerinae). Psyche (Camb.) 96:111-121. [1989.12.18]

Hölldobler, B., Palmer, J. M., Masuko, K., Brown, W. L., Jr. 1989. New exocrine glands in the legionary ants of the genus *Leptanilla* (Hymenoptera: Formicidae: Leptanillinae). Zoomorphology (Berl.) 108:255-262. [1989.01]

Hölldobler, B., Palmer, J. M., Moffett, M. W. 1990. Chemical communication in the dacetine ant *Daceton armigerum* (Hymenoptera: Formicidae). J. Chem. Ecol. 16:1207-1219. [1990.04]

Hölldobler, B., Peeters, C., Obermayer, M. 1994. Exocrine glands and the attractiveness of the ergatoid queen in the ponerine ant *Megaponera foetens*. Insectes Soc. 41:63-72. [1994]

Hölldobler, B., Stanton, R. C., Engel, H. 1976. A new exocrine gland in *Novomessor* (Hymenoptera: Formicidae) and its possible significance as a taxonomic character. Psyche (Camb.) 83:32-41. [1976.11.04]

Hölldobler, B., Stanton, R. C., Markl, H. 1978. Recruitment and food-retrieving behavior in *Novomessor* (Formicidae, Hymenoptera). I. Chemical signals. Behav. Ecol. Sociobiol. 4:163-181. [1978.12.06]

Hölldobler, B., Taylor, R. W. 1984 ("1983"). A behavioral study of the primitive ant *Nothomyrmecia macrops* Clark. Insectes Soc. 30:384-401. [1984.04]

Hölldobler, B., Traniello, J. F. A. 1980. The pygidial gland and chemical recruitment communication in *Pachycondyla* (=*Termitopone*) *laevigata*. J. Chem. Ecol. 6:883-893. [1980.09]

Hölldobler, B., Wilson, E. O. 1971 ("1970"). Recruitment trails in the harvester ant *Pogonomyrmex badius*. Psyche (Camb.) 77:385-399. [1971.10.21]

Hölldobler, B., Wilson, E. O. 1977a. The number of queens: an important trait in ant evolution. Naturwissenschaften 64:8-15. [1977.01]

Hölldobler, B., Wilson, E. O. 1977b. Weaver ants: social establishment and maintenance of territory. Science (Wash. D. C.) 195:900-902. [1977.03.04]

Hölldobler, B., Wilson, E. O. 1977c. Weaver ants. Sci. Am. 237(6):146-154. [1977.12]
Hölldobler, B., Wilson, E. O. 1978. The multiple recruitment systems of the African Weaver ant *Oecophylla longinoda* (Latreille) (Hymenoptera: Formicidae). Behav. Ecol. Sociobiol. 3: 19-60. [1978.01.27]
Hölldobler, B., Wilson, E. O. 1983a. Queen control in colonies of weaver ants (Hymenoptera: Formicidae). Ann. Entomol. Soc. Am. 76:235-238. [1983.03.15]
Hölldobler, B., Wilson, E. O. 1983b. The evolution of communal nest-weaving in ants. Am. Sci. 71:490-499. [1983.10]
Hölldobler, B., Wilson, E. O. 1986a. Soil-binding pilosity and camouflage in ants of the tribes Basicerotini and Stegomyrmecini (Hymenoptera, Formicidae). Zoomorphology (Berl.) 106: 12-20. [1986.01]
Hölldobler, B., Wilson, E. O. 1986b. Ecology and behavior of the primitive cryptobiotic ant *Prionopelta amabilis* (Hymenoptera: Formicidae). Insectes Soc. 33:45-58. [1986.06]
Hölldobler, B., Wilson, E. O. 1986c. Nest area exploration and recognition in leaf cutter ants (*Atta cephalotes*). J. Insect Physiol. 32:143-150. [1986]
Hölldobler, B., Wilson, E. O. 1990. The ants. Cambridge, Mass.: Harvard University Press, xii + 732 pp. [1990.03.28] [Date from publisher.]
Hölldobler, B., Wilson, E. O. 1992. *Pheidole nasutoides*, a new species of Costa Rican ant that apparently mimics termites. Psyche (Camb.) 99:15-22. [1992.12.04]
Hölldobler, K. 1936. Beiträge zur Kenntnis der Koloniegründung der Ameisen. Biol. Zentralbl. 56:230-248. [1936.06]
Hölldobler, K. 1944. Über die forstlich wichtigen Ameisen der nordostkarelischen Urwaldes. Z. Angew. Entomol. 30:587-622. [1944]
Hölldobler, K. 1961. Systematische Klarstellungen zur Ameisenfauna des nordostkarelischen Urwaldes. Z. Angew. Entomol. 48:186-187. [1961.12]
Holliday, M. 1903. A study of some ergatogynic ants. Zool. Jahrb. Abt. Syst. Geogr. Biol. Tiere 19:293-328. [1903.11.24]
Hollingsworth, M. J. 1960. Studies on the polymorphic workers of the army ant *Dorylus* (*Anomma*) *nigricans* Illiger. Insectes Soc. 7:17-37. [1960.06]
Hollingsworth, M. J. 1961. A gynandromorph in the weaver ant, *Oecophylla longinoda* Latr. Entomologist 94:108-111. [1961.05.12]
Holmgren, N. 1908. Über einige myrmecophile Insekten aus Bolivia und Peru. Zool. Anz. 33:337-349. [1908.08.18]
Holt, J. A., Greenslade, P. J. M. 1980 ("1979"). Ants (Hymenoptera, Formicidae) in mounds of *Amitermes laurensis* (Isoptera: Termitidae). J. Aust. Entomol. Soc. 18:349-361. [1980.04.11]
Hölzel, E. 1941. Ameisenstudien und Beobachtungen in der näheren und weiteren Umgebung von Klagenfurt und in den Karawanken. Carinthia II 131:86-120. [1941.09]
Hölzel, E. 1952. Ameisen Kärntens. Carinthia II 142:89-132. [1952.11]
Hölzel, E. 1956. Neue Ameisenbeobachtungen in Kärnten. Carinthia II 146:68-77. [1956.11]
Hölzel, E. 1966. Hymenoptera-Heterogyna: Formicidae. Cat. Faunae Austriae 16p:1-12. [1966]
Holway, D. 1995. Distribution of the Argentine ant (*Linepithema humile*) in northern California. Conserv. Biol. 9:1634-1637. [1995.12]
Hong, Y.-C. 1974. [Untitled. *Eomyrmex* Hong gen. nov. *Eomyrmex guchengziensis* Hong gen. et sp. nov.] Pp. 138-139, 147-148 in: Hong, Y.-C., Yang, T.-C., Wang, S.-T., Wang, S.-E., Li, Y.-K., Sun, M.-R., Sun, H.-C., Tu, N.-C. Stratigraphy and palaeontology of Fushun coal-field, Liaoning Province. Acta Geol. Sin. 48:113-150. [1974.11]
Hong, Y.-C. 1984 ("1983"). Fossil insects in the diatoms of Shanwang. [In Chinese.] Bull. Tianjin Inst. Geol. Miner. Resour. 8:1-12. [1984]
Hong, Y.-C., Yang, T.-C., Wang, S.-T., Wang, S.-E., Li, Y.-K., Sun, M.-R., Sun, H.-C., Tu, N.-C. 1974. Stratigraphy and paleontology of Fushun coal-field, Liaoning Province. [In Chinese and English.] Acta Geol. Sin. 48:113-150. [1974.11]

Hood, W. G., Tschinkel, W. R. 1990. Dessication resistance in arboreal and terrestrial ants. Physiol. Entomol. 15:23-25. [1990.03]

Hope, F. W. 1845. The auto-biography of John Christian Fabricius, translated from the Danish, with additional notes and observations. Trans. Entomol. Soc. Lond. 4:i-xvi. [1845.04.07] [This article is appended to the beginning of volume 4, with pagination separate from the roman pagination of the volume proper (Evenhuis et al., 1989:886).]

Hopkinson, J. 1908 ("1907"). Dates of publication of the separate parts of Gmelin's edition (13th) of the "Systema Naturae" of Linnaeus. Proc. Zool. Soc. Lond. 1907:1035-1037. [1908.06.04]

Horák, J. 1959. Nález gynandromorfa *Polyergus rufescens* Latr. (Hym., Formicidae). Cas. Cesk. Spol. Entomol. 56:202-205. [1959.04.20]

Howard, D. F., Blum, M. S., Jones, T. H., Tomalski, M. D. 1982. Behavioral responses to an alkylpyrazine from the mandibular gland of the ant *Wasmannia auropunctata*. Insectes Soc. 29:369-374. [1982.05]

Howse, P. E., Bradshaw, J. W. S. 1980. Chemical systematics of social insects with particular reference to ants and termites. Pp. 71-90 in: Bisby, F. A., Vaughan, J. G., Wright, C. A. (eds.) Chemosystematics: principles and practice. Systematics Association Special Volume 16. New York: Academic Press, xii + 449 pp. [1980]

Howse, P. E., Clement, J.-L. (eds.) 1981. Biosystematics of social insects. Systematics Association Special Volume No. 19. London: Academic Press, 346 pp. [1981]

Hôzawa, S. 1912. Ants producing sound. [In Japanese.] Dobutsugaku Zasshi (Zool. Mag.) 24:176-177. [1912.03.25]

Hsu, S.-L. 1970. Biometrical study on interspecific differences and affinities of the genus *Formica* L. (Hym. Form.). Bull. Inst. Zool. Acad. Sin. (Taipei) 9:69-81. [1970.12]

Huddleston, E. W., Fluker, S. S. 1968. Distribution of ant species in Hawaii. Proc. Hawaii. Entomol. Soc. 20:45-69. [1968.12.31]

Huddleston, E. W., Laplante, A. A., Fluker, S. S. 1968. Pictorial key of the ants of Hawaii based on the worker forms. Proc. Hawaii. Entomol. Soc. 20:71-79. [1968.12.31]

Hughes, J. 1994. Notable records of ants (Hymenoptera: Formicidae) in south-east Sutherland. Entomol. Rec. J. Var. 106:75-76. [1994.03.25]

Hung, A. C. F. 1962. Preliminary studies on the ants of Taiwan (Formosa). (I). Genus *Polyrhachis* Fr. Smith (Hymenoptera, Formicidae). Bull. Soc. Entomol. Taiwan Prov. Chung-Hsing Univ. 1(1):22-38. [1962.06]

Hung, A. C. F. 1967a. A new species and two new names of the *Polyrhachis* ants (Hymenoptera: Formicidae). Mushi 40:199-202. [1967.03.24]

Hung, A. C. F. 1967b. A revision of the ant genus *Polyrhachis* at the subgeneric level (Hymenoptera: Formicidae). Trans. Am. Entomol. Soc. 93:395-422. [1967.12.20]

Hung, A. C. F. 1969. The chromosome numbers of six species of formicine ants. Ann. Entomol. Soc. Am. 62:455-456. [1969.03.17]

Hung, A. C. F. 1970. A revision of ants of the subgenus *Polyrhachis* Fr. Smith (Hymenoptera: Formicidae: Formicinae). Orient. Insects 4:1-36. [1970.04]

Hung, A. C. F. 1971. On the taxonomic status of *Polyrhachis kirkae* Donisthorpe and its presumed mimicry (Hymenoptera: Formicidae). Entomol. News 82:43-47. [1971.05.28]

Hung, A. C. F. 1974a. A systematic study of the *Formica obscuriventris* subgroup, with notes on colony-founding in the *Formica rufa* group. Diss. Abstr. Int. B. Sci. Eng. 34:3646. [1974.02]

Hung, A. C. F. 1974b ("1973"). Induced mating in *Formica* ants (Hymenoptera: Formicidae). Entomol. News 84:310-313. [1974.05.31]

Hung, A. C. F. 1985. Isozymes of two fire ant species and their hybrid. Biochem. Syst. Ecol. 13:337-339. [1985.08.14]

Hung, A. C. F., Barlin, M. R., Vinson, S. B. 1977. Identification, distribution, and biology of fire ants in Texas. Tex. Agric. Exp. Stn. Bull. 1185:1-24. [1977.10]

Hung, A. C. F., Brown, W. L., Jr. 1966. Structure of gastric apex as a subfamily character of the Formicinae (Hymenoptera: Formicidae). J. N. Y. Entomol. Soc. 74:198-200. [1966.12.29]

Hung, A. C. F., Dowler, M. G., Vinson, S. B. 1979. Electrophoretic variants of α-glycerophosphate dehydrogenase in the fire ant *Solenopsis invicta*. Can. J. Genet. Cytol. 21:537-542. [1979.12]

Hung, A. C. F., Imai, H. T., Kubota, M. 1972. The chromosomes of nine ant species (Hymenoptera: Formicidae) from Taiwan, Republic of China. Ann. Entomol. Soc. Am. 65:1023-1025. [1972.09.15]

Hung, A. C. F., Norton, W., Vinson, S. B. 1975. Gynandromorphism in the red imported fire ant, *Solenopsis invicta* Buren (Hymenoptera: Formicidae). Entomol. News 86:45-46. [1975.10.08]

Hung, A. C. F., Vinson, S. B. 1975. Notes on the male reproductive system in ants (Hymenoptera: Formicidae). J. N. Y. Entomol. Soc. 83:192-197. [1975.09.26]

Hung, A. C. F., Vinson, S. B. 1977. Interspecific hybridization and caste specificity of protein in fire ant. Science (Wash. D. C.) 196:1458-1460. [1977.06.24]

Hung, A. C. F., Vinson, S. B. 1978. Factors affecting the distribution of fire ants in Texas (Myrmicinae: Formicidae). Southwest. Nat. 23:205-214. [1978.03.20]

Hunt, J. H. 1977. Ants. Pp. 195-198 in: Thrower, N. J. W., Bradbury, D. E. (eds.) Chile-California mediterranean scrub atlas. US/IBP synthesis series 2. Stroudsburg, Pennsylvania: Dowden, Hutchinson & Ross, Inc., xv + 237 pp. [1977]

Hunt, J. H. 1983. Foraging and morphology in ants: the role of vertebrate predators as agents of natural selection. Pp. 83-104 in: P. Jaisson (ed.) Social insects in the tropics. Volume 2. Paris: Université Paris-Nord, 252 pp. [1983.07]

Hunt, J. H., Snelling, R. R. 1975. A checklist of the ants of Arizona. J. Ariz. Acad. Sci. 10:20-23. [1975.02]

Huot, L., Francoeur, A. 1968. Étude écologique et systématique des espèces du groupe *fusca*, genre *Formica* (Formicidae, Hymenoptera) pour l'Amérique du Nord. [Abstract.] Rech. Agron. 13:18. [1968]

Huxley, C. R. 1986. Evolution of benevolent ant-plant relationships. Pp. 257-282 in: Juniper, B., Southwood, T. R. E. (eds.) Insects and the plant surface. London: Edward Arnold, viii + 360 pp. [1986]

Huxley, C. R., Cutler, D. F. (eds.) 1991. Ant-plant interactions. Oxford: Oxford University Press, xviii + 601 pp. [1991]

Huxley, J. 1930. Ants. London: E. Benn Ltd., 80 pp. [1930]

Ihering, H. von. 1894. Die Ameisen von Rio Grande do Sul. Berl. Entomol. Z. 39:321-446. [1894.10]

Ihering, H. von. 1907a. Die Cecropien und ihre Schutzameisen. [part] Bot. Jahrb. Syst. Pflanzengesch. Pflanzengeogr. 39:666-670. [1907.01.15]

Ihering, H. von. 1907b. Die Cecropien und ihre Schutzameisen. [concl.] Bot. Jahrb. Syst. Pflanzengesch. Pflanzengeogr. 39:671-714. [1907.02.19]

Ihering, H. von. 1912. Biologie und Verbreitung der brasilianischen Arten von *Eciton*. Entomol. Mitt. 1:226-235. [1912.08.01]

Illiger, K. 1802. Neue Insekten. Mag. Insektenkd. (Illiger) 1:163-208. [1802]

Illiger, K. 1807. Vergleichung der Gattungen der Hautflügler Piezata Fabr. Hymenoptera Linn. Jur. Mag. Insektenkd. (Illiger) 6:189-199. [1807]

Imai, H. T. 1966a. Chromosome observation method of ants by using a squash method. (Part 1.). [In Japanese.] Ari 3:1-2. [1966.06.25]

Imai, H. T. 1966b. The chromosome observation techniques of ants and the chromosomes of Formicinae and Myrmicinae. Acta Hymenopterol. 2:119-131. [1966.09.30]

Imai, H. T. 1966c. Chromosome observation method of ants by using a squash method. (Part 2.). [In Japanese.] Ari 4:1-2. [1966.10.20]

Imai, H. T. 1969. Karyological studies of Japanese ants. I. Chromosome evolution and species differentiation in ants. Sci. Rep. Tokyo Kyoiku Daigaku Sect. B 14:27-46. [1969.09.25]

Imai, H. T. 1971. Karyological studies of Japanese ants II. Species differentiation in *Aphaenogaster*; with special regard to their morphology, distribution and chromosomes. Mushi 44:137-151. [1971.05.05]

Imai, H. T. 1974. B-Chromosomes in the myrmicine ant, *Leptothorax spinosior*. Chromosoma (Berl.) 45:431-444. [1974.05.10]

Imai, H. T. 1986a. Modes of species differentiation and karyotype alteration in ants and mammals. Annu. Rep. Natl. Inst. Genet. Jpn. 36:58-59. [1986]

Imai, H. T. 1986b. Modes of species differentiation and karyotype alteration in ants and mammals. Pp. 87-105 in: Iwatuki, K., Raven, P. H., Bock, W. J. (eds.) Modern aspects of species. Tokyo: University of Tokyo Press, xvii + 240 pp. [1986]

Imai, H. T., Baroni Urbani, C., Kubota, M. 1984. Karyological and cytotaxonomic studies on the 94 species of Indian ants. [In Japanese.] Ari 12:2. [1984.05.01]

Imai, H. T., Baroni Urbani, C., Kubota, M., Sharma, G. P., Narasimhanna, M. H., Das, B. C., Sharma, A. K., Sharma, A., Deodikar, G. B., Vaidya, V. G., Rajasekarasetty, M. R. 1984. Karyological survey of Indian ants. Jpn. J. Genet. 59:1-32. [1984.02.25]

Imai, H. T., Brown, W. L., Jr., Kubota, M., Yong, H.-S., Tho, Y. P. 1984. Chromosome observations on tropical ants from western Malaysia. II. Annu. Rep. Natl. Inst. Genet. Jpn. 34:66-69. [1984]

Imai, H. T., Crozier, R. H., Taylor, R. W. 1977. Karyotype evolution in Australian ants. Chromosoma (Berl.) 59:341-393. [1977.02.23]

Imai, H. T., Hirai, H., Satta, Y., Shiroishi, T., Yamada, M., Taylor, R. W. 1992. Phase specific Ag-staining of nucleolar organizer regions (NORs) and kinetochores in the Australian ant *Myrmecia croslandi*. Jpn. J. Genet. 67:437-447. [1992.12.25]

Imai, H. T., Kubota, M. 1972. Karyological studies of Japanese ants (Hymenoptera, Formicidae) III. Karyotypes of nine species in Ponerinae, Formicinae and Myrmicinae. Chromosoma (Berl.) 37:193-200. [1972.05.12]

Imai, H. T., Kubota, M. 1975. Chromosome polymorphism in the ant *Pheidole nodus*. Chromosoma (Berl.) 51:391-399. [1975.08.11]

Imai, H. T., Kubota, M. 1981. Karyological survey of Indian ants - some preliminary reports. Nucleus (Calcutta) 24:93-96. [1981.08]

Imai, H. T., Kubota, M. 1989 ("1988"). Collection of ants from Lesser Sunda Islands. [Abstract.] [In Japanese.] Ari 16:4. [1989.08]

Imai, H. T., Kubota, M., Brown, W. L., Jr., Ihara, M., Tohari, M., Pranata, R. I. 1985. Chromosome observations on tropical ants from Indonesia. Annu. Rep. Natl. Inst. Genet. Jpn. 35:46-48. [1985]

Imai, H. T., Taylor, R. W. 1986. The exceptionally low chromosome number n=2 in an Australian bulldog ant *Myrmecia piliventris* Smith. Annu. Rep. Natl. Inst. Genet. Jpn. 36:59-61. [1986]

Imai, H. T., Taylor, R. W. 1989. Chromosomal polymorphisms involving telomere fusion, centromeric inactivation and centromere shift in the ant *Myrmecia* (*pilosula*) n=1. Chromosoma (Berl.) 98:456-460. [1989.12]

Imai, H. T., Taylor, R. W., Crosland, M. W. J., Crozier, R. H. 1988. Modes of spontaneous chromosomal mutation and karyotype evolution in ants with reference to the minimum interaction hypothesis. Jpn. J. Genet. 63:159-185. [1988.04.25]

Imai, H. T., Taylor, R. W., Crozier, R. H. 1994. Experimental bases for the minimum interaction theory. I. Chromosome evolution in ants of the *Myrmecia pilosula* species complex (Hymenoptera: Formicidae: Myrmeciinae). Jpn. J. Genet. 69:137-182. [1994.04.25]

Imai, H. T., Taylor, R. W., Crozier, R. H., Crosland, M. W. L., Browning, G. P. 1988. Chromosomal polymorphism in the ant *Myrmecia* (*pilosula*) n=1. Annu. Rep. Natl. Inst. Genet. Jpn. 38:82-84. [1988]

Imai, H. T., Taylor, R. W., Kubota, M., Ogata, K., Wada, M. Y. 1991 ("1990"). Notes on the remarkable karyology of the primitive ant *Nothomyrmecia macrops*, and of the related genus *Myrmecia* (Hymenoptera: Formicidae). Psyche (Camb.) 97:133-140. [1991.02.26]

Imai, H. T., Yosida, T. H. 1965. Chromosome observation in Japanese ants. Annu. Rep. Natl. Inst. Genet. Jpn. 15:64-66. [1965]

Imai, H. T., Yosida, T. H. 1966a. Chromosomes of male ants *Aphaenogaster osimensis* produced by parthenogenesis of workers. Annu. Rep. Natl. Inst. Genet. Jpn. 16:53-54. [1966]

Imai, H. T., Yosida, T. H. 1966b. Polyploid cells observed in male and queen ants of *Aphaenogaster osimensis*. Annu. Rep. Natl. Inst. Genet. Jpn. 16:54. [1966]

Imanishi, K. 1930. An example of vertical distribution, seen in ants. [In Japanese.] Kontyû 4:185-187. [1930.10.31]

Imhoff, L. 1852. Ueber eine Art afrikanischer Ameisen. Ber. Verh. Naturforsch. Ges. Basel 10:175-177. [1852.06]

Imhoff, L. 1854. On a species of African ant. Ann. Mag. Nat. Hist. (2)13:75-76. [1854.01] [Partial translation of Imhoff (1852).]

International Commission on Zoological Nomenclature. 1956. Opinion 424. Validation under the plenary powers of the specific name *rufa* Linnaeus, 1761, as published in the combination *Formica rufa* and designation under the same powers of the species so named to be the type species of the genus *Formica* Linnaeus 1758 (class Insecta, order Hymenoptera). Opin. Declar. Int. Comm. Zool. Nomencl. 14:215-242. [1956.10.12]

International Commission on Zoological Nomenclature. 1976. Opinion 1053. *Formica maxima* Moore, 1842 (Insecta: Hymenoptera): suppressed under the plenary powers. Bull. Zool. Nomencl. 32:244-245. [1976.01.30]

Ipinza, J. See: Ipinza-Regla, J. H.

Ipinza-Regla, J. See: Ipinza-Regla, J. H.

Ipinza-Regla, J. H., Capurro, S. L. 1972. *Heteroponera carinifrons* en Antofagasta. MNHN (Mus. Nac. Hist. Nat.) Not. Mens. 16(192):7. [1972.07]

Ipinza-Regla, J. H., Covarrubias Berríos, R. 1983 ("1982"). Distribución de especies de hormigas de la subfamilia Myrmicinae (Formicidae) en Chile. EOS. Rev. Esp. Entomol. 58:135-141. [1983.03.15]

Ipinza-Regla, J. H., Covarrubias Berríos, R., Ladrón de Guevara, R. F. 1983. Distribución altitudinal de Formicidae en los Andes de Chile central. Folia Entomol. Mex. 55:103-128. [1983.07]

Ipinza-Regla, J. H., Covarrubias, R. 1989. Répartition des espèces de Dolichoderinae (Hymenoptera: Formicidae) au Chili. Ann. Soc. Entomol. Fr. (n.s.)25:377-379. [1989.09.30]

Iredale, T. 1922. Book notes. Proc. Malacol. Soc. 15:78-92. [1922.12.22]

Ito, F. 1992 ("1991"). Preliminary report on queenless reproduction in a primitive ponerine ant *Amblyopone* sp. (*reclinata* group) in West Java, Indonesia. Psyche (Camb.) 98:319-322. [1992.09.04]

Ito, F. 1993. Observation of group recruitment to prey in a primitive ponerine ant, *Amblyopone* sp. (*reclinata* group) (Hymenoptera: Formicidae). Insectes Soc. 40:163-167. [1993]

Ito, F. 1994a ("1993"). Queenless reproduction in a primitive ponerine ant *Amblyopone bellii* (Hymenoptera: Formicidae) in southern India. J. N. Y. Entomol. Soc. 101:574-575. [1994.02.22]

Ito, F. 1994b. Aggressive interactions among gamergates in queenless ponerine ant *Amblyopone* sp. (*reclinata* group). P. 247 in: Lenoir, A., Arnold, G., Lepage, M. (eds.) Les insectes sociaux. 12ème congrès de l'Union Internationale pour l'Étude des Insectes Sociaux. Paris, Sorbonne, 21-27 août 1994. Paris: Université Paris Nord, xxiv + 583 pp. [1994.09]

Ito, F. 1995 ("1994"). Colony composition of two Malaysian ponerine ants, *Platythyrea tricuspidata* and *P. quadridenta*: sexual reproduction by workers and production of queens (Hymenoptera: Formicidae). Psyche (Camb.) 101:209-218. [1995.11.20]

Ito, F., Higashi, S. 1990. Temporary social parasitism in the enslaving ant species *Formica sanguinea* Latreille: an important discovery related to the evolution of dulosis in *Formica* ants. J. Ethol. 8:33-35. [1990]

Ito, T. 1912. [Untitled. *Camponotus Itoi* (*Myrmamblys*) Forel stirps *tokioensis* Ito nov. st.] Pp. 341-342 in: Forel, A. Quelques fourmis de Tokio. Ann. Soc. Entomol. Belg. 56:339-342. [1912.10.03]

Ito, T. 1914a. Formicidarum japonicarum species novae vel minus cognitae. Ann. Soc. Entomol. Belg. 58:40-45. [1914.02.06]

Ito, T. 1914b. *Crematogaster Auberti* Em. var. *Nawai* nov. var. [In Japanese.] Insect World 18:135-138. [1914.04.15]

Ito, T. 1921. On the ant *Camponotus fallax* var. *nawai*. [In Japanese.] Insect World 25:3. [1921.01.15]

Itow, T., Kobayashi, K., Kubota, M., Ogata, K., Imai, H. T., Crozier, R. H. 1984. The reproductive cycle of the queenless ant *Pristomyrmex pungens*. Insectes Soc. 31:87-102. [1984.07]

Jackson, B. D., Billen, J. P. J., Morgan, E. D. 1989. Dufour gland contents of three species of *Myrmecia* (Hymenoptera: Formicidae), primitive ants of Australia. J. Chem. Ecol. 15:2191-2205. [1989.08]

Jackson, B. D., Keegans, S. J., Morgan, E. D., Cammaerts, M.-C., Cammaerts, R. 1990. Trail pheromone of the ant *Tetramorium meridionale*. Naturwissenschaften 77:294-296. [1990.06]

Jackson, B. D., Keegans, S. J., Morgan, E. D., Clark, W. H., Blom, P. E. 1991. Chemotaxonomic study of undescribed species of *Myrmica* ant from Idaho. J. Chem. Ecol. 17:335-342. [1991.02]

Jackson, B. D., Morgan, E. D., Collingwood, C. A. 1989. The chemical secretions of *Myrmica specioides* Bondroit and *Myrmica gallieni* Bondroit (Myrmicinae). Actes Colloq. Insectes Soc. 5:315-321. [1989.06]

Jackson, D. A. 1984. Ant distributions in a Cameroonian cocoa plantation: investigation of the ant mosaic hypothesis. Oecologia (Berl.) 62:318-324. [1984.06]

Jacobson, E. 1909. Ameisen aus Java und Krakatau beobachtet und gesammelt von Herrn Edward Jacobson. II. Biologischer Theil. Notes Leyden Mus. 31:233-251. [1909.12.20] [See also Forel (1909j).]

Jacobson, E. 1912. Ameisen aus Java beobachtet und gesammelt von Edward Jacobson, bestimmt und beschrieben von Dr. A. Forel. Biologische Beobachtungen. Notes Leyden Mus. 34:113-122. [1912.04.01] [See also Forel (1912e).]

Jacobson, H. 1936. Die Ameisenfauna der Kanjerseemore. Beitrag zur Fauna Ost-Baltischer Hochmoore. Folia Zool. Hydrobiol. 9:143-165. [1936.05.23]

Jacobson, H. 1939. Die Ameisenfauna des Ostbaltischen Gebietes. Z. Morphol. Ökol. Tiere 35:389-454. [1939.06.24]

Jacobson, H. 1940a. Beitrag zur Ameisenfauna Lapplands. Zool. Anz. 129:171-176. [1940.02.15]

Jacobson, H. 1940b. Mitteilungen zur Ameisenfauna Pommerns sowie über das Vorkommen einer für Deutschland neuen Art: *Myrmica rolandi* Bondr. Zool. Anz. 131:145-150. [1940.08.15]

Jacobson, H. R., Kistner, D. H. 1992. Cladistic study, taxonomic restructuring, and revision of the myrmecophilous tribe Crematoxenini with comments on its evolution and host relationships (Coleoptera: Staphylinidae; Hymenoptera: Formicidae). Sociobiology 20:91-201. [1992]

Jaeger, E. 1933. Zur Kenntnis der Hymenoptera aculeata des Sotlatales (Jugoslavien). I. Konowia 12:98-102. [1933.08.25] [Description of *Leptothorax clypeatus* male, p. 101.]

Jaffe, K. 1984. Negentropy and the evolution of chemical recruitment in ants. J. Theor. Biol. 106:587-604. [1984.02.21]

Jaffe, K. 1985 ("1984"). Evolución de los sistemas de comunicación química en hormigas (Hymenoptera, Formicidae). Folia Entomol. Mex. 61:189-203. [1985.06.28]

Jaffe, K. 1993. El mundo de las hormigas. Baruta, Venezuela: Equinoccio (Ediciones de la Universidad Simón Bolívar), 188 pp. [1993.07]

Jaffe, K., Lattke, J. E. 1994. Ant fauna of the French and Venezuelan islands in the Caribbean. Pp. 181-190 in: Williams, D. F. (ed.) Exotic ants. Biology, impact, and control of introduced species. Boulder: Westview Press, xvii + 332 pp. [1994]

Jaffe, K., Lattke, J. E., Pérez, R. 1993. Ants on the tepuys of the Guiana Shield: a zoogeographic study. Ecotrópicos 6:22-29. [Not seen. Citation from Lattke, pers. comm.]

Jaffe, K., Lopez, M. E., Aragort, W. 1986. On the communication systems of the ants *Pseudomyrmex termitarius* and *P. triplarinus*. Insectes Soc. 33:105-117. [1986.09]

Jaffe, K., Perez, E. 1989. Comparative study of brain morphology in ants. Brain Behav. Evol. 33:25-33. [1989.04]

Jaffe, K., Romero, H., Lattke, J. E. 1989. Mirmecofauna de los tepuyes Marahuaka y Huachamakare (Territorio Federal Amazonas, Venezuela). Acta Terramaris 1:33-37. [1989.04]

Jaffe, K., Vasquez, C., Brandwijk, L., Cabrera, A. 1994. Metapleural gland secretions in ants. P. 217 in: Lenoir, A., Arnold, G., Lepage, M. (eds.) Les insectes sociaux. 12ème congrès de l'Union Internationale pour l'Étude des Insectes Sociaux. Paris, Sorbonne, 21-27 août 1994. Paris: Université Paris Nord, xxiv + 583 pp. [1994.09]

Jagodzinska, Z. 1933. Mrówki okolic Grodna. Pr. Tow. Przyj. Nauk Wilnie Wydz. Nauk Mat. Przyr. 7:273-289. [1933]

Jaisson, P. 1970. Note préliminaire sur le polymorphisme sensoriel et l'existence d'un nouveau type de sensillum chez la fourmi champignonniste: *Atta laevigata* Fred. Smith. C. R. Séances Acad. Sci. Ser. D. Sci. Nat. 271:1192-1194. [1970.10.12]

Jaisson, P., Fresneau, D., Taylor, R. W., Lenoir, A. 1992. Social organization in some primitive Australian ants. I. *Nothomyrmecia macrops* Clark. Insectes Soc. 39:425-438. [1992]

Jaisson, P., Nicolosi, F., Taylor, R. W., Fresneau, D. 1990. Phylogeny of the Formicidae and the behavioural analysis of some archaic Australian ants. Pp. 311-312 in: Veeresh, G. K., Mallik, B., Viraktamath, C. A. (eds.) Social insects and the environment. Proceedings of the 11th International Congress of IUSSI, 1990. New Delhi: Oxford & IBH Publishing Co., xxxi + 765 pp. [1990.08]

Jakubisiak, S. 1948. Mrówki okolic Przybyszewa (poludniowe Mazowsze). Studium ekologiczne. Ann. Univ. Mariae Curie-Sklodowska Sect. C Biol. 3:319-353. [1948.11.02]

Janet, C. 1893. Note sur la production des sons chez les fourmis et sur les organes qui les produisent. Ann. Soc. Entomol. Fr. 62:159-167. [1893.07.31]

Janet, C. 1894a. Étude sur les fourmis. 6e note. Sur l'appareil de stridulation de *Myrmica rubra* L. Ann. Soc. Entomol. Fr. 63:109-117. [1894.04.10]

Janet, C. 1894b. Sur les nerfs de l'antenne et les organes chordotonaux chez les fourmis. C. R. Séances Acad. Sci. 118:814-817. [1894.04.11]

Janet, C. 1894c. Sur le système glandulaire des fourmis. C. R. Séances Acad. Sci. 118:989-992. [1894.05.07]

Janet, C. 1894d. Études sur les fourmis. (Quatrième note.) *Pelodera* des glandes pharyngiennes de *Formica rufa* L. Mém. Soc. Zool. Fr. 7:45-62. [1894]

Janet, C. 1894e. Études sur les fourmis. 5e note. Sur la morphologie du squelette des segments post-thoraciques chez les Myrmicides (*Myrmica rubra* L. femelle). Mém. Soc. Acad. Archéol. Sci. Arts Dép. Oise 15:591-611. [1894]

Janet, C. 1894f. Études sur les fourmis. (Septième note.) Sur l'anatomie du petiole de *Myrmica rubra* L. Mém. Soc. Zool. Fr. 7:185-202. [1894]

Janet, C. 1895a ("1894"). Études sur les fourmis. 8e note. Sur l'organe de nettoyage tibio-tarsien de *Myrmica rubra* L., race *levinodis* Nyl. Ann. Soc. Entomol. Fr. 63:691-704. [1895.05.30]

Janet, C. 1895b. Sur les muscles des fourmis, des guêpes et des abeilles. C. R. Séances Acad. Sci. 121:610-613. [1895.11.04]

Janet, C. 1895c. Études sur les fourmis, les guêpes et les abeilles. 12me note. Structure des membranes articulaires des tendons et des muscles (*Myrmica*, *Camponotus*, *Vespa*, *Apis*). Limoges: H. Ducourtieux, 25 pp. [1895]

Janet, C. 1896. Les fourmis. Bull. Soc. Zool. Fr. 21:60-93. [1896]

Janet, C. 1897a. Études sur les fourmis, les guêpes et les abeilles. Note 13. Sur le *Lasius mixtus* l'*Antennophorus ulhmanni*, etc. Limoges: H. Ducourtieux, 62 pp. [1897]

Janet, C. 1897b. Études sur les fourmis, les guêpes et les abeilles. Note 16. Limites morphologiques des anneaux post-céphaliques et musculature des anneaux post-thoraciques chez la *Myrmica rubra*. Lille: Le Bigot Frères, 36 pp. [1897]

Janet, C. 1898a. Sur les limites morphologiques des anneaux du tégument et sur la situation des membranes articulaires chez les Hyménoptères arrivés à l'état d'imago. C. R. Séances Acad. Sci. 126:435-439. [1898.02.07]

Janet, C. 1898b. Sur une cavité du tégument servant, chez les Myrmicinae, à étaler, au contact de l'air, un produit de sécrétion. C. R. Séances Acad. Sci. 126:1168-1171. [1898.04.25]

Janet, C. 1898c. Sur un organe non décrit, servant à la fermeture du réservoir du venin et sur le mode de fonctionnement de l'aiguillon chez les fourmis. C. R. Séances Acad. Sci. 127:638-641. [1898.10.31]

Janet, C. 1898d. Études sur les fourmis, les guêpes et les abeilles. Note 17. Système glandulaire tégumentaire de la *Myrmica rubra*. Observations diverses sur les fourmis. Paris: G. Carré et C. Naud, 30 pp. [1898]

Janet, C. 1898e. Études sur les fourmis, les guêpes et les abeilles. Note 18. Aiguillon de la *Myrmica rubra*. Appareil de fermeture de la glande à venin. Paris: G. Carré et C. Naud, 27 pp. [1898]

Janet, C. 1898f. Études sur les fourmis, les guêpes et les abeilles. (19me Note.) Anatomie du corselet de la *Myrmica rubra* reine. Mém. Soc. Zool. Fr. 11:393-450. [1898]

Janet, C. 1899a. Sur le mécanisme du vol chez les insectes. C. R. Séances Acad. Sci. 128:249-253. [1899.01.30]

Janet, C. 1899b. Sur les nerfs céphaliques, les corpora allata et le tentorium de la fourmi (*Myrmica rubra* L.). Mém. Soc. Zool. Fr. 12:295-335. [1899]

Janet, C. 1902. Anatomie du gaster de la *Myrmica rubra*. Paris: G. Carré et C. Naud, 68 pp. [1902]

Janet, C. 1904. Observations sur les fourmis. Limoges: Ducourtieux et Gout, 68 pp. [1904]

Janet, C. 1905. Anatomie de la tête du *Lasius niger* reine. Limoges: Ducourtieux et Gout, 40 pp. [1905]

Janet, C. 1906. Sur un organe non décrit du thorax des fourmis ailées. C. R. Séances Acad. Sci. 143:522-524. [1906.10.15]

Janet, C. 1907. Anatomie du corselet et histolyse des muscles vibrateurs, après le vol nuptial, chez la reine de la fourmi (*Lasius niger*). Limoges: Ducourtieux et Gout, 149 pp. [1907]

Janet, C. 1909. Sur la morphologie de l'insecte. Limoges: Ducourtieux et Gout, 75 pp. [1909]

Janzen, D. H. 1966. Coevolution of mutualism between ants and acacias in Central America. Evolution 20:249-275. [1966.10.10]

Janzen, D. H. 1967. Interaction of the bull's-horn acacia (*Acacia cornigera* L.) with an ant inhabitant (*Pseudomyrmex ferruginea* F. Smith) in eastern Mexico. Univ. Kans. Sci. Bull. 47:315-558. [1967.10.11]

Janzen, D. H. 1969. Birds and the ant x acacia interaction in Central America, with notes on birds and other myrmecophytes. Condor 71:240-256. [1969.08.25]

Janzen, D. H. 1973a. Dissolution of mutualism between *Cecropia* and its *Azteca* ants. Biotropica 5:15-28. [1973.04]

Janzen, D. H. 1973b. Evolution of polygynous obligate acacia-ants in western Mexico. J. Anim. Ecol. 42:727-750. [1973.10]

Janzen, D. H. 1974. Swollen-thorn acacias of Central America. Smith. Contrib. Bot. 13:1-131. [1974.04.23]

Janzen, D. H. 1975. *Pseudomyrmex nigropilosa*: a parasite of a mutualism. Science (Wash. D. C.) 188:936-937. [1975.05.30]

Janzen, D. H. 1983. *Pseudomyrmex ferruginea* (hormiga del cornizuelo, acacia-ant). Pp. 762-764 in: Janzen, D. H. (ed.) Costa Rican natural history. Chicago: University of Chicago Press, xi + 816 pp. [1983]

Járdán, C., Gallé, L., Margóczi, K. 1993. Ant assemblage composition in a successional Hungarian sand dune area. Tiscia (Szeged) 27:9-15. [1993] [May have been published in 1994. Journal received at UC Berkeley library 9 January 1995.]

Jayasuriya, A. K., Traniello, J. F. A. 1986 ("1985"). The biology of the primitive ant *Aneuretus simoni* (Emery) (Formicidae: Aneuretinae). I. Distribution, abundance, colony structure, and foraging ecology. Insectes Soc. 32:363-374. [1986.06]

Jeanne, R. L. 1979. A latitudinal gradient in rates of ant predation. Ecology 60:1211-1224. [1979.12]

Jeffery, H. G. 1931. The Formicidae (or ants) of the Isle of Wight. Proc. Isle Wight Nat. Hist. Archaeol. Soc. 2:125-128. [1931]

Jekel, H. 1854. Fabricia entomologica. Première partie. 1 Livraison. Montmarte: published by the author, pp. 1-96. [1854]

Jell, P. A., Duncan, P. M. 1986. Invertebrates, mainly insects, from the freshwater, Lower Cretaceous, Koonwarra fossil bed (Korumburra group), South Gippsland, Victoria. Pp. 111-205 in: Jell, P. A., Roberts, J. (eds.) Plants and invertebrates from the Lower Cretaceous Koonwarra fossil bed, South Gippsland, Victoria. Mem. Assoc. Australas. Palaeontol. 3:i-x, 1-205. [1986] [*Cretacoformica explicata*, gen. et sp. nov., pp. 190-191. See also Naumann (1993).]

Jennings, D. T., Houseweart, M. W., Francoeur, A. 1986. Ants (Hymenoptera: Formicidae) associated with strip-clearcut and dense spruce-fir forests of Maine. Can. Entomol. 118:43-50. [1986.01]

Jensen, T. F., Nielson, M. G. 1982. En status over udbredelsen af myreslaegten *Camponotus* i Danmark (Hymenoptera: Formicidae). Entomol. Medd. 49:113-116. [1982]

Jensen, T. F., Skott, C. 1980. Danske myrer. Nat. Mus. (Århus) 20(1):3-29. [1980]

Jerdon, T. C. 1851. A catalogue of the species of ants found in Southern India. Madras J. Lit. Sci. 17:103-127. [1851]

Jerdon, T. C. 1854a. A catalogue of the species of ants found in Southern India. [part] Ann. Mag. Nat. Hist. (2)13:45-56. [1854.01]

Jerdon, T. C. 1854b. A catalogue of the species of ants found in Southern India. [concl.] Ann. Mag. Nat. Hist. (2)13:100-110. [1854.02]

Jerdon, T. C. 1865a. [Untitled. Introduced by: "Mr. F. Smith read the following letter from Dr. T. C. Jerdon, dated "Lahore, March 16, 1865:""] Trans. Entomol. Soc. Lond. (3)2(Proceedings):93. [1865.09.27]

Jerdon, T. C. 1865b. [Untitled. Introduced by: "Reliable information has at last been supplied by Dr. Jerdon regarding workers of one of the Indian species of *Dorylus*".] J. Asiat. Soc. Bengal Part II Phys. Sci. 34:189-190. [1865.10.23] [Reprint of letter published in Trans. Entomol. Soc. Lond. (3)2.]

Jermiin, L. S., Crozier, R. H. 1994a. The cytochrome b region in the mitochondrial DNA of the ant *Tetraponera rufoniger*: sequence divergence in Hymenoptera may be associated with nucleotide content. J. Mol. Evol. 38:282-294. [1994.03]

Jermiin, L. S., Crozier, R. H. 1994b. The phylogenetic relationship among seven ant subfamilies: a molecular study. P. 41 in: Lenoir, A., Arnold, G., Lepage, M. (eds.) Les insectes sociaux. 12ème congrès de l'Union Internationale pour l'Étude des Insectes Sociaux. Paris, Sorbonne, 21-27 août 1994. Paris: Université Paris Nord, xxiv + 583 pp. [1994.09]

Jessen, K. 1987a. Biosystematic revision of the parasitic ant genus *Epimyrma*. Pp. 41-42 in: Eder, J., Rembold, H. (eds.) Chemistry and biology of social insects. München: Verlag J. Peperny, xxxv + 757 pp. [1987]

Jessen, K. 1987b. Gastral exocrine glands in ants - functional and systematical aspects. Pp. 445-446 in: Eder, J., Rembold, H. (eds.) Chemistry and biology of social insects. München: Verlag J. Peperny, xxxv + 757 pp. [1987]

Jessen, K., Buschinger, A. 1994. Specificity of chemical communication during slave raids in dulotic ants, the genera *Epimyrma* Emery and *Myrmoxenus* Ruzsky (Hymenoptera, Formicidae). Memorabilia Zool. 48:109-114. [1994]

Jessen, K., Klinkicht, M. 1990. Hybridization in the social parasitic ant genus *Epimyrma* (Hymenoptera, Formicidae). Insectes Soc. 37:273-293. [1990.12]

Jessen, K., Maschwitz, U. 1983. Abdominaldrüsen bei *Pachycondyla tridentata* (Smith): Formicidae, Ponerinae. Insectes Soc. 30:123-133. [1983.09]

Jessen, K., Maschwitz, U., Hahn, M. 1979. Neue Abdominaldrüsen bei Ameisen. I. Ponerini (Formicidae: Ponerinae). Zoomorphologie 94:49-66. [1979.12]

Jiménez Rojas, J., Tinaut, A. 1992. Mirmecofauna de la Sierra de Loja (Granada) (Hymenoptera, Formicidae). Orsis (Org. Sist.) 7:97-111. [1992]

Johnson, C. 1986. A north Florida ant fauna (Hymenoptera: Formicidae). Insecta Mundi 1:243-246. [1986.12]

Johnson, C. 1987. Biogeography and habitats of *Ponera exotica* (Hymenoptera: Formicidae). J. Entomol. Sci. 22:358-361. [1987.10]

Johnson, C. 1988. Species identification in the eastern *Crematogaster* (Hymenoptera: Formicidae). J. Entomol. Sci. 23:314-332. [1988.11.03]

Johnson, C. 1989a. Identification and nesting sites of North American species of *Dolichoderus* Lund (Hymenoptera: Formicidae). Insecta Mundi 3:1-9. [1989.03]

Johnson, C. 1989b. Taxonomy and diagnosis of *Conomyrma insana* (Buckley) and *C. flava* (McCook) (Hymenoptera: Formicidae). Insecta Mundi 3:179-194. [1989.09]

Johnson, N. F. 1988. Midcoxal articulations and the phylogeny of the order Hymenoptera. Ann. Entomol. Soc. Am. 81:870-881. [1988.11]

Johnson, R. A. 1995 ("1994"). Distribution and natural history of the workerless inquiline ant *Pogonomyrmex anergismus* Cole (Hymenoptera: Formicidae). Psyche (Camb.) 101:257-262. [1995.11.20]

Johnson, W. F. 1914. A teratological specimen of *Myrmica rubra*. Ir. Nat. 23:94. [1914.04]

Jolivet, P. 1986a. Les fourmis et les plantes. Un exemple de coévolution. Paris: Boubée, 254 pp. [1986.02]

Jolivet, P. 1986b. Insects and plants. Parallel evolution and adaptations. New York: Brill (Flora & Fauna Publications), xii + 197 pp. [1986]

Jones, J. W. 1943. Known distribution of the shining slave maker ant *Polyergus lucidus* Mayr. Am. Midl. Nat. 29:185. [1943.02.27]

Jones, S. R., Phillips, S. A., Jr. 1985. Gynandromorphism in the ant *Pheidole dentata* Mayr (Hymenoptera: Formicidae). Proc. Entomol. Soc. Wash. 87:583-586. [1985.07.11]

Jones, T. H. 1987. Novel pyrrolizidine alkaloids from myrmecine [sic] ants native to New Zealand. Pp. 417-418 in: Eder, J., Rembold, H. (eds.) Chemistry and biology of social insects. München: Verlag J. Peperny, xxxv + 757 pp. [1987]

Jones, T. H., Blum, M. S., Fales, H. M. 1982. Ant venom alkaloids from *Solenopsis* and *Monomorium* species. Recent developments. Tetrahedron 38:1949-1958. [1982]

Jones, T. H., Blum, M. S., Fales, H. M., Brandão, C. R. F., Lattke, J. E. 1991. Chemistry of venom alkaloids in the ant genus *Megalomyrmex*. J. Chem. Ecol. 17:1897-1908. [1991.09]

Jones, T. H., Blum, M. S., Howard, R. W., McDaniel, C. A., Fales, H. M., DuBois, M. B., Torres, J. 1982. Venom chemistry of ants in the genus *Monomorium*. J. Chem. Ecol. 8:285-300. [1982.01]

Jones, T. H., Blum, M. S., Robertson, H. G. 1990. Novel dialkylpiperidines in the venom of the ant *Monomorium delagoense*. J. Nat. Prod. (Lloydia) 53:429-435. [1990.05.14]

Jones, T. H., DeVries, P. J., Escoubas, P. 1991. Chemistry of venom alkaloids in the ant *Megalomyrmex foreli* (Myrmicinae) from Costa Rica. J. Chem. Ecol. 17:2507-2518. [1991.12]

Jones, T. H., Fales, H. M. 1983. E-6-(1-pentenyl)-2H-pyran-2-one from carpenter ants (*Camponotus* spp.). Tetrahedron Lett. 24:5439-5440. [1983]

Jones, T. H., Highet, R. J., Blum, M. S., Fales, H. M. 1984. (5Z,9Z)-3-alkyl-5-methylindolizidines from *Solenopsis* (*Diplorhoptrum*) species. J. Chem. Ecol. 10:1233-1249. [1984.08]

Jones, T. H., Highet, R. J., Don, A. W., Blum, M. S. 1986. Alkaloids of the ant *Chelaner antarcticus*. J. Org. Chem. 51:2712-2716. [1986.07.11]

Jones, T. H., Laddago, A., Don, A. W., Blum, M. S. 1990. A novel (5E,9Z)-dialkylindolizidine from the ant *Monomorium smithii*. J. Nat. Prod. (Lloydia) 53:375-381. [1990.05.14]

Jones, T. H., Stahly, S. M., Don, A. W., Blum, M. S. 1988. Chemotaxonomic implications of the venom chemistry of some *Monomorium "antarcticum"* populations. J. Chem. Ecol. 14:2197-2212. [1988.12]
Jonkman, J. C. M. 1979. Distribution and densities of nests of the leaf-cutting ant *Atta vollenweideri* Forel, 1983 in Paraguay. Z. Angew. Entomol. 88:27-43. [1979.06]
Jordan, K. H. C. 1968. Die Ameisenfauna der Oberlausitz. Abh. Ber. Naturkundemus. Görlitz 43(3):1-19. [1968]
Joseph, G. 1882. Systematisches Verzeichniss der in den Tropfstein-Grotten von Krain einheimischen Arthropoden nebst Diagnosen der vom Verfasser entdeckten und bisher noch nicht beschriebenen Arten. Berl. Entomol. Z. 26:1-50. [1882.12]
Jucci, C. 1963. Sull'evoluzione del parassitismo sociale nelle formiche. Symp. Genet. Biol. Ital. 12:389-422. [1963]
Jurine, L. 1801. [Untitled.] In: Panzer, G. W. F. Nachricht von einem neuen entomolischen [sic] Werke, des Hrn. Prof. Jurine in Geneve. Intelligenzbl. Litt.-Ztg. Erlangen 1:161-165. [1801.05.30] [See also Morice & Durrant (1915).]
Jurine, L. 1807. Nouvelle méthode de classer les Hyménoptères et les Diptères. Hyménoptères. Vol. 1. Genève: Paschoud, 319 pp. [1807] [Ants pp. 269-283.]
Jusino-Atresino, R., Phillips, S. A., Jr. 1992. New ant records for Taylor Co., Texas. Southwest. Nat. 37:430-433. [1992.12]
Jusino-Atresino, R., Phillips, S. A., Jr. 1994. Impact of red imported fire ants on the ant fauna of central Texas. Pp. 259-268 in: Williams, D. F. (ed.) Exotic ants. Biology, impact, and control of introduced species. Boulder: Westview Press, xvii + 332 pp. [1994]
Kannowski, P. B. 1954. Notes on the ant *Novomessor manni* Wheeler and Creighton. Occas. Pap. Mus. Zool. Univ. Mich. 556:1-6. [1954.06.30]
Kannowski, P. B. 1956a. The ants of Ramsey County, North Dakota. Am. Midl. Nat. 56:168-185. [1956.07]
Kannowski, P. B. 1956b. Ecological distribution of certain bog ants of southeastern Michigan. [Abstract.] Proc. North Cent. Branch Entomol. Soc. Am. 11:54. [1956]
Kannowski, P. B. 1958 ("1957"). Notes on the ant *Leptothorax provancheri* Emery. Psyche (Camb.) 64:1-5. [1958.01.10]
Kannowski, P. B. 1959a. The flight activities of *Dolichoderus* (*Hypoclinea*) *taschenbergi* (Hymenoptera: Formicidae). Ann. Entomol. Soc. Am. 52:755-760. [1959.11.30]
Kannowski, P. B. 1959b. The flight activities and colony-founding behavior of bog ants in southeastern Michigan. Insectes Soc. 6:115-162. [1959.12]
Kannowski, P. B. 1962. The flight activities of formicine ants. [Abstract.] Proc. N. D. Acad. Sci. 16:34-35. [1962.07]
Kannowski, P. B. 1963. The flight activities of formicine ants. Symp. Genet. Biol. Ital. 12:74-102. [1963]
Kannowski, P. B. 1967. Colony populations of two species of *Dolichoderus* (Hymenoptera: Formicidae). Ann. Entomol. Soc. Am. 60:1246-1252. [1967.11.15]
Kannowski, P. B. 1969. Daily and seasonal periodicities in the nuptial flights of neotropical ants. I. Dorylinae. Pp. 77-83 in: Ernst, E. et al. (eds.) International Union for the Study of Social Insects. VI Congress. Bern 15-20 September 1969. Proceedings. Bern: Organizing Committee of the VI Congress IUSSI, 309 pp. [1969.09.15]
Kaplin, V. G. 1989. On the fauna and ecology of ants (Hymenoptera, Formicidae) in the Repetek Biosphere Reserve (eastern Karakum). [In Russian.] Izv. Akad. Nauk Turkm. SSR Ser. Biol. Nauk 1989(3):40-47. [1989.06.21]
Kappel, A. W. 1896. General index to the first twenty volumes of the Journal (Zoology) and the zoological portions of the Proceedings, November 1838 to 1890, of the Linnean Society. London: Longmans, Green, and Co., viii + 437 pp. [1896]
Karavaiev, V. 1906. Systematisch-Biologisches über drei Ameisen aus Buitenzorg. Z. Wiss. Insektenbiol. 2:369-376. [1906.12.30]

Karavaiev, V. 1910. Nachtrag zu meinen "Ameisen aus Transcaspien und Turkestan". Rus. Entomol. Obozr. 9:268-272. [1910.03.10]

Karavaiev, V. 1911a ("1910"). Ameisen aus Transkaspien und Turkestan. Tr. Rus. Entomol. Obshch. 39:1-72. [1911.01.13] [Corrected (Gregorian calendar) date.]

Karavaiev, V. 1911b. Ameisen aus Aegypten und dem Sudan. Rus. Entomol. Obozr. 11:1-12. [1911.05.29]

Karavaiev, V. 1912a. Ameisen aus Tunesien und Algerien, nebst einigen unterwegs in Italien gesammelten Arten. Rus. Entomol. Obozr. 12:1-22. [1912.04.20]

Karavaiev, V. 1912b. Ameisen aus dem paläarktischen Faunengebiete. Rus. Entomol. Obozr. 12:581-596. [1912.12.28]

Karavaiev, V. 1913. Through the islands of the Malayan archipelago, the Moluccas and Aru with a myrmecological aim. General impressions and natural observations. [In Russian.] Izv. Imp. Rus. Geogr. Obshch. 49:395-522. [1913]

Karavaiev, V. 1914. Eine neue Weberameise, *Polyrhachis armata* le Guillou. Biol. Centralbl. 34:440-444. [1914.07.20]

Karavaiev, V. 1916a ("1915"). Ants from Gadjatsh district of the government of Poltava and from the Province of Ferghana. [In Russian.] Rus. Entomol. Obozr. 15:496-507. [1916.03.28]

Karavaiev, V. 1916b. Scientific results of the expedition of the brothers Kuznecov (Kouznetzov) to the Arctic Ural in 1909, under the direction of H. Backlund. Formicidae. [In Russian.] Zap. Imp. Akad. Nauk Fiz.-Mat. Otd. 28(17):1-4. [1916]

Karavaiev, V. 1924. Zur Systematik der paläarktischen *Myrmecocystus* (Formicidae), nebst einigen biologischen Notizen. Konowia 3:301-308. [1924.10.31]

Karavaiev, V. 1925a. Ponerinen (Fam. Formicidae) aus dem Indo-Australischen Gebiet. Konowia 4:69-81. [1925.03.15]

Karavaiev, V. 1925b. Ponerinen (Fam. Formicidae) aus dem Indo-Australischen Gebiet. (Fortsetzung). Konowia 4:115-131. [1925.07.01]

Karavaiev, V. 1925c. Ponerinen (Fam. Formicidae) aus dem Indo-Australischen Gebiet. (Schluss). Konowia 4:276-296. [1925.10.25]

Karavaiev, V. 1926a. Über den Nestbau von *Polyrhachis* (subg. *Myrmhopla*) *tubifex* sp. n. (Fam. Formicidae). Biol. Zentralbl. 46:143-145. [1926.03]

Karavaiev, V. 1926b. Beiträge zur Ameisenfauna des Kaukasus, nebst einigen Bemerkungen über andere palaearktische Formen. Konowia 5:93-109. [1926.06.10]

Karavaiev, V. 1926c. Beiträge zur Ameisenfauna des Kaukasus, nebst einigen Bemerkungen über andere palaearktische Formen. (Fortsetzung). Konowia 5:161-169. [1926.07.28]

Karavaiev, V. 1926d. Ameisen aus dem Indo-Australischen Gebiet. Treubia 8:413-445. [1926.07]

Karavaiev, V. 1926e. Beiträge zur Ameisenfauna des Kaukasus, nebst einigen Bemerkungen über andere palaearktische Formen. (Schluss). Konowia 5:187-199. [1926.11.20]

Karavaiev, V. 1926f. Myrmekologische Fragmente. Zb. Prats Zool. Muz. 1:47-51 [= Tr. Ukr. Akad. Nauk Fiz.-Mat. Vidd. 4:65-69]. [1926]

Karavaiev, V. 1927a ("1926"). Übersicht der Ameisenfauna der Krim nebst einigen Neubeschreibungen. Konowia 5:281-303. [1927.01.28]

Karavaiev, V. 1927b. Ein Fall von lateralem Hermaphroditismus bei Ameisen und ein Fall defekter Körperbildung. Folia Myrmecol. Termit. 1:45-47. [1927.02]

Karavaiev, V. 1927c ("1926"). Myrmekologische Miszellen. Ezheg. Zool. Muz. 27:104-112. [1927.07]

Karavaiev, V. 1927d. The ant fauna of Ukraine. [In Ukrainian.] Zb. Prats Zool. Muz. 2:1-52 [= Tr. Ukr. Akad. Nauk Fiz.-Mat. Vidd. 4:247-296]. [1927]

Karavaiev, V. 1927e. Ameisen aus dem paläarktischen Gebiet. II. Zb. Prats Zool. Muz. 2:89-104 [= Tr. Ukr. Akad. Nauk Fiz.-Mat. Vidd. 4:333-348]. [1927]

Karavaiev, V. 1927f. Ameisen aus dem Indo-Australischen Gebiet. III. Zb. Prats Zool. Muz. 3: 3-52 [= Tr. Ukr. Akad. Nauk Fiz.-Mat. Vidd. 7:3-52]. [1927]

Karavaiev, V. 1928a. Ameisen aus dem Indo-Australischen Gebiet. IV. Ueber Ameisennester, hauptsächlich von *Polyrhachis*-Arten. Zb. Prats Zool. Muz. 5:129-150 [= Tr. Ukr. Akad. Nauk Fiz.-Mat. Vidd. 6:307-328]. [1928]
Karavaiev, V. 1928b. Eine anormale Thoraxbildung bei *Formica rufibarbis* F. (Fam. Formicidae). Zb. Prats Zool. Muz. 5:169-170 [= Tr. Ukr. Akad. Nauk Fiz.-Mat. Vidd. 6:473-474]. [1928]
Karavaiev, V. 1929a. Die Spinndrüsen der Weberameisen (Hym. Formicid.). Zool. Anz. 82:247-256. [1929]
Karavaiev, V. 1929b. Ameisen aus dem Indo-Australischen Gebiet. VI [sic]. Zb. Prats Zool. Muz. 7:43-58 [= Tr. Vseukr. Akad. Nauk Fiz.-Mat. Vidd. 13:41-56]. [1929]
Karavaiev, V. 1929c. Beitrag zur Ameisenfauna der Wälder in der Umgegend von Brjansk. Zb. Prats Zool. Muz. 7:59-63 [= Tr. Vseukr. Akad. Nauk Fiz.-Mat. Vidd. 13:57-61]. [1929]
Karavaiev, V. 1929d. Myrmekologische Fragmente. II. Zb. Prats Zool. Muz. 7:205-220 [= Tr. Vseukr. Akad. Nauk Fiz.-Mat. Vidd. 13:203-218]. [1929]
Karavaiev, V. 1929e. Ameisen aus dem Indo-Australischen Gebiet. VI. Zb. Prats Zool. Muz. 7:235-248 [= Tr. Vseukr. Akad. Nauk Fiz.-Mat. Vidd. 13:233-246]. [1929]
Karavaiev, V. 1930a. Ameisen von den Molukken und Neuguinea. (Ergebnisse der Sunda-Expedition der Notgemeinschaft der deutschen Wissenschaft 1929/30.) Zool. Anz. 92:206-214. [1930.10.18]
Karavaiev, V. 1930b. Beitrag zur Ameisenfauna der schwedischen Inseln Gotland und Oeland. Zb. Prats Zool. Muz. 8:5-46 [= Tr. Vseukr. Akad. Nauk Fiz.-Mat. Vidd. 15:109-150]. [1930]
Karavaiev, V. 1931a ("1930"). Myrmekologische Fragmente, III. Zool. Anz. 92:309-317. [1931.01.05]
Karavaiev, V. 1931b. Beitrag zur Ameisenfauna der Umgebung des Baikalsees. Zool. Anz. 93: 28-32. [1931.01.15]
Karavaiev, V. 1931c. Beitrag zur Ameisenfauna Jakutiens. (Auf Grund der Sammelergebnisse der Expeditionen der Wissenschaften der UdSSR., ausgeführt in den Jahren 1925 und 1926.) Zool. Anz. 94:104-117. [1931.04.20]
Karavaiev, V. 1931d. Ameisen aus Englisch-Ostafrika. Zool. Anz. 95:42-51. [1931.06.15]
Karavaiev, V. 1931e. Uebersicht der Ameisenfauna von Schweden. Zb. Prats Zool. Muz. 10:207-220 [= Tr. Vseukr. Akad. Nauk Pryr.-Tech. Vidd. 5:207-220]. [1931]
Karavaiev, V. 1932. Zwei neue Ameisen aus Aserbeidschan (Transkaukasien). Zool. Anz. 98:248-250. [1932.05.05]
Karavaiev, V. 1933a ("1932"). Ameisen aus dem Indo-Australischen Gebiet, VII. Konowia 11:305-320. [1933.03.15]
Karavaiev, V. 1933b. Ameisen aus dem Indo-Australischen Gebiet, VII. (Fortsetzung). Konowia 12:103-120. [1933.08.25]
Karavaiev, V. 1933c. Ameisen aus dem Indo-Australischen Gebiet, VII. (Schluss). Konowia 12:260-271. [1933.12.30]
Karavaiev, V. 1933d. The fauna of the family Formicidae (ants) of the Ukraine. General part. [In Ukrainian.] Zb. Prats Zool. Muz. 12:1-32. [1933]
Karavaiev, V. 1934. The fauna of the family Formicidae (ants) of the Ukraine. [In Ukrainian.] Tr. Inst. Zool. Biol. Vseukr. Akad. Nauk Ser. 1 Pr. Syst. Faun. 1934:1-164. [> 1934.09.27] [Date signed for printing ("pidpisano do druku").]
Karavaiev, V. 1935a. Neue Ameisen aus dem Indo-Australischen Gebiet, nebst Revision einiger Formen. Treubia 15:57-118. [1935.06]
Karavaiev, V. 1935b. Contribution to the ant fauna of Mariupols'kaya Province. [In Ukrainian.] Zb. Prats Zool. Muz. 16:107-111. [1935]
Karavaiev, V. 1936. The fauna of the family Formicidae (ants) of the Ukraine. Part II (conclusion). [In Ukrainian.] Tr. Inst. Zool. Biol. Ukr. Akad. Nauk Ser. 1 Pr. Syst. Faun. 1936:161-316. [> 1936.05.26] [Date signed for printing ("pidpisano do druku").]
Karavaiev, V. 1937. Ants collected in the nature reserves of Kinburnskaya Peninsula and Burkuty. [In Ukrainian.] Zb. Prats Zool. Muz. 19:171-181. [1937]

Karavajev, V. See: Karavaiev, V.
Karawaiew, W. See: Karavaiev, V.
Karawajew, W. See: Karavaiev, V.
Karpinski, J. J. 1956. Mrzówki w biocenozie Bialowieskiego Parku Narodowego. Rocz. Nauk Lesn. 14:201-221. [1956.07]
Kaschef, A. H., Mohamed, A. H. 1982. Taxonomy of ants (Fam. Formicidae) in A. R. Egypt. [Abstract.] P. 406 in: Breed, M. D., Michener, C. D., Evans, H. E. (eds.) The biology of social insects. Proceedings of the Ninth Congress of the IUSSI, Boulder, Colorado, August 1982. Boulder: Westview Press, xii + 420 pp. [1982.08]
Kaspari, M., Byrne, M. M. 1995. Caste allocation in litter *Pheidole*: lessons from plant defense theory. Behav. Ecol. Sociobiol. 37:255-263. [1995.10]
Kaspari, M., Vargo, E. L. 1995. Colony size as a buffer against seasonality: Bergmann's rule in social insects. Am. Nat. 145:610-632. [1995.04]
Kasugai, M., Takeda, S., Sakurai, H. 1983. Some observations on the microgyne form of ant *Myrmica ruginodis* Nylander (Hymenoptera, Formicidae) in Sapporo. Kontyû 51:73-79. [1983.03.25]
Keall, J. B. 1980a. Some ants recently intercepted entering New Zealand (Hymenoptera: Formicidae). N. Z. Entomol. 7:119-122. [1980.12]
Keall, J. B. 1980b. Some arthropods recently intercepted entering New Zealand in orchids from Honduras. N. Z. Entomol. 7:127-129. [1980.12]
Keall, J. B. 1981. A note on the occurrence of *Myrmecia brevinoda* (Hymenoptera: Formicidae) in New Zealand. Rec. Auckl. Inst. Mus. 18:203-204. [1981.12.18]
Keall, J. B., Somerfield, K. G. 1980. The Australian ant *Iridomyrmex darwinianus* established in New Zealand (Hymenoptera: Formicidae). N. Z. Entomol. 7:123-127. [1980.12]
Keegans, S. J., Billen, J. P. J., Morgan, E. D., Gökcen, O. A. 1993. Volatile glandular secretions of three species of New World army ants, *Eciton burchelli*, *Labidus coecus* and *Labidus praedator*. J. Chem. Ecol. 19:2705-2719. [1993.11]
Keegans, S. J., Morgan, E. D., Agosti, D., Wehner, R. 1992. What do glands tell us about species? A chemical case study of *Cataglyphis* ants. Biochem. Syst. Ecol. 20:559-572. [1992.09.22]
Keilbach, R. 1982. Bibliographie und Liste der Arten tierischer Einschlüsse in fossilen Harzen sowie ihrer Aufbewahrungsorte. Teil 1. Dtsch. Entomol. Z. (N.F.)29:129-286. [1982.03.15] [Formicidae pp. 272-281.]
Keister, M. 1963. The anatomy of the tracheal system of *Camponotus pennsylvanicus* (Hymenoptera: Formicidae). Ann. Entomol. Soc. Am. 56:336-340. [1963.05.15]
Keller, L. 1991. Queen number, mode of colony founding, and queen reproductive success in ants (Hymenoptera Formicidae). Ethol. Ecol. Evol. 3:307-316. [1991.10]
Keller, L. (ed.) 1993. Queen number and sociality in insects. Oxford: Oxford University Press, 451 pp. [1993]
Keller, L. 1995. Social life: the paradox of multiple-queen colonies. Trends Ecol. Evol. 10:355-360. [1995.09]
Keller, L., Passera, L. 1989. Size and fat content of gynes in relation to the mode of colony founding in ants (Hymenoptera; Formicidae). Oecologia (Berl.) 80:236-240. [1989.08]
Kempf, W. W. 1949. A new species and subspecies of *Procryptocerus* from Espírito Santo, Brazil (Hymenoptera, Formicidae). Rev. Entomol. (Rio J.) 20:423-426. [1949.08.31]
Kempf, W. W. 1951. A taxonomic study on the ant tribe Cephalotini (Hymenoptera: Formicidae). Rev. Entomol. (Rio J.) 22:1-244. [1951.12.31]
Kempf, W. W. 1952. A synopsis of the *pinelii*-complex in the genus *Paracryptocerus* (Hym. Formicidae). Stud. Entomol. 1:1-30. [1952.05.15]
Kempf, W. W. 1953. Uma nova espécie de *Paracryptocerus* da Colômbia, praga do cafeeiro (Hymenoptera, Formicidae). Pap. Avulsos Zool. (São Paulo) 11:79-88. [1953.02.10]
Kempf, W. W. 1954. A descoberta do primeiro macho do gênero *Thaumatomyrmex* Mayr (Hymenoptera, Formicidae). Rev. Bras. Entomol. 1:47-52. [1954.01]

Kempf, W. W. 1956. A morphological study on the male genitalia of *Paracryptocerus* (*P.*) *pusillus* (Hymenoptera: Formicidae). Rev. Bras. Entomol. 5:101-110. [1956.05.25]

Kempf, W. W. 1957. Sôbre algumas espécies de *Procryptocerus* com a descrição duma espécie nova (Hymenoptera, Formicidae). Rev. Bras. Biol. 17:395-404. [1957.09]

Kempf, W. W. 1958a. New studies of the ant tribe Cephalotini (Hym. Formicidae). Stud. Entomol. (n.s.)1:1-168. [1958.01.31]

Kempf, W. W. 1958b. Three new ants of the genus *Strumigenys* from Colombia (Hym. Formicidae). Rev. Bras. Entomol. 8:59-68. [1958.03.15]

Kempf, W. W. 1958c. Sôbre algumas formigas neotrópicas do gênero *Leptothorax* Mayr (Hymenoptera: Formicidae). An. Acad. Bras. Cienc. 30:91-102. [1958.03.31]

Kempf, W. W. 1958d. Synonymic note on ants of the genus *Paracryptocerus* Emery (Hymenoptera: Formicidae). Entomol. News 69:108-110. [1958.05.07]

Kempf, W. W. 1958e. Discovery of the ant genus *Wadeura* in Brazil (Hymenoptera: Formicidae). Rev. Bras. Entomol. 8:175-180. [1958.05.15]

Kempf, W. W. 1958f. Estudos sôbre *Pseudomyrmex*. II. (Hymenoptera: Formicidae). Stud. Entomol. (n.s.)1:433-462. [1958.11.29]

Kempf, W. W. 1958g. The ants of the tribe Dacetini in the State of Sao Paulo, Brazil, with the description of a new species of *Strumigenys*. (Hymenoptera: Formicidae). Stud. Entomol. (n.s.)1:553-560. [1958.11.29]

Kempf, W. W. 1959a. Sôbre algumas formigas Cephalotini do Museu de Oxford (Hymenoptera, Formicidae). Rev. Bras. Biol. 19:91-98. [1959.04]

Kempf, W. W. 1959b. Insecta Amapaensia. - Hymenoptera: Formicidae. Stud. Entomol. (n.s.)2:209-218. [1959.09.29]

Kempf, W. W. 1959c. A revision of the Neotropical ant genus *Monacis* Roger (Hymenoptera: Formicidae). Stud. Entomol. (n.s.)2:225-270. [1959.09.29]

Kempf, W. W. 1959d. A synopsis of the New World species belonging to the *Nesomyrmex*-group of the ant genus *Leptothorax* Mayr (Hymenoptera: Formicidae). Stud. Entomol. (n.s.)2:391-432. [1959.09.29]

Kempf, W. W. 1959e. Transmission of the Borgmeier Collection of ants. Stud. Entomol. (n.s.)2:474. [1959.09.29]

Kempf, W. W. 1959f. Two new species of *Gymnomyrmex* Borgmeier, 1954 from southern Brazil, with remarks on the genus (Hymenoptera, Formicidae). Rev. Bras. Biol. 19:337-344. [1959.10]

Kempf, W. W. 1960a. Estudo sôbre *Pseudomyrmex* I. (Hymenoptera: Formicidae). Rev. Bras. Entomol. 9:5-32. [1960.05.21]

Kempf, W. W. 1960b. *Phalacromyrmex*, a new ant genus from southern Brazil (Hymenoptera, Formicidae). Rev. Bras. Biol. 20:89-92. [1960.05]

Kempf, W. W. 1960c. *Tranopeltoides* Wheeler, a synonym of *Crematogaster* Lund (Hymenoptera, Formicidae). Entomol. News 71:173-175. [1960.07.14]

Kempf, W. W. 1960d. A review of the ant genus *Mycetarotes* Emery (Hymenoptera, Formicidae). Rev. Bras. Biol. 20:277-283. [1960.10]

Kempf, W. W. 1960e. Insecta Amapaensia. - Hymenoptera: Formicidae (segunda contribuição). Stud. Entomol. (n.s.)3:385-400. [1960.12.20]

Kempf, W. W. 1960f. Miscellaneous studies on Neotropical ants (Hymenoptera, Formicidae). Stud. Entomol. (n.s.)3:417-466. [1960.12.20]

Kempf, W. W. 1960g. Ameisen als Schaedlinge von Polyaethylen-ummantelten Kabeln. Stud. Entomol. (n.s.)3:506-507. [1960.12.20]

Kempf, W. W. 1961a. Estudos sôbre *Pseudomyrmex*. III. (Hymenoptera: Formicidae). Stud. Entomol. 4:369-408. [1961.10.10]

Kempf, W. W. 1961b. A survey of the ants of the soil fauna in Surinam (Hymenoptera: Formicidae). Stud. Entomol. 4:481-524. [1961.10.10]

Kempf, W. W. 1961c. William M. Mann, 1886-1960. Stud. Entomol. 4:547-549. [1961.10.10]

Kempf, W. W. 1961d. *Labidus coecus* as a cave ant. Stud. Entomol. 4:551-552. [1961.10.10]
Kempf, W. W. 1961e. As formigas do gênero *Pachycondyla* Fr. Smith no Brasil (Hymenoptera: Formicidae). Rev. Bras. Entomol. 10:189-204. [1961.12.26]
Kempf, W. W. 1961f. Nota preliminar sôbre a fauna das formigas de Agudos, S. P. (Hymenoptera: Formicidae). Rev. Bras. Entomol. 10:205-208. [1961.12.26]
Kempf, W. W. 1961g. Remarks on the ant genus *Irogera* Emery, with the description of a new species (Hymenoptera, Formicidae). Rev. Bras. Biol. 21:435-441. [1961.12]
Kempf, W. W. 1962a. Retoques à classificação das formigas neotropicais do gênero *Heteroponera* Mayr (Hym., Formicidae). Pap. Avulsos Zool. (São Paulo) 15:29-47. [1962.07.02]
Kempf, W. W. 1962b. Miscellaneous studies on neotropical ants. II. (Hymenoptera, Formicidae). Stud. Entomol. 5:1-38. [1962.10.20]
Kempf, W. W. 1963a. Additions to the Neotropical ant genus *Rogeria* Emery, with a key to the hitherto recorded South American species (Hym., Formicidae). Rev. Bras. Biol. 23:189-196. [1963.08]
Kempf, W. W. 1963b. A review of the ant genus *Mycocepurus* Forel, 1893 (Hymenoptera: Formicidae). Stud. Entomol. 6:417-432. [1963.12.10]
Kempf, W. W. 1963c. Nota sinonímica acêrca de formigas da tribo Cephalotini (Hymenoptera, Formicidae). Rev. Bras. Biol. 23:435-438. [1963.12]
Kempf, W. W. 1964a. Uma nova *Platythyrea* do Brasil (Hym., Formicidae). Rev. Bras. Entomol. 11:141-144. [1964.05.15]
Kempf, W. W. 1964b. Additions to the knowledge of the Cephalotini ants (Hymenoptera, Formicidae). Pap. Avulsos Zool. (São Paulo) 16:243-255. [1964.07.22]
Kempf, W. W. 1964c. A propósito de um estudo sôbre as formigas do gênero *Acanthostichus* Mayr (Hymenoptera, Formicidae). Pap. Avulsos Zool. (São Paulo) 16:263-266. [1964.07.22]
Kempf, W. W. 1964d. A revision of the Neotropical fungus-growing ants of the genus *Cyphomyrmex* Mayr. Part I: Group of *strigatus* Mayr (Hym., Formicidae). Stud. Entomol. 7: 1-44. [1964.12.10]
Kempf, W. W. 1964e. Miscellaneous studies on Neotropical ants. III. (Hymenoptera: Formicidae). Stud. Entomol. 7:45-71. [1964.12.10]
Kempf, W. W. 1964f. The ants of the genus *Anochetus* (*Stenomyrmex*) in Brazil (Hym., Formicidae). Stud. Entomol. 7:237-246. [1964.12.10]
Kempf, W. W. 1964g. On the number of ant species in the Neotropical region. Stud. Entomol. 7:481-482. [1964.12.10]
Kempf, W. W. 1964h. [Review of: Wheeler, G. C., Wheeler, J. 1963. The ants of North Dakota. Grand Forks, North Dakota: University of North Dakota Press, viii + 326 pp.] Stud. Entomol. 7:493-494. [1964.12.10]
Kempf, W. W. 1965. Nota preliminar sôbre algumas formigas neotrópicas, descritas por Frederick Smith (Hymenoptera, Formicidae). Rev. Bras. Biol. 25:181-186. [1965.07]
Kempf, W. W. 1966 ("1965"). A revision of the Neotropical fungus-growing ants of the genus *Cyphomyrmex* Mayr. Part II: Group of *rimosus* (Spinola) (Hym., Formicidae). Stud. Entomol. 8:161-200. [1966.02.28]
Kempf, W. W. 1967a ("1966"). New ants from southeastern and central Brazil (Hymenoptera, Formicidae). Stud. Entomol. 9:121-128. [1967.01.30]
Kempf, W. W. 1967b ("1966"). A synopsis of the Neotropical ants of the genus *Centromyrmex* Mayr (Hymenoptera: Formicidae). Stud. Entomol. 9:401-410. [1967.01.30]
Kempf, W. W. 1967c. Estudos sôbre *Pseudomyrmex*. IV (Hymenoptera: Formicidae). Rev. Bras. Entomol. 12:1-12. [1967.02]
Kempf, W. W. 1967d. Three new South American ants (Hym. Formicidae). Stud. Entomol. 10:353-360. [1967.12.15]
Kempf, W. W. 1967e. A new revisionary note on the genus *Paracryptocerus* Emery (Hym. Formicidae). Stud. Entomol. 10:361-368. [1967.12.15]

Kempf, W. W. 1968a. A new species of *Cyphomyrmex* from Colombia, with further remarks on the genus (Hymenoptera, Formicidae). Rev. Bras. Biol. 28:35-41. [1968.04]
Kempf, W. W. 1968b. Miscellaneous studies on Neotropical ants. IV. (Hymenoptera, Formicidae). Stud. Entomol. 11:369-415. [1968.09.30]
Kempf, W. W. 1969. Miscellaneous studies on Neotropical ants. V. (Hymenoptera, Formicidae). Stud. Entomol. 12:273-296. [1969.11.28]
Kempf, W. W. 1970a. Catálogo das formigas do Chile. Pap. Avulsos Zool. (São Paulo) 23:17-43. [1970.08.15]
Kempf, W. W. 1970b. Levantamento das formigas da mata amazônica, nos arredores de Belém do Pará, Brasil. Stud. Entomol. 13:321-344. [1970.11.30]
Kempf, W. W. 1970c. Taxonomic notes on ants of the genus *Megalomyrmex* Forel, with the description of new species (Hymenoptera, Formicidae). Stud. Entomol. 13:353-364. [1970.11.30]
Kempf, W. W. 1971. A preliminary review of the ponerine ant genus *Dinoponera* Roger (Hymenoptera: Formicidae). Stud. Entomol. 14:369-394. [1971.11.29]
Kempf, W. W. 1972a. Dedication to Fr. Thomas Borgmeier, O.F.M., Sc.D., on his 80th birthday, October 31, 1972. Stud. Entomol. 15:1-2. [1972.08.25]
Kempf, W. W. 1972b. Catálogo abreviado das formigas da região Neotropical. Stud. Entomol. 15:3-344. [1972.08.25]
Kempf, W. W. 1972c. Correção para o "Catálogo das formigas do Chile". Stud. Entomol. 15:448. [1972.08.25]
Kempf, W. W. 1972d. A study of some Neotropical ants of genus *Pheidole* Westwood. I. (Hymenoptera: Formicidae). Stud. Entomol. 15:449-464. [1972.08.25]
Kempf, W. W. 1972e. Frei Thomaz Borgmeier. Stud. Entomol. 15:503-504. [1972.08.25]
Kempf, W. W. 1972f. List of publications in Entomology by Walter W. Kempf, O.F.M. Stud. Entomol. 15:507-510. [1972.08.25]
Kempf, W. W. 1972g. A new species of the dolichoderine ant genus *Monacis* Roger, from the Amazon, with further remarks on the genus (Hymenoptera, Formicidae). Rev. Bras. Biol. 32:251-254. [1972.08]
Kempf, W. W. 1973a. Uma nova *Solenopsis* do Rio Grande do Sul, Brasil (Hymenoptera, Formicidae). Rev. Bras. Entomol. 17:29-32. [1973.05.14]
Kempf, W. W. 1973b. A revision of the Neotropical myrmicine ant genus *Hylomyrma* Forel (Hymenoptera: Formicidae). Stud. Entomol. 16:225-260. [1973.10.31]
Kempf, W. W. 1973c. A new *Zacryptocerus* from Brazil, with remarks on the generic classification of the tribe Cephalotini (Hymenoptera: Formicidae). Stud. Entomol. 16:449-462. [1973.10.31]
Kempf, W. W. 1974a. Taxonomic and faunistic notes on some Neotropical Cephalotini ants (Hymenoptera, Formicidae). Rev. Bras. Entomol. 18:67-76. [1974.08.15]
Kempf, W. W. 1974b. A review of the Neotropical ant genus *Oxyepoecus* Santschi (Hymenoptera: Formicidae). Stud. Entomol. 17:471-512. [1974.10.31]
Kempf, W. W. 1974c. A remarkable new Neotropical species in the ant genus *Odontomachus* Latreille (Hymenoptera: Formicidae). Stud. Entomol. 17:551-553. [1974.10.31]
Kempf, W. W. 1974d. [Review of: Wheeler, G. C., Wheeler, J. 1973. Ants of Deep Canyon. Riverside, Calif.: University of California, xiii + 162 pp.] Stud. Entomol. 17:558-559. [1974.10.31]
Kempf, W. W. 1975a ("1974"). Report on Neotropical Dacetine ant studies (Hymenoptera: Formicidae). Rev. Bras. Biol. 34:411-424. [1975.03.26]
Kempf, W. W. 1975b. A revision of the Neotropical ponerine ant genus *Thaumatomyrmex* Mayr (Hymenoptera: Formicidae). Stud. Entomol. 18:95-126. [1975.11.30]
Kempf, W. W. 1975c. Miscellaneous studies on neotropical ants. VI. (Hymenoptera, Formicidae). Stud. Entomol. 18:341-380. [1975.11.30]
Kempf, W. W. 1975d. Karol Lenko (1914-1975). Stud. Entomol. 18:619-621. [1975.11.30]

Kempf, W. W. 1976a. Discovery of a major worker in *Camponotus branneri* (Mann), a new combination (Hymenoptera: Formicidae). Psyche (Camb.) 83:106-111. [1976.11.04]

Kempf, W. W. 1976b. Father Thomas Borgmeier, O.F.M. (1892-1975). Stud. Entomol. 19:1-37. [1976.12.30]

Kempf, W. W. 1976c. A new species of *Strumigenys* from the lower Amazon, Brazil (Hym., Formicidae). Stud. Entomol. 19:39-44. [1976.12.30]

Kempf, W. W. 1978a. Formicidae-Hym. 111. Chave para identificação de operárias des sete subfamilias que ocorrem no Brasil. Stud. Entomol. 20:33. [1978.08.30]

Kempf, W. W. 1978b. Five new synonyms for the Argentine ant fauna (Hymenoptera, Formicidae). 112. Stud. Entomol. 20:35-38. [1978.08.30]

Kempf, W. W. 1978c. Considerações zoo-geográficas de um levantamento mirmecológico no Estado de São Paulo - Brasil. 113. Stud. Entomol. 20:39-41. [1978.08.30]

Kempf, W. W. 1978d. A preliminary zoogeographical analysis of a regional ant fauna in Latin America. 114. Stud. Entomol. 20:43-62. [1978.08.30]

Kempf, W. W., Brown, W. L., Jr. 1968. Report on some Neotropical ant studies. Pap. Avulsos Zool. (São Paulo) 22:89-102. [1968.12.04]

Kempf, W. W., Brown, W. L., Jr. 1969. Two new *Strumigenys* ants from the Amazon valley in Brasil (Hymenoptera, Formicidae). Rev. Bras. Biol. 29:17-24. [1969.04]

Kempf, W. W., Brown, W. L., Jr. 1970. Two new ants of tribe Ectatommini from Colombia (Hymenoptera: Formicidae). Stud. Entomol. 13:311-320. [1970.11.30]

Kempf, W. W., Lenko, K. 1968. Novas observações e estudos sôbre *Gigantiops destructor* (Fabricius) (Hymenoptera: Formicidae). Pap. Avulsos Zool. (São Paulo) 21:209-230. [1968.03.05]

Kempf, W. W., Lenko, K. 1976. Levantamento da formicifauna no litoral norte e ilhas adjacentes do Estado de São Paulo, Brasil. I. Subfamilias Dorylinae, Ponerinae e Pseudomyrmecinae (Hym., Formicidae). Stud. Entomol. 19:45-66. [1976.12.30]

Kennedy, C. H. 1938. *Solenopsis rosella* Kennedy, a new ant from southern Ontario. Can. Entomol. 70:232-236. [1938.11.26]

Kennedy, C. H. 1939. A plea for preservation of the south point of Pelee Island as a wild-life-study preserve. Can. Entomol. 71:84-86. [1939.05.05]

Kennedy, C. H., Dennis, C. A. 1937. New ants from Ohio and Indiana, *Formica prociliata*, *F. querquetulana*, *F. postoculata* and *F. lecontei*, (Formicidae: Hymenoptera). Ann. Entomol. Soc. Am. 30:531-544. [1937.10.14]

Kennedy, C. H., Schramm, M. M. 1933. A new *Strumigenys* with notes on Ohio species (Formicidae: Hymenoptera). Ann. Entomol. Soc. Am. 26:95-104. [1933.03.22]

Kennedy, C. H., Talbot, M. 1939. Notes on the hypogaeic ant, *Proceratium silaceum* Roger. Proc. Indiana Acad. Sci. 48:202-210. [1939]

Kermarrec, A., Febvay, G. 1985. Une glande épidermique ventrale chez les larves de *Acromyrmex octospinosus* (Reich) (Hymenoptera - Formicidae). Insectes Soc. 32:213-217. [1985.10]

Kermarrec, A., Mauléon, H., Antun, A. A. 1976. La stridulation de *Acromyrmex octospinosus* Reich. (Formicidae, Attini): biométrie de l'appareil stridulateur et analyse du signal produit. Insectes Soc. 23:29-47. [1976.03]

Kerr, W. E. 1961. Acasalamento de rainhas com vários machos em duas espécies da tribu Attini (Hymenoptera, Formicoidea). Rev. Bras. Biol. 21:45-48. [1961.06]

Kessler, K. 1868. Material for knowledge of Onega Lake and Obonega region, primarily in zoological relationships. [In Russian.] Tr. Rus. Estestvoispyt. 1:1-144. [1868] [*Cyathocephalus* (Cestoda), senior homonym of *Cyathocephalus* Emery, 1915.]

Kevan, D. K. McE. 1970. Agassiz's Nomenclatoris zoologici index universalis - a correction. J. Soc. Bibliogr. Nat. Hist. 5:286. [1970.04]

Kholová, H. 1955 ("1954"). Mravenci karlstejnska. Roc. Cesk. Spol. Entomol. 51:157-163. [1955]

Khoo, Y. H. 1990. A note on the Formicidae (Hymenoptera) from pitfall traps at Ulu Kinchin, Pahang, Malaysia. Malay. Nat. J. 43:290-293. [1990.05]

Kidd, L. N. 1975 ("1974"). *Ponera punctatissima* Roger (Hym., Formicidae) and *Neobisnius procerulus* (Grav.) (Col., Staphylinidae) in Lancashire. Entomol. Mon. Mag. 110:96. [1975.05.21]

Kieffer, J. J. 1896. Diagnose de deux espèces nouvelles de Cécidomyies (Dipt.) Bull. Soc. Entomol. Fr. 1896:236-237. [1896.06.06] [*Janetia* (Diptera), senior homonym of *Janetia* Forel, 1899.]

Kim, B.-J. 1986a. On the unrecorded species of Korean ants, *Formica fusca* on the basis of external fine features. J. Nat. Sci. (Res. Inst. Basic Sci. Won-Kwang Univ.) 5(2):15-20. [Not seen. Cited in Terayama & Choi (1991).]

Kim, B.-J. 1986b. A systematic study of ants in Is. Ullungdo of Korea on the basis of external fine features. J. Nat. Sci. (Res. Inst. Basic Sci. Won-Kwang Univ.) 5(2):84-94. [Not seen. Cited in Terayama, Choi & Kim (1992).]

Kim, B.-J. 1988. A systematic study of the subfamily Formicinae (Hym., Formicidae) from Korea. [Abstract.] Korean J. Entomol. 18:119. [1988.04.28]

Kim, B.-J., Kim, C.-W. 1983a. A review of myrmicinae ants from Korea on the basis of external fine features (Hym., Formicidae). [In Korean.] J. Natl. Acad. Sci. Repub. Korea Nat. Sci. Ser. 22:51-90. [1983.12.26]

Kim, B.-J., Kim, C.-W. 1983b. A systematic revision of the genus *Formica* in Korea on the basis of external fine features (Hym.: Formicidae). Entomol. Res. Bull. (Seoul) 9:57-67. [1983.12]

Kim, B.-J., Kim, C.-W. 1986. On the one new species, *Camponotus jejuensis* (n.sp.) from Korea (Hym., Formicidae). Korean J. Entomol. 16:139-144. [1986.10.25]

Kim, B.-J., Kim, K.-G., Ryu, D.-P., Kim, J.-H. 1995. Ants of Chindo Island in Korea (Hymenoptera, Formicidae). Korean J. Syst. Zool. 11:101-113. [1995.03]

Kim, B.-J., Kim, K.-Y. 1994. On the two new species, *Camponotus concavus* n. sp and *fuscus* n. sp from Korea (Hym., Formicidae). Korean J. Entomol. 24:285-292. [1994.10.25]

Kim, B.-J., Kim, K.-Y., Lim, K.-H. 1993. Systematic study of ants from Chejudo province. Korean J. Entomol. 23:117-141. [1993.07.25]

Kim, B.-J., Kyunghoon, I. 1992. Systematic study of Korean *Lasius* ants on the basis of electrophoretic data. [Abstract.] P. 53 in: Proceedings XIX International Congress of Entomology. Abstracts. Beijing, China, June 28 - July 4, 1992. Beijing: XIX International Congress of Entomology, xi + 730 pp. [1992.07.04]

Kim, B.-J., Lim, K.-H. 1994. Systematic study of Korean *Lasius* ants on the basis of electrophoretic data. Korean J. Entomol. 24:117-129. [1994.04.25]

Kim, B.-J., Park, J.-Y. 1994. Systematic study of Korean *Formica* ants on the basis of electrophoretic data. Korean J. Entomol. 24:131-143. [1994.04.25]

Kim, B.-J., Ryu, D.-P., Park, S.-J., Kim, J.-H. 1994. Systematic study on ants from coasts of Korean peninsula (Hym., Formicidae). Korean J. Entomol. 24:293-309. [1994.10.25]

Kim, C.-H., Choi, B.-M. 1987. On the kinds of ants (Hymenoptera: Formicidae) and vertical distribution in Jiri Mountain. [In Korean.] Korean J. Plant Prot. 26:123-132. [1987.09.30]

Kim, C.-H., Choi, B.-M., Bang, J.-R. 1992. Studies on the distribution of ants (Formicidae) in Korea (8). Ant fauna in 10 islands, Chollanam-Do. [In Korean.] Korean J. Appl. Entomol. 31:345-359. [1992.12.20]

Kim, C.-H., Choi, B.-M., Bang, J.-R. 1991. Studies on the character of ant (Formicidae) in Korea on the basis of scanning electron microscope (I). On the form-character of *Smithistruma japonica* (Ito). [In Korean.] Korean J. Appl. Entomol. 30:285-290. [1991.12.20]

Kim, K.-I., Kim, C.-H., Choi, B.-M. 1989. The ant fauna of the southern shore in Gyeongsangnamdo, Korea. [In Korean.] Kyongsang Taehak Nonmunjip (J. Kyongsang Natl. Univ.) 28(2):213-226. [1989.11.30]

King, G. B. 1901a. A check-list of the Massachusetts Formicidae, with some notes on the species. Psyche (Camb.) 9:260-262. [1901.10]

King, G. B. 1901b. Some new records of the New England Formicidae. Psyche (Camb.) 9:270-271. [1901.11]

King, G. B. 1902. Further notes on New England Formicidae. Psyche (Camb.) 9:367. [1902.07]

King, R. L. 1948. A tropical ant temporarily established in Iowa. Proc. Iowa Acad. Sci. 55:395. [1948]

King, R. L. 1955. Winged workers in the ant, *Formica obscuriventris clivia* Creighton. Proc. Iowa Acad. Sci. 62:509-513. [1955.12.15]

King, R. L., Sallee, R. M. 1952. Macropseudogynes (or pterergates?) in *Formica fossaceps* Buren (Formicidae). Proc. Iowa Acad. Sci. 59:469-474. [1952.12]

King, R. L., Sallee, R. M. 1959. *Formica fossaceps* Buren and *Formica obscuriventris clivia* Creighton as slaves of *Formica rubicunda* Emery. Proc. Iowa Acad. Sci. 66:472-473. [1959.12.18]

King, R. L., Sallee, R. M. 1965. Notes on Iowa ants. Proc. Iowa Acad. Sci. 71:484-485. [1965]

King, T. G., Green, S. C. 1995. Factors affecting the distribution of pavement ants (Hymenoptera, Formicidae) in Atlantic Coast urban fields. Entomol. News 106:224-228. [1995.12.28]

Kinomura, K., Yamauchi, K. 1992. A new workerless socially parasitic species of the genus *Vollenhovia* (Hymenoptera, Formicidae) from Japan. Jpn. J. Entomol. 60:203-206. [1992.03.25]

Kinomura, K., Yamauchi, K. 1994. Frequent occurrence of gynandromorphs in the natural population of the ant *Vollenhovia emeryi* (Hymenoptera: Formicidae). Insectes Soc. 41:273-278. [1994]

Kipyatkov, V. E. 1983. Artificial breeding of gynandromorphs in ants. [In Russian.] Priroda (Mosc.) 1983(7):113-114. [> 1983.06.15] [Date signed for printing ("podpisano k pechati"). July 1983 issue of journal.]

Kipyatkov, V. E. 1991. The world of social insects. [In Russian.] Leningrad: Izdatel'stvo Leningradskogo Universiteta, 406 pp. [> 1990.11.28] [Date signed for printing ("podpisano k pechati"). Title page dated 1991.]

Kipyatkov, V. E., Lopatina, E. B. 1994. Seasonality and biodiversity of ants in the North temperate zone. P. 30 in: Lenoir, A., Arnold, G., Lepage, M. (eds.) Les insectes sociaux. 12ème congrès de l'Union Internationale pour l'Étude des Insectes Sociaux. Paris, Sorbonne, 21-27 août 1994. Paris: Université Paris Nord, xxiv + 583 pp. [1994.09]

Kirby, W. 1819 ("1818"). A description of several new species of insects collected in New Holland by Robert Brown, Esq. F.R.S. Lib. Linn. Soc. Trans. Linn. Soc. Lond. 12:454-478. [1819.07.02] [Publication date from Raphael (1970).]

Kirby, W. 1837. Fauna boreali-americana; or the zoology of the northern parts of British America. Part four. Insects. Norwich: J. Fletcher, xxxix + 325 pp. [1837.10] [Other parts of this series were authored by J. Richardson and W. Swainson. Kirby is the sole author of Part 4. Date of publication from Sherborn (1932:cxliv). Formicidae pp. 261-263.]

Kirby, W. F. 1889 ("1888"). On the insects (exclusive of Coleoptera and Lepidoptera) of Christmas Island. Proc. Zool. Soc. Lond. 1888:546-555. [1889.04] [Date of publication from Duncan (1937).]

Kirby, W. F. 1896. Supplement to the Zoological Report. Hymenoptera. Pp. 203-209 in: Spencer, B. (ed.) Report on the work of the Horn Scientific Expedition to central Australia. Part 1. London: Dulau and Co., xviii + 220 pp. [1896.09]

Kirchner, W. 1959. Erstmaliger Nachweis von *Formica aquilonia* Yarrow, einer neuen Waldameisenart, für deutsches Gebiet. Waldhygiene 3:54. [1959]

Kiseleva, E. F. 1923. Contribution to the ant fauna of the southern Urals. [In Russian.] Izv. Tomsk. Gos. Univ. 72(5th part):1-12. [1923]

Kiseleva, E. F. 1925. On the ant fauna of the Ussuri region. [In Russian.] Izv. Tomsk. Gos. Univ. 75:73-75. [1925]

Kistner, D. H. 1972. A new genus of the staphylinid tribe Dorylomimini from Africa and its possible significance to ant phylogeny. Entomol. News 83:85-91. [1972.04.20]

Kistner, D. H. 1989. New genera and species of Aleocharinae associated with ants of the genus *Leptogenys* and their relationships (Coleoptera: Staphylinidae; Hymenoptera, Formicidae). Sociobiology 15:299-323. [1989]

Kistner, D. H., Jacobson, H. R. 1990. Cladistic analysis and taxonomic revision of the ecitophilous tribe Ecitocharini with studies of their behavior and evolution (Coleoptera, Staphylinidae, Aleocharinae). Sociobiology 17:333-480. [1990]

Klein, R. W. 1987a. Colony structures of three species of *Pseudomyrmex* (Hymenoptera: Formicidae: Pseudomyrmecinae) in Florida. Pp. 107-108 in: Eder, J., Rembold, H. (eds.) Chemistry and biology of social insects. München: Verlag J. Peperny, xxxv + 757 pp. [1987]

Klein, R. W. 1987b. A workerless inquiline in *Pseudomyrmex* (Hymenoptera: Formicidae: Pseudomyrmecinae). Pp. 623-624 in: Eder, J., Rembold, H. (eds.) Chemistry and biology of social insects. München: Verlag J. Peperny, xxxv + 757 pp. [1987]

Klimetzek, D. 1973. Die Variabilität der Standortansprüche hügelbauender Waldameisen der *Formica rufa*-Gruppe (Hymenoptera: Formicidae). Mitt. Bad. Landesver. Naturkd. Naturschutz Freibg. Breisgau 11:9-25. [1973.10.01]

Klimetzek, D. 1976. Bildschlüssel der Ameisenfauna Badens. Mitt. Bad. Landesver. Naturkd. Naturschutz Freibg. Breisgau 11:345-357. [1976.08.01]

Klimetzek, D. 1977. Die Ameisenfauna des Naturschutzgebietes "Mindelsee" (Hymenoptera: Formicidae). Beitr. Naturkd. Forsch. Südwestdtschl. 36:159-171. [1977.12.01]

Klimetzek, D. 1983. Die Verbreitung der Ameisen im Mindelsee-Gebiet. Natur-Landschaftsschutzgeb. Baden-Württ. 11:661-669. [1983]

Klimetzek, D., Kobel-Lamparski, A. 1990. Die Ameisenfauna des Naturschutzgebietes "Isteiner Klotz". Mitt. Bad. Landesver. Naturkd. Naturschutz Freibg. Breisgau 15:145-158. [1990.11.30]

Klimetzek, D., Pelz, D. R. 1992. Nest counts versus trapping in ant surveys: influence on diversity. Pp. 171-179 in: Billen, J. P. J. (ed.) Biology and evolution of social insects. Leuven: Leuven University Press, ix + 390 pp. [1992]

Klimetzek, D., Wellenstein, G. 1970. Vorkommen und Verbreitung hügelbauender Waldameisen der *Formica rufa*-Gruppe (Hymenoptera: Formicidae) in Baden-Württemberg. Allg. Forst-Jagdztg. 141:172-178. [1970.09]

Kloft, W. J., Wilkinson, R. C., Whitcomb, W. H., Kloft, E. S. 1973. *Formica integra* (Hymenoptera: Formicidae). 1. Habitat, nest construction, polygyny, and biometry. Fla. Entomol. 56:67-76. [1973.06.27]

Klotz, J. H., Mangold, J. R., Vail, K. M., Davis, L. R., Jr., Patterson, R. S. 1995. A survey of the urban pest ants (Hymenoptera: Formicidae) of peninsular Florida. Fla. Entomol. 78:109-118. [1995.03.20]

Klug, F. 1824. Entomologische Monographien. Berlin: Reimer, 242 pp. [1824]

Knauer, F. 1906 Die Ameisen. (Aus Natur und Geisteswelt Bd. 94.) Leipzig: B. G. Teubner, 156 pp. [1906]

Knechtel, W. K. 1956. Contributii la studiul formicidelor din valea Prahovei. Bul. Stiinte Acad. Repub. Pop. Rom. Sect. Biol. Stiinte Agric. 8:769-744. [1956.12]

Knechtel, W. K. 1958. Zur Färbungsvariation bei *Formica truncorum* Fab. Pp. 239-242 in: Becker, E. C. et al. (eds.) Proceedings of the Tenth International Congress of Entomology, Montreal, August 17-25, 1956. Volume 1. Ottawa: Mortimer Ltd., 941 pp. [1958.12]

Knechtel, W. K., Paraschivescu, D. 1962. Zur Kenntnis der geographischen Verbreitung der Ameisen in der Rumänischen Volksrepublik. Rev. Biol. Acad. Répub. Pop. Roum. 7:243-254. [1962]

Kneitz, G. 1972. Untersuchungen zur Verbreitung der hügelbauenden Waldameisen (Formicidae, *Formica*) in Europa unter besonderer Beachtung Skandinaviens. Pp. 367-369 in: Rafes, P. M. (ed.) XIII International Congress of Entomology, Moscow, 2-9 August, 1968. Proceedings, Volume III. Leningrad: Nauka, 494 pp. [1972.06.08]

Kneitz, G., Emmert, W. A. 1962. Waldameisenfunde (Formicidae, Gen. *Formica*) im französischen Zentralmassiv und den Pyrenäen. Waldhygiene 4:220-227. [1962]

Kneitz, G., Gernert, W., Rammoser, H. 1962. Hügelbauende Waldameisen (Formicidae, Gen. *Formica*) in den Vogesen. Waldhygiene 4:203-219. [1962]

Kneitz, G., Neumann, M. 1977. Rastermikroskopische Abbildungen von Kopfstrukturen bei *Formica polyctena* Foerster (Hymenoptera, Formicidae). Waldhygiene 12:83-95. [1977]

Knight, R. L., Rust, M. K. 1990. The urban ants of California with distribution notes of imported species. Southwest. Entomol. 15:167-178. [1990.06]

Knowlton, G. F. 1970. Ants of Curlew Valley. Proc. Utah Acad. Sci. Arts Lett. 47(1):208-212. [1970]

Knowlton, G. F., Harmston, F. C. 1939. Utah Hymenoptera. Utah Agric. Exp. Stn. Mimeogr. Ser. 200(4):1-8. [1939.03] [Ants pp. 6-8.]

Knudtson, B. K. 1978. Intrapopulation genetic similarity in *Pogonomyrmex californicus* (Buckley) (Hymenoptera: Formicidae). J. Tenn. Acad. Sci. 53:78-80. [1978.04]

Koehler, W. 1951. Fauna mrówek Pieninskiego Parku Narodowego. Warszawa: Panstwowe Wydawnictwo Rolnicze i Lesne, 55 pp. [1951.03]

Koehler, W. 1958. Nowe stanowisko *Anergates atratulus* Schenck (Hymenoptera, Formicidae) na ziemiach Polski. Pol. Pismo Entomol. 27:105-108. [1958.03]

Koen, J. H. 1988. Ant species richness of fynbos and forest ecosystems in the southern Cape. S. Afr. J. Zool. 23:184-188. [1988.07]

Kofler, A., Mildner, P. 1994. Die Pharaonenameise in Kärnten (*Monomorium pharaonis* (Linné)) (Hymenoptera: Formicidae, Myrmicinae). Carinthia II 104:427-430.

Kogure, T. 1959. Ants found in the vicinity of Lake Shikotsu, Hokkaidô. [In Japanese.] Saishû Shiiku 21:34-35. [1959.02.08]

Kohl, F. F. 1909 ("1908"). Dr. Gustav Mayr, ein Lebensbild. Verh. K-K. Zool.-Bot. Ges. Wien 58:512-528. [1909.01.20]

Kohn, M., Vlcek, M. 1984. The outdoor persistence of *Monomorium pharaonis* (Hymenoptera, Formicidae) colonies in Czechoslovakia. Acta Entomol. Bohemoslov. 81:186-189. [1984.05.30]

Kohout, R. J. 1987. Three new *Polyrhachis sexspinosa*-group species from the Philippines (Hymenoptera: Formicidae). Mem. Qld. Mus. 25:169-176. [1987.10.31]

Kohout, R. J. 1988a. New nomenclature of the Australian ants of the *Polyrhachis gab* Forel species complex (Hymenoptera: Formicidae: Formicinae). Aust. Entomol. Mag. 15:49-52. [1988.06.15]

Kohout, R. J. 1988b. A new species of *Polyrhachis* (*Polyrhachis*) from Papua New Guinea with a review of the New Guinean and Australian species (Hymenoptera: Formicidae: Formicinae). Mem. Qld. Mus. 25:417-427. [1988.11.07]

Kohout, R. J. 1988c. Nomenclatural changes and new Australian records in the ant genus *Polyrhachis* Fr. Smith (Hymenoptera: Formicidae: Formicinae). Mem. Qld. Mus. 25:429-438. [1988.11.07]

Kohout, R. J. 1989. The Australian ants of the *Polyrhachis relucens* species-group (Hymenoptera: Formicidae: Formicinae). Mem. Qld. Mus. 27:509-516. [1989.11.13]

Kohout, R. J. 1990. A review of the *Polyrhachis viehmeyeri* species-group (Hymenoptera: Formicidae: Formicinae). Mem. Qld. Mus. 28:499-508. [1990.08.31]

Kohout, R. J. 1994a. New synonymy of three Australian ants (Formicidae: Formicinae: Polyrhachis). Mem. Qld. Mus. 35:135-136. [1994.06.01]

Kohout, R. J. 1994b. *Polyrhachis lama*, a new ant from the Tibetan plateau (Formicidae: Formicinae). Mem. Qld. Mus. 35:137-138. [1994.06.01]

Kohout, R. J., Taylor, R. W. 1990. Notes on Australian ants of the genus *Polyrhachis* Fr. Smith, with a synonymic list of the species (Hymenoptera: Formicidae: Formicinae). Mem. Qld. Mus. 28:509-522. [1990.08.31]

Kohriba, O. 1963. A parasitic life of *Polyrhachis lamellidens* F. Smith. First report. [In Japanese.] Kontyû 31:200-209. [1963.09.30]
Kolossov, J. 1932. Synonymische u. systematische Bemerkungen über palaearctische Insekten. Entomol. Nachrichtenbl. Troppau 6:115-118. [1932.12]
Kondoh, M. 1961. Ants from Hakone region. [In Japanese.] Hakone Hakubutsu 1:16-27. [1961.08]
Kondoh, M. 1978. A comparison among ant communities in the anthropogenic environment. Memorabilia Zool. 29:79-92. [1978.12]
Kondoh, M. 1982. Altitudinal distribution of Japanese *Serviformica*. [Abstract.] Pp. 103-104 in: Breed, M. D., Michener, C. D., Evans, H. E. (eds.) The biology of social insects. Proceedings of the Ninth Congress of the IUSSI, Boulder, Colorado, August 1982. Boulder: Westview Press, xii + 420 pp. [1982.08]
Kondoh, M. 1987. Ant communities at Mt. Fuji, with special reference to ecocline between scoria grassland and coniferous forest. Pp. 595-596 in: Eder, J., Rembold, H. (eds.) Chemistry and biology of social insects. München: Verlag J. Peperny, xxxv + 757 pp. [1987]
Kondoh, M. 1994. The relationship between vegetation types and ant communities at the riverside of Agatsuma-gawa, Gumma prefecture, Central Japan. P. 450 in: Lenoir, A., Arnold, G., Lepage, M. (eds.) Les insectes sociaux. 12ème congrès de l'Union Internationale pour l'Étude des Insectes Sociaux. Paris, Sorbonne, 21-27 août 1994. Paris: Université Paris Nord, xxiv + 583 pp. [1994.09]
Kondoh, M., Kitazawa, Y. 1984. Ant communities on the campus of UOEH and in an adjacent natural forest. J. UOEH (Univ. Occup. Environ. Health) 6:221-234.
Königsmann, E. 1976. Das phylogenetische System der Hymenoptera. Teil 1: Einführung, Grundplanmerkmale, Schwestergruppe und Fossilfunde. Dtsch. Entomol. Z. (N.F.)23:253-279. [1976.12.10]
Königsmann, E. 1977. Das phylogenetische System der Hymenoptera. Teil 2: Symphyta. Dtsch. Entomol. Z. (N.F.)24:1-40. [1977.06.15]
Königsmann, E. 1978a. Das phylogenetische System der Hymenoptera. Teil 3: Terebrantes (Unterordnung Apocrita). Dtsch. Entomol. Z. (N.F.)25:1-55. [1978.06.01]
Königsmann, E. 1978b. Das phylogenetische System der Hymenoptera. Teil 4: Aculeata (Unterordnung Apocrita). Dtsch. Entomol. Z. (N.F.)25:365-435. [1978.11.30]
Kôno, H., Sugihara, Y. 1939. The ants of fir and pine forests in Japan. [In Japanese.] Trans. Kansai Entomol. Soc. 8:8-14. [1939.04.20]
Kostjuk, J. A. 1976. Catalogue of ant types (Hymenoptera, Formicidae) from the collection of V. A. Karavaev, preserved at the Institute of Zoology of the Academy of Sciences of the Ukrainian SSR. (Part 1). [In Russian.] Sb. Tr. Zool. Muz. (Kiev) 36:91-106. [1976]
Kotzias, H. 1930. *Formica picea* Nyl. in Schlesien. Zool. Anz. 92:56-58. [1930.11.01]
Kramer, K. U. 1950. Een verwaarloosde Nederlandse *Myrmica* vorm? Entomol. Ber. (Amst.) 13:97-98. [1950.07.01]
Kramer, K. U. 1952. Koloniestichting en sociaal parasitisme bij mieren. Vakbl. Biol. 32:65-72. [1952.04]
Kratochvíl, J. 1936a. Památce prvního naseho profesora pouzité entomologie prof. Dra Stepána Soudka. Cas. Cesk. Spol. Entomol. 33:125-131. [1936.05.15]
Kratochvíl, J. 1936b. Rozbor mravencí zvíreny Pavlovskych vrchu. Acta Soc. Sci. Nat. Moravo-Sil. 10(2):1-30. [1936]
Kratochvíl, J. 1938. Myrmekologické poznámky, 1-2. Entomol. Listy 1:161-162. [1938]
Kratochvíl, J. 1940a. Doplnek nalezist k Záleského Prodromu mravencu. Sb. Entomol. Oddel. Nár. Mus. Praze 18:241-249. [1940.12.01]
Kratochvíl, J. 1940b. Príspevky k poznání mravence *Strongylognathus Kratochvili* Silhavy. Vestn. Cesk. Zool. Spol. 8:24-46. [1940]
Kratochvíl, J. 1941a. Myrmekologické poznámky, 5. Príspevek k rozsírení vzácnejsích a sporadickych mravencu. Cas. Cesk. Spol. Entomol. 38:40-45. [1941.05.15]

Kratochvíl, J. 1941b. [Untitled. New subgenera of *Tetramorium*, and a new subspecies, *Tetramorium ferox silhavyi*, attributed to Kratochvíl.] P. 84 in: Novák, V., Sadil, J. Klíc k urcování mravencu strední Evropy se zvlástním zretelem k mravencí zvírene Cech a Moravy. Entomol. Listy 4:65-115. [1941]

Kratochvíl, J. 1941c. [Untitled. New species, *Tetramorium moravicum*, attributed to Kratochvíl.] P. 86 in: Novák, V., Sadil, J. Klíc k urcování mravencu strední Evropy se zvlástním zretelem k mravencí zvírene Cech a Moravy. Entomol. Listy 4:65-115. [1941]

Kratochvíl, J. 1941d. [Untitled. New variety, *Formica cinerera* var. *novaki*, attributed to Kratochvíl.] P. 106 in: Novák, V., Sadil, J. Klíc k urcování mravencu strední Evropy se zvlástním zretelem k mravencí zvírene Cech a Moravy. Entomol. Listy 4:65-115. [1941]

Kratochvíl, J. 1941e. [Untitled. New species, *Sysphincta fialai*, attributed to Kratochvíl.] P. 111 in: Novák, V., Sadil, J. Klíc k urcování mravencu strední Evropy se zvlástním zretelem k mravencí zvírene Cech a Moravy. Entomol. Listy 4:65-115. [1941]

Kratochvíl, J. 1941f. [Untitled. New variety, *Tetramorium staerckei* var. *gregori*, attributed to Kratochvíl.] P. 112 in: Novák, V., Sadil, J. Klíc k urcování mravencu strední Evropy se zvlástním zretelem k mravencí zvírene Cech a Moravy. Entomol. Listy 4:65-115. [1941]

Kratochvíl, J. 1944. Mravenci mohelnské reservace. Rozbor taxonomicky, faunisticko-oekologicky, sociologicky a zoogeograficky. Pp. 9-102 in: Kratochvíl, J., Novák, V., Snoflák, J. Mohelno. Soubor práci venoványch studiu vyznamne památky prírodní. 5. Hymenoptera - Aculeata. Formicidae - Apidae - Vespidae. Arch. Svazu Ochr. Prír. Domov. Moravé 6:1-155. [1944]

Kratochvíl, J. 1949. Mravenci jeseníku. Entomol. Listy 12:13-20. [1949.01.31]

Kratochvíl, J., Novák, V., Snoflák, J. 1944. Mohelno. Soubor práci venoványch studiu vyznamne památky prírodní. 5. Hymenoptera - Aculeata. Formicidae - Apidae - Vespidae. Arch. Svazu Ochr. Prír. Domov. Moravé 6:1-155. [1944]

Krausse, A. See: Krausse, A. H.

Krausse, A. H. 1910. Über Stridulationstöne bei Ameisen. Zool. Anz. 35:523-526. [1910.03.15]

Krausse, A. H. 1911a ("1909"). Über *Messor structor* Ltr. und einige andere Ameisen auf Sardinien. Bull. Soc. Entomol. Ital. 41:14-18. [1911.02.28]

Krausse, A. H. 1911b. *Formica*-Arten auf Sardinien. Wien. Entomol. Ztg. 30:168. [1911.08.08]

Krausse, A. H. 1912a. *Formica fusca* var. *glebaria* Nyl. auf Sardinien. Wien. Entomol. Ztg. 31:250. [1912.08.01]

Krausse, A. H. 1912b. Eine neue Ameisenform von Sardinien (*Pheidole pallidula* v. n. *Emeryi* m.). Int. Entomol. Z. 6:169. [1912.09.14]

Krausse, A. H. 1912c. Ueber sardische Ameisen. Arch. Naturgesch. (A)78(7):162-166. [1912.10]

Krausse, A. H. 1913. Über einige sardische Ameisen. Arch. Naturgesch. (A)79(6):39-41. [1913.09]

Krausse, A. H. 1919. Uebersicht über die Ameisen Sardiniens nebst einigen biologischen Beobachtungen. Z. Wiss. Insektenbiol. 15:96-103. [1919.12.15]

Krausse, A. H. 1922a. Myrmekologie und Phylogenie. Arch. Naturgesch. (A)88(9):79-87. [1922.09]

Krausse, A. H. 1922b. Einige myrmekologische Notizen. Z. Wiss. Insektenbiol. 17:161-163. [1922.11.15]

Krausse, A. H. 1922c. Notizen über *Leptothorax acervorum* F. (Form.). Entomol. Jahrb. 31:152-153. [1922]

Krausse, A. H. 1922d. *Formica rufa pratensis incisa* m. v. n. Entomol. Jahrb. 31:155. [1922]

Krausse, A. H. 1926a. *Formica rufa* an Weidenkätzchen. Int. Entomol. Z. 20:107-108. [1926.07.01]

Krausse, A. H. 1926b. Waldameisen-Varietäten. Int. Entomol. Z. 20:114-115. [1926.07.08]

Krausse, A. H. 1926c. *Formica rufa Aruncicola* Menozzii m.f.n. Entomol. Z. 40:336. [1926.10.08]

Krausse, A. H. 1926d. Über einige Formen der *Formica rufa* und *exsecta*. Int. Entomol. Z. 20:264. [1926.11.01]

Krausse, A. H. 1928. Einige myrmekologische Notizen. Int. Entomol. Z. 22:118-119. [1928.06.15]
Krausse, A. H. 1929. Ameisenkunde. Eine Einführung in die Systematik und Biologie der Ameisen. Stuttgart: A. Kernen, 172 pp. [1929]
Kremer, G., Berndt, K. P. 1986. Zur Morphologie normaler und gynandromorpher Pharaoameisen *Monomorium pharaonis* (L.) (Hym. Formicidae). Dtsch. Entomol. Z. (N.F.)3:177-221. [1986.10.17]
Krishna Ayyar, P. N. 1937. A new carton-building species of ant in south India. *Crematogaster dohrni artifex*, Mayr. J. Bombay Nat. Hist. Soc. 39:291-308. [1937.04.15]
Krombein, K. V. 1958. Date of publication, First supplement, Synoptic catalog of North American Hymenoptera. Proc. Entomol. Soc. Wash. 60:266. [1958.12.18]
Kroyer, H. N. 1846. Danmarks Fiske. Vol. 3. Kjobenhavn [= Copenhagen]: S. Triers, 1279 pp. [1846] [*Acantholepis* (Pisces), senior homonym of *Acantholepis* Mayr, 1861.]
Krzysztofiak, A., Krzysztofiak, L. 1989. Anomalie w budowie ciala robotnic *Myrmica ruginodis* Nyl. (Hymenoptera, Formicidae). Przegl. Zool. 33:467-469. [1989.10]
Krzysztofiak, L. 1984. Mrówki (Hymenoptera, Formicoidea) Swietokrzyskiego Parku Narodowego. Fragm. Faun. (Warsaw) 28:309-323. [1984.10.30]
Krzysztofiak, L. 1985. Rozmieszczenie i zageszczenie gniazd mrówek w Puszczy Augustowskiej (Pojezierze Mazurskie). Fragm. Faun. (Warsaw) 29:137-149. [1985.01.15]
Kubota, M. 1971. A checklist of the ants of Japan. Odawara: published by the author, 34 pp. [Not seen. Cited in Onoyama (1980) and Onoyama & Terayama (1994).]
Kubota, M. 1974. Ants. [In Japanese.] Pp. 29-32 in: Encyclopedia Genre Japonica. Volume 20. Animals. Tokyo: Shogakukan, 682 pp. [1974.06.10]
Kubota, M. 1976. Dacetine ants from Ogasawara Islands. [In Japanese.] Ari 7:4. [1976.05.01]
Kubota, M. 1977. Records of ants (1). [In Japanese.] Ari 8:4. [1977.08.01]
Kubota, M. 1983. Records of ants (3). [In Japanese.] Ari 11:7-8. [1983.08.16]
Kubota, M. 1984a. Anomalous female wings in *Vollenhovia emeryi* Wheeler. [Abstract.] [In Japanese.] Ari 12:2-3. [1984.05.01]
Kubota, M. 1984b. Records of ants (4). [In Japanese.] Ari 12:12. [1984.05.01]
Kubota, M. 1989a ("1988"). Some informations for the identification of Japanese *Pachycondyla*. [Abstract.] [In Japanese.] Ari 16:2-3. [1989.08]
Kubota, M. 1989b ("1988"). Records of ants (5). [In Japanese.] Ari 16:16. [1989.08] [Date of publication from Onoyama & Terayama (1994:20).]
Kubota, M. 1994. Geographical distribution of the indo-australian ant genus *Lordomyrma* Emery. [Abstract.] [In Japanese.] Ari 17:3. [1994.04.30]
Kubota, M., Kondoh, M. 1966. Notes on the captures of *Polyrhachis hippomanes moesta* Emery and *Pheidole indica* Mayr. [In Japanese.] Ari 3:4. [1966.06.25]
Kubota, S., Terayama, M. 1989 ("1988"). Ant fauna of Tokyo. (1) A list of ants collected at the parks. [In Japanese.] Ari 16:14-16. [1989.08] [Date of publication from Onoyama & Terayama (1994:21).]
Kugler, C. 1977 ("1976"). A new species of *Platythyrea* (Hymenoptera, Formicidae) from Costa Rica. Psyche (Camb.) 83:216-221. [1977.02.24]
Kugler, C. 1978a. A comparative study of the myrmicine sting apparatus (Hymenoptera, Formicidae). Stud. Entomol. 20:413-548. [1978.08.30]
Kugler, C. 1978b. Pygidial glands in the myrmicine ants (Hymenoptera, Formicidae). Insectes Soc. 25:267-274. [1978.09]
Kugler, C. 1979a ("1978"). Description of the ergatoid queen of *Pogonomyrmex mayri* with notes on the worker and male (Hym., Formicidae). Psyche (Camb.) 85:169-182. [1979.04.24]
Kugler, C. 1979b ("1978"). Further studies of the myrmicine sting apparatus: *Eutetramorium*, *Oxyopomyrmex*, and *Terataner* (Hymenoptera, Formicidae). Psyche (Camb.) 85:255-263. [1979.04.24]
Kugler, C. 1979c. Evolution of the sting apparatus in the myrmicine ants. Evolution 33:117-130. [1979.05.15]

Kugler, C. 1979d. Alarm and defense: a function for the pygidial gland of the myrmicine ant, *Pheidole biconstricta*. Ann. Entomol. Soc. Am. 72:532-536. [1979.07.15]

Kugler, C. 1980a. *Odontomachus cornutus* rediscovered (Hymenoptera: Formicidae: Ponerinae). J. Kansas Entomol. Soc. 53:225-229. [1980.01.02]

Kugler, C. 1980b. The sting apparatus in the primitive ants *Nothomyrmecia* and *Myrmecia*. J. Aust. Entomol. Soc. 19:263-267. [1980.12.15]

Kugler, C. 1986. Stings of ants of the tribe Pheidologetini (Myrmicinae). Insecta Mundi 1:221-230. [1986.12]

Kugler, C. 1990. The sting apparatus and phylogeny of the ants. Pp. 315-316 in: Veeresh, G. K., Mallik, B., Viraktamath, C. A. (eds.) Social insects and the environment. Proceedings of the 11th International Congress of IUSSI, 1990. New Delhi: Oxford & IBH Publishing Co., xxxi + 765 pp. [1990.08]

Kugler, C. 1991. Stings of ants of the tribe Ectatommini (Formicidae: Ponerinae). Insecta Mundi 5:153-166. [1991.12]

Kugler, C. 1992. Stings of ants of the Leptanillinae (Hymenoptera: Formicidae). Psyche (Camb.) 99:103-115. [1992.12.04]

Kugler, C. 1994. A revision of the ant genus *Rogeria* with description of the sting apparatus (Hymenoptera: Formicidae). J. Hym. Res. 3:17-89. [1994.10.15]

Kugler, C., Brown, W. L., Jr. 1982. Revisionary and other studies on the ant genus *Ectatomma*, including the description of two new species. Search Agric. (Ithaca N. Y.) 24:1-8. [1982.11]

Kugler, C., Hincapié, M. del C. 1983. Ecology of the ant *Pogonomyrmex mayri*: distribution, abundance, nest structure, and diet. Biotropica 15:190-198. [1983.09]

Kugler, J. 1981. A new species of *Cataglyphis* Förster (Hymenoptera: Formicidae) from Israel and Sinai. Isr. J. Entomol. 15:83-88. [1981.12]

Kugler, J. 1984 ("1983"). The males of *Cardiocondyla* Emery (Hymenoptera: Formicidae) with the description of the winged male of *Cardiocondyla wroughtoni* (Forel). Isr. J. Entomol. 17:1-21. [1984.04.15]

Kugler, J. 1987 ("1986"). The Leptanillinae (Hymenoptera: Formicidae) of Israel and a description of a new species from India. Isr. J. Entomol. 20:45-57. [1987.03]

Kugler, J. 1988. The zoogeography of Israel. 9. The zoogeography of social insects of Israel and Sinai. Monogr. Biol. 62:251-275. [1988]

Kugler, J., Soussan, O. 1992. Redescription and biological observations on *Leptothorax flavispinus* André (Hymenoptera: Formicidae). Isr. J. Entomol. 25-26:209-213. [1992.03.15]

Kula, E. 1985. *Formica lugubris* Zett. (Hymenoptera, Formicidae) ve smrkovych porostech lokality Alfrédka v Jeseníkách. Cas. Slezského Muz. Ser. A Vedy Prír. 34:31-42. [1985.04]

Kulmatycki, W. 1920a. Mrówki niektórych okolic Malopolski. Spraw. Kom. Fizjogr. Mater. Fizjogr. Kraju 53-54:157-172. [1920]

Kulmatycki, W. 1920b. Przycznek do fauny myrmekologicznej b. Królestwa Polskiego. Spraw. Kom. Fizjogr. Mater. Fizjogr. Kraju 53-54:189-194. [1920]

Kulmatycki, W. 1922. Przycznek do fauny mrówek Wielkopolski i Pomorza. Spraw. Kom. Fizjogr. Mater. Fizjogr. Kraju 55-56:71-86. [1922]

Kumar, R., Akotoye, N. A. K. 1972. Variation in number and structure of the Malpighian tubules in *Oecophylla longinoda* (Latreille) (Hymenoptera: Formicidae-Formicinae) workers. Bull. Entomol. Soc. Niger. 3:121-126. [1972.08]

Kumbkarni, C. G. 1965. Cytological studies in Hymenoptera. III. Cytology of parthenogenesis in the formicid ant, *Camponotus compressus*. Caryologia 18:305-312. [1965.09.10]

Kume, S. 1914. On the distribution of *Polyrhachis lamellidens*. [In Japanese.] Insect World 18: 24-25. [1914.01.15]

Kupyanskaya, A. N. 1980. Ants of the genus *Formica* Linnaeus (Hymenoptera, Formicidae) of the Soviet Far East. [In Russian.] Pp. 95-108 in: Ler, P. A. (ed.) Taxonomy of insects of the Far East. [In Russian.] Vladivostok: Akademiya Nauk SSSR, 120 pp. [> 1980.02.06] [Date signed for printing ("podpisano k pechati").]

Kupyanskaya, A. N. 1981. Rare and relict species of ants (Hymenoptera, Formicidae) in the south of the Soviet Far East. [In Russian.] Pp. 36-42 in: Bromlei, G. F., Kostenko, V. A., Yudin, V. G., Nechaev, V. A. (eds.) Rare and endangered terrestrial animals of the Far East of the USSR. [In Russian.] Vladivostok: Akademiya Nauk SSSR, 192 pp. [> 1981.12.20] [Date signed for printing ("podpisano k pechati").]

Kupyanskaya, A. N. 1982 ("1981"). Ants of the genera *Camponotus* Mayr, *Polyergus* Latr., and *Paratrechina* Motsch. (Hymenoptera, Formicidae) from the Far East of the USSR. [In Russian.] Pp. 117-124 in: Ler, P. A., Kupyanskaya, A. N., Lelej, A. S., Fedikova, V. S. (eds.) Hymenoptera of the Far East. [In Russian.] Vladivostok: Akademiya Nauk SSSR, 134 pp. [> 1982.02.15] [Date signed for printing ("podpisano k pechati").]

Kupyanskaya, A. N. 1984. The ant *Formica yessensis* Forel, 1901 (Hymenoptera, Formicidae) in southern Primorye Territory. [In Russian.] Pp. 98-112 in: Ler, P. A. (ed.) The fauna and ecology of insects in the south of the Far East. Collected scientific papers. [In Russian.] Vladivostok: Akademiya Nauk SSSR, 151 pp. [> 1984.01.06] [Date signed for printing ("podpisano k pechati").]

Kupyanskaya, A. N. 1985. Ants (Hymenoptera, Formicidae) of the Bolshoi Khekhtsir Reserve. [In Russian.] Pp. 76-84 in: Ler, P. A., Storozhenko, S. Y. (eds.) Taxonomy and ecology of arthropods from the Far East. [In Russian.] Vladivostok: Far Eastern Scientific Centre, 132 pp. [> 1984.11.23] [Date signed for printing ("podpisano k pechati"). Title page dated 1985.]

Kupyanskaya, A. N. 1986a. Ants (Hymenoptera, Formicidae) of the group *Myrmica lobicornis* Nylander from the Far East: [In Russian.] Pp. 83-90 in: Ler, P. A. (ed.) Systematics and ecology of insects from the Far East. [In Russian.] Vladivostok: Akademiya Nauk SSSR, 155 pp. [> 1986.01.16] [Date signed for printing ("podpisano k pechati").]

Kupyanskaya, A. N. 1986b. Ants (Hymenoptera, Formicidae) from the northern part of the Far East. [In Russian.] Pp. 91-102 in: Ler, P. A. (ed.) Systematics and ecology of insects from the Far East. [In Russian.] Vladivostok: Akademiya Nauk SSSR, 155 pp. [> 1986.01.16] [Date signed for printing ("podpisano k pechati").]

Kupyanskaya, A. N. 1987. Biotopic distribution, nesting and feeding area of *Liometopum orientalis* (Hymenoptera, Formicidae) in Primorye Territory. [In Russian.] Zool. Zh. 66:1321-1327. [1987.09] [September issue. Date signed for printing ("podpisano k pechati"): 18 August 1987.]

Kupyanskaya, A. N. 1988. A Far Eastern representative of the genus *Liometopum* (Hymenoptera: Formicidae). [In Russian.] Vestn. Zool. 1988(1):29-34. [1988.02] [January-February issue. Date signed for printing ("podpisano k pechati"): 26 January 1988.]

Kupyanskaya, A. N. 1989. Ants of the subgenus *Dendrolasius* Ruzsky, 1912 (Hymenoptera, Formicidae, genus *Lisius* [sic] Fabricius, 1804) of the Far Eastern USSR. [In Russian.] Entomol. Obozr. 68:779-789. [> 1989.12.08] [Date signed for printing ("podpisano k pechati"). Translated in Entomol. Rev. (Wash.) 69(4):99-111. See Kupyanskaya (1990b).]

Kupyanskaya, A. N. 1990a. Ants of the Far Eastern USSR. [In Russian.] Vladivostok: Akademiya Nauk SSSR, 258 pp. [> 1990.09.13] [Date signed for printing ("podpisano k pechati").]

Kupyanskaya, A. N. 1990b. Ants of the subgenus *Dendrolasius* Ruzsky, 1912 (Hymenoptera, Formicidae, genus *Lasius* Fabricius, 1804) from the Soviet Far East. Entomol. Rev. (Wash.) 69(4):99-111. [1990] [English translation of Kupyanskaya (1989).]

Kurian, C. 1955. Bethyloidea (Hymenoptera) from India. Agra Univ. J. Res. 4:67-155. [1955.01] [See Brown (1988a).]

Kuroiwa, T. 1926. Provisional list of the Hymenoptera collected in Loochoo Islands during the years 1905-1907. Trans. Nat. Hist. Soc. Formosa 16:138-141. [1926.08.01] [Formicidae pp. 138-139.]

Kürschner, I. 1969. Über den Bau der Antennen von *Formica pratensis* Retzius, 1783, unter besonderer Berücksichtigung der Sensillen. Beitr. Entomol. 19:273-280. [1969]

Kürschner, I. 1970. Zur Anatomie von *Formica pratensis* Retzius, 1783. Das Gelenk zwischen Thorax und Gaster einschliesslich der im Gelenkgebiet vorhandenen und der Gelenkfunktion dienenden Organe (Hymenoptera: Formicidae). Beitr. Entomol. 20:375-387. [1970]

Kürschner, I. 1971. Zur Anatomie von *Formica pratensis* Retzius, 1783. Morphologische Untersuchungen der sekretorischen Kopfdrüsen (Postpharynxdrüse, Maxillardrüse, Mandibulardrüse, Zungendrüse) und der am Kopf ausmündenden Labialdrüse. Beitr. Entomol. 21:191-210. [1971]

Kusnezov, N. See also: Kuznetsov-Ugamsky, N. N. (1923-1930)

Kusnezov, N. 1948. Instrucciones preliminares para la caza de las hormigas. Misc. Inst. Miguel Lillo 13:1-28. [1948]

Kusnezov, N. 1949a. El género *Monomorium* (Hymenoptera, Formicidae) en la Argentina. Acta Zool. Lilloana 7:423-448. [1949.12.31]

Kusnezov, N. 1949b. Sobre la reproducción de las formas sexuales en *Solenopsis patagonica* Emery (Hymenoptera, Formicidae). Acta Zool. Lilloana 8:281-290. [1949.12.31]

Kusnezov, N. 1949c. *Pogonomyrmex* del grupo *Ephebomyrmex* en la fauna de la Patagonia (Hymenoptera, Formicidae). Acta Zool. Lilloana 8:291-307. [1949.12.31]

Kusnezov, N. 1949d. El género *Cyphomyrmex* (Hymenoptera, Formicidae) en la Argentina. Acta Zool. Lilloana 8:427-456. [1949.12.31]

Kusnezov, N. 1949e. *Crematogaster* (*Neocrema*) *descolei* n. sp. (Hymenoptera, Formicidae). Acta Zool. Lilloana 8:587-590. [1949.12.31]

Kusnezov, N. 1951a. Hormigas en el Parque General San Martín (Entre Rios). Bol. Minist. Educ. Entre Ríos 32: [Not seen. Cited in Gavrilov, 1964:16).]

Kusnezov, N. 1951b. *Dorymyrmex emmaericaellus* n. sp. Folia Univ. Cochabamba 5:59-61. [1951.10]

Kusnezov, N. 1951c. Los segmentos palpales en hormigas. Folia Univ. Cochabamba 5:62-70. [1951.10]

Kusnezov, N. 1951d. "Dinergatogina" en *Oligomyrmex bruchi* Santschi (Hymenoptera Formicidae). Rev. Soc. Entomol. Argent. 15:177-181. [1951.10.20]

Kusnezov, N. 1951e. El género *Pogonomyrmex* Mayr (Hym., Formicidae). Acta Zool. Lilloana 11:227-333. [1951.11.13]

Kusnezov, N. 1951f. *Myrmelachista* en la Patagonia (Hymenoptera, Formicidae). Acta Zool. Lilloana 11:353-365. [1951.11.13]

Kusnezov, N. 1951g. *Descolemyrma ogloblini*, género y especie nuevos de la tribu Attini (Hymenoptera, Formicidae). Acta Zool. Lilloana 11:459-465. [1951.11.13]

Kusnezov, N. 1952a ("1951"). Un caso de evolución eruptiva. *Eriopheidole symbiotica* nov. gen., nov. sp. (Himenoptera, Formicidae). Mem. Mus. Entre Ríos 29:5-31. [1952.01.30]

Kusnezov, N. 1952b. La posición sistemática de la subfamilia Cerapachyinae (Hymenoptera, Formicidae). Dusenia 3:115-117. [1952.03.31]

Kusnezov, N. 1952c. Algunos datos sobre la dispersión geográfica de las hormigas (Hymenoptera, Formicidae) en la República Argentina. An. Soc. Cient. Argent. 153:230-242. [1952.06]

Kusnezov, N. 1952d ("1951"). El género *Pheidole* en la Argentina (Hymenoptera, Formicidae). Acta Zool. Lilloana 12:5-88. [1952.10.15]

Kusnezov, N. 1952e ("1951"). *Lasiophanes* Emery en la Patagonia. Acta Zool. Lilloana 12: 89-100. [1952.10.15]

Kusnezov, N. 1952f ("1951"). El género *Camponotus* en la Argentina (Hymenoptera, Formicidae). Acta Zool. Lilloana 12:183-252. [1952.10.15]

Kusnezov, N. 1952g ("1951"). Un caso de ergatandromorfismo en *Lasiophanes nigriventris* Spin. (Hymenoptera, Formicidae). Acta Zool. Lilloana 10:153-162. [1952]

Kusnezov, N. 1952h ("1951"). El género *Wasmannia* en la Argentina (Hymenoptera, Formicidae). Acta Zool. Lilloana 10:173-182. [1952]

Kusnezov, N. 1952i ("1951"). El género *Oligomyrmex* Mayr en la Argentina (Hymenoptera, Formicidae). Acta Zool. Lilloana 10:183-187. [1952]

Kusnezov, N. 1952j ("1951"). El estado real del grupo *Dorymyrmex* Mayr (Hymenoptera, Formicidae). Acta Zool. Lilloana 10:427-448. [1952]
Kusnezov, N. 1952k ("1951"). Acerca de las hormigas simbióticas del género *Martia* Forel (Hymenoptera, Formicidae). Acta Zool. Lilloana 10:717-722. [1952]
Kusnezov, N. 1953a. Die Ameisenfauna Argentiniens. Zool. Anz. 150:15-25. [1953.01]
Kusnezov, N. 1953b. Las hormigas en los Parques Nacionales de la Patagonia y los problemas relacionados. An. Mus. Nahuel Huapí 3:105-124. [1953.11]
Kusnezov, N. 1953c. Lista de las hormigas de Tucumán con descripción de dos nuevos géneros (Hymenoptera, Formicidae). Acta Zool. Lilloana 13:327-339. [1953.12]
Kusnezov, N. 1953d. El género *Synsolenopsis* Forel (Hymenoptera, Formicidae). Acta Zool. Lilloana 13:341-348. [1953.12]
Kusnezov, N. 1953e. Tendencias evolutivas de las hormigas en la parte austral de Sud América. Folia Univ. Cochabamba 6:86-210. [1953]
Kusnezov, N. 1953f. La fauna mirmecológica de Bolivia. Folia Univ. Cochabamba 6:211-229. [1953]
Kusnezov, N. 1953g. *Bisolenopsis sea*. Un nuevo género y especie de hormigas y sus relaciones con los géneros vecinos (Hymenoptera, Formicidae). Mem. Mus. Entre Ríos 31:7-44. [1953]
Kusnezov, N. 1954a. Un género nuevo de hormigas (*Paranamyrma solenopsidis* nov. gen. nov. sp.) y los problemas relacionados (Hymenoptera, Formicidae). Mem. Mus. Entre Ríos 30:7-21. [1954.04]
Kusnezov, N. 1954b. Phyletische Bedeutung der Maxillar- und Labialtaster der Ameisen. Zool. Anz. 153:28-38. [1954.07]
Kusnezov, N. 1954c. La formula palpal in las hormigas (nuevos datos y algunas interpretaciones) (Hymenoptera: Formicidae). Dusenia 5:255-258. [1954.11.30]
Kusnezov, N. 1954d. Reacciones defensivas y offensivas en las hormigas. Folia Univ. Cochabamba 7:55-81. [1954]
Kusnezov, N. 1954e. Descripción de *Pogonomyrmex marcusi* Kusnezov. Folia Univ. Cochabamba 7:82-86. [1954]
Kusnezov, N. 1955a. Evolución de las hormigas. Dusenia 6:1-34. [1955.03.31]
Kusnezov, N. 1955b. Zwei neue Ameisengattungen aus Tucuman (Argentinien). Zool. Anz. 154:268-277. [1955.06]
Kusnezov, N. 1955c. Un nuevo caracter de importancia filogenética en las hormigas (Hymenoptera, Formicidae). Dusenia 6:183-186. [1955.09.30]
Kusnezov, N. 1955d. El parasitismo social en Hymenoptera y el problema de evolución. Cienc. Invest. 11:555-558. [1955.12]
Kusnezov, N. 1955e. Ochos años en Sud America. Folia Univ. Cochabamba 8:45-93. [1955]
Kusnezov, N. 1956a. Claves para la identificación de las hormigas de la fauna argentina. Idia 104-105:1-56. [1956.09]
Kusnezov, N. 1956b. A comparative study of ants in desert regions of Central Asia and of South America. Am. Nat. 90:349-360. [1956.12.10]
Kusnezov, N. 1956c. Sobre la "reversibilidad" en la evolución orgánica. Dusenia 7:325-328. [1956.12.31]
Kusnezov, N. 1956d. Der Sexualdimorphismus bei den Ameisen. Z. Wiss. Zool. 159:319-347. [1956.12]
Kusnezov, N. 1957a. Südamerikanische Ameisengattungen (Cerapachynae [sic] und Ponerinae). Zool. Anz. 158:196-208. [1957.05]
Kusnezov, N. 1957b. Die Solenopsidinen-Gattungen von Südamerika (Hymenoptera, Formicidae). Zool. Anz. 158:266-280. [1957.06]
Kusnezov, N. 1957c. Numbers of species of ants in faunae of different latitudes. Evolution 11:298-299. [1957.09.27]
Kusnezov, N. 1957e. Nuevas especies de hormigas (Hymenoptera, Formicidae). Rev. Soc. Urug. Entomol. 2:7-18. [1957.12.01]

Kusnezov, N. 1958a. La posición sistematica del género *Rogeria*, con descripción de una nueva especie. Acta Zool. Lilloana 15:41-45. [1958.08.29]

Kusnezov, N. 1958b. *Lilidris metatarsalis* gen. et spec. nov. (Hym. Formicidae). Acta Zool. Lilloana 15:189-193. [1958.08.29]

Kusnezov, N. 1958c. La posición sistemática de las especies argentinas del genero *Leptothorax* Mayr, 1855. Acta Zool. Lilloana 16:265-271. [1958.08.29]

Kusnezov, N. 1958d. Nota sobre la sinonimia de *Linepithema* Mayr (Hym. Formicidae. Dolichoderinae). Acta Zool. Lilloana 16:273. [1958.09.25]

Kusnezov, N. 1959. Die Dolichoderinen-Gattungen von Süd-Amerika (Hymenoptera, Formicidae). Zool. Anz. 162:38-51. [1959.01]

Kusnezov, N. 1960a ("1959"). La posición sistemática del género *Metapone* Forel (Hymenoptera, Formicidae). Acta Zool. Lilloana 18:119-126. [1960.06.10]

Kusnezov, N. 1960b ("1959"). La fauna de hormigas en el oeste de la Patagonia y Tierra del Fuego. Acta Zool. Lilloana 17:321-401. [1960.06.10]

Kusnezov, N. 1960c. *Brachymyrmex physogaster* n. sp. aus Argentinien und das Problem der Physogastrie bie den Ameisen. Zool. Anz. 165:381-388. [1960.11]

Kusnezov, N. 1961 ("1960"). Zoogeografía de Formicidae en Sud América. Pp. 509-512 in: Strouhal, H., Beier, M. (eds.) XI. Internationaler Kongress für Entomologie. Wien, 17. bis 25. August 1960. Verhandlungen. Band I. Wien: Organisationskomitee des XI. Internationalen Kongresses für Entomologie, xliv + 803 pp. [1961] [Publication date ("Ende 1961") from statement in Band II.]

Kusnezov, N. 1962a. El género *Acanthostichus* Mayr (Hymenoptera, Formicidae). Acta Zool. Lilloana 18:121-138. [1962.05.23]

Kusnezov, N. 1962b. Una nueva especie del género *Brownidris* Kusnezov (Hymenoptera, Formicidae). Acta Zool. Lilloana 18:155-161. [1962.05.23]

Kusnezov, N. 1962c. El ala posterior de las hormigas. Acta Zool. Lilloana 18:367-378. [1962.05.23]

Kusnezov, N. 1962d. El vuelo nupcial de las hormigas. Acta Zool. Lilloana 18:385-442. [1962.05.23]

Kusnezov, N. 1963. Evolution der Ameisen. Symp. Genet. Biol. Ital. 12:103-121. [1963]

Kusnezov, N. 1964 ("1963"). Zoogeografía de las hormigas en Sudamérica. Acta Zool. Lilloana 19:25-186. [1964.06.10]

Kusnezov, N. 1967. La ley de la desigualdad del desarrollo. Acta Zool. Lilloana 21:123-251. [1967.11]

Kusnezov, N. 1969. Nuevas especies de hormigas. Acta Zool. Lilloana 24:33-38. [1969.07]

Kusnezov, N. 1978. Hormigas argentinas: clave para su identificación. Misc. Inst. Miguel Lillo 61:1-147 + 28 pl. [1978] [Edited by R. Golbach. Revised edition of Kusnezov (1956a).]

Kusnezov, N., Golbach, R. 1952 ("1951"). Lista de las especies argentinas de la tribu Dacetini, Hymenoptera, Formicidae. Acta Zool. Lilloana 10:423-426. [1952]

Kusnezov, N. N. (1925-1926) See: Kuznetsov-Ugamsky, N. N.

Kusnezov-Ugamsky, N. See: Kuznetsov-Ugamsky, N. N.

Kusnezow, N. (1948) See: Kusnezov, N.

Kusnezow, N. N. (1923) See: Kuznetsov-Ugamsky, N. N.

Kusnezow-Ugamsky, N. See: Kuznetsov-Ugamsky, N. N.

Kutter, H. 1916 ("1915"). Eine myrmecologische Excursion nach dem Südfuss der Alpen. Mitt. Schweiz. Entomol. Ges. 12:344-348. [1916.03]

Kutter, H. 1917. Myrmikologische Beobachtungen. Biol. Zentralbl. 37:429-437. [1917.09.30]

Kutter, H. 1919. Beiträge zur Ameisenfauna der Schweiz. Mitt. Schweiz. Entomol. Ges. 13:13-17. [1919.11]

Kutter, H. 1920. *Strongylognathus Huberi* For. r. *alpinus* Wh. eine Sklaven raubende Ameise. Biol. Zentralbl. 40:528-538. [1920.12.15]

Kutter, H. 1921. *Strongylognathus alpinus* Wh., ein neuer Sklavenräuber. Mitt. Schweiz. Entomol. Ges. 13:117-119. [1921.09.01]

Kutter, H. 1923. Der Sklavenräuber *Strongylognathus huberi* For. ssp. *alpinus* Wheeler. Rev. Suisse Zool. 30:387-424. [1923.12]

Kutter, H. 1924a. *Myrmica scabrinodis* Nyl. var. *rugulosoides* For. Schweiz. Entomol. Anz. 3:8. [1924.01.01]

Kutter, H. 1924b. *Myrmica scabrinodis* Nyl. var. *rugulosoides* For. (Schluss.) Schweiz. Entomol. Anz. 3:15. [1924.01.01]

Kutter, H. 1925. Eine neue Ameise der Schweiz. Mitt. Schweiz. Entomol. Ges. 13:409-412. [1925.07.15]

Kutter, H. 1927a. Ein myrmekologischer Streifzug durch Sizilien. Folia Myrmecol. Termit. 1: 94-104. [1927.04]

Kutter, H. 1927b. Ein myrmekologischer Streifzug durch Sizilien (Schluss). Folia Myrmecol. Termit. 1:135-136. [1927.06]

Kutter, H. 1928. A. Formicidae. Pp. 65-67 in: Schulthess, A. v. Beiträge zur Kenntnis nordafrikanischer Hymenopteren. EOS. Rev. Esp. Entomol. 4:65-92. [1928.04.28]

Kutter, H. 1931a. Mit Bananen eingeschleppte Ameisen. Mitt. Schweiz. Entomol. Ges. 15:61-64. [1931.03.15]

Kutter, H. 1931b. Verzeichnis der entomologischen Arbeiten von Prof. Dr. August Forel. Mitt. Schweiz. Entomol. Ges. 15:180-193. [1931.12.15]

Kutter, H. 1931c. *Forelophilus*, eine neue Ameisengattung. Mitt. Schweiz. Entomol. Ges. 15:193-195. [1931.12.15]

Kutter, H. 1932. Ameisen aus dem Museum zu Dresden. Mitt. Schweiz. Entomol. Ges. 15:207-210. [1932.03.15]

Kutter, H. 1933. Einige Ameisen von der Südküste von Neu-Britannien. Mitt. Schweiz. Entomol. Ges. 15:471-474. [1933.06.15]

Kutter, H. 1936. Neue Schweizerameisen. Mitt. Schweiz. Entomol. Ges. 16:722. [1936.09.15]

Kutter, H. 1945a. Ein neue Ameisengattung. Mitt. Schweiz. Entomol. Ges. 19:485-487. [1945.10.20]

Kutter, H. 1945b. Beitrag zur Kenntnis von *Strongylognathus Huberi* For. ssp. *alpinus* Wh. (Hym. Form.). Mitt. Schweiz. Entomol. Ges. 19:645-646. [1945.12]

Kutter, H. 1946. *Lasius* (*Chthonolasius*) *carniolicus* Mayr, eine neue Schweizerameise. Mitt. Schweiz. Entomol. Ges. 19:698-699. [1946.03]

Kutter, H. 1948. Beitrag zur Kenntnis der Leptanillinae (Hym. Formicidae). Eine neue Ameisengattung aus Süd-Indien. Mitt. Schweiz. Entomol. Ges. 21:286-295. [1948.08.25]

Kutter, H. 1950a. Über eine neue, extrem parasitische Ameise. 1. Mitteilung. Mitt. Schweiz. Entomol. Ges. 23:81-94. [1950.08.10]

Kutter, H. 1950b. Über zwei neue Ameisen. Mitt. Schweiz. Entomol. Ges. 23:337-346. [1950.10.15]

Kutter, H. 1950c. Über *Doronomyrmex* und verwandte Ameisen. 2. Mitteilung. Mitt. Schweiz. Entomol. Ges. 23:347-353. [1950.10.15]

Kutter, H. 1951. *Epimyrma Stumperi* Kutter (Hym. Formicid.). 2. Mitteilung. Mitt. Schweiz. Entomol. Ges. 24:153-174. [1951.07.20]

Kutter, H. 1952a. Über *Plagiolepis xene* Stärcke (Hym., Formicid.). Mitt. Schweiz. Entomol. Ges. 25:57-72. [1952.05.15]

Kutter, H. 1952b. Bemerkenswerte Missbildung bei *Leptothorax acervorum* Latr. (Hym., Formicid.). Mitt. Schweiz. Entomol. Ges. 25:354. [1952.12.17]

Kutter, H. 1956. Beiträge zur Biologie palaearktischer *Coptoformica* (Hym. Form.). Mitt. Schweiz. Entomol. Ges. 29:1-18. [1956.03.05]

Kutter, H. 1957. Zur Kenntnis schweizerischer *Coptoformica*arten (Hym. Form.). 2. Mitteilung. Mitt. Schweiz. Entomol. Ges. 30:1-24. [1957.05.30]

Kutter, H. 1958a. Einsame Ameisen. Mitt. Schweiz. Entomol. Ges. 31:177-190. [1958.10.10]

Kutter, H. 1958b. Über Modificationen bei Ameisenarbeiterinnen, welche durch den Parasitismus von Mermithiden (Nematod.) verursacht worden sind. Mitt. Schweiz. Entomol. Ges. 31:313-316. [1958.12.31]

Kutter, H. 1961a. Der individuelle Insektenpass als taxonomische Platzkarte, ein Vorschlag. Mitt. Schweiz. Entomol. Ges. 33:238-244. [1961.01.31]

Kutter, H. 1961b. Bericht über die Sammelaktion schweizerischer Waldameisen der *Formica-rufa*-Gruppe 1960/61. Schweiz. Z. Forstwes. 112:788-797. [1961.12]

Kutter, H. 1962. Bericht über die Sammelaktion schweizerischer Waldameisen der *Formica-rufa*-Gruppe 1960/61. Waldhygiene 4:193-202. [1962]

Kutter, H. 1963a. Miscellanea myrmecologica I. Mitt. Schweiz. Entomol. Ges. 35:129-137. [1963.06.30]

Kutter, H. 1963b. Ein kleiner Beitrag zur Kenntnis unserer Waldameisenfauna. Schweiz. Z. Forstwes. 114:646-653. [1963.11]

Kutter, H. 1964. Miscellanea myrmecologica II. Mitt. Schweiz. Entomol. Ges. 36:321-329. [1964.02.25]

Kutter, H. 1965a. Miscellanea myrmecologica III. Mitt. Schweiz. Entomol. Ges. 37:127-137. [1965.02.15]

Kutter, H. 1965b. *Formica nigricans* Em. (=*cordieri* Bondr.) bona species? Mitt. Schweiz. Entomol. Ges. 37:138-150. [1965.02.15]

Kutter, H. 1967a ("1966"). Einige Ergebnisse weiterer *Coptoformica*-Studien. Insectes Soc. 13:227-240. [1967.03]

Kutter, H. 1967b. Variationsstatistische Erhebungen an Weibchen von *Formica lugubris* Zett. Mitt. Schweiz. Entomol. Ges. 40:63-77. [1967.07.20]

Kutter, H. 1967c. Beschreibung neuer Sozialparasiten von *Leptothorax acervorum* F. (Formicidae). Mitt. Schweiz. Entomol. Ges. 40:78-91. [1967.07.20]

Kutter, H. 1968a. Persönliche Erinnerungen an August Forel. Mitt. Schweiz. Entomol. Ges. 40:281-289. [1968.04.15]

Kutter, H. 1968b ("1969"). Die sozialparasitischen Ameisen der Schweiz. Neujahrsbl. Naturforsch. Ges. Zür. 171:1-62. [1968.12.31]

Kutter, H. 1968c ("1967"). Liste sozialparasitischer Ameisen. Arch. Inst. Grand-Ducal Luxemb. (n.s.)33:201-210. [1968]

Kutter, H. 1970. Über den Formenreichtum bei *Myrmica lobicornis*-Arbeiterinnen (Hymenoptera, Formicidae). Mitt. Schweiz. Entomol. Ges. 43:143-146. [1970.10.15]

Kutter, H. 1971. Taxonomische Studien an schweizer Ameisen (Hymenopt., Formicidae). Mitt. Schweiz. Entomol. Ges. 43:258-271. [1971.03.15]

Kutter, H. 1973a ("1972"). Über *Xenhyboma mystes* Santschi. Mitt. Schweiz. Entomol. Ges. 45:321-324. [1973.02.28]

Kutter, H. 1973b ("1972"). *Epitritus argiolus* Emery(?), Genus und Species neu für die Schweiz. Mitt. Schweiz. Entomol. Ges. 45:325-326. [1973.02.28]

Kutter, H. 1973c ("1972"). *Leptothorax arcanus* n. sp.(?) - eine neue und rätselhafte Schmalbrustameise. Mitt. Schweiz. Entomol. Ges. 45:327-328. [1973.02.28]

Kutter, H. 1973d. Über die morphologischen Beziehungen der Gattung *Myrmica* zu ihren Satellitengenera *Sifolinia* Em., *Symbiomyrma* Arnoldi und *Sommimyrma* Menozzi (Hymenoptera, Formicidae). Mitt. Schweiz. Entomol. Ges. 46:253-268. [1973.12.28]

Kutter, H. 1973e. Zur Taxonomie der Gattung *Chalepoxenus* (Hymenoptera, Formicidae, Myrmicinae). Mitt. Schweiz. Entomol. Ges. 46:269-280. [1973.12.28]

Kutter, H. 1973f. Beitrag zur Lösung taxonomischer Probleme in der Gattung *Epimyrma* (Hymenoptera, Formicidae). Mitt. Schweiz. Entomol. Ges. 46:281-289. [1973.12.28]

Kutter, H. 1975a. Über die Waldameisenfauna der Türkei. Mitt. Schweiz. Entomol. Ges. 48:159-163. [1975.05.30]

Kutter, H. 1975b. 74. Die Ameisen (Hym. Formicidae) des Schweizerischen Nationalparkes und seiner Umgebung. Ergeb. Wiss. Unters. Schweiz. Nationalpark 14:398-414. [1975]

Kutter, H. 1976a. Beitrag zur Kenntnis der Gattung *Calyptomyrmex* (Hym. Formicidae Subf. Myrmicinae). Mitt. Schweiz. Entomol. Ges. 49:269-272. [1976.12.31]
Kutter, H. 1976b. Beitrag zur Kenntnis der Gattung *Simopone* (Hym. Formicidae Subfam. Cerapachyinae resp. Ponerinae). Mitt. Schweiz. Entomol. Ges. 49:273-276. [1976.12.31]
Kutter, H. 1977a. Zur Kenntnis der Gattung *Hylomyrma* (Hym. Formicidae, Subf. Myrmicinae). Mitt. Schweiz. Entomol. Ges. 50:85-89. [1977.03.31]
Kutter, H. 1977b. Zweiter Beitrag zur Kenntnis der Gattung *Simopone* Forel (Hym. Formicidae, Subf. Cerapachyinae resp. Ponerinae). Mitt. Schweiz. Entomol. Ges. 50:173-176. [1977.06.30]
Kutter, H. 1977c. Hymenoptera, Formicidae. Insecta Helv. Fauna 6:1-298. [1977]
Kutter, H. 1978. Hymenoptera, Formicidae. Insecta Helv. Fauna 6a. 108 pp., illustr. [1978]
Kutter, H. 1981. *Iridomyrmex humilis* Mayr (Hym., Formicidae), Gattung und Art neu für die Schweiz. Mitt. Schweiz. Entomol. Ges. 54:171-172. [1981.06.30]
Kutter, H. 1986. Über Anomalien einheimischer Formiciden. Mitt. Schweiz. Entomol. Ges. 59:229-238. [1986.12.15]
Kutter, H., Le Masne, G., Baroni Urbani, C. 1969a. [Review of: Bernard, F. 1967 ("1968"). Les fourmis (Hymenoptera Formicidae) d'Europe occidentale et septentrionale. Paris: Masson, 411 pp.] Boll. Soc. Entomol. Ital. 99-101:117-120. [1969.06.21]
Kutter, H., Le Masne, G., Baroni Urbani, C. 1970 ("1969"). Ire Conférence européenne de Myrmécologie. Sienne (Italie), 3-7 février 1969. Insectes Soc. 16:313-316. [1970.06]
Kuznecov-Ugamskij, N. N. See: Kuznetsov-Ugamsky, N. N.
Kuznetsov, G. T. 1990. Comparative analysis of ant populations (Hymenoptera, Formicidae) in altitudinal zones of central Kopetdag. Izv. Akad. Nauk Turkm. SSR Ser. Biol. Nauk 1990(3):64-67. [> 1990.09.26] [Date signed for printing ("podpisano k pechati").]
Kuznetzov-Ugamskij, N. N. See: Kuznetsov-Ugamsky, N. N.
Kuznetsov-Ugamsky, N. N. See also: Kusnezov, N. (1948-1978)
Kuznetsov-Ugamsky, N. N. 1923a. Die genetischen Elementes der Ameisenfauna des russischen Turkestan. Zool. Anz. 57:82-88. [1923.07.24]
Kuznetsov-Ugamsky, N. N. 1923b. The ant fauna of Tashkent district. [In Russian.] Tr. Turkest. Nauchn. Obshch. 1:239-558. [1923]
Kuznetsov-Ugamsky, N. N. 1925a. Zur Frage der vertikalen Verteilung der Faunenelemente Turkestans. Zool. Anz. 62:107-117. [1925.02.20]
Kuznetsov-Ugamsky, N. N. 1925b. Der Nestbau turkestanischer Ameisen als ökologischer Verbreitungsfaktor. Zool. Anz. 64:253-260. [1925.11.20]
Kuznetsov-Ugamsky, N. N. 1926a. Die Entstehung der Wüstenameisenfauna Turkestans. Zool. Anz. 65:140-160. [1926.01.05]
Kuznetsov-Ugamsky, N. N. 1926b. Neue turkestanische Ameisen. Rus. Entomol. Obozr. 20:71-77. [1926]
Kuznetsov-Ugamsky, N. N. 1926c. Contributions to the knowledge of the myrmecology of Turkestan. I. [In Russian.] Rus. Entomol. Obozr. 20:93-100. [1926]
Kuznetsov-Ugamsky, N. N. 1926d. The geographical relationships of the myrmecofauna of the Caucasus. [In Russian.] Bull. Univ. Asie Cent. 12:63-74. [1926]
Kuznetsov-Ugamsky, N. N. 1926e. Die Entstehung der Wüstenfauna von Turkestan. Rus. Zool. Zh. 6(1):81-82. [1926]
Kuznetsov-Ugamsky, N. N. 1927a. Vorläufige Uebersicht über die mittelasiatischen Formen der Gattung *Messor* (Hym., Form.). Folia Myrmecol. Termit. 1:89-94. [1927.04]
Kuznetsov-Ugamsky, N. N. 1927b. Zur Kenntnis der Gattung *Proformica* Ruzsky. Soc. Entomol. Stuttg. 42:26-28. [1927.06.01]
Kuznetsov-Ugamsky, N. N. 1927c. *Polyergus rufescens tianschanicus* subsp. nov. aus Turkestan. Soc. Entomol. Stuttg. 42:41-42. [1927.11.01]
Kuznetsov-Ugamsky, N. N. 1927d. Contributions to the knowledge of the myrmecology of Turkestan. II. [In Russian.] Rus. Entomol. Obozr. 21:33-42. [1927]

Kuznetsov-Ugamsky, N. N. 1927e. Contributions to the knowledge of the myrmecology of Turkestan. III. [In Russian.] Rus. Entomol. Obozr. 21:186-196. [1927]
Kuznetsov-Ugamsky, N. N. 1927f. On the nuptial flight of the ants. [In Russian.] Rus. Zool. Zh. 7(2):77-104. [1927]
Kuznetsov-Ugamsky, N. N. 1928a. Beiträge zur Ameisenfauna Mittelasiens. I. Die Gattung *Proformica*, Ruzsky 1903. Zool. Anz. 75:7-23. [1928.01.02]
Kuznetsov-Ugamsky, N. N. 1928b. Ants of the South Ussuri Region. [In Russian.] Zap. Vladivost. Otd. Gos. Rus. Geogr. Obshch. 1(18):1-47. [1928]
Kuznetsov-Ugamsky, N. N. 1929a. Die Ameisen des Süd-Ussuri-Gebietes. Zool. Anz. 83:16-34. [1929.06.15]
Kuznetsov-Ugamsky, N. N. 1929b. Die Ameisenfauna Daghestans. Zool. Anz. 83:34-45. [1929.06.15]
Kuznetsov-Ugamsky, N. N. 1929c. Die Gattung *Acantholepis* in Turkestan. Zool. Anz. 82:477-492. [1929]
Kuznetsov-Ugamsky, N. N. 1929d. The ants of the genus *Messor* in the fauna of Middle Asia. [In Russian.] Acta Univ. Asiae Mediae Ser. 8A Zool. 6:1-25. [1929]
Kuznetsov-Ugamsky, N. N. 1930. Contributions to the study of the agents of evolution of faunistic ensembles. [In Russian.] Dokl. Akad. Nauk SSSR 1930:367-372. [1930.06]
Kvamme, T. 1982. Atlas of the Formicidae of Norway (Hymenoptera: Aculeata). Insecta Norv. 2:1-56. [1982]
Kvamme, T., Bakke, A. 1977. *Stenamma westwoodi* (Hym., Formicidae) new to Norway. Norw. J. Entomol. 24:178-179. [1977]
Kvamme, T., Midtgaard, F. 1984. *Myrmecina graminicola* (Latreille, 1802) (Hym., Formicidae) new to the Norwegian fauna. Fauna Norv. Ser. B 31:64-65. [1984]
LaBerge, W. E. 1952. Locality records of two ants found in Kansas. J. Kansas Entomol. Soc. 25:59. [1952.05.15]
Labram, J. D., Imhoff, L. 1838. Insekten der Schweiz. 2. Basel: published by the authors (in commission of C. J. Spittler), 84 pls., 84 pp. [1838]
Labuda, M. 1971 ("1970"). Die Ameisen (Hymenoptera, Formicoidea) im mittleren Teil des Waagtals. Zb. Slov. Nár. Múz. Prír. Vedy 16(2):141-168. [1971.01]
Lachaud, J. P., Fresneau, D., Corbara, B. 1988. Mise en évidence de sous-castes comportementales chez *Amblyopone pallipes*. Actes Colloq. Insectes Soc. 4:141-147. [1988.06]
Lacordaire, T. 1848. Monographie des coléoptères subpentamères de la famille des Phytophages. Mém. Soc. R. Sci. Liège 5(tome second):1-890. [1848.05] [*Ceratobasis* (Coleoptera), senior homonym of *Ceratobasis* F. Smith, 1860.]
Laine, K. J., Niemelä, P. 1989. Nests and nest sites of red wood ants (Hymenoptera, Formicidae) in subarctic Finland. Ann. Entomol. Fenn. 55:81-88. [1989.08.15]
Lamarck, J. B. P. A. 1817. Histoire naturelle des animaux sans vertebres. Tome 4. Paris: Verdière, 603 pp. [1817.03] [Date of publication from Sherborn (1922); see also Iredale (1922:84).]
Lameere, A. 1892. Note sur les fourmis de la Belgique. Ann. Soc. Entomol. Belg. 36:61-69. [> 1892.02.06]
Lameere, A. 1902. Note sur les moeurs des fourmis du Sahara. Ann. Soc. Entomol. Belg. 46:160-169. [1920.05.01]
LaMon, B., Topoff, H. 1981. Avoiding predation by army ants: defensive behaviours of three ant species of the genus *Camponotus*. Anim. Behav. 29:1070-1081. [1981.11]
Lange, R. 1956. Experimentelle Untersuchungen über die Variabilität bei Waldameisen (*Formica rufa* L.). Z. Naturforsch. Teil B 11:538-543. [1956]
Lange, R. 1958a. Über die Variabilität der Beborstung der Waldameisen. Zugleich ein Beitrag zur Systematik der *Formica rufa*-Gruppe. Zool. Jahrb. Abt. Syst. Ökol. Geogr. Tiere 86:217-226. [1958.10.30]
Lange, R. 1958b. Die deutschen Arten der *Formica rufa*-Gruppe. Zool. Anz. 161:238-243. [1958.11]

Lange, R. 1959a. Die morphologischen Merkmale von *Formica rufa* L. und *Formica polyctena* Först., zwei für die Vermehrung wichtige Ameisenarten. Anz. Schädlingskd. 32:57-58. [1959.04]

Lange, R. 1959b. Zur Trennung von *Formica rufa* L. und *Formica polyctena* Först. (Hym., Formicidae). Der artspezifische Duft. Z. Angew. Entomol. 45:188-197. [1959.12]

Lange, R. 1960. Die systematischen Grundlagen der Waldameisenvermehrung in Deutschland. Entomophaga 5:81-86. [1960.02.24]

Lanham, U. N. 1979. Possible phylogenetic significance of complex hairs in bees and ants. J. N. Y. Entomol. Soc. 87:91-94. [1979.04.26]

Lanham, U. N. 1979. Possible phylogenetic significance of complex hairs in bees and ants. J. N. Y. Entomol. Soc. 87:91-94. [1979.04.26]

Lanne, B. S., Bergström, G., Löfqvist, J. 1988. Dufour gland alkenes from the four ant species: *F. polyctena*, *F. lugubris*, *F. truncorum*, and *F. uralensis*. Comp. Biochem. Physiol. B. Comp. Biochem. 91:729-734. [1988]

Lappano, E. R. 1958. A morphological study of larval development in polymorphic all-worker broods of the army ant *Eciton burchelli*. Insectes Soc. 5:31-66. [1958.06]

La Rivers, I. 1968. A first listing of the ants of Nevada (Hymenoptera, Formicidae). Occas. Pap. Biol. Soc. Nev. 17:1-12. [1968.02.15]

Larsson, S. G. 1943. Myrer. Dan. Fauna 49:1-190. [1943]

LaSalle, J., Gauld, I. D. 1993. Hymenoptera: their diversity, and their impact on the diversity of other organisms. Pp. 1-26 in: LaSalle, J., Gauld, I. D. (eds.) Hymenoptera and biodiversity. Oxford: CAB International, xi + 348 pp. [1993]

Latreille, P. A. 1798. Essai sur l'histoire des fourmis de la France. Brive: F. Bourdeaux, 50 pp. [1798]

Latreille, P. A. 1802a. Histoire naturelle des fourmis, et recueil de mémoires et d'observations sur les abeilles, les araignées, les faucheurs, et autres insectes. Paris: Impr. Crapelet (chez T. Barrois), xvi + 445 pp. [1802.04] [Publication date from Richards (1935:174).]

Latreille, P. A. 1802b. Histoire naturelle générale et particulière des Crustacés et des insectes. Tome 3. Familles naturelles des genres. Paris: F. Dufart, xii + 467 pp. [1802.11.06] [Publication date from Dupuis (1986:208).]

Latreille, P. A. 1802c. Description d'une nouvelle espèce de fourmi. Bull. Sci. Soc. Philomath. Paris 3:65-66. [1802]

Latreille, P. A. 1803. Histoire naturelle générale et particulière des Crustacés et des insectes. Tome 5. Paris: F. Dufart, 406 pp. [1803.04.14] [Publication date from Dupuis (1986:208).]

Latreille, P. A. 1804. Tableau méthodique des insectes. Pp. 129-200 in: Société de Naturalistes et d'Agriculteurs. Nouveau dictionnaire d'histoire naturelle. Tome 24. Paris: Déterville, 84 + 85 + 238 + 18 + 34 pp. [1804.03.07] [This appears in the third (and largest) of five individually paginated sections. Publication date from Osgood (1914:3) and Whitley (1936:189); see also Stafleu & Cowan (1976:636).]

Latreille, P. A. 1805. Histoire naturelle générale et particulière des Crustacés et des insectes. Tome 13. Paris: F. Dufart, 432 pp. [1805.03.21] [Publication date from upper bound in Dupuis (1986:209). Ants pp. 241-261.]

Latreille, P. A. 1807. Genera crustaceorum et insectorum secundum ordinem naturalem in familias disposita, iconibus exemplisque plurimus explicata. Tomus 3. Parisiis et Argentorati [= Paris and Strasbourg]: A. Koenig, 258 pp. [1807]

Latreille, P. A. 1809. Genera crustaceorum et insectorum secundum ordinem naturalem in familias disposita, iconibus exemplisque plurimus explicata. Tomus 4. Parisiis et Argentorati [= Paris and Strasbourg]: A. Koenig, 399 pp. [1809.03.20] [Publication date from Evenhuis (1994b). Ants pp. 123-132.]

Latreille, P. A. 1810. Considérations générales sur l'ordre natural des animaux composant les classes des Crustacés, des Arachnides et des Insectes; avec un tableau méthodique de leurs genres, disposés en familles. Paris: F. Schoell, 444 pp. [1810.05.23] [Publication date from Evenhuis (1994b).]

Latreille, P. A. 1817a. Fourmi. Pp. 51-100 in: Nouveau dictionnaire d'histoire naturelle. Nouvelle édition. Tome XII. Paris: Déterville, 608 pp. [1817.06.21] [Publication date from Stafleu & Cowan (1976).]

Latreille, P. A. 1817b. Insectes de l'Amérique equinoxiale, recueilliés pendant le voyage de MM. Humboldt et Bonpland. Seconde partie. [part] Pp. 97-138 in: Humboldt, F. H. A. von, Bonpland, A. 1813-1832. Voyage aux régions equinoxiales du nouveau continent fait en 1799-1804. Recueil d'observations de zoologie et d'anatomie comparée. Vol. 2. Paris: Schoell, 352 pp. [1817.12.13] [Publication date from Sherborn & Woodward (1901a:202).]

Latreille, P. A. 1818a. Oecodome. P. 50 in: Nouveau dictionnaire d'histoire naturelle. Nouvelle édition. Tome XXIII. Paris: Déterville, 612 pp. [1818.09.05] [Publication date from Stafleu & Cowan (1976).]

Latreille, P. A. 1818b. Ponère. Pp. 569-570 in: Nouveau dictionnaire d'histoire naturelle. Nouvelle édition. Tome XXVII. Paris: Déterville, 586 pp. [1818.12.26] [Publication date from Stafleu & Cowan (1976).]

Lattke, J. See: Lattke, J. E.

Lattke, J. E. 1981. Primera captura de *Anochetus striatulus* Emery (Hymenoptera, Formicidae) en Sur América. Bol. Entomol. Venez. (n.s.)2:20. [1981.11.07]

Lattke, J. E. 1985a. Hormigas nuevas para Venezuela II. (Hymenoptera: Formicidae). Bol. Entomol. Venez. (n.s.)4:51. [1985.06.28]

Lattke, J. E. 1985b. Hallazgos de hormigas nuevas para Venezuela (Hymenoptera: Formicidae). Bol. Entomol. Venez. (n.s.)4:82-84. [1985.11.30]

Lattke, J. E. 1986. New records of attines in Venezuela. Attini 17:15. [1986.06]

Lattke, J. E. 1987a ("1986"). Two new species of neotropical *Anochetus* Mayr (Hymenoptera: Formicidae). Insectes Soc. 33:352-358. [1987.03]

Lattke, J. E. 1987b ("1986"). Notes on the ant genus *Hypoclinea* Mayr, with descriptions of three new species (Hymenoptera: Formicidae). Rev. Biol. Trop. 34:259-265. [1987.08]

Lattke, J. E. 1990a. A new genus of myrmicine ants (Hymenoptera: Formicidae) from Venezuela. Entomol. Scand. 21:173-178. [1990.06]

Lattke, J. E. 1990b. Revisión del género *Gnamptogenys* Roger en Venezuela (Hymenoptera: Formicidae). Acta Terramaris 2:1-47. [1990.06]

Lattke, J. E. 1990c. Phylogenetic relationships and classification in the Ectatommini (Hymenoptera: Formicidae). Pp. 323-324 in: Veeresh, G. K., Mallik, B., Viraktamath, C. A. (eds.) Social insects and the environment. Proceedings of the 11th International Congress of IUSSI, 1990. New Delhi: Oxford & IBH Publishing Co., xxxi + 765 pp. [1990.08]

Lattke, J. E. 1990d. Ecology and phylogenetic relationships within Neotropical members of the genus *Gnamptogenys* (Hymenoptera: Formicidae). Pp. 325-326 in: Veeresh, G. K., Mallik, B., Viraktamath, C. A. (eds.) Social insects and the environment. Proceedings of the 11th International Congress of IUSSI, 1990. New Delhi: Oxford & IBH Publishing Co., xxxi + 765 pp. [1990.08]

Lattke, J. E. 1990e. Systematics and morphology. Overview. Pp. 71-74 in: Vander Meer, R. K., Jaffe, K., Cedeno, A. (eds.) Applied myrmecology: a world perspective. Boulder: Westview Press, xv + 741 pp. [1990]

Lattke, J. E. 1991a ("1990"). Una nueva especie de *Pogonomyrmex* Mayr de selva húmeda tropical (Hymenoptera: Formicidae). Rev. Biol. Trop. 38:305-309. [1991.03]

Lattke, J. E. 1991b ("1990"). The genus *Proceratium* Roger in Dominican amber (Hymenoptera: Formicidae). Caribb. J. Sci. 26:101-103. [1991.03]

Lattke, J. E. 1991c. Estudios de hormigas de Venezuela (Hymenoptera: Formicidae). Bol. Entomol. Venez. (n.s.)6:57-61. [1991.07]

Lattke, J. E. 1991d. Studies of neotropical *Amblyopone* Erichson (Hymenoptera: Formicidae). Contr. Sci. (Los Angel.) 428:1-7. [1991.09.04]

Lattke, J. E. 1992a. Revision of the *minuta*-group of the genus *Gnamptogenys* (Hymenoptera, Formicidae). Dtsch. Entomol. Z. (N.F.)39:123-129. [1992.03.06]

Lattke, J. E. 1992b. *Trichoscapa inusitata* n. sp., a remarkable dacetine ant from the Orinoco watershed (Hymenoptera Formicidae). Trop. Zool. 5:141-144. [1992.12]

Lattke, J. E. 1993. XVIII. La identificación de los géneros neotropicales: claves para su identificación. Pp. 145-170 in: Jaffe, K. El mundo de las hormigas. Baruta, Venezuela: Equinoccio (Ediciones de la Universidad Simón Bolívar), 188 pp. [1993.07]

Lattke, J. E. 1994. Phylogenetic relationships and classification of ectatommine ants (Hymenoptera: Formicidae). Entomol. Scand. 25:105-119. [1994.03]

Lattke, J. E. 1995. Revision of the ant genus *Gnamptogenys* in the New World (Hymenoptera: Formicidae). J. Hym. Res. 4:137-193. [1995.09.25]

Lauterer, P. 1969 ("1968"). Notes on the occurrence of four rare species of ants in Moravia. Zb. Slov. Nár. Múz. Prír. Vedy 14(1):95-98. [1969.11]

Lavigne, R. J. 1977. Notes on the ants of Luquillo Forest, Puerto Rico (Hymenoptera: Formicidae). Proc. Entomol. Soc. Wash. 79:216-237. [1977.04.28]

Lavigne, R. J., Tepedino, V. J. 1976. Checklist of the insects in Wyoming. I. Hymenoptera. Res. J. Univ. Wyo. Agric. Exp. Stn. 106:1-61. [1976.12]

Law, J. H., Wilson, E. O., McCloskey, J. A. 1965. Biochemical polymorphism in ants. Science (Wash. D. C.) 149:544-546. [1965.07.30]

Laza, J. H. 1982. Signos de actividad atribuibles a *Atta* (Myrmicidae, Hymenoptera) en el Mioceno de la provincia de La Pampa, Republica Argentina. Ameghiniana 19:109-124. [1982.10.15]

Leach, W. E. 1815. Entomology. Pp. 57-172 in: Brewster, D. (ed.) The Edinburgh encyclopedia. Volume 9. Edinburgh. [1815.04] [Publication date from Sherborn (1937).]

Leach, W. E. 1819. [Fam. IX. Formicadae. Leach.] Pp. 272-273 in: Samouelle, G. The entomologist's useful compendium. London: Thomas Boys, 496 pp. [1819.06] [Publication date from Sherborn (1922:cxi).]

Leach, W. E. 1825. Descriptions of thirteen species of *Formica* and three species of *Culex* found in the environs of Nice. Zool. J. Lond. 2:289-293. [1825.10]

Leclerc, J., Francoeur, A., Maire, A. 1993 ("1991"). Les fourmis (Formicidae, Hymenoptera) de trois érablières de la région de Trois-Rivières, Québec. Rev. Entomol. Qué. 36:43-49. [1993.12]

Ledoux, A. 1954. Recherches sur le cycle chromosomique de la fourmi fileuse *Oecophylla longinoda* Latr. (Hyménoptère Formicoidea). Insectes Soc. 1:149-175. [1954.06]

Ledoux, A. 1958. La construction du nid chez quelques fourmis arboricoles de France et d'Afrique tropicale. Pp. 521-528 in: Becker, E. C. (ed.) Proceedings of the Tenth International Congress of Entomology, Montreal, August 17-25, 1956. Volume 2. Ottawa: Mortimer Limited, 1055 pp. [1958.12]

Leech, H. B. 1942. The dates of publication of certain numbers of the Proceedings of the Entomological Society of British Columbia. Proc. Entomol. Soc. B. C. 38:29-36. [1942.02.07] [Provides publication date for Buckell (1928).]

Lefeber, B. A. 1995. Drie aanwinsten voor de Nederlandse Aculeatenfauna (Hymenoptera: Apidae, Formicidae). Entomol. Ber. (Amst.) 55:135. [1995.08] [*Leptothorax affinis*, new for the Netherlands.]

Legakis, A. 1983a. First contribution to the study of the ants (Hymenoptera, Formicidae) of the Zagori Region (Epirus, Greece): an annotated list of species. Entomol. Hell. 1:3-6. [1983.06.30]

Legakis, A. 1983b. Aspects of the distribution of ants in an insular mediterranean ecosystem (Cyclades Is., Greece). Rapp. P.-V. Réun. Comm. Int. Explor. Sci. Mer Méditerr. Monaco 28(8):121-122. [1983]

Legakis, A. 1985. Contribution to the zoogeography of ants (Hymenoptera, Formicidae) in the Greek Islands. Biol. Gallo-Hell. 10:253-257. [1985.07]

Legakis, A. 1987. Aspects of the zoogeography of the ants of Crete (Greece). P. 77 in: Eder, J., Rembold, H. (eds.) Chemistry and biology of social insects. München: Verlag J. Peperny, xxxv + 757 pp. [1987]

Legakis, A. 1990 ("1989"). Aspects of the distribution of ants in an insular mediterranean ecosystem (Cyclades is., Greece). Rev. Écol. Biol. Sol 26:363-369. [1990.05]

Le Guillou, E. J. F. 1842 ("1841"). Catalogue raisonné des insectes hyménoptères recueillis dans le voyage de circumnavigation des corvettes l'Astrolabe et la Zélée. Ann. Soc. Entomol. Fr. 10:311-324. [1842.01]

Leininger, H. 1927. Zur Ameisenfauna Nordbadens. Arch. Insektenkd. Oberrheingeb. Angrenz. Länder 2:131-133. [1927.02]

Leininger, H. 1951. Über Bienen, Grab-, Weg-, Faltenwespen und Ameisen aus dem badischen Oberrheingebiet (Hym. aculeata). Beitr. Naturkd. Forsch. Südwestdtschl. 10:113-136. [1951]

Le Masne, G. 1948. Observations sur les mâles ergatoïdes de la fourmi *Ponera eduardi* Forel. C. R. Séances Acad. Sci. 226:2009-2011. [1948.06.21]

Le Masne, G. 1956a. La signification des reproducteurs aptères chez la fourmi *Ponera eduardi* Forel. Insectes Soc. 3:239-259. [1956.06]

Le Masne, G. 1956b. Recherches sur les fourmis parasites. *Plagiolepis grassei* et l'évolution des *Plagiolepis* parasites. C. R. Séances Acad. Sci. 243:673-675. [1956.08.20]

Le Masne, G., Bonavita, A. 1969. La fondation des sociétés selon un type archaïque par une fourmi appartenant à une sous famille évoluée. C. R. Séances Acad. Sci. Ser. D. Sci. Nat. 269:2373-2376. [1969.12.22]

Le Moli, F. 1980. On the origin of slaves in dulotic ant societies. Boll. Zool. 47:207-212. [1980.12]

Le Moli, F., Mori, A. 1986. The aggression test as a possible taxonomic tool in the *Formica rufa* group. Aggressive Behav. 12:93-102. [1986]

Le Moli, F., Mori, A., Parmigiani, S. 1984. Studies on interspecific aggression among red wood ant species. *Formica rufa* L. vs *Formica lugubris* Zett. (Hymenoptera Formicidae). Monit. Zool. Ital. (n.s.)18:41-51. [1984.04.20]

Le Moli, F., Rosi, M. R. 1991. Ricerche corologiche sulla mirmecofauna dell'Umbria. I. L'Appennino Spoletino. Atti Soc. Ital. Sci. Nat. Mus. Civ. Stor. Nat. Milano 132:25-40. [1991.11]

Le Moli, F., Zaccone, A. 1995. Ricerche sulla mirmecofauna del Cansiglio (Prealpi Carniche). Soc. Veneziana Sci. Nat. Lav. 20:33-52. [1995.01.31] [Not seen. Cited in Zoological Record, volume 132.]

Lenko, K. 1964a. Uma nova espécie de *Gnamptogenys* de Mato Grosso (Hymenoptera, Formicidae). Pap. Avulsos Zool. (São Paulo) 16:257-261. [1964.07.22]

Lenko, K. 1964b. Sôbre um caso teratológico observado em *Paracryptocerus pusillus* (Klug) (Hymenoptera: Formicidae). Stud. Entomol. 7:201-204. [1964.12.10]

Lenko, K. 1966 ("1965"). Sôbre a ocorrência de *Stegomyrmex manni* no Estado de S. Paulo, Brasil (Hymenoptera: Formicidae). Stud. Entomol. 8:201-204. [1966.02.28]

Lenoir, A. 1972 ("1971"). Les fourmis de Touraine, leur intérêt biogéographique. Cah. Nat. (Bull. Nat. Paris.) 27:21-30. [1972.04]

Lenoir, A., Arnold, G., Lepage, M. (eds.) 1994. Les insectes sociaux. 12ème congrès de l'Union Internationale pour l'Étude des Insectes Sociaux. Paris, Sorbonne, 21-27 août 1994. Paris: Université Paris Nord, xxiv + 583 pp. [1994.09]

Lenoir, A., Dejean, A. 1994. Semi-claustral colony foundation in the formicine ants of the genus *Polyrhachis* (Hymenoptera: Formicidae). Insectes Soc. 41:225-234. [1994]

Lenoir, A., Jaisson, P. 1982. Évolution et rôle des communications antennaires chez les insectes sociaux. Pp. 157-180 in: Jaisson, P. (ed.) Social insects in the tropics. Volume 1. Paris: Université de Paris-Nord, 280 pp. [1982.06]

Leo, P., Fancello, L. 1990. Osservazioni sul genere *Leptanilla* Emery in Sardegna e riabilitazione di *L. doderoi* Emery (Hymenoptera Formicidae Leptanillinae). Boll. Soc. Entomol. Ital. 122:128-132. [1990.09.30]

Leonard, P. 1911. The honey ants of Point Loma, California. Trans. San Diego Soc. Nat. Hist. 1:85-97. [1911]

Lepeletier de Saint-Fargeau, A. 1835 ("1836"). Histoire naturelle des insectes. Hyménoptères. Tome I. Paris: Roret, 547 pp. [1835.12.26]

Leprince, D. J., Francoeur, A. 1986. Hilltop swarming by ants (Hymenoptera: Formicidae) in southwestern Quebec and interspecific competition at the swarm marker. Ann. Entomol. Soc. Am. 79:865-869. [1986.11]

Leston, D. 1971. The Ectatommini (Hymenoptera: Formicidae) of Ghana. J. Entomol. Ser. B 40:117-120. [1971.11.26]

Leston, D. 1973. The ant mosaic - tropical tree crops and the limiting of pests and diseases. Pest Artic. News Summ. 19:311-341. [1973.09]

Leston, D. 1978. A Neotropical ant mosaic. Ann. Entomol. Soc. Am. 71:649-653. [1978.07.17]

Leston, D. 1979. Dispersal by male doryline ants in West Africa. Psyche (Camb.) 86:63-77. [1979.12.28]

Letendre, M., Francoeur, A., Béique, R., Pilon, J.-G. 1971. Inventaire des fourmis de la station de biologie de l'Université de Montréal, St-Hippolyte, Québec (Hymenoptera: Formicidae). Nat. Can. (Qué.) 98:591-606. [1971.09]

Letendre, M., Huot, L. 1972. Considérations préliminaires en vue de la revision taxonomique des fourmis du groupe *microgyna*, genre *Formica* (Hymenoptera: Formicidae). Ann. Soc. Entomol. Qué. 17:117-132. [1972.09]

Letendre, M., Pilon, J.-G. 1973. La faune myrmécologique de différents peuplements forestiers caractérisant la zone des Basses Laurentides, Québec (Hymenoptera: Formicidae). Nat. Can. (Qué.) 100:195-235. [1973.06]

Leutert, W. 1962. Beitrag zur Kenntnis des Polymorphismus bei *Camponotus ligniperda* Latr. (Hym., Formicidae). Mitt. Schweiz. Entomol. Ges. 35:146-154. [1962.06.30]

Leutert, W. 1963. Systematics of ants. Nature (Lond.) 200:496-497. [1963.11.02]

Leutert, W. 1965. Phenotypic variability in worker ants of *Lasius flavus* De Geer and their progeny (Hym., Formicidae). J. Entomol. Soc. Aust. (N. S. W.) 2:40-42. [1965.12.30]

Leuthold, R. H. 1968. A tibial gland scent-trail and trail-laying behavior in the ant *Crematogaster ashmeadi* Mayr. Psyche (Camb.) 75:233-248. [1968.11.15]

Leuthold, R. H. 1969. Trail-laying from the tibial gland in *Crematogaster ashmeadi* Mayr and *Crematogaster scutellaris* Olivier (Formicidae: Myrmicinae). Pp. 149-151 in: Ernst, E. et al. (eds.) International Union for the Study of Social Insects. VI Congress. Bern 15-20 September 1969. Proceedings. Bern: Organizing Committee of the VI Congress IUSSI, 309 pp. [1969.09.15]

Lever, R. J. A. W. 1930. The female of *Ectatomma tuberculatum* (Oliv.) var. *punctigerum* Emery (Hymenoptera, Formicidae). Entomol. Mon. Mag. 66:214-215. [1930.09]

Lévieux, J. 1972a. Comportement d'alimentation et relations entre les individus chez une fourmi primitive, *Amblyopone pluto* Gotwald et Lévieux. C. R. Séances Acad. Sci. Ser. D. Sci. Nat. 275:483-485. [1972.07.24]

Lévieux, J. 1972b. Les fourmis de la savane de Lamto (Côte d'Ivoire): éléments de taxonomie. Bull. Inst. Fondam. Afr. Noire Sér. A Sci. Nat. 34:611-654. [1972.12]

Lévieux, J. 1973 ("1972"). Le rôle des fourmis dans les réseaux trophiques d'une savane préforestière de Côte-d'Ivoire. Ann. Univ. Abidjan Sér. E (Écol.) 5(1):143-240. [1973.06]

Lévieux, J. 1974 ("1973"). Étude de peuplement en fourmis terricoles d'une savane préforestière de Côte d'Ivoire. Rev. Écol. Biol. Sol 10:379-428. [1974.03]

Lévieux, J. 1982. A comparison of the ground dwelling ant populations between a Guinea savanna and an evergreen rain forest of the Ivory Coast. Pp. 48-53 in: Breed, M. D., Michener, C. D.,

Evans, H. E. (eds.) The biology of social insects. Proceedings of the Ninth Congress of the IUSSI, Boulder, Colorado, August 1982. Boulder: Westview Press, xii + 420 pp. [1982.08]

Lévieux, J. 1983. The soil fauna of tropical savannas. IV. The ants. Pp. 525-540 in: Bourliere, F. (ed.) Ecosystems of the world. Vol. 13. Amsterdam: Elsevier, xii + 730 pp. [1983]

Levings, S. C. 1983. Seasonal, annual, and among-site variation in the ground ant community of a deciduous tropical forest: some causes of patchy species distributions. Ecol. Monogr. 53:435-455. [1983.12]

Levings, S. C., Traniello, J. F. A. 1982 ("1981"). Territoriality, nest dispersion, and community structure in ants. Psyche (Camb.) 88:265-319. [1982.05.28]

Levins, R., Pressick, M. L., Heatwole, H. 1973. Coexistence patterns in insular ants. Am. Sci. 61:463-472. [1973.08]

Linnaeus, C. 1758. Systema naturae per regna tria naturae, secundum classes, ordines, genera, species, cum characteribus, differentiis, synonymis, locis. Tomus I. Editio decima, reformata. Holmiae [= Stockholm]: L. Salvii, 824 pp. [1758.01.01]

Linnaeus, C. 1761. Fauna suecica sistens animalia Sueciae regni: Mammalia, Aves, Amphibia, Pisces, Insecta, Vermes. Editio altera, auctior. Stockholmiae [=Stockholm]: L. Salvii, 48 + 578 pp. [1761]

Linnaeus, C. 1763. CXXI. Centuria insectorum quam praeside D. D. Car. von Linné proposuit Boas Johansson, Calmariensis (Upsaliae 1763. Junii 23). Pp. 384-415 in: Amoenitates academicae; seu dissertationes variae physicae, medicae, botanicae, antehac seorsim editae, nunc collectae et auctae cum tabulis aeneis. Tomus 6. Holmiae [= Stockholm]: G. Kiesewetter, 486 pp. [1763]

Linnaeus, C. 1764. Museum S:ae R:ae M:tis Ludovicae Ulricae Reginae Svecorum, Gothorum, Vandalorumque, &c. In quo animalia rariora, exotica, imprimis. Insecta & Conchilia describuntur & determinantur. Prodromi instar. Holmiae [= Stockholm]: Salvius, 8 + 720 pp. [1764]

Linnaeus, C. 1767. Systema naturae, per regna tria naturae, secundum classes, ordines, genera, species, cum caracteribus, differentiis, synonymis, locis. Tomus I. Pars 2. Editio duodecima, reformata. Holmiae [= Stockholm]: L. Salvii, pp. 533-1327. [1767.06.14] [Date of publication from Evenhuis (1994b:521).]

Linnaeus, C. 1771. Mantissa plantarum altera generum editionis VI & specierum editionis II. Holmiae [= Stockholm]: Laurentii Salvii, [6] + 143-587 pp. [1771.10] [Description of Formica quadripunctata, p. 541.]

Linsley, E. G., Usinger, R. L. 1966. Insects of the Galápagos Islands. Proc. Calif. Acad. Sci. (4)33:113-196. [1966.01.31] [Ants pp. 174-177.]

Lipski, N., Heinze, J., Hölldobler, B. 1992. Social organization of three European *Leptothorax* species (Hymenoptera, Formicidae). Pp. 287-290 in: Billen, J. P. J. (ed.) Biology and evolution of social insects. Leuven: Leuven University Press, ix + 390 pp. [1992]

Lipski, N., Heinze, J., Hölldobler, B. 1994. Untersuchungen zu Koloniestruktur und genetischem Verwandtschaftsgrad bei *Leptothorax muscorum* (Hym., Formicidae). Mitt. Dtsch. Ges. Allg. Angew. Entomol. 9:355-359. [1994.12]

Llave, P. de la. 1832. Sobre las busileras u hormigas de miel. Regist. Trimest. Colecc. Mem. Hist. Lit. Cienc. Artes 1:455-463. [1832.10]

Lloyd, H. A., Blum, M. S., Snelling, R. R., Evans, S. L. 1989. Chemistry of mandibular and Dufour's gland secretions of ants in genus *Myrmecocystus*. J. Chem. Ecol. 15:2589-2599. [1989.11]

Lloyd, H. A., Schmuff, N.R., Hefetz, A. 1986. Chemistry of the anal glands of *Bothriomyrmex syrius* Forel. Olfactory mimetism and temporary social parasitism. Comp. Biochem. Physiol. B. Comp. Biochem. 83:71-73. [1986.01.15]

Lobry de Bruyn, L. A. 1993. Ant composition and activity in naturally-vegetated and farmland environments on contrasting soils at Kellerberrin, Western Australia. Soil Biol. Biochem. 25:1043-1056. [1993.08]

Lofgren, C. S., Vander Meer, R. K. (eds.) 1986. Fire ants and leaf-cutting ants. Biology and management. Boulder: Westview Press, xv + 435 pp. [1986]

Lohmander, H. 1949. Eine neue schwedische Ameise. Myrmekologische Fragmente 1. Opusc. Entomol. 14:163-167. [1949.12.31]

Loiselle, R., Francoeur, A. 1988. Régression du dimorphisme sexuel dans le genre *Formicoxenus* et polymorphisme comparé des sexes dans la famille des Formicidae (Hymenoptera). Nat. Can. (Qué.) 115:367-378. [1988.12]

Loiselle, R., Francouer, A., Fischer, K., Buschinger, A. 1990. Variations and taxonomic significance of the chromosome numbers in the Nearctic species of the genus *Leptothorax* (s.s.) (Formicidae: Hymenoptera). Caryologia 43:321-334. [1990.10.20]

Lokay, E. 1860. Popsání hlavních druhu mravencu v Cechách zijících, s ohledem na hosti u nás dosud v mravenistích nalezené. Ziva 8:238-253. [1860]

Lokkers, C. 1986. The distribution of the weaver ant, *Ocecophylla smaragdina* (Fabricius) (Hymenoptera: Formicidae) in northern Australia. Aust. J. Zool. 34:683-687. [1986.09.30]

Lombardini, G. 1926. Formiche di Faenza. Boll. Soc. Entomol. Ital. 58:18-21. [1926.02.27]

Lombarte, A., Romero de Tejada, S., De Haro, A. 1989. Contribución al conocimiento faunístico de los formícidos de la sierra de Collserola (Barcelona). Orsis (Org. Sist.) 4:125-140. [1989]

Lomholdt, O. 1986. Myren *Pheidole anastasii* Emery, 1896 i Botanisk Have, Kobenhavn (Hymenoptera, Formicidae). Entomol. Medd. 53:58. [1986]

Lomholdt, O. 1988. Myrerne *Oecophylla smaragdina* (Fabricius, 1775) og *Camponotus vagus* (Scopoli, 1763) fundet i Danmark. Entomol. Medd. 56:72. [1988]

Lomholdt, O. 1989. Myren *Leptothorax nylanderi* (Förster, 1850) fundet i Danmark (Hymenoptera, Formicidae). Entomol. Medd. 57:142. [1989]

Lomholdt, O., Rasmussen, N. 1986. Tyvmyren *Diplorhoptrum fugax* (Latreille, 1798) fundet i Danmark (Hymenoptera, Formicidae). Entomol. Medd. 53:111-112. [1986]

Lomnicki, J. 1919. Stanowiska krajowe nieróbki czarniawej (*Anergates atratulus* Schenck.). Rozpr. Wiad. Muz. Dziedusz. Lwów 3:199-200. [1919]

Lomnicki, J. 1920. Z fauny mrówek Litwy. Rozpr. Wiad. Muz. Dziedusz. Lwów 4:99-100. [1920]

Lomnicki, J. 1922a. Przyczynek do opisu królowej mrówki: Powolnicy Europejskiej (*Sysphincta europaea* For.). Pol. Pismo Entomol. 1:3-4. [1922]

Lomnicki, J. 1922b. O mrówce zniwiarce jarowej (*Messor structor orientalis* var. *clivorum* Ruzsky) z Podola. Rozpr. Wiad. Muz. Dziedusz. Lwów 5-6:183-188. [1922]

Lomnicki, J. 1922c. Drobny przyczynek do znajomosci mrówek zniwiarek (*Messor* Forel) Podola. Rozpr. Wiad. Muz. Dziedusz. Lwów 5-6:189-190. [1922]

Lomnicki, J. 1924. O trzech gynandromorfach mrówki wscieklicy marszczystej (*Myrmica rugulosa* Nyl.). Kosmos (Warsaw) 49:817-830. [1924]

Lomnicki, J. 1925a ("1924"). Przeglad polskich gatunków rodzaju mrówki (*Formica* Linné). Pol. Pismo Entomol. 3:151-182. [< 1925.03.15] [Undated beyond the year 1925, but presumably appeared before the next issue (volume 4, part 1), which is dated 15 March 1925.]

Lomnicki, J. 1925b. Une contribution à la connaissance de la faune des fourmis des îles Baléares. Pol. Pismo Entomol. 4:1-3. [1925.03.15]

Lomnicki, J. 1925c. *Plagiolepis vindobonensis* n. sp. (Hym. Formicidae). Pol. Pismo Entomol. 4:77-79. [1925.06.15]

Lomnicki, J. 1928. Spis mrówek Lwowa i okolicy. Ksiegi Pamiatkowej (Lecia Gimn. IV Jana Dlugosza Lwowie) 50:1-10. [1928]

Lomnicki, J. 1931. Przeglad mrówek (Formicidae) Tatr polskich. Pol. Pismo Entomol. 10:97-101. [1931.06.30]

Longhurst, C., Johnson, R. A., Wood, T. G. 1979. Foraging, recruitment and predation by *Decamorium uelense* (Santschi) (Formicidae: Myrmicinae) on termites in southern Guinea Savannah, Nigeria. Oecologia (Berl.) 38:83-91. [1979.01.18]

Longino, J. T. 1988. Notes on the taxonomy of the neotropical ant genus *Thaumatomyrmex* Mayr (Hymenoptera: Formicidae). Pp. 35-42 in: Trager, J. C. (ed.) Advances in myrmecology. Leiden: E. J. Brill, xxvii + 551 pp. [1988]

Longino, J. T. 1989a. Geographic variation and community structure in an ant-plant mutualism: *Azteca* and *Cecropia* in Costa Rica. Biotropica 21:126-132. [1989.06]

Longino, J. T. 1989b. Taxonomy of the *Cecropia*-inhabiting ants in the *Azteca alfari* species group (Hymenoptera: Formicidae): evidence for two broadly sympatric species. Contr. Sci. (Los Angel.) 412:1-16. [1989.12.19]

Longino, J. T. 1991a. Taxonomy of the *Cecropia*-inhabiting *Azteca* ants. J. Nat. Hist. 25:1571-1602. [1991.12.31]

Longino, J. T. 1991b. *Azteca* ants in *Cecropia* trees: taxonomy, colony structure, and behaviour. Pp. 271-288 in: Huxley, C. R., Cutler, D. F. (eds.) Ant-plant interactions. Oxford: Oxford University Press, xviii + 601 pp. [1991]

Longino, J. T., Hanson, P. E. 1995. The ants (Formicidae). Pp. 588-620 in: Hanson, P. E., Gauld, I. D. (eds.) The Hymenoptera of Costa Rica. Oxford: Oxford University Press, xx + 893 pp. [1995]

Longino, J. T., Hartley, D. A. 1995 ("1994"). *Perissomyrmex snyderi* (Hymenoptera: Formicidae) is native to Central America and exhibits worker polymorphism. Psyche (Camb.) 101:195-202. [1995.11.20]

Longino, J. T., Nadkarni, N. M. 1990. A comparison of ground and canopy leaf litter ants (Hymenoptera: Formicidae) in a Neotropical montane forest. Psyche (Camb.) 97:81-94. [1990.10.14]

Loos-Frank, B. 1976. Electrophoretischer Nachweis von Vitellogenin in der Hämolymphe von Arbeiterinnen der Ameisen-Gattung *Formica* (Hymenoptera: Formicidae). Entomol. Ger. 3:212-226. [1976.12.29]

Loos-Frank, B. 1978. Vergleichend-elektrophoretische Darstellung der Hämolymph-Proteine verschiedener Ameisen-Arten der Gattung *Formica* (Hymenoptera: Formicidae). Entomol. Gen. 5:25-34. [1978.10.30]

Lopes, H. S. 1978. Frei Walter Entomólogo. Stud. Entomol. 20:3-15. [1978.08.30]

López, F. 1991a. Estudio morfológico y taxonómico de los grupos de especies ibéricas del género *Tetramorium* Mayr, 1855 (Hym., Formicidae). Bol. Asoc. Esp. Entomol. 15:29-52. [1991.12.30]

López, F. 1991b. Variabilidad morfológica y problemas taxonómicos en *Tetramorium caespitum* (Linné, 1758) y *Tetramorium semilaeve* André, 1881 (Hym., Formicidae). Bol. Asoc. Esp. Entomol. 15:65-78. [1991.12.30]

López, F., Martínez, M. D., Barandica, J. M. 1994. Four new species of the genus *Leptanilla* (Hymenoptera: Formicidae) from Spain - relationships to other species and ecological issues. Sociobiology 24:179-212. [1994]

López, F., Ortuño, V. M. 1992. Five teratological cases in *Tetramorium* Mayr and *Leptothorax* Mayr, and some comments on the incidence of teratologies in ants. Entomologist 111:123-133. [1992.07]

López, F., Serrano, J. M., Acosta Salmerón, F. J. 1990. Compared Iberian distribution of *Tetramorium caespitum* (Linné, 1758) and *Tetramorium semilaeve* André, 1881 (Hym., Formicidae). An. Biol. 16:53-61. [1990]

López, F., Zorrilla, J. M., Acosta Salmerón, F. J., Serrano, J. M. 1992 ("1991"). Comparative morphological study of *Tetramorium caespitum* (Linné, 1758) and *Tetramorium semilaeve* André, 1881 (Hym., Formicidae). Misc. Zool. (Barc.) 15:169-178. [1992]

López Moreno, I. R., Diaz Bitancourt, M. E. 1982. Observaciones preliminares sobre la mirmecofauna del desierto del Pinacate, Sonora, México. Folia Entomol. Mex. 54:40-41. [1982.12]

Lorber, B. E. 1981a. Le groupe *Formica rufa* (Hym. Formicidae) en Alsace. 1) Le complexe forestier de Haguenau. Entomologiste (Paris) 37:241-250. [1981.12]

Lorber, B. E. 1981b. Les difficultés taxonomiques du "Groupe *Formica rufa*" (Hym., Formicidae, Formicinae). Nouvelles données. Bull. Intér. Sect. Fr. UIEIS 1981:143-146. [1981]

Lorber, B. E. 1982. Le groupe *Formica rufa* (Hym. Formicidae) en Alsace. 2) Le petit complexe forestier de Brumath. 3) La forêt de la Hardt à Mulhouse. Entomologiste (Paris) 38:129-138. [1982.06]

Lorber, B. E. 1983. Deux formes tératologiques viables de *Myrmica* (Hym. Formicidae Myrmicinae). Entomologiste (Paris) 39:66-70. [1983.04]

Losana, M. 1834. Saggio sopra le formiche indigene del Piemonte. Mem. R. Accad. Sci. Torino 37:307-333. [1834]

Loureiro, M. C., Soares, M. A. S. 1990. Bibliografia especializada, 36. Insecta: Formicidae. Viçosa: Universidade Federal de Viçosa, 55 pp. [1990]

Lowery, B. B. 1965. The rediscovery of two rare ant species. News Bull. Aust. Entomol. Soc. 1:17. [1965.11]

Lowery, B. B. 1967. A new ant of the dacetine genus *Orectognathus* (Hymenoptera: Formicidae). J. Aust. Entomol. Soc. 6:137-140. [1967.12.31]

Lowery, B. B., Taylor, R. J. 1994. Occurrence of ant species in a range of sclerophyll forest communities at Old Chum Dam, north-eastern Tasmania. Aust. Entomol. 21:11-14. [1994.06.30]

Lowne, B. T. 1865a. Contributions to the natural history of Australian ants. Entomologist 2:275-280. [1865.09]

Lowne, B. T. 1865b. Contributions to the natural history of Australian ants. Entomologist 2:331-336. [1865.12]

Lubbock, J. 1877. On some points in the anatomy of ants. Mon. Microsc. J. 18:121-142. [1877.09]

Lubbock, J. 1879. On the anatomy of ants. [Abstract.] J. Linn. Soc. Lond. Zool. 14:738-739. [1879.09.02]

Lubbock, J. 1880. Observations on ants, bees, and wasps; with a description of a new species of honey-ant.- Part VII. Ants. J. Linn. Soc. Lond. Zool. 15:167-187. [1880.09.03]

Lubbock, J. 1881. On the anatomy of ants. Trans. Linn. Soc. Lond. Zool. (2)2:141-154. [1881.03]

Lubbock, J. 1883. Observations on ants, bees, and wasps.- Part X. With a description of a new genus of honey-ant. J. Linn. Soc. Lond. Zool. 17:41-52. [1883.04.17]

Lubin, Y. D. 1984. Changes in the native fauna of the Galápagos Islands following invasion by the little red fire ant, *Wasmannia auropunctata*. Biol. J. Linn. Soc. 21:229-242. [1984.03.12]

Lucas, H. 1849. Exploration scientifique de l'Algérie pendant les années 1840, 1841, 1842. Sciences physiques. Zoologie. III. Histoire naturelle des animaux articulés. Troisième partie - Insectes (Suite). Paris: A. Bertrand, 527 pp. [1849] [See Baker (1994) for discussion of dating. Ants pp. 299-303.]

Lucas, H. 1855. [Untitled. Introduced by "On communique, au nom de M. H. Lucas, la note suivante sur une *Formica* des plus remarquables".] Ann. Soc. Entomol. Fr. (Bull.) (3)3:liv-lv. [1855.09.12] [Note on *Myrmecocystus mexicanus*.]

Lucas, H. 1856. [Untitled. Introduced by "M. H. Lucas fait connaître la note suivante."] Ann. Soc. Entomol. Fr. (Bull.) (3)4:xix-xxi. [1856.07.23] [*Myrmica algirica* Lucas newly synonymized under *Crematogaster scutellaris* (Olivier). Bolton (1995b) misattributes this synonymy to Mayr.]

Lucas, H. 1860. Observations sur les Busileras ou Fourmis à miel du Mexique (*Myrmecocystus melligerus*). Rev. Mag. Zool. Pure Appl. (2)12:269-280. [1860.06] [Synonymy of *Myrmecocystus mexicanus* Wesmael under *M. melligerus* (Llave).]

Lucas, H. 1873a. [Untitled. Introduced by: "M. H. Lucas communique la rectification synonymique suivante".] Bull. Bimens. Soc. Entomol. Fr. 1873(1):12-13. [1873.04.19] [Reissued as Lucas (1873b).]

Lucas, H. 1873b. [Untitled. Introduced by: "M. H. Lucas communique la rectification synonymique suivante".] Ann. Soc. Entomol. Fr. (Bull.) (5)3:lxvi-lxvii. [1873.10.20] [First published as Lucas (1873a).]

Luederwaldt, H. 1918. Notas myrmecologicas. Rev. Mus. Paul. 10:29-64. [1918]
Luederwaldt, H. 1920a. Chave para determinar os Dorylineos brasileiros. Rev. Mus. Paul. 12:229-257. [1920]
Luederwaldt, H. 1920b. Neue Brasilianische Ameisen. São Paulo: Weiszflog Irmãos, 14 pp. [Not seen. Cited in Kempf (1972b:285).]
Luederwaldt, H. 1926a. Observações biologicas sobre formigas brasileiras especialmente do estado de São Paulo. Rev. Mus. Paul. 14: 185-303. [1926]
Luederwaldt, H. 1926b. Addenda á memoria "A Ilha dos Alcatrazes" do tomo XIII, 1923, da Revista do Museu Paulista. Rev. Mus. Paul. 14:397-401. [1926]
Lund, A. W. 1831a. Lettre sur les habitudes de quelques fourmis du Brésil, adressée à M. Audouin. Ann. Sci. Nat. 23:113-138. [1831.06]
Lund, A. W. 1831b. Ueber die Lebensweise einiger brasilianischer Ameisen. Notizen Geb. Natur-Heilkd. 32:97-106. [1831.11]
Lutz, H. 1986. Eine neue Unterfamilie der Formicidae (Insecta: Hymenoptera) aus dem mitteleozänen Olschiefer der "Grube Messel" bei Darmstadt (Deutschland, S-Hessen). Senckenb. Leth. 67:177-218. [1986.10.31]
Lutz, H. 1990. Systematische und palökologische Untersuchungen an Insekten aus dem Mittel-Eozän der Grube Messel bei Darmstadt. Cour. Forschungsinst. Senckenb. 124:1-165. [1990.09.28] [Formicidae pp. 43-82.]
Lynch, J. F. 1981. Seasonal, successional, and vertical segregation in a Maryland ant community. Oikos 37:183-198. [1981.09]
Lynch, J. F. 1988 ("1987"). An annotated checklist and key to the species of ants (Hymenoptera: Formicidae) of the Chesapeake Bay region. Md. Nat. 31:61-106. [1988.07.27]
Lynch, J. F., Johnson, A. K., Balinsky, E. C. 1988. Spatial and temporal variation in the abundance and diversity of ants (Hymenoptera: Formicidae) in the soil and litter layers of a Maryland forest. Am. Midl. Nat. 119:31-44. [1988.03.25]
Maavara, V. 1953. Ants of Estonian SSR. [In Estonian.] Abiks Loodusevaatl. 10:1-44. [> 1953.03.07] [Date submitted for printing ("trükkimisele antud").]
Maavara, V. 1983. Variability and differentiation in some closely related *Formica* species (Hymenoptera, Formicidae). [In Russian.] Pp. 54-78 in: Paaver, K., Sutt, T. (eds.) Problems of contemporary Darwinism. [In Russian.] Tartu: Academy of Sciences of the Estonian SSR, 190 pp. [> 1983.02.25] [Date signed for printing ("podpisano k pechati").]
Mabelis, A. A. 1987. Verspreidung en habitat van de stronkmier, *Formica truncorum* Fabricius (Hymenoptera: Formicidae). Entomol. Ber. (Amst.) 47:129-136. [1987.09]
Mabelis, A. A., Boomsma, J. J. 1984. The ant fauna of coastal and continental islands in the Netherlands. [Abstract.] P. 498 in: XVII International Congress of Entomology. Hamburg, Federal Republic of Germany, August 20-26, 1984. Abstract Volume. Hamburg: XVII International Congress of Entomology, 960 pp. [1984.08.10]
Mabelis, A. A., Boting, P. H., Dijkstra, P. J., Zaaijer, P. M. 1986. De stronkmier (*Formica truncorum* Fabricius) toch inheems! (Hymenoptera: Formicidae). Entomol. Ber. (Amst.). 46:173-175. [1986.12.01]
Mabelis, A. A., Mabelis-Jonkers, J. C. F. 1978. Vespreidung van mieren in kalkrijke gebieden van Zuid-Limburg (Hym., Formicidae). Entomol. Ber. (Amst.) 38:165-168. [1978.11.01]
MacArthur, R. H., Wilson, E. O. 1967. The theory of island biogeography. Princeton: Princeton University Press, xi + 203 pp. [1967]
MacConnell, J. G., Blum, M. S., Buren, W. F., Williams, R. N., Fales, H. M. 1976. Fire ant venoms: chemotaxonomic correlations with alkaloidal compositions. Toxicon 14:69-78. [1976]
MacDonagh, E. J. 1935 ("1934"). [Untitled. Introduced by: "El doctor Emiliano J. Mac Donagh presentó una comunicación sobre la Distribución geográfica de las hormigas cultivadoras de hongos, es decir del grupo de las 'Attinas' ".] Bol. Univ. Nac. La Plata 18(6):3-5. [1935]

MacDonagh, E. J. 1937. Sobre hormigas podadoras del extremo sur de Buenos Aires. Notas Mus. La Plata (Zool.) 2:45-52. [1937.06.11]

MacDonagh, E. J. 1939. Las hormigas "Attinae" de Patagones y rectificación de la supuest "*Oecodoma*" de Hudson. Physis 17:211-215. [1939.05.31]

MacKay, E. E., MacKay, W. P. 1985. Apoyo a la actual division generica de hormigas usando etologia comparativa (Hymenoptera, Formicidae). Folia Entomol. Mex. 61:179-188. [1985.06.28]

MacKay, W. See: MacKay, W. P.

MacKay, W. P. 1980. A new harvester ant from the mountains of southern California (Hymenoptera: Formicidae). Southwest. Nat. 25:151-156. [1980.05.15]

MacKay, W. P. 1981. A comparison of the nest phenologies of three species of *Pogonomyrmex* harvester ants (Hymenoptera: Formicidae). Psyche (Camb.) 88:25-74. [1981.12.28]

MacKay, W. P. 1982. An altitudinal comparison of oxygen consumption rates in three species of *Pogonomyrmex* harvester ants (Hymenoptera: Formicidae). Physiol. Zool. 55:367-377. [1982.10]

MacKay, W. P. 1985. *Acanthostichus sanchezorum* (Hymenoptera: Formicidae), una nueva especie de Colombia. Sociobiology 11:127-131. [1985]

MacKay, W. P. 1988. [Untitled. *Pheidole wheelerorum* W. MacKay new species.] Pp. 96-98 in: Mackay, W., Lowrie, D., Fisher, A., MacKay, E., Barnes, F., Lowrie, D. The ants of Los Alamos County, New Mexico (Hymenoptera: Formicidae). Pp. 79-131 in: Trager, J. C. (ed.) Advances in myrmecology. Leiden: E. J. Brill, xxvii + 551 pp. [1988]

MacKay, W. P. 1989. A new *Aphaenogaster* (Hymenoptera: Formicidae) from southern New Mexico. J. N. Y. Entomol. Soc. 97:47-49. [1989.03.29]

MacKay, W. P. 1991 ("1990"). *Anochetus brevidentatus*, new species, a second fossil Odontomachiti ant (Hymenoptera: Formicidae). J. N. Y. Entomol. Soc. 99:138-140. [1991.02.14]

MacKay, W. P. 1993a. The status of the ant *Leptothorax pergandei* Emery (Hymenoptera: Formicidae). Sociobiology 21:287-297. [1993]

MacKay, W. P. 1993b. A review of the New World ants of the genus *Dolichoderus* (Hymenoptera: Formicidae). Sociobiology 22:1-148. [1993]

MacKay, W. P. 1993c. Succession of ant species (Hymenoptera: Formicidae) on low-level nuclear waste sites in northern New Mexico. Sociobiology 23:1-11. [1993]

MacKay, W. P. 1995. New distributional records for the ant genus *Cardiocondyla* in the New World (Hymenoptera: Formicidae). Pan-Pac. Entomol. 71:169-172. [1995.10.26]

MacKay, W. P., Anderson, R. S. 1992 ("1991"). New distributional records for the ant genus *Ponera* (Hymenoptera: Formicidae) in North America. J. N. Y. Entomol. Soc. 99:696-699. [1992.01.07]

MacKay, W. P., Anderson, R. S. 1993. New distributional records for the ant genus *Smithistruma* (Hymenoptera: Formicidae) in southern United States. Southwest. Nat. 38:388-389. [1993.12]

MacKay, W. P., Baena, M. L. 1993. A new "horned" fungus growing ant, *Cyphomyrmex castagnei*, from Colombia. Sociobiology 23:31-37. [1993]

MacKay, W. P., Cover, S. P., Heinze, J., Holldöbler, B. 1995. Range extensions for the ant *Leptothorax pergandei* (Hymenoptera: Formicidae): a mesic forest species discovered in the Chihuahuan Desert. Proc. Entomol. Soc. Wash. 97:888. [1995.10.31]

MacKay, W. P., Elias, S. A. 1993 ("1992"). Late Quaternary ant fossils from packrat middens (Hymenoptera: Formicidae): implications for climatic change in the Chihuahuan desert. Psyche (Camb.) 99:169-184. [1993.03.22]

MacKay, W. P., Lowrie, D., Fisher, A., MacKay, E. E., Barnes, F., Lowrie, D. 1988. The ants of Los Alamos County, New Mexico (Hymenoptera: Formicidae). Pp. 79-131 in: Trager, J. C. (ed.) Advances in myrmecology. Leiden: E. J. Brill, xxvii + 551 pp. [1988]

MacKay, W. P., MacKay, E. E. 1988 ("1986"). Las hormigas de Colombia: arrieras del género *Atta* (Hymenoptera: Formicidae). Rev. Colomb. Entomol. 12(1):23-30. [1988.06]

MacKay, W. P., MacKay, E. E. 1994. *Lasius xerophilus* (Hymenoptera: Formicidae), a new ant species from White Sands National Monument, New Mexico. Psyche (Camb.) 101:37-43. [1994.06.28]

MacKay, W. P., MacKay, E. E., Perez Dominguez, J. F., Valdez Sanchez, L. I., Orozco, P. V. 1985. Las hormigas del estado de Chihuahua Mexico: el genero *Pogonomyrmex* (Hymenoptera: Formicidae). Sociobiology 11:39-54. [1985]

MacKay, W. P., Perez Dominguez, J. F., Valdez Sanchez, L. I. 1985. The army ants of the state of Chihuahua Mexico (Hymenoptera: Formicidae: Ecitoninae). Southwest. Nat. 30:611-612. [1985.11.27]

MacKay, W. P., Porter, S., Fowler, H. G., Vinson, B. 1994. A distribuição das formigas lava-pés (*Solenopsis* spp.) no Estado de Mato Grosso do Sul, Brasil (Hymenoptera: Formicidae). Sociobiology 24:307-317. [1994]

MacKay, W. P., Rebeles M., A., Arredondo B., H. C., Rodríguez R., A. D., González, D. A., Vinson, S. B. 1991. Impact of the slashing and burning of a tropical rain forest on the native ant fauna (Hymenoptera: Formicidae). Sociobiology 18:257-268. [1991]

MacKay, W. P., Van Vactor, S. 1985. New host record for the social parasite *Pogonomyrmex anergismus* (Hymenoptera: Formicidae). Proc. Entomol. Soc. Wash. 87:863. [1985.10.18]

MacKay, W. P., Vinson, S. B. 1988 ("1989"). Rediscovery of the ant *Gnamptogenys hartmani* (Hymenoptera: Formicidae) in eastern Texas. Proc. Entomol. Soc. Wash. 91:127. [1988.12.13]

MacKay, W. P., Vinson, S. B. 1989a. Two new ants of the genus *Solenopsis* (*Diplorhoptrum*) from eastern Texas (Hymenoptera: Formicidae). Proc. Entomol. Soc. Wash. 91:175-178. [1989.04.13]

MacKay, W. P., Vinson, S. B. 1989b. A guide to species identification of New World ants (Hymenoptera: Formicidae). Sociobiology 16:3-47. [1989]

MacLeay, W. 1873. Miscellanea entomologica. Trans. Entomol. Soc. N. S. W. 2:319-370. [1873] [Publication date (year) from Andrewes (1930:vii).]

Maes, J.-M., MacKay, W. P. 1993. Catálogo de las hormigas (Hymenoptera: Formicidae) de Nicaragua. Rev. Nicar. Entomol. 23:1-46. [1993.03]

Magalhães, L. E. de, Silva, V. P. da. 1971. Estudo da genitália feminina do gênero *Atta* (Hym., Formicidae). [Abstract.] Ciênc. Cult. (São Paulo) 23(supl.):280. [1971.06]

Magretti, P. 1884a. Raccolte imenotterologiche nell'Africa Orientale. Bull. Soc. Entomol. Ital. 15:241-253. [1884.04.15]

Magretti, P. 1884b. Risultati di raccolte imenotterologiche nell'Africa Orientale. Ann. Mus. Civ. Stor. Nat. 21[=(2)1]:523-636. [1884.11.15]

Maidl, F. 1922. Beiträge zur Hymenopterenfauna Dalmatiens, Montenegros und Albaniens. I. Teil: Aculeata und Chrysididae. Ann. Naturhist. Mus. Wien 35:36-106. [1922.05] [Formicidae pp. 43-45.]

Majer, J. D. 1972. The ant mosaic in Ghana cocoa farms. Bull. Entomol. Res. 62:151-160. [1972.11.14]

Majer, J. D. 1976a. The maintenance of the ant mosaic in Ghana cocoa farms. J. Appl. Ecol. 13:123-144. [1976.04]

Majer, J. D. 1976b. The ant mosaic in Ghana cocoa farms: further structural considerations. J. Appl. Ecol. 13:145-155. [1976.04]

Majer, J. D. 1983. Notes on a deformed worker of the ant genus *Iridomyrmex* (Hymenoptera: Formicidae). Aust. Entomol. Mag. 10:11-12. [1983.06]

Majer, J. D. 1984. Recolonization by ants in rehabilitated open-cut mines in northern Australia. Reclam. Reveg. Res. 2:279-298. [1984.03]

Majer, J. D. 1985. Recolonization by ants of rehabilitated mineral sand mines on North Stradbroke Island, Queensland, with particular reference to seed removal. Aust. J. Ecol. 10:31-48. [1985.03]

Majer, J. D. 1988. The Formicidae. Pp. 219-224 in: Specht, R. L. (ed.) Mediterranean-type ecosystems. A data source book. Dordrecht: Kluwer Academic Publishers, xii + 248 pp. [1988]

Majer, J. D. 1990. The abundance and diversity of arboreal ants in northern Australia. Biotropica 22:191-199. [1990.06]

Majer, J. D. 1993. Comparison of the arboreal ant mosaic in Ghana, Brazil, Papua New Guinea and Australia - its structure and influence on ant diversity. Pp. 115-141 in: LaSalle, J., Gauld, I. D. (eds.) Hymenoptera and biodiversity. Oxford: CAB International, xi + 348 pp. [1993]

Majer, J. D., Brown, K. R. 1987 ("1986"). The effects of urbanization on the ant fauna of the Swan Coastal Plain near Perth, Western Australia. J. R. Soc. West. Aust. 69:13-17. [1987.05.22]

Majer, J. D., Camer-Pesci, P. 1991. The ant mosaic in tropical Australian tree crops and native ecosystems - is there a mosaic? Biotropica 23:173-181. [1991.06]

Majer, J. D., Delabie, J. H. C. 1994. Comparison of the ant communities of annually inundated and terra firme forests at Trombetas in the Brazilian Amazon. Insectes Soc. 41:343-359. [1994]

Majer, J. D., Delabie, J. H. C., Smith, M. R. B. 1994. Arboreal ant community patterns in Brazilian cocoa farms. Biotropica 26:73-83. [1994.03]

Majer, J. D., Queiroz, M. V. B. 1990. The composition of ant communities in Brasilian Atlantic rainforest. Pp. 704-705 in: Veeresh, G. K., Mallik, B., Viraktamath, C. A. (eds.) Social insects and the environment. Proceedings of the 11th International Congress of IUSSI, 1990. New Delhi: Oxford & IBH Publishing Co., xxxi + 765 pp. [1990.08]

Majer, J. D., Queiroz, M. V. B. 1993. Distribution and abundance of ants in a Brazilian subtropical coffee plantation. P. N. G. J. Agric. For. Fish. 36(2):29-35. [1993.11]

Mallis, A. 1941. A list of the ants of California with notes on their habits and distribution. Bull. South. Calif. Acad. Sci. 40:61-100. [1941.10.25]

Malozemova, L. A. 1972. Ants of steppe forests, their distribution by habitats, and perspectives of their utilization for protection of forests (north Kazakhstan). [In Russian.] Zool. Zh. 51:57-68. [1972.01] [January issue. Date signed for printing ("podpisano k pechati"): 4 January 1972.]

Malozemova, L. A., Malozemov, Y. A. 1993. Ecological peculiarities of the ant population in forest and meadow biogeocenoses of the Visimsky Reserve. [In Russian.] Ekologiya (Mosc.) 1993(6):83-86. [1993.12] [November-December issue. Date signed for printing ("podpisano k pechati"): 11 November 1993.]

Malyshev, S. I. 1954. Ways and conditions of the origin of ants (Hymenoptera, Formicidae). [In Russian.] Dokl. Akad. Nauk SSSR (n.s.)94:1185-1188. [> 1954.02.23] [Date signed for printing ("podpisano k pechati").]

Malyshev, S. I. 1959a. Pathways and conditions of evolution of the ants (Hymenoptera, Formicidae). [In Russian.] Tr. Inst. Morfol. Zhivotn. Akad. Nauk SSSR 27:249-260. [> 1959.07.09] [Date signed for printing ("podpisano k pechati").]

Malyshev, S. I. 1959b. Hymenoptera, their origin and evolution. [In Russian.] Moskva: Sovetskaya Nauka, 291 pp. [> 1959.08.15] [Date signed for printing ("podpisano k pechati"). Origin of ants, pp. 175-234.]

Malyshev, S. I. 1960. Patterns and conditions of instinct origination in ants during the process of evolution (Hymenoptera, Formicoidea). [In Russian.] Tr. Vses. Entomol. Obshch. 47:5-52. [> 1960.01.12] [Date signed for printing ("podpisano k pechati").]

Malyshev, S. I. 1966. Genesis of the Hymenoptera and the phases of their evolution. [In Russian.] Moskva: Nauka, 329 pp. [> 1966.11.09] [Date signed for printing ("podpisano k pechati"). Origin of ants, pp. 204-266.]

Malyshev, S. I. 1968. Genesis of the Hymenoptera and the phases of their evolution. [Translated by B. Haigh, edited by O. W. Richards and B. Uvarov.] London: Methuen, 319 pp. [1968] [English translation of Malyshev (1966).]

Mamaev, B. M., Medvedev, L. N., Pravdin, F. N. 1976. Guide to insects of the European part of the USSR. [In Russian.] Moskva: Prosveshchenie, 304 pp. [> 1976.06.30] [Date signed for printing ("podpisano k pechati"). Ants pp. 243-246, including keys.]

Mamet, R. 1954. The ants (Hymenoptera Formicidae) of the Mascarene Islands. Mauritius Inst. Bull. 3:249-259. [1954.05.01]

Manabe, K. 1994a. Ants of the shrine forest in Fukuoka Prefecture (First report; ants of lowland). [Abstract.] [In Japanese.] Ari 17:6. [1994.04.30]

Manabe, K. 1994b. Ants of the shrine forest in Fukuoka Prefecture (Second report; ants of mountain zone). [Abstract.] [In Japanese.] Ari 17:8. [1994.04.30]

Mann, W. M. 1911a. Notes on the guests of some Californian ants. Psyche (Camb.) 18:27-31. [1911.02] [Stamp date in MCZ library: 1911.03.04.]

Mann, W. M. 1911b. On some Northwestern ants and their guests. Psyche (Camb.) 18:102-109. [1911.06] [Stamp date in MCZ library: 1911.07.07.]

Mann, W. M. 1912. Parabiosis in Brazilian ants. Psyche (Camb.) 19:36-41. [1912.04] [Stamp date in MCZ library: 1912.05.02.]

Mann, W. M. 1913. Literature for 1912 on the behavior of ants and myrmecophiles. J. Anim. Behav. 3:429-445. [1913.12]

Mann, W. M. 1915. A new form of a southern ant from Naushon Island, Massachusetts. Psyche (Camb.) 22:51. [1915.04] [Stamp date in MCZ library: 1915.05.27.]

Mann, W. M. 1916. The Stanford Expedition to Brazil, 1911, John C. Branner, Director. The ants of Brazil. Bull. Mus. Comp. Zool. 60:399-490. [1916.09]

Mann, W. M. 1919. The ants of the British Solomon Islands. Bull. Mus. Comp. Zool. 63:273-391. [1919.12]

Mann, W. M. 1920a. Ant guests from Fiji and the British Solomon Islands. Ann. Entomol. Soc. Am. 13:60-69. [1920.04.27]

Mann, W. M. 1920b. Additions to the ant fauna of the West Indies and Central America. Bull. Am. Mus. Nat. Hist. 42:403-439. [1920.12.20]

Mann, W. M. 1921. The ants of the Fiji Islands. Bull. Mus. Comp. Zool. 64:401-499. [1921.02]

Mann, W. M. 1922. Ants from Honduras and Guatemala. Proc. U. S. Natl. Mus. 61:1-54. [1922.06.14]

Mann, W. M. 1923. Two new ants from Bolivia. (Results of the Mulford Biological Exploration. - Entomology.). Psyche (Camb.) 30:13-18. [1923.02] [Stamp date in MCZ library: 1923.03.19.]

Mann, W. M. 1924a. Notes on Cuban ants. Psyche (Camb.) 31:19-23. [1924.02] [Stamp date in MCZ library: 1924.03.15.]

Mann, W. M. 1924b. Myrmecophiles from the western United States and Lower California. Ann. Entomol. Soc. Am. 17:87-95. [1924.04.18]

Mann, W. M. 1925a. New beetle guests of army ants. J. Wash. Acad. Sci. 15:73-77. [1925.02.19]

Mann, W. M. 1925b. Ants collected by the University of Iowa Fiji-New Zealand Expedition. Stud. Nat. Hist. Iowa Univ. 11(4):5-6. [1925.03.15]

Mann, W. M. 1926. Some new neotropical ants. Psyche (Camb.) 33:97-107. [1926.10] [Stamp date in MCZ library: 1926.11.20.]

Mann, W. M. 1929. Notes on Cuban ants of the genus *Macromischa* (Hymenoptera: Formicidae). Proc. Entomol. Soc. Wash. 31:161-166. [1929.12.26]

Mann, W. M. 1931. A new ant from Porto Rico. J. Wash. Acad. Sci. 21:440-441. [1931.10.19]

Mann, W. M. 1934. Stalking ants, savage and civilized. Natl. Geogr. Mag. 66:171-192. [1934.08]

Mann, W. M. 1935. Two new ants collected in quarantine. Psyche (Camb.) 42:35-37. [1935.03] [Stamp date in MCZ library: 1935.04.04.]

Mann, W. M. 1948. Ant hill odyssey. Boston: Little, Brown and Co., 338 pp. [1948.11]

Mantero, G. 1898. Res Ligusticae XXX. Materiali per un catalogo degli Imenotteri liguri. Parte I. Formicidi. Ann. Mus. Civ. Stor. Nat. 39[=(2)19]:146-160. [1898.08.05]

Mantero, G. 1905a. Materiali per una fauna dell'Arcipelago Toscano. Isola del Giglio. II. Ann. Mus. Civ. Stor. Nat. 41[=(3)1]:449-454. [1905.01.11]

Mantero, G. 1905b. Materiali per una fauna dell'Arcipelago Toscano. IV. Isola del Giglio. Catalogo degli Imenotteri. Parte I. [part] Ann. Mus. Civ. Stor. Nat. 42[=(3)2]:40-48. [1905.10.10]

Mantero, G. 1905c. Materiali per una fauna dell'Arcipelago Toscano. IV. Isola del Giglio. Catalogo degli Imenotteri. Parte I. [concl.] Ann. Mus. Civ. Stor. Nat. 42[=(3)2]:49-86. [1905.11.03] [List of Formicidae pp. 54-58.]

Mantero, G. 1908. Res Ligusticae XL. Materiali per un catalogo degli Imenotteri liguri. Parte V. Supplemento ai Formicidi, Crisidi, Mutillidi, Braconidi e Cinipidi. Ann. Mus. Civ. Stor. Nat. 44[=(3)4]:43-74. [1908.11.20] [Ants pp. 44-46.]

Mantero, G. 1911 ("1909"). Imenotteri dell'isola dell'Asinara raccolti dal Signor Silvio Folchini. Bull. Soc. Entomol. Ital. 41:56-83. [1911.02.28]

Marcus, H. 1944. Estudios mirmecológicos. I. Estudio comparado de la articulación mandíbular de las hormigas y termites. Acta Zool. Lilloana 2:259-284. [1944.12.29]

Marcus, H. 1945a. Estudios mirmecológicos. IV. Las articulaciones "trampas" en las mandíbulas de los Odontomachini. Rev. Agric. Cochabamba 3:13-20. [1945.06]

Marcus, H. 1945b. La diferencia en la fauna de las hormigas de Cochabamba y Liriuni. Rev. Agric. Cochabamba 3:28-29. [1945.06]

Marcus, H. 1948a. Un órgano de estridulación en hormigas y termitas. Folia Univ. Cochabamba 1:39-44. [< 1948.07.30] [Receipt date of volume 1 in the National Agricultural Library (Beltsville, Maryland): 1948.07.30.]

Marcus, H. 1948b. Los órganos de copulación de *Acromyrmex*. Folia Univ. Cochabamba 1:73-78. [< 1948.07.30] [Receipt date of volume 1 in the National Agricultural Library (Beltsville, Maryland): 1948.07.30.]

Marcus, H. 1948c. De la evolución de los órganos respiratorios en los insectos holometabolicos. Folia Univ. Cochabamba 1:83-96. [< 1948.07.30] [Receipt date of volume 1 in the National Agricultural Library (Beltsville, Maryland): 1948.07.30.]

Marcus, H. 1948d. La embriogenesis de *Aethalion*, termitas y hormigas con una comparación de la anatomía de los insectos y vertebrados. Folia Univ. Cochabamba 1:97-118. [< 1948.07.30] [Receipt date of volume 1 in the National Agricultural Library (Beltsville, Maryland): 1948.07.30.]

Marcus, H. 1948e. La variación de altura melanotica de *Camponotus conspicuus*. Folia Univ. Cochabamba 2:28-29. [1948.10]

Marcus, H. 1949a. Receptores en hormigas. Folia Univ. Cochabamba 3:35-43. [1949.08]

Marcus, H. 1949b. El órgano postantenal en Apterygota, termitas y hormigas. Folia Univ. Cochabamba 3:44-51. [1949.08]

Marcus, H. 1949c. Anatomia comparada de las articulaciones antenales de las hormigas. Folia Univ. Cochabamba 3:86-94. [1949.08]

Marcus, H. 1949d. Como las hormigas evitan el incesto. Folia Univ. Cochabamba 3:95-96. [1949.08]

Marcus, H. 1950. Como evita el incesto la *Solenopsis saevissima* var. Folia Univ. Cochabamba 4:78-79. [1950.10]

Marcus, H. 1951. Una glándula endocrina y la formación de la casta trabajadora de hormigas. Folia Univ. Cochabamba 5:75-82. [1951.10]

Marcus, H. 1951b. La conexión de la fibra muscular con la tendón en *Acromyrmex silvestrii* (Emery). Folia Univ. Cochabamba 5:115-116. [1951.10]

Marcus, H. 1953. Estudios mirmecológicos. Folia Univ. Cochabamba 6:17-68. [1953]

Marcus, H. 1954. La formación de las castas en Ecitones. Folia Univ. Cochabamba 7:3-12. [1954]

Marcus, H. 1958. Über die Polymorphie. Zool. Anz. 161:61-77. [1958.08]

Marcus, H., Marcus, E. E. 1951. Los nidos y los órganos de estridulación y de equilibrio de *Pogonomyrmex marcusi* y de *Dorymyrmex emmaericaellus* (Kusn.). Folia Univ. Cochabamba 5:117-143. [1951.10]

Mariconi, F. A. M. 1970. As saúvas. São Paulo: Agronômica Ceres, 167 pp. [Not seen. Cited in Cherrett & Cherrett (1989) and in Della Lucia (1993).]

Mariconi, F. A. M., Zamith, A. P. L., Paiva Castro, U. de, Joly, S. 1963. Nova contribuição para o conhecimento das saúvas de Piracicaba (*Atta* spp.) (Hym. - Formicidae). Rev. Agric. (Piracicaba) 38:85-93. [1963.06]

Marikovsky, P. I. 1962. Materials on the ant fauna (Formicinae) of the basin of the middle and lower reaches of the Ili River. [In Russian.] Tr. Inst. Zool. Akad. Nauk Kaz. SSR 18:161-176. [> 1962.04.14] [Date signed for printing ("podpisano k pechati").]

Marikovsky, P. I. 1963a. A new species of ant *Polyergus nigerrimus* Marik., sp. n. (Hymenoptera, Formicidae) and some features of its biology. [In Russian.] Entomol. Obozr. 42:110-114. [> 1963.03.15] [Date signed for printing ("podpisano k pechati"). Translated in Entomol. Rev. (Wash.) 42:58-59. See Marikovsky (1963b).]

Marikovsky, P. I. 1963b. A new ant, *Polyergus nigerrimus* Marik., sp. n., (Hymenoptera, Formicidae) and some features of its biology. Entomol. Rev. (Wash.) 42:58-59. [1963] [English translation of Marikovsky (1963a).]

Marikovsky, P. I. 1979. Ants of the Semireche Desert. [In Russian.] Alma Ata: Nauka, 263 pp. [> 1979.06.12] [Date signed for printing ("podpisano k pechati").]

Marikovsky, P. I., Yakushkin, V. T. 1974. The ant *Cardiocondyla uljanini* Em., 1889 and the systematic status of the "*Xenometra* parasitic ant". [In Russian.] Izv. Akad. Nauk Kaz. SSR Ser. Biol. 3:57-62. [> 1974.06.17] [Date signed for printing ("podpisano k pechati").]

Markl, H. 1962. Borstenfelder an den Gelenken als Schweresinnesorgane bei Ameisen und anderen Hymenopteren. Z. Vgl. Physiol. 45:475-569. [1962.06.28]

Markl, H. 1966. Peripheres Nervensystem und Muskulatur im Thorax der Arbeiterin von *Apis mellifica* L., *Formica polyctena* Foerster und *Vespa vulgaris* L. und der Grundplan der Innervierung des Insektenthorax. Zool. Jahrb. Abt. Anat. Ontog. Tiere 83:107-184. [1966]

Markl, H. 1973. The evolution of stridulatory communication in ants. Pp. 258-265 in: Proceedings IUSSI VIIth International Congress, London, 10-15 September, 1973. Southampton: University of Southampton, vi + 418 pp. [1973]

Markl, H., Hölldobler, B. 1978. Recruitment and food-retrieving behavior in *Novomessor* (Formicidae, Hymenoptera). II. Vibration signals. Behav. Ecol. Sociobiol. 4:183-216. [1978.12.06]

Markl, H., Hölldobler, B., Hölldobler, T. 1977. Mating behavior and sound production in harvester ants (*Pogonomyrmex*, Formicidae). Insectes Soc. 24:191-212. [1977.06]

Marsh, A. C. 1985. Forager abundance and dietary relationships in a Namib Desert ant community. S. Afr. J. Zool. 20:197-203. [1985.10]

Marsh, A. C. 1986a. Ant species richness along a climatic gradient in the Namib Desert. J. Arid Environ. 11:235-241. [1986.11]

Marsh, A. C. 1986b. Checklist, biological notes and distribution of ants in the central Namib Desert. Madoqua 14:333-344. [1986]

Martínez, M. D. 1986a. Nuevas citas para la península Ibérica de *Proceratium melinum*, *Aphaenogaster cardenai* y *Messor lobicornis* (Hym. Formicidae). Bol. Asoc. Esp. Entomol. 10:403. [1986.05]

Martínez, M. D. 1986b. Algunos datos sobre las malformaciones en hormigas (Hymenoptera. Formicidae). Pp. 1035-1041 in: Actas de las VIII Jornadas de la Asociación Española de Entomología. Sevilla: Servicio Publicaciones Universidad de Sevilla, [8] + 1280 pp. [1986.10]

Martínez, M. D. 1987. Las hormigas (Hym. Formicidae) de la Sierra de Guadarrama. Bol. Asoc. Esp. Entomol. 11:385-394. [1987.11]

Martínez, M. D., Espadaler, X. 1986. Revisión de las hormigas ibéricas de la colección M. Medina y nuevos datos de distribución (Hymenoptera, Formicidae). Pp. 1022-1034 in: Actas de las VIII Jornadas de la Asociación Española de Entomología. Sevilla: Servicio Publicaciones Universidad de Sevilla, [8] + 1280 pp. [1986.10]

Martínez, M. D., Serrano Talavera, J. M. 1985. Contribución al conocimiento de las hormigas (Hym. Formicidae) del sabinar español. Bol. Soc. Port. Entomol. Supl. 1(2):33-41. [1985]

Martinez, M. J. 1992. A new ant introduction for North America: *Pheidole teneriffana* (Forel) (Hymenoptera: Formicidae). Pan-Pac. Entomol. 68:153-154. [1992.05.14]
Martinez, M. J. 1993. The first field record for the ant *Tetramorium bicarinatum* Nylander (Hymenoptera: Formicidae) in California. Pan-Pac. Entomol. 69:272-273. [1993.10.06]
Martinez, M. J. 1995. The first record of mixed nests of *Conomyrma bicolor* (Wheeler) and *Conomyrma insana* (Buckley) (Hymenoptera: Formicidae). Pan-Pac. Entomol. 71:252. [1995.11.03]
Martínez Ibáñez, M. D. See: Martínez, M. D.
Martini, R., Raqué, K.-F. 1986. Zwei bedrohte Rossameisen-Arten in Heidelberg. Carolinea 44:171-172. [1986.12.29]
Maschwitz, U. 1974. Vergleichende Untersuchungen zur Funktion der Ameisenmetathorakaldrüse. Oecologia (Berl.) 16:303-310. [1974.09.30]
Maschwitz, U. 1975a. Old and new chemical weapons in ants. Pp. 41-45 in: Noirot, C., Howse, P. E., Le Masne, G. (eds.) Pheromones and defensive secretions in social insects. Dijon: French Section of the IUSSI, vi + 248 pp. [1975.12]
Maschwitz, U. 1975b. Old and new trends in the investigation of chemical recruitment in ants. Pp. 47-59 in: Noirot, C., Howse, P. E., Le Masne, G. (eds.) Pheromones and defensive secretions in social insects. Dijon: French Section of the IUSSI, vi + 248 pp. [1975.12]
Maschwitz, U., Dumpert, K., Schmidt, G. H. 1985. Silk pavilions of two *Camponotus* (*Karavaievia*) species from Malaysia: description of a new nesting type in ants (Formicidae: Formicinae). Z. Tierpsychol. 69:237-249. [1985.07]
Maschwitz, U., Fiala, B., Moog, J., Saw, L. G. 1991. Two new myrmecophytic associations from the Malay Peninsula: ants of the genus *Cladomyrma* (Formicidae, Camponotinae) as partners of *Saraca thaipingensis* (Caesalpiniaceae) and *Crypteronia griffithii* (Crypteroniaceae). Insectes Soc. 38:27-35. [1991]
Maschwitz, U., Hölldobler, B., Möglich, M. 1974. Tandemlaufen als Rekrutierungsverhalten bei *Bothroponera tesserinoda* Forel (Formicidae: Ponerinae). Z. Tierpsychol. 35:113-123. [1974.09]
Maschwitz, U., Jessen, K., Maschwitz, E. 1981. Foaming in *Pachycondyla*: a new defense mechanism in ants. Behav. Ecol. Sociobiol. 9:79-81. [1981.08]
Maschwitz, U., Koob, K., Schildknecht, H. 1970. Ein Beitrag zur Funktion der Metathoracaldrüse der Ameisen. J. Insect Physiol. 16:387-404. [1970.02]
Maschwitz, U., Schönegge, P. 1977. Recruitment gland of *Leptogenys chinensis*. A new type of pheromone gland in ants. Naturwissenschaften 64:589-590. [1977.11]
Maschwitz, U., Steghaus-Kovac, S. 1991. Individualismus versus Kooperation: gegensätzliche Jagd- und Rekrutierungsstrategien bei tropischen Ponerinen (Hymenoptera: Formicidae). Naturwissenschaften 78:103-113. [1991.03]
Maschwitz, U., Steghaus-Kovac, S., Gaube, R., Hänel, H. 1989. A South East Asian ponerine ant of the genus *Leptogenys* (Hym., Form.) with army ant habits. Behav. Ecol. Sociobiol. 24:305-316. [1989.05.10]
Mason, W. R. M. 1986. Standard drawing conventions and definitions for venational and other features of wings of Hymenoptera. Proc. Entomol. Soc. Wash. 88:1-7. [1986.01.15]
Masson, C. 1969. Études anatomique et électrophysiologique du deutocérébron de *Camponotus ligniperda* Latr. (Hyménoptère - Formicidae). Pp. 171-180 in: Ernst, E. et al. (eds.) International Union for the Study of Social Insects. VI Congress. Bern 15-20 September 1969. Proceedings. Bern: Organizing Committee of the VI Congress IUSSI, 309 pp. [1969.09.15]
Masson, C. 1970. Étude anatomique et fonctionnelle d'une nouvelle structure réceptrice en rapport avec l'antenne chez les fourmis. C. R. Séances Acad. Sci. Ser. D. Sci. Nat. 271:346-349. [1970.07.27]
Masson, C. 1972a. Le système antennaire chez les fourmis. I. Histologie et ultrastructure du deutocérébron. Étude comparée chez *Camponotus vagus* (Formicinae) et *Mesoponera caffraria* (Ponerinae). Z. Zellforsch. Mikrosk. Anat. 134:31-64. [1972.10.31]

Masson, C. 1972b. Organisation sensorielle des principales articulations de l'antenne de la fourmi *Camponotus vagus* Scop. (Hymenoptera, Formicidae). Z. Morphol. Tiere 73:343-359. [1972.11.08]

Masson, C., Friggi, A. 1971. Étude morphologique et topographique des sensilla coeloconica et des sensilla ampullacea de l'antenne de l'ouvrière de *Camponotus vagus* Scop. (Hym. Formicidae). C. R. Séances Acad. Sci. Ser. D. Sci. Nat. 272:618-621. [1971.02.01]

Masson, C., Gabouriaut, D., Friggi, A. 1972. Ultrastructure d'un nouveau type de récepteur olfactif de l'antenne d'insecte trouvé chez la fourmi *Camponotus vagus* Scop. (Hymenoptera, Formicinae). Z. Morphol. Tiere 72:349-360. [1972.06.30]

Masuko, K. 1982. Data on the myrmecofauna of Miyake Island. [In Japanese.] Ari 10:1-2. [1982.10.01]

Masuko, K. 1985 ("1984"). Studies on the predatory biology of oriental dacetine ants (Hymenoptera: Formicidae). I. Some Japanese species of *Strumigenys*, *Pentastruma*, and *Epitritus*, and a Malaysian *Labidogenys*, with special reference to hunting tactics in short-mandibulate forms. Insectes Soc. 31:429-451. [1985.04]

Masuko, K. 1986. Larval hemolymph feeding: a nondestructive parental cannibalism in the primitive ant *Amblyopone silvestrii* Wheeler (Hymenoptera: Formicidae). Behav. Ecol. Sociobiol. 19:249-255. [1986.09]

Masuko, K. 1987. *Leptanilla japonica*: the first bionomic information on the enigmatic ant subfamily Leptanillinae. Pp. 597-598 in: Eder, J., Rembold, H. (eds.) Chemistry and biology of social insects. München: Verlag J. Peperny, xxxv + 757 pp. [1987]

Masuko, K. 1989. Larval hemolymph feeding in the ant *Leptanilla japonica* by use of a specialized duct organ, the "larval hemolymph tap" (Hymenoptera: Formicidae). Behav. Ecol. Sociobiol. 24:127-132. [1989.02.15]

Masuko, K. 1990a. Behavior and ecology of the enigmatic ant *Leptanilla japonica* Baroni Urbani (Hymenoptera: Formicidae: Leptanillinae). Insectes Soc. 37:31-57. [1990.03]

Masuko, K. 1990b. The instars of the ant *Amblyopone silvestrii* (Hymenoptera: Formicidae). Sociobiology 17:221-244. [1990]

Masuko, K. 1993. Predation of centipedes by the primitive ant *Amblyopone silvestrii*. Bull. Assoc. Nat. Sci. Senshu Univ. 24:35-44. [1993]

Masuko, K. 1995 ("1994"). Specialized predation on oribatid mites by two species of the ant genus *Myrmecina* (Hymenoptera: Formicidae). Psyche (Camb.) 101:159-173. [1995.11.20]

Masuko, K., Kannari, T. 1980. A new local record and some notes on the behavior of *Lordomyrma azumai* (Santschi). [In Japanese.] Ari 9:5-6. [1980.07.15]

Masuko, K., Terayama, M. 1984. Ants from Izu and Ogasawara Islands. [In Japanese.] Ari 12: 7-12. [1984.05.01]

Masuko, K., Terayama, M. 1986. Ants from Izu and Ogasawara Islands (Supplement). [In Japanese.] Ari 14:7. [1986.01.31]

Mathew, R. 1981 ("1980"). Description of new species of ant (Hymenoptera: Formicidae) from the Khasi Hills, Meghalaya. Orient. Insects 14:425-427. [1981.01]

Mathew, R. 1983. Studies on house-hold insect-pests (Hymenoptera: Formicidae) of Shillong. Bull. Zool. Surv. India 5(1):125-127. [1983.06]

Mathias, M. I. C., Caetano, F. H. 1991. Corpora allata and corpora cardiaca in female ants of the species *Neoponera villosa* (Hymenoptera: Ponerinae). Morphology and histology. Rev. Bras. Biol. 51:349-354. [1991.05.31]

Mathias, M. I. C., Caetano, F. H. 1992a. Inner female genitalia histology in the ant *Neoponera villosa* (Hymenoptera: Ponerinae). Rev. Bras. Biol. 52:235-244. [1992.05.31]

Mathias, M. I. C., Caetano, F. H. 1992b. Ovarian morphology of the ants *Neoponera villosa* (Hymenoptera: Ponerinae). Rev. Bras. Biol. 52:251-257. [1992.05.31]

Mathias, M. I. C., Cruz Landim, C. da, Caetano, F. H. 1991. Ultrastructural aspects of the mandibular glands of *Neoponera villosa* workers (Hymenoptera: Ponerinae). J. Adv. Zool. 12:72-80. [1991.12]

Matsumura, S. 1898. Insects collected on Mt. Fuji. Annot. Zool. Jpn. 2:113-124. [1898.12.29] [Ants pp. 115-116.]

Matsumura, S. 1911. Erster Beitrag zur Insekten-Fauna von Sachalin. J. Coll. Agric. Tohoku Imp. Univ. 4:1-145. [1911.03.13] [Formicidae p. 99.]

Matsumura, S. 1912. Thousand insects of Japan. Supplement IV. Tokyo: Keishu-sha, 247 pp. + 14 pl. + 4 pp. (index). [1912.07.10] [Formicidae pp. 191-192. Second edition of this work was published in 1918.]

Matsumura, S., Uchida, T. 1926. Die Hymenopteren-Fauna von den Riukiu-Inseln. Insecta Matsumurana 1:32-52. [1926.07] [Ants pp. 51-52.]

Mayhé-Nunes, A. J., Caetano, F. H. 1994a. Ultramorphology of, and comparison between, the mandibular glands and mandibles of two species of *Acromyrmex* (Hymenoptera, Formicidae). Naturalia (São Paulo) 19:17-27. [1994]

Mayhé-Nunes, A. J., Caetano, F. H. 1994b. The integumentary sculpturation of the gaster of *Acromyrmex* (Hymenoptera, Formicidae) in scanning electron microscopy. Naturalia (São Paulo) 19:29-43. [1994]

Mayr, G. 1853a. Beiträge zur Insektenfauna von Siebenbürgen. Verh. Mitt. Siebenb. Ver. Naturwiss. Hermannstadt 4:141-143. [1853.08]

Mayr, G. 1853b. Einige neue Ameisen. Verh. Zool.-Bot. Ver. Wien 2:143-150. [1853]

Mayr, G. 1853c. Beiträge zur Kenntniss der Ameisen. Verh. Zool.-Bot. Ver. Wien 3:101-114. [1853]

Mayr, G. 1853d. Beschreibungen einiger neuer Ameisen. Verh. Zool.-Bot. Ver. Wien 3:277-286. [1853]

Mayr, G. 1853e. Ueber die Abtheilung der Myrmiciden, und eine neue Gattung derselben. Verh. Zool.-Bot. Ver. Wien 3:387-394. [1853]

Mayr, G. 1854. [Untitled. Introduced by: "Hr. G. Mayr theilt rücksichtlich der Synonymie der *Myrmica rubiceps* Nyl. oder *Acrocoelia ruficeps* Mayr Folgendes mit:"] Verh. Zool.-Bot. Ver. Wien 4:30-32. [1854]

Mayr, G. 1855. Formicina austriaca. Beschreibung der bisher im österreichischen Kaiserstaate aufgefundenen Ameisen, nebst Hinzufügung jener in Deutschland, in der Schweiz und in Italien vorkommenden Arten. Verh. Zool.-Bot. Ver. Wien 5:273-478. [1855]

Mayr, G. 1856. Ausflug nach Szegedin im Herbste des Jahres 1855. Verh. Zool.-Bot. Ver. Wien 6:175-178. [1856]

Mayr, G. 1857. Ungarn's Ameisen. Programm Städt. Oberrealsch. Pesth 3:5-26.

Mayr, G. 1859. Beitrag zur Ameisenfauna Russlands. Stett. Entomol. Ztg. 20:87-90. [1859.03]

Mayr, G. 1861. Die europäischen Formiciden. Nach der analytischen Methode bearbeitet. Wien: C. Gerolds Sohn, 80 pp. [1861]

Mayr, G. 1862. Myrmecologische Studien. Verh. K-K. Zool.-Bot. Ges. Wien 12:649-776. [1862]

Mayr, G. 1863a. Formicidarum index synonymicus. Verh. K-K. Zool.-Bot. Ges. Wien 13:385-460. [1863]

Mayr, G. 1863b. Beitrag zur Orismologie der Formiciden. Arch. Naturgesch. 29(1):103-118. [1863]

Mayr, G. 1864. Das Leben und Wirken der einheimischen Ameisen. Oesterr. Rev. 3(3):201-209. [1864.03]

Mayr, G. 1865. Formicidae. In: Reise der Österreichischen Fregatte "Novara" um die Erde in den Jahren 1857, 1858, 1859. Zoologischer Theil. Bd. II. Abt. 1. Wien: K. Gerold's Sohn, 119 pp. [1865.12.14] [Publication date from Higgins (1963).]

Mayr, G. 1866a. Myrmecologische Beiträge. Sitzungsber. Kais. Akad. Wiss. Wien Math.-Naturwiss. Cl. Abt. I 53:484-517. [1866.05]

Mayr, G. 1866b. Diagnosen neuer und wenig gekannter Formiciden. Verh. K-K. Zool.-Bot. Ges. Wien 16:885-908. [1866]

Mayr, G. 1867a. Adnotationes in monographiam formicidarum Indo-Neerlandicarum. Tijdschr. Entomol. 10:33-117. [1867]

Mayr, G. 1867b. Vorläufige Studien über die Radoboj-Formiciden, in der Sammlung der k. k. geologischen Reichsanstalt. Jahrb. K-K. Geol. Reichsanst. Wien 17:47-62. [1867]

Mayr, G. 1867c. On fossil insects. [Translation and abstract of Mayr (1867b).] Q. J. Geol. Soc. Lond. 23(II):7. [1867]

Mayr, G. 1867d. [Untitled. Hymenopteren.] Pp. 440-442 in: Frauenfeld, G. R. von. Zoologische Miscellen. XI. Verh. K-K. Zool.-Bot. Ges. Wien 17:425-502. [1867]

Mayr, G. 1868a. [Untitled. Introduced by: "Bei dieser Termiten-Art fanden sich Ameisen vor, welche ich meinem Freunde Dr. Mayr zur Untersuchung übergab. Dr. Mayr erkannte sie als neue Art, und sendet mir folgende Beschreibung derselben: *Prenolepis braueri* Mayr."] P. 49 in: Brauer, F. Neuropteren. In: Reise der österreichischen Fregatte "Novara" um die Erde in der Jahren 1857, 1858, 1859. Zoologischer Theil. Bd. II. Abt. 1A4. Wien: K. Gerold's Sohn, 107 pp. [1868.10.08] [Publication date from Higgins (1963).]

Mayr, G. 1868b. Formicidae novae Americanae collectae a Prof. P. de Strobel. Annu. Soc. Nat. Mat. Modena 3:161-178. [1868]

Mayr, G. 1868c. Die Ameisen des baltischen Bernsteins. Beitr. Naturkd. Preuss. 1:1-102. [1868]

Mayr, G. 1868d. *Cremastogaster Ransonneti* n. sp. Verh. K-K. Zool.-Bot. Ges. Wien 18:287-288. [1868]

Mayr, G. 1869. Die Ameisen des baltischen Bernsteins. [Summary.] Neues Jahrb. Miner. Geol. Paläontol. 1869:620-625. [1869]

Mayr, G. 1870a. Formicidae novogranadenses. Sitzungsber. Kais. Akad. Wiss. Wien Math.-Naturwiss. Cl. Abt. I 61:370-417. [1870.04]

Mayr, G. 1870b. Neue Formiciden. Verh. K-K. Zool.-Bot. Ges. Wien 20:939-996. [1870]

Mayr, G. 1872. Formicidae Borneenses collectae a J. Doria et O. Beccari in territorio Sarawak annis 1865-1867. Ann. Mus. Civ. Stor. Nat. 2:133-155. [1872.04]

Mayr, G. 1876. Die australischen Formiciden. J. Mus. Godeffroy 12:56-115. [1876]

Mayr, G. 1877a. Formicidae. [In Russian.] In: Fedchenko, A. P. Travels in Turkestan. Vol. 2, Div. 5, No. 7. [In Russian.] Izv. Imp. Obshch. Lyubit. Estestvozn. Antropol. Etnogr. Imp. Mosk. Univ. 26:i-iii, 1-20 (+1). [1877]

Mayr, G. 1877b. [Untitled. Introduced by: "Herr Professor Dr. Gustav Mayr sprach über Dr. Emery's Gruppirung der Myrmiciden:".] Verh. K-K. Zool.-Bot. Ges. Wien 27(Sitzungsber.):23-26. [1877]

Mayr, G. 1878 ("1877"). Formiciden gesammelt in Brasilien von Professor Trail. Verh. K-K. Zool.-Bot. Ges. Wien 27:867-878. [1878]

Mayr, G. 1879. Beiträge zur Ameisen-Fauna Asiens. Verh. K-K. Zool.-Bot. Ges. Wien 28:645-686. [1879]

Mayr, G. 1880. Die Ameisen Turkestan's, gesammelt von A. Fedtschenko. Tijdschr. Entomol. 23:17-40. [1880] [Translation of Mayr (1877a).]

Mayr, G. 1883a. Drei neue ost-indische Formiciden-Arten. Notes Leyden Mus. 5:245-247. [1883.10]

Mayr, G. 1883b. Az *Epitritus argiolus* Em. nevü hangya elöfordulása magyarországban. Természetr. Füz. 6:141-142. [1883]

Mayr, G. 1883c. Über das Vorkommen der *Epitritus argiolus* Em. genannten Ameise in Ungarn. Természetr. Füz. 6:196-197. [1883] [German translation of Mayr (1883b).]

Mayr, G. 1884. [Untitled. Descriptions of eight new species.] Pp. 31-38 in: Radoszkowsky, O. Fourmis de Cayenne Française. Tr. Rus. Entomol. Obshch. 18:30-39. [1884.05.13]

Mayr, G. 1886a. Ueber *Eciton-Labidus*. Wien. Entomol. Ztg. 5:33-36. [1886.02.20]

Mayr, G. 1886b. Ueber *Eciton-Labidus*. (Schluss). Wien. Entomol. Ztg. 5:115-122. [1886.05.10]

Mayr, G. 1886c. Notizen über die Formiciden-Sammlung des British Museum in London. Verh. K-K. Zool.-Bot. Ges. Wien 36:353-368. [1886.09]

Mayr, G. 1886d. Die Formiciden der Vereinigten Staaten von Nordamerika. Verh. K-K. Zool.-Bot. Ges. Wien 36:419-464. [1886.12]

Mayr, G. 1887. Südamerikanische Formiciden. Verh. K-K. Zool.-Bot. Ges. Wien 37:511-632. [1887.09]

Mayr, G. 1889. Insecta in itinere Cl. Przewalski in Asia Centrali novissime lecta. XVII. Formiciden aus Tibet. Tr. Rus. Entomol. Obshch. 24:278-280. [1889.08.24]

Mayr, G. 1892a. Nachtrag. [Appendix to Forel (1892j).] Verh. K-K. Zool.-Bot. Ges. Wien 42:317-318. [1892.09]

Mayr, G. 1892b. [Untitled. Description of worker of *Drepanognathus rugosus*.] P. 127 in: Mocsáry, A. Hymenoptera in expeditione comitis Belae Szechenyi in China et Tibet a dom. G. Kreitner et L. Lóczy anno 1879 collecta. Természetr. Füz. 15:126-131. [1892.10.31]

Mayr, G. 1893a. Ergänzende Bemerkungen zu E. Wassmann's Artikel über springende Ameisen. Wien. Entomol. Ztg. 12:23. [1893.01.31]

Mayr, G. 1893b. Formiciden von Herrn Dr. Fr. Stuhlmann in Ost-Afrika gesammelt. Jahrb. Hambg. Wiss. Anst. 10:194-201. [1893]

Mayr, G. 1895. Afrikanische Formiciden. Ann. K-K. Naturhist. Mus. Wien 10:124-154. [1895]

Mayr, G. 1896. Beiträge zur Kenntniss der Insektenfauna von Kamerun. 5. Formiciden gesammelt von Herrn Yngve Sjöstedt. Entomol. Tidskr. 17:225-252. [1896.08.26]

Mayr, G. 1897. Formiciden aus Ceylon und Singapur. Természetr. Füz. 20:420-436. [1897.06.06]

Mayr, G. 1901a ("1900"). Drei neue Formiciden aus Kamerun gesammelt von Herrn Prof. Dr. Reinhold Buchholz. Entomol. Tidskr. 21:273-279. [1901.01.31]

Mayr, G. 1901b. Südafrikanische Formiciden, gesammelt von Dr. Hans Brauns. Ann. K-K. Naturhist. Mus. Wien 16:1-30. [1901]

Mayr, G. 1902. Hymenopterologische Miscellen. Verh. K-K. Zool.-Bot. Ges. Wien 52:287-303. [1902.07.05]

Mayr, G. 1903. Hymenopterologische Miszellen. II. Verh. K-K. Zool.-Bot. Ges. Wien 53:387-403. [1903.10.10]

Mayr, G. 1904a. Hymenopterologische Miszellen. III. Verh. K-K. Zool.-Bot. Ges. Wien 54:559-598. [1904.10.15]

Mayr, G. 1904b. Formiciden aus Ägypten und dem Sudan. In: Jägerskiöld, L. A. Results of the Swedish Zoological Expedition to Egypt and the White Nile, 1901. Part 1 (no. 9). Uppsala: Library of the Royal University of Uppsala, 11 pp. [1904]

Mayr, G. 1907a. Ergebnisse der mit Subvention aus der Erbschaft Treitl unternommenen zoologischen Forschungsreise Dr. F. Werner's nach dem ägyptischen Sudan und nach Nord-Uganda. XI. Liste der von Dr. Franz Werner am oberen Nil gesammelten Ameisen nebst Beschreibung einer neuen Art. Sitzungsber. Kais. Akad. Wiss. Wien Math.-Naturwiss. Kl. Abt. I 116:387-392. [1907.03]

Mayr, G. 1907b. 2. Formicidae. Pp. 7-24 in: Sjöstedt, Y. 1907-1910. Wissenschaftliche Ergebnisse der Schwedischen Zoologischen Expedition nach dem Kilimandjaro, dem Meru und umgebenden Massaisteppen Deutsch-Ostafrikas 1905-1906. 2. Band, Abt. 8. Stockholm: K. Schwed. Akad. Wissenschaften, 316 pp. [1907.09] [This is the second of eight articles in Abteilung 8, each individually published.]

Mayr, G. 1907c. Fam. Formicidae. Pp. 117-118 in: Rechinger, K. Botanische und zoologische Ergebnisse einer wissenschaftlichen Forschungreise nach den Samoa-Inseln, dem Neuguinea-Archipel und den Salomons-Inseln von März bis Dezember 1905. I. Teil. Wien: A. Hölder, 121 pp. [1907] [Date from Stafleu & Cowan (1983). Reprinted, as pp. 313-314, in: Rechinger, K. [same title] Denkschr. Akad. Wiss. Math.-Naturwiss. Kl. 81:197-317.]

Mayr, G. 1908. Ameisen aus Tripolis und Barka. Zool. Jahrb. Abt. Syst. Geogr. Biol. Tiere 26:415-418. [1908.06.23]

Mayr, G., Aurivillius, C. 1896. Beiträge zur Kenntniss der Insektenfauna von Kamerun. 5. Formiciden gesammelt von Herrn Yngve Sjöstedt. Anhang. Beschreibung der von Dr Y. Sjöstedt heimbrachten Ameisennester. Entomol. Tidskr. 17:253-256. [1896.08.26]

Mayr, G. L. See: Mayr, G.

Mazur, S. 1983. Mrówki borów sosnowych Polski. Rozprawy naukowe i monografie 25. Warszawa: Wydawnictwo SGGW-AR, 71 pp. [1983]

Mazur, S. 1984. Structure and plasticity of ant communities under conditions of fire stress. Pp. 85-87 in: IInd Symposium of the Protection of Forest Ecosystems: organisation of biocenosis as a basis of the prevention methods in protection of forests. Rogów, 7-8 December 1981. Warszawa: Warsaw Agricultural University Press, 239 pp. [1984]

Mazur, S. 1986. Nowe stanowisko gmachówki pniowej, *Camponotus vagus* (Scopoli, 1763) (Hymenoptera, Formicidae) na Roztoczu. Przegl. Zool. 30:71-72. [1986.05]

Mazur, S. 1995. Wplyw skazenia srodowiska na epigeiczne mrówki (Formicidae) borów sosnowych Polski. Sylwan 139(7):5-14.

McAreavey, J. 1947. New species of the genera *Prolasius* Forel and *Melophorus* Lubbock (Hymenoptera, Formicidae). Mem. Natl. Mus. Vic. 15:7-27. [1947.10]

McAreavey, J. 1949. Australian Formicidae. New genera and species. Proc. Linn. ´ . N. S. W. 74:1-25. [1949.06.15]

McAreavey, J. 1956. A new species of the genus *Meranoplus*. Mem. Qld. Mus. 13:148-150. [1956.04.26]

McAreavey, J. 1957. Revision of the genus *Stigmacros* Forel. Mem. Natl. Mus. Vic. 21:7-64. [1957.08.06]

McArthur, A. J., Miller, L. R. 1994. Seed harvester ants of the Loxton area of South Australia. S. Aust. Nat. 68:54-59. [1994]

McCluskey, E. S. 1974a ("1973"). Generic diversity in phase of rhythm in formicine ants. Psyche (Camb.) 80:295-304. [1974.02.20]

McCluskey, E. S. 1974b. Generic diversity in phase of rhythm in myrmicine ants. J. N. Y. Entomol. Soc. 82:93-102. [1974.08.22]

McCluskey, E. S. 1982. Multivariate generic comparison of ant flights. [Abstract.] P. 408 in: Breed, M. D., Michener, C. D., Evans, H. E. (eds.) The biology of social insects. Proceedings of the Ninth Congress of the IUSSI, Boulder, Colorado, August, 1982. Boulder: Westview Press, xii + 420 pp. [1982.08]

McCluskey, E. S., McCluskey, D. K. 1984. Hour of mating flight in three species of ants (Hymenoptera: Formicidae). Pan-Pac. Entomol. 60:151-154. [1984.04.17]

McCluskey, E. S., Soong, S.-M. A. 1979. Rhythm variables as taxonomic characters in ants. Psyche (Camb.) 86:91-102. [1979.12.19]

McCook, H. C. 1876. Notes on the architecture and habits of *Formica pennsylvanica*, the Pennsylvania carpenter ant. Trans. Am. Entomol. Soc. 5:277-289. [1876.12]

McCook, H. C. 1877a. On the vital powers of ants. [part] Proc. Acad. Nat. Sci. Phila. 29:134-136. [1877.05.22]

McCook, H. C. 1877b. On the vital powers of ants. [concl.] Proc. Acad. Nat. Sci. Phila. 29:137. [1877.06.26]

McCook, H. C. 1877c. Mound-making ants of the Alleghenies, their architecture and habits. Trans. Am. Entomol. Soc. 6:253-296. [1877.11]

McCook, H. C. 1878a ("1877"). The agricultural ants of Texas. Proc. Acad. Nat. Sci. Phila. 29:299-304. [1878.01.15]

McCook, H. C. 1878b. Mound-making ants of the Alleghenies. Am. Nat. 12:431-445. [1878.07]

McCook, H. C. 1879a. Cutting or parasol ant, *Atta fervens*, Say. Proc. Acad. Nat. Sci. Phila. 31:33-40. [1879.04.15]

McCook, H. C. 1879b. On *Myrmecocystus mexicanus*, Wesm. Proc. Acad. Nat. Sci. Phila. 31:197-198. [1879.11.04]

McCook, H. C. 1879c. The natural history of the agricultural ant of Texas. A monograph of the habits, architecture, and structure of *Pogonomyrmex barbatus*. Philadelphia: Lippincott, 311 pp. [1879]

McCook, H. C. 1880 ("1879"). Formicariae. Pp. 182-189 in: Comstock, J. H. Report upon cotton insects. Washington, D.C.: Government Printing Office, 511 pp. [1880.05.18] [Publication date from Comstock (1881:275).]

McCook, H. C. 1881a ("1880"). Note on a new northern cutting ant, *Atta septentrionalis*. Proc. Acad. Nat. Sci. Phila. 32:359-363. [1881.02.22]

McCook, H. C. 1881b. The honey ants of the Garden of the Gods. [part] Proc. Acad. Nat. Sci. Phila. 33:17-24. [1881.05.10]

McCook, H. C. 1881c. The honey ants of the Garden of the Gods. [part] Proc. Acad. Nat. Sci. Phila. 33:25-56. [1881.05.31]

McCook, H. C. 1881d. The honey ants of the Garden of the Gods. [concl.] Proc. Acad. Nat. Sci. Phila. 33:57-77. [1881.06.14]

McCook, H. C. 1882. The honey ants of the Garden of the Gods, and the occident ants of the American plains. Philadelphia: J. B. Lippincott & Co., 188 pp. [1882]

McGurk, D. J., Frost, J., Waller, G. R., Eisenbraun, E. J., Vick, K., Drew, W. A., Young, J. 1968. Iridodial isomer variation in dolichoderine ants. J. Insect Physiol. 14:841-845. [1968.06]

McInnes, D. A., Tschinkel, W. R. 1995. Queen dimorphism and reproductive strategies in the fire ant *Solenopsis geminata* (Hymenoptera: Formicidae). Behav. Ecol. Sociobiol. 36:367-375. [1995.06.20]

McKey, D., Davidson, D. W. 1993. Ant-plant symbioses in Africa and the Neotropics: history, biogeography and diversity. Pp. 568-606 in: Goldblatt, P. (ed.) Biological relationships between Africa and South America. New Haven: Yale University Press, 630 pp. [1993]

Medeiros, M. A. de, Fowler, H. G., Bueno, O. C. 1995. Ant (Hym., Formicidae) mosaic stability in Bahian cocoa plantations: implications for management. J. Appl. Entomol. 119:411-414. [1995.08]

Medel, R. G. 1995. Convergence and historical effects in harvester ant assemblages of Australia, North America, and South America. Biol. J. Linn. Soc. 55:29-44. [1995.05.31]

Medel, R. G., Vásquez, R. A. 1994. Comparative analysis of harvester ant assemblages of Argentinian and Chilean arid zones. J. Arid Environ. 26:363-371. [1994.04]

Medina, M. 1889. [Untitled. Introduced by: "El Sr. Medina leyó la nota siguiente."] An. Soc. Esp. Hist. Nat. (Actas) 18:31. [1889.04.30]

Medina, M. 1892. [Untitled. Introduced by: "El Sr. Medina leyó la nota siguiente."] An. Soc. Esp. Hist. Nat. (Actas) (2)1[=21]:69-70. [1892.09.01]

Medina, M. 1893. Formicidos de Pozuelo de Calatrava (Ciudad-Real). An. Soc. Esp. Hist. Nat. (Actas) (2)1[=21]:104-105. [1893.01.31]

Medler, J. T. 1980. Insects of Nigeria - Check list and bibliography. Mem. Am. Entomol. Inst. 30:1-919. [1980] [Ants pp. 484-488.]

Mehlhop, P., Gardner, A. L. 1982. A rapid field technique for preparing ant chromosomes for karyotypic analysis. Stain Technol. 57:99-101. [1982.03]

Mei, M. 1984 ("1982"). Nuovi reperti de formicidi per l'Italia centrale (Hymenoptera, Formicidae). Boll. Assoc. Rom. Entomol. 37:49-58. [1984.12.15]

Mei, M. 1987. *Myrmica samnitica* n. sp.: una nuova formica parassita dell'Appennino Abruzzese (Hymenoptera, Formicidae). Fragm. Entomol. 19:457-469. [1987.10.15]

Mei, M. 1992a. Su alcune specie endogee o criptobiotiche della mirmecofauna italiana. Fragm. Entomol. 23:411-422. [1992.04.30]

Mei, M. 1992b. A survey of the socially parasitic ant genera *Epimyrma* Emery, 1915 and *Chalepoxenus* Menozzi, 1922 in Italy (Hymenoptera, Formicidae, Myrmicinae). Insectes Soc. 39:145-156. [1992]

Mei, M. 1995. Arthropoda di Lampedusa, Linosa e Pantelleria (Canale di Sicilia, Mar Mediterraneo). Hymenoptera Formicidae (con diagnosi di due nuove specie). Nat. Sicil. (4)19(suppl.):753-772.

Meinert, F. 1861. Bidrag til de danske Myrers Naturhistorie. K. Dan. Vidensk. Selsk. Skr. (5)5:273-340. [1861]

Meinwald, J., Wiemer, D. F., Hölldobler, B. 1983. Pygidial gland secretions of the ponerine ant *Rhytidoponera metallica*. Naturwissenschaften 70:46-47. [1983.01]

Melander, A. L., Carpenter, F. M. 1937. William Morton Wheeler. Ann. Entomol. Soc. Am. 30:433-437. [1937.10.14]

Menozzi, C. 1918. Primo contributo alla conoscenza della fauna mirmecologica del Modenese. Atti Soc. Nat. Mat. Modena (5)4:81-88. [1918]

Menozzi, C. 1921. Formiche dei dintorni di Sambiase di Calabria. Boll. Lab. Zool. Gen. Agrar. R. Sc. Super. Agric. 15:24-32. [1921.05.25]

Menozzi, C. 1922a. Miscellanea mirmecologica. [part] Ann. Mus. Civ. Stor. Nat. "Giacomo Doria" 49[=(3)9]:347-352. [1922.04.28]

Menozzi, C. 1922b. Miscellanea mirmecologica. [concl.] Ann. Mus. Civ. Stor. Nat. "Giacomo Doria" 49[=(3)9]:353-358. [1922.06.19]

Menozzi, C. 1922c. Contribution à la faune myrmécologique de l'Espagne. Bol. R. Soc. Esp. Hist. Nat. 22:324-332. [1922.07]

Menozzi, C. 1922d. Nota complementare per la distinzione specifica dei *Camponotus herculeanus* L. e *ligniperda* Latr. (Hym.-Formic.). Boll. Soc. Entomol. Ital. 54:141-145. [1922.12.20]

Menozzi, C. 1923a ("1922"). Nota su un nuovo genere e nuova specie di formica parassita. Atti Soc. Ital. Sci. Nat. Mus. Civ. Stor. Nat. Milano 61:256-260. [1923.02]

Menozzi, C. 1923b. Trois fourmis nouvelles (Hym.). Bull. Soc. Entomol. Fr. 1923:209-212. [1923.12.05]

Menozzi, C. 1924a. Una specie inedita di *Eciton* Latr. del sottogenere *Labidus* Jur. (Hymen: Formicidae). Boll. Soc. Entomol. Ital. 56:29-31. [1924.02.26]

Menozzi, C. 1924b. Alcune nuove formiche africane. [part] Ann. Mus. Civ. Stor. Nat. "Giacomo Doria" 51:220-224. [1924.05.07]

Menozzi, C. 1924c. Alcune nuove formiche africane. [concl.] Ann. Mus. Civ. Stor. Nat. "Giacomo Doria" 51:225-227. [1924.06.26]

Menozzi, C. 1924d. Nuova specie di *Dolichoderus* Lund del Perù. *Dolichoderus* (*Hypoclinea*) *Grandii* n. sp. Boll. Soc. Entomol. Ital. 56:119-120. [1924.10.29]

Menozzi, C. 1925a. Note staccate di mirmecologia. Boll. Soc. Entomol. Ital. 57:17-22. [1925.02.25]

Menozzi, C. 1925b. Qualche formica nuova od interessante del Deutsch. Ent. Institut di Dahlem (Form.). Entomol. Mitt. 14:368-371. [1925.10.10]

Menozzi, C. 1925c. Nouvelles fourmis des Philippines. Philipp. J. Sci. 28:439-451. [1925.11.06]

Menozzi, C. 1925d ("1924"). Res mutinenses. Formicidae (Hymenoptera). Atti Soc. Nat. Mat. Modena 55[=(6)3]:22-47. [1925]

Menozzi, C. 1926a. Formiche dell'Africa centrale. Boll. Soc. Entomol. Ital. 58:36-41. [1926.03.31]

Menozzi, C. 1926b. Zur Kenntnis der Ameisenfauna der Balearen. Zool. Anz. 66:180-182. [1926.04.05]

Menozzi, C. 1926c. Neue Ameisen aus Brasilien. Zool. Anz. 69:68-72. [1926.11.20]

Menozzi, C. 1926d. Due nuove specie di *Eciton* Latr. (Hymenoptera - Formicidae). Folia Myrmecol. Termit. 1:29-32. [1926.12]

Menozzi, C. 1926e ("1925"). Nuove formiche delle isole Filippine e di Singapore. Atti Soc. Nat. Mat. Modena 56[=(6)4]:92-103. [1926]

Menozzi, C. 1927a. Zur Kenntnis des Weibchens von *Dorylus* (*Anomma*) *nigricans* var. *molesta* Gerst. (Hymenoptera - Formicidae). Zool. Anz. 70:263-266. [1927.03.20]

Menozzi, C. 1927b. Beitrag zur Ameisenfauna des nördlichen und östlichen Spaniens. Aufzählung der von den Herren Dr. F. Haas (1914-1919) und Prof. A. Seitz (1923) gesammelten Arten. Senckenbergiana 9:89-92. [1927.06.18]

Menozzi, C. 1927c. Formiche raccolte dal Sig. H. Schmidt nei dintorni di San José di Costa Rica. Entomol. Mitt. 16:266-277. [1927.07.01]

Menozzi, C. 1927d. Formiche raccolte dal Sig. H. Schmidt nei dintorni di San José di Costa Rica (Schluss). Entomol. Mitt. 16:336-345. [1927.08.15] [Date is that of preprint, copies of which are in the Natural History Museum, London. Date on the cover of the issue (Heft 5 of volume 16) in which the article appeared is 1 September 1927.]
Menozzi, C. 1927e. Zur Erforschung des Persischen Golfes. (Beitrag nr. 12) Formicidae (Hym.). Suppl. Entomol. 16:117-119. [1927.11.10]
Menozzi, C. 1927f. Raccolte mirmecologiche dell'Africa orientale conservate nel Museo Civico di Storia Naturale "Giacomo Doria" di Genova. Parte I. Formiche raccolte dal Marchese Saverio Patrizi nella Somalia italiana ed in alcune località dell'Africa orientale inglese. Ann. Mus. Civ. Stor. Nat. "Giacomo Doria" 52:356-362. [1927.11.10]
Menozzi, C. 1927g. Risultati zoologici della Missione inviata dalla R. Società Geografica Italiana per l'esplorazione dell'Oasi di Giarabub (1926-1927). Formicidae (Hymenoptera). Ann. Mus. Civ. Stor. Nat. "Giacomo Doria" 52:379-382. [1927.12.20]
Menozzi, C. 1928a. Note sulla mirmecofauna paleartica. Boll. Lab. Zool. Gen. Agrar. R. Sc. Super. Agric. 21:126-129. [1928.05.22]
Menozzi, C. 1928b. Tre nuove formiche della Sumatra orientale. Misc. Zool. Sumatr. 30:1-5. [1928.07]
Menozzi, C. 1929a. Ricerche faunistiche nelle isole italiane dell'Egeo. Imenotteri (formiche). Arch. Zool. Ital. 13:145-146. [1929.04.24]
Menozzi, C. 1929b. Formiche di Cuba e delle Isole Canarie raccolte dal Prof. Filippo Silvestri. Boll. Lab. Zool. Gen. Agrar. R. Sc. Super. Agric. 23:1-5. [1929.05.31]
Menozzi, C. 1929c. Una nuova specie di Formica del genere *Aphaenogaster* Mayr del Nord America. Boll. Lab. Zool. Gen. Agrar. R. Sc. Super. Agric. 22:282-284. [1929.08.06]
Menozzi, C. 1929d. A new species of *Camponotus* belonging to the subgenus *Orthonotomyrmex*, Ashm. (Hymenoptera). Ann. Mag. Nat. Hist. (10)4:430-433. [1929.11]
Menozzi, C. 1929e. Revisione delle formiche del genere *Mystrium* Roger. Zool. Anz. 82:518-536. [1929]
Menozzi, C. 1929f. Formiche del Sinai raccolte dal Dr. F. S. Bodenheimer, con descrizione di una nuova specie di *Monomorium* del sottogen. *Equestrimessor*. Pp. 125-128 in: Bodenheimer, F. S., Theodor, O. Ergebnisse der Sinai-Expedition 1927 der Hebräischen Universität, Jerusalem. Leipzig: J. C. Hinrichs'schum, 142 pp. [1929]
Menozzi, C. 1930a. Formiche di Sumatra raccolte dal Prof. J. C. van der Meer Mohr. Misc. Zool. Sumatr. 47:1-5. [1930.03]
Menozzi, C. 1930b. Formiche della Somalia italiana meridionale. Mem. Soc. Entomol. Ital. 9:76-130. [1930.05.02]
Menozzi, C. 1930c. [Untitled. *Epimyrma goesswaldi*, new species, attributed to Menozzi.] Pp. 464-465 in: Gösswald, K. Die Biologie einer neuen *Epimyrma*art aus dem mittleren Maingebiet. Z. Wiss. Zool. 136:464-484. [1930.07]
Menozzi, C. 1930d. [Untitled. *Crematogaster* (*Orthocrema*) *russoi* n. sp.] Pp. 155-156 in: Menozzi, C., Russo, G. Contributo alla conoscenza della mirmecofauna della Repubblica Dominicana (Antille). Boll. Lab. Zool. Gen. Agrar. R. Sc. Super. Agric. 24:148-173. [1930.09.01]
Menozzi, C. 1930e. [Untitled. *Macromischa* (*Antillaemyrma*) *ciferrii* n. sp.] Pp. 160-161 in: Menozzi, C., Russo, G. Contributo alla conoscenza della mirmecofauna della Repubblica Dominicana (Antille). Boll. Lab. Zool. Gen. Agrar. R. Sc. Super. Agric. 24:148-173. [1930.09.01]
Menozzi, C. 1930f. [Untitled. *Camponotus* (*Tanaemyrmex*) *larvigerus* var. *maculifrons* n. var.] P. 167 in: Menozzi, C., Russo, G. Contributo alla conoscenza della mirmecofauna della Repubblica Dominicana (Antille). Boll. Lab. Zool. Gen. Agrar. R. Sc. Super. Agric. 24:148-173. [1930.09.01]

Menozzi, C. 1930g. Spedizione di S. A. R. il Duca degli Abruzzi alle sorgenti dell'Uebi Scebeli. Risultati zoologici. Formicidae (Hymenoptera). Ann. Mus. Civ. Stor. Nat. "Giacomo Doria" 55:25-26. [1930.10.14]

Menozzi, C. 1930h. Formicidae. Pp. 327-332 in: Eidmann, H. Entomologishe Ergebnisse einer Reise nach Ostasien. Verh. Zool.-Bot. Ges. Wien 79:308-335. [1930.10.31]

Menozzi, C. 1931a. Spedizione del Barone Raimondo Franchetti in Dancalia. Hymenoptera Formicidae. Ann. Mus. Civ. Stor. Nat. "Giacomo Doria" 55:154-156. [1931.03.24]

Menozzi, C. 1931b. Contribuzione alla conoscenza del "microgenton" di Costa Rica. III. Hymenoptera - Formicidae. Boll. Lab. Zool. Gen. Agrar. R. Sc. Super. Agric. 25:259-274. [1931.08.24]

Menozzi, C. 1931c. Revisione del genere *Epimyrma* Em. (Hymen. Formicidae) e descrizione di una specie inedita di questo genere. Mem. Soc. Entomol. Ital. 10:36-53. [1931.09.10]

Menozzi, C. 1931d. Qualche nuova formica di Costa Rica (Hym.). Stett. Entomol. Ztg. 92:188-202. [1931.10]

Menozzi, C. 1932a. Contributo alla conoscenza della fauna mirmecologica d'Italia. Boll. Lab. Entomol. R. Ist. Super. Agrar. Bologna 5:8-12. [1932.04.10]

Menozzi, C. 1932b. Formiche del Nord America raccolte dal Prof. F. Silvestri. Boll. Lab. Zool. Gen. Agrar. R. Sc. Super. Agric. 26:310-312. [1932.06.25]

Menozzi, C. 1932c. Missione scientifica del Prof. E. Zavattari nel Fezzan (1931). Hymenoptera-Formicidae. Boll. Soc. Entomol. Ital. 64:93-95. [1932.06.30]

Menozzi, C. 1932d. Formiche dell'Isola di Nias. Misc. Zool. Sumatr. 65:1-13. [1932.06]

Menozzi, C. 1932e. Spedizione scientifica all'Oasi di Cufra (marzo-luglio 1931). Formiche. Ann. Mus. Civ. Stor. Nat. "Giacomo Doria" 55:451-456. [1932.08.16]

Menozzi, C. 1932f. Una nuova specie de *Polyrhachis* (Hym.-Formicidae) e descrizione di tre nidi di formiche appartenenti a questo genere. Wien. Entomol. Ztg. 49:303-308. [1932.12.08]

Menozzi, C. 1932g. Raccolte mirmecologiche dell'Africa orientale conservate nel Museo Civico di Storia Naturale "Giacomo Doria" di Genova. Parte II. Formiche dell'Uganda e delle isole Sesse raccolte dal Dr. E. Bayon. [part] Ann. Mus. Civ. Stor. Nat. "Giacomo Doria" 56:93-112. [1932.12.10]

Menozzi, C. 1932h. Formiche. Pp. 103-107 in: Commissione Reale del Parco (ed.) Il Parco Nazionale del Gran Paradiso. Volume III. Torino: L. Cecchini, 153 pp. [1932]

Menozzi, C. 1933a. Raccolte mirmecologiche dell'Africa orientale conservate nel Museo Civico di Storia Naturale "Giacomo Doria" di Genova. II. Formiche dell'Uganda e delle isole Sesse raccolte dal Dr. E. Bayon. [concl.] Ann. Mus. Civ. Stor. Nat. "Giacomo Doria" 56:113-114. [1933.01.23]

Menozzi, C. 1933b. Le formiche della Palestina. Mem. Soc. Entomol. Ital. 12:49-113. [1933.06.15]

Menozzi, C. 1933c. Description préliminaire d'une espèce nouvelle de fourmi constituant un genre nouveau. Natuurhist. Maandbl. 22:146-147. [1933.12.29]

Menozzi, C. 1934. Reperti mirmecofaunistici raccolti dal Prof. L. di Caporiacco nelle oasi di Cufra e in altre località del deserto Libico. Atti Soc. Nat. Mat. Modena 65[=(6)13]:153-166. [1934]

Menozzi, C. 1935a. Formiche indo-australiane del genere *Crematogaster* Lund raccolte da W. Karawaiew. Konowia 14:103-116. [1935.03.15]

Menozzi, C. 1935b. Spedizione del Prof. Nello Beccari nella Guiana Britannica. Hymenoptera-Formicidae. Redia 21:189-203. [1935.03.23]

Menozzi, C. 1935c. Fauna Chilensis. II. (Nach Sammlungen von W. Goetsch). Le formiche del Cile. Zool. Jahrb. Abt. Syst. Ökol. Geogr. Tiere 67:319-336. [1935.11.06]

Menozzi, C. 1936a. Due nuovi Dacetini di Costa Rica e descrizione della larva di uno di essi (Hymenoptera: Formicidae). Arb. Morphol. Taxon. Entomol. Berl.-Dahl. 3:81-85. [1936.06.08]

Menozzi, C. 1936b. Nuovi contributi alla conoscenza della fauna delle Isole italiane dell'Egeo. VI. Hymenoptera - Formicidae. Boll. Lab. Zool. Gen. Agrar. R. Sc. Super. Agric. 29:262-311. [1936.06.24]

Menozzi, C. 1939a. Formiche dell'Himalaya e del Karakorum raccolte dalla Spedizione italiana comandata da S. A. R. il Duca di Spoleto (1929). Atti Soc. Ital. Sci. Nat. Mus. Civ. Stor. Nat. Milano 78:285-345. [1939.10]

Menozzi, C. 1939b. Qualche nuova formica di Sumatra. Tijdschr. Entomol. 82:175-181. [1939.11]

Menozzi, C. 1939c. Hymenoptera Formicidae. Missione Biol. Paese Borana 3:97-110. [1939]

Menozzi, C. 1940a. Contributo alla fauna della Tripolitania. Boll. Lab. Zool. Gen. Agrar. Fac. Agrar. Portici 31:244-273. [1940.02.22]

Menozzi, C. 1940b. Contribution à la faune myrmécologique du Japon. Mushi 13:11-12. [1940.05.30]

Menozzi, C. 1941. In morte di Bruno Finzi. Mem. Soc. Entomol. Ital. 20:190-191. [1941.12.22]

Menozzi, C. 1942a. Formiche dell'isola Fernando Poo e del territorio del Rio Muni (Guinea Spagnola). 24. Beitrag zu den wissenschaftlichen Ergebnissen der Forschungsreise H. Eidmann nach Spanisch-Guinea 1939 bis 1940. Zool. Anz. 140:164-182. [1942.12.15]

Menozzi, C. 1942b. Esplorazione entomologica del Parco Nazionale del Circeo. Hymenoptera-Formicidae. Salerno: M. Spadafora, 8 pp. [Not seen. Cited in Grandi (1942:201).]

Menozzi, C., Consani, M. 1952 ("1951"). Missione biologica Sagan-Omo diretta dal Prof. E. Zavattari. Hymenoptera Formicidae. Riv. Biol. Colon. 11:57-71. [1952.12]

Menozzi, C., Russo, G. 1930. Contributo alla conoscenza della mirmecofauna della Repubblica Dominicana (Antille). Boll. Lab. Zool. Gen. Agrar. R. Sc. Super. Agric. 24:148-173. [1930.09.01]

Menzel, J. G., Tautz, J. 1994. Functional morphology of the subgenual organ of the carpenter ant. Tissue Cell 26:735-746. [1994.10]

Menzel, R. 1972. The fine structure of the compound eye of *Formica polyctena* - functional morphology of a hymenopterean [sic] eye. Pp. 37-47 in: Wehner, R. (ed.) Information processing in the visual systems of arthropods. Berlin: Springer-Verlag, xi + 334 pp. [1972]

Menzel, R., Wehner, R. 1970. Augenstrukturen bei verschieden grossen Arbeiterinnen von *Cataglyphis bicolor* Fabr. (Formicidae, Hymenoptera). Z. Vgl. Physiol. 68:446-449. [1970.09.02]

Mercovich, T. C. 1958. A new species of the genus *Orectognathus*. Mem. Qld. Mus. 13:195-198. [1958.07.28]

Merickel, F. W., Clark, W. H. 1994. *Tetramorium caespitum* (Linnaeus) and *Liometopum luctuosum* W. M. Wheeler (Hymenoptera: Formicidae): new state records for Idaho and Oregon, with notes on their natural history. Pan-Pac. Entomol. 70:148-158. [1994.09.29]

Meunier, F. 1915. Über einige fossile Insekten aus den Braunkohlenschichten (Aquitanien) von Rott (Siebengebirge). Z. Dtsch. Geol. Ges. 67:205-217. [1915]

Meunier, F. 1917. Sur quelques insectes de l'Aquitainien de Rott, Sept Montagnes (Prusse rhénane). Verh. K. Akad. Wet. Amst. Tweede Sect. 20:3-17. [1917.11]

Meunier, F. 1923. Sur quelques nouveaux insectes des lignites oligocènes (Aquitanien) de Rott, Siebengebirge (Rhénanie). Proc. Sect. Sci. K. Akad. Wet. Amst. 26:605-612. [1923]

Meyne, J., Hirai, H., Imai, H. T. 1995. FISH analysis of the telomere sequences of bulldog ants (*Myrmecia*: Formicidae). Chromosoma (Berl.) 104:14-18. [1995.10]

Miller, C. D. F. 1957. Taxonomic status of *Formica subnitens* Creighton and *F. integroides* Emery, with a description of the sexuals of *F. subnitens* (Hymenoptera: Formicidae). Insectes Soc. 4:253-257. [1957.06]

Miller, E. M. 1941. Dates of receipt of the Annals in university libraries. Ann. Entomol. Soc. Am. 34:689. [1941.10.06]

Milne-Edwards, A. 1879. Description de quelques crustacés nouveaux. Bull. Soc. Philomath. Paris (7)3:103-110. [1879] [*Goniothorax* (Crustacea), senior homonym of *Goniothorax* Emery, 1896.]

Minkiewicz, R. 1939 ("1937-38"). Les sexués du *Leptothorax clypeatus* Mayr et le problème de la sexualisation somatique chez les fourmis. Pol. Pismo Entomol. 16-17:215-239. [1939.06.24]

Minozzi, C. See: Menozzi, C.

Mintzer, A. C. 1979. Colony foundation and pleometrosis in *Camponotus* (Hymenoptera: Formicidae). Pan-Pac. Entomol. 55:81-89. [1979.07.19]

Mintzer, A. C., Mintzer, C. L. 1988. Population status of the Mexican leafcutting ant, *Atta mexicana* (Formicidae), in the Sonoran desert of Arizona. Southwest. Nat. 33:250-251. [1988.06.15]

Miradoli Zatti, M. A., Pavan, M. 1957. Studi sui Formicidae. III. Nuovi reperti dell'organo ventrale nei Dolichoderinae. Boll. Soc. Entomol. Ital. 87:84-87. [1957.07.01]

Mitchell, J. D., Pierce, W. D. 1912. The ants of Victoria County, Texas. Proc. Entomol. Soc. Wash. 14:67-76. [1912.06.19]

Mitroshina, L. A. 1986. Fauna and biotopic distribution of ants in South Ustyurt. [In Russian.] Izv. Akad. Nauk Turkm. SSR Ser. Biol. Nauk 1986(5):28-32. [> 1986.11.27] [Date signed for printing ("podpisano k pechati").]

Mizutani, A. 1979. A myrmecofaunal survey at Hiyama Experiment Forest, Hokkaido University. Res. Bull. Coll. Exp. For. Coll. Agric. Hokkaido Univ. 36:509-516. [1979.07.30]

Mizutani, A. 1981. On the two forms of the ant *Myrmica ruginodis* Nylander (Hymenoptera, Formicidae) from Sapporo and its vicinity, Japan. Jpn. J. Ecol. 31:131-137. [1981.06.30]

Mizutani, A., Yamane, S. 1978. Description of the larva of the ant *Myrmica ruginodis* (Hymenoptera, Formicidae) from Japan. Kontyû 46:38-42. [1978.03.25]

Mocsáry, A. 1883. Literatura hymenopterorum. Természetr. Füz. 6:3-122. [1883]

Mocsáry, A. 1892. Hymenoptera in expeditione comitis Belae Szechenyi in China et Tibet a dom. G. Kreitner et L. Lóczy anno 1879 collecta. Természetr. Füz. 15:126-131. [1892.10.31] [Untitled contribution by G. Mayr on page 127.]

Mocsáry, S., Szépligeti, V. 1901. Hymenopterák (Hymenopteren). Pp. 121-169 in: Horváth, G. Zichy Jenó Gróf harmadik ázsiai utazásának állattani eredményei. Vol. 2. Budapest: V. Hornyánsky, xli + 470 pp. [1901] [List of ants pp. 131-132. Untitled contribution by Emery (descriptions of two new taxa) on page 159.]

Moffett, M. W. 1984. Swarm raiding in a myrmicine ant. Naturwissenschaften 71:588-589. [1984.11]

Moffett, M. W. 1985a. Behavioral notes on the Asiatic harvesting ants *Acanthomyrmex notabilis* and *A. ferox*. Psyche (Camb.) 92:165-179. [1985.11.28]

Moffett, M. W. 1985b. Revision of the genus *Myrmoteras* (Hymenoptera: Formicidae). Bull. Mus. Comp. Zool. 151:1-53. [1985.12.26]

Moffett, M. W. 1986a. Observations on *Lophomyrmex* ants from Kalimantan, Java and Malaysia. Malay. Nat. J. 39:207-211. [1986.02]

Moffett, M. W. 1986b. Trap-jaw predation and other observations on two species of *Myrmoteras* (Hymenoptera: Formicidae). Insectes Soc. 33:85-99. [1986.06]

Moffett, M. W. 1986c. Revision of the myrmicine genus *Acanthomyrmex* (Hymenoptera: Formicidae). Bull. Mus. Comp. Zool. 151:55-89. [1986.08.15]

Moffett, M. W. 1986d. Notes on the behavior of the dimorphic ant *Oligomyrmex overbecki* (Hymenoptera: Formicidae). Psyche (Camb.) 93:107-116. [1986.10.10]

Moffett, M. W. 1986e. Evidence of workers serving as queens in the genus *Diacamma* (Hymenoptera: Formicidae). Psyche (Camb.) 93:151-152. [1986.10.10]

Moffett, M. W. 1986f. Mandibles that snap: notes on the ant *Mystrium camillae* Emery. Biotropica 18:361-362. [1986.12]

Moffett, M. W. 1987 ("1986"). Behavior of the group-predatory ant *Proatta butteli* (Hymenoptera: Formicidae): an Old World relative of the attine ants. Insectes Soc. 33:444-457. [1987.06]

Möglich, M. 1975. Recruitment of *Leptothorax*. Pp. 235-242 in: Noirot, C., Howse, P. E., Le Masne, G. (eds.) Pheromones and defensive secretions in social insects. Dijon: French Section of the IUSSI, vi + 248 pp. [1975.12]

Möglich, M. 1978. Social organization of nest emigration in *Leptothorax* (Hym., Form.). Insectes Soc. 25:205-225. [1978.09]
Möglich, M. 1979. Tandem calling pheromone in the genus *Leptothorax* (Hymenoptera: Formicidae): behavioral analysis of specificity. J. Chem. Ecol. 5:35-52. [1979.01]
Möglich, M., Hölldobler, B. 1974. Social carrying behavior and division of labor during nest moving in ants. Psyche (Camb.) 81:219-236. [1974.09.26]
Möglich, M., Maschwitz, U., Hölldobler, B. 1974. Tandem calling: a new kind of signal in ant communication. Science (Wash. D. C.) 186:1046-1047. [1974.12.13]
Monastero, S. 1950. Le formiche della Sicilia. G. Sci. Nat. Econ. Palermo 47(2):1-10. [1950]
Moody, J. V., Francke, O. F. 1982. The ants (Hymenoptera, Formicidae) of western Texas. Part 1: Subfamily Myrmicinae. Grad. Stud. Tex. Tech Univ. 27:1-80. [1982.10.29]
Moody, J. V., Francke, O. F., Merickel, F. W. 1981. The distribution of fire ants, *Solenopsis* (*Solenopsis*) in western Texas (Hymenoptera: Formicidae). J. Kansas Entomol. Soc. 54:469-480. [1981.07.17]
Moore, G. F. 1842. A descriptive vocabulary of the language in common use amongst the aborigines of Western Australia. London: W. S. Orr, xiii + 171 pp. [1842] [See Ride & Taylor (1973).]
Moore, W. E. 1940. The ant fauna of swamps in the Canterbury mountains. Rec. Canterbury Mus. 4:299-303. [1940.04.09]
Morais, H. C., Benson, W. W. 1988. Recolonização de vegetação de cerrado após queimada, por formigas arborícolas. Rev. Bras. Biol. 48:459-466. [1988.08.31]
Moreira, D. D. O., Della Lucia, T. M. C. 1993. The morphology of the venom gland of *Acromyrmex subterraneus subterraneus* (Hymenoptera, Formicidae). Naturalia (São Paulo) 18:117-121. [1993]
Morgan, E. D. 1990. Exocrine gland chemistry and phylogeny of ants. Pp. 319-320 in: Veeresh, G. K., Mallik, B., Viraktamath, C. A. (eds.) Social insects and the environment. Proceedings of the 11th International Congress of IUSSI, 1990. New Delhi: Oxford & IBH Publishing Co., xxxi + 765 pp. [1990.08]
Morgan, E. D. 1992. The chemical diversity of ants: is there system in diversity? Pp. 183-194 in: Billen, J. P. J. (ed.) Biology and evolution of social insects. Leuven: Leuven University Press, ix + 390 pp. [1992]
Morgan, E. D., Agosti, D., Keegans, S. J. 1990. Chemical secretions and species discrimination in *Cataglyphis* ants (Hymenoptera; Formicidae). Pp. 402-403 in: Veeresh, G. K., Mallik, B., Viraktamath, C. A. (eds.) Social insects and the environment. Proceedings of the 11th International Congress of IUSSI, 1990. New Delhi: Oxford & IBH Publishing Co., xxxi + 765 pp. [1990.08]
Morgan, E. D., Hölldobler, B., Vaisar, T., Jackson, B. D. 1992. Contents of poison apparatus and their relation to trail-following in the ant *Daceton armigerum*. J. Chem. Ecol. 18:2161-2168. [1992.12]
Mori, A., Le Moli, F. 1993. The aggression test as a taxonomic tool: evaluation in sympatric and allopatric populations of wood-ant species. Aggressive Behav. 19:151-156. [1993]
Morice, F. D., Durrant, J. H. 1915. The authorship and first publication of the "Jurinean" genera of Hymenoptera: being a reprint of a long-lost work by Panzer, with a translation into English, an introduction, and bibliographical and critical notes. Trans. Entomol. Soc. Lond. 1914:339-436. [1915.02.27]
Morice, F. D., Durrant, J. H. 1917. Further notes on the "Jurinean" genera of Hymenoptera, correcting errors and omissions in a paper on that subject published in Trans. Ent. Soc. Lond. 1914, pp. 339-436. Trans. Entomol. Soc. Lond. 1916:432-442. [1917.09.29]
Morillo, C. 1977. Sobre la distribución de *Formica nigricans* Em., 1909 en la Península Ibérica (Hym. Formicidae). Graellsia 32:215-218. [1977.12.07]
Morillo, C. 1981. Inventario y fenología de los Formicidae de un encinar de España central. Rev. Écol. Biol. Sol 18:243-251. [1981.06]

Morisita, M. 1940. On the vertical distributions and the horizontal distributions of ants, especially reconsideration on the micro biogeographical line. [In Japanese.] Akitu 2:153-157. [1940.11.01]

Morisita, M. 1941. Notes on *Camponotus herculeanus* subsp. *vagus* var. *yessensis* Teranishi (Hymenoptera, Formicidae). [In Japanese.] Mushi 13:93-96. [1941.02.28]

Morisita, M. 1945. Ants of the southern part of Hokkaido, Japan. [In Japanese.] Mushi 16:21-28. [1945.04.10]

Morisita, M. 1984. On the ants of southern element distributed in the highland hotspring zone in Kyushu. [Abstract.] [In Japanese.] Ari 12:3-4. [1984.05.01]

Morisita, M. 1985. Some questions in *Paratrechina flavipes* and *Cardiocondyla emeryi*. [Abstract.] [In Japanese.] Ari 13:5. [1985.06.01]

Morisita, M. 1986a. Problems in the classification of some Japanese ants. [Abstract.] [In Japanese.] Ari 14:2. [1986.01.31]

Morisita, M. 1986b. Life zone arrangement of Japanese ants. [Abstract.] [In Japanese.] Ari 14:4-5. [1986.01.31]

Morisita, M. 1987. Problems on *Odontomachus* in Japan. [Abstract.] [In Japanese.] Ari 15:4-5. [1987.10.10]

Morisita, M. 1989 ("1988"). On a variation form (?) of *Monomorium chinense* Santschi. [Abstract.] [In Japanese.] Ari 16:2. [1989.08]

Morisita, M., Kubota, M., Onoyama, K., Ogata, K., Terayama, M., Kondoh, M., Imai, H. T. 1989. A guide for the identification of Japanese ants. I. Ponerinae, Cerapachyinae, Pseudomyrmecinae, Dorylinae, and Leptanillinae (Hymenoptera: Formicidae). [In Japanese.] Tokyo: Myrmecological Society of Japan, 42 pp. [1989.11.30]

Morisita, M., Kubota, M., Onoyama, K., Ogata, K., Terayama, M., Kondoh, M., Imai, H. T., Yamauchi, K., Masuko, K. 1988. A list of the ants of Japan with common Japanese names. [In Japanese.] Tokyo: Myrmecologists Society (Japan), 50 pp. [1988.03.15]

Morisita, M., Kubota, M., Onoyama, K., Ogata, K., Terayama, M., Yamauchi, K., Sonobe, R., Kondoh, M., Imai, H. T. 1991. A guide for the identification of Japanese ants. II. Dolichoderinae and Formicinae (Hymenoptera: Formicidae). [In Japanese.] Tokyo: Myrmecological Society of Japan, 56 pp. [1991.03.31]

Morisita, M., Kubota, M., Onoyama, K., Ogata, K., Terayama, M., Yamauchi, K., Sonobe, R., Yamane, S., Kondoh, M., Imai, H. T. 1992. A guide for the identification of Japanese ants. III. Myrmicinae and supplement to Leptanillinae. (Hymenoptera: Formicidae). [In Japanese.] Tokyo: Myrmecological Society of Japan, 94 pp. [1992.03.31]

Morisita, M., Onoyama, K. 1974. The ants of Kyoto Prefecture. [In Japanese.] Pp. 31-40 in: Morisita, M. (ed.) The fauna of Kyoto Prefecture. [In Japanese.] Kyoto: Kyoto Prefectural Government. [Not seen. Cited in Onoyama (1980) and Onoyama & Terayama (1994).]

Morley, B. D. W. 1938. An outline of the phylogeny of the Formicidae. Bull. Soc. Entomol. Fr. 43:190-194. [1938.10.05]

Morley, B. D. W. 1939. The phylogeny of the Cerapachyinae, Dorylinae, and Leptanillinae (Hym. Formicidae). Bull. Soc. Entomol. Fr. 44:114-118. [1939.05.27]

Morley, B. D. W. 1953. The ant world. London: Penguin Books, 190 pp. [1953]

Morley, C. 1911. Clare Island survey. Part 24. Hymenoptera. Proc. R. Ir. Acad. Sect. B 31(24): 1-18. [1911.12.07]

Morris, G. K. 1880. A new harvesting ant. Am. Nat. 14:669-670. [1880.09.21]

Morris, G. K. 1881. A new leaf cutting ant. Am. Nat. 15:100-102. [1881.02]

Morris, R. L. 1944 ("1942"). An annotated list of the ants of Indiana. Proc. Indiana Acad. Sci. 52:203-224. [1944]

Morton, S. R. 1982. Granivory in the Australian arid zone: diversity of harvester ants and structure of their communities. Pp. 257-262 in: Barker, W. R., Greenslade, P. J. M. (eds.) Evolution of the flora and fauna of arid Australia. Frewville, South Australia: Peacock Publications, 392 pp. [1982]

Morton, S. R., Christian, K. A. 1994. Ecological observations on the spinifex ant, *Ochetellus flavipes* (Kirby) (Hymenoptera: Formicidae), of Australia's northern arid zone. J. Aust. Entomol. Soc. 33:309-316. [1994.11.30]
Morton, S. R., Davidson, D. W. 1988. Comparative structure of harvester ant communities in arid Australia and North America. Ecol. Monogr. 58:19-38. [1988.03]
Moser, J. C., Blum, M. S. 1960. The Formicidae of Louisiana. Insect Cond. La. 3:48-50. [1960]
Motschoulsky, V. de See also: Victor, T. (1839).
Motschoulsky, V. de. 1860a ("1859"). Catalogue des insectes rapportés des environs du fl. Amour, depuis la Schilka jusqu'à Nikolaëvsk, examinés et énumérés. Bull. Soc. Imp. Nat. Mosc. 32(II):487-507. [> 1860.03.05] [Approved by the censor 22 February 1860 (= 5 March 1860 by the Gregorian calendar). Formicidae p. 500.]
Motschoulsky, V. de. 1860b ("1859"). Entomologie speciale. Insectes des Indes orientales, et de contrées analogues. Études entomologiques. Part 8. Helsinki: Imprim. Soc. Littér. Finnoise, 187 pp. [1860.05.03] [Copy received at the "Société impériale des Naturalistes de Moscou" on 21 April 1860 (Griffin, 1936; Blackwelder, 1949a) which would be 3 May 1860 by the Gregorian calendar. Description of *Meranoplus villosus*, p. 115.]
Motschoulsky, V. de. 1863. Essai d'un catalogue des insectes de l'île Ceylan (suite). Bull. Soc. Imp. Nat. Mosc. 36(3):1-153. [> 1863.12.24] [Allowed by the censor 12 December 1863 (= 24 December 1863 by the Gregorian calendar). Formicidae pp. 11-22 and 29.]
Motschoulsky, V. de. 1866. Catalogue des insectes reçus du Japon. Bull. Soc. Imp. Nat. Mosc. 39:163-200. [1866.03.27] [Formicidae p. 183.]
Motschulsky, V. de See: Motschoulsky, V. de.
Mrázek, M. 1908. Myrmekologické poznámky. Cas. Ceské Spol. Entomol. 5:139-146. [1908]
Muesebeck, C. F. W., Krombein, K. V., Townes, H. 1951. Date of publication, Synoptic catalog of North American Hymenoptera. Proc. Entomol. Soc. Wash. 53:232. [1951.08.28]
Muhlenberg, M., Leipold, D., Steinhauer, B. 1977. Island ecology of arthropods. II. Niches and relative abundances of Seychelles ants (Formicidae) in different habitats. Oecologia (Berl.) 29:135-144. [1977.06.24]
Mukerjee, D. 1925. Note on *Typhlopone fulva*, var. *labiata*, Emery, syn. *Dorylus labiatus*, Shuck. J. Dep. Sci. Calcutta Univ. 7(Zoology)(2nd part):1-4. [1925.02.28]
Mukerjee, D. 1927 ("1926"). Digestive and reproductive systems of the male ant *Dorylus labiatus* Shuck. J. Proc. Asiatic Soc. Bengal (n.s.)22:87-91. [1927.08]
Mukerjee, D. 1930. Report on a collection of ants in the Indian Museum, Calcutta. J. Bombay Nat. Hist. Soc. 34:149-163. [1930.03.01]
Mukerjee, D. 1933. On the anatomy of the worker of the ant *Dorylus* (*Alaopone*) *orientalis* Westw. Zool. Anz. 105:97-105. [1933.12.15]
Mukerjee, D. 1934. Entomological investigations on the spike disease of sandal (22). Formicidae (Hymen.). Indian Forest Rec. Entomol. Ser. 20(5):1-15. [1934.05.12]
Mukerjee, D., Ribeiro, S. 1925. On a collection of ants (Formicidae) from the Andaman Islands. Rec. Indian Mus. 27:205-209. [1925.05.22]
Mukerji, D. See: Mukerjee, D.
Mukherji, D. See: Mukerjee, D.
Müller, G. 1921. Due nuove formiche della regione Adriatica. Boll. Soc. Adriat. Sci. Nat. Trieste 27(2):46-49. [1921]
Müller, G. 1923a. Note mirmecologiche. Mem. Soc. Entomol. Ital. 2:65-76. [1923.08.24]
Müller, G. 1923b. Le formiche della Venezia Guilia e della Dalmazia. Boll. Soc. Adriat. Sci. Nat. Trieste 28:11-180. [1923]
Mulsant, E. 1842. Histoire naturelle des coléoptères de France. Lamellicornes. Paris: Maison, 623 pp. [1842.08.06] [Date of publication from Esben-Petersen (1922:621).]
Münch, W. 1987. Ant communities around the Federsee and the Schmiechener See: a comparative study of two nature reserves in southern Germany. P. 603 in: Eder, J., Rembold, H. (eds.) Chemistry and biology of social insects. München: Verlag J. Peperny, xxxv + 757 pp. [1987]

Münch, W., Engels, W. 1994. Vorkommen der Moor-Knotenameise *Myrmica gallienii* im Riedgürtel des Federsees (Hymenoptera: Myrmicidae). Entomol. Gen. 19:15-20. [1994.09]
Munsee, J. R. 1968. Nine species of ants (Formicidae) recently recorded from Indiana. Proc. Indiana Acad. Sci. 77:222-227. [1968]
Munsee, J. R. 1977. *Smithistruma filitalpa* W. L. Brown, an Indiana dacetine ant (Hymenoptera: Formicidae). Proc. Indiana Acad. Sci. 86:253-256. [1977]
Munsee, J. R., Schrock, J. R. 1983 ("1982"). A comparison of ant faunae on unreclaimed stripmines in Indiana. Proc. Indiana Acad. Sci. 92:257-261. [1983]
Murakami, M. 1899. *Polyrhachis lamellidens* collected at Mt. Kinpu, Kyusyu. [In Japanese.] Dobutsugaku Zasshi (Zool. Mag.) 11:208-210. [1899.06.22]
Murakami, Y., Kim, C. H. 1980. Is it true the occurrence of *Formica* (*F.*) *yessensis* Forel in Kyushu? [In Japanese.] Ari 9:7-8. [1980.07.15]
Musthak Ali, T. M. 1982. Ant fauna (Hymenoptera: Formicidae) of Bangalore with observations on their nesting and foraging habits. Thesis Abstr. Haryana Agric. Univ. 8:370-371. [1982.12]
Nachtwey, R. 1962 ("1961"). Tonerzeugung durch schwingenden Membranen bei Ameisen (*Plagiolepis*, *Leptothorax*, *Solenopsis*). Insectes Soc. 8:369-381. [1962.03]
Nachtwey, R. 1963. Das Phon-Organ der Ameisengattungen *Solenopsis*, *Leptothorax*, *Iridomyrmex* und *Lasius*. Insectes Soc. 10:43-57. [1963.06]
Nachtwey, R. 1964 ("1963"). Vergleichende Studien über das Phon-Organ der Myrmicinae, Dolichoderinae und Formicinae. Insectes Soc. 10:359-378. [1964.03]
Nadig, A. 1918. Alcune note sulla fauna dell'alta Valsesia. Formicidae. Atti Soc. Ital. Sci. Nat. Mus. Civ. Stor. Nat. Milano 56:331-341. [1918.01]
Najt, J. 1987. Le Collembole fossile *Paleosminthurus juliae* est un Hyménoptère. Rev. Fr. Entomol. (Nouv. Sér.) 9:152-154. [1987.12.15]
Nakagawa, K. 1903. A list of known species of Evaniidae, Cynipidae and Formicidae in Japan. [In Japanese.] Dobutsugaku Zasshi (Zool. Mag.) 15:400-404. [1903.12.22]
Nakano, T. 1938. The scientific achievements for the Japanese myrmecology by Mr. Teranishi. [In Japanese.] Kansai Konchu Zasshi 5:80-84. [1938.12.10]
Naora, N. 1933. Notes on some fossil insects from East-Asiatic continent, with descriptions of three new species. Entomol. World Tokyo 1:208-219. [1933.06]
Nascimento, R. R. do, Billen, J. P. J., Morgan, E. D. 1993. The exocrine secretions of the jumping ant *Harpegnathos saltator*. Comp. Biochem. Physiol. B. Comp. Biochem. 104:505-508. [1993.03.15]
Nascimento, R. R. do, Jackson, B. D., Morgan, E. D., Clark, W. H., Blom, P. E. 1993. Chemical secretions of two sympatric harvester ants, *Pogonomyrmex salinus* and *Messor lobognathus*. J. Chem. Ecol. 19:1993-2005. [1993.09]
Nasonov, N. V. 1889. Contribution to the natural history of the ants, primarily of Russia. 1. Contribution to the ant fauna of Russia. [In Russian.] Izv. Imp. Obshch. Lyubit. Estestvozn. Antropol. Etnogr. Imp. Mosk. Univ. 58:1-78. [1889]
Nasonov, N. V. 1892. Entomological studies in 1892. On the ant fauna of Russia (fauna of the near Visla River region). [In Russian.] Izv. Imp. Varsh. Univ. 5:1-14. [Not seen. Cited in Ruzsky (1905b:52), Pisarski (1975:68) and Radchenko (1994f:81).]
Naumann, I. D. 1993. The supposed Cretaceous ant *Cretacoformica explicata* Jell and Duncan. J. Aust. Entomol. Soc. 32:355-356. [1993.11.30]
Naumann, I. D., Cardale, J. C., Taylor, R. W., MacDonald, J. 1994. Type specimens of Australian Hymenoptera (Insecta) transferred from the Macleay Museum, University of Sydney, to the Australian National Insect Collection, Canberra. Proc. Linn. Soc. N. S. W. 114:69-72. [1994.05.27]
Naumann, K. 1994. An occurrence of two exotic ant (Formicidae) species in British Columbia. J. Entomol. Soc. B. C. 91:69-70. [1994.12]
Naves, M. A. 1985. A monograph of the genus *Pheidole* in Florida, USA (Hymenoptera: Formicidae). Insecta Mundi 1:53-90. [1985.03]

Nazaraw, U. I. 1994. [Untitled. *Iridomyrmex bogdassarovi* V. Nazarov, sp. nov.] P. 106 in: Nazaraw, U. I., Bagdasaraw, A. A., Ur'ew, I. I. First findings of insects (Diptera, Hymenoptera) in amber from the Belarussian Polesye region. [In Belarussian.] Vestsi Akad. Navuk Belarusi Ser. Biial. Navuk 1994(2):104-108. [> 1994.06.29] [Date signed for printing.]

Nazaraw, U. I., Bagdasaraw, A. A., Ur'ew, I. I. 1994. First findings of insects (Diptera, Hymenoptera) in amber from the Belarussian Polesye region. [In Belarussian.] Vestsi Akad. Navuk Belarusi Ser. Biial. Navuk 1994(2):104-108. [> 1994.06.29] [Date signed for printing ("padpisana w druk").]

Nefedov, N. I. 1930. Ants of the Troitsk Forest-Steppe Reserve and their distributions in elements of landscape. [In Russian.] Izv. Biol. Nauchno-Issled. Inst. Permsk. Gos. Univ. 7:259-291. [1930]

Negoro, H. 1994. Ants from Toyama Prefecture, Hokuriku, Japan. [In Japanese.] Bull. Toyama Sci. Mus. 17:35-47. [1994.03.25]

Negrobov, O. P., Uspensky, K. V. 1990. Fauna and biotopical distribution of ants in Sysert reserve. [In Russian.] Mater. Kollok. Sekts. Obshchestv. Nasek. Vses. Entomol. Obshch. 1:165-168. [> 1990.12.25] [Date signed for printing ("podpisano k pechati").]

Nelmes, E. 1938. A survey of the distribution of the wood ant (*Formica rufa*) in England, Wales, and Scotland. J. Anim. Ecol. 7:74-104. [1938.05]

Newman, E. 1850. Description of an apparently new lepidopterous insect, of the family Glaucopidae, from the Upper Amazons. Zoologist 8(Appendix):cxxii-cxxiii. [1850.12] [*Myrmecopsis* (Lepidoptera), senior homonym of *Myrmecopsis* F. Smith, 1865.]

Newman, E. 1867. *Formica herculanea* a British insect. Entomologist 3:244. [1867.03]

Nickerson, J. C., Cromroy, H. L., Whitcomb, W. H., Cornell, J. A. 1975. Colony organization and queen numbers in two species of *Conomyrma*. Ann. Entomol. Soc. Am. 68:1083-1085. [1975.11.17]

Nickerson, J. C., Harris, D. L. 1985. The Florida carpenter ant, *Camponotus abdominalis floridanus* (Buckley) (Hymenoptera: Formicidae). Fla. Dep. Agric. Consum. Serv. Div. Plant Ind. Entomol. Circ. 269:1-2. [1985.01]

Nielsen, M. G. 1987. The ant fauna (Hymenoptera: Formicidae) in northern and interior Alaska. A survey along the trans-Alaskan pipeline and a few highways. Entomol. News 98:74-88. [1987.05.08]

Nielsen, M. G. 1995. Mangrove ants in Northern Australia. [Abstract.] P. 87 in: Demeter, A., Peregovits, L. (eds.) Ecological processes: current status and perspectives. Abstracts. 7th European Ecological Congress. Budapest: Hungarian Biological Society, 294 pp. [1995.08.20]

Nielsen, M. G., Jensen, T. F. 1982. Myrefaunaen på Bornholm. Flora Fauna 88:35-36. [1982.06]

Niemeyer, H. 1976. Zur Artansprache von *Formica polyctena* (Foerst.) und *Formica rufa* (L.) in der Praxis der Ameisenhege. Z. Pflanzenkr. Pflanzenschutz 83:120-130. [1976]

Nikitin, M. I. 1979. Geographical distribution of three species of small ants common in New South Wales. Aust. Entomol. Mag. 5:101-102. [1979.04]

Nilsson, O., Douwes, P. 1987. Norrländska myror med bestämningstabell till arbetarna. Nat. Norr 6:49-90.

Nishizono, Y., Yamane, S. 1990. The genus *Aphaenogaster* (Hymenoptera, Formicidae) in Kagoshima-ken, southern Japan. [In Japanese.] Rep. Fac. Sci. Kagoshima Univ. (Earth Sci. Biol.) 23:23-40. [1990.12]

Nolan, E. J. (ed.) 1913. An index to the scientific contents of the Journal and Proceedings of the Academy of Natural Sciences of Philadelphia. Philadelphia: Academy of Natural Sciences, xiv + 1419 pp. [1913]

Nomura, K. 1935. On *Odontomachus monticola formosae* found from Tanegashima Is., Kyushu. [In Japanese.] Mushi 8:58. [1935.09.15]

Nonacs, P. 1986. Sex-ratio determination within colonies of ants. Evolution 40:199-204. [1986.01.14]

Nonacs, P. 1988. Queen number in colonies of social Hymenoptera as a kin-selected adaptation. Evolution 42:566-580. [1988.04.26]

Nonacs, P., Tobin, J. E. 1993 ("1992"). Selfish larvae: development and the evolution of parasitic behavior in the Hymenoptera. Evolution 46:1605-1620. [1993.01.13]

Norton, E. 1868a. Notes on Mexican ants. Am. Nat. 2:57-72. [1868.04]

Norton, E. 1868b. Remarks by Edward Norton. [Followed by: "The species of *Eciton* forwarded by Prof. Sumichrast may be temporarily classified as follows."] Trans. Am. Entomol. Soc. 2:44-46. [1868.06]

Norton, E. 1868c. Description of Mexican ants noticed in the American Naturalist, April, 1868. Proc. Essex Inst. (Commun.) 6:1-10. [1868.07]

Norton, E. 1875a. Report upon the collections of Hymenoptera made in portions of Nevada, Utah, Colorado, New Mexico, and Arizona, during the years 1872, 1873, and 1874. Report upon the collections of Formicidae. Pp. 729-736 in: Yarrow, H. C. (ed.) Report upon geographical and geological explorations and surveys west of the 100th meridian, in charge of First Lieut. Geo. M. Wheeler. Volume V. Zoology. Washington, D. C.: U. S. Government Printing Office, 1021 pp. [1875]

Norton, E. 1875b. Notas sobre las hormigas mexicanas. Naturaleza (Méx.) 3:179-189. [1875] [Translation, by A. Moreno, of Norton (1868a). Publication date from H. M. Smith (1942).]

Novák, O. 1877. Fauna der Cyprisschiefer des Egerer Tertiärbeckens. Sitzungsber. Kais. Akad. Wiss. Wien Math.-Naturwiss. Cl. Abt. I 76:71-96. [1877.07] [Fossil ants pp. 90-92.]

Novák, V. 1939. Príspevek k poznání mravencu severních Cech. Cas. Cesk. Spol. Entomol. 36: 38-39. [1939.03.15]

Novák, V. 1941. *Dolichoderus* (subgen. *Hypoclinea*) *quadripunctatus* var. *kratochvíli* var. nova. (Formicoidea, Dolichoderidae). Cas. Cesk. Spol. Entomol. 38:45-48. [1941.05.15]

Novák, V. 1942. Novy nález mravence *Monomorium pharaonis* L. (Myrmicidae) v Praze. Cas. Cesk. Spol. Entomol. 39:135-136. [1942.12.15]

Novák, V. 1944a. Gynandromorphus *Myrmica sabuleti*. Entomol. Listy 7:56-59. [1944.03.31]

Novák, V. 1944b. Dicephallus *Leptothorax nylanderi* r. *parvulus* Sch. Cas. Cesk. Spol. Entomol. 41:31-34. [1944.06.01]

Novák, V. 1944c. K taxonomii mravencu rodu *Bothriomyrmex* a *Leptothorax*. Pp. 103-132 in: Kratochvíl, J., Novák, V., Snoflák, J. Mohelno. Soubor práci venoványch studiu vyznamne památky prírodní. 5. Hymenoptera - Aculeata. Formicidae - Apidae - Vespidae. Arch. Svazu Ochr. Prír. Domov. Moravé 6:1-155. [1944]

Novák, V. 1947. Exotictí mravenci ve sklenících Prazské botanické zahrady. Cas. Cesk. Spol. Entomol. 44:144-146. [1947.12.01]

Novák, V. 1948. Príspevek k otázce vzniku pathologickich jedincu (pseudogyn) u mravencu z rodu *Formica*. (Predbezné sdelení). Vestn. Cesk. Zool. Spol. 12:97-131. [1948]

Novák, V., Sadil, J. 1939. Dodatek k poznání mravencu hadcové stepi u Mohelna. Cas. Cesk. Spol. Entomol. 36:52-58. [1939.11.01]

Novák, V., Sadil, J. 1941. Klíc k urcování mravencu strední Evropy se zvlástním zretelem k mravencí zvírene Cech a Moravy. Entomol. Listy 4:65-115. [1941]

Nowbahari, E., Lenoir, A., Clément, J. L., Lange, C., Bagneres, A. G., Joulie, C. 1990. Individual, geographical and experimental variation of cuticular hydrocarbons of the ant *Cataglyphis cursor* (Hymenoptera: Formicidae): their use in nest and subspecies recognition. Biochem. Syst. Ecol. 18:63-74. [1990.03.27]

Nowotny, H. 1931a. Verzeichnis der bisher in Oberschlesien aufgefundenen Ameisen. Mitt. Beuthen. Gesch.-Museumsver. 13-14:150-157. [Not seen. Cited in Pisarski (1975:68).]

Nowotny, H. 1931b. Verzeichnis der oberschlesischen Ameisen. Anhang. Beuthen. Abh. Oberschles. Heimatforsch. 6:3-10. [Not seen. Cited in Pisarski (1975:68).]

Nowotny, H. 1931c. Nachtrag zum Verzeichnis oberschlesischer Ameisen. Mitt. Beuthen. Gesch.-Museumsver. 13-14:294. [Not seen. Cited in Pisarski (1975:68).]

Nowotny, H. 1937. 2. Nachtrag zur Ameisenfauna Oberschlesiens. Z. Entomol. (Breslau) 18(2): 5-6. [1937.07.01]

Nuhn, T. P., Wright, C. G. 1979. An ecological survey of ants (Hymenoptera: Formicidae) in a landscaped suburban habitat. Am. Midl. Nat. 102:353-362. [1979.10.31]

Nylander, W. 1846a. Adnotationes in monographiam formicarum borealium Europae. Acta Soc. Sci. Fenn. 2:875-944. [1846]

Nylander, W. 1846b. Additamentum adnotationum in monographiam formicarum borealium Europae. Acta Soc. Sci. Fenn. 2:1041-1062. [1846]

Nylander, W. 1849 ("1848"). Additamentum alterum adnotationum in monographiam formicarum borealium. Acta Soc. Sci. Fenn. 3:25-48. [1849]

Nylander, W. 1851. Remarks on "Hymenopterologische Studien by Arnold Foerster, 1stes Heft, Formicariae, Aachen, 1850". Ann. Mag. Nat. Hist. (2)8:126-129. [1851.08]

Nylander, W. 1856a. [Untitled. Introduced by: "M. L. Fairmaire communique la note suivante, de M. Nylander, et la Société en décide l'impression dans le Bulletin."] Ann. Soc. Entomol. Fr. (Bull.) (3)4:xxviii. [1856.07.23]

Nylander, W. 1856b. Synopsis des Formicides de France et d'Algérie. Ann. Sci. Nat. Zool. (4)5:51-109. [1856]

Nylander, W. 1857 ("1856"). [Untitled. Introduced by: "M. L. Fairmaire communique la note suivant de M. Nylander sur les Formicides du Mont-Dore, et la Société en décide l'impression dans le Bulletin."] Ann. Soc. Entomol. Fr. (Bull.) (3)4:lxxviii-lxxix. [1857.01.28]

Obin, M. S., Vander Meer, R. K. 1989. Between and within-species recognition among imported fire ants and their hybrids (Hymenoptera: Formicidae): application to hybrid zone dynamics. Ann. Entomol. Soc. Am. 82:649-652. [1989.09]

Ochi, K. 1983. Distribution pattern of ants in pine stands, with special reference to *Monomorium nipponense* Wheeler (Hymenoptera: Formicidae). [In Japanese.] Gensei 44:1-6. [1983.12.28]

Oeser, R. 1961. Vergleichend-morphologische Untersuchungen über den Ovipositor der Hymenopteren. Mitt. Zool. Mus. Berl. 37:1-119. [1961.05.27]

Ofer, J. 1970. *Polyrhachis simplex*. The weaver ant of Israel. Insectes Soc. 17:49-82. [1970.06]

Ofer, J., Shulov, A., Noy-Meir, I. 1978. Associations of ant species in Israel: a multivariate analysis. Isr. J. Zool. 27:199-208. [1978.12]

Ogata, K. 1982. Taxonomic study of the ant genus *Pheidole* Westwood of Japan, with a description of a new species (Hymenoptera, Formicidae). Kontyû 50:189-197. [1982.06.25]

Ogata, K. 1983a. Structure and function of the male genitalia of ants. [Abstract.] [In Japanese.] Ari 11:2. [1983.08.16]

Ogata, K. 1983b. The ant genus *Cerapachys* F. Smith of Japan, with description of a new species (Hymenoptera, Formicidae). Esakia 20:131-137. [1983.12.15]

Ogata, K. 1984. Morphological notes on the wing venation of ants. [Abstract.] [In Japanese.] Ari 12:4. [1984.05.01]

Ogata, K. 1985. A generic synopsis of the poneroid complex in Japan. [Abstract.] [In Japanese.] Ari 13:3-4. [1985.06.01]

Ogata, K. 1986. Note on the ant fauna of Kyushu, Japan, with reference to the species-area relationship and species composition in different habitats. [Abstract.] [In Japanese.] Ari 14:3. [1986.01.31]

Ogata, K. 1987a. A generic synopsis of the poneroid complex of the family Formicidae in Japan (Hymenoptera). Part 1. Subfamilies Ponerinae and Cerapachyinae. Esakia 25:97-132. [1987.01.31]

Ogata, K. 1987b. Taxonomic note on the ant tribe Dacetini in Japan. [Abstract.] [In Japanese.] Ari 15:3. [1987.10.10]

Ogata, K. 1990. A new species of the ant genus *Epitritus* Emery from Japan (Hymenoptera, Formicidae). Esakia Spec. Issue 1:197-199. [1990.04.20]

Ogata, K. 1991a. Ants of the genus *Myrmecia* Fabricius: a review of the species groups and their phylogenetic relationships (Hymenoptera: Formicidae: Myrmeciinae). Syst. Entomol. 16:353-381. [1991.07]

Ogata, K. 1991b. A generic synopsis of the poneroid complex of the family Formicidae (Hymenoptera). Part II. Subfamily Myrmicinae. Bull. Inst. Trop. Agric. Kyushu Univ. 14:61-149. [1991.10]

Ogata, K. 1992a. The ant fauna of the Oriental region: an overview (Hymenoptera, Formicidae). [Abstract.] P. 52 in: Proceedings XIX International Congress of Entomology. Abstracts. Beijing, China, June 28 - July 4, 1992. Beijing: XIX International Congress of Entomology, xi + 730 pp. [1992.07.04]

Ogata, K. 1992b. The ant fauna of the Oriental region: an overview (Hymenoptera, Formicidae). Bull. Inst. Trop. Agric. Kyushu Univ. 15:55-74. [1992.12]

Ogata, K. 1994a. A taxonomic note on the subfamily Leptanillinae. [Abstract.] [In Japanese.] Ari 17:2-3. [1994.04.30]

Ogata, K. 1994b. The Oriental chaos: current state and prospect of ant taxonomy in the Oriental region. [Abstract.] [In Japanese.] Ari 18:36-37. [1994.05.30]

Ogata, K. 1994c. Ant fauna of the Nansei Islands, Japan. [Abstract.] [In Japanese.] Ari 18:39. [1994.05.30]

Ogata, K., Bolton, B. 1989. A taxonomic note on the ant *Monomorium intrudens* F. Smith (Hymenoptera, Formicidae). Jpn. J. Entomol. 57:459-460. [1989.06.25]

Ogata, K., Hirashima, Y., Miura, T., Maeta, Y., Yano, K., Ko, J.-H. 1985. Ants collected in pine forests infested by the pine needle gall midge in Korea (Hymenoptera, Formicidae). Esakia 23:159-163. [1985.11.30]

Ogata, K., Taylor, R. W. 1991. Ants of the genus *Myrmecia* Fabricius: a preliminary review and key to the named species (Hymenoptera: Formicidae: Myrmeciinae). J. Nat. Hist. 25:1623-1673. [1991.12.31]

Ogata, K., Terayama, M., Masuko, K. 1995. The ant genus *Leptanilla*: discovery of the worker-associated male of *L. japonica*, and a description of a new species from Taiwan (Hymenoptera: Formicidae: Leptanillinae). Syst. Entomol. 20:27-34. [1995.03.17]

Ogata, K., Touyama, Y., Choi, B.-M. 1994. Ant fauna of Hiroshima Prefecture, Japan. [In Japanese.] Ari 18:18-25. [1994.05.30]

Ohta, Y. 1931a. The check-list of ants from Gifu Prefecture, 1. [In Japanese.] Insect World 35: 85-89. [1931.03.15]

Ohta, Y. 1931b. The check-list of ants from Gifu Prefecture, 2. [In Japanese.] Insect World 35:115-120. [1931.04.15]

Ohta, Y. 1935a. A list of Japanese ants. [In Japanese.] [part] Insect World 39:286-289. [1935.08.15]

Ohta, Y. 1935b. A list of Japanese ants. [In Japanese.] [part] Insect World 39:329-333. [1935.09.15]

Ohta, Y. 1936a. A list of Japanese ants. [In Japanese.] [part] Insect World 40:166-169. [1936.05.15]

Ohta, Y. 1936b. A list of Japanese ants. [In Japanese.] [concl.] Insect World 40:426-433. [1936.12.15]

Ohta, Y. 1938a. Five species of ants from Kinkasan, Gifu. [part] [In Japanese.] Insect World 42:125-129. [1938.04.15]

Ohta, Y. 1938b. Five species of ants from Kinkasan, Gifu. [concl.] [In Japanese.] Insect World 42:186-188. [1938.06.15]

Ohta, Y. 1940. Two species of ants from South China. [In Japanese.] Insect World 44:332-335. [1940.11.15]

Okamoto, H. 1952. Ants from Shikoku, Japan (1). [In Japanese.] Gensei 1:9-12. [1952.12.25]

Okamoto, H. 1953. Ants from Shikoku, Japan (2). [In Japanese.] Gensei 2:39-43. [1953.12.25]

Okamoto, H. 1954. Ants from Shikoku, Japan (3). [In Japanese.] Gensei 3:43-49. [1954.12.29]

Okamoto, H. 1957. Ants from Shikoku, Japan (4). [In Japanese.] Gensei 5(2):39-43. [1957.02.16]

Okamoto, H. 1966. Ants from Shikoku, Japan (5). [In Japanese.] Gensei 16:5-8. [1966.12.23]

Okamoto, H. 1969. Ants from Shikoku, Japan (6). [In Japanese.] Gensei 19:5-10. [1969.10.30]

Okamoto, H. 1972a. Ants from Shikoku, Japan (7). [In Japanese.] Gensei 23:11-14. [1972.01.29]

Okamoto, H. 1972b. The occurrence of *Triglyphothrix striatidens* (Emery) in Kyushu. [In Japanese.] Gensei 23:14. [1972.01.29]

Okano, K. 1945. On the ants of Enoshima Is., Kanagawa Prefecture on the late spring. [In Japanese.] Kontyû Kenkyu 2(4):19-20. [Not seen. Cited in Okano (1989) and Onoyama & Terayama (1994).]

Okano, K. 1946. Some taxonomical notes on the group "*Camponotus itoi*". [In Japanese.] Kagaku (Tokyo) 16:187-188. [Not seen. Cited in Okano (1989) and Onoyama & Terayama (1994).]

Okano, K. 1947a. Some discussions on the classification of ants. [in Japanese.] Igaku Seibutsugaku 11:164-167. [Not seen. Cited in Okano (1989) and Onoyama & Terayama (1994).]

Okano, K. 1947b. On the ant "suzukikuroyamaari" described by Mr. Teranishi. [In Japanese.] Igaku Seibutsugaku 11:167-169. [Not seen. Cited in Okano (1989) and Onoyama & Terayama (1994).]

Okano, K. 1948. Delle note sulle formiche giapponesi. Kontyû Kagaku 5:91-93. [Not seen. Cited in Okano (1989) and Onoyama & Terayama (1994).]

Okano, K. 1989. The list of works concerning Formicidae (Hymenoptera) and it's related papers made by the Japanese during the years 1900-1949. Entomo Shirogane 2:1-15. [1989.12.19]

Oldham, N. J., Morgan, E. D., Gobin, B., Billen, J. P. J. 1994. First identification of a trail pheromone of an army ant (*Aenictus* species). Experientia (Basel) 50:763-765. [1994.08.15]

Oldham, N. J., Morgan, E. D., Gobin, B., Schoeters, E., Billen, J. P. J. 1994. Volatile secretions of Old World army ant *Aenictus rotundatus* and chemotaxonomic implications of army ant Dufour gland chemistry. J. Chem. Ecol. 20:3297-3305. [1994.12]

Oldham, N. J., Morgan, E. D., Hölldobler, B. 1994. The recruitment pheromones of *Aphaenogaster (=Novomessor) cockerelli* and *A. (=Novomessor) albisetosus*, two closely related myrmicine ants. P. 486 in: Lenoir, A., Arnold, G., Lepage, M. (eds.) Les insectes sociaux. 12ème congrès de l'Union Internationale pour l'Étude des Insectes Sociaux. Paris, Sorbonne, 21-27 août 1994. Paris: Université Paris Nord, xxiv + 583 pp. [1994.09]

Oliveira, M. A. de, Della Lucia, T. M. C. 1993 ("1992"). Levantamento de Formicidae de chão em áreas mineradas sob recuperação florestal de Porto Trombetas, Pará. Bol. Mus. Para. Emílio Goeldi Sér. Zool. 8:375-384. [1993]

Oliveira, P. S., Brandão, C. R. F. 1991. The ant community associated with extrafloral nectaries in the Brazilian cerrados. Pp. 198-212 in: Huxley, C. R., Cutler, D. F. (eds.) Ant-plant interactions. Oxford: Oxford University Press, xviii + 601 pp. [1991]

Oliveira, P. S., Klitzke, C., Vieira, E. 1995. The ant fauna associated with the extrafloral nectaries of *Ouratea hexasperma* (Ochnaceae) in an area of cerrado vegetation in central Brazil. Entomol. Mon. Mag. 131:77-82. [1995.03.20]

Oliver, I., Beattie, A. J. 1993. A possible method for the rapid assessment of biodiversity. Conserv. Biol. 7:562-568. [1993.09]

Olivier, A. G. 1792. Encyclopédie méthodique. Histoire naturelle. Insectes. Tome 6. (pt. 2). Paris: Panckoucke, pp. 369-704. [1792] [Publication date from Sherborn & Woodward (1906b:577). Ants pp. 469-506.]

Olsen, O. W. 1934. Notes on the North American harvesting ants of the genus *Pogonomyrmex* Mayr. Bull. Mus. Comp. Zool. 77:493-514. [1934.12]

Olson, D. M. 1991. A comparison of the efficacy of litter sifting and pitfall traps for sampling leaf litter ants (Hymenoptera, Formicidae) in a tropical wet forest, Costa Rica. Biotropica 23:166-172. [1991.06]

Onoyama, K. 1974. The fossil ants in Mizunami amber: preliminary report (Insecta: Hymenoptera: Formicidae). [In Japanese.] Bull. Mizunami Fossil Mus. 1:445-453. [1974.12.25]

Onoyama, K. 1976. A preliminary study on the ant fauna of Okinawa-Ken, with taxonomic notes (Japan; Hymenoptera: Formicidae). Pp. 121-141 in: Ikehara, S. (ed.) Ecological studies of nature conservation of the Ryukyu Islands - (II). Naha, Okinawa: University of the Ryukyus, 141 pp. [1976.03]

Onoyama, K. 1980a. An introduction to the ant fauna of Japan, with a check list (Hymenoptera, Formicidae). Kontyû 48:193-212. [1980.06.25]

Onoyama, K. 1980b. The ant fauna of the Tsushima Island, with a comparison with the ant faunas of the adjacent areas. [Abstract.] P. 32 in: Abstracts. XVI International Congress of Entomology, Kyoto, Japan 3-9 August, 1980. Kyoto: XVI International Congress of Entomology, 480pp. [1980]

Onoyama, K. 1982a. Immature stages of the harvester ant *Messor aciculatus* (Hymenoptera, Formicidae). Kontyû 50:324-329. [1982.06.25]

Onoyama, K. 1982b. The locations of type specimens of the Japanese ants described by Japanese. [In Japanese.] Ari 10:2-4. [1982.10.01]

Onoyama, K. 1982c. Which should we use - *Messor aciculatus* or *M. aciculatum*, and *Formica transkaucasica* or *F. picea*? [In Japanese.] Ari 10:5-6. [1982.10.01]

Onoyama, K. 1987. An opinion on giving Japanese names of the ants of Japan. [In Japanese.] Ari 15:6-8. [1987.10.10]

Onoyama, K. 1989a. Confirmation of the occurrence of *Myrmica rubra* (Hymenoptera: Formicidae) in Japan with taxonomic and ecological notes. Jpn. J. Entomol. 57:131-135. [1989.03.25]

Onoyama, K. 1989b. Notes on the ants of the genus *Hypoponera* in Japan (Hymenoptera: Formicidae). Edaphologia 41:1-10. [1989.09.25]

Onoyama, K. 1989c. Three ants (Hymenoptera: Formicidae) new to Hokkaido, Japan. Jpn. J. Entomol. 57:604. [1989.09.25]

Onoyama, K. 1991. A new synonym of the ant *Proceratium japonicum* (Hymenoptera, Formicidae). Jpn. J. Entomol. 59:695-696. [1991.09.25]

Onoyama, K., Terayama, M. 1994. A list of references on Japanese ants. [In Japanese.] Tokyo: Myrmecological Society of Japan, iii + 50 pp. [1994.03.25]

O'Rourke, F. J. 1940. Notes on the ant fauna of Howth, County Dublin. Ir. Nat. J. 7:30. [1940.09]

O'Rourke, F. J. 1945. A further extension of the range of *Myrmica schenki*, Emery. Entomol. Rec. J. Var. 57:85-86. [1945.07.15]

O'Rourke, F. J. 1946. The discovery of the rare ant, *Stenamma westwoodi* Westwood, in Co. Wicklow. Ir. Nat. J. 8:413-414. [1946.10]

O'Rourke, F. J. 1950a. The distribution and general ecology of the Irish Formicidae. Proc. R. Ir. Acad. Sect. B 52:383-410. [1950.07.13]

O'Rourke, F. J. 1950b. Myrmecological notes from Narvik, northern Norway. Nor. Entomol. Tidsskr. 8:47-50. [1950.09.01]

O'Rourke, F. J. 1952. A preliminary ecological classification of ant communities in Ireland. Entomol. Gaz. 3:69-72. [1952.04]

O'Rourke, F. J. 1979. The social Hymenoptera of County Wexford. Proc. R. Ir. Acad. Sect. B 79: 1-14. [1979.03.15]

Ortiz, F. J., Tinaut, A. 1986. Distribuciones geográficas notables de los formicidos en el litoral granadino (Insecta, Hymenoptera). Pp. 1051-1061 in: Actas de las VIII Jornadas de la Asociación Española de Entomología. Sevilla: Servicio Publicaciones Universidad de Sevilla, [8] + 1280 pp. [1986.10]

Ortiz, F. J., Tinaut, A. 1987. Citas nuevas o interesantes de Formícidos (Hym. Formicidae) para Andalucía. Bol. Asoc. Esp. Entomol. 11:31-34. [1987.11]

Ortiz, F. J., Tinaut, A. 1988. Formícidos del litoral granadino. Orsis (Org. Sist.) 3:145-163. [1988]

Ortiz y Sánchez, F. J. See: Ortiz, F. J.

Osgood, W. H. 1914. Dates for *Ovis canadensis*, *Ovis cervina*, and *Ovis montana*. Proc. Biol. Soc. Wash. 27:1-3. [1914.02.02] [Provides evidence on the date of publication of Latreille (1804).]

Oster, G. F., Wilson, E. O. 1978. Caste and ecology in the social insects. Princeton: Princeton University Press, xv + 352 pp. [1978]

Oswald, J. D., Penny, N. D. 1991. Genus-group names of the Neuroptera, Megaloptera and Raphidioptera of the world. Occas. Pap. Calif. Acad. Sci. 147:1-94. [1991.12.02] [Includes a discussion of the dating of Fitch (1855).]

Otto, D. 1958. Über die Homologieverhältnisse der Pharynx- und Maxillardrüsen bei Formicidae und Apidae (Hymenopt.). Zool. Anz. 161:216-226. [1958.11]

Otto, D. 1959. Statische Untersuchungen zur Systematik der Roten Waldameise (engere *Formica rufa* L.-Gruppe). Naturwissenschaften 46:458-459. [1959.07.02]

Otto, D. 1961. Zur Systematik der Waldameisenformen. Neue Methoden zur sicheren Arten- und Rassenbestimmung als Grundlage für den praktischen Einsatz der Waldameisen im Forstschutz. Arch. Forstwes. 10:531-535. [1961]

Otto, D. 1966. Kurze Übersicht über die nomenklatorischen Änderungen im Subgenus *Formica* und Angaben zur Unterscheidung der Geschlechtstiere. Entomol. Nachr. Dres. 10:121-125. [1966.09.30]

Otto, D. 1968. Zur Verbreitung der Arten der *Formica rufa* Linnaeus-Gruppe. I. Häufigkeit, geographische Verteilung und Vorzugsstandorte der Roten Waldameisen im Gebiet der Deutschen Demokratischen Republik (Hymenoptera: Formicidae). Beitr. Entomol. 18:671-692. [1968]

Otto, D., Paraschivescu, D. 1968. Zur Verbreitung der Arten der *Formica rufa* Linnaeus-Gruppe. II. Die hügelbauenden *Formica*-Arten in der Sozialistischen Republik Rumänien (Hymenoptera: Formicidae). Beitr. Entomol. 18:693-698. [1968]

Ovazza, M. 1950. Contribution à la connaissance des fourmis des Pyrénées-Orientales. Récoltes de J. Hamon. Vie Milieu 1:93-94. [1950.09]

Pace, R. 1975. Due interessanti reperti mirmecologici per i Monti Lessini e per i Monti Berici (Hymenoptera Formicidae). (XVI Contributo alla conoscenza della fauna endogea.) Boll. Soc. Entomol. Ital. 107:166-170. [1975.12.20]

Padilla, R. C., Miyazawa, E. E. 1972. Distribución geográfica de las especies de hormigas arrieras existentes en la República Mexicana. Fitofilo 67:35-36. [1972]

Pagliano, G., Scaramozzino, P. 1990 ("1989"). Elenco dei generi di Hymenoptera del mundo. Mem. Soc. Entomol. Ital. 68:1-210. [1990.05.10]

Paik, W. H. 1983. Key to the Korean insects. 1. Key to genera of Formicidae (Hymenoptera). [In Korean.] Seoul Natl. Univ. Coll. Agric. Bull. 8:69-74. [1983.12.15]

Paik, W. H. 1984. A check list of Formicidae (Hymenoptera) of Korea. [In Korean.] Korean J. Plant Prot. 23:193-195. [1984.09.30]

Paiva, R. V. S., Brandão, C. R. F. 1995. Nests, worker population, and reproductive status of workers, in the giant queenless ponerine ant *Dinoponera* Roger (Hymenoptera Formicidae). Ethol. Ecol. Evol. 7:297-312. [1995.12]

Paiva, M. R., Way, M. J., Cammel, M. 1990. Estudo preliminar sobre a distribuição das formigas nos sistemas florestais em Portugal. Bol. Soc. Port. Entomol. (4)17:197-205. [1990.05.31]

Paiva, M. R., Way, M. J., Cammell, M. 1995. Habitat related diversity of ant communities in Portugal. [Abstract.] P. 87 in: Demeter, A., Peregovits, L. (eds.) Ecological processes: current status and perspectives. Abstracts. 7th European Ecological Congress. Budapest: Hungarian Biological Society, 294 pp. [1995.08.20]

Palacio, E. E., Fernández, F. 1995. Hormigas de Colombia V: nuevos registros. Tacaya 4:6-7. [1995.11]

Palenitschko, Z. G. 1927. Zur vergleichenden Variabilität der Arten und Kasten bei den Ameisen. Z. Morphol. Ökol. Tiere 9:410-438. [1927.11.23]

Palma, R. L., Lovis, P. M., Tother, C. 1989. An annotated list of primary types of the phyla Arthropoda (except Crustacea) and Tardigrada held in the National Museum of New Zealand. Natl. Mus. N. Z. Misc. Ser. 20:1-49. [1989.09] [Formicidae p. 38.]

Palomeque, T., Chica, E., Cano, M. A., Díaz de la Guardia, R. 1988. Karyotypes, C-banding, and chromosomal location of active nucleolar organizing regions in *Tapinoma* (Hymenoptera, Formicidae). Genome 30:277-280. [1988.04]

Palomeque, T., Chica, E., Cano, M. A., Díaz de la Guardia, R. 1990. Development of silver stained structures during spermatogenesis in different genera of Formicidae. Genetica 81:51-58. [1990.04]

Palomeque, T., Chica, E., Cano, M. A., Díaz de la Guardia, R., Tinaut, A. 1989 ("1988"). Cytogenetic studies in the genera *Pheidole* and *Tetramorium* (Hymenoptera, Formicidae, Mirmicinae). Caryologia 41:289-298. [1989.02.28]

Palomeque, T., Chica, E., Díaz de la Guardia, R. 1990. Karyotype, C-banding, chromosomal location of active nucleolar organizing regions, and B-chromosomes in *Lasius niger* (Hymenoptera, Formicidae). Genome 33:267-272. [1990.04]

Palomeque, T., Chica, E., Díaz de la Guardia, R. 1993a. Karyotype evolution and chromosomal relationship between several species of the genus *Aphaenogaster* (Hymenoptera, Formicidae). Caryologia 46:25-40. [1993.05.10]

Palomeque, T., Chica, E., Díaz de la Guardia, R. 1993b. Supernumerary chromosome segments in different genera of Formicidae. Genetica 90:17-29. [1993.09]

Pamilo, P. 1982. Multiple mating in *Formica* ants. Hereditas (Lund) 97:37-45. [1982.09.01]

Pamilo, P. 1983. Genetic differentiation within subdivided populations of *Formica* ants. Evolution 37:1010-1022. [1983.09.13]

Pamilo, P. 1987. Population genetics of the *Formica rufa* group. Pp. 68-70 in: Eder, J., Rembold, H. (eds.) Chemistry and biology of social insects. München: Verlag J. Peperny, xxxv + 757 pp. [1987]

Pamilo, P. 1990. Sex allocation and queen-worker conflict in polygynous ants. Behav. Ecol. Sociobiol. 27:31-36. [1990.07.17]

Pamilo, P., Chautems, D., Cherix, D. 1992. Genetic differentiation of disjunct populations of the ants *Formica aquilonia* and *Formica lugubris* in Europe. Insectes Soc. 39:15-29. [1992]

Pamilo, P., Rosengren, R. 1983. Sex ratio strategies in *Formica* ants. Oikos 40:24-35. [1983.01]

Pamilo, P., Rosengren, R. 1984. Evolution of nesting strategies of ants: Genetic evidence from different population types of *Formica* ants. Biol. J. Linn. Soc. 21:331-348. [1984.10.10]

Pamilo, P., Rosengren, R., Vepsäläinen, K., Varvio-Aho, S.-L., Pisarski, B. 1978. Population genetics of *Formica* ants. I. Patterns of enzyme gene variation. Hereditas (Lund) 89:233-248. [1978.12.20]

Pamilo, P., Sundström, L., Fortelius, W., Rosengren, R. 1994. Diploid males and colony-level selection in *Formica* ants. Ethol. Ecol. Evol. 6:221-235. [1994.07]

Pamilo, P., Vepsäläinen, K. 1977. Heretical notes on the taxonomy of *Formica* s. str. (Hym.). Pp. 128-129 in: Velthuis, H. H. W., Wiebes, J. T. (eds.) Proceedings of the Eighth International Congress of the IUSSI, Wageningen, The Netherlands, September 5-10, 1977. Wageningen: Centre for Agricultural Publishing and Documentation, 325 pp. [1977]

Pamilo, P., Vepsäläinen, K., Rosengren, R. 1975. Low allozymic variability in *Formica* ants. Hereditas (Lund) 80:293-296. [1975.09.26]

Pamilo, P., Vepsäläinen, K., Rosengren, R. 1979. Population genetics of *Formica* ants II. Genic differentiation between species. Ann. Entomol. Fenn. 45:65-76. [1979.11.30]

Panzer, G. W. F. 1798. Fauna insectorum germanicae initia, oder Deutschlands Insecten. Heft 54. Nürnberg: Felssecker, 24 pp., 24 pls. [1798] [Publication date from Sherborn (1923:566).]

Panzer, G. W. F. 1801. Nachricht von einem neuen entomolischen [sic] Werke, des Hrn. Prof. Jurine in Geneve. Intelligenzbl. Litt.-Ztg. Erlangen 1:161-165. [1801.05.30] [See also Jurine (1801) and Morice & Durrant (1915).]

Paramonov, S. Y. 1941. V. O. Karavaiev. Zb. Prats Zool. Muz. 24:3-8.

Parapura, E. 1972. An ergatandromorph *Formica exsecta* Nyl. (Hymenoptera, Formicidae) from Poland. Bull. Acad. Pol. Sci. Sér. Biol. 20:763-767. [1972.11]

Parapura, E., Pisarski, B. 1971. Mrówki (Hymenoptera, Formicidae) Bieszczadów. Fragm. Faun. (Warsaw) 17:319-356. [1971.11.15]

Paraschivescu, D. 1960. Untersuchungen über die individuellen Standortsformen der Art *Messor structor* Latr. (Hym. Formicidae). Rev. Biol. Acad. Répub. Pop. Roum. 5:227-231. [1960]

Paraschivescu, D. 1961. Contributii la cunoasterea formicidelor din stepa si podisul Dobrogei. Stud. Cercet. Biol. Ser. Biol. Anim. 13:457-465. [1961]

Paraschivescu, D. 1963. Cercetari zoogeografice asupra formicidelor din bazinul Trotusului. Comun. Acad. Repub. Pop. Rom. 13:559-566. [1963.06]

Paraschivescu, D. 1967. Cercetari asupra faunei de Formicide din regiunea Portile de Fier. Stud. Cercet. Biol. Ser. Zool. 19:393-403. [1967]

Paraschivescu, D. 1969. Geographische Verbreitung der Formiciden in Rumänien (Hymenoptera). Pp. 221-232 in: Ernst, E. et al. (eds.) International Union for the Study of Social Insects. VI Congress. Bern 15-20 September 1969. Proceedings. Bern: Organizing Committee of the VI Congress IUSSI, 309 pp. [1969.09.15]

Paraschivescu, D. 1972a. Fauna mirmecologica din zonele saline ale României. Stud. Cercet. Biol. Ser. Zool. 24:489-495. [1972]

Paraschivescu, D. 1972b. Pozitia sistematica a speciilor *Formica pratensis* Retz. si *F. nigricans* Em. (Hym. Formicidae). Stud. Cercet. Biol. Ser. Zool. 24:527-535. [1972]

Paraschivescu, D. 1972c. Die Ameisenfauna des Naturschutzgebietes im Retezatgebirge (Südkarpaten). Waldhygiene 9:213-222. [1972]

Paraschivescu, D. 1974. Die Fauna der Formiciden in dem Gebiet um Bukarest. Trav. Mus. Hist. Nat. "Grigore Antipa" 15:297-302. [1974]

Paraschivescu, D. 1975a. Investigations upon the Formicidae belonging to the collections of the "Gr. Antipa" Museum, Bucharest. Trav. Mus. Hist. Nat. "Grigore Antipa" 16:187-192. [1975]

Paraschivescu, D. 1975b. Hymenoptera - Formicidae. Pp. 187-189 in: Ionescu, M. (ed.) Fauna. Academia Republicii Socialiste România. Grupul de cercetari complexe "Portile de Fier". Seria monografica. Bucharest: Editura Academiei Republicii Socialiste România, 316 pp. [1975]

Paraschivescu, D. 1978a. *Monomorium pharaonis* (L.) o noua specie în mirmecofauna R. S. România si importanta el economica. Nymphaea 6:459-461. [1978]

Paraschivescu, D. 1978b. Elemente balcanice în mirmecofauna R. S. România. Nymphaea 6:463-474. [1978]

Paraschivescu, D. 1982. Cercetari mirmecologice în unele localitati din Muntii Apuseni (Brad, Risculita, Cîmpeni, Abrud, Gîrda). Nymphaea 10:255-262. [1982]

Paraschivescu, D. 1993. Untersuchungen zur Ameisenfauna von Ufersandzonen im rumänischen Donaubereich. Waldhygiene 20:21-27. [1993]

Paraschivescu, D., Arcasu, C. R. 1976. Mirmecofauna vaii Crisului Repede. Nymphaea 4:161-167. [1976]

Pardo, X., Tormos, J., Sendra, A. 1986 ("1985"). Mirmecofauna de les Suredes valencianes. Misc. Zool. (Barc.) 9:251-256. [1986]

Pardo Vargas, R. 1964. Clave para identificar los Formicidae de la Provincia de Chiclayo. Rev. Peru. Entomol. 7:98-102. [1964.12]

Parfitt, E. 1880. The fauna of Devon. Hymenoptera. Section Aculeata. Rep. Trans. Devon. Assoc. Adv. Sci. Lit. Art 12:501-559. [1880] [Ants pp. 506-510, 513-518, 559.]

Parker, G. H. 1938. Biographical memoir of William Morton Wheeler 1865-1937. Biogr. Mem. Natl. Acad. Sci. 19:203-241. [1938]

Pascovici, V. D. 1979. Espèces du groupe *Formica rufa* L. de la Rep. Soc. de Roumanie et leur utilisation dans la lutte contre les ravageurs forestiers. Bull. SROP 1979(II-3):111-134. [1979]

Pascovici, V. D. 1981. Le specie del gruppo *Formica rufa* L. in Romania e la loro utilizzazione nella lotta contro gli insetti distruttori delle foreste. Collana Verde 59:165-183. [1981]

Pascovici, V. D., Ronchetti, G. 1965. Il gruppo *Formica rufa* in Romania. Collana Verde 16:297-304. [1965.09.10]

Pascual, R., Tinaut, A. 1985. Evolución anual de las mirmecocenosis en las sierras de La Alfaguara y Harana (Granada. España) (HYM. Formicidae). Bol. Soc. Port. Entomol. Supl. 1(4):31-40. [1985]

Passera, L. 1967. Inhibitions interspécifiques déclenchées par les femelles fécondes des fourmis parasites *Plagiolepis xene* St. et *Plagiolepis grassei* Le Mas. Pas. C. R. Séances Acad. Sci. Ser. D. Sci. Nat. 265:1721-1724. [1967.12.04]

Passera, L. 1969 ("1968"). Observations biologiques sur la fourmi *Plagiolepis grassei* Le Masne Passera, parasite social de *Plagiolepis pygmaea* Latr. (Hym. Formicidae). Insectes Soc. 15:327-336. [1969.03]

Passera, L. 1970 ("1969"). Biologie de la reproduction chez *Plagiolepis pygmaea* Latreille et ses deux parasites sociaux *Plagiolepis grassei* Le Masne et Passera et *Plagiolepis xene* Stärcke (Hymenoptera, Formicidae). Ann. Sci. Nat. Zool. Biol. Anim. (12)11:327-481. [1970.06]

Passera, L. 1979 ("1977"). Peuplement myrmécologique du cordon littoral du Languedoc-Roussillon. Modifications anthropiques. Vie Milieu Sér. C Biol. Terr. 27:249-265. [1979.02]

Passera, L. 1984. L'organisation sociale des fourmis. Toulouse: Editions Privat, 360 pp. [1984.09]

Passera, L. 1985. Soldier determination in ants of the genus *Pheidole*. Pp. 331-346 in: Watson, J. A. L., Okot-Kotber, B. M., Noirot, C. (eds.) Caste differentiation in social insects. Oxford: Pergamon Press, xiv + 405 pp. [1985]

Passerin d'Entrèves, P. 1983. "Faunae Ligusticae Fragmenta" e "Insectorum Liguriae species novae" di Massimiliano Spinola: note bibliografiche. Boll. Mus. Reg. Sci. Nat. Torino 1:215-226. [1983.12.23]

Pasteels, J. M., Crewe, R., Blum, M. 1970. Étude histologique et examen au microscope électronique à balayage de la glande sécrétant la phéromone de piste chez deux *Crematogaster* nord-américains (Formicidae, Myrmicinae). C. R. Séances Acad. Sci. Ser. D. Sci. Nat. 271:835-838. [1970.09.14]

Patrizi, S. 1946. Contribuzioni alla conoscenza delle formiche e dei mirmecofili dell'Africa orientale. Boll. Ist. Entomol. Univ. Studi Bologna 15:292-296. [1946.12.24]

Patrizi, S. 1947. Contribuzioni alla conoscenza delle formiche e dei mirmecofili dell'Africa orientale. II. *Microdaceton leakeyi* n. sp. (Hymenoptera-Formicidae). Boll. Ist. Entomol. Univ. Studi Bologna 16:219-221. [1947.11.15]

Patrizi, S. 1948. Contribuzioni alla conoscenza delle formiche e dei mirmecofili dell'Africa orientale. VI. *Crateropsis elmenteitae* nuovo sottogenere aberrante di *Solenopsis* Westw. (Hymenoptera Formicidae). Boll. Ist. Entomol. Univ. Studi Bologna 17:174-176. [1948.11.27]

Patton, W. H. 1879. A gall-inhabiting ant. Am. Nat. 13:126. [1879.02.04]

Patton, W. H. 1894. Habits of the leaping-ant of southern Georgia. Am. Nat. 28:618-619. [1894.07]

Pavan, M. 1955. Studi sui Formicidae. I. Contributo alla conoscenza degli organi gastrali dei Dolichoderinae. Natura (Milan) 46:135-145. [1955.09]

Pavan, M. 1959. Attivita' italiana per la lotta biologica con formiche del gruppo *Formica rufa* contro gli insetti dannosi alle foreste. Collana Verde 4:1-79. [1959.12.30]

Pavan, M. 1979. Significance of ants of the *Formica rufa* group in Italy in ecological forestry regulation. Bull. SROP 1979(II-3):161-169. [1979]

Pavan, M. 1981. Utilità delle formiche del gruppo *Formica rufa* (2a edizione aggiornata). Collana Verde 57:1-99. [1981]

Pavan, M., Ronchetti, G. 1955. Studi sulla morfologia esterna e anatomia interna dell'operaia di *Iridomyrmex humilis* Mayr e ricerche chimiche e biologiche sulla iridomirmecina. Atti Soc. Ital. Sci. Nat. Mus. Civ. Stor. Nat. Milano 94:379-477. [1955.12]

Pavan, M., Ronchetti, G. 1972. Le formiche del gruppo *Formica rufa* in Italia nell'assestamento ecologico forestale. Waldhygiene 9:223-238. [1972]

Pavan, M., Ronchetti, G., Vendegna, V. 1971. Corologia del gruppo *Formica rufa* in Italia (Hymenoptera: Formicidae). Collana Verde 30:1-93. [1971]

Pearson, B. 1981. The electrophoretic determination of *Myrmica rubra* microgynes as a social parasite: possible significance in the evolution of ant social parasites. Pp. 75-84 in: Howse, P. E., Clement, J.-L. (eds.) Biosystematics of social insects. Systematics Association Special Volume No. 19. London: Academic Press, 346 pp. [1981]

Pearson, B. 1982. The taxonomic status of morphologically-anomalous ants in the *Lasius niger/Lasius alienus* taxon. Insectes Soc. 29:95-101. [1982.06]

Pearson, B. 1983. The classification of morphologically intermediate ants in the *Lasius alienus/Lasius niger* taxon. Further observations. Insectes Soc. 30:100-105. [1983.07]

Pearson, B. 1984 ("1983"). Hybridisation between the ant species *Lasius niger* and *Lasius alienus*: the genetic evidence. Insectes Soc. 30:402-411. [1984.04]

Pearson, B., Child, A. R. 1980. The distribution of an esterase polymorphism in macrogynes of *Myrmica rubra* Latreille. Evolution 34:105-109. [1980.03.19]

Peeters, C. 1984. Could the queenless ant *Ophthalmopone berthoudi* re-evolve a reproductive caste from its fertile workers? [Abstract.] P. 515 in: XVII International Congress of Entomology. Hamburg, Federal Republic of Germany, August 20-26, 1984. Abstract Volume. Hamburg: XVII International Congress of Entomology, 960 pp. [1984.08.10]

Peeters, C. 1987. The diversity of reproductive systems in ponerine ants. Pp. 253-254 in: Eder J., Rembold H. (eds.) Chemistry and biology of social insects. München: Verlag J. Peperny, xxxv + 757 pp. [1987]

Peeters, C. 1991a. The occurrence of sexual reproduction among ant workers. Biol. J. Linn. Soc. 44:141-152. [1991.10.30]

Peeters, C. 1991b. Ergatoid queens and intercastes in ants: two distinct adult forms which look morphologically intermediate between workers and winged queens. Insectes Soc. 38:1-15. [1991]

Peeters, C. 1993. Monogyny and polygyny in ponerine ants with or without queens. Pp. 234-261 in: Keller, L. (ed.) Queen number and sociality in insects. Oxford: Oxford University Press, 451 pp. [1993]

Peeters, C., Billen, J. P. J. 1991. A novel exocrine gland inside the thoracic appendages ('gemmae') of the queenless ant *Diacamma australe*. Experientia (Basel) 47:229-231. [1991.03.15]

Peeters, C., Billen, J. P. J., Hölldobler, B. 1992. Alternative dominance mechanisms regulating monogyny in the queenless ant genus *Diacamma*. Naturwissenschaften 79:572-573. [1992.12]

Peeters, C., Crewe, R. M. 1985. Worker reproduction in the ponerine ant *Ophthalmopone berthoudi*: an alternative form of eusocial organization. Behav. Ecol. Sociobiol. 18:29-37. [1985.11]

Peeters, C., Crewe, R. M. 1986. Queenright and queenless breeding systems within the genus *Pachycondyla* (Hymenoptera: Formicidae). J. Entomol. Soc. South. Afr. 49:251-255. [1986.10]

Peeters, C., Crewe, R. M. 1987. Foraging and recruitment in ponerine ants: solitary hunting in the queenless *Ophthalmopone berthoudi* (Hymenoptera: Formicidae). Psyche (Camb.) 94:201-214. [1987.10.06]

Peeters, C., Crozier, R. H. 1989 ("1988"). Caste and reproduction in ants: not all mated egg-layers are "queens". Psyche (Camb.) 95:283-288. [1989.06.13]

Peeters, C., Higashi, S. 1989. Reproductive dominance controlled by mutilation in the queenless ant *Diacamma australe*. Naturwissenschaften 76:177-180. [1989.04]

Peeters, C., Hölldobler, B. 1992. Notes on the morphology of the sticky "doorknobs" of larvae in an Australian *Hypoponera* sp. (Formicidae; Ponerinae). Psyche (Camb.) 99:23-30. [1992.12.04]

Peeters, C., Hölldobler, B. 1995. Reproductive cooperation between queens and their mated workers: the complex life history of an ant with a valuable nest. Proc. Natl. Acad. Sci. U. S. A. 92:10977-10979. [1995.11.21] [Reproductive biology of *Harpegnathos saltator*.]

Penev, S., Ronchetti, G. 1965. Ricerche sulla distribuzione delle formiche del gruppo *Formica rufa* nella Bulgaria nord-orientale. Collana Verde 16:305-311. [1965.09.10]

Penzig, O. 1895 ("1894"). Note di biologie vegetale. I. Sopra una nuova pianta formicaria d'Africa (*Stereospermum dentatum* Rich.). Malpighia 8:466-471. [1895]

Peregrine, D. J., Mudd, A., Cherrett, J. M. 1974 ("1973"). Anatomy and preliminary chemical analysis of the post-pharyngeal glands of the leaf-cutting ant, *Acromyrmex octospinosus* (Reich) (Hym., Formicidae). Insectes Soc. 20:355-363. [1974.03]

Perfecto, I., Snelling, R. R. 1995. Biodiversity and the transformation of a tropical agroecosystem: ants in coffee plantations. Ecol. Appl. 5:1084-1097. [1995.11]

Perfecto, I., Vandermeer, J. 1994. Understanding biodiversity loss in agroecosystems: reduction of ant diversity resulting from transformation of the coffee ecosystem in Costa Rica. Entomol. (Trends Agric. Sci.) 2:7-13.

Pergande, T. 1893. On a collection of Formicidae from Lower California and Sonora, Mexico. Proc. Calif. Acad. Sci. (2)4:26-36. [1893] [Pagebase date 19 September 1893 but article said to be "in press" in the annual report for 1893 presented at the Academy meeting of 2 January 1894. Paper read by title at the Academy meeting of 6 November 1893. Whole volume issued 28 September 1894.]

Pergande, T. 1894. Formicidae of Lower California, Mexico. Proc. Calif. Acad. Sci. (2)4:161-165. [> 1894.05.17] [Article is likely to have been published shortly after the pagebase date of 17 May 1894. Whole volume issued 28 September 1894.]

Pergande, T. 1896. Mexican Formicidae. Proc. Calif. Acad. Sci. (2)5:858-896. [< 1896.01.18] [Pagebase date is 30 December 1895. Publication is likely to have occurred within a week or two of this date. Whole volume issued 18 January 1896.]

Pergande, T. 1900. Papers from the Harriman Alaska expedition. XVII. Entomological results (11): Formicidae. Proc. Wash. Acad. Sci. 2:519-521. [1900.12.20]

Perkins, R. C. L. 1891. Male and worker characters combined in the same individual of *Stenamma westwoodi*. Entomol. Mon. Mag. 27:123-124. [1891.05]

Perkins, R. C. L. 1924. The aculeate Hymenoptera of Devon. Rep. Trans. Devon. Assoc. Adv. Sci. Lit. Art 55:188-241. [1924] [Ants pp. 238-241.]

Perrault, G. H. 1976. Description des ouvrières et des soldats de *Oligomyrmex tahitiensis* Wheeler. Mise au point concernant les sexués. Nouv. Rev. Entomol. 6:303-307. [1976.11.30]

Perrault, G. H. 1981. *Proceratium deelemani*, nouvelle espèce de Kalimantan. Nouv. Rev. Entomol. 11:189-193. [1981.06.15]

Perrault, G. H. 1986a. *Gymnomyrmex villiersi*, nouvelle espèce de Guyane (Hymenoptera, Formicidae). Rev. Fr. Entomol. (Nouv. Sér.) 8:1-4. [1986.04.10]

Perrault, G. H. 1986b. *Gnamptogenys falcifera* Kempf, 1967. Description de l'ouvrière et levée d'un doute (Hymenoptera, Formicidae). Rev. Fr. Entomol. (Nouv. Sér.) 8:157-159. [1986.12.16]

Perrault, G. H. 1988a ("1987"). Notes sur des types de *Pseudomyrmex* décrits par F. Smith. I. (Hymenoptera, Formicidae). Nouv. Rev. Entomol. (n.s.)4:381-385. [1988.02.05]

Perrault, G. H. 1988b ("1987"). Les fourmis de Tahiti. Bull. Soc. Zool. Fr. 112:429-446. [1988.03]

Perrault, G. H. 1988c. *Octostrumma* [sic] *betschi*, n. sp. de Guyane Française (Hymenoptera, Formicidae). Rev. Fr. Entomol. (Nouv. Sér.) 10:303-307. [1988.12.15]

Perrault, G. H. 1993. Peuplement en fourmis de l'atoll de Fangataufa. Bull. Soc. Entomol. Fr. 98:323-338. [1993.12.01]

Perris, E. 1876. Nouvelles promenades entomologiques. [part] Ann. Soc. Entomol. Fr. (5)6:177-244. [1876.10.11] [First part of the article (pp. 171-176) was published 24 July 1876.]

Perris, E. 1878 ("1877"). Rectifications et additions à mes promenades entomologiques. Ann. Soc. Entomol. Fr. (5)7:379-386. [1878.04.10]

Perty, M. 1833. Delectus animalium articulatorum, quae in itinere per Brasiliam annis MDCCCXVII-MDCCCXX jussu et auspiciis Maximiliani Josephi I. Bavariae regis augustissimi peracto, collegerunt Dr. J. B. Spix et Dr. C. F. Ph. de Martius. Fasc. 3. Monachii [= Munich]: published by the author, pp. 125-224. [1833]

Péru, L. 1984. Individus tératologiques chez les fourmis *Leptothorax*. Actes Colloq. Insectes Soc. 1:141-149. [1984.09]

Petal, J. 1961. Materialy do znajomosci mrówek (Formicidae) Lubelszczyzny. (I-IV). Fragm. Faun. (Warsaw) 9:135-151. [1961.12.15]

Petal, J. 1963a ("1962"). *Formica forsslundi* Lohm. ssp. *strawinskii* n. ssp. Ann. Univ. Mariae Curie-Sklodowska Sect. C Biol. 17:195-202. [1963.03.16]

Petal, J. 1963b. Materialy do znajomosci mrówek (Formicidae, Hymenoptera) Lubelszczyzny (V-VI). Fragm. Faun. (Warsaw) 10:463-472. [1963.09.16]

Petal, J. 1963c. Données pour la morphologie de *Myrmica rugulosoides* For. et *Leptothorax nigrescens* Ruzsky (Hymenoptera, Formicidae). Bull. Acad. Pol. Sci. Sér. Biol. 11:379-382. [1963.10]

Petal, J. 1964 ("1963"). Fauna mrówek projektowanego rezerwatu torfowiskowego Rakowskie Bagno k. Frampola (woj. lubelskie). Ann. Univ. Mariae Curie-Sklodowska Sect. C Biol. 18:143-173. [1964.04.30]

Petal, J. 1968 ("1967"). Materialy do znajomosci mrówek (Formicidae, Hymenoptera) Lubelszczyzny. VII. Zespoly mrówek srodowisk torfowiskowych, lesnych i wydmowych okolic Libiszowa (pow. Parczew). Ann. Univ. Mariae Curie-Sklodowska Sect. C Biol. 22:117-127. [1968.11.27]

Petal, J., Pisarski, B. 1993. Formicidae. Pp. 253-262 in: Górny, M., Grüm, L. (eds.) Methods in soil zoology. Amsterdam: Elsevier, xii + 459 pp. [1993]

Petersen, B. 1968. Some novelties in presumed males of Leptanillinae (Hym., Formicidae). Entomol. Medd. 36:577-598. [1968.12.30]

Petersen-Braun, M., Buschinger, A. 1975. Entstehung und Funktion eines thorakalen Kropfes bei Formiciden-Königinnen. Insectes Soc. 22:51-66. [1975.06]

Petralia, R. S., Haut, C. F. 1986. Morphology of the labial gland system of the mature larva of the black carpenter ant, *Camponotus pennsylvanicus* (DeGeer). Proc. Iowa Acad. Sci. 93:16-20. [1986.03.25]

Petralia, R. S., Vinson, S. B. 1979. Developmental morphology of larvae and eggs of the imported fire ant, *Solenopsis invicta*. Ann. Entomol. Soc. Am. 72:472-484. [1979.07.15]

Petralia, R. S., Vinson, S. B. 1980 ("1979"). Comparative anatomy of the ventral region of ant larvae, and its relation to feeding behavior. Psyche (Camb.) 86:375-394. [1980.12.19]

Petrov, I. Z. 1986a. Distribution of species of the genus *Cataglyphis* Foerster, 1850 (Formicidae, Hymenoptera) in Yugoslavia. Arh. Biol. Nauka 38:11-12.

Petrov, I. Z. 1986b. Prilog poznavanju faune mrava (Formicidae, Hymenoptera) nekih hrastovih zajednica na Jastrepcu. Glas. Prir. Muz. Beogradu Ser. B Biol. Nauke 41:109-114. [1986]

Petrov, I. Z. 1992. Myrmecofauna (Formicidae, Hymenoptera) of Serbia - up to now investigations. Glas. Prir. Muz. Beogradu Ser. B Biol. Nauke 47:247-259. [Not seen. Cited in Zoological Record for 1994/1995.]

Petrov, I. Z., Collingwood, C. A. 1992. Survey of the myrmecofauna (Formicidae, Hymenoptera) of Yugoslavia. Arch. Biol. Sci. (Belgrade) 44:79-91.

Petrov, I. Z., Collingwood, C. A. 1993. *Formica balcanica* sp. n., a new species related to the *Formica cinerea*-group (Hymenoptera: Formicidae). Eur. J. Entomol. 90:349-354. [1993.10.25]

Petrov, I. Z., Mesaros, G. 1989 ("1988"). Prilog poznavanju faune mrava (Formicidae, Hymenoptera) Stare Planine. Acta Biol. Iugosl. Ser. G Biosist. 14(1):43-50. [1989] [Dated "1988" but last paper in this issue was accepted for publication 29 March 1989.]

Pfitzer, D. W. 1951. A new species of *Smithistruma* from Tennessee (Hymenoptera: Formicidae). J. Tenn. Acad. Sci. 26:198-200. [1951.07]

Phillips, R. A. 1921. Nests of the ant *Stenamma westwoodi* discovered in Ireland. Ir. Nat. 30:125-127. [1921.11]

Phillips, S. A., Jr., Rogers, W. M., Wester, D. B., Chandler, L. 1987. Ordination analysis of ant faunae along the range expansion front of the red imported fire ant in south-central Texas. Tex. J. Agric. Nat. Resour. 1:11-15. [1987.06]

Phillips, S. A., Jr., Vinson, S. B. 1980. Comparative morphology of glands associated with the head among castes of the red imported fire ant, *Solenopsis invicta* Buren. J. Ga. Entomol. Soc. 15:215-226. [1980.04]

Picquet, N. 1958. Contribution à l'étude des larves de Formicidae de la Côte-d'Or. Trav. Lab. Zool. Stn. Aquic. Grimaldi Fac. Sci. Dijon 23:1-48. [1958.01.30]

Piek, T. 1992. A toxinological argument in favour of the close relationship of Vespidae, Scoliidae, Tiphiidae, Mutillidae and Formicidae (Hymenoptera). Proc. Sect. Exp. Appl. Entomol. Neth. Entomol. Soc. 3:99-104. [1992]

Pierce, W. D., Gibron, J., Sr. 1962. Fossil arthropods of California. 24. Some unusual fossil arthropods from the Calico Mountains nodules. Bull. South. Calif. Acad. Sci. 61:143-151. [1962.10.24] [See Najt (1987).]

Pignalberi, C. T. 1961. Contribución al conocimiento de los formícidos de la provincia de Santa Fé. Pp. 165-173 in: Actas y trabajos del primer Congreso Sudamericano de Zoología (La Plata, 12-24 octubre 1959). Tomo III. Buenos Aires: Librart, 276 pp. [1961.09.08]

Pijoan, M. 1928. Formicidae of southern California. J. Entomol. Zool. 20:37-42. [1928.06]

Pisarski, B. 1953. Mrówki okolic Kazimierza. Fragm. Faun. (Warsaw) 6:465-500. [1953.06.25]

Pisarski, B. 1957. O wystepowaniu egzotycznych gatunków mrówek w Polsce. Fragm. Faun. (Warsaw) 7:283-288. [1957.04.15]

Pisarski, B. 1960. Dwa nowe gatunki z rodzaju *Paratrechina* Motschulsky (Hymenoptera, Formicidae). Ann. Zool. (Warsaw) 18:349-356. [1960.04.20]

Pisarski, B. 1961. Studien über die polnischen Arten der Gattung *Camponotus* Mayr (Hymenoptera, Formicidae). Ann. Zool. (Warsaw) 19:147-207. [1961.02.10]

Pisarski, B. 1962a ("1961"). Nouvelle espèce de *Cataglyphis* Först. (Formicidae) de l'Inde. Bull. Acad. Pol. Sci. Sér. Biol. 9:515-516. [1962.01]

Pisarski, B. 1962b. Materialy do znajomosci mrówek (Formicidae, Hymenoptera) Polski. I. Gatunki z podrodzaju *Coptoformica* Müll. Fragm. Faun. (Warsaw) 10:125-136. [1962.08.20]

Pisarski, B. 1962c. Notes synonymiques sur les espèces balcaniques du genre *Cardiocondyla* Emery (Hymenoptera, Formicidae). Bull. Acad. Pol. Sci. Sér. Biol. 10:331-333. [1962.08]

Pisarski, B. 1962d. Sur *Sifolinia pechi* Sams. trouvée en Pologne (Hymenoptera, Formicidae). Bull. Acad. Pol. Sci. Sér. Biol. 10:367-369. [1962.10]

Pisarski, B. 1963. Nouvelle espèce du genre *Harpagoxenus* For. de la Mongolie (Hymenoptera, Formicidae). Bull. Acad. Pol. Sci. Sér. Biol. 11:39-41. [1963.02]

Pisarski, B. 1965. Les fourmis du genre *Cataglyphis* Foerst. en Irak (Hymenoptera, Formicidae). Bull. Acad. Pol. Sci. Sér. Biol. 13:417-422. [1965.09]

Pisarski, B. 1966. Études sur les fourmis du genre *Strongylognathus* Mayr (Hymenoptera, Formicidae). Ann. Zool. (Warsaw) 23:509-523. [1966.12.10]

Pisarski, B. 1967a. Fourmis (Hymenoptera: Formicidae) d'Afghanistan récoltées par M. Dr. K. Lindberg. Ann. Zool. (Warsaw) 24:375-425. [1967.04.20]

Pisarski, B. 1967b. Ameisen (Formicidae) von Dr. J. Klapperich in Afghanistan gesammelt. Pol. Pismo Entomol. 37:47-51. [1967.08.15]

Pisarski, B. 1969a. Fourmis (Hymenoptera: Formicidae) de la Mongolie. Fragm. Faun. (Warsaw) 15:221-236. [1969.10.10]

Pisarski, B. 1969b. 175. Myrmicidae und Formicidae. Ergebnisse der zoologischen Forschungen von Dr. Z. Kaszab in der Mongolei (Hymenoptera). Faun. Abh. (Dres.) 2(29):295-316. [1969]

Pisarski, B. 1970a ("1969"). Beiträge zur Kenntnis der Fauna Afghanistans. (Sammelergebnisse von O. Jakes 1963-64, D. Povolny 1965, D. Povolny & Fr. Tenora 1966, J. Simek 1965-66, D. Povolny, J. Gaisler, Z. Sebek & Fr. Tenora 1967). Formicidae, Hym. Cas. Morav. Mus. Brne 54(Suppl.):305-326. [1970.05.15]

Pisarski, B. 1970b. Badania entomofaunistyczne Instytutu Zoologicznego PAN w Karpatach. Pol. Pismo Entomol. 40:631-635. [1970.12]

Pisarski, B. 1971a. Nouvelles espèces de fourmis (Hymenoptera, Formicidae) du sous-genre *Tanaemyrmex* Ashm. d'Iraq. Bull. Acad. Pol. Sci. Sér. Biol. 19:671-675. [1971.11]

Pisarski, B. 1971b. Les fourmis du genre *Camponotus* Mayr (Hymenoptera, Formicidae) d'Iraq. Bull. Acad. Pol. Sci. Sér. Biol. 19:727-731. [1971.11]

Pisarski, B. 1973. Struktura spoleczna *Formica* (*C.*) *exsecta* Nyl. (Hymenoptera: Formicidae) i jej wplyw na morfologie, ekologie i etologie gatunku. Warszawa: Polska Akademia Nauk, Instytut Zoologiczny, 134 pp. [1973]

Pisarski, B. 1975. Mrówki Formicoidea. Kat. Fauny Pol. 26:3-85. [1975.12]

Pisarski, B. 1981a. Zoocenologiczne podstawy ksztaltowania srodowiska przyrodniczego osiedla mieszkaniowego Bialoleka Dworska w Warszawie. I. Sklad gatunkowy i struktura fauny terenu projektowanego osiedla mieszkaniowego. Mrówki (Formicidae, Hymenoptera). Fragm. Faun. (Warsaw) 26:341-354. [1981.12.31]

Pisarski, B. 1981b. Intraspecific variations in ants of the genus *Formica* L. Pp. 17-25 in: Howse, P. E., Clement, J.-L. (eds.) Biosystematics of social insects. Systematics Association Special Volume No. 19. London: Academic Press, 346 pp. [1981]

Pisarski, B., Blum, M. S. 1988. Pheromone differentiation in *Camponotus herculeanus* and *C. ligniperdus* (Hymenoptera: Formicidae). Ann. Zool. (Warsaw) 41:527-532. [1988.05.30]

Pisarski, B., Czechowski, W. 1987. Structure and origin of ant communities of Warsaw. P. 605 in: Eder, J., Rembold, H. (eds.) Chemistry and biology of social insects. München: Verlag J. Peperny, xxxv + 757 pp. [1987]

Pisarski, B., Czechowski, W. 1991. Ant communities (Hymenoptera, Formicoidea) of moist and wet deciduous forests of Central Europe. Fragm. Faun. (Warsaw) 35:167-172. [1991.12.15]

Pisarski, B., Krzysztofiak, L. 1981. Myrmicidae und Formicidae (Hymenoptera) aus der Mongolei, II. Folia Entomol. Hung. 42:155-166. [1981.05.14]

Pisarski, B., Vepsäläinen, K. 1989. Competition hierarchies in ant communities (Hymenoptera, Formicidae). Ann. Zool. (Warsaw) 42:321-328. [1989.05.31]

Piton, L. 1935a. [Untitled. Descriptions of new taxa: *Camponotus obesus* Piton nov. spec.; *Formica cantalica* Piton nov. spec.] Pp. 68-69 in: Piton, L., Théobald, N. La faune entomologique des gisements Mio-Pliocènes du Massif Central. Rev. Sci. Nat. Auvergne (n.s.)1:65-104. [1935.09.22]

Piton, L. 1935b. [Untitled. *Formica auxillacensis* Piton nov. spec.] P. 91 in: Piton, L., Théobald, N. La faune entomologique des gisements Mio-Pliocènes du Massif Central. Rev. Sci. Nat. Auvergne (n.s.)1:65-104. [1935.09.22]

Piton, L., Théobald, N. 1935. La faune entomologique des gisements Mio-Pliocènes du Massif Central. Rev. Sci. Nat. Auvergne (n.s.)1:65-104. [1935.09.22]

Plateaux, L. 1971 ("1970"). Sur le polymorphisme social de la fourmi *Leptothorax nylanderi* (Förster). I. Morphologie et biologie comparées des castes. Ann. Sci. Nat. Zool. Biol. Anim. (12)12:373-478. [1971.03]

Plateaux, L. 1976. Hybridation expérimentale de deux espèces de fourmis *Leptothorax*. Arch. Zool. Exp. Gén. 117:255-271. [1976]

Plateaux, L. 1977. L'isolement reproductif de quelques fourmis du genre *Leptothorax*. Pp. 130-131 in: Velthuis, H. H. W., Wiebes, J. T. (eds.) Proceedings of the Eighth International Congress of the IUSSI, Wageningen, The Netherlands, September 5-10, 1977. Wageningen: Centre for Agricultural Publishing and Documentation, 325 pp. [1977]

Plateaux, L. 1978a. L'essaimage de quelques fourmis *Leptothorax*: rôles de l'éclairement et de divers autres facteurs. Effet sur l'isolement reproductif et la répartition géographique. 1re partie, I à VI. Ann. Sci. Nat. Zool. Biol. Anim. (12)20:129-164. [1978.09]

Plateaux, L. 1978b. L'essaimage de quelques fourmis *Leptothorax*: rôles de l'éclairement et de divers autres facteurs. Effet sur l'isolement reproductif et la répartition géographique. 2e partie, suite et fin. Ann. Sci. Nat. Zool. Biol. Anim. (12)20:165-192. [1978.12]

Plateaux, L. 1980 ("1979"). Polymorphisme ovarien des reines de fourmis *Leptothorax*, variations interspecifiques, infériorité d'hybrides interspécifiques. Arch. Zool. Exp. Gén. 120:381-398. [1980]

Plateaux, L. 1981. The pallens morph of the ant *Leptothorax nylanderi*: description, formal genetics, and study of populations. Pp. 63-74 in: Howse, P. E., Clément, J.-L. (eds.) Biosystematics of social insects. Systematics Association Special Volume 19. London: Academic Press, 346 pp. [1981]

Plateaux, L. 1984. L'isolement reproductif chez les fourmis *Leptothorax* (Hyménoptères, Myrmicidae). Rev. Fac. Sci. Tunis 3:215-234.

Plateaux, L. 1986. Comparaison des cycles saisonniers, des durées des sociétés et des productions des trois espèces de fourmis *Leptothorax* (*Myrafant*) du groupe *nylanderi*. Actes Colloq. Insectes Soc. 3:221-234. [1986]

Plateaux, L. 1987. Reproductive isolation in ants of the genus *Leptothorax*, subgenus *Myrafant*. Pp. 33-34 in: Eder, J., Rembold, H. (eds.) Chemistry and biology of social insects. München: Verlag J. Peperny, xxxv + 757 pp. [1987]

Plateaux, L., Genermont, J. 1981. Modification du phénotype de fourmis *Leptothorax* par le régime alimentaire. Arch. Zool. Exp. Gén. 122:47-53. [1981]

Plaza, J., Tinaut, A. 1989. Descripción de los hormigueros de *Cataglyphis rosenhaueri* (Emery, 1906) y *Cataglyphis iberica* (Emery, 1906) en diferentes biotopos de la provincia de Granada (Hym. Formicidae). Bol. Asoc. Esp. Entomol. 13:109-116. [1989.11]

Pogorevici, N. 1947. Contributiuni la studiul faunei formicidelor din România. Not. Biol. Bucur. 5:279-286. [1947]

Pohl, I. E., Kollar, V. 1832. Brasiliens vorzüglich lästige Insecten. Besonderer Abdruck aus der Reise im Innern von Brasilien von Dr. Pohl. Wien: publisher/printer not given, 20 pp. [1832]

Poldi, B. 1961. Alcuni appunti su una rara formica (*Lasius bicornis* Först.) nuova per la Sardegna. Stud. Sassar. Sez. III 9:509-516. [1961]

Poldi, B. 1979. Un nuovo *Tetramorium* dell'Anatolia (Hymenoptera, Formicidae). Entomol. Basil. 4:499-503. [1979.10.10]

Poldi, B. 1980. Imenotteri formicidi della Brughiera di Rovasenda (Piemonte). In: Quaderni sulla "Struttura delle zoocenosi terrestri". 1. La brughiera pedemontana. III. Serie AQ/1/110. Collana del programma finalizzato "Promozione della qualità dell'ambiente". Roma: Consiglio Nazionale delle Ricerche, pp. 7-19.

Poldi, B. 1989. Studi sulla palude del Busatello (Veneto - Lombardia). 23. Gli imenotteri formicidi. Mem. Mus. Civ. Stor. Nat. Verona (II Ser.) Sez. Sci. Vita (A Biol.) 7:203-218. [1989.08]

Poldi, B. 1992. A poorly known taxon: *Diplorhoptrum orbulum* (Emery 1875). Ethol. Ecol. Evol. Spec. Issue 2:91-94.

Poldi, B. 1994. *Strongylognathus pisarskii* species nova (Hymenoptera, Formicidae). Memorabilia Zool. 48:187-191. [1994]

Poldi, B. 1995. [Untitled. *Tetramorium pelagium* Poldi n. sp.] P. 765 in: Mei, M. Arthropoda di Lampedusa, Linosa e Pantelleria (Canale di Sicilia, Mar Mediterraneo). Hymenoptera Formicidae (con diagnosi di due nuove specie). Nat. Sicil. (4)19(suppl.):753-772.

Poldi, B., Mei, M., Rigato, F. 1995 ("1994"). Hymenoptera Formicidae. Checklist Specie Fauna Ital. 102:1-10.

Pollock, G. B., Rissing, S. W. 1989. Intraspecific brood raiding, territoriality, and slavery in ants. Am. Nat. 133:61-70. [1989.01]

Pontin, A. J. 1957. The *Formica fusca* group (Hym., Formicidae) on Lundy Island. Entomol. Mon. Mag. 93:36. [1957.03.11]

Pontin, A. J. 1962. Notes on the collection of British ants in the Hope Department at Oxford. Entomol. Mon. Mag. 98:63. [> 1962.05] [Precise publication dates are not available for volume 98 of Entomol. Mon. Mag.]

Pontin, A. J. 1976a. The taxonomic status of *Lasius rabaudi* (Bondroit) (Hym., Formicidae). Entomol. Mon. Mag. 111:99-100. [1976.06.30]

Pontin, A. J. 1976b. *Anergates atratulus* (Schenck) (Hym., Formicidae) in Surrey. Entomol. Mon. Mag. 111:187. [1976.11.14]

Popov, V. 1933. Two new fossil ants from Caucasus (Hymenoptera, Formicidae). Tr. Paleozool. Inst. Akad. Nauk SSSR 2:17-21. [> 1933.06.17] [Date signed for printing ("podpisano k pechati").]

Popovici-Baznosanu, A. 1937. Die Variabilität der Waldameise in Rumänien. Zool. Anz. 117:280-282. [1937.03.15]

Porter, S. D. 1992. Frequency and distribution of polygyne fire ants (Hymenoptera: Formicidae) in Florida. Fla. Entomol. 75:248-257. [1992.06.26]

Porter, S. D., Bhatkar, A., Mulder, R., Vinson, S. B., Clair, D. J. 1991. Distribution and density of polygyne fire ants (Hymenoptera: Formicidae) in Texas. J. Econ. Entomol. 84:866-874. [1991.06]

Porter, S. D., Savignano, D. A. 1990. Invasion of polygyne fire ants decimates native ants and disrupts arthropod community. Ecology 71:2095-2106. [1990.12]

Potts, R. W. L. 1948. New records of *Ponera trigona* var. *opacior* Forel (Hymenoptera: Formicidae). Pan-Pac. Entomol. 24:26. [1948.03.31]

Poulton, E. B. 1922a. Notes on some Australian ants. Biological notes by E. B. Poulton, D.Sc., M.A., F.R.S., and notes and descriptions of new forms by W. C. Crawley, B.A., F.E.S., F.R.M.S. [part] Entomol. Mon. Mag. 58:118-120. [1922.05]

Poulton, E. B. 1922b. Notes on some Australian ants. Biological notes by E. B. Poulton, D.Sc., M.A., F.R.S., and notes and descriptions of new forms by W. C. Crawley, B.A., F.E.S., F.R.M.S. [concl.] Entomol. Mon. Mag. 58:121-126. [1922.06]

Poulton, E. B. 1933. The gentle Driver ant "Baongo", discovered by Miss Vinall at Bongandanga, Belgian Congo. Proc. Entomol. Soc. Lond. 7:58. [1933.04.21] [See also Santschi, 1933c.]

Presl, J. S. 1822. Additamenta ad faunam protogaem, sistens descriptiones aliquat animalium in succino inclusorum. Pp. 191-210 in: Presl, J. S., Presl, K. B. Deliciae Pragenses, historiam naturalem spectantes. Tome 1. Pragae: Calve, 244 pp. [1822.07] [Date from Stafleu & Cowan (1983).]

Pressick, M. L., Herbst, E. 1973. Distribution of ants on St. John, Virgin Islands. Caribb. J. Sci. 13:187-197. [1973.12]

Preuss, G. 1979a. *Formica (Serviformica) transkaukasica* Nass. - Neu für Rheinland-Pfalz und Westdeutschland. Pfälz. Heim. 30:11-12. [1979.03.11]

Preuss, G. 1979b. *Strongylognathus testaceus* (Schenk). - Vorkommen in Rheinland-Pfalz. Pfälz. Heim. 30:87. [1979.09.30]

Preuss, G. 1979c. *Formica truncorum* Fabr. 1804. - Erstnachweis für die Rheinpfalz und Nachweise in Rheinland-Pfalz. Pfälz. Heim. 30:125-126. [1979.12.31]

Preuss, G. 1982. *Formica exsecta* Nylander 1846. - Vorkommen in Rheinland-Pfalz. Pfälz. Heim. 33:35. [1982.03]

Prins, A. J. 1963. A list of the ants collected in the Kruger National Park with notes on their distribution. Koedoe 6:91-108. [1963]

Prins, A. J. 1964a. A revised list of the ants collected in the Kruger National Park. Koedoe 7: 77-93. [1964]

Prins, A. J. 1964b. 'n opname van die Formicidae in Suidelike Afrika. S. Afr. Dep. Agric. Tech. Serv. Tech. Commun. 12:111-116. [1964]

Prins, A. J. 1965a. Notes on African Formicidae (Hymenoptera) - 1. J. Entomol. Soc. South. Afr. 27:153-159. [1965.02.28]

Prins, A. J. 1965b. Notes on African Formicidae (Hymenoptera) - 2. J. Entomol. Soc. South. Afr. 28:77-79. [1965.11]

Prins, A. J. 1965c. African Formicidae (Hymenoptera). Description of a new species. S. Afr. J. Agric. Sci. 8:1021-1024. [1965.12]

Prins, A. J. 1965d. Ants of the Kruger National Park. Koedoe 8:104-108. [1965]

Prins, A. J. 1967. The ants of our national parks. Koedoe 10:63-81. [1967]

Prins, A. J. 1973. African Formicidae (Hymenoptera) in the South African Museum. Description of four new species and notes on *Tetramorium* Mayr. Ann. S. Afr. Mus. 62:1-40. [1973.08]

Prins, A. J. 1978. Hymenoptera. Pp. 823-875 in: Werger, M. J. A. (ed.) Biogeography and ecology of southern Africa. The Hague: Junk, xv + 1439 pp. [1978] [Ants pp. 857-871.]

Prins, A. J. 1982. Review of *Anoplolepis* with reference to male genitalia, and notes on *Acropyga* (Hymenoptera, Formicidae). Ann. S. Afr. Mus. 89:215-247. [1982.06]

Prins, A. J. 1983. A new ant genus from southern Africa (Hymenoptera, Formicidae). Ann. S. Afr. Mus. 94:1-11. [1983.11]

Prins, A. J. 1985. Formicoidea. Pp. 443-451 in: Scholtz, C. H., Holm, E. (eds.) Insects of southern Africa. Durban: Butterworths, 502 pp. [1985]

Prins, A. J., Cillie, J. J. 1968. The ants collected in the Hluhluwe and Umfolozi Game Reserves. Lammergeyer 8:40-47. [1968.03]

Prins, A. J., Robertson, H. G., Prins, A. 1990. Pest ants in urban and agricultural areas of southern Africa. Pp. 25-33 in: Vander Meer, R. K., Jaffe, K., Cedeno, A. (eds.) Applied myrmecology: a world perspective. Boulder: Westview Press, xv + 741 pp. [1990]

Prins, A. J., Roux, A. 1989. [Untitled. Three new species of *Ocymyrmex*, attributed to Prins and Roux.] Pp. 1282, 1291-1292 in: Bolton, B., Marsh, A. C. The Afrotropical thermophilic ant genus *Ocymyrmex* (Hymenoptera: Formicidae). J. Nat. Hist. 23:1267-1308. [1898.11.09] [The Prins and Roux manuscript, cited in this paper as in press in Ann. S. Afr. Mus., was never published.]

Provancher, L. 1881a. Faune canadienne. Les insectes - Hyménoptères. (Continué de la page 180.) Nat. Can. (Qué.) 12:193-207. [1881.02]

Provancher, L. 1881b. Faune canadienne. (Continué de la page 333.) Nat. Can. (Qué.) 12:353-362. [1881.12]

Provancher, L. 1883. Petite faune entomologique du Canada et particulièrement de la province de Québec. Vol. II. Comprenant les Orthoptères, les Névroptères et les Hyménoptères. Québec: C. Darveau, vii + v + 879 pp. [with some irregularities in pagination] [1883]

Provancher, L. 1887a. Fam. X. - Les Formicides. [part] Pp. 224-228 in: Provancher, L. 1885-1889. Additions et corrections au Volume II de la Faune entomologique du Canada, traitant des Hyménoptères. Québec: C. Darveau, 477 pp. [1887.08.27]

Provancher, L. 1887b. Fam. X. - Les Formicides. [part] Pp. 229-244 in: Provancher, L. 1885-1889. Additions et corrections au Volume II de la Faune entomologique du Canada, traitant des Hyménoptères. Québec: C. Darveau, 477 pp. [1887.09.28]

Provancher, L. 1887c. Fam. X. - Les Formicides. [concl.] Pp. 245-249 in: Provancher, L. 1885-1889. Additions et corrections au Volume II de la Faune entomologique du Canada, traitant des Hyménoptères. Québec: C. Darveau, 477 pp. [1887.10.29]

Provancher, L. 1888a. Supplément aux additions aux Hyménoptères de la province de Québec. Fam X - Formicides. [part] P. 408 in: Provancher, L. 1885-1889. Additions et corrections au Volume II de la Faune entomologique du Canada, traitant des Hyménoptères. Québec: C. Darveau, 477 pp. [1888.10]

Provancher, L. 1888b. Supplément aux additions aux Hyménoptères de la province de Québec. Fam X - Formicides. [concl.] Pp. 409 in: Provancher, L. 1885-1889. Additions et corrections au Volume II de la Faune entomologique du Canada, traitant des Hyménoptères. Québec: C. Darveau, 477 pp. [1888.11]

Provancher, L. 1895. Les dernières descriptions de l'Abbé Provancher. Nat. Can. (Qué.) 22:95-97. [1895.06]

Pullen, B. E. 1961. Non-granivorous food habits of *Pheidole grallipes* Wheeler and its possible phyletic significance (Hymenoptera: Formicidae). Pan-Pac. Entomol. 37:93-96. [1961.06.14]

Pullen, B. E. 1963. Termitophagy, myrmecophagy, and the evolution of the Dorylinae (Hymenoptera, Formicidae). Stud. Entomol. 6:405-414. [1963.12.10]

Punttila, P., Haila, Y., Niemelä, J., Pajunen, T. 1994. Ant communities in fragments of old-growth taiga and managed surroundings. Ann. Zool. Fenn. 31:131-144. [1994.01.31]

Punttila, P., Haila, Y., Pajunen, T., Tukia, H. 1991. Colonisation of clearcut forests by ants in the southern Finnish taiga: a quantitative survey. Oikos 61:250-262. [1991.06]

Pusvaskyte, O. 1979. Myrmecofauna of the Lithuanian SSR. [In Russian.] Acta Entomol. Lituan. 4:99-105. [> 1979.05.07] [Date signed for printing ("pasirasyta spausdinti").]

Queiroz, M. V. B., Avilés, D. P. 1992. Chave de identificacão das subfamílias de formicídeos neotropicais. Rev. Ceres 39:189-193. [1992.04]

Quicke, D. L. J., Fitton, M. G., Tunstead, J. R., Ingram, S. N., Gaitens, P. V. 1994. Ovipositor structure and relationships within the Hymenoptera, with special reference to the Ichneumonoidea. J. Nat. Hist. 28:635-682. [1994.05.31]

Quiran, E. See: Quiran, E. M.

Quiran, E. M., Casadío, A. A. 1988. Lista preliminar anotada de los formícidos de la Provincia de La Pampa. Rev. Fac. Agron. Univ. Nac. La Pampa 3(1):99-104. [1988.07]

Quiran, E. M., Casadío, A. A. 1994. Aportes al conocimiento de las Formicidae (Hymenoptera) de La Pampa, Argentina. Rev. Soc. Entomol. Argent. 53:100. [1994.06]

Quiroz-Robledo, L., Valenzuela-González, J. 1995. A comparison of ground ant communities in a tropical rainforest and adjacent grassland in Los Tuxtlas, Veracruz, Mexico. Southwest. Entomol. 20:203-213. [1995.06]

Quispel, A. 1941a. De verspreiding van de mierenfauna in het National Park De Hoge Veluwe. [part] Ned. Bosb. Tijdschr. 14:183-201. [1941.05]

Quispel, A. 1941b. De verspreiding van de mierenfauna in het National Park De Hoge Veluwe. [concl.] Ned. Bosb. Tijdschr. 14:258-286. [1941.06]

Radchenko, A. G. 1983. *Tapinoma kinburni* (Hymenoptera, Formicidae), endemic species of Ukrainian fauna. [In Russian.] Zool. Zh. 62:1904-1907. [1983.12] [December issue. Date signed for printing ("podpisano k pechati"): 23 November 1983.]

Radchenko, A. G. 1985a. On the discovery of the rare species *Anergates atratulus* Schenck. (Hymenoptera, Formicidae) in the Kherson district of the Ukraine. [In Russian.] Vestn. Zool. 1985(1):85. [1985.02] [January-February issue. Date signed for printing ("podpisano k pechati"): 28 January 1985.]

Radchenko, A. G. 1985b. Ants of the genus *Strongylognathus* (Hymenoptera: Formicidae) in the European part of the USSR. [In Russian.] Zool. Zh. 64:1514-1523. [1985.10] [October issue. Date signed for printing ("podpisano k pechati"): 26 September 1985.]

Radchenko, A. G. 1985c. A new ant species (Hymenoptera, Formicidae) from the Ukraine. [In Russian.] Vestn. Zool. 1985(6):64-66. [1985.12] [November-December issue. Date signed for printing ("podpisano k pechati"): 29 November 1985.]

Radchenko, A. G. 1986. *Lasius meridionalis* (Hymenoptera, Formicidae) in the USSR fauna. [In Russian.] Vestn. Zool. 1986(3):72-74. [1986.06] [May-June issue. Date signed for printing ("podpisano k pechati"): 28 May 1986.]

Radchenko, A. G. 1987. On the ways of formation of the myrmecofauna of the North Pontic region. [In Russian.] Pp. 75-80 in: Savchenko, E. N., Vasiliev, V. P., Dolin, V. G. et al. (eds.) Zoogeography and biocenotic connections of the insects of the Ukraine. [In Russian.] Kiev: Naukova Dumka, 92 pp. [> 1987.11.03] [Date signed for printing ("podpisano k pechati").]

Radchenko, A. G. 1989a. Ants of the steppe Natural Reserves of the Ukraine. [In Russian.] Pp. 244-246 in: Sokolov, V. E., Syroechkovsky, E. E., Bajanov, M. A. et al. (eds.) Proceedings of the Second All-Union Conference on the problems of kadastre of animal world. Part 4. [In Russian.] Ufa: Bashkirskoe Knizhnoe Izdatelstvo, 351 pp. [> 1988.12.29] [Not seen. Citation from Radchenko (pers. comm.). Date signed for printing ("podpisano k pechati"): 29 December 1988.]

Radchenko, A. G. 1989b. The ants of the genus *Chalepoxenus* (Hymenoptera, Formicidae) of the USSR fauna. [In Russian.] Vestn. Zool. 1989(2):37-41. [1989.04] [March-April issue. Date signed for printing ("podpisano k pechati"): 24 March 1989.]

Radchenko, A. G. 1989c. Ants of the *Plagiolepis* genus of the European part of the USSR. [In Russian.] Zool. Zh. 68(9):153-156. [1989.09] [September issue. Date signed for printing ("podpisano k pechati"): 21 August 1989.]

Radchenko, A. G. 1990. Myrmecofauna of Ukraine: state of knowledge, zoogeographical uniqueness and probable ways of origin. [In Russian.] Mater. Kollok. Sekts. Obshchestv. Nasek. Vses. Entomol. Obshch. 1:190-199. [> 1990.12.25] [Date signed for printing ("podpisano k pechati").]

Radchenko, A. G. 1991a. On the discovery of *Leptothorax semenovi* Ruzs. (Hymenoptera, Formicidae) in the Crimea. [Abstract.] [In Russian.] Vestn. Zool. 1991(2):86. [1991.04] [March-April issue. Date signed for printing ("podpisano k pechati"): 26 March 1991.]

Radchenko, A. G. 1991b. Ants of the genus *Strongylognathus* (Hymenoptera, Formicidae) of the USSR fauna. [In Russian.] Zool. Zh. 70(10):84-90. [1991.10] [October issue. Date signed for printing ("podpisano k pechati"): 17 September 1991. Translated in Entomol. Rev. (Wash.) 71(2):40-48. See Radchenko (1992c).]

Radchenko, A. G. 1992a. Ants of the genus *Tetramorium* (Hymenoptera, Formicidae) of the USSR fauna. Report 1. [In Russian.] Zool. Zh. 71(8):39-49. [1992.08] [August issue. Date signed for printing ("podpisano k pechati"): 13 July 1992. Translated in Entomol. Rev. (Wash.) 72(1):129-140. See Radchenko (1993b).]

Radchenko, A. G. 1992b. Ants of the genus *Tetramorium* (Hymenoptera, Formicidae) of the USSR fauna. Report 2. [In Russian.] Zool. Zh. 71(8):50-58. [1992.08] [August issue. Date signed for printing ("podpisano k pechati"): 13 July 1992.]

Radchenko, A. G. 1992c. Ants of the genus *Strongylognathus* (Hymenoptera, Formicidae) of the USSR. Entomol. Rev. (Wash.) 71(2):40-48. [1992.08] [English translation of Radchenko (1991b).]

Radchenko, A. G. 1993a. Ants from Vietnam in the collection of the Institute of Zoology, PAS, Warsaw. I. Pseudomyrmicinae [sic], Dorylinae, Ponerinae. Ann. Zool. (Warsaw) 44:75-82. [1993.02.15]

Radchenko, A. G. 1993b. Ants of the genus *Tetramorium* (Hymenoptera, Formicidae) of the USSR. Communication 1. Entomol. Rev. (Wash.) 72(1):129-140. [1993.07] [English translation of Radchenko (1992a).]

Radchenko, A. G. 1993c. New ants of the subfamily Cerapachyinae (Hymenoptera, Formicidae) from Vietnam. Zh. Ukr. Entomol. Tov. 1(1):43-47. [1993]

Radchenko, A. G. 1993d ("1992"). Ants (Hymenoptera, Formicidae) of the Daurian Natural Reserve and adjacent territories. [In Russian.] Pp. 77-82 in: Dolin, V. G., Kotenko, A. G. (eds.) Insects of Dauria and adjacent territories. Issue 1. [In Russian.] Moskva: Central Research Laboratory of Game Management and Reserves, 143 pp. [1993] [Signed for printing ("podpisano k pechati") 20 March 1992, but actually published in 1993 (Radchenko, pers. comm.).]

Radchenko, A. G. 1994a. Taxonomic structure of the ant genus *Myrmica* (Hymenoptera, Formicidae) of Eurasia. Report 1. [In Russian.] Zool. Zh. 73(6):39-51. [1994.06] [June issue. Date signed for printing ("podpisano k pechati"): 17 May 1994. Translated in Entomol. Rev. (Wash.) 74(3):91-106. See Radchenko (1995d).]

Radchenko, A. G. 1994b. Identification table for the ants (Hymenoptera, Formicidae) of southern Siberia. [In Russian.] Pp. 95-118 in: Hymenopterous insects of Siberia and the Far East. [In Russian.] Tr. Zapov. "Daurskii" 3:1-147. [> 1994.06.10] [Date signed for printing ("podpisano k pechati").]

Radchenko, A. G. 1994c ("1993"). New species of ants of the genus *Leptothorax* (Hymenoptera, Formicidae) from southern and eastern Palearctic. [In Russian.] Zh. Ukr. Entomol. Tov. 2(2):23-34. [< 1994.06.21] [Cited in Radchenko (1994e) which has an imprimatur date of 21 June 1994.]

Radchenko, A. G. 1994d. Identification table for ants of the genus *Myrmica* (Hymenoptera, Formicidae) from central and eastern Palearctic. [In Russian.] Zool. Zh. 73(7-8):130-145.

[1994.08] [July-August issue. Date signed for printing ("podpisano k pechati"): 21 June 1994. Translated in Entomol. Rev. (Wash.) 74(3):154-169. See Radchenko (1995e).]

Radchenko, A. G. 1994e. Identification table for ants of the genus *Leptothorax* (Hymenoptera, Formicidae) from central and eastern Palearctic. [In Russian.] Zool. Zh. 73(7-8):146-158. [1994.08] [July-August issue. Date signed for printing ("podpisano k pechati"): 21 June 1994. Translated in Entomol. Rev. (Wash.) 74(2):128-142. See Radchenko (1995b).]

Radchenko, A. G. 1994f. Survey of the species of the *scabrinodis* group of the genus *Myrmica* (Hymenoptera, Formicidae) from central and eastern Palearctic. [In Russian.] Zool. Zh. 73(9):75-82. [1994.09] [September issue. Date signed for printing ("podpisano k pechati"): 10 August 1994. Translated in Entomol. Rev. (Wash.) 74(5):116-124. See Radchenko (1995f).]

Radchenko, A. G. 1994g. Taxonomic structure, distribution and species diversity centres of the ant genus *Myrmica* Latr. in the Eurasia. P. 503 in: Lenoir, A., Arnold, G., Lepage, M. (eds.) Les insectes sociaux. 12ème congrès de l'Union Internationale pour l'Étude des Insectes Sociaux. Paris, Sorbonne, 21-27 août 1994. Paris: Université Paris Nord, xxiv + 583 pp. [1994.09]

Radchenko, A. G. 1994h. Survey of the species of the *rubra*, *rugosa*, *arnoldii*, *luteola* and *schenki* groups of the genus *Myrmica* (Hymenoptera, Formicidae) from central and eastern Palearctic. [In Russian.] Zool. Zh. 73(11):72-80. [1994.11] [November issue. Date signed for printing ("podpisano k pechati"): 14 October 1994.]

Radchenko, A. G. 1994i. Survey of the species of the *lobicornis* group of the genus *Myrmica* (Hymenoptera, Formicidae) from central and eastern Palearctic. [In Russian.] Zool. Zh. 73(11):81-92. [1994.11] [November issue. Date signed for printing ("podpisano k pechati"): 14 October 1994.]

Radchenko, A. G. 1994j. New Palaearctic species of the genus *Myrmica* Latr. (Hymenoptera, Formicidae). Memorabilia Zool. 48:207-217. [1994]

Radchenko, A. G. 1995a ("1994"). A review· of the ant genus *Leptothorax* (Hymenoptera, Formicidae) of the central and eastern Palearctic. Communication 1. Subdivision into groups. Groups *acervorum* and *bulgaricus*. [In Russian.] Vestn. Zool. 1994(6):22-28. [> 1995.03.01] [November-December 1994 issue. Date signed for printing ("podpisano k pechati"): 1 March 1995.]

Radchenko, A. G. 1995b. A key to species of *Leptothorax* (Hymenoptera, Formicidae) of the central and eastern Palearctic region. Entomol. Rev. (Wash.) 74(2):128-142. [1995.06] [English translation of Radchenko (1994e).]

Radchenko, A. G. 1995c. A review of the ant genus *Leptothorax* (Hymenoptera, Formicidae) of the central and eastern Palearctic. Communication 2. Groups *tuberum*, *corticalis*, *affinis*, *clypeatus*, *alinae* and *singularis*. [In Russian.] Vestn. Zool. 1995(2-3):14-21. [> 1995.06.15] [March-June issue. Date signed for printing ("podpisano k pechati"): 15 June 1995.]

Radchenko, A. G. 1995d. Taxonomic structure of the genus *Myrmica* (Hymenoptera, Formicidae) of Eurasia. Communication 1. Entomol. Rev. (Wash.) 74(3):91-106. [1995.07] [English translation of Radchenko (1994a).]

Radchenko, A. G. 1995e. A key to species of the genus *Myrmica* (Hymenoptera, Formicidae) of the central and eastern Palearctic region. Entomol. Rev. (Wash.) 74(3):154-169. [1995.07] [English translation of Radchenko (1994d).]

Radchenko, A. G. 1995f. A review of species of *Myrmica* belonging to the group of *scabrinodus* (Hymenoptera, Formicidae) from the central and eastern Palearctic. Entomol. Rev. (Wash.) 74(5):116-124. [1995.08] [English translation of Radchenko (1994f).]

Radchenko, A. G. 1995g. Palearctic ants of the genus *Cardiocondyla* (Hymenoptera, Formicidae). Entomol. Obozr. 74:447-455. [> 1995.08.22] [Date signed for printing ("podpisano k pechati").]

Radchenko, A. G., Arakelian, G. R. 1990. Ants of the group *Tetramorium ferox* Ruzsky (Hymenoptera, Formicidae) from Crimea and the Caucasus. [In Russian.] Biol. Zh. Arm. 43:371-378. [> 1990.08.28] [Date signed for printing.]

Radchenko, A. G., Arakelian, G. R. 1991. New ant species (Hymenoptera, Formicidae) from Armenia. [In Russian.] Vestn. Zool. 1991(5):72-75. [1991.10] [September-October issue. Date signed for printing ("podpisano k pechati"): 23 September 1991.]

Radchenko, A. G., Malij, E. N. 1990 ("1989"). Zoogeographical characteristics of the myrmecofauna of the Crimea. [In Russian.] Pp. 105-113 in: Dolin, V. G., Zerova, M. D., Kononova, S. V. et al. (eds.) Ecology and taxonomy of the insects of the Ukraine. Scientific transactions. Issue 3. [In Russian.] Odessa: Vyscha Shkola, 192 pp. [1990] [Not seen. Citation from Radchenko (pers. comm.). Signed for printing ("podpisano k pechati") 12 July 1989, but actually published in 1990.]

Radchenko, A. G., Tsyubik, M. M. 1990. On the discovery of a new member of the genus *Plagiolepis* Mayr (Hymenoptera, Formicidae) in the USSR. [Abstract.] [In Russian.] Vestn. Zool. 1990(1):85. [1990.02] [January-February issue. Date signed for printing ("podpisano k pechati"): 30 January 1990.]

Radoszkowsky, O. 1876. Comte-rendu des hyménoptères recueillis en Égypte et Abyssinie en 1873. Tr. Rus. Entomol. Obshch. 12:111-150. [1876.09.27] [Ants pp. 140-141.]

Radoszkowsky, O. 1881. Hymenoptères d'Angola. J. Sci. Math. Phys. Nat. 8(31):197-221. [1881.12]

Radoszkowsky, O. 1884. Fourmis de Cayenne Française. Tr. Rus. Entomol. Obshch. 18:30-39. [1884.05.13] [See also Mayr (1884).]

Radtchenko, A. G. See: Radchenko, A. G.

Radtschenko, A. G. See: Radchenko, A. G.

Raffray, A. 1887. Pselaphides nouveaux ou peu connus. [part] Rev. Entomol. (Caen) 6:18-36. [1887.01]

Rafin, G. 1885. Ignivorous ant. Am. Nat. 19:403. [1885.03.21]

Rafinesque, C. S. 1815. Analyse de la nature, ou tableau de l'univers et des corps organisés. Palermo: J. Barravecchia, 224 pp. [1815.07] [Ants p. 124.]

Ragonot, E. L. 1887. Diagnoses of North American Phycitidae and Galleridae. Paris: privately published, 20 pp. [1887] [*Martia* (Lepidoptera), senior homonym of *Martia* Forel, 1907.]

Raignier, A. 1938. Le caractère primitif de l'instinct esclavagiste chez les Formicines (Formicidae, Hymenoptera). Arch. Néerl. Zool. 3(suppl.):167-182. [1938]

Raignier, A., Boven, J. K. A. van. 1949. La première colonie belge de la fourmi amazone (*Polyergus rufescens* Latreille) et description de trois ergatandromorphes nouveaux de cette espèce (Hymenoptera, Formicidae). Bull. Inst. R. Sci. Nat. Belg. 25(4):1-11. [1949.01]

Raignier, A., Boven, J. K. A. van. 1953 ("1952"). Quelques aspects nouveaux de la taxonomie et de la biologie des doryles africains (Hyménoptères, Formicidae). Ann. Sci. Nat. Zool. Biol. Anim. (11)14:397-403. [1953.03]

Raignier, A., Boven, J. K. A. van. 1955. Étude taxonomique, biologique et biométrique des *Dorylus* du sous-genre *Anomma* (Hymenoptera Formicidae). Ann. Mus. R. Congo Belge Nouv. Sér. Quarto Sci. Zool. 2:1-359. [1955.08]

Raignier, A., Boven, J. K. A. van, Ceusters, R. 1974. Der Polymorphismus der afrikanischen Wanderameisen unter biometrischen und biologischen Gesichtspunkten. Pp. 668-693 in: Schmidt, G. H. (ed.) Sozialpolymorphismus bei Insekten. Stuttgart: Wissenschaftliche Verlagsgesellschaft mbH, xxiv + 974 pp. [1974]

Ramdas Menon, M. G., Punjabi, G. M. 1961 ("1960"). Some observations and new records of ants from Delhi. Indian J. Entomol. 22:133-135. [1961.07]

Rammoser, H. 1965. Zur Verbreitung der hügelbauenden Waldameisen im Spessart. Waldhygiene 6:44-82. [1965]

Ramos, J. A. 1946. The insects of Mona Island (West Indies). J. Agric. Univ. P. R. 30:1-74. [1946.01] [Ants pp. 64-66, determined by M. R. Smith.]

Ramos-Elorduy de Conconi, J., MacGregor Loaeza, R., Cuadriello Aguilar, J., Sampedro Rosas, G. 1983. Quelques données sur la biologie des fourmis *Liometopum* (Dolichoderinae) du Mexique

et en particulier sur leurs rapports avec les homoptères. Pp. 125-130 in: Jaisson, P. (ed.) Social insects in the tropics. Volume 2. Paris: Université Paris-Nord, 252 pp. [1983.07]
Ranta, E., Vepsäläinen, K., As, S., Haila, Y., Pisarski, B., Tiainen, J. 1983. Island biogeography of ants (Hymenoptera, Formicidae) in four Fennoscandian archipelagoes. Acta Entomol. Fenn. 42:64. [1983]
Raphael, S. 1970. The publication dates of the Transactions of the Linnaean Society of London, Series I, 1791-1875. Biol. J. Linn. Soc. 2:61-76. [1970.03.26]
Raqué, K.-F. 1994. Die Ameisenfauna der Sandhausener Dünen. Beih. Veröff. Naturschutz Landschaftspfl. Baden-Württ. 80:241-254. [Not seen. Cited in Zoological Record, volume 132.]
Räsänen, V. 1915. Stridulationsapparate bei Ameisen besonders bei Formicidae. Acta Soc. Fauna Flora Fenn. 40(8):1-19. [1915] [The term "Formicidae" as used here refers to the subfamily Formicinae.]
Rasnitsyn, A. P. 1975. Hymenoptera Apocrita of Mesozoic. [In Russian.] Tr. Paleontol. Inst. Akad. Nauk SSSR 147:1-134. [> 1975.01.07] [Date signed for printing ("podpisano k pechati").]
Rasnitsyn, A. P. 1980. Origin and evolution of Hymenoptera. [In Russian.] Tr. Paleontol. Inst. Akad. Nauk SSSR 174:1-192. [> 1980.01.24] [Date signed for printing ("podpisano k pechati").]
Rasnitsyn, A. P. 1988. An outline of evolution of the hymenopterous insects (Order Vespida). Orient. Insects 22:115-145. [1988.03.31]
Razoumowsky, G. de. 1789. Histoire naturelle du Jorat et de ses environs; et celle des trois lacs de Neufchatel, Morat et Bienne. Tome premier. Lausanne: J. Mourer, xvi + 322 pp. [1789]
Réaumur, R. A. F. de. 1926. The natural history of ants. From an unpublished manuscript in the archives of the Academy of Sciences of Paris by René Antoine Ferchault de Réaumur. Translated and annotated by William Morton Wheeler. New York: A. Knopf, 280 pp. [1926] [The original manuscript was written between October 1742 and January 1743.]
Reder, E., Veith, H. J., Buschinger, A. 1995. Neuartige Alkaloide aus dem Giftdrüsensekret sozialparasitischer Ameisen (Myrmicinae: Leptothoracini). Helv. Chim. Acta 78:73-79. [1995.02.08]
Rees, D. M., Grundmann, A. W. 1940. A preliminary list of the ants of Utah. Bull. Univ. Utah Biol. Ser. 31(5):3-12. [1940.11.05]
Regnier, F. E., Wilson, E. O. 1968. The alarm-defense system of the ant *Acanthomyops claviger*. J. Insect Physiol. 14:955-970. [1968.07]
Regnier, F. E., Wilson, E. O. 1969. The alarm-defense system of the ant *Lasius alienus*. J. Insect Physiol. 15:893-898. [1969.05]
Regnier, F. E., Wilson, E. O. 1971. Chemical communication and "propaganda" in slave-maker ants. Science (Wash. D. C.) 172:267-269. [1971.04.16]
Reich, G. C. 1793. Kurze Beschreibung neuen, oder noch wenig bekkanten Thiere, welche Herr Le Blond der naturforschenden Gesellschaft zu Paris aus Cayenne als Geschenk überschikt hat. Mag. Thierreichs 1:128-134. [1793]
Reichensperger, A. 1912 ("1911"). Die Ameisenfauna der Rheinprovinz nebst Angaben über einige Ameisengäste. Ber. Versamml. Bot. Zool. Ver. Rheinl.-Westfal. 68:114-130. [Not seen. Cited in Stitz (1939) and Rohe & Heller (1990b).]
Reichensperger, A. 1917. Beobachtungen an Ameisen. II. Ein Beitrag zur Pseudogynen-Theorie. Z. Wiss. Insektenbiol. 13:145-152. [1917.08.31]
Reichensperger, A. 1922. Myrmekologische Beobachtungen aus Luxemburg. Bull. Mens. Soc. Nat. Luxemb. (n.s.)16:105-115. [1922.12.31]
Reichensperger, A. 1924. Das Weibchen von *Eciton quadriglume* Hal., einige neue ecitophile Histeriden und allgemeine Bemerkungen. Zool. Anz. 60:201-213. [1924.08.05]
Reichensperger, A. 1926. Das Weibchen von *Eciton mattogrossensis* Luederw. (Hym.). Entomol. Mitt. 15:401-404. [1926.10.10]

Reichensperger, A. 1930. Ein Nest und die Königin von *Eciton* (*Acamatus*) *legionis* Sm. Zool. Anz. 88:321-325. [1930.05.10]

Reichensperger, A. 1934. Beitrag zur Kenntnis von *Eciton lucanoides* Em. Zool. Anz. 106:240-245. [1934.05.20]

Reichensperger, A. 1939. Beiträge zur Kenntnis der Myrmecophilenfauna Costa Ricas und Brasiliens VII, nebst Beschreibung der Königin von *Eciton* (*Acamatus*) *pilosum*. Zool. Jahrb. Abt. Syst. Ökol. Geogr. Tiere 73:261-300. [1939.11.24]

Reid, J. A. 1941. The thorax of the wingless and short-winged Hymenoptera. Trans. R. Entomol. Soc. Lond. 91:367-446. [1941.11.25]

Reimer, N. J. 1994. Distribution and impact of alien ants in vulnerable Hawaiian ecosystems. Pp. 11-22 in: Williams, D. F. (ed.) Exotic ants. Biology, impact, and control of introduced species. Boulder: Westview Press, xvii + 332 pp. [1994]

Reimer, N., Beardsley, J. W., Jahn, G. 1990. Pest ants in the Hawaiian islands. Pp. 40-50 in: Vander Meer, R. K., Jaffe, K., Cedeno, A. (eds.) Applied myrmecology: a world perspective. Boulder: Westview Press, xv + 741 pp. [1990]

Reiskind, J. 1965. A revision of the ant tribe Cardiocondylini (Hymenoptera, Formicidae). I. The genus *Prosopidris* Wheeler. Psyche (Camb.) 72:79-86. [1965.06.25]

Reiskind, J. 1983. Request for a ruling to correct homonymy in names of the family-groups based on *Myrmecia* (Insecta) and *Myrmecium* (Arachnida). Z.N.(S.)2223. Bull. Zool. Nomencl. 40:43-44. [1983.03.29]

Restrepo, C., Espadaler, X., De Haro, A. 1986 ("1985"). Contribución al conocimiento faunístico de los formícidos del Macizo de Garraf (Barcelona). Orsis (Org. Sist.) 1:113-129. [1986]

Retana, J., Cerdá, X. 1990. Social organization of *Cataglyphis cursor* colonies (Hymenoptera, Formicidae): inter- and intraspecific comparisons. Ethology 84:105-122. [1990.02]

Retana, J., Cerdá, X. 1995. Agonistic relationships among sympatric Mediterranean ant species (Hymenoptera: Formicidae). J. Insect Behav. 8:365-380. [1995.05]

Rettenmeyer, C. W. 1963. Behavioral studies of army ants. Univ. Kans. Sci. Bull. 44:281-465. [1963.09.13]

Rettenmeyer, C. W. 1974. Description of the queen and male with some biological notes on the army ant, *Eciton rapax*. Pp. 291-302 in: Beard, R. L. (ed.) Connecticut Entomological Society 25th Anniversary Memoirs. New Haven: Connecticut Entomological Society, 322 pp. [1974]

Rettenmeyer, C. W. 1983. Checklist of army ants (arrieras) (Formicidae: Ecitoninae). P. 650 in: Janzen, D. H. (ed.) Costa Rican natural history. Chicago: University of Chicago Press, xi + 816 pp. [1983]

Rettenmeyer, C. W., Chadab-Crepet, R., Naumann, M. G., Morales, L. 1983. Comparative foraging by neotropical army ants. Pp. 59-73 in: Jaisson, P. (ed.) Social insects in the tropics. Volume 2. Paris: Université Paris-Nord, 252 pp. [1983.07]

Rettenmeyer, C. W., Watkins, J. F., II. 1978. Polygyny and monogyny in army ants (Hymenoptera: Formicidae). J. Kansas Entomol. Soc. 51:581-591. [1978.11.17]

Retzius, A. J. 1783. Caroli de Geer. Genera et species insectorum e generosissimi auctoris scriptis extraxit, digessit, Latine quoad partem reddidit, et terminologiam insectorum Linneanam addidit. Lipsiae [= Leipzig]: Cruse, 220 pp. [1783]

Reuter, O. M. 1904 ("1902-03"). *Lasius alienus* Först., funnen i Finland. Medd. Soc. Fauna Flora Fenn. 29:120-121. [1904]

Reyes, J. L. 1985. Descripción de *Messor celiae* nov. sp. (Hym., Formicidae). Bol. Asoc. Esp. Entomol. 9:255-261. [1985.05]

Reyes, J. L., Espadaler, X., Rodríguez, A. 1987. Descripción de *Goniomma baeticum* nov. sp. (Hym., Formicidae). EOS. Rev. Esp. Entomol. 63:269-276. [1987.12.15]

Reyes, J. L., Porras Castillo, A. 1985 ("1984"). Alar biometry in the taxonomy of the species *Goniomma hispanicum* and *G. baeticum*. Insectes Soc. 31:473-475. [1985.04]

Reyes, J. L., Rodríguez, A. 1987. [Untitled. *Goniomma baeticum*, Reyes y Rodríguez, nov. sp.] Pp. 269-272 in: Reyes, J. L., Espadaler, X., Rodríguez, A. Descripción de *Goniomma baeticum* nov. sp. (Hym., Formicidae). EOS. Rev. Esp. Entomol. 63:269-276. [1987.12.15]

Reyes Lopez, J. L. See: Reyes, J. L.

Reza, A. 1925. Recursos alimentícios de México de origen animal poco conocidos. Mem. Rev. Soc. Cient. "Antonio Alzate" 44:1-22. [1925.02]

Reznikova, Z. I. 1980. Stability of the structure of multispecific associations of steppe ants in different natural zones. [In Russian.] Tr. Biol. Inst. (Novosibirsk) 40:116-130. [1980] [Signed for printing ("podpisano k pechati") 19 December 1979. Title page dated 1980.]

Reznikova, Z. I., Samoshilova, N. M. 1981a. The role of ants as predators in steppe biocenoses. [In Russian.] Ekologiya (Mosc.) 12(1):69-75. [> 1981.01.13] [Date signed for printing ("podpisano k pechati"). Translated in Sov. J. Ecol. 12:55-60. See Reznikova & Samoshilova (1981b).]

Reznikova, Z. I., Samoshilova, N. M. 1981b. The role of ants as predators in steppe biocenoses. Sov. J. Ecol. 12:55-60. [1981.09] [English translation of Reznikova & Samoshilova (1981a).]

Reznikova, Z. I., Shillerova, O. A. 1979. Ecology and behavior of *Formica uralensis* in the dry steppes of Tuva. [In Russian.] Vopr. Ekol. Novosib. Gos. Univ. 5:112-131.

Richards, O. W. 1935. Notes on the nomenclature of the Aculeate Hymenoptera, with special reference to the British genera and species. Trans. Entomol. Soc. Lond. 83:143-176. [1935.06.27]

Richards, O. W. 1956. Hymenoptera. Introduction and key to families. Handb. Identif. Br. Insects 6(1):1-94. [1956.12.31]

Richards, O. W. 1959. *Lasius brunneus* (Latr.) (Hym., Formicidae) and other ants, in Buckinghamshire. Entomol. Mon. Mag. 95:83. [1959.09.10]

Richards, O. W. 1971. The thoracic spiracles and some associated structures in the Hymenoptera and their significance in classification, especially of the Aculeata. Pp. 1-13 in: Asahina, S., et al. (eds.) Entomological essays to commemorate the retirement of Professor K. Yasumatsu. Tokyo: Hokuryukan Publishing Co., vi + 389 pp. [1971]

Richards, O. W. 1977. Hymenoptera. Introduction and key to families. Handb. Identif. Br. Insects 6(1)(2nd edn.):1-100. [1977]

Ride, W. D. L., Taylor, R. W. 1973. *Formica maxima* Moore, 1842 (Insecta, Hymenoptera): proposed suppression under the plenary powers in accordance with article 23(a-b). Z.N.(S.)2023. Bull. Zool. Nomencl. 30:58-60. [1973.07.06]

Riemann, H. 1995. Zur Stechimmenfauna des Bremer Bürgerparks (Hymenoptera: Aculeata). Abh. Naturwiss. Ver. Bremen 43:45-72. [1995]

Rigato, F. 1994a. *Dacatria templaris* gen. n., sp. n. A new myrmicine ant from the Republic of Korea. Dtsch. Entomol. Z. (N.F.)41:155-162. [1994.03.04]

Rigato, F. 1994b. Revision of the myrmicine ant genus *Lophomyrmex*, with a review of its taxonomic position (Hymenoptera: Formicidae). Syst. Entomol. 19:47-60. [1994.03.12]

Rigato, F., Sciaky, R. 1989. Contributo alla conoscenza della mirmecofauna della Val Gesso (Alpi Marittime) (Hymenoptera Formicidae). Boll. Mus. Reg. Sci. Nat. Torino 7:427-442. [1989.12.12]

Rigato, F., Sciaky, R. 1991. The myrmecofauna of the Gesso Valley (Maritime Alps) (Hymenoptera Formicidae). Ethol. Ecol. Evol. Spec. Issue 1:87-89. [1991.03]

Rissing, S. W. 1983. Natural history of the workerless inquiline ant *Pogonomyrmex colei* (Hymenoptera: Formicidae). Psyche (Camb.) 90:321-332. [1983.12.09]

Rissing, S. W., Pollock, G. B. 1988. Pleometrosis and polygyny in ants. Pp. 179-222 in: Jeanne, R. L. (ed.) Interindividual behavioral variability in social insects. Boulder: Westview Press, ix + 456 pp. [1988]

Rissing, S. W., Pollock, G. B. 1989. Behavioral ecology and community organization of desert seed-harvesting ants. J. Arid Environ. 17:167-173. [1989.09]

Ritsema, C. 1874. Aanteekeningen betreffende eene kleine collectie Hymenoptera van Neder-Guinea, en beschrijving van de nieuwe soorten. Tijdschr. Entomol. 17:175-211. [1874] [Ants (Dorylinae) pp. 182-184.]

Robbins, W. W. 1910. An introduction to the study of the ants of northern Colorado. Univ. Colo. Stud. 7:215-222. [1910.06]

Robertson, H. G. 1990. Unravelling the *Camponotus fulvopilosus* species complex (Hymenoptera: Formicidae). Pp. 327-328 in: Veeresh, G. K., Mallik, B., Viraktamath, C. A. (eds.) Social insects and the environment. Proceedings of the 11th International Congress of IUSSI, 1990. New Delhi: Oxford & IBH Publishing Co., xxxi + 765 pp. [1990.08]

Robertson, H. G. 1995. Sperm transfer in the ant *Carebara vidua* F. Smith (Hymenoptera: Formicidae). Insectes Soc. 42:411-418. [1995]

Robertson, H. G., Villet, M. H. 1989a. Colony founding in the ant *Carebara vidua*: the dispelling of a myth. S. Afr. J. Sci. 85:121-122. [1989.02]

Robertson, H. G., Villet, M. H. 1989b. Mating behavior in three species of myrmicine ants (Hymenoptera: Formicidae). J. Nat. Hist. 23:767-773. [1989.07.03]

Robertson, P. L. 1968. A morphological and functional study of the venom apparatus in representatives of some major groups of Hymenoptera. Aust. J. Zool. 16:133-166. [1968.02]

Rockwood, L. L. 1973. Distribution, density and dispersion of two species of *Atta* (Hymenoptera: Formicidae) in Guanacaste Province, Costa Rica. J. Anim. Ecol. 42:803-817. [1973.10]

Rodríguez, A. 1982 ("1981"). Contribución al conocimiento de las hormigas (Hymenoptera, Formicidae) de Sierra Morena Central. Bol. Asoc. Esp. Entomol. 5:181-188. [1982.05]

Rodríguez, A., Reyes, J. L., Espadaler, X. 1983. Descripción del macho de *Messor hispanicus* Santschi, 1919 (Hym., Formicidae). Bol. Asoc. Esp. Entomol. 7:37-42. [1983.11]

Rodríguez Garza, J. A., Pozo de la Tijera, C. 1994 ("1993"). Nuevos registros de hormigas (Hymenoptera: Formicidae) para México. Rev. Biol. Trop. 41:916-917. [1994.06]

Rodríguez Garza, J. A., Prado Beltrán, E., Equihua Martínez, A. 1989. Claves para subfamilias y géneros de hormigas (Hymenoptera: Formicidae) de Nuevo León. Agrociencia 76:257-268. [1989.11]

Rodríguez González, A. See: Rodríguez, A.

Rögener, J., Pfau, J. 1994. Untersuchungen zur Ameisenfauna (Hym., Formicidae) eines Kalkmagerrasen-Gehölz-Komplexes des Halbesberges bei Witzenhausen (Werra-Meissner-Kreis). Braunschw. Naturkd. Schr. 4:553-574. [1994.10]

Roger, J. 1857. Einiges über Ameisen. Berl. Entomol. Z. 1:10-20. [1857]

Roger, J. 1859. Beiträge zur Kenntniss der Ameisenfauna der Mittelmeerländer. I. Berl. Entomol. Z. 3:225-259. [1859]

Roger, J. 1860. Die *Ponera*-artigen Ameisen. Berl. Entomol. Z. 4:278-312. [1860]

Roger, J. 1861a. Die *Ponera*-artigen Ameisen (Schluss). Berl. Entomol. Z. 5:1-54. [1861]

Roger, J. 1861b. Myrmicologische Nachlese. Berl. Entomol. Z. 5:163-174. [1861]

Roger, J. 1862a. Einige neue exotische Ameisen-Gattungen und Arten. Berl. Entomol. Z. 6:233-254. [1862.05]

Roger, J. 1862b. Beiträge zur Kenntniss der Ameisenfauna der Mittelmeerländer. II. Berl. Entomol. Z. 6:255-262. [1862.05]

Roger, J. 1862c. Synonymische Bemerkungen. 1. Ueber Formiciden. Berl. Entomol. Z. 6:283-297. [1862.05]

Roger, J. 1863a. Die neu aufgeführten Gattungen und Arten meines Formiciden-Verzeichnisses nebst Ergänzung einiger früher gegebenen Beschreibungen. Berl. Entomol. Z. 7:131-214. [1863.06]

Roger, J. 1863b. Verzeichniss der Formiciden-Gattungen und Arten. Berl. Entomol. Z. 7(Beilage): 1-65. [1863.06]

Rohe, W. 1990. Wiederfund von *Formicoxenus nitidulus* (Nylander 1846) in Rheinland-Pfalz (Hymenoptera: Formicidae). Mainz. Naturwiss. Arch. 28:137-142. [1990]

Rohe, W., Heller, G. 1990a. Vorläufige Ameisenliste (Formiciden) mit Kurzkommentar für Rheinhessen, die Pfalz und den Naheraum. Fauna Flora Rheinl.-Pfalz 5(4):803-818. [Not seen. Cited in Rohe & Heller (1990b).]

Rohe, W., Heller, G. 1990b. Vorschlag für eine Rote Liste der Ameisen in Rheinhessen, der Pfalz und dem Naheraum (Hymenoptera, Formicidae). Mainz. Naturwiss. Arch. 28:143-157. [1990]

Rohlfien, K. 1979. Aus der Geschichte der entomologischen Sammlungen des ehemaligen Deutschen Entomologischen Instituts. III. Die Hymenopterensammlung. Beitr. Entomol. 29:415-438. [1979] [Collection has 162 genera and 1360 species of Formicidae, 190 of the latter with types.]

Rojas-Fernández, P., Fragoso, C. 1994. The ant fauna (Hymenoptera: Formicidae) of the Mapimi Biosphere Reserve, Durango, Mexico. Sociobiology 24:47-75. [1994]

Romand, B. E. de. 1846. Notice sur un insecte nouveau. Ann. Soc. Entomol. Fr. (Bull.) (2)4:xxxii. [1846.07.08]

Ronchetti, G. 1961. Secretergati in popolazioni di formiche del gruppo *Formica rufa*. Collana Verde 7:61-74. [1961.04.15]

Ronchetti, G. 1963. Il gruppo *Formica rufa* in Lombardia (Italia settentrionale). Symp. Genet. Biol. Ital. 12:200-213. [1963]

Ronchetti, G. 1965. Il gruppo *Formica rufa* in Piemonte, Val d'Aosta e Liguria (Italia settentrionale). Collana Verde 16:341-354. [1965.09.10]

Ronchetti, G. 1966. Le formiche del gruppo *Formica rufa* sulle Alpi orientali italiane. Boll. Soc. Entomol. Ital. 96:123-137. [1966.10.20]

Ronchetti, G. 1978a. Distribution of ants of the *Formica rufa* group in Europe (5 maps). Distribution des fourmis du groupe *Formica rufa* en Europe (5 cartes). Verbreitung der Ameisen der Gruppe *Formica rufa* in Europa (5 Karten). Distribuzione delle formiche del gruppo *Formica rufa* in Europa (5 carte). Pavia: Istituto di Entomologia dell'Università di Pavia, 11 pp., 5 maps. [1978.09.02] [Presented at a meeting of the Organisation Internationale de Lutte Biologique contre les Animaux et les Plants Nuisible, Section Regionale Ouest Paléarctique, in Varenna, August 28 - September 2 1978.]

Ronchetti, G. 1978b. Distribution of ants of the *Formica rufa* group in Italy (5 maps). Distribution des fourmis du groupe *Formica rufa* en Italie (5 cartes). Verbreitung der Ameisen der Gruppe *Formica rufa* in Italien (5 Karten). Distribuzione delle formiche del gruppo *Formica rufa* in Italia (5 carte). Pavia: Istituto di Entomologia dell'Università di Pavia, 1 p., 5 maps. [1978.09.02]

Ronchetti, G. 1980. Distribution des fourmis du groupe *Formica rufa* en Europe (5 cartes). Deuxième édition. Pavia: Istituto di Entomologia dell'Università di Pavia, 5 maps. [1980]

Ronchetti, G. 1981. Distribuzione delle formiche del gruppo *Formica rufa* in Europa. Collana Verde 59:225-236. [1981]

Room, P. M. 1971. The relative distributions of ant species in Ghana's cocoa farms. J. Anim. Ecol. 40:735-751. [1971.10]

Room, P. M. 1975. Diversity and organization of the ground foraging ant faunas of forest, grassland and tree crops in Papua New Guinea. Aust. J. Zool. 23:71-89. [1975.02]

Roonwal, M. L. 1976. Plant-pest status of root-eating ant, *Dorylus orientalis*, with notes on taxonomy, distribution and habits (Insecta: Hymenoptera). J. Bombay Nat. Hist. Soc. 72:305-313. [1976.01.31]

Roques, L., Torossian, C. 1976 ("1975"). Étude par microscopie électronique à balayage de la morphologie céphalique des ouvrières de fourmis du groupe *Formica rufa*. Bull. Soc. Hist. Nat. Toulouse 111:322-336. [1976.01.31]

Roques, L., Torossian, C. 1977 ("1976"). Étude par microscopie électronique à balayage de la morphologie du thorax des ouvrières de fourmis du groupe *Formica rufa*. (2e note.) Bull. Soc. Hist. Nat. Toulouse 112:314-326. [1977.01]

Rosciszewski, K. 1994. *Rostromyrmex*, a new genus of myrmicine ants from peninsular Malaysia (Hymenoptera: Formicidae). Entomol. Scand. 25:159-168. [1994.06]

Rosciszewski, K., Maschwitz, U. 1994. Prey specialization of army ants of the genus *Aenictus* in Malaysia. Andrias 13:179-187. [1994.09.30]

Rosengren, M., Rosengren, R., Soderlund, V. 1980. Chromosome numbers in the genus *Formica* with special reference to the taxonomical position of *Formica uralensis* Ruzsk. and *Formica truncorum* Fabr. Hereditas (Lund) 92:321-326. [1980.06.09]

Rosengren, R., Chautems, D., Cherix, D., Fortelius, W., Keller, L. 1994. Separation of two sympatric sibling species of *Formica* L. ants by a behavioural choice test based on brood discrimination. Memorabilia Zool. 48:237-249. [1994]

Rosengren, R., Cherix, D. 1981. The pupa-carrying test as a taxonomic tool in the *Formica rufa* group. Pp. 263-281 in: Howse, P. E., Clement, J.-L. (eds.) Biosystematics of social insects. Systematics Association Special Volume No. 19. London: Academic Press, 346 pp. [1981]

Rosengren, R., Cherix, D., Pamilo, P. 1986. Insular ecology of the red wood ant *Formica truncorum* Fabr. II. Distribution, reproductive strategy and competition. Mitt. Schweiz. Entomol. Ges. 59:63-94. [1986.07.15]

Rosengren, R., Cherix, D., Poutanen, R. 1981. The pupa-carrying test as a taxonomic tool in the *Formica rufa* group. Memo. Soc. Fauna Flora Fenn. 57:14. [1981]

Rosengren, R., Pamilo, P. 1983. The evolution of polygyny and polydomy in mound-building *Formica* ants. Acta Entomol. Fenn. 42:65-77. [1983]

Rosengren, R., Pamilo, P. 1984. The interrelationship between mating behaviour, sex ratio and queen number in *Formica*. [Abstract.] P. 517 in: XVII International Congress of Entomology. Hamburg, Federal Republic of Germany, August 20-26, 1984. Abstract Volume. Hamburg: XVII International Congress of Entomology, 960 pp. [1984.08.10]

Rosengren, R., Vepsäläinen, K., Wuorenrinne, H. 1979. Distribution, nest densities and ecological significance of wood ants (the *Formica rufa* group) in Finland. Bull. SROP 1979(II-3):181-213. [1979]

Rosengren, R., Vepsäläinen, K., Wuorenrinne, H. 1981. Distribuzione, densità delle colonie ed importanza ecologica delle formiche del gruppo *Formica rufa* in Finlandia. Collana Verde 59:237-259. [1981]

Ross, E. S. 1954. [Review of: Cook, T. W. 1953. The ants of California. Palo Alto: Pacific Books, 462 pp.] Pan-Pac. Entomol. 30:14. [1954.03.30]

Ross, H. H., Rotramel, G. L., La Berge, W. E. 1971. A synopsis of common and economic Illinois ants, with keys to the genera (Hymenoptera, Formicidae). Ill. Nat. Hist. Surv. Biol. Notes 71: 1-22. [1971.01]

Ross, K. G. 1992. Strong selection on a gene that influences reproductive competition in a social insect. Nature (Lond.) 355:347-349. [1992.01.23]

Ross, K. G., Carpenter, J. M. 1991. Phylogenetic analysis and the evolution of queen number in eusocial Hymenoptera. J. Evol. Biol. 4:117-130. [1991.01]

Ross, K. G., Fletcher, D. J. C. 1985a. Genetic origin of male diploidy in the fire ant, *Solenopsis invicta* (Hymenoptera: Formicidae), and its evolutionary significance. Evolution 39:888-903. [1985.08.14]

Ross, K. G., Fletcher, D. J. C. 1985b. Comparative study of genetic and social structure in two forms of the fire ant *Solenopsis invicta* (Hymenoptera: Formicidae). Behav. Ecol. Sociobiol. 17:349-356. [1985.10]

Ross, K. G., Keller, L. 1995a. Joint influence of gene flow and selection on a reproductively important genetic polymorphism in the fire ant *Solenopsis invicta*. Am. Nat. 146:325-348. [1995.09]

Ross, K. G., Keller, L. 1995b. Ecology and evolution of social organization: insights from fire ants and other highly eusocial insects. Annu. Rev. Ecol. Syst. 26:631-656. [1995]

Ross, K. G., Robertson, J. L. 1990. Developmental stability, heterozygosity, and fitness in two introduced fire ants (*Solenopsis invicta* and *S. richteri*) and their hybrid. Heredity 64:93-103. [1990.01]

Ross, K. G., Shoemaker, D. D. 1994 ("1993"). An unusual pattern of gene flow between the two social forms of the fire ant *Solenopsis invicta*. Evolution 47:1595-1605. [1994.05.17]

Ross, K. G., Trager, J. C. 1991 ("1990"). Systematics and population genetics of fire ants (*Solenopsis saevissima* complex) from Argentina. Evolution 44:2113-2134. [1991.02.20]

Ross, K. G., Vander Meer, R. K., Fletcher, D. J. C., Vargo, E. L. 1987. Biochemical phenotypic and genetic studies of two introduced fire ants and their hybrid (Hymenoptera: Formicidae). Evolution 41:280-293. [1987.03.12]

Ross, K. G., Vargo, E. L., Fletcher, D. J. C. 1987. Comparative biochemical genetics of three fire ant species in North America, with special reference to the two social forms of *Solenopsis invicta* (Hymenoptera: Formicidae). Evolution 41:979-990. [1987.09.15]

Ross, K. G., Vargo, E. L., Fletcher, D. J. C. 1988. Colony genetic structure and queen mating frequency in fire ants of the subgenus *Solenopsis* (Hymenoptera: Formicidae). Biol. J. Linn. Soc. 34:105-117. [1988.06.17]

Ross, K. G., Vargo, E. L., Keller, L., Trager, J. C. 1993. Effect of a founder event on variation in the genetic sex-determining system of the fire ant *Solenopsis invicta*. Genetics 135:843-854. [1993.11]

Rossbach, M. H., Majer, J. D. 1983. A preliminary survey of the ant fauna of the Darling Plateau and Swan Coastal Plain near Perth, Western Australia. J. R. Soc. West. Aust. 66:85-90. [1983.12.08]

Röszler, P. 1935 ("1933-34"). Beiträge zur Kenntnis der Ameisenfauna von Siebenbürgen und Ungarn. Verh. Mitt. Siebenb. Ver. Naturwiss. Hermannstadt 83-84:72-83. [1935]

Röszler, P. 1936a. Beitrag zur Kenntnis der Ameisenfauna von Japan. Entomol. Anz. 16:121-122. [1936.06.01]

Röszler, P. 1936b. Beiträge zur Kenntnis der Ameisenfauna von Mitteleuropa. III. Teil der Arbeit: "Ein Versuch der systematischen Einteilung der mitteleuropäischen *Tetramorium*". Tijdschr. Entomol. 79:55-63. [1936.06]

Röszler, P. 1937a. Beitrag zur Kenntnis der Verbreitung der *Serviformica picea* Nyl. Entomol. Rundsch. 55:57-60. [1937.11.01]

Röszler, P. 1937b. Beitrag zur Kenntnis der Verbreitung der *Serviformica picea* Nyl. (Fortsetzung und Schluss.) Entomol. Rundsch. 55:76-77. [1937.11.15]

Röszler, P. 1937c ("1935-36"). Morphologie und Nestbau der *Serviformica picea* Nyl. Pol. Pismo Entomol. 14-15:215-226. [1937.12.04]

Röszler, P. 1937d ("1935-36"). Beiträge zur Kenntnis der Ameisenfauna von Spanien und anderer mitteleuropäischer Länder. II. Teil der Arbeit: "Ein Versuch zur systematischen Einteilung der mitteleuropäischen *Tetramorium*". Verh. Mitt. Siebenb. Ver. Naturwiss. Hermannstadt 85-86:195-208. [1937]

Röszler, P. 1942a. Myrmecologisches 1938. Tijdschr. Entomol. 85:50-71. [1942.12]

Röszler, P. 1942b ("1941-42"). Myrmekologische Mitteilungen 1939. Mitt. Arbeitsgem. Naturwiss. Sibiu-Hermannstadt 91-92:27-41. [1942]

Röszler, P. 1943. Ameisen aus Siebenbürgen mit der Beschreibung ihrer Lebensweise. Zool. Anz. 144:41-48. [1943.10.30]

Röszler, P. 1950. Die Ameisenwelt des Nagy Pietrosz, 2305 m (Ungarn) und Umgebung. Zool. Anz. 145:210-225. [1950.10]

Röszler, P. 1951. Myrmecologisches aus dem Jahre 1938. Zool. Anz. 146:88-96. [1951.02]

Roth, D. S., Perfecto, I., Rathcke, B. 1994. The effects of management systems on ground-foraging ant diversity in Costa Rica. Ecol. Appl. 4:423-436. [1994.08]

Rothney, G. A. J. 1889. Notes on Indian ants. Trans. Entomol. Soc. Lond. 1889:347-374. [1889.10.26]

Rothney, G. A. J. 1890. Notes on Indian ants. J. Bombay Nat. Hist. Soc. 5:38-64. [1890] [Reprint of Rothney (1889).]

Rothney, G. A. J. 1903. The aculeate Hymenoptera of Barrackpore, Bengal. Trans. Entomol. Soc. Lond. 1903:93-116. [1903.04.29]

Rotramel, G. L. 1967 ("1966"). The comparative morphology of the male genitalia of ants. [Abstract.] Proc. North Cent. Branch Entomol. Soc. Am. 21:96. [1967.01.21]

Ruano, F., Ballesta, M., Hidalgo, J., Tinaut, A. 1995. Mirmecocenosis del Paraje Natural Punta Entinas-El Sabinar (Almería) (Hymenoptera: Formicidae). Aspectos ecológicos. Bol. Asoc. Esp. Entomol. 19:89-107. [1995.04]

Ruano, F., Tinaut, A. 1993. Estructura del nido de *Cataglyphis floricola* Tinaut, 1993. Estudio comparado con los hormigueros de *C. iberica* (Emery, 1906) y *C. rosenhaueri* (Emery, 1906) (Hymenoptera: Formicidae). Bol. Asoc. Esp. Entomol. 17:179-189. [1993.11]

Russell, W. E. 1969 ("1968"). Miscellaneous European ant records (Hym., Formicidae). Entomol. Mon. Mag. 104:256. [1969.04.01]

Ruzsky, M. 1895. Faunistic investigations in east Russia 1. Contribution to the ant fauna of east Russia. 2. Zoological excursion in the Orenburg region in 1894. [In Russian.] Tr. Obshch. Estestvoispyt. Imp. Kazan. Univ. 28(5):1-32. [1895]

Ruzsky, M. 1896. Verzeichniss der Ameisen des östlichen Russlands und des Uralgebirges. Berl. Entomol. Z. 41:67-74. [1896.05]

Ruzsky, M. 1902a. Contribution to the ant fauna of the Turgai region. [In Russian.] Rus. Entomol. Obozr. 2:232-235. [1902.08]

Ruzsky, M. 1902b. Neue Ameisen aus Russland. Zool. Jahrb. Abt. Syst. Geogr. Biol. Tiere 17:469-484. [1902.12.18]

Ruzsky, M. 1902c. The ants of the vicinity of the Aral Sea. [In Russian.] Izv. Turkest. Otd. Imp. Rus. Geogr. Obshch. 3(1):1-24. [1902] [Summary in Zool. Zentralbl. 10:166 (10 March 1903).]

Ruzsky, M. 1902d. Material on the ant fauna of the Caucasus and the Crimea. [In Russian.] Protok. Obshch. Estestvoispyt. Imp. Kazan. Univ. 206(suppl.):1-33. [1902]

Ruzsky, M. 1902e. Formicidae. [In Russian.] Pp. 15-16 in: Report on the recorded animal life of Moscow province (No. 4). [In Russian.] Izv. Imp. Obshch. Lyubit. Estestvozn. Antropol. Etnogr. Imp. Mosk. Univ. 98(4):6-18 [= Dnev. Zool. Otd. Imp. Obshch. Lyubit. Estestvozn. Antropol. Etnogr. 3(4):6-18]. [1902]

Ruzsky, M. 1903a. A new species of ant from the Transcaspian region. [In Russian.] Rus. Entomol. Obozr. 3:36-37. [1903.02]

Ruzsky, M. 1903b. Essay on the myrmecofauna of the Kirghiz steppe. [In Russian.] Tr. Rus. Entomol. Obshch. 36:294-316. [1903.07.13]

Ruzsky, M. 1903c. Ants of the Transbaikalian region. [In Russian.] Rus. Entomol. Obozr. 3:205-207. [1903.08]

Ruzsky, M. 1904a. On ants from Archangel province. [In Russian.] Zap. Imp. Rus. Geogr. Obshch. Obshch. Geogr. 41:287-294. [1904]

Ruzsky, M. 1904b. Ants of the Dzungarian Ala Tau. [In Russian.] Izv. Imp. Tomsk. Univ. 30(5th part):1-6. [1904] [Date on cover of volume 30 is 1908, but individual articles were evidently released earlier. Dating of Ruzsky's paper is based on citation in Ruzsky (1905). The paper was apparently intended for publication in volume 24 but does not appear there.]

Ruzsky, M. 1905a. Über *Tetramorium striativentre* Mayr und *Tetr. schneideri* Emery. Zool. Anz. 29:517-518. [1905.11.14]

Ruzsky, M. 1905b. The ants of Russia. (Formicariae Imperii Rossici). Systematics, geography and data on the biology of Russian ants. Part I. [In Russian.] Tr. Obshch. Estestvoispyt. Imp. Kazan. Univ. 38(4-6):1-800. [1905]

Ruzsky, M. 1907a. List of ants from Minsk province, collected on the study trip of the Moscow student group. [In Russian.] Tr. Stud. Kruzh. Izsled. Rus. Prir. Imp. Mosk. Univ. 3:99-103. [Not seen. Cited in Ruzsky (1907c:123) and in the Zoological Record for 1907.]

Ruzsky, M. 1907b. On the ant fauna of Vilna province. [In Russian.] Tr. Stud. Kruzh. Izsled. Rus. Prir. Imp. Mosk. Univ. 3:104-105. [Not seen. Cited in Ruzsky (1907c:123) and in the Zoological Record for 1907.]

Ruzsky, M. 1907c. The ants of Russia. (Formicariae Imperii Rossici). Systematics, geography and data on the biology of Russian ants. Part II. [In Russian.] Tr. Obshch. Estestvoispyt. Imp. Kazan. Univ. 40(4):1-122 + 3 pp. [1907]

Ruzsky, M. 1912. Myrmecological notes. [In Russian.] Uch. Zap. Kazan. Vet. Inst. 29:629-636. [1912]

Ruzsky, M. 1914a ("1913"). Myrmekologische Notizen. Arch. Naturgesch. (A)79(9):58-63. [1914.05] [Heft 9 not dated, but probably published between February 1914 and May 1914, based on other dated issues of volume (A)79.]

Ruzsky, M. 1914b. Ants of Surgut district, Tobolsk province. [In Russian.] Rus. Entomol. Obozr. 14:100-105. [1914.07.14]

Ruzsky, M. 1914c. Eine neue Ameisenform aus dem europäischen Russland. Rus. Entomol. Obozr. 14:323. [1914.10.14]

Ruzsky, M. 1915a. On the ants of Tibet and the southern Gobi. On material collected on the expedition of Colonel P. K. Kozlov. [In Russian.] Ezheg. Zool. Muz. 20:418-444. [1915.10]

Ruzsky, M. 1915b. Material on Siberian myrmecology. First output. On the myrmecofauna of Tomsk province and certain other Siberian localities. (From research in 1914-1915). [In Russian.] Izv. Imp. Tomsk. Univ. 64(5th part):1-14. [1915]

Ruzsky, M. 1916. On zoological research in Yeniseisk province, work of summer of 1915. [In Russian.] Izv. Imp. Tomsk. Univ. 65(3rd part):1-21. [1916]

Ruzsky, M. 1920. Ants of Kamchatka. [In Russian.] Izv. Inst. Issled. Sib. 2:76-80. [Not seen. Cited in Dlussky (1967a:215).]

Ruzsky, M. 1923. Ants of Cheleken Island. [In Russian.] Izv. Tomsk. Gos. Univ. 72(2nd part):1-6. [1923]

Ruzsky, M. 1924. A new *Leptothorax*-ant species from Siberia. [In Russian.] Izv. Tomsk. Gos. Univ. 74:152. [1924]

Ruzsky, M. 1925a. Material on the fauna of the spa "Karachinskoe Ozero". [In Russian.] Izv. Tomsk. Gos. Univ. 75:283-290. [1925]

Ruzsky, M. 1925b. New data on the ant fauna of Siberia. [In Russian.] Rus. Entomol. Obozr. 19:41-46. [1925]

Ruzsky, M. 1926. A systematic list of the ants found in Siberia. I. Review of the species of the genera *Camponotus* (s. ext.) and *Formica* (s. str.). [In Russian.] Izv. Tomsk. Gos. Univ. 77:107-111. [1926]

Ruzsky, M. 1936. Ants of the Transbaikal region. [In Russian.] Tr. Biol. Nauchn.-Issled. Inst. Tomsk. Gos. Univ. 2:89-97. [1936]

Ruzsky, M. 1946. Ants of Tomsk province and contiguous localities. [In Russian.] Tr. Tomsk. Gos. Univ. 97:69-72. [Not seen. Cited in Radchenko (1994g:77).]

Ruzsky, M., Gordyagin, A. 1894. Some data on the ant fauna of east Russia. [In Russian.] Tr. Obshch. Estestvoispyt. Imp. Kazan. Univ. 27(2):1-33. [1894]

Ruzsky, M. D. See: Ruzsky, M.

Saaristo, M. I. 1986. The fourth record of *Sifolinia karavajevi* (Arnoldi) (Hym., Formicidae) from the Nordic countries. Not. Entomol. 66:97-98. [1986.09.23]

Saaristo, M. I. 1990. New provincial records for twelve ant species (Hymenoptera, Formicidae) from Finland. Entomol. Fenn. 1:191-192. [1990.12.03]

Saaristo, M. I. 1995. Distribution maps of the outdoor myrmicid ants (Hymenoptera, Formicidae) of Finland, with notes on their taxonomy and ecology. Entomol. Fenn. 6:153-162. [1995.12.05]

Sadil, J. V. 1939a. Nová varieta druhu *Leptothorax unifasciatus* Latr. (Hymenoptera Formicidae). Cas. Cesk. Spol. Entomol. 36:30. [1939.03.15]

Sadil, J. V. 1939b. Príspevek k soupisu ceskych mravencu. Cas. Cesk. Spol. Entomol. 36:84-85. [1939.11.01]

Sadil, J. V. 1939c. Mravenec *Messor semirufus* André var. *meridionalis* André na Slovensku (Hym., Form.). Entomol. Listy 2:40-41. [1939]

Sadil, J. V. 1939d. Mravenec *Myrmica moravica* Soudek u Prahy. Cas. Nár. Mus. (Prague) 113(2):106-114. [1939]

Sadil, J. V. 1940. Nekolik kritickych poznámek o mravenci *Myrmica moravica* Soudek. Entomol. Listy 3:102-107. [1940]

Sadil, J. V. 1945. Príspevek k poznání mravencí zvíreny Ceskomoravské vysociny. Entomol. Listy 8:11-20. [1945.01.01]

Sadil, J. V. 1952 ("1951"). A revision of the Czechoslovak forms of the genus *Myrmica* Latr. (Hym.). Sb. Entomol. Oddel. Nár. Mus. Praze 27:233-278. [1952]

Sadil, J. V. 1953a. *Epimyrma záleskyí* nov. spec. (Hym., Formicoidea). Roc. Cesk. Spol. Entomol. 50:188-196. [1953.10.01]

Sadil, J. V. 1953b. Príspevek k poznání mravencí zvíreny nasich hor (Hym., Formicoidea). Roc. Cesk. Spol. Entomol. 50:197-202. [1953.10.01]

Sadil, J. V. 1953c. Príspevek k poznání mravencu sirsího okolí Prahy (Hym., Formicoidea). Roc. Cesk. Spol. Entomol. 50:203-205. [1953.10.01]

Sadil, J. V. 1953d. Mravencí zvírena nasich vápencovych skalních stepí (Hym., Formicoidea). Roc. Cesk. Spol. Entomol. 50:206-209. [1953.10.01]

Safford, W. E. 1922. Ant acacias and acacia ants of Mexico and Central America. Annu. Rep. Smithson. Inst. 1921:381-394. [1922]

Sahlberg, J. 1913. *Ponera punctatissima* Roger, funnen i Jyväskylä-trakten. Medd. Soc. Fauna Flora Fenn. 39:68-73. [1913.02.01]

Saini, M. S. 1984. Propleural transformations with respect to the disposition of propleural suture in order Hymenoptera. J. N. Y. Entomol. Soc. 92:150-155. [1984.08.09]

Saini, M. S. 1985. Comparative study of the mesoprepectal area in Hymenoptera. J. Entomol. Res. (New Delhi) 9:132-136. [1985.12]

Saini, M. S., Dhillon, S. S. 1978a. Antenna-cleaning apparatus and its evolutionary trends in Hymenoptera. Entomon 3:77-84. [1978.06]

Saini, M. S., Dhillon, S. S. 1978b. Structural modifications pertaining to mesopostnotum and epimeropostnotal bridge in the order Hymenoptera. J. Entomol. Res. (New Delhi) 2:142-147. [1978.12]

Saini, M. S., Dhillon, S. S. 1979a ("1978"). Adaptational modifications of the first and second abdominal segments in order Hymenoptera. J. Anim. Morphol. Physiol. 25:44-53. [1979.06]

Saini, M. S., Dhillon, S. S. 1979b ("1978"). Functional modifications of the meta-tibial spurs in order Hymenoptera. J. Anim. Morphol. Physiol. 25:54-60. [1979.06]

Saini, M. S., Dhillon, S. S. 1979c. Comparative morphology of galea and lacinia in Hymenoptera. Entomon 4:149-155. [1979.06]

Saini, M. S., Dhillon, S. S. 1979d. Glossal and paraglossal transformation in order Hymenoptera. Entomon 4:355-360. [1979.12]

Saini, M. S., Dhillon, S. S. 1980. Metapleural transformations with respect to propodeum and metapostnotum in Hymenoptera. Fla. Entomol. 63:286-292. [1980.09.15]

Saini, M. S., Dhillon, S. S., Aggarwal, R. 1982. Skeletomuscular differences in the thorax of winged and non-winged forms of *Camponotus camelinus* (Smith) (Hym., Formicidae). Dtsch. Entomol. Z. (N.F.)29:447-458. [1982.11.01]

Saini, M. S., Dhillon, S. S., Kaur, S. 1981. Comparative morphology of the poison apparatus in some apocritan families of Hymenoptera. Rev. Bras. Biol. 41:371-377. [1981.05]

Saini, M. S., Dhillon, S. S., Singh, T. 1979. Positional variations and modifications relating to the protergum in Hymenoptera. J. N. Y. Entomol. Soc. 87:208-215. [1979.10.18]

Salinas, P. J. 1992. Ants of Venezuela. Catalogue and distribution. [Abstract.] P. 254 in: Proceedings XIX International Congress of Entomology. Abstracts. Beijing, China, June 28 - July 4, 1992. Beijing: XIX International Congress of Entomology, xi + 730 pp. [1992.07.04]

Samsinák, K. 1949. Jak poznáme nase mravence v prírode? Vesmir 1948-49:149-151. [1949.04.15]

Samsinák, K. 1950a. Dalsí gynandromorfní mravenec z Cech. (Gynandromorphus apud *Myrmica scabrinodis* Nyl. 1846.) Entomol. Listy 13:116-118. [1950.06.30]

Samsinák, K. 1950b. Hromadné rojení mravencu na zámku Humprechte u Sobotky. Entomol. Listy 13:164-166. [1950.12.30]

Samsinák, K. 1951. *Formica fusca* r. *lemani* Bondr. (Hym. Formic.). Cas. Cesk. Spol. Entomol. 48:122-127. [1951.11.30]

Samsinák, K. 1952. Mravenci ze Sobotecka. Cas. Cesk. Spol. Entomol. 49:69-81. [1952.10.01]

Samsinák, K. 1956. Einige interessante Ameisenarten aus dem Elbsandsteingebirge. Abh. Ber. Staatl. Mus. Tierkd. Dres. 23:9-13. [1956]

Samsinák, K. 1957. *Sifolinia pechi* n. sp. (Hymenoptera, Formicidae). Cas. Cesk. Spol. Entomol. 53:167-170. [1957]

Samsinák, K. 1964. Zur Kenntnis der Ameisenfauna der Tschechoslowakei (Hym.). Cas. Cesk. Spol. Entomol. 61:156-158. [1964.04.20]

Samsinák, K. 1965. Mravenec *Formica uralensis* Ruzsky nalezen v blízkosti nasich hranic. Ziva (n.s.)13:142. [1965.07]

Samsinák, K. 1967. *Camponotus novotnyi* sp. n., eine neue tertiäre Ameise aus Böhmen. Vestn. Ustred. Ustavu Geol. 42:365-366. [1967.09]

Samways, M. J. 1983. Community structure of ants (Hymenoptera: Formicidae) in a series of habitats associated with citrus. J. Appl. Ecol. 20:833-847. [1983.12]

Samways, M. J. 1990. Ant assemblage structure and ecological management in citrus and subtropical fruit orchards in southern Africa. Pp. 570-587 in: Vander Meer, R. K., Jaffe, K., Cedeno, A. (eds.) Applied myrmecology: a world perspective. Boulder: Westview Press, xv + 741 pp. [1990]

Samways, M. J., Lelyveld, L. J. van. 1982. Disc electrophoresis comparisons of the soluble proteins and esterases in some South African ants (Hymenoptera: Formicidae). Phytophylactica 14:7-11. [1982.03]

Samways, M. J., Nel, M., Prins, A. J. 1982. Ants (Hymenoptera: Formicidae) foraging in citrus trees and attending honeydew-producing Homoptera. Phytophylactica 14:155-157. [1982.12]

Sanders, C. J. 1970. The distribution of carpenter ant colonies in the spruce-fir forests of northwestern Ontario. Ecology 51:865-873. [1970]

Sanderson, M. W., Farr, T. H. 1960. Amber with insect and plant inclusions from the Dominican Republic. Science (Wash. D. C.) 131:1313. [1960.04.29]

Sanetra, M., Heinze, J., Buschinger, A. 1994. Enzyme polymorphism in the ant genus *Tetramorium* Mayr and its social parasites (Hymenoptera: Formicidae). Biochem. Syst. Ecol. 22:753-759. [1994.10]

Santis, L. de See: De Santis, L.

Santschi, F. 1906. A propos de moeurs parasitiques temporaires des fourmis du genre *Bothriomyrmex*. Ann. Soc. Entomol. Fr. 75:363-392. [1906.11]

Santschi, F. 1907. Fourmis de Tunisie capturées en 1906. Rev. Suisse Zool. 15:305-334. [1907.11.15]

Santschi, F. 1908. Nouvelles fourmis de l'Afrique du Nord (Égypte, Canaries, Tunisie). Ann. Soc. Entomol. Fr. 77:517-534. [1908.12.23]

Santschi, F. 1909a. Une nouvelle fourmi (Hym.) de Cuba. Bull. Soc. Entomol. Fr. 1909:309-310. [1909.12.13]

Santschi, F. 1909b. Sur la signification de la barbe des fourmis arénicoles. Rev. Suisse Zool. 17:449-458. [1909.12.30]

Santschi, F. 1909c. *Leptothorax rottenbergi* et espèces voisines. Rev. Suisse Zool. 17:459-482. [1909.12.30]

Santschi, F. 1910a. Nouvelles fourmis de Tunisie (3e note). Bull. Soc. Hist. Nat. Afr. Nord 1: 43-46. [1910.01.15]

Santschi, F. 1910b. Nouvelles fourmis de Tunisie (suite). Bull. Soc. Hist. Nat. Afr. Nord 1:61-64. [1910.02.15]

Santschi, F. 1910c ("1909"). Formicides nouveaux ou peu connus du Congo français. Ann. Soc. Entomol. Fr. 78:349-400. [1910.02.23]

Santschi, F. 1910d. Nouvelles fourmis de Tunisie (suite). Bull. Soc. Hist. Nat. Afr. Nord 1:70-72. [1910.03.15]

Santschi, F. 1910e. Contributions à la faune entomologique de la Roumanie. Formicides capturées par Mr. A. L. Montandon. Bul. Soc. Rom. Stiinte 19:648-652. [1910.09]

Santschi, F. 1910f. Mission [Gruvel et Chudeau] en Mauritanie occidentale. III. Partie zoologique. Hyménoptères. 2e partie. Actes Soc. Linn. Bordx. 64:233-234. [1910.10.08]

Santschi, F. 1910g. Nouveaux dorylines africains. Rev. Suisse Zool. 18:737-759. [1910.12.15]

Santschi, F. 1910h. Deux nouvelles fourmis du Tonkin. Naturaliste 32:283-284. [1910.12.15]

Santschi, F. 1910i. [Untitled. *Technomyrmex nigriventris*, n. sp.] P. 22 in: Forel, A. Zoologische und anthropologische Ergebnisse einer Forschungsreise im westlichen und zentralen Südafrika ausgeführt in den Jahren 1903-1905 von Dr. Leonhard Schultze. Vierter Band. Systematik und Tiergeographie. D) Formicidae. Denkschr. Med.-Naturwiss. Ges. Jena 15:1-30. [1910]

Santschi, F. 1911a. Formicides nouveaux de l'Afrique Mineure (4e note). Bull. Soc. Hist. Nat. Afr. Nord 2:11-14. [1911.01.15]

Santschi, F. 1911b. Deux nouvelles fourmis de Buenos-Ayres (Hym. Formicidae). Bull. Soc. Entomol. Fr. 1911:52-53. [1911.02.22]

Santschi, F. 1911c ("1910"). Nouvelle fourmis d'Afrique. Ann. Soc. Entomol. Fr. 79:351-369. [1911.02.22]

Santschi, F. 1911d ("1909"). Formicides récoltés par Mr. le Prof. F. Silvestri aux Etats Unis en 1908. Bull. Soc. Entomol. Ital. 41:3-7. [1911.02.28]

Santschi, F. 1911e. Nouvelles fourmis de Madagascar. Rev. Suisse Zool. 19:117-134. [1911.02]

Santschi, F. 1911f. Formicides nouveaux de l'Afrique Mineure (4e note suite). Bull. Soc. Hist. Nat. Afr. Nord 2:78-85. [1911.05.15]

Santschi, F. 1911g. Nouvelles fourmis du Congo et du Benguela. Rev. Zool. Afr. (Bruss.) 1:204-217. [1911.08.31]

Santschi, F. 1911h. Une nouvelle espèce d'*Eciton*. Berl. Entomol. Z. 56:113. [1911.08]

Santschi, F. 1911i. Formicides de diverses provenances. Ann. Soc. Entomol. Belg. 55:278-287. [1911.10.06]

Santschi, F. 1911j. Une nouvelle variété de *Formica rufa* L. (Hym. Formicidae). Bull. Soc. Entomol. Fr. 1911:349-350. [1911.11.22]

Santschi, F. 1912a. Un *Carebara américain* (Hym. Formicidae). Bull. Soc. Entomol. Fr. 1912:139-141. [1912.04.10]

Santschi, F. 1912b. Fourmis d'Afrique et de Madagascar. Ann. Soc. Entomol. Belg. 56:150-167. [1912.06.01]

Santschi, F. 1912c. Quelques nouvelles variétés de fourmis africaines. Bull. Soc. Hist. Nat. Afr. Nord 3:147-149. [1912.07.15]

Santschi, F. 1912d. Deux nouveaux *Carebara* africains (Hym. Formicidae). Bull. Soc. Entomol. Fr. 1912:284-286. [1912.08.29]

Santschi, F. 1912e. Quelques fourmis de l'Amérique australe. Rev. Suisse Zool. 20:519-534. [1912.09]

Santschi, F. 1912f. Nouvelles fourmis de Tunisie récoltées par le Dr. Normand. Bull. Soc. Hist. Nat. Afr. Nord 3:172-175. [1912.11.15]

Santschi, F. 1912g. Contributions à la faune entomologique de la Roumanie. Description d'une nouvelle espèce de Formicide. Bul. Soc. Rom. Stiinte 20:657-658. [1912.12]

Santschi, F. 1913a ("1912"). *Cremastogaster* du groupe *tricolor-Menileki* (Hym. Formicidae). Bull. Soc. Entomol. Fr. 1912:411-414. [1913.01.08]

Santschi, F. 1913b. Clé analytique des fourmis africaines du genre *Strumigenys* Sm. (Hym.). Bull. Soc. Entomol. Fr. 1913:257-259. [1913.06.11]

Santschi, F. 1913c. Glanures de fourmis africaines. Ann. Soc. Entomol. Belg. 57:302-314. [1913.11.07]

Santschi, F. 1913d. Un nouvel *Oligomyrmex* de Cochinchine (Hym. Formicidae). Bull. Soc. Entomol. Fr. 1913:457-459. [1913.12.10]
Santschi, F. 1913e. Clé dichotomique des *Oligomyrmex* africains (Hym. Formicidae). Bull. Soc. Entomol. Fr. 1913:459-460. [1913.12.10]
Santschi, F. 1913f. Une nouvelle fourmis parasite. Bull. Soc. Hist. Nat. Afr. Nord 4:229-230. [1913.12.15]
Santschi, F. 1913g. Genre nouveau et espèce nouvelle de Formicides (Hym.). Bull. Soc. Entomol. Fr. 1913:478. [1913.12.30]
Santschi, F. 1913h. Hyménoptères. Formicides. Pp. 33-43 in: André, E. et al. Mission du service géographique de l'armée pour la mesure d'un arc de méridien équatorial en Amérique du Sud. Tome 10. Fasc. 1. Insectes. Paris: Gauthier-Villars, 119 pp. [1913]
Santschi, F. 1914a ("1913"). Mélanges myrmecologiques. Ann. Soc. Entomol. Belg. 57:429-437. [1914.01.02]
Santschi, F. 1914b. Voyage de Ch. Alluaud et R. Jeannel en Afrique Orientale, 1911-1912. Résultats scientifiques. Insectes Hyménoptères. II. Formicidae. Paris: Libr. A. Schulz, pp. 41-148. [1914.02.25]
Santschi, F. 1914c. XII. Fam. Formicidae. P. 288 in: Schulthess, A. v. Hymenopteren aus Kamerun. Gesammelt von Herrn von Rothkirch, Oberleutnant der Schutztruppe. Dtsch. Entomol. Z. 1914:283-297. [1914.06.01]
Santschi, F. 1914d. Formicides de l'Afrique occidentale et australe du voyage de Mr. le Professeur F. Silvestri. Boll. Lab. Zool. Gen. Agrar. R. Sc. Super. Agric. 8:309-385. [1914.07.29]
Santschi, F. 1914e. Meddelanden från Göteborgs Musei Zoologiska Afdelning. 3. Fourmis du Natal et du Zoulouland récoltées par le Dr. I. Trägårdh. Göteb. K. Vetensk. Vitterh. Samh. Handl. 15:1-44. [1914]
Santschi, F. 1915a. Nouvelles fourmis d'Algérie, Tunisie et Syrie. Bull. Soc. Hist. Nat. Afr. Nord 6:54-63. [1915.04.25]
Santschi, F. 1915b. Deux *Cryptocerus* nouveaux (Hym. Formicidae). Bull. Soc. Entomol. Fr. 1915:207-209. [1915.08.07]
Santschi, F. 1915c. Nouvelles fourmis d'Afrique. Ann. Soc. Entomol. Fr. 84:244-282. [1915.10.27]
Santschi, F. 1916a. Description d'un nouveau Formicide (Hym.) de l'Afrique occidentale. Bull. Soc. Entomol. Fr. 1916:50-51. [1916.02.23]
Santschi, F. 1916b ("1915"). Descriptions de fourmis nouvelles d'Afrique et d'Amérique. Ann. Soc. Entomol. Fr. 84:497-513. [1916.04]
Santschi, F. 1916c. Deux nouvelles fourmis d'Australie. Bull. Soc. Entomol. Fr. 1916:174-175. [1916.06.20]
Santschi, F. 1916d. Rectifications à la nomenclature de quelques Formicides (Hym.). Bull. Soc. Entomol. Fr. 1916:242-243. [1916.11.13]
Santschi, F. 1916e. Formicides sudaméricains nouveaux ou peu connus. Physis (B. Aires) 2:365-399. [1916.12.30]
Santschi, F. 1917a ("1916"). Description d'une nouvelle reine de Formicide du genre *Aenictus* Shuckard. Ann. Soc. Entomol. Fr. 85:277-278. [1917.01.10]
Santschi, F. 1917b ("1916"). Fourmis nouvelles de la Colonie du Cap, du Natal et de Rhodesia. Ann. Soc. Entomol. Fr. 85:279-296. [1917.01.10]
Santschi, F. 1917c. Note sur *Dorylus affinis* Shuckard mâle et ses variétés. Bull. Soc. Hist. Nat. Afr. Nord 8:18-21. [1917.01.15]
Santschi, F. 1917d. *Acantholepis Frauenfeldi* Mayr et ses variétés. Bull. Soc. Hist. Nat. Afr. Nord 8:42-48. [1917.02.15]
Santschi, F. 1917e. Races et variétés nouvelles du *Messor barbarus* L. Bull. Soc. Hist. Nat. Afr. Nord 8:89-94. [1917.04.15]
Santschi, F. 1917f. Description de quelques nouvelles fourmis de la République Argentine. An. Soc. Cient. Argent. 84:277-283. [1917.12]

Santschi, F. 1918a. *Leptothorax* nouveaux de l'Afrique Mineure. Bull. Soc. Hist. Nat. Afr. Nord 9:31-38. [1918.02.15]
Santschi, F. 1918b. Nouveaux *Tetramorium* africains. Bull. Soc. Hist. Nat. Afr. Nord 9:121-132. [1918.06.15]
Santschi, F. 1918c. Nouveaux *Tetramorium* africains (Suite et fin). Bull. Soc. Hist. Nat. Afr. Nord 9:153-156. [1918.07.15]
Santschi, F. 1918d. Sous-genres et synoymies [sic] de *Cremastogaster* (Hym. Formic.). Bull. Soc. Entomol. Fr. 1918:182-185. [1918.08.27]
Santschi, F. 1918e. Cinq notes myrmécologiques. Bull. Soc. Vaudoise Sci. Nat. 52(procès-verbaux):63-64. [1918.12]
Santschi, F. 1919a. Cinq notes myrmécologiques. Bull. Soc. Vaudoise Sci. Nat. 52:325-350. [1919.03.14]
Santschi, F. 1919b. Fourmis nouvelles éthiopiennes. Rev. Zool. Afr. (Bruss.) 6:229-240. [1919.04.01]
Santschi, F. 1919c. Fourmis nouvelles du Congo. Rev. Zool. Afr. (Bruss.) 6:243-250. [1919.04.01]
Santschi, F. 1919d. Nouveaux genre et sous-genre de fourmis barbaresques (Hym.). Bull. Soc. Entomol. Fr. 1919:90-92. [1919.04.18]
Santschi, F. 1919e. Fourmis d'Espagne et des Canaries. Bol. R. Soc. Esp. Hist. Nat. 19:241-248. [1919.04]
Santschi, F. 1919f. Nouveaux formicides de la République Argentine. An. Soc. Cient. Argent. 87:37-57. [1919.06]
Santschi, F. 1919g. Trois nouvelles fourmis des Canaries. Bol. R. Soc. Esp. Hist. Nat. 19:405-407. [1919.07]
Santschi, F. 1919h. Nouvelles fourmis du Congo Belge du Musée du Congo Belge, à Tervueren. Rev. Zool. Afr. (Bruss.) 7:79-91. [1919.08.15]
Santschi, F. 1919i. Fourmis du genre *Bothriomyrmex* Emery. (Systématique et moeurs.) Rev. Zool. Afr. (Bruss.) 7:201-224. [1919.12.15]
Santschi, F. 1919j. Fam. Formicidae. P. 290 in: Vitalis de Salvaza, R. Essai d'un traité d'entomologie indochinoise. Hanoi: Impr. Minsang, xi + 308 pp. [1919]
Santschi, F. 1920a. Quelques nouveaux Camponotinae d'Indochine et Australie. Bull. Soc. Vaudoise Sci. Nat. 52:565-569. [1920.20.16]
Santschi, F. 1920b. Formicides nouveaux du Gabon, du Congo, de la Rhodesia et du Natal. Ann. Soc. Entomol. Belg. 60:6-17. [1920.04.08]
Santschi, F. 1920c. Fourmis nouvelles du Congo Belge. Rev. Zool. Afr. (Bruss.) 8:118-120. [1920.05.15]
Santschi, F. 1920d ("1919"). Formicides africains et américains nouveaux. Ann. Soc. Entomol. Fr. 88:361-390. [1920.06.30]
Santschi, F. 1920e. Quelques nouvelles fourmis de Bolivie (expédition Lizer-Delétang, 1917). An. Soc. Cient. Argent. 89:122-126. [1920.06]
Santschi, F. 1920f. Nouvelles fourmis du genre *Cephalotes* Latr. Bull. Soc. Entomol. Fr. 1920:147-149. [1920.07.20]
Santschi, F. 1920g. Cinq nouvelles notes sur les fourmis. Bull. Soc. Vaudoise Sci. Nat. 53:163-186. [1920.09.15]
Santschi, F. 1920h. Fourmis d'Indo-Chine. Ann. Soc. Entomol. Belg. 60:158-176. [1920.12.03]
Santschi, F. 1920i. Études sur les maladies et les parasites du cacaoyer et d'autres plantes cultivées à S. Thomé. X. Fourmis de S. Thomé. Extrait des Mémoires publiés par la Société Portugaise des Sciences Naturelles. Lisbonne: Imprimerie de la Librairie Ferin, 4 pp. [1920] [See also Santschi, 1922f.]
Santschi, F. 1921a. Notes sur les fourmis paléarctiques. II. Fourmis d'Asie Mineure récoltées par M. H. Gadeau de Kerville. Bol. R. Soc. Esp. Hist. Nat. 21:110-116. [1921.02]

Santschi, F. 1921b. Notes sur les fourmis paléarctiques. I. Quelques fourmis du nord de l'Afrique et des Canaries. Mem. R. Soc. Esp. Hist. Nat., Tomo del Cincuentenario:424-436. [1921.03.15]

Santschi, F. 1921c. Quelques nouveaux Formicides africains. Ann. Soc. Entomol. Belg. 61:113-122. [1921.04.02]

Santschi, F. 1921d. Formicides nouveaux de l'Afrique du Nord. Bull. Soc. Hist. Nat. Afr. Nord 12:68-77. [1921.04.15]

Santschi, F. 1921e. Nouvelles fourmis paléarctiques. 3ème. note. Bol. R. Soc. Esp. Hist. Nat. 21:165-170. [1921.04]

Santschi, F. 1921f. Retouches aux sous-genres de *Camponotus*. Ann. Soc. Entomol. Belg. 61:310-312. [1921.09.02]

Santschi, F. 1921g. Ponerinae, Dorylinae et quelques autres formicides néotropiques. Bull. Soc. Vaudoise Sci. Nat. 54:81-103. [1921.11.16]

Santschi, F. 1921h. Quelques nouveaux *Cryptocerus* de l'Argentine et pays voisins. An. Soc. Cient. Argent. 92:124-128. [1921.09]

Santschi, F. 1922a. [Untitled. Descriptions of new *Crematogaster* taxa by Santschi.] Pp. 153-158 in: Wheeler, W. M. Ants of the American Museum Congo Expedition. II. The ants collected by the American Museum Congo Expedition. Bull. Am. Mus. Nat. Hist. 45:39-269. [1922.02.10]

Santschi, F. 1922b. Quelques nouvelles variétés de fourmis paléarctiques. Bull. Soc. Hist. Nat. Afr. Nord 13:66-68. [1922.03.15]

Santschi, F. 1922c. Myrmicines, dolichodérines et autres formicides néotropiques. Bull. Soc. Vaudoise Sci. Nat. 54:345-378. [1922.07.15]

Santschi, F. 1922d. *Camponotus* néotropiques. Ann. Soc. Entomol. Belg. 62:97-124. [1922.08.01]

Santschi, F. 1922e. Description de nouvelles fourmis de l'Argentine et pays limitrophes. An. Soc. Cient. Argent. 94:241-262. [1922.12]

Santschi, F. 1922f. Études sur les maladies et les parasites du cacaoyer et d'autres plantes cultivées à S. Thomé. X. Fourmis de S. Thomé. Mém. Soc. Port. Sci. Nat. Sér. Zool. 2:60-63. [1922] [Appeared earlier in the form of a separate; see Santschi, 1920i.]

Santschi, F. 1923a. Notes sur les fourmis paléarctiques. 4ème note. Bol. R. Soc. Esp. Hist. Nat. 23:133-137. [1923.03]

Santschi, F. 1923b. Revue des fourmis du genre *Brachymyrmex* Mayr. An. Mus. Nac. Hist. Nat. B. Aires 31:650-678. [1923.04.02]

Santschi, F. 1923c. *Solenopsis* et autres fourmis néotropicales. Rev. Suisse Zool. 30:245-273. [1923.06]

Santschi, F. 1923d. *Pheidole* et quelques autres fourmis néotropiques. Ann. Soc. Entomol. Belg. 63:45-69. [1923.07.25]

Santschi, F. 1923e. Descriptions de nouveaux Formicides éthiopiens et notes diverses. I. Rev. Zool. Afr. (Bruss.) 11:259-295. [1923.10.30]

Santschi, F. 1923f. *Messor* et autres fourmis paléarctiques. Rev. Suisse Zool. 30:317-336. [1923.12]

Santschi, F. 1923g. Descriptions de quelques nouvelles fourmis du Brésil. Rev. Mus. Paul. 13:1255-1264. [1923]

Santschi, F. 1924a. Revue du genre *Plectroctena* F. Smith. Rev. Suisse Zool. 31:155-173. [1924.05]

Santschi, F. 1924b. Descriptions de nouveaux Formicides africains et notes diverses. II. Rev. Zool. Afr. (Bruss.) 12:195-224. [1924.06.01]

Santschi, F. 1924c. Fourmis d'Indochine. Opusc. Inst. Sci. Indoch. 3:95-117 [= Faune Entomol. Indoch. Fr. 8:95-117]. [1924.09]

Santschi, F. 1925a. Revision du genre *Acromyrmex* Mayr. Rev. Suisse Zool. 31:355-398. [1925.03]

Santschi, F. 1925b ("1924"). Nouvelles fourmis brésiliennes. Ann. Soc. Entomol. Belg. 64:5-20. [1925.04]
Santschi, F. 1925c ("1924"). Révision des *Myrmicaria* d'Afrique. Ann. Soc. Entomol. Belg. 64:133-176. [1925.04]
Santschi, F. 1925d. Nouveaux Formicides brésiliens et autres. Bull. Ann. Soc. Entomol. Belg. 65:221-247. [1925.07.31]
Santschi, F. 1925e. Fourmis des provinces argentines de Santa Fe, Catamarca, Santa Cruz, Córdoba et Los Andes. Comun. Mus. Nac. Hist. Nat. "Bernardino Rivadavia" 2:149-168. [1925.09.24]
Santschi, F. 1925f. Contribution à la faune myrmécologique de la Chine. Bull. Soc. Vaudoise Sci. Nat. 56:81-96. [1925.11.30]
Santschi, F. 1925g. Fourmis d'Espagne et autres espèces paléarctiques (Hymenopt.). EOS. Rev. Esp. Entomol. 1:339-360. [1925.12.31] [This paper includes an unavailable name, *Messor barbarus* st. *sordidus* v. *tingitanus*, used first here and attributed to Emery, but Santschi appears to be responsible for the description.]
Santschi, F. 1925h. Formicidae. Mission Rohan-Chabot 4(3):159-168. [1925]
Santschi, F. 1926a. Trois notes myrmécologiques. Ann. Soc. Entomol. Fr. 95:13-28. [1926.03.31]
Santschi, F. 1926b. Description de nouveaux Formicides éthiopiens (IIIme partie). Rev. Zool. Afr. (Bruss.) 13:207-267. [1926.04.01]
Santschi, F. 1926c. Nouvelles notes sur les *Campontous*. Rev. Suisse Zool. 33:597-618. [1926.07]
Santschi, F. 1926d. Deux nouvelles fourmis parasites de l'Argentine. Folia Myrmecol. Termit. 1: 6-8. [1926.10]
Santschi, F. 1926e. Quelques fourmis nord-africaines. Bull. Soc. Hist. Nat. Afr. Nord 17:229-236. [1926.11.15]
Santschi, F. 1926f. Travaux scientifiques de l'Armée d'Orient (1916-1918). Fourmis. Bull. Mus. Natl. Hist. Nat. 32:286-293. [1926.11.25]
Santschi, F. 1927a ("1926"). Deux notices sur les fourmis. Bull. Ann. Soc. Entomol. Belg. 66:327-330. [1927.01.10]
Santschi, F. 1927b. A propos du *Tetramorium caespitum* L. Folia Myrmecol. Termit. 1:52-58. [1927.02]
Santschi, F. 1927c. Notes myrmécologiques. Bull. Soc. Entomol. Fr. 1927:126-128. [1927.05.20]
Santschi, F. 1927d. Revision des *Messor* du groupe *instabilis* Sm. (Hymenopt.). Bol. R. Soc. Esp. Hist. Nat. 27:225-250. [1927.05]
Santschi, F. 1927e. Révision myrmécologique. Bull. Ann. Soc. Entomol. Belg. 67:240-248. [1927.11.10]
Santschi, F. 1928a. Formicidae (Fourmis). Insects Samoa 5:41-58. [1928.02.25]
Santschi, F. 1928b. Nouvelles fourmis de Chine et du Turkestan Russe. Bull. Ann. Soc. Entomol. Belg. 68:31-46. [1928.03.03]
Santschi, F. 1928c. Fourmis de îles Fidji. Rev. Suisse Zool. 35:67-74. [1928.05]
Santschi, F. 1928d. Descriptions de nouvelles fourmis éthiopiennes (quatrième note). Rev. Zool. Bot. Afr. 16:54-69. [1928.06.01]
Santschi, F. 1928e. Nouvelles fourmis d'Australie. Bull. Soc. Vaudoise Sci. Nat. 56:465-483. [1928.08.30]
Santschi, F. 1928f. Descriptions de nouvelles fourmis éthiopiennes (suite). Rev. Zool. Bot. Afr. 16:191-213. [1928.09.15]
Santschi, F. 1928g. Quelques nids de fourmis du Muséum d'Histoire Naturelle de Paris. Ann. Sci. Nat. Zool. (10)11:247-259. [1928.09]
Santschi, F. 1928h. Fourmis de Sumatra, récoltées par Mr. J. B. Corporaal. Tijdschr. Entomol. 71:119-140. [1928.12.31]
Santschi, F. 1929a ("1928"). Sur quelques nouvelles fourmis du Brésil (Hym. Form.). Dtsch. Entomol. Z. 1928:414-416. [1929.01.10]
Santschi, F. 1929b. Étude sur les *Cataglyphis*. Rev. Suisse Zool. 36:25-70. [1929.03]

Santschi, F. 1929c. Fourmis du Sahara central récoltées par la Mission du Hoggar (février-mars 1928). Bull. Soc. Hist. Nat. Afr. Nord 20:97-108. [1929.04.15]
Santschi, F. 1929d. Nouvelles fourmis de la République Argentine et du Brésil. An. Soc. Cient. Argent. 107:273-316. [1929.04]
Santschi, F. 1929e. Fourmis du Maroc, d'Algérie et de Tunisie. Bull. Ann. Soc. Entomol. Belg. 69:138-165. [1929.06.15]
Santschi, F. 1929f. Mélange myrmécologique. Wien. Entomol. Ztg. 46:84-93. [1929.09.15]
Santschi, F. 1929g. Note additionnelle aux fourmis du Sahara central récoltées par la Mission du Hoggar (février-mars 1928). Bull. Soc. Hist. Nat. Afr. Nord 20:164-166. [1929.11.15]
Santschi, F. 1929h. Révision du genre *Holcoponera* Mayr. Zool. Anz. 82:437-477. [1929]
Santschi, F. 1930a. Description de Formicides éthiopiens nouveaux ou peu connus. V. Bull. Ann. Soc. Entomol. Belg. 70:49-77. [1930.03.08]
Santschi, F. 1930b. Résultats de la Mission scientifique suisse en Angola, 1928-1929. Formicides de l'Angola. Rev. Suisse Zool. 37:53-81. [1930.03]
Santschi, F. 1930c. Trois notes myrmécologiques. Bull. Ann. Soc. Entomol. Belg. 70:263-270. [1930.12.01]
Santschi, F. 1930d. Un nouveau genre de fourmi parasite sans ouvrières de l'Argentine. Rev. Soc. Entomol. Argent. 3:81-83. [1930.12.26]
Santschi, F. 1930e. Quelques fourmis de Cuba et du Brésil. Bull. Soc. Entomol. Egypte 14:75-83. [1930]
Santschi, F. 1931a. Inventa entomologica itineris Hispanici et Maroccani, quod a. 1926 fecerunt Harald et Håkan Lindberg. Fourmis du Bassin Méditerranéen occidental et du Maroc récoltées par MM. Lindberg. Comment. Biol. 3(14):1-13. [1931.02]
Santschi, F. 1931b. *Engramma taylori*, n. sp. [Translated by H. Howard.] P. 42 in: Taylor, J. S. A note on the fauna of mangrove. Entomol. Rec. J. Var. 43:41-42 + pl.II. [1931.03.15]
Santschi, F. 1931c. Notes sur le genre *Myrmica* (Latreille). Rev. Suisse Zool. 38:335-355. [1931.05]
Santschi, F. 1931d. Fourmis de Cuba et de Panama. Rev. Entomol. (Rio J.) 1:265-282. [1931.09.05]
Santschi, F. 1931e. La reine du *Dorylus fulvus* Westw. Bull. Soc. Hist. Nat. Afr. Nord 22:401-408. [1931.11.15]
Santschi, F. 1931f. Contribution à l'étude des fourmis de l'Argentine. An. Soc. Cient. Argent. 112:273-282. [1931.11]
Santschi, F. 1932a. Deux cas de parasitisme social chez les fourmis. Nature (Paris) 60:457-461. [1932.05.15]
Santschi, F. 1932b. Formicides sud-africains. Pp. 381-392 in: Jeannel, R. (ed.) Société Entomologique de France. Livre du centenaire. Paris: Société Entomologique de France, xii + 729 pp. [1932.06.30]
Santschi, F. 1932c. Notes sur les fourmis du Sahara. Bull. Mus. Natl. Hist. Nat. (2)4:516-520. [1932.08.02]
Santschi, F. 1932d. Résultats scientifiques du voyage aux Indes orientales néerlandaises de LL. AA. RR. le Prince et la Princesse Léopold de Belgique. Hymenoptera. Formicidae. Mém. Mus. R. Hist. Nat. Belg. 4:11-29. [1932.10.15]
Santschi, F. 1932e. Liste de fourmis d'Espagne recueilliés par Mr. J. M. Dusmet. Bol. Soc. Entomol. Esp. 15:69-74. [1932.10.30]
Santschi, F. 1932f. Études sur quelques *Attomyrma* paléarctiques. Mitt. Schweiz. Entomol. Ges. 15:338-346. [1932.12.15]
Santschi, F. 1932g. Quelques fourmis inédites de l'Amérique centrale et Cuba. Rev. Entomol. (Rio J.) 2:410-414. [1932.12.28]
Santschi, F. 1932h. Fourmis du Portugal. Mem. Estud. Mus. Zool. Univ. Coimbra Sér. 1. Zool. Sist. 59:1-3. [1932]

Santschi, F. 1933a. Voyage de MM. L. Chopard et A. Méquignon aux Açores. V. Fourmis. Ann. Soc. Entomol. Fr. 102:21-22. [1933.03.30]

Santschi, F. 1933b. Contribution à l'étude des fourmis de l'Afrique tropicale. Bull. Ann. Soc. Entomol. Belg. 73:95-108. [1933.03.30]

Santschi, F. 1933c. [Untitled. *Anomma titan*, Sants., *vinalli*, var. n.] P. 58 in: Poulton, E. B. The gentle Driver ant "Baongo", discovered by Miss Vinall at Bongadanga, Belgian Congo. Proc. Entomol. Soc. Lond. 7:58. [1933.04.21]

Santschi, F. 1933d. Étude sur le sous-genre *Aphaenogaster* Mayr. Rev. Suisse Zool. 40:389-408. [1933.07]

Santschi, F. 1933e. Formicides des collections de S. A. R. le Prince Léopold de Belgique. Voyage aux Indes orientales, 1932. Bull. Mus. R. Hist. Nat. Belg. 9(27):1-3. [1933.09]

Santschi, F. 1933f. Fourmis de la République Argentine en particulier du territoire de Misiones. An. Soc. Cient. Argent. 116:105-124. [1933.09]

Santschi, F. 1933g. Sur l'origine de la nervure cubitale chez les Formicides. Mitt. Schweiz. Entomol. Ges. 15:557-566. [1933.12.15]

Santschi, F. 1934a. Deux nouveaux *Crematogaster* intéressants. Bull. Soc. Vaudoise Sci. Nat. 58:187-191. [1934.02.25]

Santschi, F. 1934b. Mission J. de Lépiney au Soudan Français 1933-1934. (Huitième note.) Fourmis. Bull. Soc. Sci. Nat. Maroc 14:33-34. [1934.03.31]

Santschi, F. 1934c. Fourmis de Misiones et du Chaco argentin. Rev. Soc. Entomol. Argent. 6: 23-34. [1934.05.30]

Santschi, F. 1934d. Fourmis d'une croisière. Bull. Ann. Soc. Entomol. Belg. 74:273-282. [1934.08.25]

Santschi, F. 1934e. Contribution aux *Solenopsis* paléarctiques. Rev. Suisse Zool. 41:565-592. [1934.10]

Santschi, F. 1934f. Fourmis du Sahara central. Mém. Soc. Hist. Nat. Afr. Nord 4:165-177. [1934] [Reprinting of Santschi (1929c and 1929g).]

Santschi, F. 1935a. Fourmis du Musée du Congo Belge. Rev. Zool. Bot. Afr. 27:254-285. [1935.09.25]

Santschi, F. 1935b. Hymenoptera. I. Formicidae. Mission Sci. Omo 2:255-277. [1935]

Santschi, F. 1936a. Étude sur les fourmis du genre *Monomorium* Mayr. Bull. Soc. Sci. Nat. Maroc 16:32-64. [1936.03.31]

Santschi, F. 1936b. Contribution à l'étude des fourmis de l'Amérique du Sud. Rev. Entomol. (Rio J.) 6:196-218. [1936.07.15]

Santschi, F. 1936c. Liste et descriptions de fourmis du Maroc. Bull. Soc. Sci. Nat. Maroc 16:198-210. [1936.09.30]

Santschi, F. 1936d. Fourmis nouvelles ou intéressantes de la République Argentine. Rev. Entomol. (Rio J.) 6:402-421. [1936.10.30]

Santschi, F. 1936e ("1935"). Hyménoptères. Fourmis. Pp. 81-82 in: Dalloni, M. et al. Mission au Tibesti. 2e volume. Zoologie. Mém. Acad. Sci. Inst. Fr. (2)62:41-92. [1936]

Santschi, F. 1937a. Glanure de fourmis éthiopiennes. Bull. Ann. Soc. Entomol. Belg. 77:47-66. [1937.02.28]

Santschi, F. 1937b. Résultats entomologiques d'un voyage au Cameroun. Formicides récoltés par Mr. le Dr. F. Zumpt. Mitt. Schweiz. Entomol. Ges. 17:93-104. [1937.03.15]

Santschi, F. 1937c. Les sexués du genre *Anillidris* Santschi. Bull. Soc. Entomol. Fr. 42:68-70. [1937.04.21]

Santschi, F. 1937d. Résultats de la Mission scientifique suisse en Angola (2me voyage) 1932-1933. Fourmis angolaises. Rev. Suisse Zool. 44:211-250. [1937.04]

Santschi, F. 1937e. Contribution à l'étude des *Crematogaster* paléarctiques. Mém. Soc. Vaudoise Sci. Nat. 5:295-317. [1937.06.30]

Santschi, F. 1937f. Note sur *Acromyrmex subterraneus* Forel (Hym. Formicidae). Rev. Entomol. (Rio J.) 7:230-233. [1937.07.27]

Santschi, F. 1937g. Fourmis du Congo et régions limitrophes. Rev. Zool. Bot. Afr. 30:71-85. [1937.10.30]
Santschi, F. 1937h. Fourmis du Japon et de Formose. Bull. Ann. Soc. Entomol. Belg. 77:361-388. [1937.10.30]
Santschi, F. 1938a ("1937"). Quelques nouvelles fourmis d'Egypte. Bull. Soc. Entomol. Egypte 21:28-44. [1938.03.31]
Santschi, F. 1938b. Notes sur quelques *Ponera* Latr. Bull. Soc. Entomol. Fr. 43:78-80. [1938.04.15]
Santschi, F. 1938c. Mission Robert Ph. Dollfus en Égypte (décembre 1927 - mars 1929). XV. Hymenoptera: Formicidae. Mém. Inst. Egypte 37:253. [1938]
Santschi, F. 1939a. Contribution au sous-genre *Alaopone* Emery. Rev. Suisse Zool. 46:143-154. [1939.01]
Santschi, F. 1939b. Fourmis de Rhodesia et du Congo. Bull. Ann. Soc. Entomol. Belg. 79:237-246. [1939.04.24]
Santschi, F. 1939c. Trois notes sur quelques fourmis du Musée Royal d'Histoire Naturelle de Belgique. Bull. Mus. R. Hist. Nat. Belg. 15(14):1-15. [1939.04]
Santschi, F. 1939d. Notes sur des *Camponotus* et autres fourmis de l'Afrique Mineure. Bull. Soc. Sci. Nat. Maroc 19:66-87. [1939.06.30]
Santschi, F. 1939e. Études et descriptions de fourmis néotropiques. Rev. Entomol. (Rio J.) 10:312-330. [1939.09.04]
Santschi, F. 1939f. Résultats scientifiques des croisières du navire-école belge, "Mercator". XIV. Formicidae s. lt. Mém. Mus. R. Hist. Nat. Belg. (2)15:159-167. [1939.12.31]
Santschi, F. 1941. Quelques fourmis japonaises inédites. Mitt. Schweiz. Entomol. Ges. 18:273-279. [1941.01.15]
Sartori, M., Cherix, D. 1983. Histoire de l'étude des insectes sociaux en Suisse à travers l'oeuvre d'Auguste Forel. Bull. Soc. Entomol. Fr. 88:66-74. [1983.07.27]
Satchell, J. E., Collingwood, C. A. 1955. The wood ants of the English Lake District. North West. Nat. (n.s.)3:23-29. [1955.03]
Satoh, T. 1989. Comparisons between two apparently distinct forms of *Camponotus nawai* Ito (Hymenoptera: Formicidae). Insectes Soc. 36:277-292. [1989.09]
Satoh, T., Masuko, K., Matumoto, T. 1994. DNA polymorphism of the ants, *Camponotus nawai* complex. [Abstract.] [In Japanese.] Ari 18:33-34. [1994.05.30]
Saulcy, F. de. 1874. Species des paussides, clavigérides, psélaphides & scydménides de l'Europe & des pays circonvoisins. Bull. Soc. Hist. Nat. Dep. Moselle Metz 13:1-132. [1874] [*Pheidole jordanica* n. sp., page 17.]
Saunders, E. 1880. Synopsis of the British Heterogyna and fossorial Hymenoptera. Trans. Entomol. Soc. Lond. 1880:201-304. [1880.12.31] [Date from Wheeler, G. (1912).]
Saunders, E. 1883. Notes on British ants. Entomol. Mon. Mag. 20:16-17. [1883.06]
Saunders, E. 1886a. The male of *Formicoxenus nitidulus*, Nyl. Entomol. Mon. Mag. 23:42. [1886.07]
Saunders, E. 1886b. *Ponera punctatissima*, Rog., at Bromley, Kent. Entomol. Mon. Mag. 23:68. [1886.08]
Saunders, E. 1890. Aculeate Hymenoptera collected by J. J. Walker, Esq., R.N., F.L.S., at Gibraltar and in North Africa. (Part I - Heterogyna). Entomol. Mon. Mag. 26:201-205. [1890.08]
Saunders, E. 1896. The Hymenoptera Aculeata of the British islands. A descriptive account of the families, genera, and species indigenous to Great Britain and Ireland, with notes as to habits, localities, habitats, etc. London: L. Reeve & Co., 391 pp. [1896] [Ants pp. 18-42.]
Saunders, E. 1905. [Untitled. Postscript to Arnold (1905).] Entomol. Mon. Mag. 41:212. [1905.09]
Saunders, W. W. 1842 ("1841"). Descriptions of two hymenopterous insects from northern India. Trans. Entomol. Soc. Lond. 3:57-58. [1842.01.27] [Date from Wheeler, G. (1912).]
Savage, T. S. 1847. On the habits of the "Drivers" or visiting ants of West Africa. Trans. Entomol. Soc. Lond. 5:1-15. [1847.06.30] [Date from Wheeler, G. (1912).]

Savage, T. S. 1849. The driver ants of western Africa. Proc. Acad. Nat. Sci. Phila. 4:195-200. [1849.08]
Say, T. 1831. Descriptions of new species of North American insects, found in Louisiana by Joseph Barabino. New Harmony, Indiana: School Press, 19 pp. [1831.03] [Contains the original description of *Formica mellea* [*Camponotus melleus*], pp. 14-15. See also Scudder (1899).]
Say, T. 1836. Descriptions of new species of North American Hymenoptera, and observations on some already described. Boston J. Nat. Hist. 1:209-305. [1836.05] [Ants pp. 286-294.]
Schaefer, M. 1988. 22. Ord.: Hymenoptera, Hautflügler. Pp. 304-325 in: Brohmer, P. Fauna von Deutschland. 17. Auflage. Heidelberg: Quelle & Meyer, x + 586 pp. [1988] [Ants pp. 318-319.]
Schaefer, M. 1992. Brohmer. Fauna von Deutschland. 18. Auflage. Heidelberg: Quelle & Meyer, xiii + 704 pp. [1992] [Ants pp. 390-391.]
Scheffrahn, R. H., Gaston, L. K., Sims, J. J., Rust, M. K. 1984. Defensive ecology of *Forelius foetidus* and ts chemosystematic relationship to *F.* (=*Iridomyrmex*) *pruinosus* (Hymenoptera: Formicidae: Dolichoderinae). Environ. Entomol. 13:1502-1506. [1984.12]
Schembri, S. P., Collingwood, C. A. 1981. A revision of the myrmecofauna of the Maltese Islands (Hymenoptera, Formicidae). Ann. Mus. Civ. Stor. Nat. "Giacomo Doria" 83:417-442. [1981.09.01]
Schembri, S. P., Collingwood, C. A. 1995. The myrmecofauna of the Maltese Islands. Remarks and additions (Hymenoptera Formicidae). Boll. Soc. Entomol. Ital. 127:153-158. [1995.10.31]
Schenck, C. F. 1852. Beschreibung nassauischer Ameisenarten. Jahrb. Ver. Naturkd. Herzogthum Nassau Wiesb. 8:1-149. [1852]
Schenck, C. F. 1853a. Die nassauischen Ameisen-Species. Stett. Entomol. Ztg. 14:157-163. [1853.05]
Schenck, C. F. 1853b. Die nassauischen Ameisen-Species. (Fortsetzung.) Stett. Entomol. Ztg. 14:185-198. [1853.06]
Schenck, C. F. 1853c. Die nassauischen Ameisen-Species. (Fortsetzung.) Stett. Entomol. Ztg. 14:225-232. [1853.07]
Schenck, C. F. 1853d. Die nassauischen Ameisen-Species. (Fortsetzung.) Stett. Entomol. Ztg. 14:296-301. [1853.09]
Schenck, C. F. 1854. Berichtigung der Druckfehler in der Beschreibung: Nassauische Ameisen. Stett. Entomol. Ztg. 15:63-64. [1854.02]
Schenck, C. F. 1855. Ueber die im Heft VIII. *Eciton testaceum* genannte Ameise. Jahrb. Ver. Naturkd. Herzogthum Nassau Wiesb. 10:150. [1855]
Schenck, C. F. 1856. Systematische Eintheilung der nassauischen Ameisen nach Mayr. Jahrb. Ver. Naturkd. Herzogthum Nassau Wiesb. 11:90-94. [1856]
Schenck, C. F. 1861. Zusätze und Berichtigungen zu der Beschreibung der nassauischen Grabwespen (Heft XII), Goldwespen (Heft XI), Bienen (Heft XIV) und Ameisen (Heft VIII und XI). Jahrb. Ver. Naturkd. Herzogthum Nassau Wiesb. 16:137-206. [1861]
Scherdlin, P. 1909. Les fourmis d'Alsace. Ann. Soc. Entomol. Belg. 53:107-112. [1909.04.02]
Scheurer, S. 1984. Erstnachweis des Higieneschädlings *Tapinoma melanocephalum* (Hymenoptera, Formicidae) in der DDR. Angew. Parasitol. 25:96-99. [1984.05]
Schilliger, E., Baroni Urbani, C. 1985. Morphologie de l'organe de stridulation et sonogrammes comparés chez les ouvrières de deux espèces de fourmis moissonneuses du genre *Messor* (Hymenoptera, Formicidae). Bull. Soc. Vaudoise Sci. Nat. 77:377-384. [1985.12.15]
Schilling, P. S. 1830. Bericht über die entomologische Section der Schlesischen Gesellschaft für vaterländische Cultur im Jahre 1829. Uebers. Arb. Veränd. Schles. Ges. Vaterl. Kult. 1829:52-56. [1830]
Schilling, P. S. 1839. Bemerkungen über die in Schlesien und der Grafschaft Glatz vorgefundenen Arten der Ameisen. Uebers. Arb. Veränd. Schles. Ges. Vaterl. Kult. 1838:51-56. [1839]

Schimmer, F. 1909 ("1908"). Beitrag zur Ameisenfauna des Leipziger Gebietes. Sitzungsber. Naturforsch. Ges. Leipz. 35:21-30. [1909.11.01]

Schkaff, B. 1923. IV. Hymenoptera. Contributions à la faune des fourmis de l'état Tchéco-Slovaque. Cas. Cesk. Spol. Entomol. 20:74-76. [1923.06.01]

Schkaff, B. 1924. Formiche di Constantinopoli. Boll. Soc. Entomol. Ital. 56:90-96. [1924.06.15]

Schkaff, B. 1925a ("1924"). Contributions à la faune des fourmis de la Tchécoslovaquie. Suite. Cas. Cesk. Spol. Entomol. 21:109-110. [1925.02.15]

Schkaff, B. 1925b. Fourmis de Macédoine, récoltées par le Professeur J. Komarek. Bull. Soc. Entomol. Fr. 1925:273-276. [1925.12.20]

Schmidt, G. See: Schmidt, G. H.

Schmidt, G. H. 1961 ("1960"). Sinnesorgane bei Ameisenlarven (Form. Hym. Ins.). Pp. 403-407 in: Strouhal, H., Beier, M. (eds.) XI. Internationaler Kongress für Entomologie. Wien, 17. bis 25. August 1960. Verhandlungen. Band I. Wien: Organisationskomitee des XI. Internationalen Kongresses für Entomologie, xliv + 803 pp. [1961] [Publication date ("Ende 1961") from statement in Band II.]

Schmidt, G. H. (ed.) 1974a. Sozialpolymorphismus bei Insekten. Probleme der Kastenbildung im Tierreich. Stuttgart: Wissenschaftliche Verlagsgesellschaft mbH, xxiv + 974 pp. [1974]

Schmidt, G. H. 1974b. Polymorphismus, Arbeitsteilung, Kastenbildung. Pp. 1-59 in: Schmidt, G. H. (ed.) Sozialpolymorphismus bei Insekten. Probleme der Kastenbildung im Tierreich. Stuttgart: Wissenschaftliche Verlagsgesellschaft mbH, xxiv + 974 pp. [1974]

Schmidt, G. H., Gürsch, E. 1970. Zur Struktur des Spinnorgans einiger Ameisenlarven (Hymenoptera, Formicidae). Z. Morphol. Tiere 67:172-182. [1970.09.28]

Schmidt, G. H., Gürsch, E. 1971. Analyse der Spinnbewegungen der Larve von *Formica pratensis* Retz. (Form. Hym. Ins.). Z. Tierpsychol. 28:19-32. [1971.01]

Schmidt, G. H., Pohlmann, G. 1960. Die mikroskopische und submikroskopische Struktur der Puppenkokons von Waldameisen (Form. Hym. Ins.). Biol. Zentralbl. 79:337-342. [1960.06]

Schmidt, J. O. 1986. Chemistry, pharmacology, and chemical ecology of ant venoms. Pp. 425-508 in: Piek, T. (ed.) Venoms of the Hymenoptera. London: Academic Press, xi + 570 pp. [1986]

Schmidt, J. O., Blum, M. S., Overal, W. L. 1986. Comparative enzymology of venoms from stinging Hymenoptera. Toxicon 24:907-921. [1986]

Schmidt, J. O., Schmidt, P. J., Snelling, R. R. 1986. *Pogonomyrmex occidentalis*, an addition to the ant fauna of Mexico, with notes on other species of harvester ants from Mexico (Hymenoptera: Formicidae). Southwest. Nat. 31:395-396. [1986.09.11]

Schmitz, E. 1911. Etwas über die Ameisen Palästinas. Heilige Land 55:237-240. [1911]

Schmitz, H. 1906. Das Leben der Ameisen und ihrer Gäste. Regensburg: G. J. Manz, 190 pp. [1906]

Schmitz, H. 1915. De Nederlandsche mieren en haar gasten. Maastricht: C. Goffin, 146 + iv pp. [1915]

Schmitz, H. 1932. In memoriam. P. Erich Wassmann S. J. Tijdschr. Entomol. 75:1-57. [1932.06.15]

Schmitz, H. 1950. Formicidae quaedam a cl. A. Stärcke determinatae, quas in Lusitania collegit. Brotéria Ser. Cienc. Nat. 19:12-16. [1950]

Schmitz, H. 1955. Ein Verzeichnis portugiesischer Ameisen (Formicidae, Hymenoptera). Brotéria Ser. Cienc. Nat. 24:27-37. [1955]

Schneider, J. S. 1909. Hymenoptera aculeata im arktischen Norwegen. Tromso Mus. Aarsh. 29:81-160. [1909.01.30] [Formicidae pp. 94-101.]

Schneirla, T. C. 1956. The army ants. Annu. Rep. Smithson. Inst. 1955:379-406. [1956]

Schneirla, T. C. 1957. A comparison of species and genera in the ant subfamily Dorylinae with respect to functional pattern. Insectes Soc. 4:259-298. [1957.06]

Schneirla, T. C. 1958. The behavior and biology of certain Nearctic army ants. Last part of the functional season, southeastern Arizona. Insectes Soc. 5:215-255. [1958.06]

Schneirla, T. C. 1965. Cyclic functions in genera of legionary ants (subfamily Dorylinae). Pp. 336-338 in: Freeman, P. (ed.) XIIth International Congress of Entomology, London, 8-16 July, 1964. Proceedings. London: XIIth International Congress of Entomology, 842 pp. [1965]

Schneirla, T. C. 1971. Army ants. A study in social organization. (Edited by H. R. Topoff.) San Francisco: W. H. Freeman & Co., xx + 349 pp. [1971]

Schneirla, T. C., Gianutsos, R. R., Pasternak, B. S. 1968. Comparative allometry in the larval broods of three army-ant genera, and differential growth as related to colony behavior. Am. Nat. 102:533-554. [1968.12.31]

Schneirla, T. C., Reyes, A. Y. 1966. Raiding and related behaviour in two surface-adapted species of the old world doryline ant, *Aenictus*. Anim. Behav. 14:132-148. [1966.01]

Schneirla, T. C., Reyes, A. Y. 1969. Emigrations and related behaviour in two surface-adapted species of the old-world doryline ant, *Aenictus*. Anim. Behav. 17:87-103. [1969.02]

Schoenherr, C. J. 1833. Genera et species Curculionidum, cum synonymia hujus familiae. Tomus 1, pars 1. Paris: Roret, xv + 381 pp. [1833] [*Acamatus* (Coleoptera), senior homonym of *Acamatus* Emery, 1894.]

Schoeters, E. 1990. Morphology of the glandular system in the leaf-cutting ants *Atta sexdens sexdens* (L., 1758) and *Atta sexdens rubropilosa* Forel, 1908 (Hymenoptera: Formicidae). [Abstract.] Belg. J. Zool. 120(Suppl. 1):52-53. [1990.11]

Schoeters, E., Billen, J. P. J. 1990. Morphology of the venom gland in relation to worker size in leaf-cutting ants (Formicidae, Attini). Actes Colloq. Insectes Soc. 6:249-252. [1990.06]

Schoeters, E., Billen, J. P. J. 1991. Morphologie des glandes pro- et postpharyngiennes chez *Atta sexdens* (Hymenoptera: Formicidae). Actes Colloq. Insectes Soc. 7:153-160. [1991.06]

Schoeters, E., Billen, J. P. J. 1992. Morphological and ultrastructural study of the metapleural gland in *Diacamma* (Hymenoptera, Formicidae). Pp. 239-247 in: Billen, J. P. J. (ed.) Biology and evolution of social insects. Leuven: Leuven University Press, ix + 390 pp. [1992]

Schoeters, E., Billen, J. P. J. 1993a. Anatomy and fine structure of the metapleural gland in *Atta* (Hymenoptera, Formicidae). Belg. J. Zool. 123:67-75. [1993.06]

Schoeters, E., Billen, J. P. J. 1993b. Glandes coxales de *Pachycondyla obscuricornis* (Formicidae, Ponerinae). Actes Colloq. Insectes Soc. 8:183-186.

Schoeters, E., Billen, J. P. J. 1994a. The intramandibular gland, a novel exocrine structure in ants (Insecta, Hymenoptera). Zoomorphology (Berl.) 114:125-131. [1994.06]

Schoeters, E., Billen, J. P. J. 1994b. The venom gland in formicine ants: an ontogenetic and ultrastructural study. P. 218 in: Lenoir, A., Arnold, G., Lepage, M. (eds.) Les insectes sociaux. 12ème congrès de l'Union Internationale pour l'Étude des Insectes Sociaux. Paris, Sorbonne, 21-27 août 1994. Paris: Université Paris Nord, xxiv + 583 pp. [1994.09]

Schoeters, E., Billen, J. P. J. 1995. Morphology and ultrastructure of the convoluted gland in the ant *Dinoponera australis* (Hymenoptera: Formicidae). Int. J. Insect Morphol. Embryol. 24:323-332. [1995.07]

Scholz, E. J. R. 1924. *Formica exsecta* Nyl. var. *sudetica* nov. var. Neue Beitr. Syst. Insektenkd. 3:48. [1924.12.30]

Schön, A. 1911. Bau und Entwicklung des tibialen Chordotonalorgans bei der Honigbiene und bei Ameisen. Zool. Jahrb. Abt. Anat. Ontog. Tiere 31:439-472. [1911.04.03]

Schönitzer, K., Lawitzky, G. 1986. Rasterelektronenmikroskopische Untersuchung zur vergleichenden Morphologie der Antennenputzapparate bei Formicidae, Tiphiidae und Mutillidae (Insecta, Hymenoptera). [Abstract.] Verh. Dtsch. Zool. Ges. 79:189. [1986]

Schönitzer, K., Lawitzky, G. 1987. A phylogenetic study of the antenna cleaner in Formicidae, Mutillidae, and Tiphiidae (Insecta, Hymenoptera). Zoomorphology (Berl.) 107:273-285. [1987.12]

Schrank, F. von P. 1837. Kritische Revisionen und Ergänzungen zu Schrank's "Enumeratio Insectorum Austriae, Fauna boica u.s.w." Faunus (N.F.)1:5-19. [1837] [*Formica malabarica*, page 16.]

Schrottky, C. 1910. Nomenklaturfragen. Dtsch. Entomol. Natl.-Bibl. 1:69-70. [1910.11.01]

Schug, A. 1966. Untersuchungen über Grössenvariabilität des männlichen Kopulationsapparates bei verschiedenen *Formica*-Arten. Zool. Anz. 177:390-399. [1966.12]

Schultz, T. R. 1995 ("1994"). [Review of: Bolton, B. 1994. Identification guide to the ant genera of the world. Cambridge, Mass.: Harvard University Press, 222 pp.] Psyche (Camb.) 101:203-208. [1995.11.20]

Schultz, T. R., Meier, R. 1994. The phylogeny of the fungus-growing ants (Myrmicinae: Attini) and their fungi. P. 177 in: Lenoir, A., Arnold, G., Lepage, M. (eds.) Les insectes sociaux. 12ème congrès de l'Union Internationale pour l'Étude des Insectes Sociaux. Paris, Sorbonne, 21-27 août 1994. Paris: Université Paris Nord, xxiv + 583 pp. [1994.09]

Schultz, T. R., Meier, R. 1995. A phylogenetic analysis of the fungus-growing ants (Hymenoptera: Myrmicinae: Attini) based on morphological characters of the larvae. Syst. Entomol. 20:337-370. [1995.12.15]

Schulz, A. 1991a. Die Ameisenfauna (Hym.: Formicidae) des Setzberges in der Wachau (Niederösterreich). Z. Arbeitsgem. Österr. Entomol. 43:55-61. [1991.07]

Schulz, A. 1991b. *Tetramorium semilaeve* (Hym.: Formicidae, Myrmicinae) und *Bothriomyrmex gibbus* (Hym.: Formicidae, Dolichoderinae) neu für Österreich sowie über die Verbreitung von *Leptothorax sordidulus* (Hym.: Formicidae, Myrmicinae). Z. Arbeitsgem. Österr. Entomol. 43:120-122. [1991.12]

Schulz, A. 1994a. *Aphaenogaster graeca* nova species (Hym: Formicidae) aus dem Olymp-Gebirge (Griechenland) und eine Gliederung der Gattung *Aphaenogaster*. Beitr. Entomol. 44:417-429. [1994.06.24]

Schulz, A. 1994b. *Epimyrma birgitae* nova species, eine sozialparasitische Ameisenart (Hym.: Formicidae) auf Teneriffa (Kanarische Inseln, Spanien). Beitr. Entomol. 44:431-440. [1994.06.24]

Schulz, W. A. 1906. Spolia hymenopterologica. Paderborn: Junfermannsche Buchhandlung, 355 pp. [1906]

Schumann, K. 1888. Einige neue Ameisenpflanzen. Jahrb. Wiss. Bot. 19:357-421. [1888]

Schumann, R. D. 1993 ("1992"). Peculiar appendages in male pupae of *Leptothorax subditivus* (Wheeler) (Hymenoptera: Formicidae). Psyche (Camb.) 99:185-188. [1993.03.22]

Schweigger, A. F. 1819. Beobachtungen auf naturhistorischen Reisen. Anatomisch-physiologische Untersuchungen über Corallen; nebst einem Anhange, Bemerkungen über den Bernstein enthaltend. Berlin: Reimer, 127 pp. [1819]

Scopoli, J. A. 1763. Entomologia carniolica exhibens insecta Carnioliae indigena et distributa in ordines, genera, species, varietates. Methodo Linnaeana. Vindobonae [=Vienna]: J. Trattner, xxxvi + 420 pp. [1763.06.23] [Date of publication from Evenhuis (1994b:544). Ants pp. 312-314.]

Scott, H. 1933. Formicidae. Pp. 102-109 in: Benson, R. B., Bequaert, J., Schulthess, A. v., Scott, H. Entomological expedition to Abyssinia, 1926-7. Hymenoptera, III: Tenthredinidae, Formicidae, Mutillidae, Scoliidae, Masaridae, Vespidae. Ann. Mag. Nat. Hist. (10)12:97-120. [1933.07]

Scudder, S. H. 1877a. Appendix to Mr. George M. Dawson's report. The insects of the Tertiary beds at Quesnel. Rep. Prog. Geol. Surv. Can. 1875-1876:266-280. [1877.04]

Scudder, S. H. 1877b. The first discovered traces of fossil insects in the American Tertiaries. Bull. U. S. Geol. Geogr. Surv. Territ. 3:741-762. [1877.08.15]

Scudder, S. H. 1878. The fossil insects of the Green River shales. Bull. U. S. Geol. Geogr. Surv. Territ. 4:747-776. [1878.12] [Date of publication from Evenhuis (1994b).]

Scudder, S. H. 1891. Index to the known fossil insects of the world, including myriapods and arachnids. Bull. U. S. Geol. Surv. 71:1-744. [1891] [Hymenoptera pp. 682-734.]

Scudder, S. H. 1899. An unknown tract on American insects by Thomas Say. Psyche (Camb.) 8:306-308. [1899.01] [See Say (1831).]

Scupola, A. 1994a. Un caso di ginandromorfismo in *Myrmica* Latr. (Hymenoptera Formicidae). Boll. Soc. Entomol. Ital. 125:252-254. [1994.01.31]

Scupola, M. 1994b ("1991"). Contributo alla mirmecofauna italiana (Hymenoptera Formicidae). Boll. Mus. Civ. Stor. Nat. Verona 18:133-136. [1994.06.30]

Seevers, C. H. 1965. The systematics, evolution and zoogeography of staphylinid beetles associated with army ants (Coleoptera, Staphylinidae). Fieldiana Zool. 47:137-351. [1965.03.22]

Seifert, B. 1982a. *Hypoponera punctatissima* (Roger) (Hymenoptera, Formicidae) - eine interessante Ameisenart in menschlichen Siedlungsgebieten. Entomol. Nachr. Ber. 26:173-175. [1982.08.15]

Seifert, B. 1982b. *Lasius* (*Chthonolasius*) *jensi* n. sp. - eine neue temporär sozialparasitische Erdameise aus Mitteleuropa (Hymenoptera, Formicidae). Reichenbachia 20:85-96. [1982.09.27]

Seifert, B. 1982c. Die Ameisenfauna (Hymenoptera, Formicidae) einer Rasen-Wald-Catena im Leutratal bei Jena. Abh. Ber. Naturkundemus. Görlitz 56(6):1-18. [1982.11.15]

Seifert, B. 1983. The taxonomical and ecological status of *Lasius myops* Forel (Hymenoptera, Formicidae) and first description of its males. Abh. Ber. Naturkundemus. Görlitz 57(6):1-16. [1983.06.01]

Seifert, B. 1984a. A method for differentiation of the female castes of *Tapinoma ambiguum* Emery and *Tapinoma erraticum* (Latr.) and remarks on their distribution in Europe north of the Mediterranean region. Faun. Abh. (Dres.) 11:151-155. [1984.03.01]

Seifert, B. 1984b. Firm evidence for synonymy of *Myrmica rugulosoides* Forel, 1915 and *Myrmica scabrinodis* Nylander, 1846. Abh. Ber. Naturkundemus. Görlitz 58(6):1-8. [1984.11.30]

Seifert, B. 1984c. Nachweis einer im Freiland aufgetretenen Bastardierung von *Leptothorax nigriceps* Mayr und *Leptothorax unifasciatus* (Latr.) mittels einer multiplen Diskriminanzanalyse. Abh. Ber. Naturkundemus. Görlitz 58(7):1-8. [1984.12.20]

Seifert, B. 1986. Vergleichende Untersuchungen zur Habitatwahl von Ameisen (Hymenoptera: Formicidae) im mittleren und südlichen Teil der DDR. Abh. Ber. Naturkundemus. Görlitz 59(5):1-124. [1986.01.30]

Seifert, B. 1987. *Myrmica georgica* n. sp., a new ant from Transcaucasia and North Kazakhstan (U.S.S.R.) (Hymenoptera, Formicidae, Myrmicinae). Reichenbachia 24:183-187. [1987.05.20]

Seifert, B. 1988a. A revision of the European species of the ant subgenus *Chthonolasius* (Insecta, Hymenoptera, Formicidae). Entomol. Abh. Staatl. Mus. Tierkd. Dres. 51:143-180. [1988.07.20]

Seifert, B. 1988b. A taxonomic revision of the *Myrmica* species of Europe, Asia Minor, and Caucasia (Hymenoptera, Formicidae). Abh. Ber. Naturkundemus. Görlitz 62(3):1-75. [1988.10.12]

Seifert, B. 1989. *Camponotus herculeanus* (Linné, 1758) und *Camponotus ligniperda* (Latr., 1802) - Determination der weiblichen Kasten, Verbreitung und Habitatwahl in Mitteleuropa. Entomol. Nachr. Ber. 33:127-133. [1989.06.15]

Seifert, B. 1990. Supplementation to the revision of European species of the ant subgenus *Chthonolasius* Ruzsky, 1913 (Hymenoptera: Formicidae). Doriana 6(271):1-13. [1990.11.15]

Seifert, B. 1991a. The phenotypes of the *Formica rufa* complex in East Germany. Abh. Ber. Naturkundemus. Görlitz 65(1):1-27. [1991.04.25]

Seifert, B. 1991b. *Lasius platythorax* n. sp., a widespread sibling species of *Lasius niger* (Hymenoptera: Formicidae). Entomol. Gen. 16:69-81. [1991.06.30]

Seifert, B. 1992a. *Formica nigricans* Emery, 1909 - an ecomorph of *Formica pratensis* Retzius, 1783 (Hymenoptera, Formicidae). Entomol. Fenn. 2:217-226. [1992.01.08]

Seifert, B. 1992b. A taxonomic revision of the Palaearctic members of the ant subgenus *Lasius* s.str. (Hymenoptera: Formicidae). Abh. Ber. Naturkundemus. Görlitz 66(5):1-67. [1992.08.07]

Seifert, B. 1993a. Rote Liste der Ameisen (Formicidae) Sachsen-Anhalts, Thüringens und Sachsens. Entomol. Nachr. Ber. 37:243-246. [1993.12.30]

Seifert, B. 1993b. Taxonomic description of *Myrmica microrubra* n sp. - a social parasitic ant so far known as the microgyne of *Myrmica rubra* (L.). Abh. Ber. Naturkundemus. Görlitz 67(5):9-12. [1993.12.31]

Seifert, B. 1994 ("1993"). Die freilebenden Ameisenarten Deutschlands (Hymenoptera: Formicidae) und Angaben zu deren Taxonomie und Verbreitung. Abh. Ber. Naturkundemus. Görlitz 67(3):1-44. [1994.01.10]

Seifert, B. 1995. Two new Central European subspecies of *Leptothorax nylanderi* (Förster, 1850) and *Leptothorax sordidulus* Müller, 1923 (Hymenoptera: Formicidae). Abh. Ber. Naturkundemus. Görlitz 68(7):1-18. [1995.12.31]

Seima, F. A. 1964. On the fauna and ecology of ants in the Oka Preserve. [In Russian.] Zool. Zh. 43:1404-1408. [1964.09] [September issue. Date signed for printing ("podpisano k pechati"): 18 August 1964.]

Sellenschlo, U. 1991a. Die Braunrote Blütenameise *Monomorium floricola* (Jerdon, 1851) erstmalig in Deutschland nachgewiesen (Hym., Myrmicidae). Prakt. Schädlingsbekämpfer 43:96-107. [1991.05]

Sellenschlo, U. 1991b. Braunrote Blütenameise, *Monomorium floricola* (Jerdon, 1851) (Hym., Myrmicidae) erstmalig nach Deutschland eingeschleppt. Anz. Schädlingskd. Pflanzenschutz Umweltschutz 64:111-115. [1991.09]

Sellenschlo, U. 1993. *Crematogaster scutellaris* (Oliv.) (Hym., Myrmicidae) nach Norddeutschland eingeschleppt. Anz. Schädlingskd. Pflanzenschutz Umweltschutz 66:105-107. [1993.09]

Sellenschlo, U. 1994. Erfahrungen mit der Hälterung der braunroten Blütenameise *Monomorium floricola* (Jerdon, 1851) (Hym., Myrmicidae) und Beschreibung der Männchen. Mitt. Dtsch. Ges. Allg. Angew. Entomol. 9:343-345. [1994.12]

Sellenschlo, U. 1995. Schuppen- und Stachelameisen in Deutschland. Fachgerechte Bestimmung ist wichtig. Prakt. Schädlingsbekämpfer 47:19-22. [1995.03]

Seppä, P. 1992. Genetic relatedness of worker nestmates in *Myrmica ruginodis* (Hymenoptera: Formicidae) populations. Behav. Ecol. Sociobiol. 30:253-260. [1992.04.21]

Seppä, P., Pamilo, P. 1994. Sociogenetic organization and gene flow in red ants. P. 79 in: Lenoir, A., Arnold, G., Lepage, M. (eds.) Les insectes sociaux. 12ème congrès de l'Union Internationale pour l'Étude des Insectes Sociaux. Paris, Sorbonne, 21-27 août 1994. Paris: Université Paris Nord, xxiv + 583 pp. [1994.09]

Seppä, P., Pamilo, P. 1995. Gene flow and population viscosity in *Myrmica* ants. Heredity 74:200-209. [1995.02]

Sergi, G. 1892. Ricerche su alcuni organi de senso nelle antenne delle formiche. Bull. Soc. Entomol. Ital. 24:18-25. [1892.06.30]

Shapley, H. 1920a. Note on pterergates in the Californian harvester ant. Psyche (Camb.) 27:72-74. [1920.08] [Stamp date in MCZ library: 1920.09.18.]

Shapley, H. 1920b. Preliminary report on pterergates in *Pogonomyrmex californicus*. Proc. Natl. Acad. Sci. U. S. A. 6:687-690. [1920.12.15]

Sharp, D. 1893. On stridulation in ants. Trans. Entomol. Soc. Lond. 1893:199-213. [1893.06.06]

Sharp, D. 1899. The Cambridge Natural History. Volume 6. Insects, Part II. Hymenoptera continued (Tubulifera and Aculeata), Coleoptera, Strepsiptera, Lepidoptera, Diptera, Aphaniptera, Thysanoptera, Hemiptera, Anpolura. London: Macmillan & Co., xii + 626 pp. [1899] [Ants in Chapter IV (pp. 131-183).]

Sharplin, J. 1966. An annotated list of the Formicidae (Hymenoptera) of central and southern Alberta. Quaest. Entomol. 2:243-252. [1966.07.04]

Shattuck, S. O. 1985. Illustrated key to ants associated with western spruce budworm. U. S. Dep. Agric. Agric. Handb. 632:1-36. [1985.05]

Shattuck, S. O. 1987. An analysis of geographic variation in the *Pogonomyrmex occidentalis* complex (Hymenoptera: Formicidae). Psyche (Camb.) 94:159-179. [1987.10.06]

Shattuck, S. O. 1990. Revision of the dolichoderine ant genus *Turneria* (Hymenoptera: Formicidae). Syst. Entomol. 15:101-117. [1990.01]

Shattuck, S. O. 1991. Revision of the dolichoderine ant genus *Axinidris* (Hymenoptera: Formicidae). Syst. Entomol. 16:105-120. [1991.01]

Shattuck, S. O. 1992a. Review of the dolichoderine ant genus *Iridomyrmex* Mayr with descriptions of three new genera (Hymenoptera: Formicidae). J. Aust. Entomol. Soc. 31:13-18. [1992.02.28]

Shattuck, S. O. 1992b. Higher classification of the ant subfamilies Aneuretinae, Dolichoderinae and Formicinae (Hymenoptera: Formicidae). Syst. Entomol. 17:199-206. [1992.04]

Shattuck, S. O. 1992c. Generic revision of the ant subfamily Dolichoderinae (Hymenoptera: Formicidae). Sociobiology 21:1-181. [1992]

Shattuck, S. O. 1993a. Revision of the *Iridomyrmex purpureus* species-group (Hymenoptera: Formicidae). Invertebr. Taxon. 7:113-149. [1993.04.27]

Shattuck, S. O. 1993b. Revision of the *Iridomyrmex calvus* species-group (Hymenoptera: Formicidae). Invertebr. Taxon. 7:1303-1325. [1993.11.17]

Shattuck, S. O. 1994. Taxonomic catalog of the ant subfamilies Aneuretinae and Dolichoderinae (Hymenoptera: Formicidae). Univ. Calif. Publ. Entomol. 112:i-xix, 1-241. [1994.01]

Shattuck, S. O. 1995. Generic-level relationships within the ant subfamily Dolichoderinae (Hymenoptera: Formicidae). Syst. Entomol. 20:217-228. [1995.10.07]

Shattuck, S. O., McArthur, A. J. 1995. Generic placements of Australian ants described by W. F. Erichson (Hymenoptera: Formicidae). J. Aust. Entomol. Soc. 34:121-123. [1995.05.31]

Shen, L., Hsu, C., Wang, H., Wu, J. 1992. A comparitive [sic] study of esterase isoenzymes in selected ants (Hymenoptera: Formicidae). [Abstract.] P. 254 in: Proceedings XIX International Congress of Entomology. Abstracts. Beijing, China, June 28 - July 4, 1992. Beijing: XIX International Congress of Entomology, xi + 730 pp. [1992.07.04]

Sherborn, C. D. 1908. On the date of publication of Frederick Dixon's "Geology of Sussex." Geol. Mag. 1908:286-287. [1908.06]

Sherborn, C. D. 1914. On the contents of the parts and dates of publication of C. W. Hahn and G. A. W. Herrich-Schaeffer, 'Die Wanzigartigen Insecten,' 1831-1853. Ann. Mag. Nat. Hist. (8)13:365. [1914.03]

Sherborn, C. D. 1922. Index animalium sive index nominum quae ab A.D. MDCCLVIII generibus et speciebus animalium imposita sunt. Sectio secunda a kalendis ianuariis, MDCCCI usque ad finem decembris MDCCCL. Part I. London: British Museum (Natural History), cxxxi + 128 pp. [1922]

Sherborn, C. D. 1923. On the dates of G. W. F. Panzer's "Fauna Insect. German.", 1792-1844. Ann. Mag. Nat. Hist. (9)11:566-568. [1923.04]

Sherborn, C. D. 1932. Index animalium sive index nominum quae ab A.D. MDCCLVIII generibus et speciebus animalium imposita sunt. Sectio secunda a kalendis ianuariis, MDCCCI usque ad finem decembris, MDCCCL. Epilogue, additions to bibliography, additions and corrections, and index to trivialia. London: British Museum (Natural History), vii + cxxxiii-cxlvii + 1098 pp. [1932.06]

Sherborn, C. D. 1934. Dates of publication of catalogues of natural history (post 1850) issued by the British Museum. Ann. Mag. Nat. Hist. (10)13:308-312. [1934.02]

Sherborn, C. D. 1937. Brewster's Edinburgh encyclopedia. Issued in 18 vols. from 18-- to 1830. J. Soc. Bibliogr. Nat. Hist. 1:112. [1937.12.08]

Sherborn, C. D., Woodward, B. B. 1901a. Bibliographical notes. XXVII. - The dates of Humboldt and Bonpland's "Voyage". J. Bot. 39:202-206. [1901.06]

Sherborn, C. D., Woodward, B. B. 1901b. Dates of publication of the zoological and botanical portions of some French voyages. Part II. [part] Ann. Mag. Nat. Hist. (7)8:161-164. [1901.08]

Sherborn, C. D., Woodward, B. B. 1901c. Dates of publication of the zoological and botanical portions of some French voyages. Part II. [part] Ann. Mag. Nat. Hist. (7)8:333-336. [1901.10]

Sherborn, C. D., Woodward, B. B. 1906a. Notes on the dates of publication of the natural history portions of some French voyages. "Voyage autour du Monde...sur...La Coquille pendent..1822-25...par L. J. Duperry." - a correction. Ann. Mag. Nat. Hist. (7)17:335-336. [1906.03]
Sherborn, C. D., Woodward, B. B. 1906b. On the dates of publication of the natural history portions of the "Encyclopedie Méthodique." Ann. Mag. Nat. Hist. (7)17:577-582. [1906.06]
Shindo, M. 1979. Ants of the Ogasawara Islands. [In Japanese.] Nat. Insects (Konchu Shizen) 14(10):24-28. [1979.08]
Shoemaker, D. D., Ross, K. G., Arnold, M. L. 1994a. Macro- and microgeographic structure of a hybrid zone formed between the two introduced fire ants *Solenopsis invicta* and *Solenopsis richteri*. P. 39 in: Lenoir, A., Arnold, G., Lepage, M. (eds.) Les insectes sociaux. 12ème congrès de l'Union Internationale pour l'Étude des Insectes Sociaux. Paris, Sorbonne, 21-27 août 1994. Paris: Université Paris Nord, xxiv + 583 pp. [1994.09]
Shoemaker, D. D., Ross, K. G., Arnold, M. L. 1994b. Development of RAPD markers in two introduced fire ants, *Solenopsis invicta* and *S. richteri*, and their application to the study of a hybrid zone. Mol. Ecol. 3:531-539. [1994.12]
Shuckard, W. E. 1838. Description of a new species of *Myrmica* which has been found in houses both in the Metropolis and Provinces. Mag. Nat. Hist. (2)2:626-627. [1838.11]
Shuckard, W. E. 1840a. Monograph of the Dorylidae, a family of the Hymenoptera Heterogyna. Ann. Nat. Hist. 5:188-201. [1840.05]
Shuckard, W. E. 1840b. Monograph of the Dorylidae, a family of the Hymenoptera Heterogyna. (Continued from p. 201.) Ann. Nat. Hist. 5:258-271. [1840.06]
Shuckard, W. E. 1840c. Monograph of the Dorylidae, a family of the Hymenoptera Heterogyna. (Concluded from p. 271.) Ann. Nat. Hist. 5:315-328. [1840.07]
Shuckard, W. E. 1840d. Appendix to Mr. Shuckard's monograph of the Dorylidae, containing a description of two new species of *Labidus*. Ann. Nat. Hist. 5:396-398. [1840.08]
Shuckard, W. E. 1841. Differences of neuters in ants. Ann. Mag. Nat. Hist. 7:525-526. [1841.08]
Shyamalanath, S. 1978. Anatomy and histology of some systems in the adult and pupa of the male of the Old World doryline ant *Aenictus gracilis* Emery (Hymenoptera: Formicidae). [Abstract.] Diss. Abstr. Int. B. Sci. Eng. 39:1618-B. [1978.10]
Shyamalanath, S., Forbes, J. 1980. Digestive system and associated organs in the adult and pupal male doryline ant *Aenictus gracilis* Emery (Hymenoptera: Formicidae). J. N. Y. Entomol. Soc. 88:15-28. [1980.06.27]
Shyamalanath, S., Forbes, J. 1984 ("1983"). Anatomy and histology of the male reproductive system in the adult and pupa of the doryline ant, *Aenictus gracilis* Emery (Hymenoptera: Formicidae). J. N. Y. Entomol. Soc. 91:377-393. [1984.03.14]
Signoret, V. 1847. Description de deux hémiptères-homoptères, tribu des octicelles, group des cicadides. Ann. Soc. Entomol. Fr. (2)5:293-296. [1847.08.20] [*Cephaloxys* (Hemiptera), senior homonym of *Cephaloxys* F. Smith, 1865.]
Silhavy, V. 1935. Mravenci z okolí Trebíce, záp. Slovenska a Kutné Hory. Veda Prír. 16:96-97. [1935.04.15]
Silhavy, V. 1936. Slození mravencí zvíreny hadcové stepi u Mohelna. (Predbezne sdeleni.) Príroda (Brno) 29:95-97. [1936]
Silhavy, V. 1937a. Muzeme pokládati mravence *Lasius flavus* var. *flavo-myops* za samostatnou systematickou jednotku? Cas. Cesk. Spol. Entomol. 34:59-61. [1937.01.20]
Silhavy, V. 1937b. *Strongylognathus kratochvili* n. sp., novy preaglacialni mravenec z Moravy. Sb. Prírodoved. Klubu Trebici 1:5-12. [1937]
Silhavy, V. 1938. Mravenci hadcové stepi u Mohelna. Sb. Prírodoved. Klubu Trebici 2:3-24. [1938]
Silhavy, V. 1939a. Die Ameisenfauna des Bezirkes von Trebíc. Entomol. Rundsch. 56:367-370. [1939.09.01]
Silhavy, V. 1939b. Poznámky k mravencí zvírene dolního toku Oslavy. Entomol. Listy 2:38-39. [1939]

Simberloff, D. S., Wilson, E. O. 1969. Experimental zoogeography of islands: the colonization of empty islands. Ecology 50:278-296. [1969]

Simberloff, D. S., Wilson, E. O. 1970. Experimental zoogeography of islands. A two year record of colonization. Ecology 51:934-937. [1970]

Singleton, J. 1971. Male genitalia compared in four species of the ant genus *Formica* L. Proc. N. D. Acad. Sci. 24:18-35. [1971]

Sjöstedt, Y. 1908. 4. Akaziengallen und Ameisen auf den Ostafrikanischen Steppen. Pp. 97-118 in: Sjöstedt, Y. 1907-1910. Wissenschaftliche Ergebnisse der Schwedischen Zoologischen Expedition nach dem Kilimandjaro, dem Meru und umgebenden Massaisteppen Deutsch-Osatafrikas 1905-1906. 2 Band, Abt. 8. Stockholm: K. Schwed. Akad. Wissenschaften, 316 pp. [1908.08] [This is the fourth of eight articles in Abteilung 8, each individually published.]

Skaff, B. See: Schkaff, B.

Skaife, S. H. 1962. The distribution of the Argentine ant *Iridomyrmex humilis* Mayr. Ann. Cape Prov. Mus. 2:297-298. [1962.08]

Skinner, G. J. 1987. Ants of the British Isles. Shire Nat. Hist. Ser. 21:1-24. [1987]

Skott, C. 1970. Myrefaunaen på Laeso. Flora Fauna 76:29-30. [1970.03]

Skott, C. 1971. Nye danske fund af myren *Ponera punctatissima* Roger (Hym., Formicidae). Entomol. Medd. 39:44-47. [1971]

Skott, C. 1973. Nye danske fund af myrerne *Stenamma westwoodi* Westw. og *Myrmecina graminicola* Latr. (Hym., Formicidae). Flora Fauna 79:11. [1973.03]

Skwarra, E. 1926. Mitteilung über das Vorkommen einer für Deutschland neuen Ameisenart *Formica uralensis* Ruzsky in Ostpreussen. Entomol. Mitt. 15:305-315. [1926.07.01]

Skwarra, E. 1929a. Die Ameisenfauna des Zehlaubruches. Beiträge zur Fauna des Zehlau-Hochmoores in Ostpreussen, IV. Schr. Phys.-Ökon. Ges. Königsb. 66(2):3-174. [1929]

Skwarra, E. 1929b. *Formica fusca-picea* Nyl. als Moorameise. Zool. Anz. 82:46-55. [1929]

Skwarra, E. 1930. Ameisen und Ameisenpflanzen im Staate Veracruz (Mexiko). Wanderversamml. Dtsch. Entomol. 4:160-170. [1930.12]

Skwarra, E. 1934a. Ökologie der Lebensgemeinschaften mexikanischer Ameisenpflanzen. Z. Morphol. Ökol. Tiere 29:306-373. [1934.12.17]

Skwarra, E. 1934b. Ökologische Studien über Ameisen und Ameisenpflanzen in Mexiko. Königsberg: published by author (printer: R. Leupold), 153 pp. [1934]

Smith, D. R. 1972. Two species of imported fire ants in the United States (Hymenoptera: Formicidae). U. S. Dep. Agric. Coop. Econ. Insect Rep. 22:103-104. [1972.03.03]

Smith, D. R. 1973a. Dr. Marion Russell Smith, a bibliography. Proc. Entomol. Soc. Wash. 75: 88-95. [1973.04.26]

Smith, D. R. 1973b. Notice of name changes in ants and sawflies (Hymenoptera: Formicidae, Tenthredinidae). U. S. Dep. Agric. Coop. Econ. Insect Rep. 23:498. [1973.07.27]

Smith, D. R. 1979. Superfamily Formicoidea. Pp. 1323-1467 in: Krombein, K. V., Hurd, P. D., Smith, D. R., Burks, B. D. (eds.) Catalog of Hymenoptera in America north of Mexico. Volume 2. Apocrita (Aculeata). Washington, D.C.: Smithsonian Institution Press, pp. i-xvi, 1199-2209. [1979]

Smith, D. R. 1983. Obituary. Dr. Marion Russell Smith, 1894-1981. Proc. Entomol. Soc. Wash. 85:628-630. [1983.08.10]

Smith, D. R. 1991. Ants (Formicidae, Hymenoptera). Pp. 297-309, 633-649 in: Gorham, J. R. (ed.) Insect and mite pests in food. U. S. Dep. Agric. Agric. Handb. 655:i-vii, 1-767. [1991.02]

Smith, D. R., Lavigne, R. J. 1973. Two new species of ants of the genera *Tapinoma* Foerster and *Paratrechina* Motschoulsky from Puerto Rico (Hymenoptera: Formicidae). Proc. Entomol. Soc. Wash. 75:181-187. [1973.07.10]

Smith, E. L. 1970. Evolutionary morphology of the external insect genitalia: 2. Hymenoptera. Ann. Entomol. Soc. Am. 63:1-27. [1970.01.15]

Smith, F. 1851. List of the specimens of British animals in the collection of the British Museum. Part VI. Hymenoptera, Aculeata. London: British Museum (Natural History), 134 pp. [1851.09.20] [Date of publication from Sherborn (1934). Ants pp. 1-6, 115-119.]

Smith, F. 1852. Descriptions of some hymenopterous insects captured in India, with notes on their economy, by Ezra T. Downes, Esq., who presented them to the Honourable the East India Company. Ann. Mag. Nat. Hist. (2)9:44-50. [1852.01] [Ants pp. 44-45.]

Smith, F. 1853 ("1854"). Monograph of the genus *Cryptocerus,* belonging to the group Cryptoceridae - family Myrmicidae - division Hymenoptera Heterogyna. Trans. Entomol. Soc. Lond. (2)2:213-228. [1853.12.26] [Date from Wheeler, G. (1912).]

Smith, F. 1855a. Essay on the genera and species of British Formicidae. [part] Trans. Entomol. Soc. Lond. (2)3:95-112. [1855.01.09] [Date from Wheeler, G. (1912).]

Smith, F. 1855b. Essay on the genera and species of British Formicidae. [concl.] Trans. Entomol. Soc. Lond. (2)3:113-135. [1855.04.23] [Date from Wheeler, G. (1912).]

Smith, F. 1855c. Descriptions of some species of Brazilian ants belonging to the genera *Pseudomyrma, Eciton* and *Myrmica* (with observations on their economy by Mr. H. W. Bates). Trans. Entomol. Soc. Lond. (2)3:156-169. [1855.08.06] [Date from Wheeler, G. (1912).]

Smith, F. 1855d. Hymenoptera. Notes on the new species of British Aculeate Hymenoptera. Entomol. Annu. 1855:64-77 [Entomol. Annu. 1855(2nd edition):87-100]. [1855]

Smith, F. 1857a. Catalogue of the hymenopterous insects collected at Sarawak, Borneo; Mount Ophir, Malacca; and at Singapore, by A. R. Wallace. [part] J. Proc. Linn. Soc. Lond. Zool. 2:42-88. [1857.11.02] [Date of publication from Kappel (1896:v). Ants pp. 52-83.]

Smith, F. 1857b. Notes on aculeate Hymenoptera, with some observations on their economy. Entomol. Annu. 1858:34-46. [1857.12.19] [Ants pp. 38-40.]

Smith, F. 1858a. Catalogue of hymenopterous insects in the collection of the British Museum. Part VI. Formicidae. London: British Museum (Natural History), 216 pp. [1858.03.27] [Date of publication from Donisthorpe (1932c) and Sherborn (1934).]

Smith, F. 1858b. Revision of an essay on the British Formicidae published in the Transactions of the Society. Trans. Entomol. Soc. Lond. (2)4:274-284. [1858.04.05]

Smith, F. 1859a. Catalogue of hymenopterous insects collected by Mr. A. R. Wallace at the islands of Aru and Key. [part] J. Proc. Linn. Soc. Lond. Zool. 3:132-158. [1859.02.01] [Date of publication from Kappel (1896:v). Ants pp. 135-150.]

Smith, F. 1859b ("1858"). Catalogue of British fossorial Hymenoptera, Formicidae, and Vespidae, in the collection of the British Museum. London: British Museum (Natural History), 236 pp. [1859.04.09] [Date of publication from Sherborn (1934). Ants pp. 2-37, 223-225.]

Smith, F. 1859c. Catalogue of hymenopterous insects in the collection of the British Museum. Part VII. Dorylidae and Thynnidae. London: British Museum (Natural History), 76 pp. [1859.07.23] [Date of publication from Sherborn (1934).]

Smith, F. 1860a. Descriptions of new species of hymenopterous insects collected by Mr. A. R. Wallace at Celebes. J. Proc. Linn. Soc. Lond. Zool. 5(17b)(suppl. to vol. 4):57-93. [1860.07.18] [Date of publication from Kappel (1896:vi). Ants pp. 68-75.]

Smith, F. 1860b. Catalogue of hymenopterous insects collected by Mr. A. R. Wallace in the islands of Bachian, Kaisaa, Amboyna, Gilolo, and at Dory in New Guinea. J. Proc. Linn. Soc. Lond. Zool. 5(17b)(suppl. to vol. 4):93-143. [1860.07.18] [Date of publication from Kappel (1896:vi). Ants pp. 94-114.]

Smith, F. 1860c. Descriptions of new genera and species of exotic Hymenoptera. J. Entomol. 1: 65-84. [1860.10] [Ants pp. 65-79.]

Smith, F. 1861a. Descriptions of some new species of ants from the Holy Land, with a synonymic list of others previously described. J. Proc. Linn. Soc. Lond. Zool. 6:31-35. [1861.11.01] [Date of publication from Kappel (1896:vi).]

Smith, F. 1861b. Catalogue of hymenopterous insects collected by Mr. A. R. Wallace in the islands of Ceram, Celebes, Ternate, and Gilolo. [part] J. Proc. Linn. Soc. Lond. Zool. 6:36-48. [1861.11.01] [Date of publication from Kappel (1896:vi). This part deals entirely with ants.]

Smith, F. 1862a. Catalogue of hymenopterous insects collected by Mr. A. R. Wallace in the islands of Ceram, Celebes, Ternate, and Gilolo. [concl.] J. Proc. Linn. Soc. Lond. Zool. 6:49-66. [1862.03.01] [Date of publication from Kappel (1896:vi). Ants pp. 49-50.]

Smith, F. 1862b. Descriptions of new species of aculeate Hymenoptera, collected at Panama by R. W. Stretch, Esq., with a list of described species, and the various localities where they have previously occurred. Trans. Entomol. Soc. Lond. (3)1:29-44. [1862.04.07] [Date from Wheeler, G. (1912). Ants pp. 29-35.]

Smith, F. 1862c. Descriptions of new species of Australian Hymenoptera, and of a species of *Formica* from New Zealand. Trans. Entomol. Soc. Lond. (3)1:53-62. [1862.05.29] [Date from Wheeler, G. (1912).]

Smith, F. 1862d. A list of the genera and species belonging to the family Cryptoceridae, with descriptions of new species; also a list of the species of the genus *Echinopla*. Trans. Entomol. Soc. Lond. (3)1:407-416. [1862.11.17] [Date from Wheeler, G. (1912).]

Smith, F. 1863a. Catalogue of hymenopterous insects collected by Mr. A. R. Wallace in the islands of Mysol, Ceram, Waigiou, Bouru and Timor. J. Proc. Linn. Soc. Lond. Zool. 7:6-48. [1863.03.04] [Date of publication from Kappel (1896:vi). Ants pp. 12-24.]

Smith, F. 1863b. Observations on ants of equatorial Africa. Trans. Entomol. Soc. Lond. (3)1:470-473. [1863.06.08]

Smith, F. 1863c. Notes on the geographic distribution of the aculeate Hymenoptera collected by Mr. A. R. Wallace in the Eastern Archipelago. J. Proc. Linn. Soc. Lond. Zool. 7:109-131. [1863.10.29] [Date of publication from Kappel (1896:vi).]

Smith, F. 1864. Notes on Hymenoptera. Entomol. Annu. 1864:108-117. [1864] [Ants pp. 111-112.]

Smith, F. 1865a. Descriptions of new species of hymenopterous insects from the islands of Sumatra, Sula, Gilolo, Salwatty, and New Guinea, collected by Mr. A. R. Wallace. J. Proc. Linn. Soc. Lond. Zool. 8:61-94. [1865.01.13] [Date of publication from Kappel (1896:vi). Ants pp. 68-77.]

Smith, F. 1865b. Observations on the genus *Dorylus*, and upon a new genus of Apidae. Entomol. Mon. Mag. 2:3-5. [1865.06]

Smith, F. 1865c. Notes on British Formicidae. Entomol. Mon. Mag. 2:28-30. [1865.07]

Smith, F. 1865d. British species of ants at Bournemouth. Entomologist 2:303-305. [1865.10]

Smith, F. 1865e. Notes on Hymenoptera. Entomol. Annu. 1865:81-96. [1865] [Ants pp. 85-89.]

Smith, F. 1866. Notes on Hymenoptera. Entomol. Annu. 1866:122-137. [1866] [Ants pp. 124-130.]

Smith, F. 1867. Descriptions of new species of Cryptoceridae. Trans. Entomol. Soc. Lond. (3)5:523-528. [1867.12.30]

Smith, F. 1868. Notes on Hymenoptera. Entomol. Annu. 1868:81-96. [1868] [Ants pp. 92-95.]

Smith, F. 1869. Notes on Hymenoptera. Entomol. Annu. 1869:65-82. [1869] [Ants pp. 69-72.]

Smith, F. 1871a. A catalogue of the Aculeate Hymenoptera and Ichneumonidae of India and the Eastern Archipelago. With introductory remarks by A. R. Wallace. [part] J. Linn. Soc. Lond. Zool. 11:285-348. [1871.10.16] [Ants pp. 302-336.]

Smith, F. 1871b. A catalogue of the aculeate Hymenoptera and Ichneumonidae of India and the Eastern Archipelago. [concl.] J. Linn. Soc. Lond. Zool. 11:349-415. [1871.12.20]

Smith, F. 1871c. A catalogue of British Hymenoptera; Aculeata. London: Entomological Society of London, viii + 44 pp. [1871] [Ants pp. 1-4.]

Smith, F. 1871d. Notes on various species of Apidae, Formicidae, Fossores, and Vespidae; with observations on some parasites of the latter. Entomol. Annu. 1871:55-70. [1871] [Ants pp. 58-61.]

Smith, F. 1873. [Untitled. Introduced by: "Mr. F. Smith exhibited a further collection of ants sent by Mr. G. A. James Rothney, from Calcutta."] Trans. Entomol. Soc. Lond. 1873:viii-ix. [1873.05.20] [Date from Wheeler, G. (1912).]

Smith, F. 1874a. [Untitled. Introduced by: "Mr. F. Smith exhibited (1) a hermaphrodite ant, *Myrmica laevinodis*."] Trans. Entomol. Soc. Lond. 1874:iv. [1874.04.27] [Date from Wheeler, G. (1912).]

Smith, F. 1874b. Descriptions of new species of Tenthredinidae, Ichneumonidae, Chrysididae, Formicidae, &c. of Japan. Trans. Entomol. Soc. Lond. 1874:373-409. [1874.06] [Date from Wheeler, G. (1912). Ants pp. 402-407, 409.]

Smith, F. 1874c. On hermaphroditism in ants. Entomol. Annu. 1874:147-148. [1874]

Smith, F. 1875. Descriptions of new species of Indian aculeate Hymenoptera, collected by Mr. G. R. James Rothney, member of the Entomological Society. Trans. Entomol. Soc. Lond. 1875:33-51. [1875.05.29] [Date from Wheeler, G. (1912). Ants pp. 34-35.]

Smith, F. 1876a. Preliminary notice of new species of Hymenoptera, Diptera, and Forficulidae collected in the island of Rodriguez by the naturalists accompanying the Transit-of-Venus expedition. Ann. Mag. Nat. Hist. (4)17:447-451. [1876.06] [Ants pp. 447-448.]

Smith, F. 1876b. Descriptions of new species of hymenopterous insects of New Zealand, collected by C. M. Wakefield, Esq., principally in the neighbourhood of Canterbury. Trans. Entomol. Soc. Lond. 1876:473-487. [1876.11.02] [Date from Wheeler, G. (1912). Ants pp. 480-482.]

Smith, F. 1876c. Descriptions of three new species of Hymenoptera (Formicidae) from New Zealand. Trans. Entomol. Soc. Lond. 1876:489-492. [1876.12.23] [Date from Wheeler, G. (1912).]

Smith, F. 1876d. Descriptions of new species of Cryptoceridae, belonging to the genera *Cryptocerus*, *Meranoplus*, and *Cataulacus*. Trans. Entomol. Soc. Lond. 1876:603-612. [1876.12.23] [Date from Wheeler, G. (1912).]

Smith, F. 1877a. VIII. Hymenoptera and Diptera. Pp. 82-84 in: Günther, A. Account of the zoological collection made during the visit of H. M. S. "Peterel" to the Galapagos Islands. Proc. Zool. Soc. Lond. 1877:64-93. [1877.02.06]

Smith, F. 1877b. Descriptions of new species of the genera *Pseudomyrma* and *Tetraponera*, belonging to the family Myrmicidae. Trans. Entomol. Soc. Lond. 1877:57-72. [1877.07.02] [Date from Wheeler, G. (1912).]

Smith, F. 1878a. Descriptions of new species of hymenopterous insects from New Zealand, collected by Prof. Hutton, at Otago. Trans. Entomol. Soc. Lond. 1878:1-7. [1878.04.24] [Date from Wheeler, G. (1912). Ants p. 6.]

Smith, F. 1878b. Scientific results of the Second Yarkand Mission; based upon the collections and notes of the late Ferdinand Stoliczka, Ph.D. Hymenoptera. Calcutta: Superintendent of Government Printing (Government of India), 22 pp. [1878] [Ants pp. 9-13.]

Smith, F. 1879a. Descriptions of new species of aculeate Hymenoptera collected by the Rev. Thos. Blackburn in the Sandwich Islands. J. Linn. Soc. Lond. Zool. 14:674-685. [1879.05.20] [Ants pp. 675-676.]

Smith, F. 1879b. Descriptions of new species of Hymenoptera in the collection of the British Museum. London: British Museum (Natural History), xxi + 240 pp. [1879.10.11] [Date of publication from Sherborn (1934). Ants pp. 228-229.]

Smith, F. 1879c. The collections from Rodriguez. Hymenoptera, Diptera, and Neuroptera. Pp. 534-540 in: Hooker, J. D., Günther, A. (eds.) An account of the petrological, botanical, and zoological collections made in Kerguelen's land and Rodriguez during the Transit of Venus expeditions, carried out by order of Her Majesty's Government in the years 1874-75. Philos. Trans. R. Soc. Lond. 168(extra volume):i-ix,1-579. [1879]

Smith, F. 1941. A list of the ants of Washington state. Pan-Pac. Entomol. 17:23-28. [1941.02.21] [Author is Smith, Falconer.]

Smith, F. 1942. Polymorphism in *Camponotus* (Hymenoptera-Formicidae). J. Tenn. Acad. Sci. 17:367-373. [1942.10] [Author is Smith, Falconer.]

Smith, H. M. 1942. The publication dates of "La Naturaleza". Lloydia 5:95-96. [1942.04.22]

Smith, J. H., Atherton, D. O. 1944. Seed-harvesting and other ants in the tobacco-growing districts of North Queensland. Qld. J. Agric. Sci. 1(3):33-61. [1944.09]

Smith, M. R. 1916a. Observations on ants in South Carolina (Hym.). Entomol. News 27:279-280. [1916.06.03]

Smith, M. R. 1916b. Notes on South Carolina ants (Hym., Hem.). Entomol. News 27:468. [1916.12.06]

Smith, M. R. 1918. A key to the known species of South Carolina ants, with notes (Hym.). Entomol. News 29:17-29. [1918.01.12]

Smith, M. R. 1922. Some ants noted to infest houses in Mississippi during the summer and fall of 1921. J. Econ. Entomol. 15:113-114. [1922.02]

Smith, M. R. 1923a. Two new Mississippi ants of the subgenus *Colobopsis*. Psyche (Camb.) 30:82-88. [1923.04] [Stamp date in MCZ library: 1923.06.07.]

Smith, M. R. 1923b. Two new varieties of ants (Hymen.: Formicidae). Entomol. News 34:306-308. [1923.12.15]

Smith, M. R. 1924a. An annotated list of the ants of Mississippi (Hym.). Entomol. News 35:47-54. [1924.02.07]

Smith, M. R. 1924b. An annotated list of the ants of Mississippi (Hym.) (continued from page 54). Entomol. News 35:77-85. [1924.03.03]

Smith, M. R. 1924c. An annotated list of the ants of Mississippi (Hym.) (continued from page 85). Entomol. News 35:121-127. [1924.04.02]

Smith, M. R. 1924d. A new species of ant from Kansas (Hym.: Formicidae). Entomol. News 35:250-253. [1924.07.03]

Smith, M. R. 1927a. A contribution to the biology and distribution of one of the legionary ants, *Ection schmitti* Emery. Ann. Entomol. Soc. Am. 20:401-404. [1927.10.15]

Smith, M. R. 1927b. An additional annotated list of ants of Mississippi, with a description of a new species of *Pheidole* (Hym.: Formicidae). Entomol. News 38:308-314. [1927.12.14]

Smith, M. R. 1928a. The biology of *Tapinoma sessile* Say, an important house-infesting ant. Ann. Entomol. Soc. Am. 21:307-330. [1928.08.18]

Smith, M. R. 1928b. An additional annotated list of the ants of Mississippi. With a description of a new species of *Aphaenogaster* (Hym.: Formicidae). Entomol. News 39:242-246. [1928.10.08]

Smith, M. R. 1928c. An additional annotated list of the ants of Mississippi. With a description of a new species of *Aphaenogaster* (Hym.: Formicidae) (continued from page 246). Entomol. News 39:275-279. [1928.11.02]

Smith, M. R. 1929a ("1928"). Observations and remarks on the slave-making raids of three species of ants found at Urbana, Illinois. J. N. Y. Entomol. Soc. 36:323-333. [1929.01.09]

Smith, M. R. 1929b. Two introduced ants not previously known to occur in the United States. J. Econ. Entomol. 22:241-243. [1929.02]

Smith, M. R. 1929c. Descriptions of five new North American ants, with biological notes. Ann. Entomol. Soc. Am. 22:543-551. [1929.10.12]

Smith, M. R. 1930a. A list of Florida ants. Fla. Entomol. 14:1-6. [1930.03]

Smith, M. R. 1930b. A description of the male of *Proceratium croceum* Emery, with remarks. Ann. Entomol. Soc. Am. 23:390-392. [1930.06.28]

Smith, M. R. 1930c. Another imported ant. Fla. Entomol. 14:23-24. [1930.06]

Smith, M. R. 1930d. Description of three new North American ants, with biological notes. Ann. Entomol. Soc. Am. 23:564-568. [1930.10.18]

Smith, M. R. 1931a. An additional annotated list of the ants of Mississippi (Hym.: Formicoidea). Entomol. News 42:16-24. [1931.01.13]

Smith, M. R. 1931b. Is *Eciton mexicanum* F. Smith really *Eciton pilosus* F. Smith? J. N. Y. Entomol. Soc. 39:295-298. [1931.09.30]

Smith, M. R. 1931c. A revision of the genus *Strumigenys* of America, north of Mexico, based on a study of the workers (Hymn.: Formicidae). Ann. Entomol. Soc. Am. 24:686-710. [1931.12.19]

Smith, M. R. 1932. An additional annotated list of the ants of Mississippi (Hym.: Formicidae). Entomol. News 43:157-160. [1932.06.03]

Smith, M. R. 1933. Additional species of Florida ants, with remarks. Fla. Entomol. 17:21-26. [1933.08]
Smith, M. R. 1934a. A list of the ants of South Carolina. J. N. Y. Entomol. Soc. 42:353-361. [1934.10.02]
Smith, M. R. 1934b. Three new North American ants. Ann. Entomol. Soc. Am. 27:384-387. [1934.10.10]
Smith, M. R. 1934c. Dates on which the immature or mature sexual phases of ants have been observed (Hymen.: Formicoidea). Entomol. News 45:247-251. [1934.11.09]
Smith, M. R. 1934d. Ponerine ants of the genus *Euponera* in the United States. Ann. Entomol. Soc. Am. 27:557-564. [1934.12.20]
Smith, M. R. 1934e. Dates on which the immature or mature sexual phases of ants have been observed (Hymen.: Formicoidea) (continued from page 251). Entomol. News 45:264-267. [1934.12.20]
Smith, M. R. 1934f. Two new North American ants. Psyche (Camb.) 41:211-213. [1934.12] [Stamp date in MCZ library: 1935.03.07.]
Smith, M. R. 1935a. Two new species of North American *Strumigenys* (Formicidae: Hymenoptera). Ann. Entomol. Soc. Am. 28:214-216. [1935.06.13]
Smith, M. R. 1935b. A list of the ants of Oklahoma (Hymen.: Formicidae). Entomol. News 46:235-241. [1935.11.08]
Smith, M. R. 1935c. A list of the ants of Oklahoma (Hymen.: Formicidae) (continued from page 241). Entomol. News 46:261-264. [1935.12.27]
Smith, M. R. 1936a. Consideration of the fire ant *Solenopsis xyloni* as an important southern pest. J. Econ. Entomol. 29:120-122. [1936.02]
Smith, M. R. 1936b. Distribution of the Argentine ant in the United States and suggestions for its control or eradication. U. S. Dep. Agric. Circ. 387:1-39. [1936.05]
Smith, M. R. 1936c. A list of the ants of Texas. J. N. Y. Entomol. Soc. 44:155-170. [1936.06.25]
Smith, M. R. 1936d. Ants of the genus *Ponera* in America, north of Mexico. Ann. Entomol. Soc. Am. 29:420-430. [1936.09.26]
Smith, M. R. 1937 ("1936"). The ants of Puerto Rico. J. Agric. Univ. P. R. 20:819-875. [1937.01]
Smith, M. R. 1938a. A study of the North American ants of the genus *Xiphomyrmex* Forel. J. Wash. Acad. Sci. 28:126-130. [1938.03.15]
Smith, M. R. 1938b. Notes on the legionary ants (*Eciton*, subgenus *Acamatus*) with a record of new specific synonymy. Proc. Entomol. Soc. Wash. 40:157-160. [1938.06.24]
Smith, M. R. 1939a. A new species of North American *Ponera*, with an ergatandrous form (Hymenoptera: Formicidae). Proc. Entomol. Soc. Wash. 41:76-78. [1939.03.24]
Smith, M. R. 1939b. The North American ants of the genus *Harpagoxenus* Forel, with the description of a new species (Hymenoptera: Formicidae). Proc. Entomol. Soc. Wash. 41:165-172. [1939.05.27]
Smith, M. R. 1939c. Notes on *Leptothorax* (*Mychothorax*) *hirticornis* Emery, and description of a related new species (Formicidae). Proc. Entomol. Soc. Wash. 41:176-180. [1939.05.27]
Smith, M. R. 1939d. A study of the subspecies of *Odontomachus haematoda* (L.) of the United States (Hymenoptera: Formicidae). J. N. Y. Entomol. Soc. 47:125-130. [1939.06.08]
Smith, M. R. 1939e. Ants of the genus *Macromischa* Roger in the United States (Hymenoptera: Formicidae). Ann. Entomol. Soc. Am. 32:502-509. [1939.09.20]
Smith, M. R. 1939f. Notes on *Formica* (*Neoformica*) *moki* Wheeler, with description of a new subspecies (Hymenoptera: Formicidae). Ann. Entomol. Soc. Am. 32:581-584. [1939.09.20]
Smith, M. R. 1940a. The discovery of the worker caste of an inquilinous ant, *Epipheidole inquilina* Wheeler. Proc. Entomol. Soc. Wash. 42:104-109. [1940.05.28]
Smith, M. R. 1940b. The identity of the ant *Camponotus* (*Myrmentoma*) *caryae* (Fitch). Proc. Entomol. Soc. Wash. 42:137-141. [1940.10.31]
Smith, M. R. 1941. Two new species of *Aphaenogaster* (Hymenoptera: Formicidae). Great Basin Nat. 2:118-121. [1941.11.29]

Smith, M. R. 1942a. A new, apparently parasitic ant. Proc. Entomol. Soc. Wash. 44:59-61. [1942.04.30]
Smith, M. R. 1942b. The males of two North American cerapachyine ants. Proc. Entomol. Soc. Wash. 44:62-64. [1942.04.30]
Smith, M. R. 1942c. The legionary ants of the United States belonging to *Eciton* subgenus *Neivamyrmex* Borgmeier. Am. Midl. Nat. 27:537-590. [1942.06.30]
Smith, M. R. 1942d. The relationship of ants and other organisms to certain scale insects on coffee in Puerto Rico. J. Agric. Univ. P. R. 26:21-27. [1942.06]
Smith, M. R. 1943a ("1942"). A new North American *Solenopsis* (*Diplorhoptrum*) (Hymenoptera: Formicidae). Proc. Entomol. Soc. Wash. 45:209-211. [1943.01.05]
Smith, M. R. 1943b. Ants of the genus *Tetramorium* in the United States with the description of a new species. Proc. Entomol. Soc. Wash. 45:1-5. [1943.01.30]
Smith, M. R. 1943c. *Pheidole* (*Macropheidole*) *rhea* Wheeler, a valid species (Hymenoptera: Formicidae). Proc. Entomol. Soc. Wash. 45:5-9. [1943.01.30]
Smith, M. R. 1943d. The first record of *Leptothorax*, subgenus *Goniothorax* Emery, in the United States, with the description of a new species. Proc. Entomol. Soc. Wash. 45:154-156. [1943.05.03]
Smith, M. R. 1943e. A generic and subgeneric synopsis of the male ants of the United States. Am. Midl. Nat. 30:273-321. [1943.10.22]
Smith, M. R. 1943f. A new male legionary ant from the Mojave Desert, California. Lloydia 6:196-197. [1943.11.10]
Smith, M. R. 1944a. Ants of the genus *Cardiocondyla* Emery in the United States. Proc. Entomol. Soc. Wash. 46:30-41. [1944.03.13]
Smith, M. R. 1944b. Ants of the genus *Thaumatomyrmex* Mayr with the description of a new Panamanian species (Hymenoptera: Formicidae). Proc. Entomol. Soc. Wash. 46:97-99. [1944.05.03]
Smith, M. R. 1944c. A key to the genus *Acanthognathus* Mayr, with the description of a new species (Hymenoptera: Formicidae). Proc. Entomol. Soc. Wash. 46:150-152. [1944.07.10]
Smith, M. R. 1944d. Additional ants recorded from Florida, with descriptions of two new subspecies. Fla. Entomol. 27:14-17. [1944.07]
Smith, M. R. 1944e. The genus *Lachnomyrmex*, with the description of a second species (Hymenoptera: Formicidae). Proc. Entomol. Soc. Wash. 46:225-228. [1944.11.21]
Smith, M. R. 1944f. A second species of *Glamyromyrmex* Wheeler (Hymenoptera: Formicidae). Proc. Entomol. Soc. Wash. 46:254-256. [1944.12.30]
Smith, M. R. 1946. A second species of *Stegomyrmex*, and the first description of a *Stegomyrmex* worker (Hymenoptera: Formicidae). Rev. Entomol. (Rio J.) 17:286-289. [1946.08.26]
Smith, M. R. 1947a. Ants of the genus *Cryptocerus* F., in the United States. Proc. Entomol. Soc. Wash. 49:29-40. [1947.01.15]
Smith, M. R. 1947b ("1946"). Ants of the genus *Apsychomyrmex* Wheeler (Hymenoptera: Formicidae). Rev. Entomol. (Rio J.) 17:468-473. [1947.01.20]
Smith, M. R. 1947c. A new and extraordinary *Pheidole* from New Guinea (Hymenoptera, Formicidae). Proc. Entomol. Soc. Wash. 49:73-75. [1947.02.28]
Smith, M. R. 1947d. A new species of *Metapone* Forel from New Guinea (Hymenoptera, Formicidae). Proc. Entomol. Soc. Wash. 49:75-77. [1947.02.28]
Smith, M. R. 1947e. A new species of *Megalomyrmex* from Barro Colorado Island, Canal Zone (Hymenoptera, Formicidae). Proc. Entomol. Soc. Wash. 49:101-103. [1947.04.15]
Smith, M. R. 1947f. A generic and subgeneric synopsis of the United States ants, based on the workers. Am. Midl. Nat. 37:521-647. [1947.07.27]
Smith, M. R. 1947g. A study of *Polyergus* in the United States, based on the workers (Hymenoptera: Formicidae). Am. Midl. Nat. 38:150-161. [1947.08.20]
Smith, M. R. 1947h. Notes on *Pheidole* (*Decapheidole*) and the description of a new species (Hymenoptera: Formicidae). Rev. Entomol. (Rio J.) 18:193-196. [1947.08.26]

Smith, M. R. 1947i. A new genus and species of ant from Guatemala (Hymenoptera, Formicidae). J. N. Y. Entomol. Soc. 55:281-284. [1947.12.16]
Smith, M. R. 1948. A new species of *Myrmecina* from California (Hymenoptera, Formicidae). Proc. Entomol. Soc. Wash. 50:238-240. [1948.12.29]
Smith, M. R. 1949a ("1948"). A new genus and species of ant from India (Hymenoptera: Formicidae). J. N. Y. Entomol. Soc. 56:205-208. [1949.01.26]
Smith, M. R. 1949b. A new species of *Probolomyrmex* from Barro Colorado Island, Canal Zone (Hymenoptera, Formicidae). Proc. Entomol. Soc. Wash. 51:38-40. [1949.02.21]
Smith, M. R. 1949c. On the status of *Cryptocerus* Latreille and *Cephalotes* Latreille (Hymenoptera: Formicidae). Psyche (Camb.) 56:18-21. [1949.05.17]
Smith, M. R. 1949d. A new species of *Camponotus*, subg. *Colobopsis* from Mexico (Hymenoptera: Formicidae). J. N. Y. Entomol. Soc. 57:177-181. [1949.09.12]
Smith, M. R. 1949e. A new *Leptothorax* commonly inhabiting the canyon live oak of California (Hymenoptera: Formicidae). Psyche (Camb.) 56:112-115. [1949.10.31]
Smith, M. R. 1950a. On the status of *Leptothorax* Mayr and some of its subgenera. Psyche (Camb.) 57:29-30. [1950.06.16]
Smith, M. R. 1950b. [Review of: Creighton, W. S. 1950. The ants of North America. Bull. Mus. Comp. Zool. 104:1-585.] Proc. Entomol. Soc. Wash. 52:274-275. [1950.10.25]
Smith, M. R. 1950c. Order Hymenoptera. Ants. Family Formicidae. Pp. 261-300 in: Weinman, C. J. (ed.) Pest control technology. Entomological section. New York: National Pest Control Association, vii + 365 + [50] pp. [1950]
Smith, M. R. 1951a. [Review of: Creighton, W. S. 1950. The ants of North America. Bull. Mus. Comp. Zool. 104:1-585.] Sci. Mon. 72:63. [1951.01]
Smith, M. R. 1951b. A new species of *Stenamma* from North Carolina (Hymenoptera, Formicidae). Proc. Entomol. Soc. Wash. 53:156-158. [1951.06.15]
Smith, M. R. 1951c. Family Formicidae. Pp. 778-875 in: Muesebeck, C. F., Krombein, K. V., Townes, H. K. (eds.) Hymenoptera of America north of Mexico. Synoptic catalogue. U. S. Dep. Agric. Agric. Monogr. 2:1-1420. [1951.07.11] [Publication date from Muesebeck, Krombein & Townes (1951).]
Smith, M. R. 1951d. Two new ants from western Nevada (Hymenoptera, Formicidae). Great Basin Nat. 11:91-96. [1951.12.29]
Smith, M. R. 1952a. The correct name for the group of ants formerly known as *Pseudomyrma* (Hymenoptera). Proc. Entomol. Soc. Wash. 54:97-98. [1952.04.25]
Smith, M. R. 1952b. On the collection of ants made by Titus Ulke in the Black Hills of South Dakota in the early nineties. J. N. Y. Entomol. Soc. 60:55-63. [1952.04.30]
Smith, M. R. 1952c. North American *Leptothorax* of the *tricarinatus-texanus* complex (Hymenoptera: Formicidae). J. N. Y. Entomol. Soc. 60:96-106. [1952.06.23]
Smith, M. R. 1953a. A revision of the genus *Romblonella* W. M. Wheeler (Hymenoptera: Formicidae). Proc. Hawaii. Entomol. Soc. 15:75-80. [1953.03.27]
Smith, M. R. 1953b. *Dolichoderus granulatus* Pergande, a synonym (Hymenoptera, Formicidae). Proc. Entomol. Soc. Wash. 55:211. [1953.11.13]
Smith, M. R. 1953c. A new species of *Probolomyrmex*, and the first description of a *Probolomyrmex* male (Hymenoptera, Formicidae). J. N. Y. Entomol. Soc. 61:127-129. [1953.11.20]
Smith, M. R. 1953d. A new *Metapone* from the Micronesian Islands (Hymenoptera, Formicidae). J. N. Y. Entomol. Soc. 61:135-137. [1953.11.20]
Smith, M. R. 1953e. A new *Pheidole* (subg. Ceratopheidole) from Utah (Hymenoptera: Formicidae). J. N. Y. Entomol. Soc. 61:143-146. [1953.11.20]
Smith, M. R. 1953f. A new *Romblonella* from Palau, and the first description of a *Romblonella* male (Hymenoptera, Formicidae). J. N. Y. Entomol. Soc. 61:163-167. [1953.11.20]
Smith, M. R. 1953g. *Pogonomyrmex salinus* Olsen, a synonym of *Pogonomyrmex occidentalis* (Cress.) (Hymenoptera: Formicidae). Bull. Brooklyn Entomol. Soc. 48:131-132. [1953.11.30]

Smith, M. R. 1954a. A new name for *Martia* Forel (Hymenoptera). Bull. Brooklyn Entomol. Soc. 49:17. [1954.02.17]

Smith, M. R. 1954b ("1953"). A new *Camponotus* in California apparently inhabiting live oak, *Quercus* sp. (Hymenoptera, Formicidae). J. N. Y. Entomol. Soc. 61:211-214. [1954.02.20]

Smith, M. R. 1954c. Ants of the Bimini Island Group, Bahamas, British West Indies (Hymenoptera, Formicidae). Am. Mus. Novit. 1671:1-16. [1954.06.28]

Smith, M. R. 1954d. Concerning type locality and type fixation of the North American ant, *Myrmica emeryana* Forel. Bull. Brooklyn Entomol. Soc. 49:138-140. [1954.12.03]

Smith, M. R. 1955a. An unusual ant collection record. Bull. Brooklyn Entomol. Soc. 50:28. [1955.02.21]

Smith, M. R. 1955b. *Acanthostichus* (*Ctenopyga*) *townsendi* (Ashm.), a synonym of *Acanthostichus texanus* Forel (Hymenoptera: Formicidae). Bull. Brooklyn Entomol. Soc. 50:48-50. [1955.04.15]

Smith, M. R. 1955c. Remarks concerning the types of five species of ants described by Roger or Forel (Hymenoptera, Formicidae). Bull. Brooklyn Entomol. Soc. 50:98-99. [1955.11.04]

Smith, M. R. 1955d. Ants of the genus *Pheidole*, subgenus *Hendecapheidole* (Hymenoptera, Formicidae). Proc. Entomol. Soc. Wash. 57:301-305. [1955.12.22]

Smith, M. R. 1955e. The correct taxonomic status of *Pheidole* (*Pheidolacanthinus*) *brevispinosa* Donisthorpe (Hymenoptera, Formicidae). Proc. Entomol. Soc. Wash. 57:305. [1955.12.22]

Smith, M. R. 1956a. A key to the workers of *Veromessor* Forel of the United States and the description of a new subspecies (Hymenoptera, Formicidae). Pan-Pac. Entomol. 32:36-38. [1956.03.14]

Smith, M. R. 1956b. A list of the species of *Romblonella* including two generic transfers (Hymenoptera, Formicidae). Bull. Brooklyn Entomol. Soc. 51:18. [1956.03.19]

Smith, M. R. 1956c. A further contribution to the taxonomy and biology of the inquiline ant, *Leptothorax diversipilosus* Smith (Hymenoptera, Formicidae). Proc. Entomol. Soc. Wash. 58:271-275. [1956.11.23]

Smith, M. R. 1956d. [Review of: Borgmeier, T. 1955. Die Wanderameisen der neotropischen Region. Stud. Entomol. 3:1-720.] Proc. Entomol. Soc. Wash. 58:275-276. [1956.11.23]

Smith, M. R. 1957a ("1956"). New synonymy of a New Guinea ant (Hymenoptera, Formicidae). Proc. Entomol. Soc. Wash. 58:347. [1957.01.31]

Smith, M. R. 1957b. Revision of the genus *Stenamma* Westwood in America north of Mexico (Hymenoptera, Formicidae). Am. Midl. Nat. 57:133-174. [1957.01]

Smith, M. R. 1958a ("1957"). New synonymy of a North American ant, *Aphaenogaster macrospina* M. R. Smith (Hymenoptera: Formicidae). Bull. Brooklyn Entomol. Soc. 52:113. [1958.01.29]

Smith, M. R. 1958b ("1957"). A contribution to the taxonomy, distribution and biology of the vagrant ant, *Plagiolepis alluaudi* Emery (Hymenoptera, Formicidae). J. N. Y. Entomol. Soc. 65:195-198. [1958.06.27]

Smith, M. R. 1958c. Family Formicidae. Pp. 108-162 in: Krombein, K. V. (ed.) Hymenoptera of America north of Mexico. Synoptic catalogue. First supplement. U. S. Dep. Agric. Agric. Monogr. 2(suppl. 1):1-305. [1958.10.08] [Publication date from Krombein (1958).]

Smith, M. R. 1959. An illustrated key for the recognition of the imported fire ant and closely related species. U. S. Dep. Agric. Coop. Econ. Insect Rep. 9:221-223. [1959.03.27]

Smith, M. R. 1961a ("1960"). Notes on the synonymy of a North American ant (Hymenoptera: Formicidae). Proc. Entomol. Soc. Wash. 62:251-252. [1961.01.04]

Smith, M. R. 1961b. A study of New Guinea ants of the genus *Aphaenogaster* Mayr (Hymenoptera, Formicidae). Acta Hymenopterol. 1:213-238. [1961.05.31]

Smith, M. R. 1962a. A remarkable new *Stenamma* from Costa Rica, with pertinent facts on other Mexican and Central American species (Hymenoptera: Formicidae). J. N. Y. Entomol. Soc. 70:33-38. [1962.03.14]

Smith, M. R. 1962b. A new species of exotic *Ponera* from North Carolina (Hymenoptera, Formicidae). Acta Hymenopterol. 1:377-382. [1962.09.30]

Smith, M. R. 1963a. [Review of: Gregg, R. E. 1963. The ants of Colorado, with reference to their ecology, taxonomy, and geographic distribution. Boulder: University of Colorado Press, xvi + 792 pp.] Ann. Entomol. Soc. Am. 56:721. [1963.09.15]

Smith, M. R. 1963b. Notes on the leaf-cutting ants, *Atta* spp., of the United States and Mexico (Hymenoptera: Formicidae). Proc. Entomol. Soc. Wash. 65:299-302. [1963.12.12]

Smith, M. R. 1963c. A new species of *Aphaenogaster* (*Attomyrma*) from the western United States (Hymenoptera: Formicidae). J. N. Y. Entomol. Soc. 71:244-246. [1963.12.23]

Smith, M. R. 1964. [Review of: Wheeler, G. C., Wheeler, J. 1963. The ants of North Dakota. Grand Forks, North Dakota: University of North Dakota Press, viii + 326 pp.] Am. Midl. Nat. 71:249-250. [1964.01.15]

Smith, M. R. 1965. House-infesting ants of the eastern United States. Their recognition, biology, and economic importance. U. S. Dep. Agric. Tech. Bull. 1326:1-105. [1965.05]

Smith, M. R. 1967a. Family Formicidae. Pp. 343-374 in: Krombein, K. V., Burks, B. D. (eds.) Hymenoptera of America north of Mexico. Synoptic catalog. Second supplement. U. S. Dep. Agric. Agric. Monogr. 2(suppl. 2):1-584. [1967.02]

Smith, M. R. 1967b. Theodore Pergande - Early student of ants. Entomol. News 78:117-122. [1967.05.05]

Smith, M. R. 1969. [Review of: Cole, A. C., Jr. 1968. *Pogonomyrmex* harvester ants. A study of the genus in North America. Knoxville: University of Tennessee Press, x + 222 pp.] Bull. Entomol. Soc. Am. 15:157-158. [1969.06.16]

Smith, M. R., Haug, G. W. 1931. An ergatandrous form in *Ponera opaciceps* Mayr. Ann. Entomol. Soc. Am. 24:507-509. [1931.10.17]

Smith, M. R., Morrison, W. A. 1916. South Carolina ants (Hym.). Entomol. News 27:110-111. [1916.03.01]

Smith, M. R., Wing, M. W. 1955 ("1954"). Redescription of *Discothyrea testacea* Roger, a little-known North American ant, with notes on the genus (Hymenoptera: Formicidae). J. N. Y. Entomol. Soc. 62:105-112. [1955.05.05]

Smith, R. G. 1973. The skeleto-musculature of the thorax (mesosoma) of the queen and worker wood ant. Pp. 362-371 in: Proceedings IUSSI VIIth International Congress, London, 10-15 September, 1973. Southampton: University of Southampton, vi + 418 pp. [1973]

Snelling, R. R. 1963. The United States species of fire ants of the genus *Solenopsis*, subgenus *Solenopsis* Westwood, with synonymy of *Solenopsis aurea* Wheeler (Hymenoptera: Formicidae). Calif. Dep. Agric. Bur. Entomol. Occ. Pap. 3:1-15. [1963.01.14]

Snelling, R. R. 1965a. Studies on California ants. 1. *Leptothorax hirticornis* Emery, a new host and descriptions of the female and ergatoid male (Hymenoptera: Formicidae). Bull. South. Calif. Acad. Sci. 64:16-21. [1965.04.26]

Snelling, R. R. 1965b. Studies on California ants. 2. *Myrmecina californica* M. R. Smith (Hymenoptera; Formicidae). Bull. South. Calif. Acad. Sci. 64:101-105. [1965.06.30]

Snelling, R. R. 1966. The female of *Eucryptocerus placidus* (F. Smith) (Hymenoptera: Formicidae). Bull. South. Calif. Acad. Sci. 65:37-40. [1966.03.31]

Snelling, R. R. 1967. Studies on California ants. 3. The taxonomic status of *Proceratium californicum* Cook (Hymenoptera: Formicidae). Contr. Sci. (Los Angel.) 124:1-10. [1967.05.31]

Snelling, R. R. 1968a. Taxonomic notes on some Mexican cephalotine ants (Hymenoptera: Formicidae). Contr. Sci. (Los Angel.) 132:1-10. [1968.02.20]

Snelling, R. R. 1968b. Studies on California ants. 4. Two species of *Camponotus* (Hymenoptera: Formicidae). Proc. Entomol. Soc. Wash. 70:350-358. [1968.12.18]

Snelling, R. R. 1968c. A new species of *Eurhopalothrix* from El Salvador (Hymenoptera: Formicidae). Contr. Sci. (Los Angel.) 154:1-4. [1968.12.31]

Snelling, R. R. 1969a. The repository of the T. W. Cook ant types (Hymenoptera: Formicidae). Bull. South. Calif. Acad. Sci. 68:57-58. [1969.03.17]

Snelling, R. R. 1969b. Taxonomic notes on the *Myrmecocystus melliger* complex (Hymenoptera: Formicidae). Contr. Sci. (Los Angel.) 170:1-9. [1969.06.30]

Snelling, R. R. 1969c. Notes on the systematics and dulosis of some western species of *Formica*, subgenus *Raptiformica* (Hymenoptera: Formicidae). Proc. Entomol. Soc. Wash. 71:194-197. [1969.06.30]

Snelling, R. R. 1970. Studies on California ants, 5. Revisionary notes on some species of *Camponotus*, subgenus *Tanaemyrmex* (Hymenoptera: Formicidae). Proc. Entomol. Soc. Wash. 72:390-397. [1970.09.25]

Snelling, R. R. 1971a. Studies on California ants. 6. Three new species of *Myrmecocystus* (Hymenoptera: Formicidae). Contr. Sci. (Los Angel.) 214:1-16. [1971.05.28]

Snelling, R. R. 1971b. A new species of *Simopelta* from Costa Rica (Hymenoptera: Formicidae). Bull. South. Calif. Acad. Sci. 70:16-17. [1971.08.27]

Snelling, R. R. 1973a. Two ant genera new to the United States (Hymenoptera: Formicidae). Contr. Sci. (Los Angel.) 236:1-8. [1973.01.10]

Snelling, R. R. 1973b. The ant genus *Conomyrma* in the United States (Hymenoptera: Formicidae). Contr. Sci. (Los Angel.) 238:1-6. [1973.01.29]

Snelling, R. R. 1973c. Studies on California ants. 7. The genus *Stenamma* (Hymenoptera: Formicidae). Contr. Sci. (Los Angel.) 245:1-38. [1973.06.28]

Snelling, R. R. 1974. Studies on California ants. 8. A new species of *Cardiocondyla* (Hymenoptera: Formicidae). J. N. Y. Entomol. Soc. 82:76-81. [1974.08.22]

Snelling, R. R. 1975. Descriptions of new Chilean ant taxa (Hymenoptera: Formicidae). Contr. Sci. (Los Angel.) 274:1-19. [1975.12.23]

Snelling, R. R. 1976. A revision of the honey ants, genus *Myrmecocystus* (Hymenoptera: Formicidae). Nat. Hist. Mus. Los Angel. Cty. Sci. Bull. 24:1-163. [1976.08.05]

Snelling, R. R. 1979a. Three new species of the Palaeotropical arboreal ant genus *Cataulacus* (Hymenoptera: Formicidae). Contr. Sci. (Los Angel.) 315:1-8. [1979.07.31]

Snelling, R. R. 1979b. *Aphomomyrmex* and a related new genus of arboreal African ants (Hymenoptera: Formicidae). Contr. Sci. (Los Angel.) 316:1-8. [1979.07.31]

Snelling, R. R. 1981. Systematics of social Hymenoptera. Pp. 369-453 in: Hermann, H. R. (ed.) Social insects. Volume 2. New York: Academic Press, xiii + 491 pp. [1981]

Snelling, R. R. 1982a ("1981"). The taxonomy and distribution of some North American *Pogonomyrmex* and descriptions of two new species (Hymenoptera: Formicidae). Bull. South. Calif. Acad. Sci. 80:97-112. [1982.08.16]

Snelling, R. R. 1982b. A revision of the honey ants, genus *Myrmecocystus*, first supplement (Hymenoptera: Formicidae). Bull. South. Calif. Acad. Sci. 81:69-86. [1982.10.13]

Snelling, R. R. 1985. [Untitled. *Camponotus bakeri* Wheeler 1904 (elevated from subspecies of *C. hyatti*).] P. 24 in: Miller, S. E. The California Channel Islands - past, present and future: an entomological perspective. Pp. 3-28 in: Menke, A. S., Miller, D. R. (eds.) Entomology of the California Channel Islands. Proceedings of the first symposium. Santa Barbara: Santa Barbara Museum of Natural History, 178 pp. [1985]

Snelling, R. R. 1986. New synonymy in Caribbean ants of the genus *Leptothorax* (Hymenoptera: Formicidae). Proc. Entomol. Soc. Wash. 88:154-156. [1986.01.15]

Snelling, R. R. 1988. Taxonomic notes on Nearctic species of *Camponotus*, subgenus *Myrmentoma* (Hymenoptera: Formicidae). Pp. 55-78 in: Trager, J. C. (ed.) Advances in myrmecology. Leiden: E. J. Brill, xxvii + 551 pp. [1988]

Snelling, R. R. 1992a. Two unusual new myrmicine ants from Cameroon (Hymenoptera: Formicidae). Psyche (Camb.) 99:95-101. [1992.12.04]

Snelling, R. R. 1992b. A newly adventive ant of the genus *Pheidole* in southern California (Hymenoptera: Formicidae). Bull. South. Calif. Acad. Sci. 91:121-125. [1992.12.17]

Snelling, R. R. 1995a. Systematics of Nearctic ants of the genus *Dorymyrmex* (Hymenoptera: Formicidae). Contr. Sci. (Los Angel.) 454:1-14. [1995.07.27]

Snelling, R. R. 1995b. [Review of: Bolton, B. 1994. Identification guide to the ant genera of the world. Cambridge, Mass.: Harvard University Press, 222 pp.] Ann. Entomol. Soc. Am. 88:593. [1995.07]

Snelling, R. R., Buren, W. F. 1985. Description of a new species of slave-making ant in the *Formica sanguinea* group (Hymenoptera: Formicidae). Great Lakes Entomol. 18:69-78. [1985.06.03]

Snelling, R. R., Cover, S. P. 1992. Description of a new *Proceratium* from Mexico (Hymenoptera: Formicidae). Psyche (Camb.) 99:49-53. [1992.12.04]

Snelling, R. R., George, C. D. 1979. The taxonomy, distribution and ecology of California desert ants. Report to California Desert Plan Program, Bureau of Land Management, U.S. Dept. Interior. 335 + 89 pp. [1979.08.01] [Copies of this work were sent to "selected institutional and personal research libraries".]

Snelling, R. R., Hunt, J. H. 1975. The ants of Chile (Hymenoptera: Formicidae). Rev. Chil. Entomol. 9:63-129. [1976.05]

Snelling, R. R., Longino, J. T. 1992. Revisionary notes on the fungus-growing ants of the genus *Cyphomyrmex, rimosus* group (Hymenoptera: Formicidae: Attini). Pp. 479-494 in: Quintero, D., Aiello, A. (eds.) Insects of Panama and Mesoamerica: selected studies. Oxford: Oxford University Press, xxii + 692 pp. [1992]

Snodgrass, R. E. 1910. The thorax of the Hymenoptera. Proc. U. S. Natl. Mus. 39:37-91. [1910.10.25]

Snodgrass, R. E. 1941. The male genitalia of Hymenoptera. Smithson. Misc. Collect. 99(14):1-86. [1941.01.14]

Snyder, L. E. 1992. The genetics of social behavior in a polygynous ant. Naturwissenschaften 79:525-527. [1992.11]

Snyder, L. E. 1993. Non-random behavioural interactions among genetic subgroups in a polygynous ant. Anim. Behav. 46:431-439. [1993.09]

Snyder, L. E., Herbers, J. M. 1991. Polydomy and sexual allocation ratios in the ant *Myrmica punctiventris*. Behav. Ecol. Sociobiol. 28:409-415. [1991.07.02]

Snyder, T. E., Graf, J. E., Smith, M. R. 1961. William M. Mann. 1886-1960. Proc. Entomol. Soc. Wash. 63:68-73. [1961.04.12]

Sokolowski, A., Wisniewski, J. 1975. Teratologische Untersuchungen an Ameisen-Arbeiterinnen aus der *Formica rufa*-Gruppe (Hymenoptera: Formicidae). Insectes Soc. 22:117-134. [1975.12]

Sokolowski, A., Wisniewski, J. 1978 ("1977"). Mikroteratologische morphologische Veränderungen der Ameisenarbeiterinnen aus der *Formica rufa* gruppe (Hym., Formicidae). Bull. Soc. Amis Sci. Lett. Poznan Sér. D Sci. Biol. 17:227-234. [1978.04]

Somfai, E. 1959. Hangya alkatúak Formicoidea. Magyarorsz. Allatvilága 43:1-79. [1959.02.11]

Sommer, F., Cagniant, H. 1988. Peuplements de fourmis des Albères Orientales (Pyrénées-Orientales, France) (Première partie). Vie Milieu 38:189-200. [1988.10]

Sommer, F., Cagniant, H. 1989 ("1988"). Étude des peuplements de fourmis des Albères Orientales (Pyrénées-Orientales, France) (Seconde partie). Vie Milieu 38:321-329. [1989.01]

Sommer, K., Hölldobler, B. 1992. Coexistence and dominance among queens and mated workers in the ant *Pachycondyla tridentata*. Naturwissenschaften 79:470-472. [1992.10]

Sommer, K., Hölldobler, B., Jessen, K. 1994. The unusual social organization of the ant *Pachycondyla tridentata* (Formicidae, Ponerinae). J. Ethol. 12:175-185. [1994.12]

Sonan, J. 1912. Studies on *Polyrhachis dives* F. Smith. [In Japanese.] Insect World 16:436-440. [1912.11.15]

Sonan, J. 1940. M. Yanagihara's collection from Daito-Islands, Okinawa: Hymenoptera. Trans. Nat. Hist. Soc. Formosa 30:369-375. [1940.10.15] [Formicidae p. 375.]

Sonobe, R. 1973. Ant fauna of the Sesoko Island, Okinawa. Sesoko Mar. Sci. Lab. Tech. Rep. 2:15-16. [1973.12]

Sonobe, R. 1974a. On the occurrence of pseudogyne of *Formica japonica* Motschulsky (Hymenoptera, Formicidae) in Japan. Kontyû 42:401-403. [1974.12.25]

Sonobe, R. 1974b. Formicidae of Japan. (1) *Proceratium* Roger. [In Japanese.] Ari 6:2-3. [1974.12.30]

Sonobe, R. 1974c. Slave ants of *Formica sanguinea* and *Polyergus samurai* in Japan. [In Japanese.] Ari 6:3-4. [1974.12.30]

Sonobe, R. 1976. Formicidae of Japan. (2) *Myrmica* Latreille and *Manica* Jurine. [In Japanese.] Ari 7:1-2. [1976.05.01]

Sonobe, R. 1977a. Ant fauna of Miyagi Prefecture, Japan. Jpn. J. Ecol. 27:111-116. [1977.06.30]

Sonobe, R. 1977b. Formicidae of Japan. (3) *Formica* Linnaeus. [In Japanese.] Ari 8:1-2. [1977.08.01]

Sonobe, R. 1985. On Japanese *Myrmica* ants. [Abstract.] [In Japanese.] Ari 13:4-5. [1985.06.01]

Sonobe, R., Dlussky, G. M. 1977. On two ant species of the genus *Formica* (Hymenoptera, Formicidae) from Japan. Kontyû 45:23-25. [1977.03.25]

Sörensen, U., Schmidt, G. H. 1983. Die hügelbauenden Waldameisen in Waldgebieten der Bredstedter Geest (Schleswig-Holstein) (Genus *Formica*, Insecta). Z. Angew. Zool. 70:285-319. [1983]

Soroker, V., Hefetz, A., Cojocaru, M., Billen, J., Franke, S., Francke, W. 1995. Structural and chemical ontogeny of the postpharyngeal gland in the desert ant *Cataglyphis niger*. Physiol. Entomol. 20:323-329. [1995.12]

Soudek, S. 1921. K variabilite mravencu *Leptothorax tuberum* Nyl. Cas. Cesk. Spol. Entomol. 18:1-5. [1921]

Soudek, S. 1922a. Príspevky k poznání moravskych mravencu. II. Sb. Klubu Prírodoved. Brne 4: 3-6. [1922]

Soudek, S. 1922b. Mravenci. Soustava, zemepisné rozsírení, oekologie a urcovací klíc mravencu zijících na území Ceskoslovenské republiky. Praha: Ceskoslovenská Spolecnost Entomologická, 100 pp. [1922]

Soudek, S. 1923a ("1922"). Príspevek ku poznání mravencu Moravy. I. Cas. Morav. Zemského Mus. Brne 20-21:44-52. [1923]

Soudek, S. 1923b ("1922"). *Myrmica moravica* n. sp., relikt fauny praeglaciální. Cas. Morav. Zemského Mus. Brne 20-21:107-134. [1923]

Soudek, S. 1925a. Four new European ants. Entomol. Rec. J. Var. 37:33-37. [1925.03.15]

Soudek, S. 1925b. Dalmatstí mravenci (Formicidae). Cas. Cesk. Spol. Entomol. 22:12-17. [1925.05.20]

Soudek, S. 1925c ("1924"). *Bothriomyrmex meridionalis gibbus* n. ssp., novy mravenec z Moravy. Cas. Morav. Zemského Mus. Brne 22:216-232. [1925]

Soudek, S. 1927. Príspevek ku poznání moravskych mravencu. III. Cas. Morav. Zemského Mus. Brne 25:234-236. [1927]

Soudek, S. 1931. Mravenci "Hádu", jizního vybezku Moravského Krasu. (Faunisticky rozbor). Zpr. Kom. Prírodoved. Vyzk. Moravy Slez. Odd. Zool. 19:1-30. [1931]

Soulié, J. 1956. La nidification chez les espèces françaises du genre *Cremastogaster* Lund (Hymenoptera - Formicoidea). Insectes Soc. 3:93-105. [1956.02] [Reprinted as Soulié (1973b).]

Soulié, J. 1961a. [Untitled. Descriptions of new taxa: *Cremastogaster* (*Acrocelia*) *skounensis* Soulié; *Cremastogaster* (*Acrocelia*) *vandeli* Soulié; *Cremastogaster* (*Acrocelia*) *ledouxi* Soulié.] Pp. 99-105 in: Fromantin, J., Soulié, J. Notes systématiques sur le genre *Cremastogaster* avec description de trois espèces nouvelles du Cambodge. Bull. Soc. Hist. Nat. Toulouse 96:87-112. [1961.05.30]

Soulié, J. 1961b ("1960"). Des considérations écologiques peuvent-elles apporter une contribution à la connaisance du cycle biologique des colonies de *Cremastogaster* (Hymenoptera - Formicoidea)? Insectes Soc. 7:283-295. [1961.06]

Soulié, J. 1961c. Les nids et le comportement nidificateur des fourmis du genre *Cremastogaster* d'Europe, d'Afrique du Nord et d'Asie du Sud-Est. Insectes Soc. 8:213-297. [1961.12]

Soulié, J. 1962a. Fourmis des Hautes-Pyrénées. Bull. Soc. Hist. Nat. Toulouse 97:35-37. [1962.05.30] [Reprinted as Soulié (1973d).]

Soulié, J. 1962b. La fondation et le développement des colonies chez quelques espèces de fourmis du genre *Cremastogaster* Lund. Insectes Soc. 9:181-195. [1962.12]

Soulié, J. 1963 ("1962"). Recherches écologiques sur quelques espèces de fourmis du genre *Cremastogaster* de l'ancien monde (Europe, Afrique du Nord, Asie du Sud-Est). Ann. Sci. Nat. Zool. Biol. Anim. (12)4:669-826. [1963.06]

Soulié, J. 1964. Sur la répartition géographique des genres de la tribu des Cremastogastrini. Bull. Soc. Hist. Nat. Toulouse 99:397-409. [1964.12.30] [Reprinted as Soulié (1973e).]

Soulié, J. 1965. Contribution à la systématique des fourmis de la tribu des "Cremastogastrini" (Hymenoptera - Formicoidea - Myrmicidae). Ann. Univ. Abidjan Sér. Sci. 1:69-83. [1965] [Reprinted as Soulié (1973a).]

Soulié, J. 1967. Un nid de *Camponotus* (*Orthonotomyrmex*) *vividus* récolté en Basse Côte (Côte d'Ivoire) (Hymenoptera Formicoidea). Bull. Soc. Hist. Nat. Toulouse 103:7-18. [1967.06.30] [Reprinted as Soulié (1973c).]

Soulié, J. 1973a ("1972"). Contribution à la systématique des fourmis de la tribu des "Cremastogastrini" (Hymenoptera - Formicoidea - Myrmicidae). Ann. Univ. Abidjan Sér. E (Écol.) 5(2):5-19. [1973.03] [Reprinted from Soulié (1965).]

Soulié, J. 1973b ("1972"). La nidification chez les espèces françaises du genre *Cremastogaster* Lund (Hymenoptera - Formicoidea). Ann. Univ. Abidjan Sér. E (Écol.) 5(2):43-55. [1973.03] [Reprinted from Soulié (1956).]

Soulié, J. 1973c ("1972"). Un nid de *Camponotus* (*Orthonotomyrmex*) *vividus* récolté en Basse Côte (Côte d'Ivoire) (Hymenoptera Formicoidea). Ann. Univ. Abidjan Sér. E (Écol.) 5(2):57-67. [1973.03] [Reprinted from Soulié (1967).]

Soulié, J. 1973d ("1972"). Fourmis des Hautes-Pyrénées. Ann. Univ. Abidjan Sér. E (Écol.) 5(2):85-87. [1973.03] [Reprinted from Soulié (1962a).]

Soulié, J. 1973e ("1972"). Sur la répartition géographique des genres de la tribu des Cremastogastrini dans la faune éthiopienne et malgache. Ann. Univ. Abidjan Sér. E (Écol.) 5(2):89-99. [1973.03] [Reprinted from Soulié (1964).]

Soulié, J. 1973f ("1972"). La répartition des genres de fourmis de la tribu des "Cremastogastrini" dans la faune éthiopienne et malgache. Hymenoptera - Formicoidea - Myrmicidae. Ann. Univ. Abidjan Sér. E (Écol.) 5(2):101-121. [1973.03] [Reprinted from Soulié & Dicko, 1965.]

Soulié, J., Dicko, L. D. 1965. La répartition des genres de fourmis de la tribu des "Cremastogastrini" dans la faune éthiopienne et malgache. Hymenoptera - Formicoidea - Myrmicidae. Ann. Univ. Abidjan Sér. Sci. 1:85-106. [1965] [Reprinted as Soulié (1973f).]

Souza Paula, H. 1956. Ocorrência de saúvas no Estado de Paraná. Bol. Fitosannit. 6:153-158. [1956]

Soyunov, O. 1990. Landscape distribution of ants in the deserts of Turkmenistan. [In Russian.] Mater. Kollok. Sekts. Obshchestv. Nasek. Vses. Entomol. Obshch. 1:201-203. [> 1990.12.25] [Date signed for printing ("podpisano k pechati").]

Spahr, U. 1987. Ergänzungen und Berichtigungen zu R. Keilbach's Bibliographie und Liste der Bernsteinfossilien - Ordnung Hymenoptera. Stuttg. Beitr. Naturkd. Ser. B (Geol. Paläontol.) 127:1-121. [1987.03.16] [Formicidae pp. 41-64, 91-96.]

Spangler, H. G., Rettenmeyer, C. W. 1966. The function of the ammochaetae or psammophores of harvester ants, *Pogonomyrmex* spp. J. Kansas Entomol. Soc. 39:739-745. [1966.12.02]

Sparks, S. D. 1941. Surface anatomy of ants. Ann. Entomol. Soc. Am. 34:572-579. [1941.10.06]

Spinola, M. 1806. Insectorum Liguriae species novae aut rariores, quae in agro ligustico nuper detexit, descripsit et iconibus illustravit Maximilianus Spinola, adjecto catalogo specierum auctoribus jam enumeratarum, quae in eadam regione passim occurrent. Tom. I. Fasc. 1. Genova: Y. Gravier, xvii + 160 pp. [1806.10.21] [Date of publication from Passerin d'Entrèves (1983).]

Spinola, M. 1808. Insectorum Liguriae species novae aut rariores, quae in agro ligustico nuper detexit, descripsit et iconibus illustravit Maximilianus Spinola, adjecto catalogo specierum auctoribus jam enumeratarum, quae in eadam regione passim occurrent. Tom. II. Fasc. 4. Genova: Y. Gravier, pp. 207-262. [1808.03.17] [Date of publication from Passerin d'Entrèves (1983).]

Spinola, M. 1839 ("1838"). Compte rendu des Hyménoptères recueillis par M. Fischer pendant son voyage en Égypte. Ann. Soc. Entomol. Fr. 7:437-546. [1839.01]

Spinola, M. 1851a. Insectos. Orden 7. Himenopteros. Pp. 153-569 in: Gay, C. Historia fisica y politica de Chile. Zoologia. Tomo 6. Paris: Maulde & Renon, 572 pp. [1851]

Spinola, M. 1851b. Compte rendu des Hyménoptères inédits provenants du voyage entomologique de M. Ghiliani dans le Para en 1846. Extrait des Mémoires de l'Académie des Sciences de Turin (2)13:3-78. [1851] [This separate preceeds the paper published in 1853 in Mem. R. Accad. Sci. Torino. See Vecht (1975).]

Spinola, M. 1853. Compte rendu des Hyménoptères inédits provenants du voyage entomologique de M. Ghiliani dans le Para en 1846. Mem. R. Accad. Sci. Torino (2)13:19-94. [1853]

Spooner, G. M. 1968. Records of three uncommon *Formica* species (Hymenoptera: Formicidae). Entomol. Mon. Mag. 104:130-131. [1968.11.13]

Spooner, G. M. 1969 ("1968"). *Strongylognathus testaceus* (Hymenoptera: Formicidae) in South Devon. Entomol. Mon. Mag. 104:251. [1969.04.01]

Spooner, G. M. 1972. *Strongylognathus*, *Solenopsis* and *Anergates* (Hymenoptera: Formicidae) in South Devon. Entomol. Mon. Mag. 108:11. [1972.09.27]

Stafleu, F. A., Cowan, R. S. 1976. Taxonomic literature. A selective guide to botanical publications and collections, with dates, commentaries and types. Volume I. A-G. Utrecht: Bohn, Scheltema & Holkema, xl + 1136 pp. [1976]

Stafleu, F. A., Cowan, R. S. 1983. Taxonomic literature. A selective guide to botanical publications and collections, with dates, commentaries and types. Volume IV. P-Sak. Utrecht: Bohn, Scheltema & Holkema, ix + 1214 pp. [1983]

Stäger, R. 1917. Beitrag zur Kenntnis stengelbewohnender Ameisen in der Schweiz. Rev. Suisse Zool. 25:95-109. [1917.05]

Stäger, R. 1928. *Anergates atratulus* Schenck am Mittelmeer (Hym., Formic.). Z. Wiss. Insektenbiol. 23:159-162. [1928.07.25]

Stäger, R. 1935. Siedelungsverhältnisse bei den Ameisen der Hochalp. Verh. Schweiz. Naturforsch. Ges. 116:349-350. [1935]

Stainforth, T. 1903. *Camponotus herculaneus* [sic] at Hull. Naturalist (Leeds) 1903:456. [1903.12.01]

Stärcke, A. 1926. Nieuwe Nederlandsche Formiciden (benevens enkele systematische opmerkingen). Entomol. Ber. (Amst.) 7:86-97. [1926.05.01]

Stärcke, A. 1927. Beginnende Divergenz bei *Myrmica lobicornis* Nyl. Tijdschr. Entomol. 70: 73-84. [1927.08]

Stärcke, A. 1928. Iets over de verspreiding van onze miersoorten. Natura (Amst.) 1928:258-265. [1928.12.15]

Stärcke, A. 1930. Verzeichnis der bis jetzt von der Insel Palau Berhala bekannt gewordenen Ameisen. Treubia 12:371-381. [1930.12]

Stärcke, A. 1933a. Contribution à l'étude de la faune népenthicole. Art. III. Un nouveau *Camponotus* de Bornéo, habitant les tiges creuses de *Nepenthes*, récolté par J. P. Schuitemaker. Natuurhist. Maandbl. 22:29-31. [1933.03.31]

Stärcke, A. 1933b. [Untitled. Introduced by: "De heer Stärcke spreekt over de larven der Dolichoderinen".] Tijdschr. Entomol. 76:xxvi-xxxi. [1933.05]

Stärcke, A. 1934. Un nouveau sous-genre de *Camponotus* (Hymenoptera Formicidae) de la Malaisie avec description de formes nouvelles, récoltées par M. F. G. Nainggolan. Zool. Meded. (Leiden) 17:20-30. [1934.08.01]

Stärcke, A. 1935a. Zoologie. Formicidae. Wiss. Ergeb. Niederl. Exped. Karakorum 1:260-269. [1935.01.15]

Stärcke, A. 1935b. Het probleem van den mikroergaat. Tijdschr. Entomol. 78:xxv-xxxi. [1935.06]

Stärcke, A. 1936. Retouches sur quelques fourmis d'Europe. I. *Plagiolepis xene* nov. sp. et *Pl. vindobonensis* Lomnicki. Entomol. Ber. (Amst.) 9:277-279. [1936.11.01]

Stärcke, A. 1937. Retouches sur quelques fourmis d'Europe. II. *Lasius* groupe *umbratus* Nylander. Tijdschr. Entomol. 80:38-72. [1937.05]

Stärcke, A. 1938. [Untitled. Introduced by: "De heer Stärcke doet eene mededeeling over gedragen ontwikkeling van enkele Javaansche mieren tijdens hun verblijf hier te lande."] Tijdschr. Entomol. 81:xxxiii-xli. [1938.05]

Stärcke, A. 1939. Le faisceau des ocelles et la fonction qu'il nous suggère, avec un appendice sur l'antenne larvale des fourmis. Tijdschr. Entomol. 82:xix-xxvii. [1939.05]

Stärcke, A. 1941. Hersenganglion van *Strumigenys*; koloniestichting van *Polyrhachis bicolor*. Tijdschr. Entomol. 84:ii-xv. [1941.05]

Stärcke, A. 1942a. Ants, collected by Dr. C. F. Engelhard at Stockmarknes (Island Hadsekoy, Vesterale Archipelago circa 68°50' Lat. N. Norway) and between 64° and 66° Lat. N. on the western coast of Norway, July 1932. Entomol. Ber. (Amst.) 11:21-23. [1942.05.01]

Stärcke, A. 1942b. Formicides de l'île de Port-Croz (Hyères) récoltées par le docteur Chassagne. Entomol. Ber. (Amst.) 11:23 [1942.05.01]

Stärcke, A. 1942c. Definities van species (soort), subspecies (ras, stirps), variëteit en aberratie. Entomol. Ber. (Amst.) 11:40-48. [1942.05.01]

Stärcke, A. 1942d. Drie nog onbeschreven Europeesche miervormen. Tijdschr. Entomol. 85:xxiv-xxix. [1942.12]

Stärcke, A. 1943. Onze verdreven boschmieren. Levende Nat. 48:1-7. [1943.05.001]

Stärcke, A. 1944a. Mieren-tabel 1926. Herziene 2e druk. Natuurhist. Mandbl. 33:6-8. [1944.01.28]

Stärcke, A. 1944b. Retouches sur quelques fourmis d'Europe. III. Autres *Lasius*. Entomol. Ber. (Amst.) 11:153-158. [1944.02.25]

Stärcke, A. 1944c. Determineertabel voor de werksterkaste der Nederlandsche mieren. Herziene 2e druk. (Vervolg). Natuurhist. Maandbl. 33:23-24. [1944.03.31]

Stärcke, A. 1944d ("1943"). Nederlandsche adventief-mieren. Tijdschr. Entomol. 86:xvii-xxii. [1944.04.07]

Stärcke, A. 1944e. Determineertabel voor de werksterkaste der Nederlandsche mieren. Herziene 2e druk. (Vervolg). Natuurhist. Maandbl. 33:29-32. [1944.04.28]

Stärcke, A. 1944f. Determineertabel voor de werksterkaste der Nederlandsche mieren. Herziene 2e druk. (Vervolg). Natuurhist. Maandbl. 33:37-38. [1944.05.31]

Stärcke, A. 1944g. Determineertabel voor de werksterkaste der Nederlandsche mieren. Herziene 2e druk. (Vervolg). Natuurhist. Maandbl. 33:43-46. [1944.06.30]

Stärcke, A. 1944h. Determineertabel voor de werksterkaste der Nederlandsche mieren. Herziene 2e druk. (Vervolg). Natuurhist. Maandbl. 33:55-56. [1944.08]

Stärcke, A. 1944i. Determineertabel voor de werksterkaste der Nederlandsche mieren. Herziene 2e druk. (Vervolg). Natuurhist. Maandbl. 33:58-60. [1944.10]

Stärcke, A. 1944j. Determineertabel voor de werksterkaste der Nederlandsche mieren. Herziene 2e druk. (Vervolg). Natuurhist. Maandbl. 33:62-65. [1944.11.30]

Stärcke, A. 1944k. Determineertabel voor de werksterkaste der Nederlandsche mieren. Herziene 2e druk. (Slot). Natuurhist. Maandbl. 33:72-76. [1944.12.29]

Stärcke, A. 1947. Die boreale vorm van de roode boschmier (*Formica rufa rufa* Nyl.) op de Hooge Veluwe. Entomol. Ber. (Amst.) 12:144-146. [1947.06.07]

Stärcke, A. 1949 ("1948"). Contribution to the biology of *Myrmica schencki* Em. (Hym., Form.). Tijdschr. Entomol. 91:25-71. [1949.12.20]

Stärcke, A. 1951. Critical commentary upon M. V. Brian's and A. D. Brian's "Observations". Entomol. Ber. (Amst.) 13:324-327. [1951.09.01]

Stawarski, I. 1961. Nowe stanowiska rzadkich gatunków mrówek (Hym., Formicidae). Pol. Pismo Entomol. 31:135-138. [1961.06]

Stefani, R. 1970 ("1968"). Contributo alla conoscenza dei formicidi cavernicoli dell Sardegna. Rend. Semin. Fac. Sci. Univ. Cagliari 38(suppl.):1-5. [1970.01]

Steghaus-Kovac, S., Maschwitz, U. 1993. Predation on earwigs: a novel diet specialization within the genus *Leptogenys* (Formicidae: Ponerinae). Insectes Soc. 40:337-340. [1993]

Steghaus-Kovac, S., Maschwitz, U. 1994. Resource partitioning in Malayan *Leptogenys* species (Formicidae: Ponerinae). P. 288 in: Lenoir, A., Arnold, G., Lepage, M. (eds.) Les insectes sociaux. 12ème congrès de l'Union Internationale pour l'Étude des Insectes Sociaux. Paris, Sorbonne, 21-27 août 1994. Paris: Université Paris Nord, xxiv + 583 pp. [1994.09]

Stein, M. B., Thorvilson, H. G. 1989. Ant species sympatric with the red imported fire ant in southeastern Texas. Southwest. Entomol. 14:225-231. [1989.09]

Stephens, J. F. 1829a. The nomenclature of British insects; being a compendious list of such species as are contained in the systematic catalogue of British insects, and forming a guide to their classification. London: Baldwin & Cradock, 68 pp. [1829.06.01] [Date of publication from Blackwelder (1949b).]

Stephens, J. F. 1829b. A systematic catalogue of British insects: being an attempt to arrange all the hitherto discovered indigenous insects in accordance with their natural affinities. London: Baldwin & Cradock, xxxiv + 416 + 388 pp. [1829.07.15] [Date of publication from Blackwelder (1949b). First use of "Formicidae" (Part I, page 356).]

Stiller, V. 1937. *Strumigenys baudieri* Em. aus Ungarn (Form.). Entomol. Nachrichtenbl. Troppau 11:175. [1937.12]

Stitz, H. 1909. Eine neue afrikanische *Dichthadia*. Zool. Anz. 35:231-232. [1909.12.21]

Stitz, H. 1910. Westafrikanische Ameisen. I. Mitt. Zool. Mus. Berl. 5:125-151. [1910.08]

Stitz, H. 1911a. Australische Ameisen. (Neu-Guinea und Salomons-Inseln, Festland, Neu-Seeland). Sitzungsber. Ges. Naturforsch. Freunde Berl. 1911:351-381. [1911.10.10]

Stitz, H. 1911b. Formicidae. Wiss. Ergeb. Dtsch. Zent.-Afr.-Exped. 3:375-392. [1911]

Stitz, H. 1912. Ameisen aus Ceram und Neu-Guinea. Sitzungsber. Ges. Naturforsch. Freunde Berl. 1912:498-514. [1912.11.12]

Stitz, H. 1913. Ameisen aus Brasilien, gesammelt von Ule. (Hym.). Dtsch. Entomol. Z. 1913:207-212. [1913.03.31]

Stitz, H. 1914. Die Ameisen (Formicidae) Mitteleuropas, insbesondere Deutschlands. Pp. 1-111 in: Schröder, C. (ed.) Die Insekten Mitteleuropas insbesondere Deutschlands. Band II, Hymenopteren, 2. Teil. Stuttgart: Franckh'sche Verlagshandlung, 256 pp. [1914]

Stitz, H. 1916. Formiciden. Ergeb. Zweiten Dtsch. Zent.-Afr. Exped. 1:369-405. [1916.10.14]

Stitz, H. 1917. Ameisen aus dem westlichen Mittelmeergebiet und von den Kanarischen Inseln. Mitt. Zool. Mus. Berl. 8:333-353. [1917.04]

Stitz, H. 1923. Hymenoptera, VII. Formicidae. Beitr. Kennt. Land- Süsswasserfauna Dtsch.-Südwestafr. 2:143-167. [1923]

Stitz, H. 1925a. Zur Kenntnis estländischer Hochmoorameisen. Beitr. Kunde Estlands 10:136-139. [1925.03]

Stitz, H. 1925b ("1924"). [Untitled. *Myrmicaria gracilis* v. *simplex* n. var.] P. 170 in: Santschi, F. Révision des Myrmicaria d'Afrique. Ann. Soc. Entomol. Belg. 64:133-176. [1925.04]

Stitz, H. 1925c. Ameisen von den Philippinen, den malayischen und ozeanischen Inseln. Sitzungsber. Ges. Naturforsch. Freunde Berl. 1923:110-136. [1925.12.19]

Stitz, H. 1930a. Entomologische Ergebnisse der Deutsch-Russischen Alai-Pamir Expedition 1928 (1). 5. Hymenoptera III. Formicidae. Mitt. Zool. Mus. Berl. 16:238-240. [1930.06.30]

Stitz, H. 1930b. XXXI. Fam. Formicidae. Ameisen. Pp. 521-563 in: Schmiedeknect, O. (ed.) Die Hymenopteren Nord- und Mitteleuropas. Zweite Auflage. Jena: G. Fischer, x + 1062 pp. [1930]

Stitz, H. 1932a. Formicidae [of the Wollebaek Galapagos Expedition]. Nyt Mag. Naturvidensk. 71:367-372. [1932.10.19]

Stitz, H. 1932b. Formicidae der Deutschen Limnologischen Sunda-Expedition. Arch. Hydrobiol. Suppl.-Bd. 9(Tropische Binnengewässer 2):733-737. [1932]

Stitz, H. 1933. Neue Ameisen des Hamburger Museums (Hym. Form.). Mitt. Dtsch. Entomol. Ges. 4:67-75. [1933.05]

Stitz, H. 1934. Schwedisch-chinesische wissenschaftliche Expedition nach den nordwestlichen Provinzen Chinas, unter Leitung von Dr. Sven Hedin und Prof. Sü Ping-chang. Insekten gesammelt vom schwedischen Arzt der Expedition Dr. David Hummel 1927-1930. 25. Hymenoptera. 3. Formicidae. Ark. Zool. 27A(11):1-9. [1934.01.26]

Stitz, H. 1937. Einige Ameisen aus Mexiko. Sitzungsber. Ges. Naturforsch. Freunde Berl. 1937:132-136. [1937.11.03]

Stitz, H. 1938. Neue Ameisen aus dem indo-malayischen Gebiet. Sitzungsber. Ges. Naturforsch. Freunde Berl. 1938:99-122. [1938.12.28]

Stitz, H. 1939. Die Tierwelt Deutschlands und der angrenzenden Meersteile nach ihren Merkmalen und nach ihrer Lebensweise. 37. Theil. Hautflüger oder Hymenoptera. I: Ameisen oder Formicidae. Jena: G. Fischer, 428 pp. [1939]

Stitz, H. 1940. Die Arthropodenfauna von Madeira nach den Ergebnissen der Reise von Prof. Dr. O. Lundblad Juli-August 1935. XXIV. Hymenoptera: Formicidae. Ark. Zool. 32B(5):1-2. [1940.02.12]

Stoll, O. 1898. Zur Kenntniss der geographischen Verbreitung der Ameisen. Mitt. Schweiz. Entomol. Ges. 10:120-126. [1898.06]

Stolpe, H. 1882. Förteckning öfver svenska myror. Preliminärt meddelande. Entomol. Tidskr. 3:127-151. [1882]

Stradling, D. J., Powell, R. J. 1986. The cloning of more highly productive fungal strains: a factor in the speciation of fungus-growing ants. Experientia (Basel) 42:962-964. [1986.08.15]

Stradling, D. J., Powell, R. J. 1987. The cloning of selected fungal strains; a factor in the speciation of attine ants. Pp. 625-626 in: Eder, J., Rembold, H. (eds.) Chemistry and biology of social insects. München: Verlag J. Peperny, xxxv + 757 pp. [1987]

Strand, E. 1911. Ein bisher unbekanntes Dorylidenweibchen aus Kamerun. Jahrb. Nassau. Ver. Naturkd. 64:118-120. [1911]

Strand, E. 1935. Miscellanea nomenclatorica zoologica et palaeontologica. VIII. Folia Zool. Hydrobiol. 8:176. [1935.08.30]

Street, M. D., Donovan, G. R., Baldo, B. A., Sutherland, S. 1994. Immediate allergic reactions to *Myrmecia ant* stings: immunochemical analysis of *Myrmecia* venoms. Clin. Exp. Allergy 24:590-597. [1994.06]

Strickland, A. H. 1945. A survey of the arthropod soil and litter fauna of some forest reserves and cacao estates in Trinidad, British West Indies. J. Anim. Ecol. 14:1-11. [1945.05] [Ants pp. 6-8.]

Stuart, R. J., Alloway, T. M. 1982. Territoriality and the origin of slave raiding in leptothoracine ants. Science (Wash. D. C.) 215:1262-1263. [1982.03.05]

Stuart, R. J., Alloway, T. M. 1985. Behavioural evolution and domestic degeneration in obligatory slave-making ants (Hymenoptera: Formicidae: Leptothoracini). Anim. Behav. 33:1080-1088. [1985.11]

Stuart, R. J., Alloway, T. M. 1988. Aberrant yellow ants: North American *Leptothorax* species as intermediate hosts of cestodes. Pp. 537-545 in: Trager, J. C. (ed.) Advances in myrmecology. Leiden: E. J. Brill, xxvii + 551 pp. [1988]

Stuart, R. J., Bell, P. D. 1981 ("1980"). Stridulation by workers of the ant, *Leptothorax muscorum* (Nylander) (Hymenoptera: Formicidae). Psyche (Camb.) 87:199-210. [1981.08.31]

Stuart, R. J., Francoeur, A., Loiselle, R. 1987. Lethal fighting among dimorphic males of the ant, *Cardiocondyla wroughtonii*. Naturwissenschaften 74:548-549. [1987.11]

Stuart, R. J., Page, R. E. 1991. Genetic component to division of labor among workers of a leptothoracine ant. Naturwissenschaften 78:375-377. [1991.08]

Stumper, R. 1918a. Zur Kenntnis des Polymorphismus der Formiciden. Bull. Mens. Soc. Nat. Luxemb. 28:18-24. [1918.01.31]

Stumper, R. 1918b. *Formicoxenus nitidulus* Nyl. I. Beitrag. Biol. Zentralbl. 38:160-179. [1918.05.10]

Stumper, R. 1934. [Review of: Wasmann, E. 1934. Die Ameisen, die Termiten und ihre Gäste. Regensburg: G. J. Manz, xviii + 148 pp.] Bull. Mens. Soc. Nat. Luxemb. (n.s.)28:76-77. [1934.10.28]

Stumper, R. 1939. Kurze Zusammenstellung der einheimischen Ameisen. Bull. Mens. Soc. Nat. Luxemb. (n.s.)33:82-87. [1939.05.20]

Stumper, R. 1951. *Teleutomyrmex schneideri* Kutter (Hym. Formicid.). II. Mitteilung. Über die Lebensweise der neuen Schmarotzerameise. Mitt. Schweiz. Entomol. Ges. 24:129-152. [1951.07.20]

Stumper, R. 1952. Données quantitatives sur la sécrétion d'acide formique par les fourmis. C. R. Séances Acad. Sci. 234:149-152. [1952.01.09]

Stumper, R. 1953a ("1952"). Études myrmécologiques. XI. Fourmis luxembourgeoises. Bull. Soc. Nat. Luxemb. (n.s.)46:122-130. [1953.11.18]

Stumper, R. 1953b. Quelques aspects nouveaux du parasitisme social chez les fourmis. Arch. Inst. Grand-Ducal Luxemb. (n.s.) 20:171-174. [1953]

Stumper, R. 1955. Sur le développement relatif des organes visuels et antennaires chez les fourmis. C. R. Séances Acad. Sci. 240:1485-1487. [1955.04.04]

Stumper, R. 1957a. Sur quelques caractéristiques physiques des fourmis. C. R. Séances Acad. Sci. 244:127-130. [1957.01.09]

Stumper, R. 1957b ("1955"). Sur l'éthologie de la fourmi à miel *Proformica nasuta* Nyl. (Étude myrmécologique LXXVIII.) Bull. Soc. Nat. Luxemb. 60:87-97. [1957.02.20]

Stumper, R. 1960. Die Giftsekretion der Ameisen. Naturwissenschaften 47:457-463. [1960.10]

Stumper, R., Kutter, H. 1950. Sur le stade ultime du parasitisme social chez les fourmis, atteint par *Teleutomyrmex Schneideri* (subtrib. nov.; gen. nov.; spec. nov. Kutter). C. R. Séances Acad. Sci. 231:876-878. [1950.10.30]

Stumper, R., Kutter, H. 1951. Sur l'éthologie du nouveau myrmécobionte *Epimyrma Stumperi* (nov. spec. Kutter). C. R. Séances Acad. Sci. 233:983-985. [1951.10.29]

Stumper, R., Kutter, H. 1952. Sur un type nouveau de myrmécobiose réalisé par *Plagiolepis Xene* (Staercke). C. R. Séances Acad. Sci. 234:1482-1485. [1952.04.07]

Sturm, J. 1826. Catalog meiner Insecten-Sammlung. 1. Käfer. Nürnberg: Verfasser, viii + 207 pp. [1826] [*Myrmex* (Coleoptera), senior homonym of *Myrmex* Guérin-Méneville, 1844.]

Sturtevant, A. H. 1925. Notes on the ant fauna of oak galls in the Woods Hole region. Psyche (Camb.) 32:313-314. [1925.12] [Stamp date in MCZ library: 1926.01.26.]

Sturtevant, A. H. 1931. Ants collected on Cape Cod, Massachusetts. Psyche (Camb.) 38:73-79. [1931.09] [Stamp date in MCZ library: 1931.10.10.]

Sudd, J. H. 1967. An introduction to the behaviour of ants. London: E. Arnold, 200 pp. [1967]

Sudd, J. H., Franks, N. R. 1987. The behavioural ecology of ants. New York: Chapman & Hall, x + 206 pp. [1987]

Suehiro, A. 1960. Insects and other arthropods from Midway Atoll. Proc. Hawaii. Entomol. Soc. 17:289-298. [1960.08.08]

Sugerman, B. B. 1979. Additions to the list of insects and other arthropods from Kwajalein Atoll (Marshall Islands). Proc. Hawaii. Entomol. Soc. 23:147-151. [1979.05.31]

Sugihara, Y. 1933. Hymenoptera-Fauna in Province Tosa. Formicidae. [In Japanese and English.] Kansai Konchu Zasshi 1:79-86. [1933.10.07]

Sumichrast, F. 1868. Notes on the habits of certain species of Mexican Hymenoptera presented to the American Entomological Society. No. 1. On the habits of the Mexican species of the genus *Eciton* Latr. Trans. Am. Entomol. Soc. 2:39-44. [1868.06]

Sundström, L. 1993. Genetic population structure and sociogenetic organisation in *Formica truncorum* (Hymenoptera; Formicidae). Behav. Ecol. Sociobiol. 33:345-354. [1993.11.22]

Sundström, L. 1994. Sex ratio bias, relatedness asymmetry and queen mating frequency in ants. Nature (Lond.) 367:266-268. [1994.01.20]

Sundström, L. 1995. Sex allocation and colony maintenance in monogyne and polygyne colonies of *Formica truncorum* (Hymenoptera: Formicidae): the impact of kinship and mating structure. Am. Nat. 146:182-201. [1995.08]

Suñer i Escriche, D. 1988. Primera referencia de *Stenamma westwoodi* Westwood, 1840 (Hym. Formicidae) a les comarques gironines (Catalunya). Sci. Gerundensis 13:143-147. [Not seen. Cited in Zoological Record for 1991/92.]

Suñer, D., Gómez, C., Espadaler, X. 1991. Poblaciones meridionales de *Lasius flavus* (Fabr.) y *L. myops* Forel: estudio biométrico (Hymenoptera, Formicidae). Orsis (Org. Sist.) 6:101-108. [1991]

Suzzoni, J. P., Cagniant, H. 1975. Étude histologique des voies génitales chez l'ouvrière et la reine de *Cataglyphis cursor* Fonsc. (Hyménoptère Formicidae, Formicinae). Arguments en faveur d'une parthénogenèse thélytoque chez cette espèce. Insectes Soc. 22:83-92. [1975.06]

Suzzoni, J. P., Kenne, M., Dejean, A. 1994. The ecology and distribution of *Myrmicaria opaciventris*. Pp. 133-150 in: Williams, D. F. (ed.) Exotic ants. Biology, impact, and control of introduced species. Boulder: Westview Press, xvii + 332 pp. [1994]

Sveum, P. 1979. Notes on the distribution of some Norwegian ant species (Hymenoptera, Formicidae). Fauna Norv. Ser. B 26:10-11. [1979]

Swainson W., Shuckard, W. E. 1840. On the history and natural arrangement of insects. London: Longman, Brown, Green & Longman's, 406 pp. [1840.12] [Publication date from Sherborn (1922:cxxi).]

Sweeney, R. C. H. 1950a. Two teratological specimens of *Myrmica ruginodis* Nylander (Hym., Formicidae) from Denmark. Entomol. Mon. Mag. 86:5. [1950.02.15]

Sweeney, R. C. H. 1950b. Identification of British ants (Hym.: Formicidae) with keys to the genera and species. Entomol. Gaz. 1:64-83. [1950.04]

Swezey, O. H. 1915. A note on "*Technomyrmex albipes*." Proc. Hawaii. Entomol. Soc. 3:56. [1915.07]

Swezey, O. H. 1927. Notes and exhibitions. *Tetramorium tonganum* Mayr. Proc. Hawaii. Entomol. Soc. 6:367-368. [1927.10]

Swezey, O. H. 1942. Insects of Guam - 1. Hymenoptera. Formicidae of Guam. Bull. Bernice P. Bishop Mus. 172:175-183. [1942.06.01]

Swezey, O. H. 1944 ("1943"). Notes and exhibitions. *Cardiocondyla emeryi* Forel. Proc. Hawaii. Entomol. Soc. 12:25. [1944.08.16]

Swezey, O. H. 1945. Insects associated with orchids. Proc. Hawaii. Entomol. Soc. 12:343-403. [1945.06.25]

Sykes, W. H. 1835. Descriptions of new species of Indian ants. Trans. Entomol. Soc. Lond. 1:99-107. [1835.10.02] [Date from Wheeler, G. (1912).]

Szabó, J. 1909. De duabus speciebus novis Formicidarum generis *Epitritus* Em. Arch. Zool. (Budapest) 1:27-28. [1909.12.30]

Szabó, J. 1910a. Formicides nouveaux ou peu connus des collections du Musée National Hongrois. [part] Ann. Hist.-Nat. Mus. Natl. Hung. 8:364-368. [1910.06.20]

Szabó, J. 1910b. Faunánk egy új hangya-nemérol. Allatt. Közl. 9:182-184. [1910.12.15]

Szabó, J. 1910c. Formicides nouveaux ou peu connus des collections du Musée National Hongrois. [concl.] Ann. Hist.-Nat. Mus. Natl. Hung. 8:369. [1910.12.30]

Szabó, J. 1910d. Uj hangya Uj-Guineából. Rovartani Lapok 17:186. [1910.12.31]

Szabó, J. 1911. A *Camponotus ligniperda* noi ivarkészülékének szerkezete. Allatt. Közl. 10:83-96. [1911.06.10]

Szabó, J. 1913. A hangyák nöstényeinek önálló államalapitásáról. Rovartani Lapok 20:186-190. [1913.12.15]

Szabó, J. 1914. Magyarország rabszolgatartó és élosködo hangyái. Allatt. Közl. 13:93-107. [1914.06.10]

Szabó-Patay, J. 1926. Trois *Orectognathus* nouveaux de la collection du Musée National Hongrois. Ann. Hist.-Nat. Mus. Natl. Hung. 24:348-351. [1926.11.23]

Taber, S. W. 1988. The gyne of the harvester ant, *Pogonomyrmex texanus* (Hymenoptera: Formicidae). J. Kansas Entomol. Soc. 61:244-246. [1988.06.02]

Taber, S. W. 1990. Cladistic phylogeny of the North American species complexes of *Pogonomyrmex* (Hymenoptera: Formicidae). Ann. Entomol. Soc. Am. 83:307-316. [1990.05]

Taber, S. W., Cokendolpher, J. C. 1988. Karyotypes of a dozen ant species from the southwestern U.S.A. (Hymenoptera: Formicidae). Caryologia 41:93-102. [1988.10.31]

Taber, S. W., Cokendolpher, J. C., Francke, O. F. 1987. Scanning electron microscopic study of North American *Pogonomyrmex* (Hymenoptera: Formicidae). Proc. Entomol. Soc. Wash. 89:512-526. [1987.07.23]

Taber, S. W., Cokendolpher, J. C., Francke, O. F. 1988. Karyological study of North American *Pogonomyrmex* (Hymenoptera: Formicidae). Insectes Soc. 35:47-60. [1988.09]

Taber, S. W., Francke, O. F. 1986. A bilateral gynandromorph of the western harvester ant, *Pogonomyrmex occidentalis* (Hymenoptera: Formicidae). Southwest. Nat. 31:274-276. [1986.05.22]

Tafuri, J. F. 1957 ("1955"). Growth and polymorphism in the larva of the army ant (*Eciton* (*E.*) *hamatum* Fabricius). J. N. Y. Entomol. Soc. 63:21-41. [1957.03.08]

Takechi, F. 1960. A list of ants unrecorded from Mt. Ishizuchi and Omogo Valley, Iyo, Shikoku (Hymenoptera: Formicidae). Trans. Shikoku Entomol. Soc. 6:91. [1960.05.30]

Talbot, M. 1934. Distribution of ant species in the Chicago region with reference to ecological factors and physiological toleration. Ecology 15:416-439. [1934.10.26]

Talbot, M. 1948. A comparison of two ants of the genus *Formica*. Ecology 29:316-325. [1948.07]

Talbot, M. 1953. Ants of an old-field community on the Edwin S. George Reserve, Livingston County, Michigan. Contr. Lab. Vertebr. Biol. Univ. Mich. 63:1-13. [1953.04]

Talbot, M. 1957. Populations of ants in a Missouri woodland. Insectes Soc. 4:375-384. [1957.10]

Talbot, M. 1963. Local distribution and flight activities of four species of ants of the genus *Acanthomyops* Mayr. Ecology 44:549-557. [1963]

Talbot, M. 1968 ("1967"). Slave-raids of the ant *Polyergus lucidus* Mayr. Psyche (Camb.) 74:299-313. [1968.06.12]

Talbot, M. 1972 ("1971"). Flights of the ant *Formica dakotensis* Emery. Psyche (Camb.) 78:169-179. [1972.02.23]

Talbot, M. 1973. Five species of the ant genus *Acanthomyops* (Hymenoptera: Formicidae) at the Edwin S. George Reserve in southern Michigan. Great Lakes Entomol. 6:19-22. [1973.05.04]

Talbot, M. 1976a ("1975"). Habitats and populations of the ant *Stenamma diecki* Emery in southern Michigan. Great Lakes Entomol. 8:241-244. [1976.03.01]

Talbot, M. 1976b ("1975"). A list of the ants (Hymenoptera: Formicidae) of the Edwin S. George Reserve, Livingston County, Michigan. Great Lakes Entomol. 8:245-246. [1976.03.01]

Talbot, M. 1977 ("1976"). The natural history of the workerless ant parasite *Formica talbotae*. Psyche (Camb.) 83:282-288. [1977.08.29]

Talbot, M. 1979. Social parasitism amoung ants at the E. S. George Reserve in southern Michigan. Great Lakes Entomol. 12:87-89. [1979.05.02]

Talbot, M. 1985. The slave-making ant *Formica gynocrates* (Hymenoptera: Formicidae). Great Lakes Entomol. 18:103-112 [1985.07.01]

Talbot, M., Kennedy, C. H. 1940. The slave-making ant, *Formica sanguinea subintegra* Emery, its raids, nuptual flights and nest structure. Ann. Entomol. Soc. Am. 33:560-577. [1940.09.30]

Tanaka, M. 1974a. Description of a new species of the ant of the genus *Probolomyrmex* Mayr from Malaysia (Hymenoptera, Formicidae). Entomol. Rev. Jpn. 26:35-37. [1974.06.30]
Tanaka, M. 1974b. A new species of the ant genus *Ponera* from Yaku Island (Hymenoptera, Formicidae). Entomol. Rev. Jpn. 27:32-36. [1974.12.31]
Tang, C., Li, S. 1982. Hymenoptera: Formicidae. Pp. 371-374 in: Academy of Science Survey Team. Insects of Tibet. Vol. 2. Beijing: Academy of Science Publishing House, ix + 508 pp. [1982.04]
Tang, J., Li, S., Chen, Y. 1992. The first discovery of *Leptanilla* in China and the description of a new species. [In Chinese.] J. Zhejiang Agric. Univ. 18:107-108.
Tanquary, M. 1912 ("1911"). A preliminary list of the ants of Illinois. Trans. Ill. Acad. Sci. 4:137-142. [1912.01.25]
Tarbinsky, Y. S. 1970. A new species of ant of the genus *Proformica* (Hymenoptera, Formicidae) from Tian Shan. [In Russian.] Zool. Zh. 49:309-311. [1970.02] [February issue. Date signed for printing ("podpisano k pechati"): 29 January 1970.]
Tarbinsky, Y. S. 1971. Role of ants as entomophages in the fruit-growing biocoenoses of Tien Shan forests. [In Russian.] Entomol. Issled. Kirg. 7:35-53. [> 1971.02.23] [Date signed for printing ("podpisano k pechati").]
Tarbinsky, Y. S. 1976. The ants of Kirghizia. [In Russian.] Frunze: Ilim, 217 pp. [> 1976.06.04] [Date signed for printing ("podpisano k pechati").]
Tarbinsky, Y. S. 1983. Genesis of the myrmecofauna of Tien Shan. [In Russian.] Entomol. Issled. Kirg. 16:31-58. [> 1983.09.20] [Date signed for printing ("podpisano k pechati").]
Tarbinsky, Y. S. 1985. Morphological-functional peculiarities of ants of the Tien Shan fauna. [In Russian.] Entomol. Issled. Kirg. 18:39-47. [> 1985.04.22] [Date signed for printing ("podpisano k pechati").]
Taschenberg, E. L. 1866. Die Hymenopteren Deutschlands nach ihren Gattungen und theilweise nach ihren Arten. Leipzig: Eduard Kummer, 277 pp. [1866] [Dallas (1867:410) cites a publication date of October 1865. Ants pp. 230-245.]
Taylor, J. S. 1931. A note on the fauna of mangrove. Entomol. Rec. J. Var. 43:41-42. [1931.03.15]
Taylor, R. W. 1958. Original material of the ant *Discothyrea antarctica* Emery, 1894, in New Zealand. N. Z. Entomol. 2(3):17. [1958.12]
Taylor, R. W. 1959a. The Australian ant *Iridomyrmex darwinianus* (Forel) recorded from New Zealand. N. Z. Entomol. 2(4):18-19. [1959.12]
Taylor, R. W. 1959b. A note on the status of the ant species *Monomorium* (*Notomyrmex*) *smithii* Forel (Hymenoptera: Formicidae). N. Z. Entomol. 2(4):20-21. [1959.12]
Taylor, R. W. 1960. Taxonomic notes on the ants *Ponera leae* Forel and *Ponera norfolkensis* (Wheeler) (Hymenoptera - Formicidae). Pac. Sci. 14:178-180. [1960.04]
Taylor, R. W. 1961. Notes and new records of exotic ants introduced into New Zealand. N. Z. Entomol. 2(6):28-37. [1961.12]
Taylor, R. W. 1962a. New Australian dacetine ants of the genera *Mesostruma* Brown and *Codiomyrmex* Wheeler (Hymenoptera-Formicidae). Breviora 152:1-10. [1962.01.15]
Taylor, R. W. 1962b. The ants of the Three Kings Islands. Rec. Auckl. Inst. Mus. 5:251-254. [1962.11.06]
Taylor, R. W. 1964. Taxonomy and parataxonomy of some fossil ants (Hymenoptera-Formicidae). Psyche (Camb.) 71:134-141. [1964.12.30]
Taylor, R. W. 1965a. New Melanesian ants of the genera *Simopone* and *Amblyopone* (Hymenoptera-Formicidae) of zoogeographic significance. Breviora 221:1-11. [1965.05.07]
Taylor, R. W. 1965b. The Australian ants of the genus *Pristomyrmex*, with a case of apparent character displacement. Psyche (Camb.) 72:35-54. [1965.06.25]
Taylor, R. W. 1965c. A new species of the ant genus *Dacetinops* from Sarawak. Breviora 237:1-4. [1965.12.15]
Taylor, R. W. 1965d. A monographic revision of the rare tropicopolitan ant genus *Probolomyrmex* Mayr (Hymenoptera: Formicidae). Trans. R. Entomol. Soc. Lond. 117:345-365. [1965.12.31]

Taylor, R. W. 1966a ("1965"). A second African species of the dacetine ant genus *Codiomyrmex*. Psyche (Camb.) 72:225-228. [1966.01.18]

Taylor, R. W. 1966b ("1965"). Notes on the Indo-Australian ants of genus *Simopone* Forel (Hymenoptera-Formicidae). Psyche (Camb.) 72:287-290. [1966.05.10]

Taylor, R. W. 1967a. A monographic revision of the ant genus *Ponera* Latreille (Hymenoptera: Formicidae). Pac. Insects Monogr. 13:1-112. [1967.05.30]

Taylor, R. W. 1967b. Entomological survey of the Cook Islands and Niue. 1 - Hymenoptera-Formicidae. N. Z. J. Sci. 10:1092-1095. [1967.12]

Taylor, R. W. 1968a. Nomenclature and synonymy of the North American ants of the genera *Ponera* and *Hypoponera* (Hymenoptera: Formicidae). Entomol. News 79:63-66. [1968.03.25]

Taylor, R. W. 1968b. The Australian workerless inquiline ant, *Strumigenys xenos* Brown (Hymenoptera - Formicidae) recorded from New Zealand. N. Z. Entomol. 4(1):47-49. [1968.03]

Taylor, R. W. 1968c. Notes on the Indo-Australian basicerotine ants (Hymenoptera: Formicidae). Aust. J. Zool. 16:333-348. [1968.05]

Taylor, R. W. 1968d. A supplement to the revision of Australian *Pristomyrmex* species (Hymenoptera: Formicidae). J. Aust. Entomol. Soc. 7:63-66. [1968.06.30]

Taylor, R. W. 1968e. A new Malayan species of the ant genus *Epitritus*, and a related new genus from Singapore (Hymenoptera: Formicidae). J. Aust. Entomol. Soc. 7:130-134. [1968.12.31]

Taylor, R. W. 1969. The identity of *Dorylozelus mjobergi* Forel (Hymenoptera: Formicidae). J. Aust. Entomol. Soc. 8:131-133. [1969.12.31]

Taylor, R. W. 1970a. Notes on some Australian and Melanesian basicerotine ants (Hymenoptera: Formicidae). J. Aust. Entomol. Soc. 9:49-52. [1970.04.30]

Taylor, R. W. 1970b. Characterization of the Australian endemic ant genus *Peronomyrmex* Viehmeyer (Hymenoptera: Formicidae). J. Aust. Entomol. Soc. 9:209-211. [1970.12.31]

Taylor, R. W. 1971. The ants (Hymenoptera: Formicidae) of the Kermadec Islands. N. Z. Entomol. 5:81-82. [1971.12]

Taylor, R. W. 1972a. Studies on the Australian ponerine army ant, *Onychomyrmex*. P. 372 in: Rafes, P. M. (ed.) XIII International Congress of Entomology, Moscow, 2-9 August, 1968. Proceedings, Volume III. Leningrad: Nauka, 494 pp. [> 1972.06.08] [Date signed for printing ("podpisano k pechati").]

Taylor, R. W. 1972b. Biogeography of insects of New Guinea and Cape York Peninsula. Pp. 213-230 in: Walker, D. (ed.) Bridge and barrier: the natural and cultural history of Torres Strait. Research School of Pacific Studies, Publication BG/3. Canberra: Australian National University, xxii + 437 pp. [1972]

Taylor, R. W. 1973. Ants of the Australian genus *Mesostruma* Brown (Hymenoptera: Formicidae). J. Aust. Entomol. Soc. 12:24-38. [1973.03.31]

Taylor, R. W. 1974. Superfamily Formicoidea. P. 111 in: CSIRO (ed.) The insects of Australia. Supplement 1974. Carlton, Victoria: Melbourne University Press, viii + 146 pp. [1974]

Taylor, R. W. 1976a. The ants of Rennell and Bellona Islands. Nat. Hist. Rennell Isl. Br. Solomon Isl. 7:73-90. [1976.12.15]

Taylor, R. W. 1976b. La faune terrestre de l'Île de Sainte-Hélène. 7. Superfam. Formicoidea. Ann. Mus. R. Afr. Cent. Sér. Octavo Sci. Zool. 215:192-199. [1976.12]

Taylor, R. W. 1977. New ants of the Australasian genus *Orectognathus*, with a key to the known species (Hymenoptera: Formicidae). Aust. J. Zool. 25:581-612. [1977.08.05]

Taylor, R. W. 1978a. *Nothomyrmecia macrops*: a living-fossil ant rediscovered. Science (Wash. D. C.) 201:979-985. [1978.09.15]

Taylor, R. W. 1978b. A taxonomic guide to the ant genus *Orectognathus* (Hymenoptera: Formicidae). CSIRO Div. Entomol. Rep. 3:1-11. [1978.10]

Taylor, R. W. 1979 ("1978"). Melanesian ants of the genus *Amblyopone* (Hymenoptera: Formicidae). Aust. J. Zool. 26:823-839. [1979.01.24]

Taylor, R. W. 1980a ("1979"). New Australian ants of the genus *Orectognathus*, with summary description of the twenty-nine known species (Hymenoptera: Formicidae). Aust. J. Zool. 27:773-788. [1980.02.15]

Taylor, R. W. 1980b. Australian and Melanesian ants of the genus *Eurhopalothrix* Brown and Kempf - notes and new species (Hymenoptera: Formicidae). J. Aust. Entomol. Soc. 19:229-239. [1980.09.26]

Taylor, R. W. 1980c ("1979"). Notes on the Russian endemic ant genus *Aulacopone* Arnoldi (Hymenoptera: Formicidae). Psyche (Camb.) 86:353-361. [1980.12.19]

Taylor, R. W. 1980d. The rare Fijian ant *Myrmecina (=Archaeomyrmex) cacabau* (Mann) rediscovered (Hymenoptera: Formicidae). N. Z. Entomol. 7:122-123. [1980.12]

Taylor, R. W. 1983. Descriptive taxonomy: past, present and future. Pp. 93-134 in: Highley, E., Taylor, R. W. (eds.) Australian systematic entomology: a bicentenary perspective. Melbourne: CSIRO, vii + 147 pp. [1983]

Taylor, R. W. 1985. The ants of the Papuasian genus *Dacetinops* (Hymenoptera: Formicidae: Myrmicinae). Pp. 41-67 in: Ball, G. E. (ed.) Taxonomy, phylogeny and zoogeography of beetles and ants: a volume dedicated to the memory of Philip Jackson Darlington, Jr., 1904-1983. Ser. Entomol. (Dordr.) 33:1-514. [1985]

Taylor, R. W. 1986. The quadrinominal infrasubspecific names of Australian ants (Hymenoptera: Formicidae). Gen. Appl. Entomol. 18:33-37. [1986.10.10]

Taylor, R. W. 1987a. A checklist of the ants of Australia, New Caledonia and New Zealand (Hymenoptera: Formicidae). CSIRO Div. Entomol. Rep. 41:1-92. [1987.06]

Taylor, R. W. 1987b. A checklist of the ants of Australia, New Caledonia and New Zealand (Hymenoptera: Formicidae). First supplement, 10 July, 1987. CSIRO Div. Entomol. Rep. 41(Suppl.1):1-5. [1987.07]

Taylor, R. W. 1988. The nomenclature and distribution of some Australian and New Caledonian ants of the genus *Leptogenys* Roger (=*Prionogenys* Emery, n. syn.) (Hymenoptera: Formicidae: Ponerinae). Gen. Appl. Entomol. 20:33-37. [1988.07.31]

Taylor, R. W. 1989a. The nomenclature and distribution of some Australian ants of the genus *Polyrhachis* Fr Smith (Hymenoptera: Formicidae: Formicinae). J. Aust. Entomol. Soc. 28:23-27. [1989.02.28]

Taylor, R. W. 1989b. Australasian ants of the genus *Leptothorax* Mayr (Hymenoptera: Formicidae: Myrmicinae). Mem. Qld. Mus. 27:605-610. [1989.11.13]

Taylor, R. W. 1990a. [Untitled. A key to the ant genera of Australia, New Caledonia, and New Zealand.] Pp. 55-60 in: Hölldobler, B., Wilson, E. O. The ants. Cambridge, Mass.: Harvard University Press, xii + 732 pp. [1990.03.28]

Taylor, R. W. 1990b. [Untitled. Anomalomyrmini Taylor tribe n., *Anomalomyrma* Taylor gen. n., *Protanilla* Taylor gen. n.] Pp. 278-279 in: Bolton, B. The higher classification of the ant subfamily Leptanillinae (Hymenoptera: Formicidae). Syst. Entomol. 15:267-282. [1990.07]

Taylor, R. W. 1990c. New Asian ants of the tribe Basicerotini, with an on-line computer interactive key to the twenty-six known Indo-Australian species (Hymenoptera: Formicidae: Myrmicinae). Invertebr. Taxon. 4:397-425. [1990.10.12]

Taylor, R. W. 1990d. The nomenclature and distribution of some Australian and New Caledonian ants of the genus *Meranoplus* Fr. Smith (Hymenoptera: Formicidae: Myrmicinae). Gen. Appl. Entomol. 22:31-40. [1990.10.31]

Taylor, R. W. 1991a ("1990"). Notes on the ant genera *Romblonella* and *Willowsiella*, with comments on their affinities, and the first descriptions of Australian species (Hymenoptera: Formicidae: Myrmicinae). Psyche (Camb.) 97:281-296. [1991.02.26]

Taylor, R. W. 1991b. Nomenclature and distribution of some Australasian ants of the Myrmicinae (Hymenoptera: Formicidae). Mem. Qld. Mus. 30:599-614. [1991.08.01]

Taylor, R. W. 1991c. *Myrmecia croslandi* sp. n., a karyologically remarkable new Australian jack-jumper ant (Hymenoptera: Formicidae: Myrmeciinae). J. Aust. Entomol. Soc. 30:288. [1991.11.29]

Taylor, R. W. 1991d. Formicidae. Pp. 980-989 in: CSIRO (ed.) The insects of Australia. Second edition. Volume II. Carlton, Victoria: Melbourne University Press, pp. i-vi, 543-1137. [1991]

Taylor, R. W. 1992a. Nomenclature and distribution of some Australian and New Guinean ants of the subfamily Formicinae (Hymenoptera: Formicidae). J. Aust. Entomol. Soc. 31:57-69. [1992.02.28]

Taylor, R. W. 1992b. *Rhoptromyrmex rawlinsoni* sp. nov., a new apparently workerless parasitic ant from Anak Krakatau, Indonesia (Hymenoptera: Formicidae: Myrmicinae). Mem. Natl. Mus. Vic. 53:125-128. [1992.05.30]

Taylor, R. W. 1994. The sociobiology of *Nothomyrmecia macrops*: considerations regarding the evolution of eusociality among ants. P. 60 in: Lenoir, A., Arnold, G., Lepage, M. (eds.) Les insectes sociaux. 12ème congrès de l'Union Internationale pour l'Étude des Insectes Sociaux. Paris, Sorbonne, 21-27 août 1994. Paris: Université Paris Nord, xxiv + 583 pp. [1994.09]

Taylor, R. W., Beaton, C. D. 1970. Insect systematics and the scanning electron microscope. Search (Sydney) 1:347-348. [1970.12]

Taylor, R. W., Brown, D. R. 1985. Formicoidea. Zool. Cat. Aust. 2:1-149, 306-348. [1985]

Taylor, R. W., Brown, W. L., Jr. 1978. *Smithistruma kempfi* species nov. Pilot Regist. Zool. Card No. 35. [1978.02.25]

Taylor, R. W., Lowery, B. B. 1972. The New Guinean species of the ant genus *Orectognathus* Fr. Smith (Hymenoptera: Formicidae). J. Aust. Entomol. Soc. 11:306-310. [1972.12.31]

Taylor, R. W., Wilson, E. O. 1962 ("1961"). Ants from three remote oceanic islands. Psyche (Camb.) 68:137-144. [1962.03.13]

Tegelström, H., Nilsson, G., Wyöni, P.-I. 1983. Lack of species differences in isoelectric focused proteins in the *Formica rufa* group (Hymenoptera, Formicidae). Hereditas (Lund) 98:161-165. [1983.06.01]

Teranishi, C. 1915a. Notes on *Camponotus marginatus* var. *quadrinotus* Forel. [In Japanese.] Insect World 19:56-59. [1915.02.15]

Teranishi, C. 1915b. Ant-fauna in the neighborhood of Osaka. [In Japanese.] Insect World 19:194-198. [1915.05.15]

Teranishi, C. 1915c. Further note on *Camponotus marginatus* var. *quadrinotatus* Forel. [In Japanese.] Insect World 19:502-503. [1915.12.15]

Teranishi, C. 1915d. A new species of Formicidae from Japan. Entomol. Mag. Kyoto 1:137-138. [1915.12]

Teranishi, C. 1916. Note on *Tetramorium caespitum* Linné. [In Japanese.] Insect World 20:444-448. [1916.11.15]

Teranishi, C. 1917. Rare ants collected by Mr. Mitsuharu Azuma. [In Japanese.] Seikô 1:8-10. [Not seen. Cited in Okano (1989) and Onoyama & Terayama (1994).]

Teranishi, C. 1924. Three interesting Hymenoptera occurring in Hokkaido and the mainland. [In Japanese.] Insect World 28:52-54. [1924.02.15]

Teranishi, C. 1925a. Notes on the scientific name of *Aphaenogaster aciculata* Smith. [In Japanese and English (part).] Dobutsugaku Zasshi (Zool. Mag.) 37:289-294. [1925.07.15]

Teranishi, C. 1925b. Ants collected by a certain person. [In Japanese.] Kagaku Noggo 5(5):53-55. [Not seen. Cited in Okano (1989) and Onoyama & Terayama (1994).]

Teranishi, C. 1927a. Notes on Japanese ants. 1. Aberrant forms. [In Japanese.] Dobutsugaku Zasshi (Zool. Mag.) 39:88-94. [1927.02.15]

Teranishi, C. 1927b. Ants found in a green-house of Tennoji Botanical Garden, Osaka. [In Japanese.] Kontyû 2:51-52. [1927.04.05]

Teranishi, C. 1927c. New localities for *Polyergus samurai* Yano. [In Japanese.] Kontyû 2:53. [1927.04.05]

Teranishi, C. 1927d. Notes on Japanese ants. 2. Larvae with glutinous tubercles. [In Japanese.] Dobutsugaku Zasshi (Zool. Mag.) 39:297-300. [1927.07.15]

Teranishi, C. 1927e. On the distribution of *Tetramorium guineense* (Fabricius) in Japan. [In Japanese.] Kontyû 2:123-125. [1927.09.05]

Teranishi, C. 1927f. Observations on the habits of *Leptothorax congruus* F. Smith. [In Japanese.] Kontyû 2:155-176. [1927.11.05]
Teranishi, C. 1927g. On a form beta-female found in ant world. [In Japanese.] Kagaku Noggo 8(6):10-13. [Not seen. Cited in Okano (1989) and Onoyama & Terayama (1994).]
Teranishi, C. 1928 ("1927"). On *Paratrechina longicornis* (Latreille) and *Monomorium pharaonis* (Linnaeus). [In Japanese.] Kontyû 2:241-242. [1928.05.20]
Teranishi, C. 1929a. Okinawan ants intruding into dwelling. [In Japanese.] Kontyû 3:41-42. [1929.03.25]
Teranishi, C. 1929b. Japanese ants, their behavior and distribution. I. [In Japanese.] Dobutsugaku Zasshi (Zool. Mag.) 41:239-251. [1929.06.15]
Teranishi, C. 1929c. Japanese ants, their behavior and distribution. II. [In Japanese.] Dobutsugaku Zasshi (Zool. Mag.) 41:312-332. [1929.07.15]
Teranishi, C. 1930a. Ants of the Palaearctic part of Japan (1). [In Japanese.] Trans. Kansai Entomol. Soc. 1:17-26. [1930.11.01]
Teranishi, C. 1930b. On the discovery of *Monomorium pharaonis* Linnaeus in Osaka. [In Japanese.] Trans. Kansai Entomol. Soc. 1:77-78. [1930.11.01]
Teranishi, C. 1930c. Winged ants attracted to the light-trap settled in the Nawa Entomological Laboratory. [In Japanese.] Trans. Kansai Entomol. Soc. 1:87-89. [1930.11.01]
Teranishi, C. 1931. Ants collected by Mr. K. Kobayashi on the island of Shikotan in the Southern Kuriles. Trans. Kansai Entomol. Soc. 2:28-29. [1931.12.13]
Teranishi, C. 1932. A list of the ants of Sakhalin. [In Japanese.] Trans. Kansai Entomol. Soc. 3: 49-54. [1932.11.12]
Teranishi, C. 1933a. Japanese ants, their behavior and distribution (III). [In Japanese.] Trans. Kansai Entomol. Soc. 4:77-80. [1933.11.12]
Teranishi, C. 1933b. A list of ants found at the sand dune of Tottori. [In Japanese.] Trans. Kansai Entomol. Soc. 4:84-85. [1933.11.12]
Teranishi, C. 1934a. Formicid found in the bottom-mud of Lake Ichibishinae, Kunashiri. [In Japanese.] Rikusuigaku Zasshi (Jpn. J. Limnol.) 3:111. [1934.05.10]
Teranishi, C. 1934b. *Formica exsecta* var. *fukaii* Wheeler and its related species. [In Japanese.] Kansai Konchu Zasshi 2:37-39. [1934.07.31]
Teranishi, C. 1935. Note on *Stigmatomma silvestrii* Wheeler. [In Japanese.] Kansai Konchu Zasshi 2:79-80. [1935.05.15]
Teranishi, C. 1936a ("1935"). An ergatogyne of *Pristomyrmex pungens* Mayr. [In Japanese.] Trans. Kansai Entomol. Soc. 6:40. [1936.02.10]
Teranishi, C. 1936b. Insects of Jehol (VII). Orders: Coleoptera (II) & Hymenoptera (I). Family Formicidae. [In Japanese.] Rep. First Sci. Exped. Manchoukuo (5)1(11)(article 60):1-12. [1936.12.30] [Citation on title page of article is given as "Section V, Division I, Part XI, Article 60". Each article in this series is individually paginated.]
Teranishi, C. 1940. Works of Cho Teranishi. Memorial Volume. [In Japanese, English and German, in part.] Osaka: Kansai Entomological Society, 312 + (posthumous section) 95 pp. [< 1940.08.07] [Additional bibliographical details in Anonymous (1969). For a complete listing of all nineteen titles in the posthumous section see Onoyama & Terayama (1994: 38-39).]
Teranishi, C. 1944. Myrmicinae species of Japan. [In Japanese.] Osaka: publisher unknown, 30 pp. [Not seen. Cited in Okano (1989) and Onoyama & Terayama (1994).]
Terayama, M. 1981. Distribution of ants of the Nansei Archipelago. (I) Ants in the Amani Islands. [In Japanese.] Nat. Insects (Konchu Shizen) 16(8):34-36. [1981.07]
Terayama, M. 1982a. Regional differences of the ant fauna of the Nansei Archipelago based on the quantitative method. I. Analysis using Nomura-Simpson's coefficient. Bull. Biogeogr. Soc. Jpn. 37:1-5. [1982.12.20]

Terayama, M. 1982b. Regional differences of the ant fauna of the Nansei Archipelago based on the quantitative method. II. Analysis using harmonity index of taxa. Bull. Biogeogr. Soc. Jpn. 37: 7-10. [1982.12.20]

Terayama, M. 1983. Biogeographic study of the ant fauna of the Izu and the Ogasawara Islands. Bull. Biogeogr. Soc. Jpn. 38:93-103. [1983.12.20]

Terayama, M. 1984. A new species of the army ant genus *Aenictus* from Taiwan (Insecta; Hymenoptera; Formicidae). Bull. Biogeogr. Soc. Jpn. 39:13-16. [1984.12.20]

Terayama, M. 1985a. Ant community of mangrove forest in the Ryukyu Islands, Japan - species composition, distribution, and biomass. [Abstract.] [In Japanese.] Ari 13:1-2. [1985.06.01]

Terayama, M. 1985b. Taxonomy, number of species, and distribution of the ant subfamilies Ponerinae, Cerapachyinae, Dorylinae, Leptanillinae, Dolichoderinae, and Formicinae of Japan. [Abstract.] [In Japanese.] Ari 13:4. [1985.06.01]

Terayama, M. 1985c. New records of some ants from the Nansei Islands, Japan. [In Japanese.] Ari 13:8. [1985.06.01]

Terayama, M. 1985d. Two new species of the ant genus *Myrmecina* (Insecta; Hymenoptera; Formicidae) from Japan and Taiwan. Edaphologia 32:35-40. [1985.06.25]

Terayama, M. 1985e. Two new species of the genus *Acropyga* (Hymenoptera, Formicidae) from Taiwan and Japan. Kontyû 53:284-289. [1985.06.25]

Terayama, M. 1985f. Descriptions of a new species of the genus *Proceratium* Roger from Taiwan (Hymenoptera, Formicidae). Kontyû 53:406-408. [1985.09.25]

Terayama, M. 1985g. Structure of communities of ants in the Japanese Islands: S/G ratio, area and species richness. [Abstract.] Zool. Sci. (Tokyo) 2:1003. [1985.12.15]

Terayama, M. 1986. Two new ants of the genus *Ponera* (Hymenoptera, Formicidae) from Taiwan. Kontyû 54:591-595. [1986.12.25]

Terayama, M. 1987a. A new species of *Amblyopone* (Hymenoptera, Formicidae) from Japan. Edaphologia 36:31-33. [1987.03.25]

Terayama, M. 1987b. Records of *Formica* (*Formica*) *yessensis* Forel from Taiwan. [In Japanese.] Ari 15:5-6. [1987.10.10]

Terayama, M. 1988. A record of the social parasitic ant *Strongylognathus koreanus* Pisarski, 1965 (Hymenoptera, Formicidae) from Japan. Kontyû 56:458. [1988.06.25]

Terayama, M. 1989a. The ant tribe Odontomachini (Hymenoptera: Formicidae) from Taiwan, with a description of a new species. Edaphologia 40:25-29. [1989.03.25]

Terayama, M. 1989b. The ant tribe Amblyoponini (Hymenoptera, Formicidae) of Taiwan with description of a new species. Jpn. J. Entomol. 57:343-346. [1989.06.25]

Terayama, M. 1989c ("1988"). Ant fauna of Saitama Prefecture, Japan. [In Japanese.] Ari 16:4-13. [1989.08] [Date of publication from Onoyama & Terayama (1994:40).]

Terayama, M. 1990a. A list of Ponerinae of Taiwan (Hymenoptera; Formicidae). [In Chinese.] Bull. Toho Gakuen 4:25-50.

Terayama, M. 1990b. Discovery of worker caste in *Trachymesopus darwinii* (Forel, 1893). Jpn. J. Entomol. 58:897-898. [1990.12.25]

Terayama, M. 1991a. Species-area relations of ant communities. [In Japanese.] Bull. Toho Gakuen 6:1-16. [1991.02.10]

Terayama, M. 1991b. The subgenus *Paramyrmamblys* of the genus *Camponotus* (Insecta: Hymenoptera: Formicidae) from Japan, with a description of a new species. Bull. Biogeogr. Soc. Jpn. 46:165-169. [1991.12.20]

Terayama, M. 1992a. Structure of ant communities in East Asia: regional differences, species-area relations, and diversities. [Abstract.] P. 52 in: Proceedings XIX International Congress of Entomology. Abstracts. Beijing, China, June 28 - July 4, 1992. Beijing: XIX International Congress of Entomology, xi + 730 pp. [1992.07.04]

Terayama, M. 1992b. The present status and prospect of the classification of Japanese Formicidae (Insecta, Hymenoptera). [In Japanese.] Nat. Insects (Konchu Shizen) 27(11):21-25. [1992.10]

Terayama, M. 1992c. Structure of ant communities in east Asia. 1. Regional differences and species richness. [In Japanese.] Bull. Biogeogr. Soc. Jpn. 47:1-31. [1992.12.20]

Terayama, M. 1993. Structure of ant communities in east Asia. II. Species and nest densities. [In Japanese.] Bull. Biogeogr. Soc. Jpn. 48:51-57. [1993.07.31]

Terayama, M. 1994a. A new record of *Acropyga* sp. from Japan. [In Japanese.] Ari 17:10. [1994.04.30]

Terayama, M. 1994b. On the Japanese species of ants changed the scientific names recently. [In Japanese.] Ari 17:20. [1994.04.30]

Terayama, M. 1994c. Ants of Okushiri-tô Island, Hokkaido. [In Japanese.] Ari 18:28-29. [1994.05.30]

Terayama, M. 1994d. Ant fauna of Saitama Prefecture, Japan (Supplement). [In Japanese.] Ari 18:30. [1994.05.30]

Terayama, M. 1995. A new species of the ant genus *Acanthomyrmex* (Hymenoptera, Formicidae) from Thailand. Jpn. J. Entomol. 63:551-555. [1995.09.25]

Terayama, M., Choi, B.-M. 1991. Four newly recorded species of Formicidae (Insecta, Hymenoptera) from Korea. Edaphologia 45:63-64. [1991.06.25]

Terayama, M., Choi, B.-M. 1994. Ant faunas of Taiwan, Korea, and Japan. [Abstract.] [In Japanese.] Ari 18:36. [1994.05.30]

Terayama, M., Choi, B.-M., Kim, C.-H. 1992. A check list of ants from Korea, with taxonomic notes. [In Japanese.] Bull. Toho Gakuen 7:19-54. [1992.03.10]

Terayama, M., Inoue, N. 1994. Ants collected by members of the Soil Zoological Expedition to Taiwan, 1988. [In Japanese.] Ari 18:25-28. [1994.05.30]

Terayama, M., Kihara, A. 1994. Distribution maps of Japanese ants. [In Japanese.] Tokyo: Myrmecological Society of Japan, vi + 63 pp. [1994.10.31]

Terayama, M., Kubota, M. 1994. Phylogeny and taxonomy of the genus *Trigonogaster*. [Abstract.] [In Japanese.] Ari 18:39. [1994.05.30]

Terayama, M., Kubota, S. 1989. The ant tribe Dacetini (Hymenoptera, Formicidae) of Taiwan, with descriptions of three new species. Jpn. J. Entomol. 57:778-792. [1989.12.25]

Terayama, M., Kubota, S. 1993. The army ant genus *Aenictus* (Hymenoptera: Formicidae) from Thailand and Viet Nam, with descriptions of three new species. Bull. Biogeogr. Soc. Jpn. 48: 68-72. [1993.12.30]

Terayama, M., Kubota, S. 1994. Ant fauna of Tokyo. (2) Ants from Aogashima Island, the Izu Island. [In Japanese.] Ari 17:11. [1994.04.30]

Terayama, M., Kubota, S., Sakai, H., Kawazoe, A. 1988. Rediscovery of *Cerapachys sauteri* Forel, 1913 (Insecta: Hymenoptera: Formicidae) from Taiwan, with notes on the Taiwanese species of the genus *Cerapachys*. Bull. Biogeogr. Soc. Jpn. 43:35-38. [1988.12.20]

Terayama, M., Kubota, S., Sakai, H., Kawazoe, A. 1994. Ant fauna of Taiwan. [Abstract.] [In Japanese.] Ari 17:3-4. [1994.04.30]

Terayama, M., Lin, C.-C., Wu, W.-J. 1995. The ant genera *Epitritus* and *Kyidris* from Taiwan (Hymenoptera: Formicidae). Proc. Jpn. Soc. Syst. Zool. 53:85-89. [1995.06.25]

Terayama, M., Miyano, S., Kurozumi, T. 1994. Ant fauna (Insecta: Hymenoptera: Formicidae) of the northern Mariana Islands, Micronesia. Nat. Hist. Res. Spec. Issue 1:231-236. [1994.03]

Terayama, M., Murata, K. 1987. Relation between ant communities and vegetations of Toshima Island, the Izu Islands. [In Japanese.] Bull. Biogeogr. Soc. Jpn. 42:57-63. [1987.12.20]

Terayama, M., Murata, K. 1990. Effects of area and fragmentation of forests for nature conservation: analysis by ant communities. [In Japanese.] Bull. Biogeogr. Soc. Jpn. 45:11-17. [1990.12.20]

Terayama, M., Ogata, K. 1988. Two new species of the ant genus *Probolomyrmex* (Hymenoptera, Formicidae) from Japan. Kontyû 56:590-594. [1988.09.25]

Terayama, M., Ogata, K., Choi, B.-M. 1994. Distribution records of Japanese ants in each prefecture. [In Japanese.] Ari 18:5-17. [1994.05.30]

Terayama, M., Sakai, H., Kubota, S., Takamine, H. 1994. Descriptions of the female and male in *Probolomyrmex longinodus* Terayama & Ogata. [In Japanese.] Ari 17:12-13. [1994.04.30]

Terayama, M., Satoh, T. 1990a. A new species of the genus *Camponotus* from Japan, with notes on two known forms of the subgenus *Myrmamblys* (Hymenoptera, Formicidae). Jpn. J. Entomol. 58:405-414. [1990.06.25]

Terayama, M., Satoh, T. 1990b. Taxonomic notes on two Japanese species of Formicidae (Hymenoptera). Jpn. J. Entomol. 58:532. [1990.09.25]

Terayama, M., Satoh, T. 1990c. *Camponotus (Myrmamblys) ogasawarensis* sp. nov. from the Ogasawara Islands, Japan (Insecta, Hymenoptera, Formicidae). Bull. Biogeogr. Soc. Jpn. 45:117-121. [1990.12.20]

Terayama, M., Watanabe, Y. 1994. Ant fauna of the Zhongyang Mountains in Taiwan. [Abstract.] [In Japanese.] Ari 18:32. [1994.05.30]

Terayama, M., Yamaguchi, T., Hasegawa, E. 1993. Ergatoid queens of slave-making ant *Polyergus samurai* Yano (Hymenoptera, Formicidae). Jpn. J. Entomol. 61:511-514. [1993.09.25]

Terayama, M., Yamane, S. 1989. The army ant genus *Aenictus* (Hymenoptera, Formicidae) from Sumatra, with descriptions of three new species. Jpn. J. Entomol. 57:597-603. [1989.09.25]

Terayama, M., Yamane, S. 1991. A new ant of the genus *Podomyrma* (Hymenoptera, Formicidae) from Sumatra, Indonesia. Proc. Jpn. Soc. Syst. Zool. 44:69-72. [1991.06.25]

Terron, G. 1968 ("1967"). Description des castes de *Tetraponera anthracina* Santschi (Hym., Formicidae, Promyrmicinae). Insectes Soc. 14:339-348. [1968.06]

Terron, G. 1969. Description de *Tetraponera ledouxi* espèce nouvelle du Cameroun, parasite temporaire de *Tetraponera anthracina* Santschi (Hym. Formicidae, Promyrmicinae). Bull. Inst. Fondam. Afr. Noire Sér. A Sci. Nat. 31:629-642. [1969.07]

Terron, G. 1970 ("1969"). Mise en évidence du parasitisme temporaire de *Tetraponera anthracina* Santschi par *Tetraponera ledouxi* nov. spec. (Hym. Formicidae, Promyrmicinae). Ann. Fac. Sci. Univ. Féd. Cameroun 3:113-115. [1970.01]

Terron, G. 1971. Description des castes de *Tetraponera nasuta* Bernard (Hym. Formicidae, Promyrmicinae). Ann. Fac. Sci. Univ. Féd. Cameroun 6:73-84. [1971.04]

Terron, G. 1972. Observations sur les mâles ergatoïdes et les mâles ailés chez une fourmie du genre *Technomyrmex* Mayr (Hym., Formicidae, Dolichoderinae). Ann. Fac. Sci. Univ. Féd. Cameroun 10:107-120. [1972.04]

Terron, G. 1974. Découverte au Cameroun de deux espèces nouvelles du genre *Prionopelta* Mayr (Hym., Formicidae). Ann. Fac. Sci. Univ. Féd. Cameroun 17:105-119. [1974.03]

Terron, G. 1977. Évolution des colonies de *Tetraponera anthracina* Santschi (Formicidae Pseudomyrmecinae) avec reines. Bull. Biol. Fr. Belg. 111:115-181. [1977.09]

Terron, G. 1981. Deux nouvelles espèces éthiopiennes pour le genre *Proceratium* (Hym.: Formicidae). Ann. Fac. Sci. Yaoundé 28:95-103. [1981.04]

Tewary, R. N. See: Tiwari, R. N.

Théobald, N. 1935a. [Untitled. *Lasius crispus* N. Théobald nov. spec.] P. 68 in: Piton, L., Théobald, N. La faune entomologique des gisements Mio-Pliocènes du Massif Central. Rev. Sci. Nat. Auvergne (n.s.)1:65-104. [1935.09.22]

Théobald, N. 1935b. [Untitled. Descriptions of new taxa: *Lasius chambonensis* N. Théobald nov. spec.; *Formica pitoni* N. Théobald nov. spec.] Pp. 82-83 in: Piton, L., Théobald, N. La faune entomologique des gisements Mio-Pliocènes du Massif Central. Rev. Sci. Nat. Auvergne (n.s.)1:65-104. [1935.09.22]

Théobald, N. 1935c. [Untitled. *Formica maculipennis* N. Théobald nov. spec.] P. 91 in: Piton, L., Théobald, N. La faune entomologique des gisements Mio-Pliocènes du Massif Central. Rev. Sci. Nat. Auvergne (n.s.)1:65-104. [1935.09.22]

Théobald, N. 1937a. Les insectes fossiles des terrains oligocènes de France. Nancy: G. Thomas, 473 pp. [1937.02.24] [Publication date from Evenhuis (1994b:550).]

Théobald, N. 1937b. Notes complémentaire sur les insectes fossiles oligocènes des gypses d'Aix-en-Provence. Bull. Mens. Soc. Sci. Nancy (n.s.5)6:157-178. [1937.06]

Theobald-Ley, S., Horstmann, K. 1990. Die Ameisenfauna (Hymenoptera, Formicidae) von Windwurfflächen und angrenzenden Waldhabitaten im Nationalpark Bayerischer Wald. Waldhygiene 18:93-118. [1990]

Thomas, J. A., Elmes, G. W., Wardlaw, J. C., Woyciechowski, M. 1989. Host specificity among *Maculinea* butterflies in *Myrmica* ant nests. Oecologia (Berl.) 79:452-457. [1989.06]

Thome, G., Thome, H. See: Tohmé, G., Tohmé, H.

Thompson, C. B. 1913. A comparative study of the brains of three genera of ants, with special reference to the mushroom bodies. J. Comp. Neurol. 23:515-572. [1913.12]

Thompson, C. R. 1981. *Solenopsis* (*Diplorhoptrum*) (Hymenoptera: Formicidae) of Florida. [Abstract.] Diss. Abstr. Int. B. Sci. Eng. 41:3306. [1981.03]

Thompson, C. R. 1982. A new *Solenopsis* (*Diplorhoptrum*) species from Florida (Hym.: Formicidae). J. Kansas Entomol. Soc. 55:485-488. [1982.07.20]

Thompson, C. R. 1989. The thief ants, *Solenopsis molesta* group, of Florida (Hymenoptera: Formicidae). Fla. Entomol. 72:268-283. [1989.06.30]

Thompson, C. R., Johnson, C. 1989. Rediscovered species and revised key to the Florida thief ants (Hymenoptera: Formicidae). Fla. Entomol. 72:697-698. [1989.12.22]

Timberlake, P. H. 1925. New records of Hawaiian ants. Proc. Hawaii. Entomol. Soc. 6:7-8. [1925.08]

Timberlake, P. H. 1926. Hymenoptera. Pp. 17-43 in: Bryan, E. H., Jr. and collaborators. Insects of Hawaii, Johnston Island and Wake Island. Bull. Bernice P. Bishop Mus. 31:1-94. [1926]

Timmins, C. J., Stradling, D. J. 1993. Horse dung: a new or old habitat for *Hypoponera punctatissima* (Roger) (Hymenoptera: Formicidae)? Entomologist 112:217-218. [1993.10]

Tinaut, A. 1979. Estudio de la mirmecofauna de los Borreguiles del S. Juan (Sierra Nevada, Granada) (Hym. Formicidae). Bol. Asoc. Esp. Entomol. 3:173-183. [1979.11]

Tinaut, A. 1981 ("1980"). *Rossomyrmex minuchae* nov. sp. (Hym. Formicidae) encontrada en Sierra Nevada, España. Bol. Asoc. Esp. Entomol. 4:195-203. [1981.05]

Tinaut, A. 1982. Evolución anual de la mirmecocenosis de un encinar. Bol. Estac. Cent. Ecol. 11:49-56. [1982]

Tinaut, A. 1983 ("1982"). Descripción de una nueva especie de *Leptothorax* Mayr, 1855, del sur de la Península Ibérica (Hymenoptera, Formicidae). EOS. Rev. Esp. Entomol. 58:319-325. [1983.03.15]

Tinaut, A. 1985. Descripción de los sexuados de *Messor lusitanicus* Santschi, 1929 (Hymenoptera, Formicidae). Nouv. Rev. Entomol. (n.s.)2:85-90. [1985.05.07]

Tinaut, A. 1986 ("1985"). Descripción del macho de *Aphaenogaster cardenai* Espadaler, 1981 (Hymenoptera, Formicidae). Misc. Zool. (Barc.) 9:245-249. [1986]

Tinaut, A. 1987a. Descripción de *Leptothorax pardoi* nov. sp. (Hym. Formicidae). EOS. Rev. Esp. Entomol. 63:315-320. [1987.12.15]

Tinaut, A. 1987b. *Leptanilla revelierei* Emery, 1870 en la península ibérica. Nueva cita de la subfamilia Leptanillinae Emery, 1910 (Hymenoptera. Formicidae). EOS. Rev. Esp. Entomol. 63:321-323. [1987.12.15]

Tinaut, A. 1988. Nuevo hallazgo de *Epitritus argiolus* Emery, 1869 en la Península Ibérica. (Hym. Formicidae). Nouv. Rev. Entomol. (n.s.)5:48. [1988.04.20]

Tinaut, A. 1989a. Descripción del macho de *Leptothorax baeticus* Emery, 1924 (Hymenoptera, Formicidae). Nouv. Rev. Entomol. (n.s.)6:51-55. [1989.06.22]

Tinaut, A. 1989b. Nueva cita de *Messor celiae* Reyes, 1985 (Hym. Formicidae). Bol. Asoc. Esp. Entomol. 13:454. [1989.11]

Tinaut, A. 1989c. Contribución al estudio de los formícidos de la región del estrecho de Gibraltar y su interés biogeográfico (Hym., Formicidae). Graellsia 45:19-29. [1989]

Tinaut, A. 1990a. Descripción del macho de *Formica subrufa* Roger, 1859 y creación de un nuevo subgénero (Hymenoptera: Formicidae). EOS. Rev. Esp. Entomol. 65:281-291. [1990.06]

Tinaut, A. 1990b. *Teleutomyrmex kutteri*, spec. nov. A new species from Sierra Nevada (Granada, Spain). Spixiana 13:201-208. [1990.07.31]

Tinaut, A. 1990c. Taxonomic situation of the genus *Cataglyphis* Förster, 1850 in the Iberian peninsula. II. New position for *C. viatica* (Fabricius, 1787) and redescription of *C. velox* Santschi, 1929 stat. n. (Hymenoptera, Formicidae). EOS. Rev. Esp. Entomol. 66:49-59. [1990.12]

Tinaut, A. 1990d ("1988"). El género *Amblyopone* Erichson en la Península Ibérica (Hymenoptera, Formicidae). Misc. Zool. (Barc.) 12:189-193. [1990]

Tinaut, A. 1991a ("1990"). Situación taxonómica del género *Cataglyphis* Förster, 1850 en la Península Ibérica. III. El grupo de *C. velox* Santschi, 1929 y descripción de *Cataglyphis humeya* sp. n. (Hymenoptera, Formicidae). EOS. Rev. Esp. Entomol. 66:215-227. [1991.03]

Tinaut, A. 1991b. Contribución al conocimiento de los formícidos del Parque Nacional de Doñana (Hymenoptera, Formicidae). Bol. Asoc. Esp. Entomol. 15:57-63. [1991.12.30]

Tinaut, A. 1993. *Cataglyphis floricola* nov. sp. new species for the genus *Cataglyphis* Förster, 1850 (Hymenoptera, Formicidae) in the Iberian Peninsula. Mitt. Schweiz. Entomol. Ges. 66:123-134. [1993.06]

Tinaut, A. 1994 ("1993"). Nueva cita de *Camponotus universitatis* Forel, 1890 en Francia (Hymenoptera: Formicidae). Nouv. Rev. Entomol. (n.s.)10:381. [1994.03.04]

Tinaut, A. 1995 ("1994"). Nueva especie de *Leptothorax* (Mayr, 1855) del grupo *laurae* Emery, 1884 *Leptothorax crepuscularis* n. sp. (Hymenoptera, Formicidae). Zool. Baetica 5:89-98. [1995]

Tinaut, A., Espadaler, X. 1987. Description [sic] del macho de *Myrmica aloba* Forel, 1909 (Hymenoptera, Formicidae). Nouv. Rev. Entomol. (n.s.)4:61-69. [1987.05.29]

Tinaut, A., Espadaler, X., Jiménez, J. J. 1992. *Camponotus universitatis* Forel, 1891, en la Península Ibérica. Descripción de sus sexuados (Hymenoptera, Formicidae). Nouv. Rev. Entomol. (n.s.)9:233-238. [1992.12.04]

Tinaut, A., Heinze, J. 1992. Wing reduction in ant queens from arid habitats. Naturwissenschaften 79:84-85. [1992.02]

Tinaut, A., Jiménez Rojas, J. 1991 ("1990"). Redescripción de *Aphaenogaster striativentris* Forel, 1895 y consideraciones sobre su polimorfismo (Hymenoptera, Formicidae). EOS. Rev. Esp. Entomol. 66:117-126. [1991.03]

Tinaut, A., Jiménez Rojas, J., Pascual, R. 1995 ("1994"). Estudio de la mirmecofauna de los bosques de *Quercus* Linneo 1753 de la provincia de Granada (Hymenoptera; Formicidae). Ecología (Madr.) 8:429-438. [1995]

Tinaut, A., Ortiz, F. J. 1988. Descripción del macho de *Monomorium algiricum* (Bernard 1955) y consideraciones sobre el valor taxonómico del genero *Epixenus* Emery, 1908 (Hymenoptera, Formicidae). Bol. Asoc. Esp. Entomol. 12:165-174. [1988.11]

Tinaut, A., Ortiz, F. J. 1988b. Introducción al conocimiento de las hormigas de la provincia de Almería (Hymenoptera: Formicidae). Publ. Inst. Estud. Almer. Bol. (Cienc.) 8:223-231. [1988.12]

Tinaut, A., Pascual, R. 1986. Confirmación de la presencia de *Aphaenogaster splendida* (Roger, 1859) en la Península Ibérica (Hymenoptera, Formicidae). Nouv. Rev. Entomol. (n.s.)3:189-192. [1986.10.20]

Tinaut, A., Plaza, J. L. 1990 ("1989"). Situación taxonómica del género *Cataglyphis* Förster, 1850 en la Península Ibérica. I. Las especies del subgénero *Cataglyphis* Förster (Hym. Formicidae). EOS. Rev. Esp. Entomol. 65:189-199. [1990.01]

Tinaut, A., Ruano, F. 1992. Braquipterismo y apterismo en formícidos. Morfología y biometría en las hembras de especies ibéricas de vida libre (Hymenoptera: Formicidae). Graellsia 48:121-131. [1992.12]

Tinaut, A., Ruano, F. 1994. Contribución al conocimiento de los formícidos de la Sierra de la Estrella (Portugal) (Hymenoptera: Formicidae). Bol. Asoc. Esp. Entomol. 18:97-99. [1994.10]

Tinaut, A., Ruano, F., Fernandez Escudero, I. 1995 ("1994"). Descripción del macho del género *Rossomyrmex* Arnoldi, 1928 (Hymenoptera, Formicidae). Nouv. Rev. Entomol. (n.s.)11:347-351. [1995.04.20]

Tinaut, A., Ruano, F., Hidalgo, J., Ballesta, M. 1995 ("1994"). Mirmecocenosis del sistema de dunas del Paraje Natural Punta Entinas-El Sabinar (Almería) (Hymenoptera, Formicidae). Aspectos taxonómicos, functionales y biogeográficos. Graellsia 50:71-84. [1995.07.04]
Tinaut, J. A. See: Tinaut, A.
Tinaut Ranera, J. A. See: Tinaut, A.
Tiwari, R. N. 1994. Two new species of a little known genus *Myrmecina* Curtis from Kerala, India. Rec. Zool. Surv. India 94:151-158.
Tiwari, R. N., Guha, D. K. 1976. A new record of *Polyrhachis* (*Campomyrma*) *hauxwelli* Bingham (Hymenoptera: Formicidae) from India. Newsl. Zool. Surv. India 2:210. [1976.10]
Tiwari, R. N., Jonathan, J. K. 1986a. A new species of *Liomyrmex* Mayr from Andaman Islands (Hymenoptera: Formicidae). Rec. Zool. Surv. India 83:87-90. [1986.07]
Tiwari, R. N., Jonathan, J. K. 1986b. A new species of *Metapone* Forel from Nicobar Islands (Hymenoptera: Formicidae: Myrmicinae). Rec. Zool. Surv. India 83:149-153. [1986.07]
Tjan, K. N., Imai, H. T., Kubota, M., Brown, W. L., Jr., Gotwald, W. H., Jr., Yong, H.-S., Leh, C. 1986. Chromosome observations of Sarawak ants. Annu. Rep. Natl. Inst. Genet. Jpn. 36:57. [1986]
Tobin, J. E. 1994. Ants as primary consumers: diet and abundance in Formicidae. Pp. 279-307 in: Hunt, J. H., Nalepa, C. A. (eds.) Nourishment and evolution in insect societies. Boulder: Westview Press, xii + 449 pp. [1994]
Tobin, J. E. 1995. Ecology and diversity of tropical forest canopy ants. Pp. 129-147 in: Lowman, M. D., Nadkarni, N. (eds.) Forest canopies. San Diego: Academic Press, xix + 624 pp. [1995]
Togashi, I. 1991. Ants (Hymenoptera, Formicidae) lured to benzyl acetate in bucket traps in Ishikawa Prefecture, Japan. [In Japanese.] Bull. Biogeogr. Soc. Jpn. 46:161-164. [1991.12.20]
Togashi, I. 1993. Ants at Kaga side of Mt. Hakusan. [In Japanese.] Bull. Biogeogr. Soc. Jpn. 48: 63-67. [1993.12.30]
Togashi, I. 1994. Ant fauna of Noto Peninsula, Ishikawa Prefecture. [In Japanese.] Bull. Biogeogr. Soc. Jpn. 49:47-50. [1994.11.20]
Tohmé, G. 1969a. Description d'espèces nouvelles de fourmis au Liban (Hymenoptera, Formicoidea). Publ. Univ. Liban. Sect. Sci. Nat. 7:1-15. [1969.08.30]
Tohmé, G. 1969b. Essaimage et premiers stades du développement du *Messor ebeninus* (Forel) (Hym. - Formicoidea - Myrmecidae [sic]). Pp. 287-293 in: Ernst, E. et al. (eds.) International Union for the Study of Social Insects. VI Congress. Bern 15-20 September 1969. Proceedings. Bern: Organizing Committee of the VI Congress IUSSI, 309 pp. [1969.09.15]
Tohmé, G. 1971 ("1970"). Description de *Messor ebeninus* (Forel) (Hymenoptera: Formicoidea - Myrmecidae). Bull. Soc. Entomol. Egypte 54:569-577. [1971.12]
Tohmé, G. 1975. Écologie, biologie de la reproduction et éthologie de *Messor ebeninus* Forel (Hymenoptera, Formicoidea-Myrmicidae). Bull. Biol. Fr. Belg. 109:171-251. [1975.12]
Tohmé, G., Tohmé, H. 1981. Les fourmis du genre *Messor* en Syrie. Position systématique. Description de quelques ailés et de formes nouvelles. Répartition géographique. Ecol. Mediterr. 7(1):139-153. [1981]
Tohmé, H. 1981. Écologie et biologie de la reproduction de la fourmi *Acantholepis frauenfeldi*, Mayr (Hymenoptera, Formicoidae, Formicinae). Publ. Univ. Liban. Sect. Sci. Nat. 12:1-197. [1981.01.30]
Tohmé, H., Tohmé, G. 1976 ("1975"). Description des castes d'*Acantholepis frauenfeldi*, Mayr et des differents stades larvaires (Hymenoptera, Formicoidea: Formicinae). Bull. Soc. Entomol. Egypte 59:131-142. [1976.12]
Tohmé, H., Tohmé, G. 1980a ("1979"). Le genre *Epixenus* Emery (Hymenoptera, Formicidae, Myrmicinae) et ses principaux représentants au Liban et en Syrie. Bull. Mus. Natl. Hist. Nat. Sect. A Zool. Biol. Écol. Anim. (4)1:1087-1108. [1980.04.18]
Tohmé, H., Tohmé, G. 1980b. Les fourmis du genre *Solenopsis* en Syrie. Description de deux nouvelles sous-espèces et d'ailés inédits. Notes biogéographiques et systématiques (1). Rev. Fr. Entomol. (Nouv. Sér.) 2:129-137. [1980.09.22]

Tohmé, H., Tohmé, G. 1980c. Contribution à l'étude systématique et biologique de *Acantholepis syriaca* André (Hymenoptera, Formicidae, Formicinae). Bull. Mus. Natl. Hist. Nat. Sect. A Zool. Biol. Écol. Anim. (4)2:517-524. [1980.10.31]

Tohmé, H., Tohmé, G. 1981. Contribution à l'étude systématique et biologique de *Bothriomyrmex syrius* (Forel), Formicoidea, Dolichoderinae [Hym.]. Bull. Soc. Entomol. Fr. 86:97-103. [1981.09.11]

Tohmé, H., Tohmé, G. 1985. Contribution à l'étude systématique et bioécologique de *Cataglyphis frigida* (André) (Hymenoptera, Formicidae, Formicinae). Rev. Fr. Entomol. (Nouv. Sér.) 7: 83-88. [1985.06.14]

Tokunaga, M. 1934. The ants observed in seabord of Kii Province. [In Japanese.] Kansai Konchu Zasshi 2:40-43. [1934.07.31]

Tomotake, M. E. M., Caetano, F. H. 1994. Digestive tract morphology of *Acanthostichus serratulus* and *Cylindromyrmex brasiliensis* (Hymenoptera: Formicidae). P. 533 in: Lenoir, A., Arnold, G., Lepage, M. (eds.) Les insectes sociaux. 12ème congrès de l'Union Internationale pour l'Étude des Insectes Sociaux. Paris, Sorbonne, 21-27 août 1994. Paris: Université Paris Nord, xxiv + 583 pp. [1994.09]

Tomotake, M. E. M., Mathias, M. I. C., Yabuki, A. T., Caetano, F. H. 1992. Scanning electron microscopy of mandibular glands of workers and queens of the ants *Pachycondyla striata* (Hymenoptera: Ponerinae). J. Adv. Zool. 13:1-6. [1992.12]

Tonapi, G. T. 1958. A comparative study of spiracular structure and mechanisms in some Hymenoptera. Trans. R. Entomol. Soc. Lond. 110:489-520. [1958.12.19]

Topoff, H. 1971. Polymorphism in army ants related to division of labor and colony cyclic behavior. Am. Nat. 105:529-548. [1971.12.31]

Topoff, H. 1990. The evolution of slave-making behavior in the parasitic ant genus *Polyergus*. Ethol. Ecol. Evol. 2:284-287. [1990.09]

Topoff, H., Zimmerli, E. 1992 ("1991"). *Formica wheeleri*: Darwin's predatory slave-making ant? Psyche (Camb.) 98:309-317. [1992.09.04]

Torgerson, R. L., Akre, R. D. 1970. Interspecific responses to trail and alarm pheromones by New World army ants. J. Kansas Entomol. Soc. 43:395-404. [1970.11.13]

Torossian, C. 1971. Faune secondaire des galles de Cynipidae: I. Étude systématique des fourmis et des principaux arthropods recoltés dans les galles. Insectes Soc. 18:135-154. [1971.12]

Torossian, C. 1977. Les fourmis du groupe *Formica rufa* de la Cerdagne. Bull. Soc. Hist. Nat. Toulouse 113:255-260. [1977.09]

Torre-Grossa, J. P., Febvay, G., Kermarrec, A. 1982 ("1981"). Larval instars of the worker caste in the attine ant, *Acromyrmex octospinosus* (Hymenoptera: Formicidae). Colemania 1:141-147. [1982.05.11]

Torres, J. A. 1984a. Niches and coexistence of ant communities in Puerto Rico: repeated patterns. Biotropica 16:284-295. [1984.12]

Torres, J. A. 1984b. Diversity and distribution of ant communities in Puerto Rico. Biotropica 16:296-303. [1984.12]

Torres, J. A., Snelling, R. R. 1995 ("1992"). Los himenópteros de Isla de Mona. Acta Cient. (P. R.) 6:87-102. [1995] [Ants pp. 93-96.]

Trager, J. C. 1984a ("1983"). A new *Paratrechina* (Hymenoptera: Formicidae) from Machu Picchu, Peru. Fla. Entomol. 66:482-486. [1984.02.28]

Trager, J. C. 1984b. A revision of the genus *Paratrechina* (Hymenoptera: Formicidae) of the continental United States. Sociobiology 9:49-162. [1984]

Trager, J. C. 1988a. A revision of *Conomyrma* (Hymenoptera: Formicidae) from the southeastern United States, especially Florida, with keys to the species. Fla. Entomol. 71:11-29. [1988.03.18] [Errata in Fla. Entomol. 71:219. (1988.06.20)]

Trager, J. C. (ed.) 1988b. Advances in myrmecology. Leiden: E. J. Brill, xxvii + 551 pp. [1988]

Trager, J. C. 1988c. George C. Wheeler - An appreciation. Pp. xvii-xxvii in: Trager, J. C. (ed.) Advances in myrmecology. Leiden: E. J. Brill, xxvii + 551 pp. [1988]

Trager, J. C. 1991. A revision of the fire ants, *Solenopsis geminata* group (Hymenoptera: Formicidae: Myrmicinae). J. N. Y. Entomol. Soc. 99:141-198. [1991.05.29]

Trager, J. C., Johnson, C. 1985. A slave-making ant in Florida: *Polyergus lucidus* with observations on the natural history of its host *Formica archboldi* (Hymenoptera: Formicidae). Fla. Entomol. 68:261-266. [1985.06.10]

Trager, J. C., Johnson, C. 1988. The ant genus *Leptogenys* (Hymenoptera: Formicidae, Ponerinae) in the United States. Pp. 29-34 in: Trager, J. C. (ed.) Advances in myrmecology. Leiden: E. J. Brill, xxvii + 551 pp. [1988]

Traniello, J. F. A. 1978. Caste in a primitive ant: absence of age polyethism in *Amblyopone*. Science (Wash. D. C.) 202:770-772. [1978.11.17]

Traniello, J. F. A. 1982. Population structure and social organization in the primitive ant *Amblyopone pallipes* (Hymenoptera: Formicidae). Psyche (Camb.) 89:65-80. [1982.12.17]

Traniello, J. F. A. 1987. Comparative foraging ecology of North Temperate ants: the role of worker size and cooperative foraging in prey selection. Insectes Soc. 34:118-130. [1987.12]

Traniello, J. F. A. 1989. Foraging strategies of ants. Annu. Rev. Entomol. 34:191-210. [1989]

Traniello, J. F. A., Hölldobler, B. 1984. Chemical communication during tandem running in *Pachycondyla obscuricornis* (Hymenoptera: Formicidae). J. Chem. Ecol. 10:783-794. [1984.08]

Traniello, J. F. A., Jayasuriya, A. K. 1981a. The sternal gland and recruitment communication in the primitive ant *Aneuretus simoni*. Experientia (Basel) 37:46-47. [1981.01.15]

Traniello, J. F. A., Jayasuriya, A. K. 1981b. Chemical communication in the primitive ant *Aneuretus simoni*: the role of the sternal and pygidial glands. J. Chem. Ecol. 7:1023-1033. [1981.11]

Traniello, J. F. A., Jayasuriya, A. K. 1986 ("1985"). The biology of the primitive ant *Aneuretus simoni* (Emery) (Formicidae: Aneuretinae). II. The social ethogram and division of labor. Insectes Soc. 32:375-388. [1986.06]

Travan, J. 1990. Bestandsaufnahme der Waldameisennester im Staatswald Oberbayerns. Teil I: Flachlandforstämter. Waldhygiene 18:119-142. [1990]

Trivers, R. L., Hare, H. 1976. Haplodipoidy and the evolution of the social insects. Science (Wash. D. C.) 191:249-263. [1976.01.23]

Trojan, P. 1994. Bohdan Pisarski (1928-1992). Memorabilia Zool. 48:3-7. [1994]

Troppmair, H. 1979. Verbreitungstypen brasilianischer Blattschneider-Ameisen der Gattung *Atta*. Biogeographica 16:159-161. [1979]

Tryon, E. H., Jr. 1986. The striped earwig, and ant predators of sugarcane rootstock borer, in Florida citrus. Fla. Entomol. 69:336-343. [1986.08.08] [Page 340: nomen nudum, "*Conomyrma edeni* Buren".]

Tryon, H. 1886 ("1885"). Notes on Queensland ants. Proc. R. Soc. Qld. 2:146-162. [1886.06]

Tryon, H. 1900. Harvesting ants. Qld. Agric. J. 7:71-79. [1900.07.01]

Tschinkel, W. R. 1987. Relationship between ovariole number and spermathecal sperm count in ant queens: a new allometry. Ann. Entomol. Soc. Am. 80:208-211. [1987.03]

Tsuji, K. 1995. Reproductive conflicts and levels of selection in the ant *Pristomyrmex pungens*: contextual analysis and partitioning of covariance. Am. Nat. 146:586-607. [1995.10]

Tsuji, K., Furukawa, T., Kinomura, K., Takamine, H., Yamauchi, K. 1991. The caste system of the dolichoderine ant *Technomyrmex albipes* (Hymenoptera: Formicidae): morphological description of queens, workers and reproductively active intercastes. Insectes Soc. 38:413-422. [1991]

Tsuji, K., Yamauchi, K. 1995. Production of females by parthenogenesis in the ant *Cerapachys biroi*. Insectes Soc. 42:333-336. [1995]

Tsuji, N., Yamauchi, K., Yamamura, N. 1994. A mathematical model for wing dimorphism in male *Cardiocondyla* ants. J. Ethol. 12:19-24. [1994.06]

Tsyubik, M. M., Radchenko, A. G. 1988. Zoogeographical aspects of the study of the myrmecofauna of the Carpathian and Transcarpathian. [In Russian.] Pp. 45-52 in: Fodor, S. S.,

Komendar, V. I., Korchinsky, A. V. et al. (eds.) Questions of the protection and rational use of the plant and animal world of the Ukrainian Carpathian. [In Russian.] Uzhgorod: Uzhgorogskoe Otdelenie MOIP, 192 pp. [1988] [Not seen. Citation from Radchenko (pers. comm.). Signed for printing ("podpisano k pechati") 19 November 1987, but actually published in 1988.]

Tuff, D. W. 1975. An anomalous wing in a male doryline ant *Labidus coecus* (Latreille) (Hymenoptera: Formicidae). Southwest. Nat. 19:449-452. [1975.01.20]

Tulloch, G. S. 1929. The proper use of the terms parapsides and parapsidal furrows. Psyche (Camb.) 36:376-382. [1929.12] [Stamp date in MCZ library: 1930.03.13.]

Tulloch, G. S. 1930a. An unusual nest of *Pogonomyrmex*. Psyche (Camb.) 37:61-70. [1930.03] [Stamp date in MCZ library: 1930.06.10.]

Tulloch, G. S. 1930b. Thoracic modifications accompanying the development of subaptery and aptery in the genus *Monomorium*. Psyche (Camb.) 37:202-206. [1930.09] [Stamp date in MCZ library: 1930.11.20.]

Tulloch, G. S. 1932. A gynergate of *Myrmecia*. Psyche (Camb.) 39:48-51. [1932.06] [Stamp date in MCZ library: 1932.07.01.]

Tulloch, G. S. 1934. Vestigial wings in *Diacamma* (Hymenoptera: Formicidae). Ann. Entomol. Soc. Am. 27:273-277. [1934.06.29]

Tulloch, G. S. 1935. Morphological studies of the thorax of the ant. Entomol. Am. (n.s.)15:93-131. [1935.07.01]

Tulloch, G. S. 1936. The metasternal glands of the ant, *Myrmica rubra*, with special reference to the Golgi bodies and the intracellular canaliculi. Ann. Entomol. Soc. Am. 29:81-84. [1936.03.31]

Tulloch, G. S. 1942. The thoracic structure of pseudogynes of *Formica sanguinea* Latreille (Hymenoptera, Formicidae). Bull. Brooklyn Entomol. Soc. 37:21-23. [1942.03.26]

Tulloch, G. S. 1946. The thoracic structure of worker ants of the genus *Pheidologeton*. Bull. Brooklyn Entomol. Soc. 41:92-93. [1946.07.29]

Tulloch, G. S., Shapiro, J. E., Hershenov, B. 1963 ("1962"). The ultrastructure of the metasternal glands of ants. Bull. Brooklyn Entomol. Soc. 57:91-101. [1963.03.07]

Turner, G. 1897. Notes upon the Formicidae of Mackay, Queensland. Proc. Linn. Soc. N. S. W. 22:129-144. [1897.09.17]

Tutt, J. W. 1905. Types of the genera of the agdistid, alucitid and orneodid plume moths. Entomol. Rec. J. Var. 17:34-37. [1905.02.15] [*Wheeleria* (Lepidoptera), senior homonym of *Wheeleria* Forel, 1905.]

Tynes, J. S., Hutchins, R. E. 1964. Studies of plant-nesting ants in east central Mississippi. Am. Midl. Nat. 72:152-156. [1964.06.29]

Übler, E., Kern, F., Bestmann, H. J., Hölldobler, B., Attygalle, A. B. 1995. Trail pheromones of two formicine ants, *Camponotus silvicola* and *C. rufipes* (Hymenoptera, Formicidae). Naturwissenschaften 82:523-525. [1995.11]

Uchida, T. 1925. A list of known species of Korean Hymenoptera which I collected in 1922 and their geographical distributions. [part] [In Japanese.] Insect World 29:328-337. [1925.10.15] [Formicidae pp. 334-335.]

Uchida, T. 1936. Some Hymenoptera from Daisetsu Mountain. [In Japanese.] Biogeographica (Tokyo) 1:63-74. [1936.06.05] [Ants pp. 71-72.]

Urich, F. W. 1895. Notes on some fungus-growing ants in Trinidad. J. Trinidad Field Nat. Club 2:175-182. [1895.04]

Urich, F. W. 1911. Miscellaneous notes. Circ. Board Agric. Trin. 3:15-25. [1911.08.09] [Includes "Preliminary list of Trinidad ants", pp. 17-18.]

Vail, K., Davis, L., Wojcik, D., Koehler, P., Williams, D. 1994. Structure-invading ants of Florida. SP 164. Gainesville: University of Florida, 15 pp. [1994.09]

Valcurone Dazzini, M. See: Dazzini Valcurone, M.

Valentine, E. W., Walker, A. K. 1991. Annotated catalogue of New Zealand Hymenoptera. DSIR Plant Protection Report No. 4. Auckland: DSIR Plant Protection, 84 pp. [1991.05] [Ants pp. 35-38, 61, 62.]

Van Boven, J. See: Boven, J. K. A. van.

Van Boven, J. K. A. See: Boven, J. K. A. van.

Vandel, A. 1926. Fourmis françaises rares ou peu connues. Bull. Soc. Entomol. Fr. 1926:196-198. [1926.12.30]

Vandel, A. 1927. Modifications déterminées par un nématode du genre *Mermis* chez les ouvrières et les soldats de la fourmi *Pheidole pallidula* Nyl. Bull. Biol. Fr. Belg. 61:38-48. [1927.03.15]

Vandel, A. 1928 ("1927"). Observations sur les moeurs d'une fourmi parasite: *Epimyrma vandeli* Santschi. Bull. Soc. Entomol. Fr. 1927:289-295. [1928.01.15]

Vandel, A. 1930a. La production d'intercastes, chez la fourmi, *Pheidole pallidula*, sous l'action de parasites du genre *Mermis*. C. R. Séances Acad. Sci. 190:770-772. [1930.03.31]

Vandel, A. 1930b. La production d'intercastes chez la fourmi *Pheidole pallidula* sous l'action de parasites du genre *Mermis*. 1. Étude morphologique des individus parasités. Bull. Biol. Fr. Belg. 64:457-494. [1930.11.03]

Vandel, A. 1931. Étude d'un gynandromorphe (dinergatandromorphe) de *Pheidole pallidula* Nyl. (Hyménoptères - Formicidés). Bull. Biol. Fr. Belg. 65:114-129. [1931.11.15]

Van der Have, T. M., Boomsma, J. J., Menken, S. B. J. 1988. Sex-investment ratios and relatedness in the monogynous ant *Lasius niger* (L.). Evolution 42:160-172. [1988.01.12]

Vander Meer, R. K. 1986. Chemical taxonomy as a tool for separating *Solenopsis* spp. Pp. 316-326 in: Lofgren, C. S., Vander Meer, R. K. (eds.) Fire ants and leaf cutting ants: biology and management. Boulder: Westview Press, xv + 435 pp. [1986]

Vander Meer, R. K., Jaffe, K., Cedeno, A. (eds.) 1990. Applied myrmecology: a world perspective. Boulder: Westview Press, xv + 741 pp. [1990]

Vander Meer, R. K., Lofgren, C. S. 1988. Use of chemical characters in defining populations of fire ants (*Solenopsis saevissima* complex) (Hymenoptera: Formicidae). Fla. Entomol. 71:323-332. [1988.09.30]

Vander Meer, R. K., Lofgren, C. S. 1989. Biochemical and behavioral evidence for hybridization between fire ants, *Solenopsis invicta* and *Solenopsis richteri* (Hymenoptera: Formicidae). J. Chem. Ecol. 15:1757-1765. [1989.06]

Vander Meer, R. K., Lofgren, C. S. 1990. Chemotaxonomy applied to fire ant systematics in the United States and South America. Pp. 75-84 in: Vander Meer, R. K., Jaffe, K., Cedeno, A. (eds.) Applied myrmecology: a world perspective. Boulder: Westview Press, xv + 741 pp. [1990]

Vander Meer, R. K., Lofgren, C. S., Alvarez, F. M. 1985. Biochemical evidence for hybridization in fire ants. Fla. Entomol. 68:501-506. [1985.09.20]

Van der Wiel, P. See: Wiel, P. van der.

Van Loon, A. J., Boomsma, J. J., Andrasfalvy, A. 1990. A new polygynous *Lasius* species (Hymenoptera: Formicidae) from central Europe. I. Description and general biology. Insectes Soc. 37:348-362. [1990.12]

Vanni, S., Bartolozzi, L., Whitman-Mascherini, S. 1986 ("1985"). Cataloghi del Museo Zoologico "La Specola" dell'Università di Firenze. II. Insecta Hymenoptera: tipi. Atti Soc. Toscana Sci. Nat. Mem. Ser. B 92:119-131. [1986] [Ant types (Menozzi) p. 128.]

Van Pelt, A. F. 1948. A preliminary key to the worker ants of Alachua County, Florida. Fla. Entomol. 30:57-67. [1948.01]

Van Pelt, A. F. 1956. The ecology of the ants of the Welaka Reserve, Florida (Hymenoptera: Formicidae). Am. Midl. Nat. 56:358-387. [1956.10]

Van Pelt, A. F. 1958. The ecology of the ants of the Welaka Reserve, Florida (Hymenoptera: Formicidae). Part II. Annotated list. Am. Midl. Nat. 59:1-57. [1958.01]

Van Pelt, A. F. 1962. Concerning the sparse ant fauna of Mt. Mitchell. J. Elisha Mitchell Sci. Soc. 78:138-141. [1962.11]

Van Pelt, A. F. 1963. High altitude ants of the southern Blue Ridge. Am. Midl. Nat. 69:205-223. [1963.01.17]

Van Pelt, A. F. 1983. Ants of the Chisos Mountains, Texas (Hymenoptera: Formicidae). Southwest. Nat. 28:137-142. [1983.05.20]

Vasconcelos, H. L., Cherrett, J. M. 1995. Changes in leaf-cutting ant populations (Formicidae: Attini) after the clearing of mature forest in Brazilian Amazonia. Stud. Neotrop. Fauna Environ. 30:107-113. [1995.06]

Vashkevich, A. F. 1924a. A new form of ant from Semirechensk Oblast. [In Russian.] Izv. Tomsk. Gos. Univ. 74:144-145. [1924]

Vashkevich, A. F. 1924b. On the ant fauna of north Tobolsk province. [In Russian.] Izv. Tomsk. Gos. Univ. 74:146-149. [1924]

Vashkevich, A. F. 1926. A new form of ant from the Ili river valley. [In Russian.] Izv. Tomsk. Gos. Univ. 77:118. [1926]

Vasilev, I. 1987 ("1984"). The ants (Formicidae, Hymenoptera) from the valley of the River Racene. [In Bulgarian.] God. Sofii. Univ. "Kliment Okhridski" Biol. Fak. Kn. 1 Zool. 78: 74-79. [> 1987.06.18] [Date signed for printing.]

Vasilev, I., Eftimov, M. 1975. The ants Formicidae (Hymenoptera) from the mountain of Lozen. [In Bulgarian.] God. Sofii. Univ. Biol. Fak. Kn. 1 Zool. Fiziol. Biohim. Zhivotn. 67:121-128. [> 1975.10.14] [Date signed for printing.]

Vecht, J. van der. 1957. On some Hymenoptera from the collection of Guérin-Méneville in the Leiden Museum. Zool. Meded. (Leiden) 35:21-31. [1957.01.23]

Vecht, J. van der. 1975. The date of publication of M. Spinola's paper on the Hymenoptera collected by V. Ghiliani in Para, with notes on the Eumenidae described in this work. Entomol. Ber. (Amst.) 35:60-63. [1975.04.01]

Vepsäläinen, K., Pisarski, B. 1981. The taxonomy of the *Formica rufa* group: chaos before order. Pp. 27-35 in: Howse, P. E., Clement, J.-L. (eds.) Biosystematics of social insects. Systematics Association Special Volume 19. London: Academic Press, 346 pp. [1981]

Vepsäläinen, K., Pisarski, B. 1982a. The structure of urban ant communities along the geographical gradient from north Finland to Poland. Pp. 155-168 in: Luniak, M., Pisarski, B. (eds.) Animals in urban environment. Proceedings of the Symposium on the occasion of the 60th anniversary of the Institute Zoology of the Polish Academy of Sciences. Wroclaw: Polska Akademia Nauk, 175 pp. [1982.04]

Vepsäläinen, K., Pisarski, B. 1982b. Assembly of island ant communities. Ann. Zool. Fenn. 19:327-335. [1982.12.30]

Verhaagh, M. 1990. The Formicidae of the rain forest in Panguana, Peru: the most diverse local ant fauna ever recorded. Pp. 217-218 in: Veeresh, G. K., Mallik, B., Viraktamath, C. A. (eds.) Social insects and the environment. Proceedings of the 11th International Congress of IUSSI, 1990. New Delhi: Oxford & IBH Publishing Co., xxxi + 765 pp. [1990.08]

Verhaagh, M. 1994a. *Pachycondyla luteola* (Hymenoptera, Formicidae), an inhabitant of *Cecropia* trees in Peru. Andrias 13:215-224. [1994.09.30]

Verhaagh, M. 1994b. Neue Fundstellen einiger Ameisen in Südwestdeutschland. Carolinea 52:115-118. [1994.12.30]

Verhaagh, M., Rosciszewski, K. 1994. Ants (Hymenoptera, Formicidae) of forest and savanna in the Biosphere Reserve Beni, Bolivia. Andrias 13:199-214. [1994.09.30]

Vernalha, M. M. 1952. Algumas formigas que ocorrem no Estado do Paraná (Hym. Formicidae). Subfamiliás: Dorylinae, Ponerinae, Dolichoderinae, Pseudomyrminae, Myrmicinae e Formicinae. Arq. Biol. Tecnol. (Curitiba) 7:43-51. [1952]

Vernier, R. 1992. Recherche écofaunistique sur les fourmis du genre *Formica* L. de la tourbière du Cachot (Jura Neuchâtelois) et hauts-marais voisins (Hymenoptera, Formicidae). 1. Liste des espèces et leurs biotopes préférentiels. Bull. Soc. Neuchâtel. Sci. Nat. 115:61-82. [1992]

Vesselinov, G. See: Wesselinoff, G. D.

Viana, M. J., Haedo Rossi, J. A. 1957. Primer hallazgo en el hemisferio sur de Formicidae extinguidos y catálogo mundial de los Formicidae fósiles. Primera parte. Ameghiniana 1(1/2):108-113. [1957.01.22]

Vick, K. W., Drew, W. A., Eisenbraun, E. J., McGurk, D. J. 1969. Comparative effectiveness of aliphatic ketones in eliciting alarm behavior in *Pogonomyrmex barbatus* and *P. comanche*. Ann. Entomol. Soc. Am. 62:380-381. [1969.03.17]

Vick, K. W., Drew, W. A., Young, J., Eisenbraun, E. J. 1969. Chemotaxonomic studies of ants: free amino acids. Can. Entomol. 101:1207-1213. [1969.11.05]

Vick, K. W., Drew, W. A., Young, J., McGurk, K. J., Eisenbraun, E. J. 1969. Chemotaxonomic studies of ants: volatile compounds. Can. Entomol. 101:879-882. [1969.08.29]

Victor, T. See also: Motschoulsky, V. de (1860-1866).

Victor, T. 1839. Insectes du Caucase et des provinces transcaucasiennes. Bull. Soc. Imp. Nat. Mosc. 12:44-68. [1839] [Page 47: description of *Formica caduca* (= *Messor caducus*).]

Viehmeyer, H. 1904. Experimente zu Wasmanns *Lomechusa*-Pseudogynen-Theorie und andere biologische Beobachtungen an Ameisen. Allg. Z. Entomol. 9:334-344. [1904.09.15]

Viehmeyer, H. 1906. Beiträge zur Ameisenfauna des Königreiches Sachsen. Sitzungsber. Abh. Naturwiss. Ges. "Isis" Dres. 1906:55-69. [1906]

Viehmeyer, H. 1908. Zur Koloniegründung der parasitischen Ameisen. Biol. Centralbl. 28:18-32. [1908.01.01]

Viehmeyer, H. 1909. Bilder aus dem Ameisenleben. Naturwiss. Bibliothek (Höller, K., Ulmer, G., eds.). Leipzig: Quelle & Meyer, viii + 159 pp. [Not seen. Cited in Zoological Record for 1909.]

Viehmeyer, H. 1910a. Bemerkungen zu Wasmanns neuester Arbeit: Über den Ursprung des sozialen Parasitismus, der Sklaverei und der Myrmecophilie bei den Ameisen. Zool. Anz. 35:450-457. [1910.02.15]

Viehmeyer, H. 1910b. Ontogenetische und phylogenetische Betrachtungen über die parasitische Koloniegründung von *Formica sanguinea*. Biol. Centralbl. 30:569-580. [1910.09.01]

Viehmeyer, H. 1910c. Ueber eine erst in den letzen Jahren in Sachsen aufgefundene Ameise: *Harpagoxenus sublevis* (Nyl.). Korrespondenzbl. Beil. Dtsch. Entomol. Z. "Iris" 1910:40. [1910.10.01]

Viehmeyer, H. 1911a. Hochzeitsflug und Hybridation bei den Ameisen. Dtsch. Entomol. Natl.-Bibl. 2:28-30. [1911.02.01]

Viehmeyer, H. 1911b. Morphologie und Phylogenie von *Formica sanguinea*. Zool. Anz. 37:427-441. [1911.05.02]

Viehmeyer, H. 1912a. Über die Verbreitung und die geflügelten Weibchen von *Harpagoxenus sublevis* Nyl. (Hym., Form.). Entomol. Mitt. 1:193-197. [1912.07.01]

Viehmeyer, H. 1912b. Ameisen aus Deutsch Neuguinea gesammelt von Dr. O. Schlaginhaufen. Nebst einem Verzeichnisse der papuanischen Arten. Abh. Ber. K. Zool. Anthropol.-Ethnogr. Mus. Dres. 14:1-26. [1912.11.05]

Viehmeyer, H. 1913. Ameisen aus dem Kopal von Celebes. Stett. Entomol. Ztg. 74:141-155. [1913.07.01]

Viehmeyer, H. 1914a. Ameisen aus Perak, Bali und Ceram (Hym.) (Freiburger Molukken-Expedition), gesammelt von E. Streesemann. Entomol. Mitt. 3:112-116. [1914.04.01]

Viehmeyer, H. 1914b ("1913"). Neue und unvollständig bekannte Ameisen der alten Welt. Arch. Naturgesch. (A)79(12):24-60. [1914.04] [Heft 12 is not dated, but the publication dates of adjacent issues suggest that Heft 12 was published in March or April 1914. Received at UC Berkeley library 28 April 1914.]

Viehmeyer, H. 1914c. Papuanische Ameisen. Dtsch. Entomol. Z. 1914:515-535. [1914.09.30]

Viehmeyer, H. 1914d. Mayr's Gattung *Ischnomyrmex* (Hym.) nebst Beschreibung einiger neuer Arten aus anderen Gattungen. Zool. Jahrb. Abt. Syst. Geogr. Biol. Tiere 37:601-612. [1914.12.10]

Viehmeyer, H. 1916a ("1915"). Ameisen von Singapore. Beobachtet und gesammelt von H. Overbeck. Arch. Naturgesch. (A)81(8):108-168. [1916.04]
Viehmeyer, H. 1916b. Ameisen von den Philippinen und anderer Herkunft (Hym.). Entomol. Mitt. 5:283-291. [1916.11.11]
Viehmeyer, H. 1916c ("1915"). Zur sächsischen Ameisenfauna. Sitzungsber. Abh. Naturwiss. Ges. "Isis" Dres. 1915:61-64. [1916]
Viehmeyer, H. 1917. Anomalien am Skelette der Ameisen (Hym.). Entomol. Mitt. 6:66-72. [1917.03.27]
Viehmeyer, H. 1920 ("1918"). Anleitung zum Sammeln von Ameisen. Arch. Naturgesch. (A)84(9):160-170. [1920.04]
Viehmeyer, H. 1921. Die mitteleuropäischen Beobachtungen von *Harpagoxenus sublevis* Mayr. Biol. Zentralbl. 41:269-278. [1921.06.01]
Viehmeyer, H. 1922. Neue Ameisen. Arch. Naturgesch. (A)88(7):203-220. [1922.06]
Viehmeyer, H. 1923. Wissenschaftliche Ergebnisse der mit Unterstützung der Akademie der Wissenschaften in Wien aus der Erbschaft Treitl von F. Werner unternommenen zoologischen Expedition nach dem anglo-ägyptischen Sudan (Kordofan) 1914. VII. Hymenoptera A. Formicidae. Denkschr. Akad. Wiss. Wien Math.-Naturwiss. Kl. 98:83-94. [1923]
Viehmeyer, H. 1924a ("1923"). Polymorphismus und Ernährung bei den Ameisen. Arch. Naturgesch. (A)89(12):1-12. [1924.03]
Viehmeyer, H. 1924b. Formiciden der australischen Faunenregion. Entomol. Mitt. 13:219-229. [1924.08.09]
Viehmeyer, H. 1924c. Formiciden der australischen Faunenregion. (Fortsetzung). Entomol. Mitt. 13:310-319. [1924.11.07]
Viehmeyer, H. 1925a. Formiciden der australischen Faunenregion. (Fortsetzung.) Entomol. Mitt. 14:25-39. [1925.01.15]
Viehmeyer, H. 1925b. Formiciden der australischen Faunenregion. (Schluss.) Entomol. Mitt. 14:139-149. [1925.04.10]
Vierbergen, G., Scheven, J. 1995. Nine new species and a new genus of Dominican amber ants of the tribe (Cephalotini Hymenoptera: Formicidae) [sic]. Creat. Res. Soc. Q. 32:158-170. [1995.12]
Viereck, H. L. 1903. Hymenoptera of Beulah, New Mexico. [part] Trans. Am. Entomol. Soc. 29: 56-87. [1903.01] [Ants pp. 72-74.]
Villet, M. H. 1989. A syndrome leading to ergatoid queens in ponerine ants (Hymenoptera: Formicidae). J. Nat. Hist. 23:825-832. [1989.07.03]
Villet, M. H. 1990. Qualitative relations of egg size, egg production and colony size in some ponerine ants (Hymenoptera: Formicidae). J. Nat. Hist. 24:1321-1331. [1990.08.31]
Villet, M. H. 1991a. Colony foundation in *Plectroctena mandibularis* F. Smith, and the evolution of ergatoid queens in *Plectroctena* (Hymenoptera: Formicidae). J. Nat. Hist. 25:979-983. [1991.09.30]
Villet, M. H. 1991b. Reproduction and division of labour in *Platythyrea* cf. *cribrinodis* (Gerstaecker 1858) (Hymenoptera Formicidae): comparisons of individuals, colonies and species. Trop. Zool. 4:209-231. [1991.12]
Villet, M. H. 1992a. The social biology of *Hagensia havilandi* (Forel 1901) (Hymenoptera Formicidae), and the origin of queenlessness in ponerine ants. Trop. Zool. 5:195-206. [1992.12]
Villet, M. H. 1992b. Exploring the biology of obligate queenlessness: social organization in Platythyreine ants. Pp. 291-294 in: Billen, J. P. J. (ed.) Biology and evolution of social insects. Leuven: Leuven University Press, ix + 390 pp. [1992]
Villet, M. H., Crewe, R. M., Duncan, F. D. 1991. Evolutionary trends in the reproductive biology of ponerine ants (Hymenoptera: Formicidae). J. Nat. Hist. 25:1603-1610. [1991.12.31]
Villet, M. H., Peeters, C. P., Crewe, R. M. 1984. The occurrence of a pygidial gland in four genera of ponerine ants (Hymenoptera: Formicidae). J. Ga. Entomol. Soc. 19:413-416. [1984.07]

Villet, M. H., Wildman, M. H. 1991. Division of labour in the obligately queenless ant *Pachycondyla* (=*Bothroponera*) *krugeri* Forel 1910 (Hymenoptera Formicidae). Trop. Zool. 4:233-250. [1991.12]

Viswanath, B. N., Belvadi, V. V., Reddy, D. N. R. 1981. The sub-family Dorylinae (Hymenoptera: Formicidae) - a review. Pp. 205-217 in: Veeresh, G. K. (ed.) Progress in soil biology and ecology in India. Hebbal, Bangalore: University of Agricultural Sciences, [11] + 351 pp. [1981]

Vogelsanger, E. 1938. Eine für die Schweiz neue Ameisenart, *Formica uralensis* Ruzsky. Mitt. Schweiz. Entomol. Ges. 17:231-232. [1938.06.15]

Vogrin, V. 1955. Prilog fauni Hymenoptera - Aculeata Jugoslavije. Zast. Bilja 31(suppl.):1-74. [1955] [Formicidae pp. 15-21.]

Von Sicard, N. A. E., Anderson, M. 1986. The gross morphology of the stinging and non-stinging states of the ant *Tetramorium caespitum* L. (Hymenoptera, Formicidae, Myrmicinae). Experientia (Basel) 42:30-31. [1986.01.15]

Vowles, D. M. 1955. The structure and connexions of the corpora pedunculata in bees and ants. Q. J. Microsc. Sci. 96:239-255. [1955.06]

Vysoky, V. 1987. Mravenci vrchu Trabice, Deblík a okolí Církvic (Hymenoptera, Formicoidea). Severocesk. Prír. Príl. 1987:69-73. [1987]

Vysoky, V. 1988. Príspevek k poznání mravencu Ceského stredohorí (Hymenoptera: Formicoidea) pruzkum SPR Rac, s príhlédnutím k faune okolních kopcu Rovny a Jedovina. Zpr. Stud. Krajského Muz. Teplicích 17:23-38. [1988]

Vysoky, V. 1990. Mravenci vrchu Kalvárie a okolí (Hymenoptera: Formicoidea). Fauna Bohemiae Septentr. 14/15:41-45. [1990]

Vysoky, V. 1994a. Príspevek k poznání mravencu SPR Novozámecky rybník. Fauna Bohemiae Septentr. 19:155-160. [1994]

Vysoky, V. 1994b. Mravenci zjistení na území ZOO Ustí nad Labem. Fauna Bohemiae Septentr. 19:161-165. [1994]

Vysoky, V., Werner, P. 1987. Poznámky k vyskytu mravencu na naucné stezce Pod Vysokym Ostrym (Hymenoptera: Formicidae). Fauna Bohemiae Septentr. 12:103-106. [1987]

Vysoky, V., Werner, P. 1988. Príspevek k poznání mravencu Terezínské kotliny (Hymenoptera, Formicidae). Fauna Bohemiae Septentr. 13:133-138. [1988]

Vysoky, V., Werner, P. 1991. Rozsírení mravencu rodu *Camponotus* Mayr, 1861 v severozápadních Cechách (Hym., Formicidae). Myrmekol. Listy 1991:2-6. [1991.10.15]

Walckenaer, C. A. 1802. Faune parisienne, insectes. Ou histoire abrégée des insectes des environs de Paris, classés d'après le système de Fabricius. Vol. 2. Paris: Dentu, 438 pp. [1802] [Published between October and December 1802, according to Richards (1935:174). Ants pp. 157-167.]

Waldkircher, G., Maschwitz, U. 1994. Silk nests of *Camponotus* species in tropical rain forest of South East Asia. P. 543 in: Lenoir, A., Arnold, G., Lepage, M. (eds.) Les insectes sociaux. 12ème congrès de l'Union Internationale pour l'Étude des Insectes Sociaux. Paris, Sorbonne, 21-27 août 1994. Paris: Université Paris Nord, xxiv + 583 pp. [1994.09]

Walker, F. 1859. Characters of some apparently undescribed Ceylon insects. [part] Ann. Mag. Nat. Hist. (3)4:370-376. [1859.11]

Walker, F. 1860. Characters of some apparently undescribed Ceylon insects. [part] Ann. Mag. Nat. Hist. (3)5:304-311. [1860.04]

Walker, F. 1861. List of Ceylon insects. Pp. 442-463 in: Tennent, J. E. Sketches of the natural history of Ceylon. London: Longman, Green, Longman, and Roberts, xxiii + 500 pp. [1861] [Ants p. 454.]

Walker, F. 1871. A list of Hymenoptera collected by J. K. Lord, Esq. in Egypt, in the neighbourhood of the Red Sea, and in Arabia, with descriptions of the new species. London: E. W. Janson, vi + 59 pp. [1871]

Walsh, B. D. 1863 ("1862"). On the genera of Aphidae found in the United States. Proc. Entomol. Soc. Phila. 1:294-311. [1863.01.12] [Publication date from F. M. Brown (1964:307).]

Walther, J. R. 1979a. Vergleichende morphologische Betrachtung der antennalen Sensillenfelder einiger ausgewählter Aculeata (Insecta, Hymenoptera). Z. Zool. Syst. Evolutionsforsch. 17: 30-56. [1979.03]

Walther, J. R. 1979b. Morphologie und Feinstruktur der Sinnesorgane auf den Geisselantennen von *Formica rufa* L. (Hymenoptera, Formicidae). Verh. Dtsch. Zool. Ges. 72:313. [1979]

Walther, J. R. 1980. Morphology and fine structure of the sense organs on the flagellum of *Formica rufa* L. (Hymenoptera; Formicidae). [Abstract.] P. 437 in: Abstracts. XVI International Congress of Entomology, Kyoto, Japan 3-9 August, 1980. Kyoto: XVI International Congress of Entomology, 480 pp. [1980]

Walther, J. R. 1981. Cuticular sense organs as characters in phylogenetic research. Mitt. Dtsch. Ges. Allg. Angew. Entomol. 3:146-150. [1981.10]

Walther, J. R. 1982. The sense organs on the flagella of ants (Formicoidea). [Abstract.] Pp. 399-400 in: Breed, M. D., Michener, C. D., Evans, H. E. (eds.) The biology of social insects. Proceedings of the Ninth Congress of the IUSSI, Boulder, Colorado, August 1982. Boulder: Westview Press, xii + 420 pp. [1982.08]

Walther, J. R. 1983. Antennal patterns of sensilla of the Hymenoptera - a complex character of phylogenetic reconstruction. Verh. Naturwiss. Ver. Hambg. (N.F.)26:373-392. [1983]

Walther, J. R. 1984. The antennal sensilla of the Plathythyreini (Ponerinae) in comparison to those of other ants (Hym., Formicoidea). [Abstract.] P. 28 in: XVII International Congress of Entomology. Hamburg, Federal Republic of Germany, August 20-26, 1984. Abstract Volume. Hamburg: XVII International Congress of Entomology, 960 pp. [1984.08.10]

Walther, J. R. 1985. The antennal pattern of sensilla of ants (Formicoidea, Hymenoptera). Mitt. Dtsch. Ges. Allg. Angew. Entomol. 4:173-176. [1985.08]

Walther, J. R. 1994. The antennal pattern of sensilla of *Nothomyrmecia macrops* in comparison to those of other ants from the Myrmeciinae and Ponerinae. P. 360 in: Lenoir, A., Arnold, G., Lepage, M. (eds.) Les insectes sociaux. 12ème congrès de l'Union Internationale pour l'Étude des Insectes Sociaux. Paris, Sorbonne, 21-27 août 1994. Paris: Université Paris Nord, xxiv + 583 pp. [1994.09]

Wang, C. 1995. [Untitled. *Leptothorax reduncus*, new combination, attributed to Wang.] P. 243 in: Bolton, B. A new general catalogue of the ants of the world. Cambridge, Mass.: Harvard University Press, 504 pp. [1995.10]

Wang, C., Wu, J. 1991. Taxonomic studies on the genus *Polyrhachis* Mayr of China (Hymenoptera, Formicidae). [In Chinese.] For. Res. 4:596-601. [1991.12]

Wang, C., Wu, J. 1992a. A new species of the ant genus *Acropyga* Roger (Hymenoptera: Formicidae) of China. [In Chinese.] Sci. Silvae Sin. 28:226-229. [1992.05]

Wang, C., Wu, J. 1992b. Two new species of genus *Camponotus* Mayr (Hymenoptera: Formicidae) from China. [Abstract.] P. 56 in: Proceedings XIX International Congress of Entomology. Abstracts. Beijing, China, June 28 - July 4, 1992. Beijing: XIX International Congress of Entomology, xi + 730 pp. [1992.07.04]

Wang, C., Wu, J. 1992c. Ants of the Jianfengling forest region in Hainan Province (Hymenoptera: Formicidae). [In Chinese.] Sci. Silvae Sin. 28:561-564. [1992.11]

Wang, C., Wu, J. 1994. Second revisionary studies on genus *Camponotus* Mayr of China (Hymenoptera: Formicidae). J. Beijing For. Univ. (Engl. Ed.) 3(1):23-34. [1994.06]

Wang, C., Xiao, G., Wu, J. 1989a. Taxonomic studies on the genus *Camponotus* Mayr in China (Hymenoptera, Formicidae). [part] [In Chinese.] For. Res. 2:221-228. [1989.06]

Wang, C., Xiao, G., Wu, J. 1989b. Taxonomic studies on the genus *Camponotus* Mayr in China (Hymenoptera, Formicidae). [concl.] [In Chinese.] For. Res. 2:321-328. [1989.08]

Wang, M. 1992. Hymenoptera: Formicidae. Pp. 677-682 in: Huang, F. (ed.) Insects of Wuling Mountains area, southwestern China. [In Chinese.] Beijing: Science Press, x + 777 pp. [1992.12]

Wang, M. 1993a. Taxonomic study of the ant tribe Odontomachini in China (Hymenoptera: Formicidae). [In Chinese.] Sci. Treatise Syst. Evol. Zool. 2:219-230. [1993.04]

Wang, M. 1993b. Two new species and three new records of Myrmicinae from China (Hymenoptera: Formicidae). [In Chinese.] Sinozoologia 10:433-436. [1993.05]

Wang, M. 1993c. Hymenoptera: Formicidae. Pp. 740-748 in: Huang, C.-M. (ed.) Animals of Longqi Mountain. [In Chinese.] Beijing: China Forestry Publishing House, 1105 pp. [1993.08]

Wang, M., Wu, J. 1988. [Untitled. Descriptions of new taxa: *Tetramorium reduncum* Wang et Wu, new species; *Tetramorium crepum* Wang et Wu, new species.] Pp. 268, 269-270 in: Wang, M., Xiao, G., Wu, J. Taxonomic studies on the genus *Tetramorium* Mayr in China (Hymenoptera, Formicidae). [In Chinese.] For. Res. 1:264-274. [1988.06]

Wang, M., Xiao, G. 1988. [Untitled. Descriptions of new taxa: *Tetramorium repletum* Wang et Xiao, new species; *Tetramorium jiangxiense* Wang et Xiao, new species.] Pp. 266, 269 in: Wang, M., Xiao, G., Wu, J. Taxonomic studies on the genus *Tetramorium* Mayr in China (Hymenoptera, Formicidae). [In Chinese.] For. Res. 1:264-274. [1988.06]

Wang, M., Xiao, G., Wu, J. 1988. Taxonomic studies on the genus *Tetramorium* Mayr in China (Hymenoptera, Formicidae). [In Chinese.] For. Res. 1:264-274. [1988.06]

Wang, W.-I. 1980. Phylum Arthropoda. Pp. 59-153 in: Paleontological atlas of Northeast China. 2. Mesozoic and Cenozoic volume. [In Chinese.] Beijing: Geological Publishing House, [14] + 403 pp. [1980.10] [Ants pp. 152-153.]

Ward, P. S. 1980a. A systematic revision of the *Rhytidoponera impressa* group (Hymenoptera: Formicidae) in Australia and New Guinea. Aust. J. Zool. 28:475-498. [1980.08.26]

Ward, P. S. 1980b. Genetic variation and population differentiation in the *Rhytidoponera impressa* group, a species complex of ponerine ants (Hymenoptera: Formicidae). Evolution 34:1060-1076. [1980.12.26]

Ward, P. S. 1981a. Ecology and life history of the *Rhytidoponera impressa* group. I. Habitats, nest sites, and foraging behavior. Psyche (Camb.) 88:89-108. [1981.12.28]

Ward, P. S. 1981b. Ecology and life history of the *Rhytidoponera impressa* group. II. Colony origin, seasonal cycles, and reproduction. Psyche (Camb.) 88:109-126. [1981.12.28]

Ward, P. S. 1983a. Genetic relatedness and colony organization in a species complex of ponerine ants. I. Genotypic and phenotypic composition of colonies. Behav. Ecol. Sociobiol. 12:285-299. [1983.07]

Ward, P. S. 1983b. Genetic relatedness and colony organization in a species complex of ponerine ants. II. Patterns of sex ratio investment. Behav. Ecol. Sociobiol. 12:301-307. [1983.07]

Ward, P. S. 1984. A revision of the ant genus *Rhytidoponera* (Hymenoptera: Formicidae) in New Caledonia. Aust. J. Zool. 32:131-175. [1984.03.09]

Ward, P. S. 1985a. Taxonomic congruence and disparity in an insular ant fauna: *Rhytidoponera* in New Caledonia. Syst. Zool. 34:140-151. [1985.07.18]

Ward, P. S. 1985b. The Nearctic species of the genus *Pseudomyrmex* (Hymenoptera: Formicidae). Quaest. Entomol. 21:209-246. [1985]

Ward, P. S. 1986. Functional queens in the Australian greenhead ant, *Rhytidoponera metallica* (Hymenoptera: Formicidae). Psyche (Camb.) 93:1-12. [1986.10.10]

Ward, P. S. 1987. Distribution of the introduced Argentine ant (*Iridomyrmex humilis*) in natural habitats of the lower Sacramento Valley and its effects on the indigenous ant fauna. Hilgardia 55(2):1-16. [1987.04]

Ward, P. S. 1988. Mesic elements in the western Nearctic ant fauna: taxonomic and biological notes on *Amblyopone*, *Proceratium*, and *Smithistruma* (Hymenoptera: Formicidae). J. Kansas Entomol. Soc. 61:102-124. [1988.03.04]

Ward, P. S. 1989a. Systematic studies on pseudomyrmecine ants: revision of the *Pseudomyrmex oculatus* and *P. subtilissimus* species groups, with taxonomic comments on other species. Quaest. Entomol. 25:393-468. [1989.12]

Ward, P. S. 1989b. Genetic and social changes associated with ant speciation. Pp. 123-148 in: Breed, M. D., Page, R. E. (eds.) The genetics of social evolution. Boulder: Westview Press, viii + 213 pp. [1989]

Ward, P. S. 1990. The ant subfamily Pseudomyrmecinae (Hymenoptera: Formicidae): generic revision and relationship to other formicids. Syst. Entomol. 15:449-489. [1990.11]

Ward, P. S. 1991. Phylogenetic analysis of pseudomyrmecine ants associated with domatia-bearing plants. Pp. 335-352 in: Huxley, C. R., Cutler, D. F. (eds.) Ant-plant interactions. Oxford: Oxford University Press, xviii + 601 pp. [1991]

Ward, P. S. 1992. Ants of the genus *Pseudomyrmex* (Hymenoptera: Formicidae) from Dominican amber, with a synopsis of the extant Antillean species. Psyche (Camb.) 99:55-85. [1992.12.04]

Ward, P. S. 1993. Systematic studies on *Pseudomyrmex* acacia-ants (Hymenoptera: Formicidae: Pseudomyrmecinae). J. Hym. Res. 2:117-168. [1993.11.17] [Errata in J. Hym. Res. 3:309-310. (1994.10.15)]

Ward, P. S. 1994. *Adetomyrma*, an enigmatic new ant genus from Madagascar (Hymenoptera: Formicidae), and its implications for ant phylogeny. Syst. Entomol. 19:159-175. [1994.08.01]

Ward, P. S. 1995. Ants on the run. [Review of: Gotwald, W. H., Jr. 1995. Army ants: the biology of social predation. Ithaca, New York: Cornell University Press, xviii + 302 pp.] Science (Wash. D. C.) 270:319-320. [1995.10.13]

Ward, P. S., Taylor, R. W. 1981. Allozyme variation, colony structure and genetic relatedness in the primitive ant *Nothomyrmecia macrops* Clark (Hymenoptera: Formicidae). J. Aust. Entomol. Soc. 20:177-183. [1981.08.24]

Ware, A. B. 1994. Factors eliciting stridulation by the ponerine ant *Streblognathus aethiopicus* Smith (Hymenoptera: Formicidae). Afr. Entomol. 2:31-36. [1994.03]

Warming, E. 1894 ("1893"). Om et par af Myrer beboede Traeer. Vidensk. Medd. Naturhist. Foren. Kbh. (5)5:173-187. [1894]

Warren, L. O., Rouse, E. P. 1969. The ants of Arkansas. Univ. Arkansas Agric. Exp. Stn. Bull. 742:1-68. [1969.06]

Wasmann, E. 1884. *Monomorium Pharaonis* in Aachen. Nat. Offenbarung 30:572-573. [1884.09]

Wasmann, E. 1888a. Diebsameisen und Gastameisen. Nat. Offenbarung 34:321-331. [1888.06]

Wasmann, E. 1888b. Diebsameisen und Gastameisen. Schluss. Nat. Offenbarung 34:321-331. [1888.09]

Wasmann, E. 1889a. Die sklavenhaltenden Ameisen. Nat. Offenbarung 35:1-11. [1889.01]

Wasmann, E. 1889b. Die sklavenhaltenden Ameisen. Fortsetzung und Schluss. Nat. Offenbarung 35:471-486. [1889.08]

Wasmann, E. 1889c. Die Wechselbeziehungen zwischen Pflanzen und Ameisen im tropischen Amerika. Nat. Offenbarung 35:487-489. [1889.08]

Wasmann, E. 1889d. Ein kleiner Beitrag zur niederländischen Ameisenfauna. Tijdschr. Entomol. 32:19. [1889]

Wasmann, E. 1891a ("1890"). Einige neue Hermaphroditen von *Myrmica scabrinodis* und *laevinodis*. Stett. Entomol. Ztg. 51:298-299. [1891.01]

Wasmann, E. 1891b ("1890"). Ueber die verschiedenen Zwischenformen von Weibchen und Arbeiterinnen bei Ameisen. Stett. Entomol. Ztg. 51:300-309. [1891.01]

Wasmann, E. 1891c. Verzeichnis der Ameisen und Ameisengäste von Holländisch Limburg. Tijdschr. Entomol. 34:39-64. [1891]

Wasmann, E. 1891d. Die zusammengesetzten Nester und gemischten Kolonien der Ameisen. Ein Beitrag zur Biologie, Psychologie und Entwicklungsgeschichte der Ameisengesellschaften. Münster in Westfalen: Aschendorffschen Buchdruckerei, vii + 262 pp. [1891]

Wasmann, E. 1892. Einiges über springende Ameisen. Wien. Entomol. Ztg. 11:316-317. [1892.12.25]

Wasmann, E. 1894. Kritisches Verzeichniss der myrmekophilen und termitophilen Arthropoden. Berlin: F. L. Dames, xv + 231 pp. [1894]

Wasmann, E. 1895a. Die Ameisen- und Termitengäste von Brasilien. Verh. K-K. Zool.-Bot. Ges. Wien 45:137-179. [1895.04]

Wasmann, E. 1895b. Die ergatogynen Formen bei den Ameisen und ihre Erklärung. Biol. Centralbl. 15:606-622 [1895.08.15]

Wasmann, E. 1895c. Die ergatogynen Formen bei den Ameisen und ihre Erklärung. (Schluss.) Biol. Centralbl. 15:625-646. [1895.09.01]

Wasmann, E. 1897a. Bemerkungen über einige Ameisen von Madagascar. Zool. Anz. 20:249-250. [1897.07.19]

Wasmann, E. 1897b. Über ergatoide Weibchen und Pseudogynen bei den Ameisen. Zool. Anz. 20:251-253. [1897.07.19]

Wasmann, E. 1899. Zur Kenntniss der bosnischen Myrmekophilen und Ameisen. Wiss. Mitt. Bosnien Herceg. 6:767-772. [1899]

Wasmann, E. 1901. Neues über die zusammengesetzten Nester und gemischten Kolonien der Ameisen. [part] Allg. Z. Entomol. 6:369-371. [1901.12.15]

Wasmann, E. 1902a. Neues über die zusammengesetzten Nester und gemischten Kolonien der Ameisen. [part] Allg. Z. Entomol. 7:1-5. [1902.01.01]

Wasmann, E. 1902b. Neues über die zusammengesetzten Nester und gemischten Kolonien der Ameisen. [part] Allg. Z. Entomol. 7:167-173. [1902.05.01]

Wasmann, E. 1902c. Biologische und phylogenetische Bemerkungen über die Dorylinengäste der alten und der neuen Welt, mit specieller Berücksichtigung ihrer Convergenzerscheinungen. Verh. Dtsch. Zool. Ges. 1902:86-98. [1902.05.22]

Wasmann, E. 1902d. Neues über die zusammengesetzten Nester und gemischten Kolonien der Ameisen. [part] Allg. Z. Entomol. 7:206-208. [1902.06.01]

Wasmann, E. 1902e. Zur Ameisenfauna von Helgoland. Dtsch. Entomol. Z. 1902:63-64. [1902.07]

Wasmann, E. 1902f. Neues über die zusammengesetzten Nester und gemischten Kolonien der Ameisen. [part] Allg. Z. Entomol. 7:293-298. [1902.08.15] [Contains footnoted descriptions by Forel (pp. 294-295, 296) and Emery (p. 297).]

Wasmann, E. 1904a. Neue Beiträge zur Kenntniss der Paussiden, mit biologischen und phylogenetischen Bemerkungen. Notes Leyden Mus. 25:1-82. [1904.10] [Pp. 72-73: *Pheidole megacephala impressiceps*, new subspecies.]

Wasmann, E. 1904b. Zur Kenntniss der Gäste der Treiberameisen und ihrer Wirthe am obern Congo, nach den Sammlungen und Beobachtungen von P. Herm. Kohl, C.S.S.C. bearbeitet. Zool. Jahrb. Suppl. 7:611-682. [1904]

Wasmann, E. 1905a. Ursprung und Entwickelung der Sklaverei bei den Ameisen. Biol. Centralbl. 25:117-127. [1905.02.15]

Wasmann, E. 1905b. Ursprung und Entwickelung der Sklaverei bei den Ameisen. (Fortsetzung.) Biol. Centralbl. 25:129-144. [1905.03.01]

Wasmann, E. 1905c. Ursprung und Entwickelung der Sklaverei bei den Ameisen. (Fortsetzung.) Biol. Centralbl. 25:161-169. [1905.03.15]

Wasmann, E. 1905d. Ursprung und Entwickelung der Sklaverei bei den Ameisen. (Fortsetzung.) Biol. Centralbl. 25:193-216. [1905.04.01]

Wasmann, E. 1905e. Ursprung und Entwickelung der Sklaverei bei den Ameisen. (Fortsetzung.) Biol. Centralbl. 25:256-270. [1905.04.15]

Wasmann, E. 1905f. Berichtigungen zu Note I dieses Bandes. Notes Leyden Mus. 25:110. [1905.04.15]

Wasmann, E. 1905g. Ursprung und Entwickelung der Sklaverei bei den Ameisen. (Schluss.) Biol. Centralbl. 25:273-292. [1905.04.15]

Wasmann, E. 1905h. Nochmals zur Frage über die temporär gemischten Kolonien und den Ursprung der Sklaverei bei den Ameisen. Biol. Centralbl. 25:644-653. [1905.10.01]

Wasmann, E. 1906. Zur Kenntniss der Ameisen und Ameisengäste von Luxemburg. Arch. Inst. Grand-Ducal Luxemb. (n.s.)1(1/2):104-124. [1906.06]

Wasmann, E. 1909a. Über den Ursprung des sozialen Parasitismus, der Sklaverei und der Myrmekophilie bei den Ameisen. Biol. Centralbl. 29:587-604. [1909.10.01]

Wasmann, E. 1909b. Über den Ursprung des sozialen Parasitismus, der Sklaverei und der Myrmekophilie bei den Ameisen. (Fortsetzung.) Biol. Centralbl. 29:619-637. [1909.10.15]

Wasmann, E. 1909c. Über den Ursprung des sozialen Parasitismus, der Sklaverei und der Myrmekophilie bei den Ameisen. (Fortsetzung.) Biol. Centralbl. 29:651-663. [1909.11.01]

Wasmann, E. 1909d. Über den Ursprung des sozialen Parasitismus, der Sklaverei und der Myrmekophilie bei den Ameisen. (Schluss.) Biol. Centralbl. 29:683-703. [1909.11.15]

Wasmann, E. 1909e. Zur Kenntniss der Ameisen und Ameisengäste von Luxemburg. III. Verzeichniss der Ameisen von Luxemburg, mit biologischen Notizen. Arch. Inst. Grand-Ducal Luxemb. (n.s.)4(3/4):1-114. [1909]

Wasmann, E. 1910. Nachträge zum sozialen Parasitismus und der Sklaverei bei den Ameisen. [part] Biol. Centralbl. 30:515-524. [1910.08.01]

Wasmann, E. 1911a. Zur Kenntnis der Termiten und Termitengäste vom belgischen Congo. [part] Rev. Zool. Afr. (Bruss.) 1:91-117. [1911.04.10]

Wasmann, E. 1911b. Ein neuer Paussus aus Ceylon, mit einer Uebersicht über die Paussidenwirte. Tijdschr. Entomol. 54:195-207. [1911.12.31]

Wasmann, E. 1913 ("1912"). The ants and their guests. Annu. Rep. Smithson. Inst. 1912:455-474. [1913]

Wasmann, E. 1915a. Revision der Gattung *Aenictonia* Wasm. Entomol. Mitt. 4:26-35. [1915.03.08]

Wasmann, E. 1915b. Zwei für Holland neue Ameisen, mit anderen Bemerkungen über Ameisen und deren Gäste aus Süd-Limburg. Tijdschr. Entomol. 58:150-162. [1915.03.15]

Wasmann, E. 1915c. Eine neue *Pseudomyrma* aus der Ochsenhorndornakazie in Mexiko, mit Bemerkungen über Ameisen in Akaziendornen und ihre Gäste. Ein kritischer Beitrag zur Pflanzen-Myrmekophilie. Tijdschr. Entomol. 58:296-325. [1915.10.15]

Wasmann, E. 1915d. *Anergatides Kohli*, eine neue arbeiterlose Schmarotzerameise vom oberen Kongo (Hym., Form.). Entomol. Mitt. 4:279-288. [1915.12.27]

Wasmann, E. 1915e. Das Gesellschaftsleben der Ameisen. Das Zusammenleben von Ameisen verschiedener Arten und von Ameisen und Termiten. Gesammelte Beiträge zur sozialen Symbiose bei den Ameisen. I. Band. Münster in Westfalen: Aschendorffsche Verlagsbuchhandlung, xviii + 413 pp. [1915]

Wasmann, E. 1916a. Nachtrag zu "Eine neue *Pseudomyrma* aus der Ochsenhorndornakazie in Mexiko." Tijdschr. Entomol. 58(suppl.):125-131. [1916.03.01]

Wasmann, E. 1916b. Neue dorylophile Staphyliniden Afrikas. Entomol. Mitt. 5:134-147. [1916.07.15]

Wasmann, E. 1917. Neue Anpassungstypen bei Dorylinengästen Afrikas (Col. Staphylinidae). Z. Wiss. Zool. 117:257-360. [1917.06.11]

Wasmann, E. 1918a. Ueber *Solenopsis geminata saevissima* Sm. und ihre Gäste. Entomol. Bl. Biol. Syst. Käfer 14:69-76. [1918.03.15]

Wasmann, E. 1918b. Ueber die von v. Rothkirch, 1912, in Kamerun gesammelten Myrmekophilen. Entomol. Mitt. 7:135-149. [1918.08.15]

Wasmann, E. 1931. *Acromyrmex Bucki*, n. sp. (Hym., Formicidae). Rev. Entomol. (Rio J.) 1:106. [1931.04.25]

Wasmann, E. 1934. Die Ameisen, die Termiten und ihre Gäste. Regensburg: G. J. Manz, xviii + 148 pp. [1934.10] [Reviewed by Stumper (1934) in October 1934.]

Watanabe, F. 1915. Ants of Okazaki, Aichi Prefecture. [In Japanese.] Insect World 19:480. [1915.11.15]

Waterhouse, F. H. 1937. List of the dates of delivery of the sheets of the 'Proceedings' of the Zoological Society of London, from commencement in 1830 to 1859 inclusive. Proc. Zool. Soc. Lond. Ser. A 107:78-83. [1937.04.15]

Watkins, J. F., II. 1964. Laboratory experiments on the trail following of army ants of the genus *Neivamyrmex* (Formicidae: Dorylinae). J. Kansas Entomol. Soc. 37:22-28. [1964.03.26]

Watkins, J. F., II. 1968. The rearing of the army ant male, *Neivamyrmex harrisi* (Haldeman) from larvae collected from a nest of *N. wheeleri* (Emery). Am. Midl. Nat. 80:273-276. [1968.07.25]

Watkins, J. F., II. 1969 ("1968"). A new species of *Neivamyrmex* (Hymenoptera: Formicidae) from Louisiana. J. Kansas Entomol. Soc. 41:528-531. [1969.01.10]

Watkins, J. F., II. 1971. A taxonomic review of *Neivamyrmex moseri*, *N. pauxillus*, and *N. leonardi*, including new distribution records and original descriptions of queens of the first two species. J. Kansas Entomol. Soc. 44:93-103. [1971.02.26]

Watkins, J. F., II. 1972. The taxonomy of *Neivamyrmex texanus*, n. sp., *N. nigrescens* and *N. californicus* (Formicidae: Dorylinae), with distribution map and keys to the species of *Neivamyrmex* of the United States. J. Kansas Entomol. Soc. 45:347-372. [1972.07.18]

Watkins, J. F., II. 1973. *Neivamyrmex baylori*, n. sp. (Formicidae: Dorylinae) from Waco, Texas, U.S.A. J. Kansas Entomol. Soc. 46:430-433. [1973.07.31]

Watkins, J. F., II. 1974. *Neivamyrmex angulimandibulatus*, new species (Formicidae: Dorylinae) from Cordoba, Mexico. Southwest. Nat. 19:309-312. [1974.09.23]

Watkins, J. F., II. 1975a. *Neivamyrmex cornutus*, n. sp. (Formicidae: Dorylinae) from Oaxaca, Mexico. J. Kansas Entomol. Soc. 48:92-95. [1975.01.22]

Watkins, J. F., II. 1975b. *Neivamyrmex digitistipus,* n. sp. (Formicidae: Dorylinae) from Costa Rica. Tex. J. Sci. 26:203-206. [1975.02]

Watkins, J. F., II. 1975c. *Neivamyrmex quadratooccipututs*, n. sp. (Formicidae: Dorylinae) from El Salvador. Tex. J. Sci. 26:207-211. [1975.02]

Watkins, J. F., II. 1975d. The relationship of *Neivamyrmex fuscipennis* to *N. macropterus* (Dorylinae: Formicidae). Southwest. Nat. 20:85-90. [1975.05.15]

Watkins, J. F., II. 1976. The identification and distribution of New World army ants (Dorylinae: Formicidae). Waco, Texas: Baylor University Press, 102 pp. [1976]

Watkins, J. F., II. 1977a. The species and subspecies of *Nomamyrmex* (Dorylinae: Formicidae). J. Kansas Entomol. Soc. 50:203-214. [1977.04.29]

Watkins, J. F., II. 1977b. *Neivamyrmex nyensis*, n. sp. (Formicidae: Dorylinae) from Nye County, Nevada, U.S.A. Southwest. Nat. 22:421-425. [1977.10.15]

Watkins, J. F., II. 1982. The army ants of Mexico (Hymenoptera: Formicidae: Ecitoninae). J. Kansas Entomol. Soc. 55:197-247. [1982.04.20]

Watkins, J. F., II. 1985. The identification and distribution of the army ants of the United States of America (Hymenoptera, Formicidae, Ecitoninae). J. Kansas Entomol. Soc. 58:479-502. [1985.07.31]

Watkins, J. F., II. 1986. *Neivamyrmex chamelensis*, n. sp. (Hymenoptera: Formicidae: Ecitoninae) from Jalisco, Mexico. J. Kansas Entomol. Soc. 59:361-366. [1986.05.28]

Watkins, J. F., II. 1990a ("1988"). The army ants (Formicidae: Ecitoninae) of the Chamela Biological Station in Jalisco, Mexico. Folia Entomol. Mex. 77:379-393. [1990.04.30]

Watkins, J. F., II. 1990b. *Neivamyrmex crassiscapus*, n. sp. (Hymenoptera: Formicidae: Ecitoninae) from Mexico. J. Kansas Entomol. Soc. 63:348-350. [1990.05.31]

Watkins, J. F., II. 1994 ("1993"). *Neivamyrmex curvinotus*, n. sp. (Hymenoptera: Formicidae: Ecitoninae) from South America. J. Kansas Entomol. Soc. 66:411-413. [1994.03.29]

Watkins, J. F., II., Cole, T. W., Baldridge, R. S. 1967. Laboratory studies on interspecies trail following and trail preference of army ants (Dorylinae). J. Kansas Entomol. Soc. 40:146-151. [1967.04.28]

Watkins, J. F., II., Coody, C. J. 1986. The taxonomy of *Neivamyrmex graciellae* (Mann) (Hymenoptera: Formicidae: Ecitoninae) including an original description of the queen and field observations. Southwest. Nat. 31:256-259. [1986.05.22]

Watson, E. M. 1945. The dates of publication of the Journal of the West Australian Natural History Society, the Journal of the Natural History and Science Society of Western Australia and the

Journal of the Royal Society of Western Australia. J. R. Soc. West. Aust. 29:174-175. [1945.07.20]

Weber, N. A. 1934a. Notes on neotropical ants, including the descriptions of new forms. Rev. Entomol. (Rio J.) 4:22-59. [1934.04.10]

Weber, N. A. 1934b. A new *Strumigenys* from Illinois (Hymenoptera: Formicidae). Psyche (Camb.) 41:63-65. [1934.06] [Stamp date in MCZ library: 1934.08.30.]

Weber, N. A. 1935. The biology of the thatching ant, *Formica rufa obscuripes* Forel, in North Dakota. Ecol. Monogr. 5:165-206. [1935.04]

Weber, N. A. 1937. The biology of the fungus-growing ants. Part I. New forms. Rev. Entomol. (Rio J.) 7:378-409. [1937.10.11]

Weber, N. A. 1938a. New ants from stomachs of *Bufo marinus* L. and *Typhlops reticulatus* (L). Ann. Entomol. Soc. Am. 31:207-210. [1938.06.29]

Weber, N. A. 1938b. The biology of the fungus-growing ants. Part IV. Additional new forms. Part V. The Attini of Bolivia. Rev. Entomol. (Rio J.) 9:154-206. [1938.09.27]

Weber, N. A. 1938c. The food of the giant toad, *Bufo marinus* (L.), in Trinidad and British Guiana with special reference to the ants. Ann. Entomol. Soc. Am. 31:499-503. [1938.12.23]

Weber, N. A. 1939a. New ants of rare genera and a new genus of ponerine ants. Ann. Entomol. Soc. Am. 32:91-104. [1939.03.20]

Weber, N. A. 1939b. Description of new North American species and subspecies of *Myrmica* Latreille (Hym.: Formicidae). Lloydia 2:144-152. [1939.07.20]

Weber, N. A. 1939c. Tourist ants. Ecology 20:442-446. [1939.07]

Weber, N. A. 1940a. Ants on a Nile River steamer. Ecology 21:292-293. [1940.04]

Weber, N. A. 1940b. The biology of the fungus-growing ants. Part VI. Key to *Cyphomyrmex*, new Attini and a new guest ant. Rev. Entomol. (Rio J.) 11:406-427. [1940.06.28]

Weber, N. A. 1940c. Rare ponerine genera in Panama and British Guiana (Hym.: Formicidae). Psyche (Camb.) 47:75-84. [1940.09] [Stamp date in MCZ library: 1940.11.14.]

Weber, N. A. 1941a. Four new genera of Ethiopian and Neotropical Formicidae. Ann. Entomol. Soc. Am. 34:183-194. [1941.04.07]

Weber, N. A. 1941b. The biology of the fungus-growing ants. Part VII. The Barro Colorado Island, Canal Zone, species. Rev. Entomol. (Rio J.) 12:93-130. [1941.07.31]

Weber, N. A. 1941c. The rediscovery of the queen of *Eciton* (*Labidus*) *coecum* Latr. (Hym.: Formicidae). Am. Midl. Nat. 26:325-329. [1941.09.30]

Weber, N. A. 1942a. A biocoenose of papyrus heads (*Cyperus papyrus*). Ecology 23:115-119. [1942.01]

Weber, N. A. 1942b. New doryline, cerapachyine and ponerine ants from the Imatong Mountains, Anglo-Egyptian Sudan. Proc. Entomol. Soc. Wash. 44:40-49. [1942.03.31]

Weber, N. A. 1942c. The genus *Thaumatomyrmex* Mayr with description of a Venezuelan species (Hym.: Formicidae). Bol. Entomol. Venez. 1:65-71. [1942.08.30]

Weber, N. A. 1943a. The queen of a British Guiana *Eciton* and a new ant garden *Solenopsis* (Hymenoptera: Formicidae). Proc. Entomol. Soc. Wash. 45:88-91. [1943.04.30]

Weber, N. A. 1943b. New ants from Venezuela and neighboring countries. Bol. Entomol. Venez. 2:67-78. [1943.06.30]

Weber, N. A. 1943c. Parabiosis in Neotropical "ant gardens". Ecology 24:400-404. [1943.07]

Weber, N. A. 1943d. The ants of the Imatong Mountains, Anglo-Egyptian Sudan. Bull. Mus. Comp. Zool. 93:263-389. [1943.12]

Weber, N. A. 1944a. The tree ants (*Dendromyrmex*) of South and Central America. Ecology 25:117-120. [1944.01]

Weber, N. A. 1944b. The neotropical coccid-tending ants of the genus *Acropyga* Roger. Ann. Entomol. Soc. Am. 37:89-122. [1944.04.10]

Weber, N. A. 1945. The biology of the fungus-growing ants. Part VIII. The Trinidad, B. W. I., species. Rev. Entomol. (Rio J.) 16:1-88. [1945.08.30]

Weber, N. A. 1946a. Two common ponerine ants of possible economic significance, *Ectatomma tuberculatum* (Olivier) and *E. ruidum* Roger. Proc. Entomol. Soc. Wash. 48:1-16. [1946.01.31]

Weber, N. A. 1946b. Dimorphism in the African *Oecophylla* worker and an anomaly (Hym.: Formicidae). Ann. Entomol. Soc. Am. 39:7-10. [1946.03.29]

Weber, N. A. 1946c. The biology of the fungus-growing ants. Part IX. The British Guiana species. Rev. Entomol. (Rio J.) 17:114-172. [1946.08.26]

Weber, N. A. 1947a. Binary anterior ocelli in ants. Biol. Bull. (Woods Hole) 93:112-113. [1947.10]

Weber, N. A. 1947b. A revision of the North American ants of the genus *Myrmica* Latreille with a synopsis of the Palearctic species. I. Ann. Entomol. Soc. Am. 40:437-474. [1947.11.21]

Weber, N. A. 1947c. Lower Orinoco River fungus-growing ants (Hymenoptera: Formicidae, Attini). Bol. Entomol. Venez. 6:143-161. [1947.12.31]

Weber, N. A. 1948a. A revision of the North American ants of the genus *Myrmica* Latreille with a synopsis of the Palearctic species. II. Ann. Entomol. Soc. Am. 41:267-308. [1948.10.19]

Weber, N. A. 1948b. Studies on the fauna of Curaçao, Aruba, Bonaire and the Venezuelan islands: No. 14. Ants from the Leeward Group and some other Caribbean localities. Natuurwet. Studiekring Suriname Ned. Antillen 5:78-86. [1948]

Weber, N. A. 1949a. New African ants of the genera *Cerapachys*, *Phryacaces*, and *Simopone*. Am. Mus. Novit. 1396:1-9. [1949.01.21]

Weber, N. A. 1949b. New ponerine ants from equatorial Africa. Am. Mus. Novit. 1398:1-9. [1949.01.25]

Weber, N. A. 1949c. The functional significance of dimorphism in the African ant, *Oecophylla*. Ecology 30:397-400. [1949.08.12]

Weber, N. A. 1949d. A new Panama *Eciton* (Hymenoptera, Formicidae). Am. Mus. Novit. 1441: 1-8. [1949.12.22]

Weber, N. A. 1950a. The African species of the genus *Oligomyrmex* Mayr (Hymenoptera, Formicidae). Am. Mus. Novit. 1442:1-19. [1950.01.04]

Weber, N. A. 1950b. Ants from Saipan, Marianas Islands. Entomol. News 61:99-102. [1950.07.19]

Weber, N. A. 1950c. A revision of the North American ants of the genus *Myrmica* Latreille with a synopsis of the Palearctic species. III. Ann. Entomol. Soc. Am. 43:189-226. [1950.08.08]

Weber, N. A. 1950d. New Trinidad Myrmicinae, with a note on Basiceros Schulz (Hymenoptera, Formicidae). Am. Mus. Novit. 1465:1-6. [1950.08.16]

Weber, N. A. 1950e. A survey of the insects and related arthropods of Arctic Alaska. Part I. Trans. Am. Entomol. Soc. 76:147-206. [1950.10.11]

Weber, N. A. 1952a. Studies on African Myrmicinae, I (Hymenoptera, Formicidae). Am. Mus. Novit. 1548:1-32. [1952.04.07]

Weber, N. A. 1952b. Biological notes on Dacetini (Hymenoptera, Formicidae). Am. Mus. Novit. 1554:1-7. [1952.05.01]

Weber, N. A. 1952c. Observations on Baghdad ants. Baghdad: The Trading and Printing Co., 30 pp. [1952]

Weber, N. A. 1958a. Nomenclatural notes on *Proatta* and *Atta* (Hym.: Formicidae). Entomol. News 69:7-13. [1958.01.18]

Weber, N. A. 1958b. Nomenclatural changes in *Trachymyrmex* (Hym.: Formicidae). Entomol. News 69:49-55. [1958.02.21]

Weber, N. A. 1958c ("1957"). The nest of an anomalous colony of the arboreal ant *Cephalotes atratus*. Psyche (Camb.) 64:60-69. [1958.05.01]

Weber, N. A. 1958d. Synonymies and types of *Apterostigma* (Hym: Formicidae). Entomol. News 69:243-251. [1958.11.12]

Weber, N. A. 1958e. Some attine synonyms and types (Hymenoptera, Formicidae). Proc. Entomol. Soc. Wash. 60:259-264. [1958.12.18]

Weber, N. A. 1958f. Evolution in fungus-growing ants. Pp. 459-473 in: Becker, E. C. et al. (eds.) Proceedings of the Tenth International Congress of Entomology, Montreal, August 17-25, 1956. Volume 2. Ottawa: Mortimer Ltd., 1055 pp. [1958.12]

Weber, N. A. 1963. Argentine myrmecology. Entomol. News 74:205-208. [1963.10.02]

Weber, N. A. 1964. Termite prey of some African ants. Entomol. News 75:197-204. [1964.09.24]

Weber, N. A. 1965. Note on the European pavement ant, *Tetramorium caespitum*, in the Philadelphia area (Hymenoptera: Formicidae). Entomol. News 76:137-139. [1965.05.05]

Weber, N. A. 1966a. The subgenus *Cyphomannia* Weber 1938 of *Cyphomyrmex* Mayr 1862, reinstated, and systematic notes (Hymenoptera: Formicidae). Entomol. News 77:166-168. [1966.06.14]

Weber, N. A. 1966b. Fungus-growing ants. Science (Wash. D. C.) 153:587-604. [1966.08.05]

Weber, N. A. 1966c. Development of pigmentation in the pupa and callow of *Trachymyrmex septentrionalis* (Hymenoptera: Formicidae). Entomol. News 77:241-246. [1966.11.15]

Weber, N. A. 1967. Synonyms of *Trachymyrmex bugnioni* Forel and *Trachymyrmex diversus* Mann (Hymenoptera: Formicidae). Proc. Entomol. Soc. Wash. 69:273-274. [1967.10.13]

Weber, N. A. 1968a. Tobago Island fungus-growing ants (Hymenoptera: Formicidae). Entomol. News 79:141-145. [1968.06.07]

Weber, N. A. 1968b. The Panamanian *Atta* species (Hymenoptera: Formicidae). Proc. Entomol. Soc. Wash. 70:348-350. [1968.12.18]

Weber, N. A. 1969a. A comparative study of the nests, gardens and fungi of the fungus-growing ants, Attini. Pp. 299-307 in: Ernst, E. et al. (eds.) International Union for the Study of Social Insects. VI Congress. Bern 15-20 September 1969. Proceedings. Bern: Organizing Committee of the VI Congress IUSSI, 309 pp. [1969.09.15]

Weber, N. A. 1969b. Ecological relations of three *Atta* species in Panama. Ecology 50:141-147. [1969]

Weber, N. A. 1970. Northern extent of attine ants (Hymenoptera: Formicidae). Proc. Entomol. Soc. Wash. 72:414-415. [1970.09.25]

Weber, N. A. 1972. Gardening ants, the attines. Philadelphia: American Philosophical Society, xvii + 146 pp. [1972]

Weber, N. A. 1982. Fungus ants. Pp. 255-363 in: Hermann, H. R. (ed.) Social insects. Volume 4. New York: Academic Press, xiii + 385 pp. [1982]

Weber, N. A., Anderson, J. L. 1950. Studies on central African ants of the genus *Pseudolasius* Emery (Hymenoptera, Formicidae). Am. Mus. Novit. 1443:1-7. [1950.01.04]

Weddell, H. A. 1850. Additions à la flore de l'Amérique du Sud (Suite). [part] Ann. Sci. Nat. Bot. (3)13:257-268. [1850.05] [Page 263: *Myrmica triplarina* (now *Pseudomyrmex triplarinus*), new species.]

Wehner, R. 1981. Verhaltensphysiologische, funktionsmorphologische und taxonomische Studien an zwei Arten der *Cataglyphis albicans*-Gruppe (Formicidae, Hymenoptera). Mitt. Dtsch. Ges. Allg. Angew. Entomol. 3:214-217. [1981.10]

Wehner, R. 1983. Taxonomie, Funktionsmorphologie und Zoogeographie der saharischen Wüstenameise *Cataglyphis fortis* (Forel 1902) stat. nov. (Insecta: Hymenoptera: Formicidae). Senckenb. Biol. (n.s.)64:89-132. [1983.11.30]

Wehner, R. 1986. Visuelle Orientierung bei Wüstenameisen: Art- und kastenspezifische Mechanismen. Jahrb. Akad. Wiss. Lit. Mainz 1986:107-112. [1986]

Wehner, R. 1990. On the brink of introducing sensory ecology: Felix Santschi (1872-1940) - Tabib-en-Neml. Behav. Ecol. Sociobiol. 27:295-306. [1990.10.18]

Wehner, R., Harkness, R. D., Schmid-Hempel, P. 1983. Foraging strategies in individually searching ants, *Cataglyphis bicolor* (Hymenoptera: Formicidae). Information processing in animals. Volume 1. Stuttgart: G. Fischer, 79 pp. [1983]

Wehner, R., Wehner, S., Agosti, D. 1994. Patterns of biogeographic distribution within the *bicolor* species group of the North African desert ant, *Cataglyphis* Foerster 1850 (Insecta: Hymenoptera: Formicidae). Senckenb. Biol. 74:163-191. [1994.12.21]

Weidner, H. 1972. Die Entomologischen Sammlungen des Zoologischen Instituts und Zoologischen Museums der Universität Hamburg. Mitt. Hambg. Zool. Mus. Inst. 68:107-134. [1972.04] [Hymenopteran types (including Formicidae) pp. 131-134.]

Weir, J. S. 1957. The functional anatomy of the mid-gut of larvae of the ant, *Myrmica*. Q. J. Microsc. Sci. 98:499-506. [1957.12]

Welch, R. C. 1991. *Hypoponera punctatissima* (Roger) (Hymenoptera: Formicidae) outdoors in a rural Northamptonshire garden. Entomol. Rec. J. Var. 103:97-98. [1991.03.25]

Welch, R. C., Greatorex-Davies, J. N. 1990. *Lasius brunneus* Lat. (Hymenoptera: Formicidae) rediscovered in Monks Wood National Nature Reserve, Cambs. Entomol. Rec. J. Var. 102:291-292. [1990.11.15]

Wellenius, O. H. 1904 ("1902-03"). För Finland nya eller sällsynta myror. Medd. Soc. Fauna Flora Fenn. 29:124. [1904]

Wellenius, O. H. 1949. Iter entomologicum et botanicum ad insulas Madeiram et Azores anno 1938 a Richard Frey, Ragnar Storå et Carl Cedercreutz factum. No. 19. Die Formiciden von den Azoren und Madeira. Comment. Biol. 8(19):1-4. [1949.10]

Wellenius, O. H. 1955. Entomologische Ergebnisse der finnländischen Kanaren-Expedition 1947-1951. No. 10. Formicidae Insularum Canariensium. Systematik, Ökologie und Verbreitung der Kanarischen Formiciden. Comment. Biol. 15(8):1-20. [1955.03]

Wellenstein, G. 1967. Zur Frage der Standortansprüche hügelbauender Waldameisen (*F. rufa*-Gruppe). Z. Angew. Zool. 54:139-166. [1967]

Wengris, J. 1962. Mrówki (Hymenoptera, Formicidae) rezerwatu torfowiskowego Redykajny pod Olsztynem. Zesz. Nauk. Wyz. Szk. Roln. Olszt. 14:93-103. [1962.10]

Wengris, J. 1964 ("1963"). Mrówki (Hymenoptera, Formicidae) rezerwatu torfowiskowego Mszar (woj. olsztynskie). Zesz. Nauk. Wyz. Szk. Roln. Olszt. 16:411-423. [1964.01]

Wengrisówna, J. 1933. Mrówki okolic Trok i Wilna. Pr. Tow. Przyj. Nauk Wilnie Wydz. Nauk Mat. Przyr. 7:387-408. [1933]

Wengrisówna, J. 1939. Nowe gatunki mrówek dla fauny Wilenszczyzny. Pr. Tow. Przyj. Nauk Wilnie Wydz. Nauk Mat. Przyr. 13:131-136. [1939]

Wenner, A. M. 1959. The ants of Bidwell Park, Chico, California. Am. Midl. Nat. 62:174-183. [1959.07]

Werner, P. 1989. Formicoidea. Pp. 153-156 in: Sedivy, J. (ed.) Enumeratio insectorum Bohemoslovakiae. Checklist of Czechoslovak insects. III (Hymenoptera). Acta Faun. Entomol. Mus. Natl. Pragae 19:1-194. [1989.11.30]

Werner, P. 1992. Fylogeneze mravencu (Hymenoptera, Formicidae). Myrmekol. Listy 1992:3-9. [1992.12.01]

Werringloer, A. 1932. Die Sehorgane und Sehzentren der Dorylinen nebst Untersuchungen über die Facettenaugen der Formiciden. Z. Wiss. Zool. 141:432-524. [1932.04]

Wesmael, C. 1838. Sur une nouvelle espèce de fourmi du Mexique. Bull. Acad. R. Belg. 5:766-771. [1838]

Wesselinoff, G. D. 1936. Die Ameise *Ponera coarctata* Latr. in Bulgarien. Izv. Bulg. Entomol. Druzh. 9:131-134. [1936.05.01]

Wesselinoff, G. D. 1967. Hügelbauende Waldameisen in den Losen-, Plana-, Werila-, Witoscha- und Ljulingebirgen. Waldhygiene 7:106-116. [1967]

Wesselinoff, G. D. 1973. Die hügelbauenden Waldameisen Bulgariens. Waldhygiene 10:103-117. [1973]

Wesselinoff, G. D. 1981. Diffusione e protezione delle formiche del gruppo *Formica rufa* in Bulgaria. Collana Verde 59:315-318. [1981]

Wesson, L. G. 1935. A new species of ant from Tennessee (Hymen.: Formicidae). Entomol. News 46:208-210. [1935.10.10]

Wesson, L. G. 1937. A slave-making *Leptothorax* (Hymen.: Formicidae). Entomol. News 48:125-129. [1937.05.13]

Wesson, L. G. 1939. *Leptothorax manni* Wesson synonymous with *L. pergandei* Emery (Hymenoptera: Formicidae). Entomol. News 50:180. [1939.06.16]

Wesson, L. G. 1940a. Observations on *Leptothorax duloticus*. Bull. Brooklyn Entomol. Soc. 35: 73-83. [1940.06.07]

Wesson, L. G. 1940b. A gynandromorph of *Aphaenogaster fulva* subsp. *aquia* Buckley (Hymenoptera: Formicidae). Entomol. News 51:241-242. [1940.11.08]

Wesson, L. G. 1949. *Strumigenys venatrix* Wesson and Wesson synonymous with *S. talpa* Weber. Psyche (Camb.) 56:21. [1949.05.17]

Wesson, L. G., Wesson, R. G. 1939. Notes on *Strumigenys* from southern Ohio, with descriptions of six new species. Psyche (Camb.) 46:91-112. [1939.09] [Stamp date in MCZ library: 1939.10.16.]

Wesson, L. G., Wesson, R. G. 1940. A collection an ants from southcentral Ohio. Am. Midl. Nat. 24:89-103. [1940.07.31]

Westhoff, V., Westhoff de Joncheere, J. N. 1942. Verspreiding en nestoecologie van de mieren in de Nederlandsche bosschen. Tijdschr. Plantenziekten 48:138-212. [1942.10]

Westwood, J. O. 1829. [Untitled.] P. 344 in: Stephens, J. F. A systematic catalogue of British insects: being an attempt to arrange all hitherto discovered indigenous insects in accordance with their natural affinities. Part II. Insecta Haustellata. London: Baldwin & Cradock, 388 pp. [1829.07.15] [*Orthonotus* Westwood (Hemiptera), senior homonym of *Orthonotus* Ashmead, 1905.]

Westwood, J. O. 1835. [Untitled. Introduced by: "Specimens were exhibited, partly from the collection of the Rev. F. W. Hope, and partly from that of Mr. Westwood, of various Hymenopterous insects, which Mr. Westwood regarded as new to science."] Proc. Zool. Soc. Lond. 3:68-72. [1835.09.02] [Date of publication from Waterhouse (1937).]

Westwood, J. O. 1839a. An introduction to the modern classification of insects; founded on the natural habits and corresponding organisation of the different families. Volume 2. Part XI. London: Longman, Orme, Brown, Green and Longmans, pp. 193-224. [1839.03] [Date from Griffin (1932) and Blackwelder (1949a). Ants pp. 216-224.]

Westwood, J. O. 1839b. An introduction to the modern classification of insects; founded on the natural habits and corresponding organisation of the different families. Volume 2. Part XII. London: Longman, Orme, Brown, Green and Longmans, pp. 225-256. [1839.04] [Date from Griffin (1932) and Blackwelder (1949a). Ants pp. 225-236.]

Westwood, J. O. 1840a. Synopsis of the genera of British insects. [Synopsis sheet G, pp. 81-96] In: Westwood, J. O. An introduction to the modern classification of insects; founded on the natural habits and corresponding organisation of the different families. Volume 2. Part XV. London: Longman, Orme, Brown, Green and Longmans, pp. 353-400. [1840.01] [Date from Griffin (1932) and Blackwelder (1949a). Ants p. 83.]

Westwood, J. O. 1840b. Observations on the genus *Typhlopone*, with descriptions of several exotic species of ants. Ann. Mag. Nat. Hist. 6:81-89. [1840.10]

Westwood, J. O. 1842. Monograph of the hymenopterous group, Dorylides. Pp. 73-80 in: Westwood, J. O. Arcana entomologica; or illustrations of new, rare, and interesting insects. Volume 1, No. 5. London: W. Smith, pp. 65-80, pl. 17-20. [1842.01.01] [Date from Bradley (1975).]

Westwood, J. O. 1845. Description of a new species of the Hymenopterous genus *Aenictus*, belonging to the Dorylidae. J. Proc. Entomol. Soc. Lond. 1840-1846:85. [1845.11.29] [Date inferred from Wheeler, G. (1912:762).]

Westwood, J. O. 1847a. Descriptions of a new Dorylideous insect from South Africa, belonging to the genus *Aenictus*. Trans. Entomol. Soc. Lond. 4:237-238. [1847.03.01] [Date from Wheeler, G. (1912).]

Westwood, J. O. 1847b. Description of the "Driver" ants, described in the preceding article. Trans. Entomol. Soc. Lond. 5:16-18. [1847.06.30] [Date from Wheeler, G. (1912).]

Westwood, J. O. 1854. Contributions to fossil entomology. Q. J. Geol. Soc. Lond. 10:378-396. [1854]
Wetterer, J. K. 1994. Nourishment and evolution in fungus-growing ants and their fungi. Pp. 309-328 in: Hunt, J. H., Nalepa, C. A. (eds.) Nourishment and evolution in insect societies. Boulder: Westview Press, xii + 449 pp. [1994]
Weyrauch, W. 1933. Über unterscheidende Geschlechtsmerkmale. 2. Beitrag. Die Variabilität der Körperlänge bei den Camponotinen. Z. Morphol. Ökol. Tiere 27:384-400. [1933.08.02]
Weyrauch, W. 1943 ("1942"). Las hormigas cortadoras de hojas del Valle de Chanchamayo. Bol. Direcc. Agric. Ganad. (Lima) 15:204-259. [1943.03]
Weyrauch, W. K. See: Weyrauch, W.
Wheeler, D. E. 1985 ("1984"). Behavior of the ant, *Procryptocerus scabriusculus* (Hymenoptera: Formicidae), with comparisons to other cephalotines. Psyche (Camb.) 91:171-192. [1985.04.22]
Wheeler, D. E. 1986. Developmental and physiological determinants of caste in social Hymenoptera: evolutionary implications. Am. Nat. 128:13-34. [1986.07]
Wheeler, D. E. 1991. The developmental basis of worker caste polymorphism in ants. Am. Nat. 138:1218-1238. [1991.11]
Wheeler, D. E. 1994. Nourishment in ants: patterns in individuals and societies. Pp. 245-278 in: Hunt, J. H., Nalepa, C. A. (eds.) Nourishment and evolution in insect societies. Boulder: Westview Press, xii + 449 pp. [1994]
Wheeler, D. E., Crichton, E. G., Krutzsch, P. H. 1990. Comparative ultrastructure of ant spermatozoa (Formicidae: Hymenoptera). J. Morphol. 206:343-350. [1990.12]
Wheeler, D. E., Hölldobler, B. 1986 ("1985"). Cryptic phargmosis: the structural modifications. Psyche (Camb.) 92:337-353. [1986.04.27]
Wheeler, D. E., Krutzsch, P. H. 1992. Internal reproductive system in adult males of the genus *Camponotus* (Hymenoptera: Formicidae: Formicinae). J. Morphol. 211:307-317. [1992.03]
Wheeler, D. E., Krutzsch, P. H. 1994. Ultrastructure of the spermatheca and its associated gland in the *Crematogaster opuntiae* (Hymenoptera, Formicidae). Zoomorphology (Berl.) 114:203-212. [1994.10]
Wheeler, G. 1912. On the dates of the publications of the Entomological Society of London. Trans. Entomol. Soc. Lond. 1911:750-767. [1912.02.10]
Wheeler, G. C. 1928. The larva of *Leptanilla* (Hym.: Formicidae). Psyche (Camb.) 35:85-91. [1928.06] [Stamp date in MCZ library: 1928.09.14.]
Wheeler, G. C. 1935. The larva of *Allomerus* (Hym.: Formicidae). Psyche (Camb.) 42:92-98. [1935.06] [Stamp date in MCZ library: 1935.08.13.]
Wheeler, G. C. 1938. Are ant larvae apodous? Psyche (Camb.) 45:139-145. [1938.12] [Stamp date in MCZ library: 1939.03.01.]
Wheeler, G. C. 1943. The larvae of the army ants. Ann. Entomol. Soc. Am. 36:319-332. [1943.07.01]
Wheeler, G. C. 1949 ("1948"). The larvae of the fungus-growing ants. Am. Midl. Nat. 40:664-689. [1949.02.24]
Wheeler, G. C. 1950. Ant larvae of the subfamily Cerapachyinae. Psyche (Camb.) 57:102-113. [1950.12.29]
Wheeler, G. C. 1954 ("1953"). [Review of: Cook, T. W. 1953. The ants of California. Palo Alto: Pacific Books, 462 pp.] Ann. Entomol. Soc. Am. 46:618-619. [1954.02.08]
Wheeler, G. C. 1955. [Review of: Wilson, E. O. 1955. A monographic revision of the ant genus *Lasius*. Bull. Mus. Comp. Zool. 113:1-201.] Q. Rev. Biol. 30:399. [1955.12]
Wheeler, G. C. 1956. Myrmecological orthoepy and onomatology. Grand Forks, North Dakota: published by the author (printed by University of North Dakota Press), 22 pp. [1956]
Wheeler, G. C. 1965. History of myrmecopedology. Bull. Entomol. Soc. Am. 11:85. [1965.06.15]

Wheeler, G. C. 1972. [Fascimile reprint and translation of: Emery, C. 1899. Intorno alle larve di alcune formiche. Mem. R. Accad. Sci. Ist. Bologna (5)8:1-10, 2 pl.] Reno, Nevada: Desert Research Institute, University of Nevada System, 24 pp. [1972]

Wheeler, G. C. 1975. Myrmecological orthoepy and onomatology: 1974 supplement. Reno, Nevada: published by the author, 1 p. [1975]

Wheeler, G. C., Wheeler, E. W. 1930. Two new ants from Java. Psyche (Camb.) 37:193-201. [1930.09] [Stamp date in MCZ library: 1930.11.20.]

Wheeler, G. C., Wheeler, E. W. 1934. New forms of *Aphaenogaster treatae* Forel from the southern United States (Hym.: Formicidae). Psyche (Camb.) 41:6-12. [1934.03] [Stamp date in MCZ library: 1934.04.07.]

Wheeler, G. C., Wheeler, E. W. 1944. The ants of North Dakota. N. D. Hist. Q. 11:231-271. [1944.10]

Wheeler, G. C., Wheeler, J. 1951. The ant larvae of the subfamily Dolichoderinae. Proc. Entomol. Soc. Wash. 53:169-210. [1951.08.28]

Wheeler, G. C., Wheeler, J. 1952a. The ant larvae of the subfamily Ponerinae - Part I. Am. Midl. Nat. 48:111-144. [1952.07]

Wheeler, G. C., Wheeler, J. 1952b. The ant larvae of the myrmicine tribe Crematogastrini. J. Wash. Acad. Sci. 42:248-262. [1952.08]

Wheeler, G. C., Wheeler, J. 1952c. The ant larvae of the subfamily Ponerinae - Part II. Am. Midl. Nat. 48:604-672. [1952.11]

Wheeler, G. C., Wheeler, J. 1953a ("1952"). The ant larvae of the myrmicine tribe Myrmicini. Psyche (Camb.) 59:105-125. [1953.01.30]

Wheeler, G. C., Wheeler, J. 1953b. The ant larvae of the myrmicine tribe Pheidolini (Hymenoptera, Formicidae). Proc. Entomol. Soc. Wash. 55:49-84. [1953.04.27]

Wheeler, G. C., Wheeler, J. 1953c. The ant larvae of the subfamily Formicinae. Ann. Entomol. Soc. Am. 46:126-171. [1953.05.14]

Wheeler, G. C., Wheeler, J. 1953d. The ant larvae of the myrmicine tribes Melissotarsini, Metaponini, Myrmicariini, and Cardiocondylini. J. Wash. Acad. Sci. 43:185-189. [1953.06]

Wheeler, G. C., Wheeler, J. 1953e. The ant larvae of the subfamily Formicinae. Part II. Ann. Entomol. Soc. Am. 46:175-217. [1953.08.03]

Wheeler, G. C., Wheeler, J. 1954a ("1953"). The ant larvae of the myrmicine tribe Pheidologetini. Psyche (Camb.) 60:129-147. [1954.04.27]

Wheeler, G. C., Wheeler, J. 1954b. The ant larvae of the myrmicine tribes Cataulacini and Cephalotini. J. Wash. Acad. Sci. 44:149-157. [1954.05]

Wheeler, G. C., Wheeler, J. 1954c. The ant larvae of the myrmicine tribe Myrmecinini (Hymenoptera). Proc. Entomol. Soc. Wash. 56:126-138. [1954.06.15]

Wheeler, G. C., Wheeler, J. 1954d. The ant larvae of the myrmicine tribes Meranoplini, Ochetomyrmicini and Tetramoriini. Am. Midl. Nat. 52:443-452. [1954.12.26]

Wheeler, G. C., Wheeler, J. 1955a ("1954"). The ant larvae of the myrmicine tribes Basicerotini and Dacetini. Psyche (Camb.) 61:111-145. [1955.01.26]

Wheeler, G. C., Wheeler, J. 1955b. The ant larvae of the myrmicine tribe Leptothoracini. Ann. Entomol. Soc. Am. 48:17-29. [1955.05.18]

Wheeler, G. C., Wheeler, J. 1955c. The ant larvae of the myrmicine tribe Solenopsidini. Am. Midl. Nat. 54:119-141. [1955.08.27]

Wheeler, G. C., Wheeler, J. 1956. The ant larvae of the subfamily Pseudomyrmecinae (Hymenoptera: Formicidae). Ann. Entomol. Soc. Am. 49:374-398. [1956.08.30]

Wheeler, G. C., Wheeler, J. 1957a. The larva of the ant genus *Dacetinops* Brown and Wilson. Breviora 78:1-4. [1957.06.21]

Wheeler, G. C., Wheeler, J. 1957b. The larva of *Simopelta* (Hymenoptera: Formicidae). Proc. Entomol. Soc. Wash. 59:191-194. [1957.09.18]

Wheeler, G. C., Wheeler, J. 1957c ("1956"). *Veromessor lobognathus* in North Dakota (Hymenoptera: Formicidae). Psyche (Camb.) 63:140-145. [1957.11.18]

Wheeler, G. C., Wheeler, J. 1959a. *Veromessor lobgnathus*: second note (Hymenoptera: Formicidae). Ann. Entomol. Soc. Am. 52:176-179. [1959.03.26]
Wheeler, G. C., Wheeler, J. 1959b. The larva of *Paramyrmica* (Hymenoptera: Formicidae). J. Tenn. Acad. Sci. 34:219-220. [1959.10]
Wheeler, G. C., Wheeler, J. 1960a. The ant larvae of the subfamily Myrmicinae. Ann. Entomol. Soc. Am. 53:98-110. [1960.02.05]
Wheeler, G. C., Wheeler, J. 1960b. Supplementary studies on the larvae of the Myrmicinae (Hymenoptera: Formicidae). Proc. Entomol. Soc. Wash. 62:1-32. [1960.04.27]
Wheeler, G. C., Wheeler, J. 1961 ("1960"). Techniques for the study of ant larvae. Psyche (Camb.) 67:87-94. [1961.07.07]
Wheeler, G. C., Wheeler, J. 1963. The ants of North Dakota. Grand Forks, North Dakota: University of North Dakota Press, viii + 326 pp. [1963]
Wheeler, G. C., Wheeler, J. 1964a. The ant larvae of the subfamily Cerapachyinae: supplement. Proc. Entomol. Soc. Wash. 66:65-71. [1964.07.06]
Wheeler, G. C., Wheeler, J. 1964b. The ant larvae of the subfamily Ponerinae: supplement. Ann. Entomol. Soc. Am. 57:443-462. [1964.07.15]
Wheeler, G. C., Wheeler, J. 1964c. The ant larvae of the subfamily Dorylinae: supplement. Proc. Entomol. Soc. Wash. 66:129-137. [1964.10.22]
Wheeler, G. C., Wheeler, J. 1965a. *Veromessor lobognathus*: third note (Hymenoptera: Formicidae). J. Kansas Entomol. Soc. 38:55-61. [1965.02.17]
Wheeler, G. C., Wheeler, J. 1965b. The ant larvae of the subfamily Leptanillinae (Hymenoptera, Formicidae). Psyche (Camb.) 72:24-34. [1965.06.25]
Wheeler, G. C., Wheeler, J. 1966. Ant larva of the subfamily Dolichoderinae: supplement. Ann. Entomol. Soc. Am. 59:726-732. [1966.07.15]
Wheeler, G. C., Wheeler, J. 1967. *Veromessor lobognathus*: fourth note (Hymenoptera: Formicidae). J. Kansas Entomol. Soc. 40:238-241. [1967.04.28]
Wheeler, G. C., Wheeler, J. 1968a. The ant larvae of the subfamily Formicinae (Hymenoptera: Formicidae): supplement. Ann. Entomol. Soc. Am. 61:205-222. [1968.01.15]
Wheeler, G. C., Wheeler, J. 1968b. The rediscovery of *Manica parasitica* (Hymenoptera: Formicidae). Pan-Pac. Entomol. 44:71-72. [1968.06.28]
Wheeler, G. C., Wheeler, J. 1969. The larva of *Acanthognathus* (Hymenoptera: Formicidae). Psyche (Camb.) 76:110-113. [1969.12.12]
Wheeler, G. C., Wheeler, J. 1970a. The natural history of *Manica* (Hymenoptera: Formicidae). J. Kansas Entomol. Soc. 43:129-162. [1970.05.07]
Wheeler, G. C., Wheeler, J. 1970b. Ant larvae of the subfamily Formicinae: second supplement. Ann. Entomol. Soc. Am. 63:648-656. [1970.05.15]
Wheeler, G. C., Wheeler, J. 1971a ("1970"). The larva of *Apomyrma* (Hymenoptera: Formicidae). Psyche (Camb.) 77:276-279. [1971.05.31]
Wheeler, G. C., Wheeler, J. 1971b. Ant larvae of the subfamily Ponerinae: second supplement. Ann. Entomol. Soc. Am. 64:1197-1217. [1971.11.15]
Wheeler, G. C., Wheeler, J. 1971c. The larvae of the ant genus *Bothroponera* (Hymenoptera: Formicidae). Proc. Entomol. Soc. Wash. 73:386-394. [1971.12.17]
Wheeler, G. C., Wheeler, J. 1971d. Ant larvae of the subfamily Myrmeciinae (Hymenoptera: Formicidae). Pan-Pac. Entomol. 47:245-256. [1971.12.30]
Wheeler, G. C., Wheeler, J. 1972a. The subfamilies of Formicidae. Proc. Entomol. Soc. Wash. 74:35-45. [1972.03.28]
Wheeler, G. C., Wheeler, J. 1972b. Ant larvae of the subfamily Myrmicinae: second supplement on the tribes Myrmicini and Pheidolini. J. Ga. Entomol. Soc. 7:233-246. [1972.10]
Wheeler, G. C., Wheeler, J. 1973a. The ant larvae of six tribes: second supplement (Hymenoptera: Formicidae: Myrmicinae). J. Ga. Entomol. Soc. 8:27-39. [1973.01]
Wheeler, G. C., Wheeler, J. 1973b. Ant larvae of four tribes: second supplement (Hymenoptera: Formicidae: Myrmicinae). Psyche (Camb.) 80:70-82. [1973.09.07]

Wheeler, G. C., Wheeler, J. 1973c. The ant larvae of the tribes Basicerotini and Dacetini: second supplement (Hymenoptera: Formicidae: Myrmicinae). Pan-Pac. Entomol. 49:207-214. [1973.10.19]

Wheeler, G. C., Wheeler, J. 1973d. Supplementary studies on ant larvae: Cerapachyinae, Pseudomyrmecinae and Myrmicinae. Psyche (Camb.) 80:204-211. [1973.12.06]

Wheeler, G. C., Wheeler, J. 1973e. Ants of Deep Canyon. Riverside, Calif.: University of California, xiii + 162 pp. [1973]

Wheeler, G. C., Wheeler, J. 1974a. Ant larvae of the subfamily Formicinae: third supplement. J. Ga. Entomol. Soc. 9:59-64. [1974.01]

Wheeler, G. C., Wheeler, J. 1974b ("1973"). Ant larvae of the subfamily Dolichoderinae: second supplement (Hymenoptera: Formicidae). Pan-Pac. Entomol. 49:396-401. [1974.04.16]

Wheeler, G. C., Wheeler, J. 1974c. Ant larvae of the subfamily Dorylinae: second supplement (Hymenoptera: Formicidae). J. Kansas Entomol. Soc. 47:166-172. [1974.04.26]

Wheeler, G. C., Wheeler, J. 1974d. Ant larvae of the myrmicine tribe Attini: second supplement (Hymenoptera: Formicidae). Proc. Entomol. Soc. Wash. 76:76-81. [1974.04.30]

Wheeler, G. C., Wheeler, J. 1974e. Supplementary studies on ant larvae: *Teratomyrmex*. Psyche (Camb.) 81:38-41. [1974.05.21]

Wheeler, G. C., Wheeler, J. 1974f. Supplementary studies on ant larvae: *Simopone* and *Turneria*. J. N. Y. Entomol. Soc. 82:103-105. [1974.08.22]

Wheeler, G. C., Wheeler, J. 1974g. Ant larvae of the subfamily Ponerinae: third supplement (Hymenoptera: Formicidae). Proc. Entomol. Soc. Wash. 76:278-281. [1974.11.07]

Wheeler, G. C., Wheeler, J. 1976a. Supplementary studies on ant larvae: Ponerinae. Trans. Am. Entomol. Soc. 102:41-64. [1976.04.27]

Wheeler, G. C., Wheeler, J. 1976b. Ant larvae: review and synthesis. Mem. Entomol. Soc. Wash. 7:1-108. [1976]

Wheeler, G. C., Wheeler, J. 1977a. Supplementary studies on ant larvae: Myrmicinae. Trans. Am. Entomol. Soc. 103:581-602. [1977.12.23]

Wheeler, G. C., Wheeler, J. 1977b. North Dakota ants updated. Reno, Nevada: published by the authors, 27 pp. [1977]

Wheeler, G. C., Wheeler, J. 1978a. *Brachymyrmex musculus*, a new ant in the United States. Entomol. News 89:189-190. [1978.09.30]

Wheeler, G. C., Wheeler, J. 1978b. Mountain ants of Nevada. Great Basin Nat. 38:379-396. [1978.12]

Wheeler, G. C., Wheeler, J. 1979. Larvae of the social Hymenoptera. Pp. 287-338 in: Hermann, H. R. (ed.) Social insects. Volume 1. New York: Academic Press, xv + 437 pp. [1979]

Wheeler, G. C., Wheeler, J. 1980. Supplementary studies on ant larvae: Ponerinae, Myrmicinae and Formicinae. Trans. Am. Entomol. Soc. 106:527-545. [1980.12.31]

Wheeler, G. C., Wheeler, J. 1981. Nest of *Formica propinqua* (Hymenoptera: Formicidae). Great Basin Nat. 41:389-392. [1981.12]

Wheeler, G. C., Wheeler, J. 1982a. Air sacs in ants (Hymenoptera: Formicidae). Entomol. News 93:25-26. [1982.03.31]

Wheeler, G. C., Wheeler, J. 1982b. Supplementary studies on ant larvae: Formicinae (Hymenoptera: Formicidae). Psyche (Camb.) 89:175-181. [1982.12.17]

Wheeler, G. C., Wheeler, J. 1983a. Supplementary studies on ant larvae: Myrmicinae. Trans. Am. Entomol. Soc. 108:601-610. [1983.01.11]

Wheeler, G. C., Wheeler, J. 1983b. The superstructure of ant nests (Hymenoptera: Formicidae). Trans. Am. Entomol. Soc. 109:159-177. [1983.10.27]

Wheeler, G. C., Wheeler, J. 1984a. The larvae of the army ants (Hymenoptera: Formicidae): a revision. J. Kansas Entomol. Soc. 57:263-275. [1984.05.18]

Wheeler, G. C., Wheeler, J. 1984b. Myrmecological orthoepy and onomatology. (2nd edition.) Grand Forks, North Dakota: published by the authors, 20 pp. [1984]

Wheeler, G. C., Wheeler, J. 1985a. A checklist of Texas ants. Prairie Nat. 17:49-64. [1985.06.01]

Wheeler, G. C., Wheeler, J. 1985b. A simplified conspectus of the Formicidae. Trans. Am. Entomol. Soc. 111:255-264. [1985.07.30]
Wheeler, G. C., Wheeler, J. 1986a ("1985"). The larva of *Dinoponera* (Hymenoptera: Formicidae: Ponerinae). Psyche (Camb.) 92:387-391. [1986.04.27]
Wheeler, G. C., Wheeler, J. 1986b ("1985"). The larva of *Proatta* (Hymenoptera: Formicidae). Psyche (Camb.) 92:447-450. [1986.04.27]
Wheeler, G. C., Wheeler, J. 1986c. Supplementary studies of ant larvae: Ponerinae. Trans. Am. Entomol. Soc. 112:85-94. [1986.06.27]
Wheeler, G. C., Wheeler, J. 1986d. Supplementary studies on ant larvae: Formicinae (Hymenoptera: Formicidae). J. N. Y. Entomol. Soc. 94:331-341. [1986.07.02]
Wheeler, G. C., Wheeler, J. 1986e. Ten-year supplement to "Ant larvae: review and synthesis". Proc. Entomol. Soc. Wash. 88:684-702. [1986.10.14]
Wheeler, G. C., Wheeler, J. 1986f. Supplementary studies on ant larvae: Myrmicinae (Hymenoptera: Formicidae). J. N. Y. Entomol. Soc. 94:489-499. [1986.10.16]
Wheeler, G. C., Wheeler, J. 1986g. The ants of Nevada. Los Angeles: Natural History Museum of Los Angeles County, vii + 138 pp. [1986]
Wheeler, G. C., Wheeler, J. 1987a ("1986"). Young larvae of *Eciton* (Hymenoptera: Formicidae: Dorylinae). Psyche (Camb.) 93:341-349. [1987.03.17]
Wheeler, G. C., Wheeler, J. 1987b. A checklist of the ants of South Dakota. Prairie Nat. 19:199-208. [1987.09.30]
Wheeler, G. C., Wheeler, J. 1988a. An additional use for ant larvae (Hymenoptera: Formicidae). Entomol. News 99:23-24. [1988.02.16]
Wheeler, G. C., Wheeler, J. 1988b ("1987"). Young larvae of *Veromessor pergandei* (Hymenoptera: Formicidae: Myrmicinae). Psyche (Camb.) 94:303-307. [1988.07.25]
Wheeler, G. C., Wheeler, J. 1988c. The larva of *Notostigma* (Hymenoptera: Formicidae: Formicinae). J. N. Y. Entomol. Soc. 96:355-358. [1988.08.12]
Wheeler, G. C., Wheeler, J. 1988d. A checklist of the ants of Montana. Psyche (Camb.) 95:101-114. [1988.12.20]
Wheeler, G. C., Wheeler, J. 1988e. A checklist of the ants of Wyoming (Hymenoptera: Formicidae). Insecta Mundi 2:231-239. [1988.12]
Wheeler, G. C., Wheeler, J. 1989a ("1988"). Notes on ant larvae: Myrmicinae. Trans. Am. Entomol. Soc. 114:319-327. [1989.01.25]
Wheeler, G. C., Wheeler, J. 1989b. Notes on ant larvae: Ponerinae. J. N. Y. Entomol. Soc. 97: 50-55. [1989.03.29]
Wheeler, G. C., Wheeler, J. 1989c ("1988"). The larva of *Leptanilla japonica*, with notes on the genus (Hymenoptera: Formicidae: Leptanillinae). Psyche (Camb.) 95:185-189. [1989.06.13]
Wheeler, G. C., Wheeler, J. 1989d. On the trail of the ant, *Veromessor lobognathus*. Prairie Nat. 21:119-124. [1989.09.28]
Wheeler, G. C., Wheeler, J. 1989e. A checklist of the ants of Oklahoma. Prairie Nat. 21:203-210. [1989.12.27]
Wheeler, G. C., Wheeler, J. 1989f. Revised techniques for the study of ant larvae. Insecta Mundi 3:307-311. [1989.12]
Wheeler, G. C., Wheeler, J. 1990a ("1989"). Notes on ant larvae. Trans. Am. Entomol. Soc. 115:457-473. [1990.01.15]
Wheeler, G. C., Wheeler, J. 1990b. Larvae of the formicine ant genus *Polyrhachis*. Trans. Am. Entomol. Soc. 116:753-767. [1990.09]
Wheeler, G. C., Wheeler, J. 1990c. Insecta: Hymenoptera Formicidae. Pp. 1277-1294 in: Dindal, D. L. (ed.) Soil biology guide. New York: Wiley Interscience, xviii + 1349 pp. [1990]
Wheeler, G. C., Wheeler, J. 1991a. The larva of *Blepharidatta* (Hymenoptera: Formicidae). J. N. Y. Entomol. Soc. 99:132-137. [1991.02.14]
Wheeler, G. C., Wheeler, J. 1991b. Instars of three ant species. Psyche (Camb.) 98:89-99. [1991.08.16]

Wheeler, G. C., Wheeler, J. 1991c. Notes on ant larvae 1989-1991. Insecta Mundi 5:167-173. [1991.12]

Wheeler, G. C., Wheeler, J., Galloway, T. D., Ayre, G. L. 1989. A list of the ants of Manitoba. Proc. Entomol. Soc. Manit. 45:34-49. [1989]

Wheeler, G. C., Wheeler, J., Kannowski, P. B. 1994. Checklist of ants of Michigan (Hymenoptera: Formicidae). Great Lakes Entomol. 26:297-310. [1994]

Wheeler, G. C., Wheeler, J., Taylor, R. W. 1980. The larval and egg stages of the primitive ant *Nothomyrmecia macrops* Clark (Hymenoptera: Formicidae). J. Aust. Entomol. Soc. 19:131-137. [1980.07.24]

Wheeler, J. 1968. Male genitalia and the taxonomy of *Polyergus* (Hymenoptera: Formicidae). Proc. Entomol. Soc. Wash. 70:156-164. [Erratum, Proc. Entomol. Soc. Wash. 70:254.] [1968.06.29]

Wheeler, J. W., Avery, J., Olubajo, O., Shamim, M. T., Storm, C. B., Duffield, R. M. 1982. Alkylpyrazines from Hymenoptera. Isolation, identification and synthesis of 5-methyl-3-n-propyl-2-(1-butenyl)pyrazine from *Aphaenogaster* ants (Formicidae). Tetrahedron 38:1939-1948. [1982]

Wheeler, J. W., Blum, M. S. 1973. Alkylpyrazine alarm pheromones in ponerine ants. Science (Wash. D. C.) 182:501-503. [1973.11.02]

Wheeler, J. W., Evans, S. L., Blum, M. S., Torgerson, R. S. 1975. Cyclopentyl ketones: identification and function in *Azteca* ants. Science (Wash. D. C.) 187:254-255. [1975.01.24]

Wheeler, W. M. 1900a. The female of *Eciton sumichrasti* Norton, with some notes on the habits of Texan Ecitons. Am. Nat. 34:563-574. [1900.07.25]

Wheeler, W. M. 1900b. A study of some Texan Ponerinae. Biol. Bull. (Woods Hole) 2:1-31. [1900.10]

Wheeler, W. M. 1900c. The habits of *Ponera* and *Stigmatomma*. Biol. Bull. (Woods Hole) 2:43-69. [1900.11]

Wheeler, W. M. 1901a. The compound and mixed nests of American ants. Part I. Observations on a new guest ant. Am. Nat. 35:431-448. [1901.06.28]

Wheeler, W. M. 1901b. Notices biologiques sur les fourmis Mexicaines. Ann. Soc. Entomol. Belg. 45:199-205. [1901.07.03]

Wheeler, W. M. 1901c. The compound and mixed nests of American ants. Part II. The known cases of social symbiosis among American ants. Am. Nat. 35:513-539. [1901.07.31]

Wheeler, W. M. 1901d. *Microdon* larvae in *Pseudomyrma* nests. Psyche (Camb.) 9:222-224. [1901.07]

Wheeler, W. M. 1901e. The compound and mixed nests of American ants. Part II (continued). Am. Nat. 35:701-724. [1901.09.12]

Wheeler, W. M. 1901f. The compound and mixed nests of American ants. Part III. Symbiogenesis and psychogenesis. Am. Nat. 35:791-818. [1901.10.14]

Wheeler, W. M. 1901g. The parasitic origin of macroërgates among ants. Am. Nat. 35:877-886. [1901.11.18]

Wheeler, W. M. 1902a. A new agricultural ant from Texas, with remarks on the known North American species. Am. Nat. 36:85-100. [1902.02.13]

Wheeler, W. M. 1902b. *Formica fusca* Linn. subsp. *subpolita* Mayr, var. *perpilosa*, n. var. Mem. Rev. Soc. Cient. "Antonio Alzate" 17:141-142. [> 1902.05.07] [The issue of volume 17 in which this article appears is dated "Abril 1902", but Wheeler signs off his article with date "May 7th 1902".]

Wheeler, W. M. 1902c. [Translation of Emery's "An analytical key to the genera of the family Formicidae, for the identification of the workers."] Am. Nat. 36:707-725. [1902.09.20]

Wheeler, W. M. 1902d. New agricultural ants from Texas. Psyche (Camb.) 9:387-393. [1902.09]

Wheeler, W. M. 1902e. An American *Cerapachys*, with remarks on the affinities of the Cerapachyinae. Biol. Bull. (Woods Hole) 3:181-191. [1902.10]

Wheeler, W. M. 1902f. The occurrence of *Formica cinerea* Mayr and *Formica rufibarbis* Fabricius in America. Am. Nat. 36:947-952. [1902.12.24]

Wheeler, W. M. 1902g. A consideration of S. B. Buckley's "North American Formicidae." Trans. Tex. Acad. Sci. 4:17-31. [1902]

Wheeler, W. M. 1903a. *Erebomyrma*, a new genus of hypogaeic ants from Texas. Biol. Bull. (Woods Hole) 4:137-148. [1903.02]

Wheeler, W. M. 1903b. Ethological observations on an American ant (*Leptothorax Emersoni* Wheeler). [part] J. Psychol. Neurol. 2:31-47. [1903.04]

Wheeler, W. M. 1903c. A decad of Texan Formicidae. Psyche (Camb.) 10:93-111. [1903.06] [Stamp date in MCZ library: 1906.06.30.]

Wheeler, W. M. 1903d. A revision of the North American ants of the genus *Leptothorax* Mayr. Proc. Acad. Nat. Sci. Phila. 55:215-260. [1903.07.08]

Wheeler, W. M. 1903e. Ethological observations on an American ant (*Leptothorax Emersoni* Wheeler). [concl.] J. Psychol. Neurol. 2:64-78. [1903.07]

Wheeler, W. M. 1903f. The North American ants of the genus *Stenamma* sensu stricto. Psyche (Camb.) 10:164-168. [1903.08] [Stamp date in MCZ library: 1903.08.22.]

Wheeler, W. M. 1903g. Extraordinary females in three species of *Formica*, with remarks on mutation in the Formicidae. Bull. Am. Mus. Nat. Hist. 19:639-651. [1903.11.21]

Wheeler, W. M. 1903h. Some new gynandromorphous ants, with a review of the previously recorded cases. Bull. Am. Mus. Nat. Hist. 19:653-683. [1903.12.05]

Wheeler, W. M. 1903i. The origin of female and worker ants from the eggs of parthenogenetic workers. Science (N. Y.) (n.s.)18:830-833. [1903.12.25]

Wheeler, W. M. 1903j. Some notes on the habits of *Ceraphachys augustae*. Psyche (Camb.) 10:205-209. [1903.12] [Stamp date in MCZ library: 1904.01.04.]

Wheeler, W. M. 1904a. Three new genera of inquiline ants from Utah and Colorado. Bull. Am. Mus. Nat. Hist. 20:1-17. [1904.01.14]

Wheeler, W. M. 1904b. Dr. Castle and the Dzierzon theory. Science (N. Y.) (n.s.)19:587-591. [1904.04.08]

Wheeler, W. M. 1904c. The American ants of the subgenus *Colobopsis*. Bull. Am. Mus. Nat. Hist. 20:139-158. [1904.04.23]

Wheeler, W. M. 1904d. A crustacean-eating ant (*Leptogenys elongata* Buckley). Biol. Bull. (Woods Hole) 6:251-259. [1904.05]

Wheeler, W. M. 1904e. Ants from Catalina Island, California. Bull. Am. Mus. Nat. Hist. 20:269-271. [1904.08.02]

Wheeler, W. M. 1904f. The ants of North Carolina. Bull. Am. Mus. Nat. Hist. 20:299-306. [1904.09.08]

Wheeler, W. M. 1904g. On the pupation of ants and the feasibility of establishing the Guatemalan Kelep, or Cotton-Weevil Ant, in the United States. Science (N. Y.) (n.s.)20:437-440. [1904.09.30]

Wheeler, W. M. 1904h. Social parasitism among ants. Am. Mus. J. 4:74-75. [1904.10.01]

Wheeler, W. M. 1904i. A new type of social parasitism among ants. Bull. Am. Mus. Nat. Hist. 20:347-375. [1904.10.11]

Wheeler, W. M. 1904j. Some further comments on the Guatemalan boll weevil ant. Science (N. Y.) (n.s.)20:766-768. [1904.12.02]

Wheeler, W. M. 1905a. An interpretation of the slave-making instincts in ants. Bull. Am. Mus. Nat. Hist. 21:1-16. [1905.02.14]

Wheeler, W. M. 1905b. Ants from Catalina Island, Cal. Bull. South. Calif. Acad. Sci. 4:60-62. [1905.05.18] [Reprint of Wheeler (1904e) with some additions and emendations.]

Wheeler, W. M. 1905c. The ants of the Bahamas, with a list of the known West Indian species. Bull. Am. Mus. Nat. Hist. 21:79-135. [1905.06.30]

Wheeler, W. M. 1905d. New species of *Formica*. Bull. Am. Mus. Nat. Hist. 21:267-274. [1905.09.28]

Wheeler, W. M. 1905e. How the queens of the parasitic and slave-making ants establish their colonies. Am. Mus. J. 5:144-148. [1905.09.29]

Wheeler, W. M. 1905f. Some remarks on temporary social parasitism and the phylogeny of slavery among ants. Biol. Centralbl. 25:637-644. [1905.10.01]

Wheeler, W. M. 1905g. The North American ants of the genus *Dolichoderus*. Bull. Am. Mus. Nat. Hist. 21:305-319. [1905.11.11]

Wheeler, W. M. 1905h. The North American ants of the genus *Liometopum*. Bull. Am. Mus. Nat. Hist. 21:321-333. [1905.11.14]

Wheeler, W. M. 1905i. Dr. O. F. Cook's "Social organization and breeding habits of the cotton-protecting kelep of Guatemala." Science (N. Y.) (n.s.)21:706-710. [1905.12.01]

Wheeler, W. M. 1905j. An annotated list of the ants of New Jersey. Bull. Am. Mus. Nat. Hist. 21:371-403. [1905.12.09]

Wheeler, W. M. 1905k. Worker ants with vestiges of wings. Bull. Am. Mus. Nat. Hist. 21:405-408. [1905.12.09]

Wheeler, W. M. 1905l. Ants from the summit of Mount Washington. Psyche (Camb.) 12:111-114. [1905.12] [Stamp date in MCZ library: 1906.01.06.]

Wheeler, W. M. 1906a. The habits of the tent-building ant (*Cremastogaster lineolata* Say). Bull. Am. Mus. Nat. Hist. 22:1-18. [1906.01.25]

Wheeler, W. M. 1906b. On certain tropical ants introduced into the United States. Entomol. News 17:23-26. [1906.01]

Wheeler, W. M. 1906c. The kelep excused. Science (N. Y.) (n.s.)23:348-350. [1906.03.02]

Wheeler, W. M. 1906d. The queen ant as a psychological study. Pop. Sci. Mon. 68:291-299. [1906.04]

Wheeler, W. M. 1906e. On the founding of colonies by queen ants, with special reference to the parasitic and slave-making species. Bull. Am. Mus. Nat. Hist. 22:33-105. [1906.05.15]

Wheeler, W. M. 1906f. New ants from New England. Psyche (Camb.) 13:38-41. [1906.06] [Stamp date in MCZ library: 1906.06.30.]

Wheeler, W. M. 1906g. Fauna of New England. 7. List of the Formicidae. Occas. Pap. Boston Soc. Nat. Hist. 7:1-24. [1906.07]

Wheeler, W. M. 1906h. The ants of Japan. Bull. Am. Mus. Nat. Hist. 22:301-328. [1906.09.17]

Wheeler, W. M. 1906i. The ants of the Grand Cañon. Bull. Am. Mus. Nat. Hist. 22:329-345. [1906.09.17]

Wheeler, W. M. 1906j. The ants of the Bermudas. Bull. Am. Mus. Nat. Hist. 22:347-352. [1906.09.29]

Wheeler, W. M. 1906k. Concerning *Monomorium destructor* Jerdon. Entomol. News 17:265. [1906.09]

Wheeler, W. M. 1906l. An ethological study of certain maladjustments in the relations of ants to plants. Bull. Am. Mus. Nat. Hist. 22:403-418. [1906.12.17]

Wheeler, W. M. 1907a. The polymorphism of ants, with an account of some singular abnormalities due to parasitism. Bull. Am. Mus. Nat. Hist. 23:1-93. [1907.01.15]

Wheeler, W. M. 1907b. A collection of ants from British Honduras. Bull. Am. Mus. Nat. Hist. 23:271-277. [1907.03.30]

Wheeler, W. M. 1907c. Notes on a new guest-ant, *Leptothorax glacialis*, and the varieties of *Myrmica brevinodis* Emery. Bull. Wis. Nat. Hist. Soc. 5:70-83. [1907.04]

Wheeler, W. M. 1907d. The fungus-growing ants of North America. Bull. Am. Mus. Nat. Hist. 23:669-807. [1907.09.30]

Wheeler, W. M. 1907e. On certain modified hairs peculiar to the ants of arid regions. Biol. Bull. (Woods Hole) 13:185-202. [1907.09]

Wheeler, W. M. 1907f. The origin of slavery among ants. Pop. Sci. Mon. 71:550-559. [1907.12]

Wheeler, W. M. 1908a. The ants of Porto Rico and the Virgin Islands. Bull. Am. Mus. Nat. Hist. 24:117-158. [1908.02.07]

Wheeler, W. M. 1908b. The ants of Jamaica. Bull. Am. Mus. Nat. Hist. 24:159-163. [1908.02.07]

Wheeler, W. M. 1908c. Ants from Moorea, Society Islands. Bull. Am. Mus. Nat. Hist. 24:165-167. [1908.02.07]
Wheeler, W. M. 1908d. Ants from the Azores. Bull. Am. Mus. Nat. Hist. 24:169-170. [1908.02.07]
Wheeler, W. M. 1908e. [Untitled. Introduced by: "Professor Wheeler gave an interesting talk on the genus *Formica* and said in part as follows:".] J. N. Y. Entomol. Soc. 16:56-58. [1908.03]
Wheeler, W. M. 1908f. The polymorphism of ants. Ann. Entomol. Soc. Am. 1:39-69. [1908.04.14] [Publication date from Miller (1941).]
Wheeler, W. M. 1908g. Honey ants, with a revision of the American Myrmecocysti. Bull. Am. Mus. Nat. Hist. 24:345-397. [1908.05.09]
Wheeler, W. M. 1908h. The ants of Texas, New Mexico and Arizona. (Part I.) Bull. Am. Mus. Nat. Hist. 24:399-485. [1908.05.20]
Wheeler, W. M. 1908i. The ants of Casco Bay, Maine, with observations on two races of *Formica sanguinea* Latreille. Bull. Am. Mus. Nat. Hist. 24:619-645. [1908.09.25]
Wheeler, W. M. 1908j. A European ant (*Myrmica levinodis*) introduced into Massachusetts. J. Econ. Entomol. 1:337-339. [1908.12.15]
Wheeler, W. M. 1908k. Comparative ethology of the European and North American ants. J. Psychol. Neurol. 13:404-435. [1908]
Wheeler, W. M. 1909a. A small collection of ants from Victoria, Australia. J. N. Y. Entomol. Soc. 17:25-29. [1909.03]
Wheeler, W. M. 1909b. Ants collected by Prof. F. Silvestri in Mexico. Boll. Lab. Zool. Gen. Agrar. R. Sc. Super. Agric. 3:228-238. [1909.04.02]
Wheeler, W. M. 1909c. Ants collected by Professor Filippo Silvestri in the Hawaiian Islands. Boll. Lab. Zool. Gen. Agrar. R. Sc. Super. Agric. 3:269-272. [1909.04.08]
Wheeler, W. M. 1909d. Ants of Formosa and the Philippines. Bull. Am. Mus. Nat. Hist. 26:333-345. [1909.06.24]
Wheeler, W. M. 1909e. A decade of North American Formicidae. J. N. Y. Entomol. Soc. 17: 77-90. [1909.06]
Wheeler, W. M. 1909f. A new honey ant from California. J. N. Y. Entomol. Soc. 17:98-99. [1909.09]
Wheeler, W. M. 1909g. Observations on some European ants. J. N. Y. Entomol. Soc. 17:172-187. [1909.12]
Wheeler, W. M. 1909h. The ants of Isle Royale, Michigan. Pp. 325-328 in: Adams, C. C. (ed.) An ecological survey of Isle Royale, Lake Superior. Lansing, Michigan: Board of Geological Survey, iv + 468 pp. [1909] [Repeats the description of *Formica adamsi* n. sp. (which first appeared in Wheeler, 1909e).]
Wheeler, W. M. 1910a. Small artificial ant-nests of novel patterns. Psyche (Camb.) 17:73-75. [1910.04] [Stamp date in MCZ library: 1910.04.09.]
Wheeler, W. M. 1910b. Ants: their structure, development and behavior. New York: Columbia University Press, xxv + 663 pp. [1910.06] [Reviewed by Brues (1910) in the June issue of Psyche.]
Wheeler, W. M. 1910c. The effects of parasitic and other kinds of castration in insects. J. Exp. Zool. 8:377-438. [1910.06]
Wheeler, W. M. 1910d. Three new genera of myrmicine ants from tropical America. Bull. Am. Mus. Nat. Hist. 28:259-265. [1910.07.30]
Wheeler, W. M. 1910e. An aberrant *Lasius* from Japan. Biol. Bull. (Woods Hole) 19:130-137. [1910.07]
Wheeler, W. M. 1910f. A new species of *Aphomomyrmex* from Borneo. Psyche (Camb.) 17:131-135. [1910.08] [Stamp date in MCZ library: 1910.08.23.]
Wheeler, W. M. 1910g. The North American ants of the genus *Camponotus* Mayr. Ann. N. Y. Acad. Sci. 20:295-354. [1910.12.30]

Wheeler, W. M. 1910h. The North American forms of *Lasius umbratus* Nylander. Psyche (Camb.) 17:235-243. [1910.12] [Stamp date in MCZ library: 1911.01.02.]

Wheeler, W. M. 1910i. The North American forms of *Camponotus fallax* Nylander. J. N. Y. Entomol. Soc. 18:216-232. [1910.12]

Wheeler, W. M. 1910j. Family Formicidae. Pp. 655-663 in: Smith, J. B. Annual report of the New Jersey State Museum, including a report of the insects of New Jersey, 1909. Trenton: MacCrellish & Quigley, 888 pp. [1910]

Wheeler, W. M. 1911a. Additions to the ant-fauna of Jamaica. Bull. Am. Mus. Nat. Hist. 30: 21-29. [1911.05.26]

Wheeler, W. M. 1911b. Ants collected in Grenada, W. I. by Mr. C. T. Brues. Bull. Mus. Comp. Zool. 54:167-172. [1911.05]

Wheeler, W. M. 1911c. Three formicid names which have been overlooked. Science (N. Y.) (n.s.)33:858-860. [1911.06.02]

Wheeler, W. M. 1911d. A new *Camponotus* from California. J. N. Y. Entomol. Soc. 19:96-98. [1911.06]

Wheeler, W. M. 1911e. Two fungus-growing ants from Arizona. Psyche (Camb.) 18:93-101. [1911.06] [Stamp date in MCZ library: 1911.07.07.]

Wheeler, W. M. 1911f. The ant-colony as an organism. J. Morphol. 22:307-325. [1911.06]

Wheeler, W. M. 1911g. A list of the type species of the genera and subgenera of Formicidae. Ann. N. Y. Acad. Sci. 21:157-175. [1911.10.17]

Wheeler, W. M. 1911h. Literature for 1910 on the behavior of ants, their guests and parasites. J. Anim. Behav. 1:413-429. [1911.12]

Wheeler, W. M. 1911i. Descriptions of some new fungus-growing ants from Texas, with Mr. C. G. Hartman's observations on their habits. J. N. Y. Entomol. Soc. 19:245-255. [1911.12]

Wheeler, W. M. 1911j. *Lasius* (*Acanthomyops*) *claviger* in Tahiti. J. N. Y. Entomol. Soc. 19:262. [1911.12]

Wheeler, W. M. 1911k. Three new ants from Mexico and Central America. Psyche (Camb.) 18:203-208. [1911.12] [Stamp date in MCZ library: 1912.01.01.]

Wheeler, W. M. 1912a. The ants of Guam. J. N. Y. Entomol. Soc. 20:44-48. [1912.03]

Wheeler, W. M. 1912b. Notes on a mistletoe ant. J. N. Y. Entomol. Soc. 20:130-133. [1912.06]

Wheeler, W. M. 1912c. New names for some ants of the genus *Formica*. Psyche (Camb.) 19:90. [1912.06] [Stamp date in MCZ library: 1912.08.01.]

Wheeler, W. M. 1912d. Notes about ants and their resemblance to man. Natl. Geogr. Mag. 23:731-766. [1912.08]

Wheeler, W. M. 1912e. Additions to our knowledge of the ants of the genus *Myrmecocystus* Wesmael. Psyche (Camb.) 19:172-181. [1912.12] [Stamp date in MCZ library: 1913.01.29.]

Wheeler, W. M. 1912f. The male of *Eciton vagans* Olivier. Psyche (Camb.) 19:206-207. [1912.12] [Stamp date in MCZ library: 1913.01.29.]

Wheeler, W. M. 1913a. Corrections and additions to "List of type species of the genera and subgenera of Formicidae". Ann. N. Y. Acad. Sci. 23:77-83. [1913.05.29]

Wheeler, W. M. 1913b. The ants of Cuba. Bull. Mus. Comp. Zool. 54:477-505. [1913.05]

Wheeler, W. M. 1913c. [Untitled. Description of *Iridomyrmex humilis* Mayr.] Pp. 27-29 in: Newell, W., Barber, T. C. The Argentine ant. U. S. Dep. Agric. Bur. Entomol. Bull. 122:1-98. [1913.06.26]

Wheeler, W. M. 1913d. Ants collected in Georgia by Dr. J. C. Bradley and Mr. W. T. Davis. Psyche (Camb.) 20:112-117. [1913.06] [Stamp date in MCZ library: 1913.08.21.]

Wheeler, W. M. 1913e. Ants collected in the West Indies. Bull. Am. Mus. Nat. Hist. 32:239-244. [1913.07.09]

Wheeler, W. M. 1913f. Zoological results of the Abor Expedition, 1911-1912, XVII. Hymenoptera, II: Ants (Formicidae). Rec. Indian Mus. 8:233-237. [1913.09]

Wheeler, W. M. 1913g. Observations on the Central American acacia ants. Pp. 109-139 in: Jordan, K., Eltringham, H. (eds.) 2nd International Congress of Entomology, Oxford, August 1912. Volume II, Transactions. London: Hazell, Watson & Viney Ltd., 489 pp. [1913.10.06]

Wheeler, W. M. 1913h. A solitary wasp (*Aphilanthops frigidus* F. Smith) that provisions its nest with queen ants. J. Anim. Behav. 3:374-387. [1913.10]

Wheeler, W. M. 1913i. A revision of the ants of the genus *Formica* (Linné) Mayr. Bull. Mus. Comp. Zool. 53:379-565. [1913.10]

Wheeler, W. M. 1914a. Gynandromorphous ants described during the decade 1903-1913. Am. Nat. 48:49-56. [1914.01]

Wheeler, W. M. 1914b. *Formica exsecta* in Japan. Psyche (Camb.) 21:26-27. [1914.02] [Stamp date in MCZ library: 1914.03.16.]

Wheeler, W. M. 1914c. Ants collected by W. M. Mann in the state of Hidalgo, Mexico. J. N. Y. Entomol. Soc. 22:37-61. [1914.03]

Wheeler, W. M. 1914d. Note on the habits of *Liomyrmex*. Psyche (Camb.) 21:75-76. [1914.04] [Stamp date in MCZ library: 1914.07.07.]

Wheeler, W. M. 1914e. The American species of *Myrmica* allied to *M. rubida* Latreille. Psyche (Camb.) 21:118-122. [1914.08] [Stamp date in MCZ library: 1914.09.17.]

Wheeler, W. M. 1914f. New and little known harvesting ants of the genus *Pogonomyrmex*. Psyche (Camb.) 21:149-157. [1914.10] [Stamp date in MCZ library: 1914.11.07.]

Wheeler, W. M. 1915a. *Neomyrma* versus *Oreomyrma*. A correction. Psyche (Camb.) 22:50. [1915.04] [Stamp date in MCZ library: 1915.05.27.]

Wheeler, W. M. 1915b. Some additions to the North American ant-fauna. Bull. Am. Mus. Nat. Hist. 34:389-421. [1915.06.04]

Wheeler, W. M. 1915c. *Paranomopone*, a new genus of ponerine ants from Queensland. Psyche (Camb.) 22:117-120. [1915.08] [Stamp date in MCZ library: 1915.08.24.]

Wheeler, W. M. 1915d. On the presence and absence of cocoons among ants, the nest-spinning habits of the larvae and the significance of black cocoons among certain Australian species. Ann. Entomol. Soc. Am. 8:323-342. [1915.10.11]

Wheeler, W. M. 1915e. The Australian honey-ants of the genus *Leptomyrmex* Mayr. Proc. Am. Acad. Arts Sci. 51:255-286. [1915.11]

Wheeler, W. M. 1915f. Two new genera of myrmicine ants from Brasil. Bull. Mus. Comp. Zool. 59:483-491. [1915.11]

Wheeler, W. M. 1915g. A new bog-inhabiting variety of *Formica fusca* L. Psyche (Camb.) 22:203-206. [1915.12] [Stamp date in MCZ library: 1916.01.04.]

Wheeler, W. M. 1915h. Hymenoptera. [In "Scientific notes on an expedition into the north-western regions of South Australia".] Trans. R. Soc. S. Aust. 39:805-823. [1915.12]

Wheeler, W. M. 1915i ("1914"). The ants of the Baltic Amber. Schr. Phys.-Ökon. Ges. Königsb. 55:1-142. [1915]

Wheeler, W. M. 1916a. The Australian ants of the genus *Onychomyrmex*. Bull. Mus. Comp. Zool. 60:45-54. [1916.01]

Wheeler, W. M. 1916b. Four new and interesting ants from the mountains of Borneo and Luzon. Proc. N. Engl. Zool. Club 6:9-18. [1916.02.10]

Wheeler, W. M. 1916c. Ants collected in British Guiana by the expedition of the American Museum of Natural History during 1911. Bull. Am. Mus. Nat. Hist. 35:1-14. [1916.02.21]

Wheeler, W. M. 1916d. The marriage-flight of a bull-dog ant (*Myrmecia sanguinea* F. Smith). J. Anim. Behav. 6:70-73. [1916.02]

Wheeler, W. M. 1916e. [Review of: Donisthorpe, H. 1915. British ants, their life-history and classification. Plymouth: Brendon & Son Ltd., xv + 379 pp.] Science (N. Y.) (n.s.)43:316-318. [1916.03.03]

Wheeler, W. M. 1916f. Ants collected in Trinidad by Professor Roland Thaxter, Mr. F. W. Urich, and others. Bull. Mus. Comp. Zool. 60:323-330. [1916.03]

Wheeler, W. M. 1916g. Some new formicid names. Psyche (Camb.) 23:40-41. [1916.04] [Stamp date in MCZ library: 1916.04.21.]
Wheeler, W. M. 1916h. Notes on some slave-raids of the western Amazon ant (*Polyergus breviceps* Emery). J. N. Y. Entomol. Soc. 24:107-118. [1916.06]
Wheeler, W. M. 1916i. Two new ants from Texas and Arizona. Proc. N. Engl. Zool. Club 6:29-35. [1916.10.18]
Wheeler, W. M. 1916j. Note on the Brazilian fire-ant, *Solenopsis saevissima* F. Smith. Psyche (Camb.) 23:142-143. [1916.10] [Stamp date in MCZ library: 1916.11.13.]
Wheeler, W. M. 1916k. An anomalous blind worker ant. Psyche (Camb.) 23:143-145. [1916.10] [Stamp date in MCZ library: 1916.11.13.]
Wheeler, W. M. 1916l. An Indian ant introduced into the United States. J. Econ. Entomol. 9:566-569. [1916.12.09]
Wheeler, W. M. 1916m. *Prodiscothyrea*, a new genus of ponerine ants from Queensland. Trans. R. Soc. S. Aust. 40:33-37. [1916.12.23]
Wheeler, W. M. 1916n. The Australian ants of the genus *Aphaenogaster* Mayr. Trans. R. Soc. S. Aust. 40:213-223. [1916.12.23]
Wheeler, W. M. 1916o. Questions of nomenclature connected with the ant genus *Lasius* and its subgenera. Psyche (Camb.) 23:168-173. [1916.12] [Stamp date in MCZ library: 1917.01.19.]
Wheeler, W. M. 1916p. Ants carried in a floating log from the Brazilian mainland to San Sebastian Island. Psyche (Camb.) 23:180-183. [1916.12] [Stamp date in MCZ library: 1917.01.19.]
Wheeler, W. M. 1916q. A phosphorescent ant. Psyche (Camb.) 23:173-174. [1916.12] [Stamp date in MCZ library: 1917.01.19.]
Wheeler, W. M. 1916r. Formicoidea. Formicidae. Pp. 577-601 in: Viereck, H. L. Guide to the insects of Connecticut. Part III. The Hymenoptera, or wasp-like insects, of Connecticut. Conn. State Geol. Nat. Hist. Surv. Bull. 22:1-824. [1916]
Wheeler, W. M. 1917a. The mountain ants of western North America. Proc. Am. Acad. Arts Sci. 52:457-569. [1917.01]
Wheeler, W. M. 1917b. The phylogenetic development of subapterous and apterous castes in the Formicidae. Proc. Natl. Acad. Sci. U. S. A. 3:109-117. [1917.02.15]
Wheeler, W. M. 1917c. The North American ants described by Asa Fitch. Psyche (Camb.) 24: 26-29. [1917.02] [Stamp date in MCZ library: 1917.03.12.]
Wheeler, W. M. 1917d. A new Malayan ant of the genus *Prodiscothyrea*. Psyche (Camb.) 24: 29-30. [1917.02] [Stamp date in MCZ library: 1917.03.12.]
Wheeler, W. M. 1917e. The ants of Alaska. Bull. Mus. Comp. Zool. 61:13-23. [1917.03]
Wheeler, W. M. 1917f. The Australian ant-genus *Myrmecorhynchus* (Ern. André) and its position in the subfamily Camponotinae. Trans. R. Soc. S. Aust. 41:14-19. [1917.12.24]
Wheeler, W. M. 1917g. Jamaican ants collected by Prof. C. T. Brues. Bull. Mus. Comp. Zool. 61:457-471. [1917.12]
Wheeler, W. M. 1917h. The temporary social parasitism of *Lasius subumbratus* Viereck. Psyche (Camb.) 24:167-176. [1917.12] [Stamp date in MCZ library: 1918.01.30.]
Wheeler, W. M. 1917i. Notes on the marriage flights of some Sonoran ants. Psyche (Camb.) 24:177-180. [1917.12] [Stamp date in MCZ library: 1918.01.30.]
Wheeler, W. M. 1917j. The pleometrosis of *Myrmecocystus*. Psyche (Camb.) 24:180-182. [1917.12] [Stamp date in MCZ library: 1918.01.30.]
Wheeler, W. M. 1917k. A list of Indiana ants. Proc. Indiana Acad. Sci. 26:460-466. [1917]
Wheeler, W. M. 1918a. The Australian ants of the ponerine tribe Cerapachyini. Proc. Am. Acad. Arts Sci. 53:215-265. [1918.01]
Wheeler, W. M. 1918b. Ants collected in British Guiana by Mr. C. William Beebe. J. N. Y. Entomol. Soc. 26:23-28. [1918.03]
Wheeler, W. M. 1918c. A study of some ant larvae, with a consideration of the origin and meaning of the social habit among insects. Proc. Am. Philos. Soc. 57:293-343. [1918.08.08]

Wheeler, W. M. 1918d. The ants of the genus *Opisthopsis* Emery. Bull. Mus. Comp. Zool. 62:341-362. [1918.11]
Wheeler, W. M. 1919a. Two gynandromorphous ants. Psyche (Camb.) 26:1-8. [1919.02] [Stamp date in MCZ library: 1919.03.18.]
Wheeler, W. M. 1919b. A new subspecies of *Aphaenogaster treatae* Forel. Psyche (Camb.) 26:50. [1919.04] [Stamp date in MCZ library: 1919.04.22.]
Wheeler, W. M. 1919c. The parasitic Aculeata, a study in evolution. Proc. Am. Philos. Soc. 58: 1-40. [1919.06.11]
Wheeler, W. M. 1919d. Expedition of the California Academy of Sciences to the Galapagos Islands, 1905-1906. XIV. The ants of the Galapagos Islands. Proc. Calif. Acad. Sci. (4)2(2):259-297. [1919.06.16]
Wheeler, W. M. 1919e. Expedition of the California Academy of Sciences to the Galapagos Islands, 1905-1906. XV. The ants of Cocos Island. Proc. Calif. Acad. Sci. (4)2(2):299-308. [1919.06.16]
Wheeler, W. M. 1919f. The ants of Borneo. Bull. Mus. Comp. Zool. 63:43-147. [1919.07]
Wheeler, W. M. 1919g. The ant genus *Lordomyrma* Emery. Psyche (Camb.) 26:97-106. [1919.08] [Stamp date in MCZ library: 1919.09.10.]
Wheeler, W. M. 1919h. A new paper-making *Crematogaster* from the southeastern United States. Psyche (Camb.) 26:107-112. [1919.08] [Stamp date in MCZ library: 1919.09.10.]
Wheeler, W. M. 1919i. The ants of Tobago Island. Psyche (Camb.) 26:113. [1919.08] [Stamp date in MCZ library: 1919.09.10.]
Wheeler, W. M. 1919j. The ants of the genus *Metapone* Forel. Ann. Entomol. Soc. Am. 12:173-191. [1919.10.21]
Wheeler, W. M. 1919k. A singular neotropical ant (*Pseudomyrma filiformis* Fabricius). Psyche (Camb.) 26:124-131. [1919.10] [Stamp date in MCZ library: 1919.11.10.]
Wheeler, W. M. 1920. The subfamilies of Formicidae, and other taxonomic notes. Psyche (Camb.) 27:46-55. [1920.06] [Stamp date in MCZ library: 1920.06.12.]
Wheeler, W. M. 1921a. Professor Emery's subgenera of the genus *Camponotus* Mayr. Psyche (Camb.) 28:16-19. [1921.02] [Stamp date in MCZ library: 1921.04.25.]
Wheeler, W. M. 1921b. A new case of parabiosis and the "ant gardens" of British Guiana. Ecology 2:89-103. [1921.04.26]
Wheeler, W. M. 1921c. Chinese ants. Bull. Mus. Comp. Zool. 64:529-547. [1921.04]
Wheeler, W. M. 1921d. Observations on army ants in British Guiana. Proc. Am. Acad. Arts Sci. 56:291-328. [1921.06]
Wheeler, W. M. 1921e. Chinese ants collected by Prof. C. W. Howard. Psyche (Camb.) 28:110-115. [1921.08] [Stamp date in MCZ library: 1921.09.19.]
Wheeler, W. M. 1921f. The *Tachigalia* ants. Zoologica (N. Y.) 3:137-168. [1921.12.24]
Wheeler, W. M. 1922a. Ants of the American Museum Congo expedition. A contribution to the myrmecology of Africa. I. On the distribution of the ants of the Ethiopian and Malagasy regions. Bull. Am. Mus. Nat. Hist. 45:13-37. [1922.02.10]
Wheeler, W. M. 1922b. Ants of the American Museum Congo expedition. A contribution to the myrmecology of Africa. II. The ants collected by the American Museum Congo Expedition. Bull. Am. Mus. Nat. Hist. 45:39-269. [1922.02.10]
Wheeler, W. M. 1922c. Observations on *Gigantiops destructor* Fabricius and other leaping ants. Biol. Bull. (Woods Hole) 42:185-201. [1922.04]
Wheeler, W. M. 1922d. Ants of the genus *Formica* in the tropics. Psyche (Camb.) 29:174-177. [1922.08] [Stamp date in MCZ library: 1922.10.10.]
Wheeler, W. M. 1922e. The ants of Trinidad. Am. Mus. Novit. 45:1-16. [1922.09.07]
Wheeler, W. M. 1922f. A new genus and subgenus of Myrmicinae from tropical America. Am. Mus. Novit. 46:1-6. [1922.09.07]
Wheeler, W. M. 1922g. Neotropical ants of the genera *Carebara*, *Tranopelta* and *Tranopeltoides*, new genus. Am. Mus. Novit. 48:1-14. [1922.10.16]

Wheeler, W. M. 1922h. Ants of the American Museum Congo expedition. A contribution to the myrmecology of Africa. Contents. Introduction. Bull. Am. Mus. Nat. Hist. 45:1-11. [1922.10.25]

Wheeler, W. M. 1922i. Ants of the American Museum Congo expedition. A contribution to the myrmecology of Africa. VII. Keys to the genera and subgenera of ants. Bull. Am. Mus. Nat. Hist. 45:631-710. [1922.10.25]

Wheeler, W. M. 1922j. Ants of the American Museum Congo expedition. A contribution to the myrmecology of Africa. VIII. A synonymic list of the ants of the Ethiopian region. Bull. Am. Mus. Nat. Hist. 45:711-1004. [1922.10.25]

Wheeler, W. M. 1922k. Ants of the American Museum Congo expedition. A contribution to the myrmecology of Africa. IX. A synonymic list of the ants of the Malagasy region. Bull. Am. Mus. Nat. Hist. 45:1005-1055. [1922.10.25]

Wheeler, W. M. 1922l. Social life among the insects. Lecture IV - Ants, their development, castes, nesting and feeding habits. Sci. Mon. 15:385-404. [1922.11]

Wheeler, W. M. 1922m. Social life among the insects. Lecture IV - Ants, their development, castes, nesting and feeding habits. II. Sci. Mon. 15:527-541. [1922.12]

Wheeler, W. M. 1922n. Formicidae from Easter Island and Juan Fernandez. Pp. 317-319 in: Skottsberg, C. (ed.) 1921-1940. The natural history of Juan Fernandez and Easter Island. Vol. III. Zoology. Uppsala: Almqvist & Wiksells, 688 pp. [1922] [The entire work, with 61 articles, spans 20 years. Exact dates of publication are given only for later articles (pp. 419 et seq.).]

Wheeler, W. M. 1923a. Wissenschaftliche Ergebnisse der schwedischen entomologischen Reise des Herrn Dr. A. Roman in Amazonas 1914-1915. 7. Formicidae. Ark. Zool. 15(7):1-6. [1923.01.31]

Wheeler, W. M. 1923b. Social life among the insects. Lecture V. Parasitic ants and ant guests. Sci. Mon. 16:5-33. [1923.01]

Wheeler, W. M. 1923c. Chinese ants collected by Professor S. F. Light and Professor A. P. Jacot. Am. Mus. Novit. 69:1-6. [1923.04.20]

Wheeler, W. M. 1923d. Report on the ants collected by the Barbados-Antigua Expedition from the University of Iowa in 1918. Stud. Nat. Hist. Iowa Univ. 10(3):3-9. [1923.08.01]

Wheeler, W. M. 1923e. The occurrence of winged females in the ant genus *Leptogenys* Roger, with descriptions of new species. Am. Mus. Novit. 90:1-16. [1923.10.16]

Wheeler, W. M. 1923f. Ants of the genera *Myopias* and *Acanthoponera*. Psyche (Camb.) 30:175-192. [1923.12] [Stamp date in MCZ library: 1924.02.11.]

Wheeler, W. M. 1923g. Social life among the insects. New York: Harcourt, Brace and Co., vii + 375 pp. [1923]

Wheeler, W. M. 1924a. Hymenoptera of the Siju Cave, Garo Hills, Assam. I. *Triglyphothrix striadens* Emery as a cave ant. Rec. Indian Mus. 26:123-124. [1924.01.31]

Wheeler, W. M. 1924b. The Formicidae of the Harrison Williams Galapagos Expedition. Zoologica (N. Y.) 5:101-122. [1924.02.27]

Wheeler, W. M. 1924c. Ants of Krakatau and other islands in the Sunda Strait. Treubia 5:239-258. [1924.02]

Wheeler, W. M. 1924d. A gynandromorph of *Tetramorium guineense* Fabr. Psyche (Camb.) 31:136-137. [1924.08] [Stamp date in MCZ library: 1924.09.22.]

Wheeler, W. M. 1924e. On the ant-genus *Chrysapace* Crawley. Psyche (Camb.) 31:224-225. [1924.10] [Stamp date in MCZ library: 1924.11.25.]

Wheeler, W. M. 1925a. Neotropical ants in the collections of the Royal Museum of Stockholm. Ark. Zool. 17A(8):1-55. [1925.02.07]

Wheeler, W. M. 1925b. Zoological results of the Swedish Expedition to Central Africa 1921. Insecta. 10. Formicidae. Ark. Zool. 17A(25):1-3. [1925.03.16]

Wheeler, W. M. 1925c. L'évolution des insectes sociaux. Rev. Sci. (Paris) 63:548-557. [1925.08.22]

Wheeler, W. M. 1925d. The finding of the queen of the army ant *Eciton hamatum* Fabricius. Biol. Bull. (Woods Hole) 49:139-149. [1925.09]

Wheeler, W. M. 1925e. A new guest-ant and other new Formicidae from Barro Colorado Island, Panama. Biol. Bull. (Woods Hole) 49:150-181. [1925.09]

Wheeler, W. M. 1925f. Obituary. Carlo Emery. Entomol. News 36:318-320. [1925.12.11]

Wheeler, W. M. 1926a. Ants of the Balearic Islands. Folia Myrmecol. Termit. 1:1-6. [1926.10]

Wheeler, W. M. 1926b. A gynandromorph of *Tetramorium guineense* Fabr. Pp. 44-45 in: Bryan, E. H., Jr. and collaborators. Insects of Hawaii, Johnston Island and Wake Island. Bull. Bernice P. Bishop Mus. 31:1-94. [1926] [The same article appears in Wheeler (1924d).]

Wheeler, W. M. 1926c. Les sociétés d'insectes: leur origine, leur évolution. Paris: Gaston Doin & Co., xii + 468 pp. [1926]

Wheeler, W. M. 1926d. The natural history of ants. From an unpublished manuscript in the archives of the Academy of Sciences of Paris by René Antoine Ferchault de Réaumur. Translated and annotated by William Morton Wheeler. New York: A. Knopf, 280 pp. [1926]

Wheeler, W. M. 1927a. The occurrence of *Formica fusca* L. in Sumatra. Psyche (Camb.) 34: 40-41. [1927.02] [Stamp date in MCZ library: 1927.04.04.]

Wheeler, W. M. 1927b. Burmese ants collected by Professor G. E. Gates. Psyche (Camb.) 34: 42-46. [1927.02] [Stamp date in MCZ library: 1927.04.04.]

Wheeler, W. M. 1927c. Ants of the genus *Amblyopone* Erichson. Proc. Am. Acad. Arts Sci. 62: 1-29. [1927.02]

Wheeler, W. M. 1927d. Chinese ants collected by Professor S. F. Light and Professor N. Gist Gee. Am. Mus. Novit. 255:1-12. [1927.03.12]

Wheeler, W. M. 1927e. A few ants from China and Formosa. Am. Mus. Novit. 259:1-4. [1927.03.18]

Wheeler, W. M. 1927f. The physiognomy of insects. Q. Rev. Biol. 2:1-36. [1927.03]

Wheeler, W. M. 1927g. The ants of the Canary Islands. Proc. Am. Acad. Arts Sci. 62:93-120. [1927.04]

Wheeler, W. M. 1927h. Ants collected by Professor F. Silvestri in Indochina. Boll. Lab. Zool. Gen. Agrar. R. Sc. Super. Agric. 20:83-106. [1927.05.16]

Wheeler, W. M. 1927i. The ants of Lord Howe Island and Norfolk Island. Proc. Am. Acad. Arts Sci. 62:121-153. [1927.05]

Wheeler, W. M. 1927j. The occurrence of the pavement ant (*Tetramorium caespitum* L.) in Boston. Psyche (Camb.) 34:164-165. [1927.08] [Stamp date in MCZ library: 1928.04.09.]

Wheeler, W. M. 1928a. *Zatapinoma*, a new genus of ants from India. Proc. N. Engl. Zool. Club 10:19-23. [1928.01.30]

Wheeler, W. M. 1928b. Mermis parasitism and intercastes among ants. J. Exp. Zool. 50:165-237. [1928.02.05]

Wheeler, W. M. 1928c. Ants collected by Professor F. Silvestri in China. Boll. Lab. Zool. Gen. Agrar. R. Sc. Super. Agric. 22:3-38. [1928.03.05]

Wheeler, W. M. 1928d. Ants collected by Professor F. Silvestri in Japan and Korea. Boll. Lab. Zool. Gen. Agrar. R. Sc. Super. Agric. 22:96-125. [1928.03.16]

Wheeler, W. M. 1928e. A new species of *Probolomyrmex* from Java. Psyche (Camb.) 35:7-9. [1928.03] [Stamp date in MCZ library: 1928.09.14.]

Wheeler, W. M. 1928f. Ants of Nantucket Island, Mass. Psyche (Camb.) 35:10-11. [1928.03] [Stamp date in MCZ library: 1928.09.14.]

Wheeler, W. M. 1928g. The evolution of ants. Pp. 210-224 in: Mason, F. (ed.) Creation by evolution. New York: Macmillan, xx + 392 pp. [1928]

Wheeler, W. M. 1928h. The social insects: their origin and evolution. New York: Harcourt, Brace and Co., xviii + 378 pp. [1928]

Wheeler, W. M. 1928i. Foibles of insects and men. New York: A. Knopf, xxvi + 217 pp. [1928] [A series of papers, mostly reprintings of earlier articles, but the paper on *Leptothorax emersoni* (pp. 83-126) is rewritten and expanded.]

Wheeler, W. M. 1929a. The identity of the ant genera *Gesomyrmex* Mayr and *Dimorphomyrmex* Ernest André. Psyche (Camb.) 36:1-12. [1929.03] [Stamp date in MCZ library: 1929.04.15.]

Wheeler, W. M. 1929b. Three new genera of ants from the Dutch East Indies. Am. Mus. Novit. 349:1-8. [1929.04.29]

Wheeler, W. M. 1929c. Two Neotropical ants established in the United States. Psyche (Camb.) 36:89-90. [1929.06] [Stamp date in MCZ library: 1929.07.09.]

Wheeler, W. M. 1929d. Note on *Gesomyrmex*. Psyche (Camb.) 36:91-92. [1929.06] [Stamp date in MCZ library: 1929.07.09.]

Wheeler, W. M. 1929e. The ant genus *Rhopalomastix*. Psyche (Camb.) 36:95-101. [1929.06] [Stamp date in MCZ library: 1929.07.09.]

Wheeler, W. M. 1929f. A *Camponotus* mermithergate from Argentina. Psyche (Camb.) 36:102-106. [1929.06] [Stamp date in MCZ library: 1929.07.09.]

Wheeler, W. M. 1929g. Some ants from China and Manchuria. Am. Mus. Novit. 361:1-11. [1929.07.20]

Wheeler, W. M. 1929h. Ants collected by Professor F. Silvestri in Formosa, the Malay Peninsula and the Philippines. Boll. Lab. Zool. Gen. Agrar. R. Sc. Super. Agric. 24:27-64. [1929.10.22]

Wheeler, W. M. 1929i. Two interesting Neotropical myrmecophytes (*Cordia nodosa* and *C. alliodora*). Pp. 342-353 in: Jordan, K., Horn, W. (eds.) Fourth International Congress of Entomology, Ithaca, August 1928. Volume II. Transactions. Naumburg a. Saale: G. Pätz, vii + 1037 pp. [1929.12] [Nomen nudum: *Allomerus octoarticulatus* var. *demerarae*, page 343.]

Wheeler, W. M. 1929j. Part II. Formicidae. Pp. 29-39 in: Wheeler, W. M., Bequaert, J. C. Amazonian myrmecophytes and their ants. Zool. Anz. 82:10-39. [1929]

Wheeler, W. M. 1930a. [Review of: Forel, A. 1928. The social world of the ants compared with that of man. Translated by C. K. Ogden. Two volumes. London: G. P. Putnam's Sons, Ltd., xlv + 551 pp (vol. 1) and xx + 444 pp. (vol. 2).] J. Soc. Psychol. 1:170-177. [1930.02]

Wheeler, W. M. 1930b. Formosan ants collected by Dr. R. Takahashi. Proc. N. Engl. Zool. Club 11:93-106. [1930.03.07]

Wheeler, W. M. 1930c. The ant *Prenolepis imparis* Say. Ann. Entomol. Soc. Am. 23:1-26. [1930.03.21]

Wheeler, W. M. 1930d. A second note on *Gesomyrmex*. Psyche (Camb.) 37:35-40. [1930.03] [Stamp date in MCZ library: 1930.06.10.]

Wheeler, W. M. 1930e. Two new genera of ants from Australia and the Philippines. Psyche (Camb.) 37:41-47. [1930.03] [Stamp date in MCZ library: 1930.06.10.]

Wheeler, W. M. 1930f. Two mermithergates of *Ectatomma*. Psyche (Camb.) 37:48-54. [1930.03] [Stamp date in MCZ library: 1930.06.10.]

Wheeler, W. M. 1930g. A new parasitic *Crematogaster* from Indiana. Psyche (Camb.) 37:55-60. [1930.03] [Stamp date in MCZ library: 1930.06.10.]

Wheeler, W. M. 1930h. A new *Emeryella* from Panama. Proc. N. Engl. Zool. Club 12:9-13. [1930.04.03]

Wheeler, W. M. 1930i. [Untitled. *Azteca xanthochroa* Roger subsp. *salti* Wheeler, subsp. nov.] Pp. 114-115 in: Wheeler, W. M., Darlington, P. J. Ant-tree notes from Rio Frio, Colombia. Psyche (Camb.) 37:107-117. [1930.06] [Stamp date in MCZ library: 1930/08/13.]

Wheeler, W. M. 1930j. Philippine ants of the genus *Aenictus* with descriptions of the females of two species. J. N. Y. Entomol. Soc. 38:193-212. [1930.07.12]

Wheeler, W. M. 1930k. A list of the known Chinese ants. Peking Nat. Hist. Bull. 5:53-81. [1930.09]

Wheeler, W. M. 1930l. Demons of the dust: a study in insect behavior. New York: W. W. Norton, xviii + 378 pp. [1930]

Wheeler, W. M. 1931a. Neotropical ants of the genus *Xenomyrmex* Forel. Rev. Entomol. (Rio J.) 1:129-139. [1931.07.15]

Wheeler, W. M. 1931b. New and little-known ants of the genera *Macromischa*, *Creosomyrmex* and *Antillaemyrmex*. Bull. Mus. Comp. Zool. 72:1-34. [1931.07]

Wheeler, W. M. 1931c. Concerning some ant gynandromorphs. Psyche (Camb.) 38:80-85. [1931.09] [Stamp date in MCZ library: 1931.10.10.]
Wheeler, W. M. 1931d. The ant *Camponotus* (*Myrmepomis*) *sericeiventris* Guérin and its mimic. Psyche (Camb.) 38:86-98. [1931.09] [Stamp date in MCZ library: 1931.10.10.]
Wheeler, W. M. 1932a. A list of the ants of Florida with descriptions of new forms. J. N. Y. Entomol. Soc. 40:1-17. [1932.03.15]
Wheeler, W. M. 1932b. Some attractions of the field study of ants. Sci. Mon. 34:397-402. [1932.05]
Wheeler, W. M. 1932c. An extraordinary ant-guest from the Philippines (*Aenictoteras Chapmani*, gen. et sp. nov.). Pp. 301-310 in: Jeannel, R. (ed.) Société Entomologique de France. Livre du centenaire. Paris: Société Entomologique de France, xii + 729 pp. [1932.06.30]
Wheeler, W. M. 1932d. An Australian *Leptanilla*. Psyche (Camb.) 39:53-58. [1932.09] [Stamp date in MCZ library: 1932.12.09.]
Wheeler, W. M. 1932e. Ants of the Marquesas Islands. Bull. Bernice P. Bishop Mus. 98:155-163. [1932.11.20]
Wheeler, W. M. 1932f. How the primitive ants of Australia start their colonies. Science (N. Y.) 76:532-533. [1932.12.09]
Wheeler, W. M. 1932g. Ants from the Society Islands. Bull. Bernice P. Bishop Mus. 113:13-19. [1932.12.23] [Separate issued on the above date. Whole volume published 1935.]
Wheeler, W. M. 1933a. The Templeton Crocker Expedition of the California Academy of Sciences, 1932. No. 6. Formicidae of the Templeton Crocker Expedition. Proc. Calif. Acad. Sci. (4)21:57-64. [1933.03.22]
Wheeler, W. M. 1933b. Mermis parasitism in some Australian and Mexican ants. Psyche (Camb.) 40:20-31. [1933.03] [Stamp date in MCZ library: 1933.06.30.]
Wheeler, W. M. 1933c. New ants from China and Japan. Psyche (Camb.) 40:65-67. [1933.06] [Stamp date in MCZ library: 1933.08.14.]
Wheeler, W. M. 1933d. A second parasitic *Crematogaster*. Psyche (Camb.) 40:83-86. [1933.06] [Stamp date in MCZ library: 1933.08.14.]
Wheeler, W. M. 1933e. A new *Myrmoteras* from Java. Proc. N. Engl. Zool. Club 13:73-75. [1933.07.17]
Wheeler, W. M. 1933f. A new species of *Ponera* and other records of ants from the Marquesas Islands. Bull. Bernice P. Bishop Mus. 114:141-144. [1933.08.30] [Separate issued on the above date. Whole volume published 1935.]
Wheeler, W. M. 1933g. Three obscure genera of ponerine ants. Am. Mus. Novit. 672:1-23. [1933.11.06]
Wheeler, W. M. 1933h. An ant new to the fauna of the Hawaiian Islands. Proc. Hawaii. Entomol. Soc. 8:275-278. [1933.11]
Wheeler, W. M. 1933i. Colony founding among the ants, with an account of some primitive Australian species. Cambridge, Mass.: Harvard University Press, viii + 179 pp. [1933]
Wheeler, W. M. 1934a. Formicidae of the Templeton Crocker Expedition, 1933. Proc. Calif. Acad. Sci. (4)21:173-181. [1934.04.20]
Wheeler, W. M. 1934b. An Australian ant of the genus *Leptothorax* Mayr. Psyche (Camb.) 41: 60-62. [1934.06] [Stamp date in MCZ library: 1934.08.30.]
Wheeler, W. M. 1934c. A second revision of the ants of the genus *Leptomyrmex* Mayr. Bull. Mus. Comp. Zool. 77:69-118. [1934.06]
Wheeler, W. M. 1934d. Contributions to the fauna of Rottnest Island, Western Australia. No. IX. The ants. J. R. Soc. West. Aust. 20:137-163. [1934.10.05]
Wheeler, W. M. 1934e. Some aberrant species of *Camponotus* (*Colobopsis*) from the Fiji Islands. Ann. Entomol. Soc. Am. 27:415-424. [1934.10.10]
Wheeler, W. M. 1934f. Ants from the islands off the west coast of Lower California and Mexico. Pan-Pac. Entomol. 10:132-144. [1934.10.17]

Wheeler, W. M. 1934g. Neotropical ants collected by Dr. Elisabeth Skwarra and others. Bull. Mus. Comp. Zool. 77:157-240. [1934.11]

Wheeler, W. M. 1934h. Some ants from the Bahama Islands. Psyche (Camb.) 41:230-232. [1934.12] [Stamp date in MCZ library: 1935.03.07.]

Wheeler, W. M. 1934i. Revised list of Hawaiian ants. Occas. Pap. Bernice P. Bishop Mus. 10(21):1-21. [1934]

Wheeler, W. M. 1935a. Two new genera of myrmicine ants from Papua and the Philippines. Proc. N. Engl. Zool. Club 15:1-9. [1935.02.06]

Wheeler, W. M. 1935b. New ants from the Philippines. Psyche (Camb.) 42:38-52. [1935.03] [Stamp date in MCZ library: 1935.04.04.]

Wheeler, W. M. 1935c. Myrmecological notes. Psyche (Camb.) 42:68-72. [1935.03] [Stamp date in MCZ library: 1935.04.04.]

Wheeler, W. M. 1935d. Ants of the genera *Belonopelta* Mayr and *Simopelta* Mann. Rev. Entomol. (Rio J.) 5:8-19. [1935.04.12]

Wheeler, W. M. 1935e. The Australian ant genus *Mayriella* Forel. Psyche (Camb.) 42:151-160. [1935.09] [Stamp date in MCZ library: 1936.01.36.]

Wheeler, W. M. 1935f. Ants of the genus *Acropyga* Roger, with description of a new species. J. N. Y. Entomol. Soc. 43:321-329. [1935.10.01]

Wheeler, W. M. 1935g. Check list of the ants of Oceania. Occas. Pap. Bernice P. Bishop Mus. 11(11):1-56. [1935]

Wheeler, W. M. 1936a. The Australian ant genus *Froggattella*. Am. Mus. Novit. 842:1-11. [1936.04.13]

Wheeler, W. M. 1936b. Binary anterior ocelli in ants. Biol. Bull. (Woods Hole) 70:185-192. [1936.04]

Wheeler, W. M. 1936c. Ants from Hispaniola and Mona Island. Bull. Mus. Comp. Zool. 80:195-211. [1936.09]

Wheeler, W. M. 1936d. A singular *Crematogaster* from Guatemala. Psyche (Camb.) 43:40-48. [1936.09] [Stamp date in MCZ library: 1936.11.25.]

Wheeler, W. M. 1936e. Ecological relations of ponerine and other ants to termites. Proc. Am. Acad. Arts Sci. 71:159-243. [1936.10]

Wheeler, W. M. 1936f. Notes on some aberrant Indonesian ants of the subfamily Formicinae. Tijdschr. Entomol. 79:217-221. [1936.11]

Wheeler, W. M. 1936g. Ants from the Society, Austral, Tuamotu and Mangareva Islands. Occas. Pap. Bernice P. Bishop Mus. 12(18):3-17. [1936.12.31]

Wheeler, W. M. 1937a. Additions to the ant-fauna of Krakatau and Verlaten Island. Treubia 16: 21-24. [1937.05]

Wheeler, W. M. 1937b. Ants mostly from the mountains of Cuba. Bull. Mus. Comp. Zool. 81:439-465. [1937.05]

Wheeler, W. M. 1937c. Mosaics and other anomalies among ants. Cambridge, Mass.: Harvard University Press, 95 pp. [1937]

Wheeler, W. M. 1938. Ants from the caves of Yucatan. Pp. 251-255 in: Pearse, A. S. Fauna of the caves of Yucatan. Carnegie Inst. Wash. Publ. 491:1-304. [1938.06.15]

Wheeler, W. M. 1942. Studies of Neotropical ant-plants and their ants. Bull. Mus. Comp. Zool. 90:1-262. [1942.10]

Wheeler, W. M., Bailey, I. W. 1920. The feeding habits of pseudomyrmine and other ants. Trans. Am. Philos. Soc. (2)22:235-279. [1920]

Wheeler, W. M., Bequaert, J. C. 1929. Amazonian myrmecophytes and their ants. Zool. Anz. 82: 10-39. [1929]

Wheeler, W. M., Chapman, J. W. 1922. The mating of *Diacamma*. Psyche (Camb.) 29:203-211. [1922.12] [Stamp date in MCZ library: 1923.01.27.]

Wheeler, W. M., Chapman, J. W. 1925. The ants of the Philippine Islands. Part I, Dorylinae and Ponerinae. Philipp. J. Sci. 28:47-73. [1925.09.21]

Wheeler, W. M., Creighton, W. S. 1934. A study of the ant genera *Novomessor* and *Veromessor*. Proc. Am. Acad. Arts Sci. 69:341-387. [1934.09]

Wheeler, W. M., Darlington, P. J., Jr. 1930. Ant-tree notes from Rio Frio, Colombia. Psyche (Camb.) 37:107-117. [1930.06] [Stamp date in MCZ library: 1930.08.13.]

Wheeler, W. M., Gaige, F. M. 1920. *Euponera gilva* (Roger), a rare North American ant. Psyche (Camb.) 27:69-72. [1920.08] [Stamp date in MCZ library: 1920.09.18.]

Wheeler, W. M., Long, W. H. 1901. The males of some Texan Ecitons. Am. Nat. 35:157-173. [1901.03.15]

Wheeler, W. M., Mann, W. M. 1914. The ants of Haiti. Bull. Am. Mus. Nat. Hist. 33:1-61. [1914.02.20]

Wheeler, W. M., Mann, W. M. 1916. The ants of the Phillips Expedition to Palestine during 1914. Bull. Mus. Comp. Zool. 60:167-174. [1916.02]

Wheeler, W. M., Mann, W. M. 1942a. [Untitled. *Pseudomyrma picta* Stitz var. *heterogyna* Wheeler and Mann, var. nov.] Pp. 172-173 in: Wheeler, W. M. Studies of Neotropical ant-plants and their ants. Bull. Mus. Comp. Zool. 90:1-262. [1942.10]

Wheeler, W. M., Mann, W. M. 1942b. [Untitled. *Pseudomyrma triplarina* (Weddell) var. *rurrenabaquensis* Wheeler & Mann, var. nov.] Pp. 188-189 in: Wheeler, W. M. Studies of Neotropical ant-plants and their ants. Bull. Mus. Comp. Zool. 90:1-262. [1942.10]

Wheeler, W. M., Mann, W. M. 1942c. [Untitled. *Allomerus decemarticulatus* Mayr subsp. *novemarticulatus* Wheeler & Mann, subsp. nov.] P. 199 in: Wheeler, W. M. Studies of Neotropical ant-plants and their ants. Bull. Mus. Comp. Zool. 90:1-262. [1942.10]

Wheeler, W. M., Mann, W. M. 1942d. [Untitled. *Allomerus decemarticulatus* Mayr subsp. *octoarticulatus* Mayr var. *exsanguis* Wheeler and Mann, var. nov.] P. 200 in: Wheeler, W. M. Studies of Neotropical ant-plants and their ants. Bull. Mus. Comp. Zool. 90:1-262. [1942.10]

Wheeler, W. M., Mann, W. M. 1942e. [Untitled. *Allomerus decemarticulatus* Mayr subsp. *octoarticulatus* Mayr var. *angulatus* Wheeler & Mann, var. nov.] P. 201 in: Wheeler, W. M. Studies of Neotropical ant-plants and their ants. Bull. Mus. Comp. Zool. 90:1-262. [1942.10]

Wheeler, W. M., Mann, W. M. 1942f. [Untitled. *Allomerus decemarticulatus* Mayr subsp. *octoarticulatus* Mayr var. *melanoticus* Wheeler & Mann, var. nov.] Pp. 202-203 in: Wheeler, W. M. Studies of Neotropical ant-plants and their ants. Bull. Mus. Comp. Zool. 90:1-262. [1942.10]

Wheeler, W. M., Mann, W. M. 1942g. [Untitled. *Azteca brevicornis* Mayr var. *boliviana* Wheeler & Mann, var. nov.] P. 225 in: Wheeler, W. M. Studies of Neotropical ant-plants and their ants. Bull. Mus. Comp. Zool. 90:1-262. [1942.10]

Wheeler, W. M., Mann, W. M. 1942h. [Untitled. *Azteca ulei* Forel var. *gagatina* Wheeler and Mann var. nov.] P. 246 in: Wheeler, W. M. Studies of Neotropical ant-plants and their ants. Bull. Mus. Comp. Zool. 90:1-262. [1942.10]

Wheeler, W. M., Mann, W. M. 1942i. [Untitled. *Myrmelachista* (*Decamera*) *schumanni* Emery var. *cordincola* Wheeler & Mann, var. nov.] P. 255 in: Wheeler, W. M. Studies of Neotropical ant-plants and their ants. Bull. Mus. Comp. Zool. 90:1-262. [1942.10]

Wheeler, W. M., McClendon, J. F. 1903. Dimorphic queens in an American ant (*Lasius latipes* Walsh). Biol. Bull. (Woods Hole) 4:149-163. [1903.03]

Whelden, R. M. 1957. Notes on the anatomy of *Rhytidoponera convexa* Mayr ("*violacea*" Forel) (Hymenoptera, Formicidae). Ann. Entomol. Soc. Am. 50:271-282. [1957.05.15]

Whelden, R. M. 1958a. Additional notes on *Rhytidoponera convexa* Mayr (Hymenoptera, Formicidae). Ann. Entomol. Soc. Am. 51:80-84. [1958.01.20]

Whelden, R. M. 1958b ("1957"). Notes on the anatomy of the Formicidae. I. *Stigmatomma pallipes* (Haldeman). J. N. Y. Entomol. Soc. 65:1-21. [1958.03.25]

Whelden, R. M. 1960. The anatomy of *Rhytidoponera metallica* F. Smith (Hymenoptera: Formicidae). Ann. Entomol. Soc. Am. 53:793-808. [1960.12.21]

Whelden, R. M. 1963a. Anatomy of adult queen and workers of army ants *Eciton burchelli* Westw. and *E. hamatum* Fabr. (Hymenoptera: Formicidae). J. N. Y. Entomol. Soc. 71:14-30. [1963.03.21]

Whelden, R. M. 1963b. The anatomy of the adult queen and workers of the army ants *Eciton burchelli* Westwood and *Eciton hamatum* Fabricus [sic] (continued). J. N. Y. Entomol. Soc. 71:90-115. [1963.06.21]

Whelden, R. M. 1963c. The anatomy of the adult queen and workers of the army ants *Eciton burchelli* Westwood and *Eciton hamatum* Fabricus [sic] (continued). J. N. Y. Entomol. Soc. 71:158-178. [1963.08.30]

Whelden, R. M. 1963d. The anatomy of the adult queen and workers of the army ants *Eciton burchelli* Westwood and *Eciton hamatum* Fabricus [sic] (continued). J. N. Y. Entomol. Soc. 71:246-261. [1963.12.23]

Whelden, R. M., Haskins, C. P. 1954 ("1953"). Cytological and histological studies in the Formicidae. I. Chromosome morphology and the problem of sex determination. Ann. Entomol. Soc. Am. 46:579-595. [1954.02.08]

Whitcomb, W. H., Denmark, H. A., Bhatkar, A. P., Greene, G. L. 1972. Preliminary studies on the ants of Florida soybean fields. Fla. Entomol. 55:129-142. [1972.09.11]

Whitcomb, W. H., Denmark, H. A., Buren, W. F., Carroll, J. F. 1972. Habits and present distribution in Florida of the exotic ant, *Pseudomyrmex mexicanus* (Hymenoptera: Formicidae). Fla. Entomol. 55:31-33. [1972.03.01]

White, A. 1846. Insects. In: Richardson, J., Gray, J. E. (eds.) The zoology of the voyage of H.M.S. Erebus & Terror, under the command of Captain Sir James Clark Ross, R.N., F.R.S., during the years 1839 to 1843. Vol. 2 [part]. London: E. W. Janson, pp. 1-24, pl. 1-6. [1846]

White, W. F. 1884. Ants and their ways. With illustrations, and an appendix giving a complete list of genera and species of the British ants. London: Religious Tract Society, 279 pp. [1884]

White, W. F. 1895. Ants and their ways. With illustrations, and an appendix giving a complete list of genera and species of the British ants. Second edition. London: Religious Tract Society, 255 pp. [1895]

Whitehead, P. F. 1994. Rural breeding populations of *Hypoponera punctatissima* (Roger) (Hym., Formicidae) in Worcestershire. Entomol. Mon. Mag. 130:194. [1994.11.30]

Whitford, W. G. 1978. Structure and seasonal activity of Chihuahua desert ant communities. Insectes Soc. 25:79-88. [1978.03]

Whitford, W. G., Gentry, J. B. 1981. Ant communities of southeastern longleaf pine plantations. Environ. Entomol. 10:183-185. [1981.04]

Whiting, J. H., Jr., Black, H. L., Jorgensen, C. D. 1989. A scanning electron microscopy study of the mouthparts of *Paraponera clavata* (Hymenoptera: Formicidae). Pan-Pac. Entomol. 65:302-309. [1989.08.03]

Whiting, P. W. 1938. Anomalies and cast [sic] determination in ants. J. Hered. 29:189-193. [1938.05]

Whitley, G. 1936. Ichthyological genotypes: some supplementary remarks. Aust. Zool. 8:189-192. [1936.06.29] [Relevant to dating Latreille (1804).]

Wiel, P. van der. 1926. *Formica pressilabris* Nyl. in Nederland. Entomol. Ber. (Amst.) 7:106. [1926.07.01]

Wiel, P. van der. 1927. Een nest van *Ponera punctatissima* Rog. Entomol. Ber. (Amst.) 7:175-176. [1927.01.01]

Wiernasz, D. C., Yencharis, J., Cole, B. J. 1995. Size and mating success in males of the western harvester ant, *Pogonomyrmex occidentalis* (Hymenoptera: Formicidae). J. Insect Behav. 8:523-531. [1995.07]

Wilkinson, R. C., Bhatkar, A. P., Whitcomb, W. H., Kloft, W. J. 1980. *Formica integra* (Hymenoptera: Formicidae) 3. Trial introduction into Florida. Fla. Entomol. 63:142-146. [1980.04.30]

Willey, R. B., Brown, W. L., Jr. 1983. New species of the ant genus *Myopias* (Hymenoptera: Formicidae: Ponerinae). Psyche (Camb.) 90:249-285. [1983.12.09]
Williams, D. F. (ed.) 1994. Exotic ants. Biology, impact, and control of introduced species. Boulder: Westview Press, xvii + 332 pp. [1994]
Williams, F. X. 1937. [Untitled. Introduced by: "Dr. F. X. Williams exhibited winged female specimens of the endemic ponerine ant *Pseudocryptopone swezeyi*".] Proc. Hawaii. Entomol. Soc. 9:375. [1937.09]
Williams, F. X. 1945. The aculeate wasps of New Caledonia, with natural history notes. Proc. Hawaii. Entomol. Soc. 12:407-452. [1945.06.25] [Account of *Myrmecia apicalis*, page 449.]
Williams, F. X. 1946. *Stigmatomma* (*Fulakora*) *zwaluwenbergi*, a new species of ponerine ant from Hawaii. Proc. Hawaii. Entomol. Soc. 12:639-640. [1946.05.28]
Williams, G. C. 1956. Records of an established infestation of *Tapinoma melanocephalum* F. (Hym., Formicidae) in Great Britain. Entomol. Mon. Mag. 92:329-330. [1956.12.14]
Williams, M. W., Williams, C. S., DeWitt, G. R. 1965. Temperature and subspecies variation on the oxygen consumption of the desert ant, *Pogonomyrmex barbatus*. Life Sci. 4:603-606. [1965.03]
Wilson, E. O. 1951a. Variation and adaptation in the imported fire ant. Evolution 5:68-79. [1951.03.26]
Wilson, E. O. 1951b ("1950"). A new *Leptothorax* from Alabama (Hymenoptera: Formicidae). Psyche (Camb.) 57:128-130. [1951.04.23]
Wilson, E. O. 1952a. Notes on *Leptothorax bradleyi* Wheeler and *L. wheeleri* M. R. Smith (Hymenoptera: Formicidae). Entomol. News 63:67-71. [1952.03.19]
Wilson, E. O. 1952b. O complexo *Solenopsis saevissima* na America do Sul (Hymenoptera: Formicidae). Mem. Inst. Oswaldo Cruz Rio J. 50:49-59. [1952.03]
Wilson, E. O. 1952c. The *Solenopsis saevissima* complex in South America (Hymenoptera: Formicidae). Mem. Inst. Oswaldo Cruz Rio J. 50:60-68. [1952.03]
Wilson, E. O. 1953a. On Flanders' hypothesis of caste determination in ants. Psyche (Camb.) 60:15-20. [1953.06.26]
Wilson, E. O. 1953b. The origin and evolution of polymorphism in ants. Q. Rev. Biol. 28:136-156. [1953.06]
Wilson, E. O. 1953c. Origin of the variation in the imported fire ant. Evolution 7:262-263. [1953.09.30]
Wilson, E. O. 1954a ("1953"). The ecology of some North American dacetine ants. Ann. Entomol. Soc. Am. 46:479-495. [1954.02.08]
Wilson, E. O. 1954b. A new interpretation of the frequency curves associated with ant polymorphism. Insectes Soc. 1:75-80. [1954.03]
Wilson, E. O. 1955a. A monographic revision of the ant genus *Lasius*. Bull. Mus. Comp. Zool. 113:1-201. [1955.03]
Wilson, E. O. 1955b. Ecology and behavior of the ant *Belonopelta deletrix* Mann (Hymenoptera: Formicidae). Psyche (Camb.) 62:82-87. [1955.07.19]
Wilson, E. O. 1955c. Division of labor in a nest of the slave-making ant *Formica wheeleri* Creighton. Psyche (Camb.) 62:130-133. [1955.11.08]
Wilson, E. O. 1955d. The status of the ant genus *Microbolbos* Donisthorpe. Psyche (Camb.) 62:136. [1955.11.08]
Wilson, E. O. 1956. Feeding behavior in the ant *Rhopalothrix biroi* Szabó. Psyche (Camb.) 63: 21-23. [1956.10.11]
Wilson, E. O. 1957a. The discovery of cerapachyine ants on New Caledonia, with the description of new species of *Phyracaces* and *Sphinctomyrmex*. Breviora 74:1-9. [1957.05.01]
Wilson, E. O. 1957b. The *tenuis* and *selenophora* groups of the ant genus *Ponera* (Hymenoptera: Formicidae). Bull. Mus. Comp. Zool. 116:355-386. [1957.05]
Wilson, E. O. 1958a. The beginnings of nomadic and group-predatory behavior in the ponerine ants. Evolution 12:24-36. [1958.03.31]

Wilson, E. O. 1958b. The fire ant. Sci. Am. 198(3):36-41. [1958.03]
Wilson, E. O. 1958c. Studies on the ant fauna of Melanesia. I. The tribe Leptogenyini. II. The tribes Amblyoponini and Platythyreini. Bull. Mus. Comp. Zool. 118:101-153. [1958.04]
Wilson, E. O. 1958d ("1957"). The organization of a nuptial flight of the ant *Pheidole sitarches* Wheeler. Psyche (Camb.) 64:46-50. [1958.05.01]
Wilson, E. O. 1958e ("1957"). Sympatry of the ants *Conomyrma bicolor* (Wheeler) and *C. pyramica* (Roger). Psyche (Camb.) 64:76. [1958.05.01]
Wilson, E. O. 1958f. Observations on the behavior of the cerapachyine ants. Insectes Soc. 5:129-140. [1958.06]
Wilson, E. O. 1958g. Studies on the ant fauna of Melanesia III. *Rhytidoponera* in western Melanesia and the Moluccas. IV. The tribe Ponerini. Bull. Mus. Comp. Zool. 119:303-371. [1958.08]
Wilson, E. O. 1958h. Character displacement and species criteria. Pp. 125-128 in: Becker, E. C. et al. (eds.) Proceedings of the Tenth International Congress of Entomology, Montreal, August 17-25, 1956. Volume 1. Ottawa: Mortimer Ltd., 941 pp. [1958.12]
Wilson, E. O. 1959a. Adaptive shift and dispersal in a tropical ant fauna. Evolution 13:122-144. [1959.03.31]
Wilson, E. O. 1959b ("1958"). Patchy distributions of ant species in New Guinea rain forests. Psyche (Camb.) 65:26-38. [1959.05.22]
Wilson, E. O. 1959c. Studies on the ant fauna of Melanesia V. The tribe Odontomachini. Bull. Mus. Comp. Zool. 120:483-510. [1959.05]
Wilson, E. O. 1959d. Studies on the ant fauna of Melanesia. VI. The tribe Cerapachyini. Pac. Insects 1:39-57. [1959.07.15]
Wilson, E. O. 1959e. Some ecological characteristics of ants in New Guinea rain forests. Ecology 40:437-447. [1959.07]
Wilson, E. O. 1960a ("1959"). Communication by tandem running and ant genus *Cardiocondyla*. Psyche (Camb.) 66:29-34. [1960.09.08]
Wilson, E. O. 1960b ("1959"). William M. Mann. Psyche (Camb.) 66:55-59. [1960.12.28]
Wilson, E. O. 1961. The nature of the taxon cycle in the Melanesian ant fauna. Am. Nat. 95:169-193. [1961.06.16]
Wilson, E. O. 1962a. The Trinidad cave ant *Erebomyrma* (=*Spelaeomyrmex*) *urichi* (Wheeler), with a comment on cavernicolous ants in general. Psyche (Camb.) 69:62-72. [1962.07.01]
Wilson, E. O. 1962b. Behavior of *Daceton armigerum* (Latreille), with a classification of self-grooming movements in ants. Bull. Mus. Comp. Zool. 127:403-421. [1962.08.28]
Wilson, E. O. 1962c. The ants of Rennell and Bellona Islands. Nat. Hist. Rennell Isl. Br. Solomon Isl. 4:13-23. [1962]
Wilson, E. O. 1963a. Social modifications related to rareness in ant species. Evolution 17:249-253. [1963.06.21]
Wilson, E. O. 1963b. The social biology of ants. Annu. Rev. Entomol. 8:345-368. [1963]
Wilson, E. O. 1964a. The true army ants of the Indo-Australian area (Hymenoptera: Formicidae: Dorylinae). Pac. Insects 6:427-483. [1964.11.10]
Wilson, E. O. 1964b. The ants of the Florida Keys. Breviora 210:1-14. [1964.11.12]
Wilson, E. O. 1965a. Trail sharing in ants. Psyche (Camb.) 72:2-7. [1965.06.25]
Wilson, E. O. 1965b. Chemical communication in the social insects. Science (Wash. D. C.) 149:1064-1071. [1965.09.03]
Wilson, E. O. 1965c. A consistency test for phylogenies based on contemporaneous species. Syst. Zool. 14:214-220. [1965.10.12]
Wilson, E. O. 1968. The ergonomics of caste in the social insects. Am. Nat. 102:41-66. [1968.02.27]
Wilson, E. O. 1971. The insect societies. Cambridge, Mass.: Harvard University Press, x + 548 pp. [1971.01.01] [Date from publisher.]

Wilson, E. O. 1973. The ants of Easter Island and Juan Fernández. Pac. Insects 15:285-287. [1973.07.20]

Wilson, E. O. 1974a. The soldier of the ant, *Camponotus* (*Colobopsis*) *fraxinicola*, as a trophic caste. Psyche (Camb.) 81:182-188. [1974.05.21]

Wilson, E. O. 1974b. The population consequences of polygyny in the ant *Leptothorax curvispinosus*. Ann. Entomol. Soc. Am. 67:781-786. [1974.09.16]

Wilson, E. O. 1975a. Sociobiology. The new synthesis. Cambridge, Mass.: Harvard University Press, 697 pp. [1975.01.01] [Date from publisher.]

Wilson, E. O. 1975b. *Leptothorax duloticus* and the beginnings of slavery in ants. Evolution 29:108-119. [1975.03.31]

Wilson, E. O. 1975c. Slavery in ants. Sci. Am. 232(6):32-36. [1975.06]

Wilson, E. O. 1975d. Enemy specification in the alarm-recruitment system of an ant. Science (Wash. D. C.) 190:798-800. [1975.11.21]

Wilson, E. O. 1976a. The organization of colony defense in the ant *Pheidole dentata* Mayr (Hymenoptera: Formicidae). Behav. Ecol. Sociobiol. 1:63-81. [1976.02.06]

Wilson, E. O. 1976b. A social ethogram of the Neotropical arboreal ant *Zacryptocerus varians* (Fr. Smith). Anim. Behav. 24:354-363. [1976.05]

Wilson, E. O. 1976c. Behavioral discretization and number of castes in an ant species. Behav. Ecol. Sociobiol. 1:141-154. [1976.08.05]

Wilson, E. O. 1976d. Which are the most prevalent ant genera? Stud. Entomol. 19:187-200. [1976.12.30]

Wilson, E. O. 1977a. [Review of: Wheeler, G. C., Wheeler, J. 1976. Ant larvae: review and synthesis. Mem. Entomol. Soc. Wash. 7:1-108.] Science (Wash. D. C.) 195:975. [1977.03.11]

Wilson, E. O. 1977b ("1976"). The first workerless parasite in the ant genus *Formica* (Hymenoptera: Formicidae). Psyche (Camb.) 83:277-281. [1977.08.29]

Wilson, E. O. 1978. Division of labor in fire ants based on physical castes (Hymenoptera: Formicidae: Solenopsis). J. Kansas Entomol. Soc. 51:615-636. [1978.11.17]

Wilson, E. O. 1979. The evolution of caste systems in social insects. Proc. Am. Philos. Soc. 123:204-210. [1979.08]

Wilson, E. O. 1981a. Epigenesis and the evolution of social systems. J. Hered. 72:70-77. [1981.04]

Wilson, E. O. 1981b. Communal silk-spinning by larvae of *Dendromyrmex* tree-ants (Hymenoptera: Formicidae). Insectes Soc. 28:182-190. [1981.09]

Wilson, E. O. 1984a. The relation between caste ratios and division of labor in the ant genus *Pheidole* (Hymenoptera: Formicidae). Behav. Ecol. Sociobiol. 16:89-98. [1984.11]

Wilson, E. O. 1984b. Tropical social parasites in the ant genus *Pheidole*, with an analysis of the anatomical parasitic syndrome (Hymenoptera: Formicidae). Insectes Soc. 31:316-334. [1984.12]

Wilson, E. O. 1985a. Ants of the Dominican amber (Hymenoptera: Formicidae). 1. Two new myrmicine genera and an aberrant *Pheidole*. Psyche (Camb.) 92:1-9. [1985.06.28]

Wilson, E. O. 1985b. Ants of the Dominican amber (Hymenoptera: Formicidae). 2. The first fossil army ants. Psyche (Camb.) 92:11-16. [1985.06.28]

Wilson, E. O. 1985c. Ants of the Dominican amber (Hymenoptera: Formicidae). 3. The subfamily Dolichoderinae. Psyche (Camb.) 92:17-37. [1985.06.28]

Wilson, E. O. 1985d. The sociogenesis of insect colonies. Science (Wash. D. C.) 228:1489-1495. [1985.06.28]

Wilson, E. O. 1985e. Invasion and extinction in the West Indian ant fauna: evidence from the Dominican amber. Science (Wash. D. C.) 229:265-267. [1985.07.19]

Wilson, E. O. 1985f. Ants from the Cretaceous and Eocene amber of North America. Psyche (Camb.) 92:205-216. [1985.11.28]

Wilson, E. O. 1985g. The principles of caste evolution. Fortschr. Zool. 31:307-324. [1985]

Wilson, E. O. 1986a ("1985"). Ants of the Dominican amber (Hymenoptera: Formicidae). 4. A giant ponerine in the genus *Paraponera*. Isr. J. Entomol. 19:197-200. [1986.02]

Wilson, E. O. 1986b. Caste and division of labor in *Erebomyrma*, a genus of dimorphic ants (Hymenoptera: Formicidae: Myrmicinae). Insectes Soc. 33:59-69. [1986.06]

Wilson, E. O. 1986c. The defining traits of fire ants and leaf-cutting ants. Pp. 1-9 in: Lofgren, C. S., Vander Meer, R. K. (eds.) Fire ants and leaf-cutting ants. Biology and management. Boulder: Westview Press, xv + 435 pp. [1986]

Wilson, E. O. 1987a. Causes of ecological success: the case of the ants. The Sixth Tansley Lecture. J. Anim. Ecol. 56:1-9. [1987.02]

Wilson, E. O. 1987b. The earliest known ants: an analysis of the Cretaceous species and an inference concerning their social organization. Paleobiol. 13:44-53. [1987.03.13]

Wilson, E. O. 1987c ("1986"). The organization of floor evacuation in the ant genus *Pheidole* (Hymenoptera: Formicidae). Insectes Soc. 33:458-469. [1987.06]

Wilson, E. O. 1987d. The arboreal ant fauna of Peruvian Amazon forests: a first assessment. Biotropica 19:245-251. [1987.09]

Wilson, E. O. 1988a. The current status of ant taxonomy. Pp. 3-10 in: Trager, J. C. (ed.) Advances in myrmecology. Leiden: E. J. Brill, xxvii + 551 pp. [1988]

Wilson, E. O. 1988b. The biogeography of the West Indian ants (Hymenoptera: Formicidae). Pp. 214-230 in: Liebherr, J. K. (ed.) Zoogeography of Caribbean insects. Ithaca, New York: Cornell University Press, ix + 285 pp. [1988]

Wilson, E. O. 1989. *Chimaeridris*, a new genus of hook-mandibled myrmicine ants from tropical Asia (Hymenoptera: Formicidae). Insectes Soc. 36:62-69. [1989.06]

Wilson, E. O. 1990. Success and dominance in ecosystems: the case of the social insects. Excellence in ecology, 2. Oldendorf/Luhe: Ecology Institute, xxi + 104 pp. [1990.09]

Wilson, E. O., Brown, W. L., Jr. 1953. The subspecies concept and its taxonomic application. Syst. Zool. 2:97-111. [1953.09]

Wilson, E. O., Brown, W. L., Jr. 1955. Revisionary notes on the *sanguinea* and *neogagates* groups of the ant genus *Formica*. Psyche (Camb.) 62:108-129. [1955.11.08]

Wilson, E. O., Brown, W. L., Jr. 1956. New parasitic ants of the genus *Kyidris*, with notes on ecology and behavior. Insectes Soc. 3:439-454. [1956.09]

Wilson, E. O., Brown, W. L., Jr. 1958a. The worker caste of the parasitic ant *Monomorium metoecus* Brown and Wilson, with notes on behavior. Entomol. News 69:33-38. [1958.02.21]

Wilson, E. O., Brown, W. L., Jr. 1958b. Recent changes in the introduced population of the fire ant *Solenopsis saevissima* (Fr. Smith). Evolution 12:211-218. [1958.06.30]

Wilson, E. O., Brown, W. L., Jr. 1967. [Untitled. Descriptions of new taxa: Sphecomyrminae Wilson and Brown, new subfamily; *Sphecomyrma* Wilson and Brown, new genus; *Sphecomyrma freyi* Wilson and Brown, new species.] Pp. 6-10 in: Wilson, E. O., Carpenter, F. M., Brown, W. L., Jr. The first Mesozoic ants, with the description of a new subfamily. Psyche (Camb.) 74:1-19. [1967.08.18]

Wilson, E. O., Brown, W. L., Jr. 1985 ("1984"). Behavior of the cryptobiotic predaceous ant *Eurhopalothrix heliscata*, n. sp. (Hymenoptera: Formicidae: Basicerotini). Insectes Soc. 31:408-428. [1985.04]

Wilson, E. O., Carpenter, F. M., Brown, W. L., Jr. 1967a. The first Mesozoic ants, with the description of a new subfamily. Psyche (Camb.) 74:1-19. [1967.08.18]

Wilson, E. O., Carpenter, F. M., Brown, W. L., Jr. 1967b. The first Mesozoic ants. Science (Wash. D. C.) 157:1038-1040. [1967.09.01]

Wilson, E. O., Eisner, T., Wheeler, G. C., Wheeler, J. 1956. *Aneuretus simoni* Emery, a major link in ant evolution. Bull. Mus. Comp. Zool. 115:81-99. [1956.08]

Wilson, E. O., Fagen, R. M. 1974. On the estimation of total behavioral repertories in ants. J. N. Y. Entomol. Soc. 82:106-112. [1974.08.22]

Wilson, E. O., Farish, D. J. 1973. Predatory behaviour in the ant-like wasp *Methocha stygia* (Say) (Hymenoptera: Tiphiidae). Anim. Behav. 21:292-295. [1973.05]

Wilson, E. O., Francoeur, A. 1974. Ants of the *Formica fusca* group in Florida. Fla. Entomol. 57:115-116. [1974.06.28]
Wilson, E. O., Hölldobler, B. 1980. Sex differences in cooperative silk-spinning by weaver ant larvae. Proc. Natl. Acad. Sci. U. S. A. 77:2343-2347. [1980.04]
Wilson, E. O., Hölldobler, B. 1985. Caste-specific techniques of defense in the polymorphic ant *Pheidole embolopyx* (Hymenoptera: Formicidae). Insectes Soc. 32:3-22. [1985.07]
Wilson, E. O., Hölldobler, B. 1986. Ecology and behavior of the Neotropical cryptobiotic ant *Basiceros manni* (Hymenoptera: Formicidae: Basicerotini). Insectes Soc. 33:70-84. [1986.06]
Wilson, E. O., Hölldobler, B. 1988. Dense heterarchies and mass communication as the basis of organization in ant colonies. Trends Ecol. Evol. 3:65-68. [1988.03]
Wilson, E. O., Hunt, G. L. 1967. Ant fauna of Futuna and Wallis Islands, stepping stones to Polynesia. Pac. Insects 9:563-584. [1967.11.20]
Wilson, E. O., Pavan, M. 1960 ("1959"). Glandular sources and specificity of some chemical releasers of social behavior in dolichoderine ants. Psyche (Camb.) 66:70-76. [1960.12.28]
Wilson, E. O., Regnier, F. E., Jr. 1971. The evolution of the alarm-defense system in the formicine ants. Am. Nat. 105:279-289. [1971.06.30]
Wilson, E. O., Simberloff, D. S. 1969. Experimental zoogeography of islands. Defaunation and monitoring techniques. Ecology 50:267-278. [1969]
Wilson, E. O., Taylor, R. W. 1964. A fossil ant colony: new evidence of social antiquity. Psyche (Camb.) 71:93-103. [1964.08.24]
Wilson, E. O., Taylor, R. W. 1967a. An estimate of the potential evolutionary increase in species density in the Polynesian ant fauna. Evolution 21:1-10. [1967.03.30]
Wilson, E. O., Taylor, R. W. 1967b. The ants of Polynesia (Hymenoptera: Formicidae). Pac. Insects Monogr. 14:1-109. [1967.08.15]
Wing, M. W. 1939. An annotated list of the ants of Maine (Hymenoptera: Formicidae). Entomol. News 50:161-165. [1939.06.16]
Wing, M. W. 1947. A proposed bibliography and catalogue of world Formicidae. Rev. Entomol. (Rio J.) 18:467-468. [1947.12.31]
Wing, M. W. 1949. A new *Formica* from northern Maine, with a discussion of its supposed type of social parasitism (Hymenoptera: Formicidae). Can. Entomol. 81:13-17. [1949.02.28]
Wing, M. W. 1950. An address list of the myrmecologists of the world. Rev. Entomol. (Rio J.) 21:417-432. [1950.12.30]
Wing, M. W. 1968a. Taxonomic revision of the Nearctic genus *Acanthomyops* (Hymenoptera: Formicidae). Mem. Cornell Univ. Agric. Exp. Stn. 405:1-173. [1968.03]
Wing, M. W. 1968b. A taxonomic revision of the Nearctic genus *Acanthomyops* Mayr (Hymenoptera: Formicidae). [Abstract.] Diss. Abstr. B. Sci. Eng. 28:3934. [1968.03]
Winter, U. 1974 ("1972"). Sozialparasiten der *Leptothorax*-Gruppe (Hym.; Formicidae) aus der Umgebung des Tennengebirges (Österreich). Z. Arbeitsgem. Österr. Entomol. 24:124-126. [1974.02]
Winter, U., Buschinger, A. 1983. The reproductive biology of a slavemaker ant, *Epimyrma ravouxi*, and a degenerate slavemaker, *E. kraussei* (Hymenoptera: Formicidae). Entomol. Gen. 9:1-15. [1983.11.30]
Winter, U., Buschinger, A. 1986. Genetically mediated queen polymorphism and caste determination in the slave-making ant, *Harpagoxenus sublaevis* (Hymenoptera: Formicidae). Entomol. Gen. 11:125-137. [1986.05.30]
Wisniewski, J. 1969. Inwentaryzacja mrowisk z grupy *Formica rufa* w Borach Niemodlinskich. Pr. Kom. Nauk Roln. Kom. Nauk Lesn. 28:383-397. [1969.12]
Wisniewski, J. 1970. Die Verbreitung der Pharaoameise - *Monomorium pharaonis* L. (Hymenoptera, Formicidae) in Polen. Pol. Pismo Entomol. 40:565-568. [1970.12]
Wisniewski, J. 1975a. Zmiennosc osobnicza zabkowania zuwaczek robotnic niektórych gatunków mrówek z rodzaju *Formica* L (Hym., Formicidae). Pr. Kom. Nauk Roln. Kom. Nauk Lesn. 40:143-148. [1975.06]

Wisniewski, J. 1975b. Osobnicza zmiennosc budowy morfologicznej pseudogin mrówki cmawej - *Formica polyctena* Först. (Hym., Formicidae). Pr. Kom. Nauk Roln. Kom. Nauk Lesn. 40:149-151. [1975.06]

Wisniewski, J. 1976. The occurrence rate of ants from the *Formica rufa*-group in various phytosociologic associations. Oecologia (Berl.) 25:193-198. [1976.09.24]

Wisniewski, J. 1977. Osobnicza oraz patologiczna zmiennosc uzylkowania skrzydel mrówek *Formica polyctena* Först. (Hym., Formicidae). Pr. Kom. Nauk Roln. Kom. Nauk Lesn. 44:157-162. [1977.03]

Wisniewski, J. 1978. Dalsze badania zmiennosci teratologicznej robotnic mrówek z grupy *Formica rufa* (Hym., Formicidae). Pr. Kom. Nauk Roln. Kom. Nauk Lesn. 46:153-166. [1978.03]

Wisniewski, J. 1979. Aktueller Stand der Forschungen über Ameisen aus der *Formica rufa*-Gruppe (Hym., Formicidae) in Polen. Bull. SROP 1979(II-3):285-301. [1979]

Wisniewski, J. 1980. Teratologische Untersuchungen an Ameisenweibchen und -Männchen aus der *Formica rufa* gruppe (Hym., Formicidae). Bull. Soc. Amis Sci. Lett. Poznan Sér. D Sci. Biol. 20:149-159. [1980.09]

Wisniewski, J. 1981a. Gynandromorfy *Formica rufa* L. i *F. pratensis* Retz. (Hymenoptera, Formicidae). Pr. Kom. Nauk Roln. Kom. Nauk Lesn. 52:181-184. [1981.05]

Wisniewski, J. 1981b. Situazione attuale delle ricerche sulle formiche del gruppo *Formica rufa* (Hymenoptera, Formicidae) in Polonia. Collana Verde 59:327-338. [1981]

Wisniewski, J., Kapyszewska, E., Zielinska, G. 1981. Mrówki z grupy *Formica rufa* (Hym., Formicidae) w lasach gospodarczych Slowinskiego Parku Narodowego. Pr. Kom. Nauk Roln. Kom. Nauk Lesn. 52:185-193. [1981.05]

Wisniewski, J., Kraszewska, M., Wybranska Sadkowska, T. 1983 ("1982"). Fauna mrówek z grupy *Formica rufa* (Hym., Formicidae) w tzw. "Królestwie Mrówek" Wolinskiego Parku Narodowego. Pr. Kom. Nauk Roln. Kom. Nauk Lesn. 54:167-172. [1983.02]

Wisniewski, J., Moskaluk, A. 1975. Mrówki z grupy *Formica rufa* (Hym., Formicidae) w Karkonoskim Parku Narodowym. Pr. Kom. Nauk Roln. Kom. Nauk Lesn. 40:153-156. [1975.06]

Wisniewski, J., Sokolowski, A. 1983a. Stanowiska pseudogin *Formica polyctena* Först. (Hym., Formicidae) w Polsce. Pr. Kom. Nauk Roln. Kom. Nauk Lesn. 56:145-149. [1983.06]

Wisniewski, J., Sokolowski, A. 1983b. Rzadko spotykane zmiany teratologiczne robotnic mrówek z grupy *Formica rufa* (Hym., Formicidae). Pr. Kom. Nauk Roln. Kom. Nauk Lesn. 56:151-156. [1983.06]

Wisniewski, J., Sokolowski, A. 1987. Missbildungen bei Waldameisen-Arbeiterinnen aus der *Formica rufa*-Gruppe (Hymenoptera, Formicidae). Waldhygiene 17:113-128. [1987]

Wojcik, D. P. 1994. Impact of the red imported fire ant on native ant species in Florida. Pp. 269-281 in: Williams, D. F. (ed.) Exotic ants. Biology, impact, and control of introduced species. Boulder: Westview Press, xvii + 332 pp. [1994]

Wojcik, D. P., Banks, W. A., Buren, W. F. 1975. First report of *Pheidole moerens* in Florida (Hymenoptera: Formicidae). U. S. Dep. Agric. Coop. Econ. Insect Rep. 25:906. [1975.12]

Wojcik, D. P., Buren, W. F., Grissell, E. E., Carlysle, T. 1976. The fire ants (*Solenopsis*) of Florida (Hymenoptera: Formicidae). Fla. Dep. Agric. Consum. Serv. Div. Plant Ind. Entomol. Circ. 173:1-4. [1976.11]

Wolcott, G. N. 1936. Insectae Borinquenses. A revised annotated check-list of the insects of Puerto Rico. J. Agric. Univ. P. R. 20:1-627. [1936.07.10] [Publication date from Wolcott (1941). Ants pp. 539-556.]

Wolcott, G. N. 1941. A supplement to "Insectae Borinquenses". J. Agric. Univ. P. R. 25:33-158. [1941.07] [Ants pp. 148-150.]

Wolcott, G. N. 1951 ("1948"). The insects of Puerto Rico. Hymenoptera. J. Agric. Univ. P. R. 32:749-975. [1951.11.29] [Ants pp. 810-839.]

Wolf, H. 1954. Über westdeutsche *Myrmica*-Arten. Dtsch. Entomol. Z. (N.F.)1:121-123. [1954.08.31]

Wolf, H. 1957 ("1956"). *Plagiolepis xene* Staerke [sic], eine für Deutschland neue Ameise (Hym. Form. Formicinae). Mitt. Dtsch. Entomol. Ges. 16:13. [1957.03.15]

Wolf, K. 1915. Studien über palaearktische Formiciden. I. Ber. Naturwiss.-Med. Ver. Innsb. 35: 37-52. [1915]

Wood, W. F., Chong, B. 1975. Alarm pheromones of the East African acacia symbionts: *Crematogaster mimosae* and *C. negriceps* [sic]. J. Ga. Entomol. Soc. 10:332-334. [1975.10]

Woyciechowski, M. 1985. Mrówki (Hymenoptera, Formicidae) Malych Pienin - Karpaty. Acta Zool. Cracov. 28:283-296. [1985.06.30]

Woyciechowski, M. 1993. Ants (Hymenoptera, Formicidae) of the glades in the Tatra Mts (the Carpathians). Tiscia (Szeged) 27:17-22. [1993] [May have been published in 1994. Journal received at UC Berkeley Library on 9 January 1995.]

Woyciechowski, M., Miszta, A. 1977 ("1976"). Spatial and seasonal structure of ant communities in a mountain meadow. Ekol. Pol. 24:577-592. [1977.04]

Wray, D. L. 1950. Insects of North Carolina. Second supplement. Raleigh: North Carolina Department of Agriculture, 59 pp. [1950.06] [Ants pp. 36-37.]

Wray, D. L. 1967. Insects of North Carolina. Third supplement. Raleigh: North Carolina Department of Agriculture, 181 pp. [1967.06] [Ants pp. 112-117.]

Wroughton, R. C. 1892. Our ants. Part I. J. Bombay Nat. Hist. Soc. 7:13-60. [1892.06.01]

Wroughton, R. C. 1892. Our ants. Part II. J. Bombay Nat. Hist. Soc. 7:175-203. [1892.10.01]

Wu, C. F. 1941. Catalogus insectorum Sinensium. Volume VI. Peiping [= Beijing]: Yenching University, x + 334 pp. [1941.05.16] [Formicoidea: pp. 141-204.]

Wu, J. 1990. Taxonomic studies on the genus *Formica* L. of China (Hymenoptera: Formicidae). [In Chinese.] For. Res. 3:1-8. [1990.02]

Wu, J., Wang, C. 1989a. [Untitled. Descriptions of new taxa: *Camponotus pseudolendus* Wu et Wang; *Camponotus chongqingensis* Wu et Wang; *Camponotus largiceps* Wu et Wang.] Pp. 225-227, 227-228 in: Wang, C., Xiao, G., Wu, J. Taxonomic studies on the genus *Camponotus* Mayr in China (Hymenoptera, Formicidae). [part] [In Chinese.] For. Res. 2:221-228. [1989.06]

Wu, J., Wang, C. 1989b. [Untitled. Descriptions of new taxa: *Camponotus pseudoirritans* Wu et Wang; *Camponotus anningensis* Wu et Wang.] Pp. 321-322, 323 in: Wang, C., Xiao, G., Wu, J. Taxonomic studies on the genus *Camponotus* Mayr in China (Hymenoptera, Formicidae). [concl.] [In Chinese.] For. Res. 2:321-328. [1989.08]

Wu, J., Wang, C. 1990. A taxonomic study on the genus *Tetraponera* Smith in China (Hymenoptera: Formicidae). [In Chinese.] Sci. Silvae Sin. 26:515-518. [1990.11]

Wu, J., Wang, C. 1992a. Taxonomic study of genus *Prenolepis* Mayr (Hymenoptera: Formicidae) of China. [Abstract.] P. 56 in: Proceedings XIX International Congress of Entomology. Abstracts. Beijing, China, June 28 - July 4, 1992. Beijing: XIX International Congress of Entomology, xi + 730 pp. [1992.07.04]

Wu, J., Wang, C. 1992b. A revisionary list of Chinese ants (Hymenoptera, Formicidae). [Abstract.] P. 262 in: Proceedings XIX International Congress of Entomology. Abstracts. Beijing, China, June 28 - July 4, 1992. Beijing: XIX International Congress of Entomology, xi + 730 pp. [1992.07.04]

Wu, J., Wang, C. 1992c. Hymenoptera: Formicidae. [In Chinese.] Pp. 1301-1320 in: Peng, J., Liu, Y. (eds.) Iconography of forest insects in Hunan, China. Hunan: Hunan Scientific and Technical Publishing House, 1473 pp. [1992]

Wu, J., Wang, C. 1994. A new genus of ants from Yunnan, China (Hymenoptera: Formicidae: Formicinae). J. Beijing For. Univ. (Engl. Ed.) 3(1):35-38. [1994.06]

Wu, J., Wang, C. 1995. The ants of China. [In Chinese.] Beijing: China Forestry Publishing House, 214 pp. [1995.06]

Wu, J., Xiao, G. 1987. A new species of the genus *Gnamptogenys* from China (Hymenoptera: Formicidae). [In Chinese.] Sci. Silvae Sin. 23:303-305. [1987.08]

Wu, J., Xiao, G. 1989. A new species of the genus *Vollenhovia* from China (Hymenoptera: Formicidae). [In Chinese.] Entomotaxonomia 11:239-241. [1989.06]

Wuorenrinne, H. 1975. Über die Notwendigkeit und die Möglichkeiten des Waldameisenschutzes in Finnland. Waldhygiene 11:48-50. [1975]

Würmli, M. 1969. Due interessanti reperti mirmecologici per la fauna d'Italia (Hymenoptera, Formicidae). Boll. Soc. Entomol. Ital. 99-101:208-210. [1969.12.20]

Xiao, G., Wang, C. 1989a. [Untitled. Descriptions of new taxa: *Camponotus spanis* Xiao et Wang; *Camponotus fuscivillosus* Xiao et Wang; *Camponotus rubidus* Xiao et Wang.] Pp. 224-225, 227 in: Wang, C., Xiao, G., Wu, J. Taxonomic studies on the genus *Camponotus* Mayr in China (Hymenoptera, Formicidae). [part] [In Chinese.] For. Res. 2:221-228. [1989.06]

Xiao, G., Wang, C. 1989b. [Untitled. Descriptions of new taxa: *Camponotus jianghuaensis* Xiao et Wang; *Camponotus helvus* Xiao et Wang.] Pp. 321, 322-323 in: Wang, C., Xiao, G., Wu, J. Taxonomic studies on the genus *Camponotus* Mayr in China (Hymenoptera, Formicidae). [concl.] [In Chinese.] For. Res. 2:321-328. [1989.08]

Xu, C., Wang, H., Shen, L., Wu, J. 1992. A comparative study of esterase isoenzymes in selected ants (Hymenoptera: Formicidae). [In Chinese.] For. Res. 5(1):22-25. [1992.02]

Xu, Z. 1994a. A taxonomic study of the ant subfamily Dorylinae in China (Hymenoptera Formicidae). [In Chinese.] J. Southwest For. Coll. 14:115-122. [1994.06]

Xu, Z. 1994b. A taxonomic study of the ant genus *Brachyponera* Emery in southwestern China (Hymenoptera Formicidae Ponerinae). [In Chinese.] J. Southwest For. Coll. 14:181-185. [1994.09]

Xu, Z. 1994c. A taxonomic study of the ant genus *Lepisiota* Santschi from southwestern China (Hymenoptera Formicidae Formicinae). [In Chinese.] J. Southwest For. Coll. 14:232-237. [1994.12]

Xu, Z. 1995a. A taxonomic study of the ant genus *Dolichoderus* Lund in China (Hymenoptera Formicidae Dolichoderinae). [In Chinese.] J. Southwest For. Coll. 15:33-39. [1995.03]

Xu, Z. 1995b. Two new species of the ant genus *Prenolepis* from Yunnan China (Hymenoptera: Formicidae). Zool. Res. 16:337-341.

Xu, Z., Zheng, Z. 1994. New species and new record species of the genus *Tetramorium* Mayr (Hymenoptera: Formicidae) from southwestern China. [In Chinese.] Entomotaxonomia 16:285-290. [1994.12]

Xu, Z., Zheng, Z. 1995. Two new species of the ant genera *Recurvidris* Bolton and *Kartidris* Bolton (Hymenoptera: Formicidae: Myrmicinae) from southwestern China. Entomotaxonomia 17:143-146. [1995.06]

Yamane, S., Harada, Y., Yano, M. 1985. Ant fauna of Tanega-shima Island, the northern Ryukyus (Hymenoptera, Formicidae). Mem. Kagoshima Univ. Res. Cent. S. Pac. 6:166-173. [1985.10.31]

Yamauchi, K. 1968. Additional notes on the ecological distribution of ants in Sapporo and the vicinity. J. Fac. Sci. Hokkaido Univ. Ser. VI. Zool. 16:382-395. [1968.09]

Yamauchi, K. 1969. Ants. Pp. 28-31 in: Tamura, H., Nakamura, Y., Yamauchi, K., Fujikawa, T. An ecological survey of soil fauna in Hidaka-Mombetsu, southern Hokkaido. J. Fac. Sci. Hokkaido Univ. Ser. VI. Zool. 17:17-57. [1969.09]

Yamauchi, K. 1979 ("1978"). Taxonomical and ecological studies on the ant genus *Lasius* in Japan (Hymenoptera: Formicidae). I. Taxonomy. Sci. Rep. Fac. Educ. Gifu Univ. (Nat. Sci.) 6:147-181. [1979.03.20]

Yamauchi, K. 1980 ("1979"). Taxonomical and ecological studies on the ant genus *Lasius* in Japan (Hymenoptera: Formicidae). II. Geographical distribution, habitat and nest site preferences and nest structure. Sci. Rep. Fac. Educ. Gifu Univ. (Nat. Sci.) 6:420-433. [1980.03.10]

Yamauchi, K., Hayashida, K. 1968. Taxonomic studies on the genus *Lasius* in Hokkaido, with ethological and ecological notes (Formicidae, Hymenoptera). I. The subgenus *Dendrolasius* or Jet Black Ants. J. Fac. Sci. Hokkaido Univ. Ser. VI. Zool. 16:396-412. [1968.09]

Yamauchi, K., Hayashida, K. 1970. Taxonomic studies on the genus *Lasius* in Hokkaido, with ethological and ecological notes (Formicidae, Hymenoptera). II. The subgenus *Lasius*. J. Fac. Sci. Hokkaido Univ. Ser. VI. Zool. 17:501-519. [1970.09]

Yamauchi, K., Ogata, K. 1995. Social structure and reproductive systems of tramp versus endemic ants (Hymenoptera: Formicidae) of the Ryukyu Islands. Pac. Sci. 49:55-68. [1995.01]

Yang, X., Wang, J. 1994. Two kinds of ant chromosomes in Heilongjiang. [In Chinese.] Zool. Res. 15(2):93-96. [1994.05] [Karyotypes of *Formica truncicola* and *Camponotus herculeanus*.]

Yano, M. 1909. Notes on ants. [In Japanese.] Hakubutsu Tomo 9:79-80. [Not seen. Cited in Okano (1989) and Onoyama & Terayama (1994).]

Yano, M. 1910a. On the ants of Japan. [In Japanese.] Dobutsugaku Zasshi (Zool. Mag.) 22:416-425. [1910.08.15]

Yano, M. 1910b. On *Camponotus ligniperdus* and *Formica japonica*. [In Japanese.] Hakubutsu Tomo 10:51-53. [Not seen. Cited in Okano (1989) and Onoyama & Terayama (1994).]

Yano, M. 1911a. The genus *Polyrhachis* of Japan. [In Japanese.] Dobutsugaku Zasshi (Zool. Mag.) 23:249-256. [1911.05.15]

Yano, M. 1911b. A new slave-making ant from Japan. Psyche (Camb.) 18:110-112. [1911.06] [Stamp date in MCZ library: 1911.07.07.]

Yano, M. 1911c. On the ant called "kumaari" in Japan. [In Japanese.] Hakubutsu Tomo 11:58-59. [Not seen. Cited in Okano (1989) and Onoyama & Terayama (1994).]

Yano, M. 1911d. Habitats of *Polyrhachis lamellidens*. [In Japanese.] Hakubutsu Tomo 11:83. [Not seen. Cited in Okano (1989) and Onoyama & Terayama (1994).]

Yano, M. 1911e. On *Strumigenys* ants in Japan. [In Japanese.] Hakubutsu Tomo 11:83-84. [Not seen. Cited in Okano (1989) and Onoyama & Terayama (1994).]

Yano, M. 1912a. On the slave-making ants and its allies in Japan. [In Japanese.] Dobutsugaku Zasshi (Zool. Mag.) 24:121-130. [1912.03.25]

Yano, M. 1912b. The number of species of ants. [In Japanese.] Dobutsugaku Zasshi (Zool. Mag.) 24:362-363. [1912.06.15]

Yano, M. 1912c. Ants of Japan. [In Japanese.] Rigakukai 9:649-654. [1912.09] [Not seen. Cited in Okano (1989:5).]

Yano, M. 1912d. The scientific names of ants appeared in Japanese books of insects. [In Japanese.] Dobutsugaku Zasshi (Zool. Mag.) 24:592-595. [1912.10.15]

Yano, M. 1926a. On distribution of *Pheidole*. [In Japanese.] Kontyû 1:129-131. [1926.11.25]

Yano, M. 1926b. On habitat of *Polyergus rufescens samurai*. [In Japanese.] Kontyû 1:134. [1926.11.25]

Yano, M. 1931. On the ants taken at Shiroyama, Kagoshima-Ken. [In Japanese.] Shiseki Meisho Tennen Kinenbutsu 6:273-277. [Not seen. Cited in Onoyama (1976) and Onoyama & Terayama (1994).]

Yano, M. 1932. Family Formicidae. [In Japanese.] Pp. 328-340 in: Esaki, T. et al. Nippon konchu zukan. Iconographia insectorum japonicorum. Editio prima. Tokyo: Hokuryukan, [5] + 97 + 123 + 15 + 2241 + 24 + 6 pp. [1932.06.26]

Yarrow, I. H. H. 1954a. The British ants allied to *Formica fusca* L. (Hym., Formicidae). Trans. Soc. Br. Entomol. 11:229-244. [1954.07.14]

Yarrow, I. H. H. 1954b. *Formica exsecta* Nylander (Hym., Formicidae) in the British Isles. Entomol. Mon. Mag. 90:183-185. [1954.10.06]

Yarrow, I. H. H. 1954c. Application for the re-examination and re-phrasing of the decision taken by the International Commission regarding the name of the type species of "*Formica*" Linnaeus, 1758 (Class Insecta, Order Hymenoptera). Bull. Zool. Nomencl. 9:313-318. [1954.12.30]

Yarrow, I. H. H. 1955a. The British ants allied to *Formica rufa* L. (Hym., Formicidae). Trans. Soc. Br. Entomol. 12:1-48. [1955.03.22]

Yarrow, I. H. H. 1955b. The British *Strongylognathus* (Hym., Formicidae). J. Soc. Br. Entomol. 5:114-115. [1955.03.22]

Yarrow, I. H. H. 1955c. The type species of the ant genus *Myrmica* Latreille. Proc. R. Entomol. Soc. Lond. Ser. B 24:113-115. [1955.06.22]

Yarrow, I. H. H. 1956. *Formica rufa* L. (Hym., Formicidae) in Oxfordshire. Entomol. Mon. Mag. 92:8. [1956.02.13]

Yarrow, I. H. H. 1967. On the Formicidae of the Azores. Bol. Mus. Munic. Funchal 21:24-32. [1967.12]

Yarrow, I. H. H. 1968. *Sifolinia laurae* Emery 1907, a workerless parasitic ant new to Britain (Hymenoptera, Formicidae). Entomologist 101:236-240. [1968.10.04]

Yasumatsu, K. 1935. Further notes on the hymenopterous fauna of the Yaeyama Group. Annot. Zool. Jpn. 15:33-45. [1935.06.15] [Ants pp. 34, 36.]

Yasumatsu, K. 1940a. Contributions to the hymenopterous fauna of Inner Mongolia and North China. Trans. Sapporo Nat. Hist. Soc. 16:90-95. [1940.02] [Ants p. 91.]

Yasumatsu, K. 1940b. Some ants from Sanko Province in Manchuria. [In Japanese.] Mushi 13:42. [1940.03.30]

Yasumatsu, K. 1940c. Matériaux pour servir à la faune myrmécologique des îles de Yaeyama. Mushi 13:67-70. [1940.03.30]

Yasumatsu, K. 1940d. Beiträge zur Kenntnis der Ameisenfauna Mikronesiens. I. Die Ameisengattung *Anochetus* Mayr der Karolinen. Annot. Zool. Jpn. 19:312-315. [1940.12.10]

Yasumatsu, K. 1940e. Ants from Kyusyu. [In Japanese.] Kyodo Shizen 5:1-5. [Not seen. Cited in Okano (1989) and Onoyama & Terayama (1994).]

Yasumatsu, K. 1941a. On the ants of the genus *Dolichoderus* of Angaran element from the Far East (Hymenoptera, Formicidae). [In Japanese.] Kontyû 14:177-183. [1941.01.25]

Yasumatsu, K. 1941b. *Camponotus vagus yessoensis* in Mt. Hikosan, Kyusyu. [In Japanese.] Mushi 13:96. [1941.02.28]

Yasumatsu, K. 1941c. Ants collected by Mr. H. Takahasi in Hingan (Hsingan) North Province, North Manchuria (Hymenoptera, Formicidae). [In Japanese.] Trans. Nat. Hist. Soc. Formosa 31:182-185. [1941.04]

Yasumatsu, K. 1941d. Some ants from the Linshoten Islands. [In Japanese.] Kontyû 15:22-23. [1941.07.30]

Yasumatsu, K. 1941e. On the new habitat of *Tapinoma indicum* in Japan. [In Japanese.] Mushi 14:11. [1941.10.29]

Yasumatsu, K. 1941f. On the new habitat of *Formica exsecta fukaii*. [In Japanese.] Mushi 14:14. [1941.10.29]

Yasumatsu, K. 1942. On the vertical distributions of ants on Mt. Hikosan, Kyushu. [In Japanese.] Mushi 14:102. [1942.05.27]

Yasumatsu, K. 1950. Discovery of an ant of the genus *Lordomyrma* Emery in eastern Asia (Hym.). Insecta Matsumurana 17:73-79. [1950.04]

Yasumatsu, K. 1960. The occurrence of the subfamily Leptanillinae in Japan (Hymenoptera, Formicidae). Esakia 1:17-20. [1960.01.20]

Yasumatsu, K. 1962. Notes on synonymies of five ants widely spread in the Orient (Hym.: Formicidae). Mushi 36:93-97. [1962.03.15]

Yasumatsu, K., Brown, W. L., Jr. 1951b. Revisional notes on *Camponotus herculeanus* Linné and close relatives in Palearctic regions (Hymenoptera: Formicidae). J. Fac. Agric. Kyushu Univ. 10:29-44. [1951.10.30]

Yasumatsu, K., Brown, W. L., Jr. 1957. A second look at the ants of the *Camponotus herculeanus* group in eastern Asia. J. Fac. Agric. Kyushu Univ. 11:45-51. [1957.08.30]

Yasumatsu, K., Hirashima, Y. 1965. Aculeate Hymenoptera collected by Lepidopterological Society of Japan expedition to Formosa in 1961. Spec. Bull. Lepid. Soc. Jpn. 1:175-177. [1965.12.30] [Ants p. 175.]

Yasumatsu, K., Murakami, Y. 1960. A revision of the genus *Stenamma* of Japan (Hymenoptera, Formicidae, Myrmicinae). Esakia 1:27-31. [1960.01.20]

Yasuno, M. 1963. The study of the ant population in the grassland at Mt. Hakkôda. I. The distribution and nest abundance of ants in the grassland. Ecol. Rev. 16:83-91. [1963.03]

Yensen, N. P., Clark, W. H., Francoeur, A. 1977. A checklist of Idaho ants (Hymenoptera: Formicidae). Pan-Pac. Entomol. 53:181-187. [1977.11.28]

Yensen, N., Yensen, E., Yensen, D. 1980. Intertidal ants from the Gulf of California, Mexico. Ann. Entomol. Soc. Am. 73:266-269. [1980.05.15]

York, A. 1994. The long-term effects of fire on forest ant communities: management implications for the conservation of biodiversity. Mem. Qld. Mus. 36:231-239. [1994.06.30]

Yoshioka, H. 1939a. A list of the ants of Gumma Prefecture. [In Japanese.] Trans. Kansai Entomol. Soc. 8:64-69. [1939.04.20]

Yoshioka, H. 1939b. A new ant of the genus *Dolichoderus* from Japan. Trans. Kansai Entomol. Soc. 8:70-71. [1939.04.20]

Young, A. M. 1983. Patterns of distribution and abundance of ants (Hymenoptera: Formicidae) in three Costa Rican cocoa farm localities. Sociobiology 8:51-76. [1983]

Young, A. M. 1986. Notes on the distribution and abundance of ground- and arboreal-nesting ants (Hymenoptera: Formicidae) in some Costa Rican cacao habitats. Proc. Entomol. Soc. Wash. 88:550-571. [1986.07.31]

Young, J., Howell, D. E. 1964. Ants of Oklahoma. Misc. Publ. Okla. Agric. Exp. Stn. 71:1-42. [1964.01]

Zakharov, A. A. 1972. Intraspecific relations in ants. [In Russian.] Moskva: Nauka, 216 pp. [> 1972.11.15] [Date signed for printing ("podpisano k pechati").]

Zakharov, A. A. 1975. Evolution of the social mode of life in ants. [In Russian.] Zool. Zh. 54:861-872. [1975.06] [June issue. Date signed for printing ("podpisano k pechati"): 26 May 1975.]

Zakharov, A. A. 1994. Ant population structure on the islands of Tonga and Western Samoa. [In Russian.] Pp. 94-142 in: Puzatchenko, Y. G., Golovatch, S. I., Dlussky, G. M., Diakonov, K. N., Zakharov, A. A., Korganova, G. A. Animal populations of the islands of southwestern Oceania (ecogeographic studies). [In Russian.] Moskva: Nauka, 254 pp. [> 1993.12.07] [Signed for printing ("podpisano k pechati") 7 December 1993. Copyright date is 1994. English summary on pages 249-250.]

Zálesky, M. 1932a. *Formica exsecta pressilabris* Nyl. v Podk. Rusi. Cas. Cesk. Spol. Entomol. 29:52-53. [1932.05.30]

Zálesky, M. 1932b. Dva dalsí rody podceledi Camponotinae Mayr (Form.) v Podkarpatské Rusi. Cas. Cesk. Spol. Entomol. 29:53-54. [1932.05.30]

Zálesky, M. 1932c. *Camponotus fallax* Nyl. a jiní mravenci toho rodu v Podk. Rusi. Cas. Cesk. Spol. Entomol. 29:55-58. [1932.05.30]

Zálesky, M. 1936. Dva pro Ceskoslovensko noví mravenci. Cas. Cesk. Spol. Entomol. 33:67-69. [1936.01.21]

Zálesky, M. 1937a. Ergatopseudogyny a monstrositní pseudogynoidy dvou nasich mravencu (Formicidae). Príroda (Brno) 30:220-221. [1937]

Zálesky, M. 1937b. K poznání mravencu (Formicidae) okolí Prostejova a útesu celechovického. Vestn. Klubu Prírodoved. Prostejove 25:61-64. [1937]

Zálesky, M. 1938a. Generacní monstrosity mravence *Myrmica rubra*. Cas. Cesk. Spol. Entomol. 35:33-34. [1938.03.15]

Zálesky, M. 1938b. Hnízda mravencu *Formica exsecta exsecta* a *F. exsecta pressilabris*. Cas. Cesk. Spol. Entomol. 35:35. [1938.03.15]

Zálesky, M. 1938c. Prvodelnice protoergate jako kasta nasich mravencu (Formicidae). Príroda (Brno) 31:184. [1938]

Zálesky, M. 1939a. Prodromus naseho blanokrídlého hmyzu. Prodromus Hymenopterorum patriae nostrae. Pars III. Formicoidea. Sb. Entomol. Oddel. Nár. Mus. Praze 17:191-240. [1939.12.01]

Zálesky, M. 1939b. *Sysphincta europea europea*. Príroda (Brno) 32:38. [1939]

Zálesky, M. 1939c. Melanismus mravence *Formica sanguinea*. Príroda (Brno) 32:38. [1939]

Zálesky, M. 1939d. Pseudogyny mravence *Formica rufa rufa* L. Príroda (Brno) 32:38. [1939]

Zálesky, M. 1939e. Pseudogyna *Leptothorax nylanderi nylanderi* (Formicidae). Príroda (Brno) 32:176. [1939]

Zalessky, Y. M. 1949. A new Tertiary ant. [In Russian.] Sov. Geol. 40:50-54. [*Lasius tertiarius* nov. sp., male, Miocene deposits, Ukraine. See also Dlussky (1981:74).]

Zayas, L. 1975. Nuevas hormigas (Hymenoptera: Formicidae) para Isla de Pinos. Misc. Zool. (Havana) 1:4. [1975.03.13]

Zdobnitzky, W. 1910. Beitrag zur Ameisenfauna Mährens. Z. Mähr. Landesmus. 10:113-125. [1910]

Zetterstedt, J. W. 1838. Insecta Lapponica. Sectio secunda. Hymenoptera. Lipsiae [= Leipzig]: L. Voss, pp. 317-475. [1838]

Zhigul'skaya, Z. A. 1991. A new species of the ant genus *Myrmica* from the upper Kolyma Basin. [In Russian.] Zool. Zh. 70(5):58-62. [1991.05] [May issue. Date signed for printing ("podpisano k pechati"): 26 April 1991. Translated in Entomol. Rev. (Wash.) 70(5):161-165. See Zhigul'skaya (1992).]

Zhigul'skaya, Z. A. 1992. A new species of the ant genus *Myrmica* (Hymenoptera, Formicidae) from the upper Kolyma River. Entomol. Rev. (Wash.) 70(5):161-165. [1992.01] [English translation of Zhigul'skaya (1991).]

Zhigul'skaya, Z. A., Kipyatkov, V. E., Kipyatkova, T. A. 1992. Seasonal developmental cycle of the ant, *Myrmica aborigenica* sp. nov., in the upper reaches of Kolyma river. [In Russian.] Zool. Zh. 71(5):72-82. [1992.05] [May issue. Date signed for printing ("podpisano k pechati"): 10 April 1992.]

Zhizhilashvili, T. I. 1964. Ecological-geographical characteristics of the myrmecofauna of the steppe zone of Georgia. [In Russian.] Soobshch. Akad. Nauk Gruz. SSR 34:651-657. [> 1964.05.20] [Date signed for printing.]

Zhizhilashvili, T. I. 1965 ("1966"). Towards a study of the myrmecofauna of the steppe zone of Georgia. [In Russian.] Mater. Faune Gruz. 1:59-77. [> 1965.12.22] [Date signed for printing.]

Zhizhilashvili, T. I. 1967. Data on the myrmecofauna of the Borzhomya-Bakurianya forests. [In Russian.] Mater. Faune Gruz. 2:50-70. [> 1967.06.22] [Date signed for printing.]

Zhizhilashvili, T. I. 1973. Towards a study of the myrmecofauna (Hymenoptera - Formicidae) of Kartlya. [In Georgian.] Mater. Faune Gruz. 3:177-185. [> 1973.06.20] [Date signed for printing.]

Zhizhilashvili, T. I. 1974a. On the ant fauna (Hymenoptera, Formicidae) of Samtskhe-Trialetya and Dzhavakhetya. [In Russian.] Mater. Faune Gruz. 4:191-220. [> 1974.11.11] [Date signed for printing.]

Zhizhilashvili, T. I. 1974b. Ecological-faunistic study on the myrmecofauna (Hymenoptera, Formicidae) of Kolkhidskaya lowland. [In Russian.] Mater. Faune Gruz. 4:221-241. [> 1974.11.11] [Date signed for printing.]

Zíkan, W., Wygodzinsky, P. 1948. Catalogo dos tipos de insetos do Instituto de Ecologia e Experimentacão Agricolas. Bol. Serv. Nac. Pesqui. Agron. 4:1-93. [1948.05] [Ants pp. 70-73.]

Zimmerman, E. C. 1941. Argentine ant in Hawaii. Proc. Hawaii. Entomol. Soc. 11:108. [1941.07]

Zimmerman, E. C. 1953. Notes and exhibitions. *Anoplolepis longipes* (Jerdon). Proc. Hawaii. Entomol. Soc. 15:10. [1953.03.27]

Zimmermann, S. 1935 ("1934"). Beitrag zur Kenntnis der Ameisenfauna Süddalmatiens. Verh. Zool.-Bot. Ges. Wien 84:1-65. [1935.02.20]

Zimsen, E. 1964. The type material of I. C. Fabricius. Copenhagen: Munksgaard, 656 pp. [1964] [Ants pp. 423-428.]

Zittser, T. V., Reshetnikov, A. N. 1982. A method of making squashed preparations of ant chromosomes (Formicidae). [In Russian.] Tsitologiya 24:838-839. [> 1982.06.24] [Date signed for printing ("podpisano k pechati").]

Zolessi, L. C. de, Abenante, Y. P. de. 1974 ("1973"). Nidificación y mesoetología de *Acromyrmex* en el Uruguay III. *Acromyrmex* (*A.*) *hispidus* Santschi, 1925 (Hymenoptera: Formicidae). Rev. Biol. Urug. 1:151-165. [1974.05]

Zolessi, L. C. de, Abenante, Y. P. de. 1977 ("1975"). Estudio comparativo de la genitalia del macho de las especies de *Acromyrmex* del Uruguay. Rev. Biol. Urug. 3:73-86. [1977.06]

Zolessi, L. C. de, Abenante, Y. P. de. 1980. Estado actual de los estudios bioecológicos y morfológicos de las especies del género *Acromyrmex* (Myrmicinae: Attini, Mayr) en la Rep. O. del Uruguay. Jorn. Cienc. Nat. Resúm. 1:103-104. [1980.09.29]

Zolessi, L. C. de, Abenante, Y. P. de. 1983. Estado actual de los estudios bioecológicos y morfológicos de las especies del género *Acromyrmex* (Myrmicinae: Attini, Mayr) en la Republica del Uruguay. Attini 14:3-4. [1983.05]

Zolessi, L. C. de, Abenante, Y. P. de, González, L. A. 1978 ("1976"). Descripción y observaciones bioetológicas sobre una nueva especie de *Brachymyrmex* (Hymenoptera: Formicidae). Rev. Biol. Urug. 4:21-44. [1978.09]

Zolessi, L. C. de, Abenante, Y. P. de, Philippi, M. E. 1983. Insectos del Uruguay. Formicoidea: Hymenoptera. Lista ilustrada de los formícidos del Uruguay. [Abstract.] Jorn. Cienc. Nat. Resúm. Comun. 3:10. [1983.09.18]

Zolessi, L. C. de, Abenante, Y. P. de, Philippi, M. E. 1988 ("1987"). Lista sistemática de las especies de Formícidos del Uruguay. Comun. Zool. Mus. Hist. Nat. Montev. 11(165):1-9. [1988]

Zolessi, L. C. de, Abenante, Y. P. de, Philippi, M. E. 1989. Catálogo sistemático de las especies de Formícidos del Uruguay (Hymenoptera: Formicidae). Montevideo: ORCYT Unesco, 40 + ix pp. [1989]

Zolessi, L. C. de, González, L. A. 1974. Nidificación y mesoetología de *Acromyrmex* en el Uruguay, II. *Acromyrmex* (*Acromyrmex*) *lobicornis* (Emery, 1887) (Hymenoptera: Formicidae). Rev. Biol. Urug. 2:37-57. [1974.10]

Zolessi, L. C. de, González, L. A. 1979 ("1978"). Observaciones sobre el género *Acromyrmex* en el Uruguay. IV. *A.* (*Acromyrmex*) *lundi* (Guérin, 1838) (Hymenoptera: Formicidae). Rev. Fac. Humanid. Cienc. Ser. Cienc. Biol. 1:9-28. [1979.05]

Zorilla, J. M., Serrano, J. M., Casado, M. A., Acosta Salmerón, F. J., Pineda, F. D. 1986. Structural characteristics of an ant community during succession. Oikos 47:346-354. [1986.11]

Zylberberg, L., Jeantet, A.-Y., Delage-Darchen, B. 1974. Particularités structurales de l'intima cuticulaire des glandes post-pharyngiennes des fourmis. J. Microsc. (Paris) 21:331-342. [1974.12]

Also available
University of California Publications in Entomology

Vol. 102. E. Gorton Linsley and John A. Chemsak. *The Cerambycidae of North America, Part VII, No. 1: Taxonomy and Classification of the Subfamily Lamiinae, Tribes Parmenini through Acanthoderini.* ISBN 0-520-09690-8

Vol. 108. Howell V. Daly. *Bees of the New Genus Ctenoceratina in Africa South of the Sahara (Hymenoptera: Apoidea).* ISBN 0-520-09725-4

Vol. 109. John C. Luhman. *A Taxonomic Revision of Nearctic Endasys Foerster 1868 (Hymenoptera: Ichneumonidae, Gelinae).* ISBN 0-520-09757-2

Vol. 110. John D. Pinto. *The Taxonomy of North American Epicauta (Coleoptera: Meloidae), with a Revision of the Nominate Subgenus and a Survey of Courtship Behavior.* ISBN 0-520-09764-5

Vol. 111. John W. Brown and Jerry A. Powell. *Systematics of the Chrysoxena Group of Genera (Lepidoptera: Tortricidae: Euliini).* ISBN 0-520-09765-3

Vol. 112. Steven O. Shattuck. *Taxonomic Catalog of the Ant Subfamilies Aneuretinae and Dolichoderinae (Hymenoptera: Formicidae).* ISBN 0-520-09787-4

Vol. 113. Robert E. Dietz IV. *Systematics and Biology of the Genus Macrocneme Hübner (Lepidoptera: Ctenuchidae).* ISBN 0-520-09780-7.

Vol. 114. E. Gorton Linsley and John A. Chemsak. *The Cerambycidae of North America, Part VII, No. 2: Taxonomy and Classification of the Subfamily Lamiinae, Tribes Acanthocinini through Hemilophini.* ISBN 0-520-09795-5.

Vol. 115. Jerry A. Powell, ed. *Biosystematic Studies of Conifer-feeding Choristoneura (Lepidoptera: Tortricidae) in the Western United States.* ISBN 0-520-09796-3

University of California Press
Berkeley, California 94720

ISBN 0-520-09814-5